Lecture Notes in Computer Science 11660

More information about this series at http://www.springer.com/series/7411

Olga Galinina · Sergey Andreev ·
Sergey Balandin · Yevgeni Koucheryavy (Eds.)

Internet of Things, Smart Spaces, and Next Generation Networks and Systems

19th International Conference, NEW2AN 2019
and 12th Conference, ruSMART 2019
St. Petersburg, Russia, August 26–28, 2019
Proceedings

 Springer

Editors
Olga Galinina (iD)
Tampere University
Tampere, Finland

Sergey Andreev (iD)
Tampere University
Tampere, Finland

Sergey Balandin (iD)
FRUCT Oy
Helsinki, Finland

Yevgeni Koucheryavy (iD)
Tampere University
Tampere, Finland

ISSN 0302-9743　　　　　　　ISSN 1611-3349　(electronic)
Lecture Notes in Computer Science
ISBN 978-3-030-30858-2　　　ISBN 978-3-030-30859-9　(eBook)
https://doi.org/10.1007/978-3-030-30859-9

LNCS Sublibrary: SL5 – Computer Communication Networks and Telecommunications

This Springer imprint is published by the registered company Springer Nature Switzerland AG
The registered company address is: Gewerbestrasse 11, 6330 Cham, Switzerland

Preface

We welcome you to the joint proceedings of the 19th NEW2AN (Next Generation Teletraffic and Wired/Wireless Advanced Networks and Systems) and 12th Conference on the Internet of Things and Smart Spaces ruSMART (Are You Smart) held in St. Petersburg, Russia, during August 26–28, 2019.

Originally, the NEW2AN conference was launched by ITC (International Teletraffic Congress) in St. Petersburg in June 1993 as an ITC-Sponsored Regional International Teletraffic Seminar. The first edition was entitled "Traffic Management and Routing in SDH Networks" and held by R&D LONIIS. In 2002, the event received its current name, the NEW2AN. In 2008, NEW2AN acquired a new companion in Smart Spaces, ruSMART, hence boosting interaction between researchers, practitioners, and engineers across different areas of ICT. From 2012, the scope of ruSMART conference has been extended to cover the Internet of the Things and related aspects.

Presently, NEW2AN and ruSMART are well-established conferences with a unique cross-disciplinary mixture of telecommunications-related research and science. NEW2AN/ruSMART are accompanied by outstanding keynotes from universities and companies across Europe, USA, and Russia.

The 19th NEW2AN technical program addressed various aspects of next-generation data networks. This year, special attention was given to advanced wireless networking and applications. In particular, the authors demonstrated novel and innovative approaches to performance and efficiency analysis of 5G and beyond systems, employed game-theoretical formulations, advanced queuing theory, and stochastic geometry. It is also worth mentioning the rich coverage of the Internet of Things, cyber security, optics, signal processing, as well as business aspects.

The 12th Conference on the Internet of Things and Smart Spaces, ruSMART 2019, provided a forum for academic and industrial researchers to discuss new ideas and trends in the emerging areas of the Internet of Things and smart spaces that create new opportunities for fully customized applications and services. The conference brought together leading experts from top affiliations around the world. This year, we saw good participation from representatives of various players in the field, including academic teams and industrial companies, particularly representatives of Russian R&D centers, which have a good reputation for high-quality research and business in innovative service creation and applications development.

We would like to thank the Technical Program Committee members of both conferences, as well as the associated reviewers, for their hard work and important contribution to the conference. This year, the conference program met the highest quality criteria with an acceptance ratio of around 35%.

The current edition of the conferences was organized in cooperation with National Instruments, IEEE Communications Society Russia Northwest Chapter, YL-Verkot OY, Open Innovations Association FRUCT, Tampere University, Peter the Great St. Petersburg Polytechnic University, Peoples' Friendship University of Russia

(RUDN University), The National Research University Higher School of Economics (HSE), St. Petersburg State University of Telecommunications, and Popov Society. The conference was held within the framework of the RUDN University Program 5-100.

We also wish to thank all of those who contributed to the organization of the conferences. In particular, we are grateful to Roman Kovalchukov for his substantial work on the compilation of camera-ready papers.

We believe that the 19th NEW2AN and 12th ruSMART conferences delivered an informative, high-quality, and up-to-date scientific program. We also hope that participants enjoyed both technical and social conference components, the Russian hospitality, and the beautiful city of St. Petersburg.

August 2019

Olga Galinina
Sergey Andreev
Sergey Balandin
Yevgeni Koucheryavy

Organization

NEW2AN and ruSMART Technical Program Committee

Torsten Braun	University of Bern, Switzerland
Paulo Carvalho	University of Minho, Portugal
Chrysostomos Chrysostomou	Frederick University, Cyprus
Roman Dunaytsev	The Bonch-Bruevich Saint-Petersburg State University of Telecommunications, Russia
Dieter Fiems	Ghent University, Belgium
Alexey Frolov	Skolkovo Institute of Science and Technology, Russia
Ivan Ganchev	University of Limerick, Ireland
Jiri Hosek	Brno University of Technology, Czech Republic
Alexey Kashevnik	SPIIRAS, Russia
Joaquim Macedo	University of Minho, Portugal
Ninoslav Marina	University of Information Science and Technology, Macedonia
Aleksandr Ometov	Tampere University, Finland
Pavel Masek	Brno University of Technology, Czech Republic
Edison Pignaton de Freitas	Federal University of Rio Grande do Sul, Brazil
Andrey Kucheryavy	The Bonch-Bruevich Saint Petersburg State University of Telecommunications, Russia

NEW2AN and ruSMART Publicity Chair

Nikita Tafintsev	Tampere University, Finland

Contents

Next Generation Wired/Wireless Advanced Networks and Systems

New Generation of Smart Services

Proactive Context-Aware IoT-Enabled Waste Management

Orsola Fejzo[1]([⊠]), Arkady Zaslavsky[2,3], Saguna Saguna[1], and Karan Mitra[1]

[1] Luleå University of Technology, Skellefteå , Sweden
`orsfej-8@student.ltu.se`
[2] Deakin University, Melbourne, Australia
[3] ITMO University, Saint Petersburg, Russia

Abstract. Exploiting future opportunities and avoiding problematic upcoming events is the main characteristic of a proactively adapting system, leading to several benefits such as uninterrupted and efficient services. In the era when IoT applications are a tangible part of our reality, with interconnected devices almost everywhere, there is potential to leverage the diversity and amount of their generated data in order to act and take proactive decisions in several use cases, smart waste management as such. Our work focuses in devising a system for proactive adaptation of behavior, named ProAdaWM. We propose a reasoning model and system architecture that handles waste collection disruptions due to severe weather in a sustainable and efficient way using decision theory concepts. The proposed approach is validated by implementing a system prototype and conducting a case study.

Keywords: Proactive adaptation · Reasoning model · Smart cities · IoT-enabled Waste Management

1 Introduction

Recent deployment of Internet of Things (IoT) infrastructure, with its substantial usage of sensors and actuators that turn everyday objects into smart connected devices, is paving the way for the emergence of many smart cities applications [14]. The vast amount of sensor generated data enables stakeholders and systems to decide and act proactively by exploiting future opportunities or avoiding undesired events in data-driven applications [4]. IoT-enabled Waste Management (WM) systems benefit from real-time data of sensor equipped containers and GPS assisted collection vehicles to offer advanced system capabilities for monitoring, routing and planning operations [1]. However, we believe that the potential of proactive adaptation of connected devices has not been fully integrated in WM.

Collection disruptions due to lack of preparedness towards severe weather events constitute a serious problem for the WM companies, as it has been

© Springer Nature Switzerland AG 2019
O. Galinina et al. (Eds.): NEW2AN 2019/ruSMART 2019, LNCS 11660, pp. 3–15, 2019.
https://doi.org/10.1007/978-3-030-30859-9_1

reported also in [6]. Consider the following motivating scenario based on a collection interruption notice[1]: *"Oscar is a truck driver of the WM company. That January he witnessed collection delays since some residential streets were not quite safe to access. Once the snow is cleared out, the collection crews have to work extra hours to remove all the uncollected waste. Sometimes collections might even be skipped completely leaving great amounts of uncollected waste in the residents' houses."*

Given this, a context aware system that deals with weather impediments in a proactive way is essential for effective and efficient WM solutions. Such system enables adaptation by providing useful information based on its tasks and goals through reasoning methods such as rule-based, probabilistic reasoning, fuzzy logic, and other machine learning techniques just to name a few [10].

Differently from existing approaches, our work contributes with a generic reasoning model and system architecture for proactive adaptation for cases of problematic weather in the context of smart WM. The system, named ProAd-aWM (Proactive Adaptation for Smart Waste Management), utilizes decision theory concepts for reasoning in order to provide a smooth and efficient service; it makes use of data coming from IoT platforms regarding smart bins, and Open Data API-s for weather information.

The rest of the paper is structured as follows: Sect. 2 presents the literature review; the reasoning model of ProAdaWM is introduced in Sect. 3. Section 4 describes the architecture and components of the system. Finally, Sects. 5 and 6 present respectively the results and conclusions.

2   Related Work

The term *Proactive Computing* was introduced by Tennenhouse in 2000; a proactive system is characterized by the ability to mitigate the effects of undesired future events or to identify and take advantage of future opportunities [4] in order to provide a smooth uninterrupted service [13]; such system satisfies the pattern "detect - forecast - decide - act" [4]. Most of the recent works tackle the decision phase by determining cost/utility functions for different adaptation alternatives [3,13], treating it as a Markov Decision Process [3,4], or an adaptive fuzzy system [12]. Nowadays, the detection phase is enabled by the numerous sensors deployed in IoT systems which allow smart objects to take autonomous decisions about their behavior.

As observed in the survey presented in [1], there is an emergence in WM solutions based on Information and Communication Technologies (ICT). Leveraging the IoT potential the authors in [2] propose a routing model that copes with unexpected hindering factors such as traffic congestion or road construction. The work in [9] presents a heuristic for cost-saving collection of underground bins equipped with level sensors; model parameters are tuned in order to cope with a highly dynamic environment. The authors highlight the need for a manual

---

[1] https://www.luton.gov.uk/seasonal/winter/Pages/Bin-collections-during-severe-weather-conditions.aspx.

parameter change in case of external changes such as start of a holiday period or changing weather conditions. In [15] it is proved that a distinctive fill-up rate for waste containers selection is more relevant than mean values. The authors in [5] propose a model with configurable parameters such as "oversize risk" and "optimal replenishment level" to minimize the covered distance and the number of utilized collection vehicles. Even though most of the above works report cost-savings and more efficient collections in a highly dynamic environment, they do not explicitly consider unexpected events such as severe weather conditions.

In our work, we try to address this problem by creating a reasoning model that utilizes the findings of the previous related works, benefits from IoT generated data and proactively selects bins for anticipated collection in case of severe weather events with the aim of avoiding unsuccessful and inefficient collection trips. Our model is not a substitute of the aforementioned works, but rather an extension for providing a more robust service.

3   ProAdaWM Context Modeling and Reasoning

ProAdaWM reasoning model is designed to give daily proactive ·recommendations for appropriately rescheduling the waste collection day when faced with problematic weather events by using decision theory concepts. The system initially checks whether there are any severe weather forecasts and any collections taking place during those days. If yes, the reasoning algorithm decides whether it is more efficient to advance the collection day for a certain area or not. Afterwards, if areas are proactively selected for a new collection day a "Routing Engine" is notified to schedule the necessary resources and the affected citizens are notified. To enable such adaptation capabilities, a context and a reasoning decision model is designed.

3.1   Context Model

The necessary context model is built based on ORM (Object Role Modeling); it is a suitable technique for modeling context with relationships and, combined with a database, provides data retrieval operations that perform relatively quickly [10]. Context modeling was performed by identifying the following entities, their main attributes and relationships:

- **Weather**: includes forecasts of the snow precipitation levels in mm and any issued weather warnings regarding ice and snowfall.
- **Area**: comprises a physical area with accessible streets whose waste bins are emptied at the same day through a predefined schedule.
- **Segment**: a street within an area characterized by the same features throughout its whole length. Main attributes include: street inclination, cleaning priority in case of snow and waste generation rate.
- **Smart Bin**: it belongs to a unique user; it is part of a segment and an area. They are characterized by their current waste level, the maximum volume of waste that they can support as well was the waste type.

– **User**: it is identified by a unique ID in the system, a preferred contact method for cases of rescheduling and the smart bins it is the owner of.

3.2   Context Reasoning

The model should be able to reason under uncertainty about the state of its entities; be aware of the its goals and the degree of usefulness it obtains from a certain outcome of its actions. From the plethora of reasoning techniques, a Bayesian Network (BN) provides suitable capabilities to handle incomplete information, deduce new knowledge from it and be extended with theoretical concepts to handle decision making for adaptation; this formalism is called a *Decision Network* [11]. A BN has the advantages of either being built by expert knowledge, learnt by data-sets, or both. Hence, it does not pose any limitations if large data-sets are not available.

Based on the concepts of decision theory, the ProAdaWM model uses a Bayesian Network to reason about street accessibility during severe winter weather events and Utility Theory to reason about advancing the collection day to avoid unnecessary or inefficient trips. For instance, based on the motivating scenario, the model should reason on the state of a certain segment during snowfall, if it will be suitable for access or not, and then determine how beneficial it is to empty the bins of that segment ahead of time, or in the predefined schedule day.

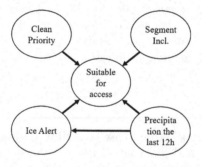

Fig. 1. Bayesian network topology for determining segment access suitability.

Bayesian Network Modeling. Through the BN topology presented in Fig. 1 the ProAdaWM system determines the probability that a certain segment will be suitable for access by the waste collection vehicle under certain conditions. Interviews were conducted with Skellefteå municipality and online reports for waste collection disruptions were consulted to determine the most relevant variables on which street access during winter weather depends:

Table 1. Notations.

	Meaning			Meaning
b	A smart bin		cw	Current waste amount
s	A segment		ew	Expected waste amount
a	An area		r	Waste generation rate
EU	Expected Utility		k_s	Segment Weight Coefficient
$maxC$	Maximum capacity (volume)		t_0	Anticipated collection day
sa	Suitable access		t_{sch}	Scheduled collection day

Algorithm 1: ProAdaWM Reasoning Algorithm

Data: t_0, $affectedAreas$, $weatherEvent$
Result: $selectedAreas$
1 initialize empty array for $selectedAreas$;
2 **foreach** a in $affectedAreas$ **do**
3   EU_{a,t_0}, $EU_{a,t_{sch}}$ = 0;
4   get $segments_a$, $maxC_a$;
5   **foreach** s in $segments_a$ **do**
6   get $maxC_s$;
7   k_s = $maxC_s$ / $maxC_a$;
8          $P_s(sa_s = true)$, $P_s(sa_s = false)$ =
           calculateSuitableAccessProbabilities(s, $weatherEvent$);
9   cw_s, ew_s = calculateWasteAmounts(s, t_0, t_{sch}, r_s);
10   EU_{s,t_0} = computeUtility(cw_s, $maxC_s$);
11         $EU_{s,t_{sch}}$ = computeUtility(ew_s, $maxC_s$);
12   EU_{a,t_0} = EU_{a,t_0} + k_s * $currU_s$;
13         $EU_{a,t_{sch}}$ = $EU_{a,t_{sch}}$ + k_s * $expU_s$;
14   **end**
15   **if** EU_{a,t_0} > $EU_{a,t_{sch}}$ **then**
16   add a to $selectedAreas$;
17   **end**
18   **return** $selectedAreas$;
19 **end**

- **Clean Priority:** The collection trucks require suitable conditions to start and stop the vehicle frequently, an aspect that can compromise the driver's and citizens' safety or cause property damage if the road is not cleared properly. The proposed model considers three different priority levels, where a greater level signifies lower priority, which in turn signifies lesser chances for a segment to be in good conditions for access during a severe weather event.
- **Street Inclination:** Operating a heavy collection vehicle, varying between 25 tonnes to 35 tonnes, becomes riskier in winter conditions as street inclination increases. Each segment is characterized by its inclination level given in percentage.

- **Snow precipitation the past 12 h:** Snowfall can quickly lead to impacts on road networks as it affects the visibility distance, pavement friction, lane obstruction, and as a consequence increases the travel time and the vehicle crash risk[2].
- **Ice Warnings:** These alerts are issued by meteorological systems whenever there is an increased risk for disruptions due to black ice on the streets which has a direct effect on drivers' safety.

Querying the BN explained above, will yield a probability value related to the state of the desired node. For instance, the query: $P(SuitableAccess = true|CleanPriority = Level1, SnowPrecipitation = s0\_5, StreetSteepness = little\_to\_gentle)$ will give the probability that node "Suitable Access" is in state "Yes" given the evidence regarding the other nodes. That probability will subsequently be used in computing the expected utilities for the decision phase.

Utility Functions. Our reasoning model aims to avoid unsuccessful trips of the collection vehicle that lead to overflown bins. Given this, the utility, which represents the satisfaction degree, is directly dependent on the amount of collected or uncollected waste in an area or segment.

Equation 1 shows how the expected waste amount is computed for a segment (refer to Table 1 for notations). Waste generation patterns vary depending on seasons, holidays, demographic and economic factors [6]. It is assumed here that the generation rate will be specific for a segment, as well as specific for a certain time of the year, to capture characteristics that depend on seasons or holidays. For instance, there might be a higher generation rate during Christmas holidays in an area with many stores.

$$ew_s = cw_s + r_{s,t_0} * (t_{sch} - t_0) \tag{1}$$

More collected waste represents a higher utility for visiting the segment for collection, whereas more uncollected waste signifies higher dissatisfaction. The latter is represented with negative values in the proposed Eq. 2 for the computation of segment utility. The model considers as uncollected waste, the amount of waste present in an inaccessible segment ($sa_s = false$).

$$U(w_s, maxC_s, sa_s) = \begin{cases} min(\frac{w_s}{maxC_s} * 100, 100) & \text{if } sa_s = True \\ -1 * min(\frac{w_s - maxC_s}{maxC_s} * 100, 100) & \text{if } sa_s = False \end{cases} \tag{2}$$

$$EU_{s,t} = P(sa_s|true) * U(ew_s, maxC_s, sa_s) + P(sa_s|false) * U(ew_s, maxC_s, sa_s) \tag{3}$$

The uncertainty factor in our model is related with the state of a segment to be suitable for access by the collection vehicle or not. The latter is introduced in the expected utility formula (Eq. 3) by the following: $P(sa_s)$; its value is inferred by the proposed Bayesian Network as explained in Algorithm 1.

---

[2] https://ops.fhwa.dot.gov/weather/q1_roadimpact.htm.

The principle of **Maximum Expected Utility** (MEU) states that a rational agent picks the action that maximizes its expected utility [11], for example the reasoning model determines that a segment has a higher expected utility to be collected in advance rather than during the weather event. However, adaptation on segment level is not suitable for the case when the waste management company requires to serve the whole area at the same day. For this reason, it is introduced in Eq. 4 the area *Score* variable, which represents the aggregated expected utility of all its segments in a specific point in time.

$$Score_{a,t} = \sum_{s=1}^{S_a} EU_{s,t} * k_s, \text{ where } k_s = \frac{maxC_s}{maxC_a} \tag{4}$$

At the end of the reasoning process, the ProAdaWM model will select for anticipated collection the areas with a higher score during the day before the severe weather event.

4   ProAdaWM System Architecture

ProAdaWM is an independent module, running on cloud infrastructure, that can be integrated to any IoT platform, to provide smart city services. Its architecture and components are shown in Fig. 2. Communication with external components such as the Weather Open Data API, IoT Platform, or the service consumers takes place via RESTful API-s through the HTTP protocol.

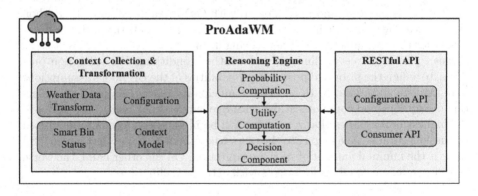

Fig. 2. ProAdaWM architecture and components.

The **Context Collection & Transformation** (CCT) component retrieves and transforms to the necessary format weather data and information related to the fill level of the smart bins. The *Configuration* sub-component is responsible for storing reasoning and data transformation parameters. For instance the area and segment characteristics of a city are part of the configuration data. The *Context Model* contains the blueprint of the relationships between the context entities; it offers tools for low-level context retrieval and manipulation.

The **Reasoning Engine** (RE) conducts the reasoning logic explained in Sect. 3.2. The decision process can either be a scheduled procedure, or it can be triggered by a daily request from the routing engine. The *Probability Computation* (PC) contains tools and libraries to perform Bayesian inference. The expected utilities are computed by the *Utility Computation* sub-component which takes as input the computed probabilities from the PC and configuration data from CCT. The *Decision Component* (DC) forwards the reasoning result to the routing engine and notifies the affected users based on the response of the routing engine.

ProAdaWM system offers a set of RESTful web services for entity **configuration**; BN configuration with custom probabilities; and for **consumer** services such as visualization, citizen notification, and for the routing engine.

5   Validation and Results

5.1   Implementation

To validate the proposed reasoning model a prototype was implemented based on the previously presented architecture. The module is cloud-based; it contains a web server implemented with Flask Microframework[3]. The CCT module implements an interface to retrieve weather information from the Swedish Meteorological and Hydrological Institute (SMHI) open data API[4]. Flask-SQLAlchemy[5] was utilized to create the context model blueprint, it also offers a set of functionalities to access the PostgreSQL database where the configurations are stored.

The BN is modeled and implemented with GeNIe Modeler[6] and the Python Wrappers it offers through its SMILE engine[7]. Table 2 presents the posterior probability for the "Suitable Access" node being in the state "Yes" for different cases of clean priority, street inclination and weather conditions. For validation purposes, to assign the prior and posterior probabilities of the BN nodes assumptions were made in accordance with observed municipality reports about waste collection disruptions and communications with Skellefteå Municipality. For instance, the highest probabilities that a segment will be in good conditions to be visited by the collection vehicle belong to the cases when there is no Ice Alert issued, and there is the minimal amount of snow precipitation. On the other hand, the worse cases for access are present for streets with extreme inclination.

5.2   Case Study

The next validation step is to analyze how the model behaves in different scenarios through a case study. Experiments are conducted with real data provided

---

[3] http://flask.pocoo.org/docs/1.0/.
[4] http://opendata.smhi.se/apidocs/.
[5] https://flask-sqlalchemy.palletsprojects.com/en/2.x/.
[6] https://www.bayesfusion.com/genie/.
[7] https://support.bayesfusion.com/docs/Wrappers/.

Table 2. Suitable access probability cases.

Clean priority	Street inclination	Snow precipitation forecast	Ice alert	Suitable access probability
Level 1	Little to gentle	Up to 5 mm	No	94%
Level 1	Little to gentle	5 mm to 20 mm	-	86%
Level 1	Little to gentle	More than 20 mm	-	75%
Level 2	Little to gentle	5 mm to 20 mm	-	83%
Level 2	Little to gentle	More than 20 mm	-	70%
Level 3	Little to gentle	5 mm to 20 mm	-	78%
Level 3	Little to gentle	More than 20 mm	-	65%
Level 1, 2, 3	Extreme to excessive	Up to 5 mm	-	40%
Level 1, 2, 3	Extreme to excessive	More than 20 mm	-	20%

by Skellefteå Municipality regarding the collection process and the bin types, as well as assumed data for other parameters. We consider an area with no segment inclination, the majority of which have a Level 3 cleaning priority. It is assumed the decision process takes place 3 days before the unwanted weather event (WeC), which can be characterised by a precipitation level forecast up to 5 mm, from 5 to 20 mm and more than 20 mm in the previous 12 h. Three different collection methods are evaluated: for private bins of 140l placed in every household, and shared bins among many houses, 1 bin of 660l for 5 households (Shared 1) and 1 bin of 660l for 10 households (Shared 2). The collection truck is dedicated to one area only; it starts and returns to the same collection point which is situated 6.3 km from the residential area. We consider average waste generation rates as reported in [7] for Europe and Central Asia. Unsafe streets, that will be skipped by the vehicle, are assigned randomly. To asses how effective and efficient the ProAdaWM model is regarding fuel efficiency, we evaluate the ratio of average consumed fuel amount per liter of collected waste (AFW) when proactive reasoning is applied, against the AFW without proactive reasoning.

$$AFW = AvgFuelPerKm * RouteLen \div CollectedWasteVol \qquad (5)$$

$$FER = \begin{cases} \dfrac{AFW_{t_0}}{AFW_{t_{sch}}} & \text{if } EU_{a,t_0} > EU_{a,t_{sch}} \\[3mm] \dfrac{AFW_{t_{sch}}}{AFW_{t_0}} & \text{if } EU_{a,t_0} \leqslant EU_{a,t_{sch}} \end{cases} \qquad (6)$$

The Fuel Efficiency Ratio (FER) is computed as the ratio between the AFW of the anticipated collection day over the AFW of the usual scheduled day, if the reasoning anticipates the collection; if the collection day remains unchanged, FER is the inverted value as shown in Eq. 6.

5.3 Results and Discussion

Figures 3a, b and c show how the expected utility evolves based on the average fill level of the bins in for the three different weather cases. The points below the identity function (grey line) represent cases where the reasoning algorithm recommends to reschedule the collection day before the unwanted weather event.

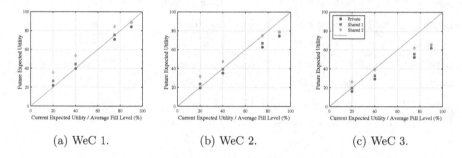

(a) WeC 1.   (b) WeC 2.   (c) WeC 3.

Fig. 3. Utility evolution.

The more severe the weather event, the more the chances that the collection will take place in advance; this is an expected result of the reasoning algorithm. For lower average fill levels, in most of the cases, especially for the shared bins, the collection is not rescheduled, this is also an expected behavior, since it is not efficient to prioritize bins with less waste amount. For weather cases with more than 5 mm of snowfall in the last 12 h the reasoning algorithm suggests to reschedule all the areas with an average fill level of more than 75% to avoid overflown bins, hence the closer to the fill-up level, more are the chances that a segment is selected for an anticipated collection.

However, the chart in Fig. 3c shows that advanced collection should take place for an extreme snowfall (more than 20 mm of precipitation) for most of the collection types even will low average fill levels. This might be disputable and non-efficient for some municipalities. Nevertheless, the areas recommended for a new collection date can be prioritized based on their utilities, hence bins of an area with expected utility of 20 will never be emptied before the bins of an area with higher expected utility.

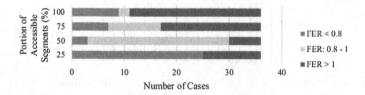

Fig. 4. FER for different ratio of accessible segments.

For the conducted case experiments, the best *FER* was obtained for *WeC 3* with an average fill level of 90% as shown in Table 3. The worst *FER* case is observed for the case when the algorithm does not anticipate the collection day and only 25% of the segments are accessible in the area with average fill level of 75%.

Table 3. Computed Fuel Efficiency Ratio from worst to best case.

Anticipated collection	Average fill level	Weather case	Accessibility	Collection type	FER
No	75%	WeC 1	25%	Shared 1	3.387
Yes	75%	WeC 2	100%	Shared 1	1.134
Yes	75%	WeC 3	50%	Private	0.818
Yes	90%	WeC 3	25%	Private	0.273

The chart in Fig. 4 gives the count of three different *FER* ranges, as explained in the legend, for different cases of street accessibility, namely when 25%, 50%, 75%, 100% of the segments are in good conditions for access. It is observed that most cases of a more efficient collection are present for the cases when less than half of the segments of an area are accessible; meaning that if the weather event's impact was not as severe as predicted, rescheduling the areas in advance, as suggested by the reasoning algorithm, leads to inefficient fuel consumption. However, if the city is easily impacted by snowfall, the proposed reasoning algorithm recommends more efficient collections.

6   Conclusions and Future Work

This paper introduced a reasoning model, ProAdaWM, for achieving proactive adaptation in a smart city context. The approach was validated by implementing a system prototype which gives instructions for adaptation via RESTful API-s. In addition, a case study was conducted to evaluate the performance and effectiveness of the reasoning model for waste operations. Results show that the system can be beneficial for cities that lack robust preparedness towards severe winter weather.

Even though our model considers seasonal and segment specific waste generation rates, for future work, it can be improved by predicting waste amounts from real-time data rather than estimating them with average generation rates. Moreover, the BN can be enhanced by learning it from real data-sets and introducing more variables to it, like temperature, month of the year or more segment characteristics.

Acknowledgement. This research was funded by the PERCCOM Erasmus Mundus Joint Master Program of the European Union [8]. Part of this study has been carried out in the scope of the project bIoTope, which is co-funded by the European Commission under Horizon-2020 program, contract number H2020-ICT- 2015/688203-bIoTope. The research was also supported by ITMO University, Russia and Luleå University of Technology, Sweden.

References

1. Anagnostopoulos, T., et al.: Challenges and opportunities of waste management in IoT-enabled smart cities: a survey. IEEE Trans. Sustain. Comput. **2**(3), 275–289 (2017). https://doi.org/10.1109/TSUSC.2017.2691049
2. Anagnostopoulos, T.V., Zaslavsky, A.: Effective waste collection with shortest path semi-static and dynamic routing. In: Balandin, S., Andreev, S., Koucheryavy, Y. (eds.) NEW2AN 2014. LNCS, vol. 8638, pp. 95–105. Springer, Cham (2014). https://doi.org/10.1007/978-3-319-10353-2_9
3. Bousdekis, A., Papageorgiou, N., Magoutas, B., Apostolou, D., Mentzas, G.: Enabling condition-based maintenance decisions with proactive event-driven computing. Comput. Ind. **100**, 173–183 (2018). https://doi.org/10.1016/j.compind. 2018.04.019
4. Engel, Y., Etzion, O., Feldman, Z.: A basic model for proactive event-driven computing. In: Proceedings of the 6th ACM International Conference on Distributed Event-Based Systems - DEBS 2012, pp. 107–118 (2012). https://doi.org/10.1145/ 2335484.2335496
5. Faccio, M., Persona, A., Zanin, G.: Waste collection multi objective model with real time traceability data. Waste Manage. **31**(12), 2391–2405 (2011). https://doi. org/10.1016/j.wasman.2011.07.005
6. Johnson, N.E., et al.: Patterns of waste generation: a gradient boosting model for short-term waste prediction in New York City. Waste Manage. **62**, 3–11 (2017). https://doi.org/10.1016/j.wasman.2017.01.037
7. Kaza, S., Yao, L.C., Bhada-Tata, P., Van Woerden, F.: What a Waste 2.0. The World Bank (2018)
8. Klimova, A., Rondeau, E., Andersson, K., Porras, J., Rybin, A., Zaslavsky, A.: An international Master's program in green ICT as a contribution to sustainable development. J. Cleaner Prod. **135**, 223–239 (2016). https://doi.org/10.1016/j. jclepro.2016.06.032
9. Mes, M., Schutten, M., Rivera, A.P.: Inventory routing for dynamic waste collection. Waste Manage. **34**(9), 1564–1576 (2014). https://doi.org/10.1016/j.wasman. 2014.05.011
10. Perera, C., Zaslavsky, A., Christen, P., Georgakopoulos, D.: Context aware computing for the Internet of Things: a survey. IEEE Commun. Surv. Tutorials **16**(1), 414–454 (2014)
11. Russell, S., Norvig, P.: Artificial Intelligence: A Modern Approach, 3rd edn. Prentice Hall Press, Upper Saddle River (2009)
12. Vainio, A.M., Valtonen, M., Vanhala, J.: Proactive fuzzy control and adaptation methods for smart homes. IEEE Intell. Syst. **23**(2), 42–49 (2008). https://doi.org/ 10.1109/MIS.2008.33
13. Vansyckel, S., Schafer, D., Schiele, G., Becker, C.: Configuration management for proactive adaptation in pervasive environments. In: International Conference on Self-Adaptive and Self-Organizing Systems, SASO, pp. 131–140 (2013). https:// doi.org/10.1109/SASO.2013.28

14. Zanella, A., Bui, N., Castellani, A., Vangelista, L., Zorzi, M.: Internet of things for smart cities. IEEE Internet Things J. **1**(1), 22–32 (2014). https://doi.org/10.1109/jiot.2014.2306328
15. Zsigraiova, Z., Semiao, V., Beijoco, F.: Operation costs and pollutant emissions reduction by definition of new collection scheduling and optimization of MSW collection routes using GIS. The case study of Barreiro, Portugal. Waste Manage. **33**(4), 793–806 (2013). https://doi.org/10.1016/j.wasman.2012.11.015

Investigation of the IoT Device Lifetime with Secure Data Transmission

Ievgeniia Kuzminykh[1,2(✉)] ⓘ, Anders Carlsson[1] ⓘ,
Maryna Yevdokymenko[2] ⓘ, and Volodymyr Sokolov[3] ⓘ

[1] Blekinge Institute of Technology,
Campus Grasvik, 371 41 Karlskrona, Sweden
`ievgeniia.kuzminykh@bth.se`
[2] Kharkiv National University of Radio Electronics,
Nauki Avenue 14, Kharkiv, Ukraine
[3] Borys Grinchenko Kyiv University, Kyiv, Ukraine

Abstract. This paper represents the approach for estimation of the lifetime of the IoT end devices. The novelty of this approach is in the taking into account not only the energy consumption for data transmission, but also for ensuring the security by using the encryption algorithms. The results of the study showed the effect of using data encryption during transmission on the device lifetime depending on the key length and the principles of the algorithm used.

Keywords: IoT · Lifetime · Power consumption · Cryptographic algorithms · Energy efficiency

1 Introduction

Nowadays, in the conditions of intensive growth of information and development of communication technologies the concept of the Internet of Things (IoT) takes one of the leading positions. This is due, primarily, an increase of the number of smart devices that are constantly being introduced in all spheres of human activities. Thus, the leading infocommunication companies (Cisco, Google, Microsoft) predict that by the end of 2020 more than 30 billion different devices will be connected to the Internet [1, 2]. These include RFID tags, contactless smart cards, global system for mobile communications (GSM), tools for automated process control systems (SCADA), wireless sensors, logistics tools, internet banking systems, public transportation, car anti-theft or key systems, airport luggage tracking, etc.

At the same time, one of the key challenges when introducing the concept of the IoT is the transmitting of data and protecting of data in the context of limited IoT device resources. In other words, today the use of IoT system should be characterized not only by the efficiency of data transmission from the point of view of the technical characteristics of devices, the network, the transmission technology that is used, etc. but also, from the point of view of the data protection since in most cases it is personal data and leakage can lead to the compromising of the entire IoT system.

© Springer Nature Switzerland AG 2019
O. Galinina et al. (Eds.): NEW2AN 2019/ruSMART 2019, LNCS 11660, pp. 16–27, 2019.
https://doi.org/10.1007/978-3-030-30859-9_2

However, due to the fact that most of the IoT devices have a limited energy resource, so-called lifetime, it is necessary to use low-power consuming technologies both for data transmission and for its protection. In this paper an up-to-date approach is proposed that allows to ensure secure data transfer and to estimate the maximum lifetime of an IoT device.

2 Review of Existing Approaches for Evaluation of the IoT Device Lifetime

During analysis of existing solutions for estimating the IoT device lifetime it should be noted that at this stage in the development of the IoT concept there are no solutions that would simultaneously take into account the energy consumption for data transmission and for ensuring data protection.

The studies presented in the papers [3–6], aimed at estimating the lifetime of a device only during data transfer depending on the technologies used. There are also works [7–10] in which aspects of the impact on the IoT device lifetime are considered from the point of view of the structural organization of the IoT system itself like topology construction and topology control. In articles [11–13] the main attention is paid to the analysis of the operating modes of the IoT device itself like active mode, sleep mode, data gathering, etc., which also allow us to estimate the lifetime of the device depending on the period of each mode.

In terms of security of data transmission there are many works also [14–19] which are aimed at studying the applied traditional encryption algorithms in the IoT devices. At the same time, there is a constant development towards the implementing such new algorithms and techniques [20–25] as lightweight algorithms, various hybrid secret key transmission schemes.

Based on the analysis of existing solutions we can distinguish three main areas that can affect the lifetime of an IoT device:

- technical characteristics of IoT devices (operating modes, critical transmission range, signal level, low resource consumption, wireless adapter, hardware, CPU, memory);
- functional characteristics of IoT devices (supported wireless communication standards, data transfer protocols used, the nature and frequency of data transmitted, the ability to use encryption and other security methods);
- structural characteristics of the IoT system (network topology, heterogeneity of IoT devices, etc.).

The review shows that today the question of the limited resources of IoT devices is important and it is almost impossible to use traditional infocommunication technologies both for data transmission and for data protection in IoT systems. Therefore, this stage of implementing of IoT systems is characterized by a variety of solutions in this area that also proves the importance and relevance of research on the IoT devices lifetime.

3 Analysis of Requirements for Modern IoT Systems

With the growing number of IoT devices and applications associated with them, the heterogeneity of the existing equipment which this concept is implemented by, as well as the development of information and communication technologies in general, the design and operation of IoT systems become more and more demanding. Thus, according to recommendation Y.2060 [26] the following high-level requirements are put forward IoT systems:

- *Identification-based connection:* it is required to ensure that the connection between an item and the IoT is established based on the identifier of this item. In addition, this includes the requirement that presumably heterogeneous identifiers of different things be processed on the basis of a unified approach.
- *Interoperability:* requires interoperability between heterogeneous and distributed systems in order to provide and consume the most diverse types of information and services.
- *Autonomous network organization:* IoT functions related to network management are required to support autonomous network organization (including methods and/or mechanisms for automatic control, automatic configuration, self-healing, automatic optimization, and automatic protection) to adapt to various applications areas, different data transmission media and a large number of devices of various types.
- *Provision of autonomous services:* it is required that services can be provided through the automatic collection, transmission and processing of data of things based on the rules set by operators or customized by subscribers. Autonomous services may depend on automated joint processing and data mining techniques.
- *Location-based capabilities:* IoT is required to provide location-based capabilities. Communications and services related to anything will depend on the information about the location of things and/or users. Location information is required to be measured and tracked automatically. Communications and location-based services may be limited by laws and regulations and must comply with safety requirements.
- *Security:* in IoT everything has a connection which leads to serious security threats, such as threats to the confidentiality, authenticity and integrity of both data and services. One of the most important examples of security requirements in IoT is the need to combine different principles and methods of ensuring security, related to a variety of user devices and networks.
- *Privacy protection:* IoT is required to ensure privacy protection. Many things have owners and users. Measurement data of things may contain personal information about their owners or users. IoT is required to protect privacy during transmission, accumulation, storage, mining, and data processing. Protection of privacy should not be an obstacle to authenticating the source of the data.
- *High-quality and high security services* related to the human body, it is required that in the IoT supported by high-quality and high security services related to the human body. Different countries have different laws and regulations regarding these services.

- *Automatic configuration:* IoT is required to provide the ability to automatically configure, allowing you to quickly create, build, or acquire configuration-based semantics for seamless integration and the interaction of attached things with applications, as well as the satisfaction of application requirements.
- *Manageability:* IoT is required to maintain manageability to ensure normal network operation. As a rule, IoT applications work automatically without the participation of people, but the whole process of their work should be manageable by the relevant parties.

However, due to the fact that mainly IoT devices operate from an autonomous power supply, and in case this power source is an ordinary battery with energy that is not recharged during operation, then at a certain time it is discharged and the IoT device stops working. And since, in the general case, all IoT devices are autonomous, a moment comes when the IoT system can no longer solve the tasks assigned to it. The time from the start of operation of the IoT system to this critical point is called the lifetime or battery life of the IoT.

When solving practical problems in IoT systems there are two main tasks related to the lifetime indicator:

1. Estimation of the assumed lifetime of IoT devices, and, accordingly, of the entire IoT system, with given input characteristics of the hardware and algorithmic mechanisms of functioning.
2. Increasing the lifetime by applying a number of methods and algorithms.

This paper proposes a solution of the first task related to estimation of the lifetime of the end IoT device with focus on a secure data transmission.

4 Estimation of Power Consumption and Lifetime of IoT Device

As it is known, an IoT device can be considered working as long as it is capable to accurately read the values from the sensors, perform the necessary calculations and transmit data to the main node where the application of this IoT system is installed. On the stage of design and deployment of an IoT system it is important to estimate in advance the approximate operating time of each node until it is necessary to replace its batteries. To do this, it is important to understand what factors affect the lifetime of its battery. In particular, it is well known that the power consumption of individual IoT devices depends on the following factors that need to be taken into account when deploying the IoT system:

1. Hardware characteristics (battery capacity, power consumption of the microcontroller, transceiver, sensors and other electronic components).
2. The frequency of data collection and data transmission depending on the application.
3. The protocols of the physical and data link layers which determine the mechanisms for controlling access to the environment where the IoT devices operate.

4. The network topology that determines the amount of information passing through each end and intermediate IoT device.
5. The routing protocol for information exchange between IoT devices which adds additional service traffic.
6. The security mechanisms for encryption and decryption of transmitted data.

Based on these factors, we formalize the above statements in the form of a method for calculating the lifetime of an IoT device. To do this, in this paper we assume that an IoT device is an end device that is designed to read data from its own sensors, encrypt them and then transfer them through the network to IoT application. Then, the frequency of reading the values from the sensors is one of the key parameters that determine the lifetime of the IoT device since the stay of the node in low-power consumption mode directly depends on it.

For clarity, we consider the cycle of the end IoT device on Fig. 1.

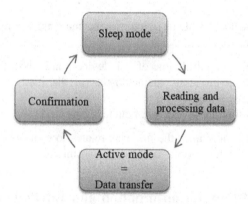

Fig. 1. IoT device operation cycle

Then, according to [26], the traditional power consumption Equation for an IoT end device is:

$$P_{IoT} = \frac{P_T \cdot t_T + P_A \cdot t_A + P_S \cdot (t_c - t_T - t_A)}{t_c} \tag{1}$$

where t_c – the duration of one cycle of the IoT device (sec);

P_T – average power during data transfer and subsequent reception confirmation (mW);

t_T – time spent on data transmission and reception of confirmation (sec);

P_A – power consumption in an active mode (reading) (mW);

t_A – the total time spent on reading data (for example, from the sensors), its processing and preparation for the transfer (sec);

P_S – power consumption in a sleep mode (mW).

In Eq. (1), the quantities P_A and P_S are constant and are determined by the features of the specific hardware implementation of IoT device. Time t_A depends both on the characteristics of the processor used and on the complexity of implementation for the data processing algorithm. Time t_T as well as power P_T depend on the wireless transmission standard used, the size of the transmitted data, and the probability of transmission errors.

Time t_T spent on the data transmission and receiving the confirmation is equal to:

$$t_T = t_{wait} + t_{ch} + t_{tr} + t_{conf}, \qquad (2)$$

where t_{ch} is a time of listening of the channel that determines its occupancy is constant and equal to 8 symbol periods or 128 μs; t_{conf} is a time to transfer confirmation (s); t_{wait} is a waiting time for data transfer (s) which is calculated as:

$$t_{wait} = W \cdot H, \qquad (3)$$

and depends on random time interval W, after which determines the channel occupancy, and which is an integer selected randomly each time, for example, in all editions of the IEEE 802.15.4 standard for the 2.4 GHz frequency band, one symbol period is $W = 3$ μs for the best case and for the worst case $W = 7$ μs; H is a constant equal to the period of 20 symbols.

Time spent on data transfer t_{tr} can be calculated by the following Equation:

$$t_{tr} = \frac{\left(D_{packet} + D_{field}\right) \cdot 8}{C}, \qquad (4)$$

where D_{packet} is a packet size, bytes; D_{field} is a size of the service fields of the packet, bytes; C is a channel data transfer rate, kbps.

The average power P_T during data transfer and subsequent receiving of confirmation will be equal to:

$$P_T = E \cdot D_{packet} \cdot F, \qquad (5)$$

where E is an energy consumed for transmission of one bit, and depending on the particular characteristics of the transceiver (J); F is a frequency of formation of the outgoing packets.

5   Calculation of the IoT Device Lifetime Taking into Account the Time for Data Encryption and Decryption

The lifetime of an IoT device depends not only on active mode and sleep mode but also on the security mechanism used such as encryption and decryption algorithms. Then, for lifetime estimation that is spent for providing data security we choose such metrics as the encryption time, the decryption time, the throughput of encryption, the throughput of decryption, the CPU process time, the power consumption and the memory utilization.

Thus, in order to calculate the lifetime of the IoT device, we introduce into the Eq. (1) additional variables that are responsible for ensuring the secure transmission of data. Then the Eq. (1) takes the following form:

$$P_{IoT} = \frac{P_T \cdot t_T + P_A \cdot t_A + TP_{Sec} + P_S \cdot (t_c - t_T - t_A - t_{Sec})}{t_c} \qquad (6)$$

where TP_{Sec} – the total power consumption for data encryption and decryption (mW); t_{Sec} – the total time spent on data encryption and decryption (sec) which depend on type of cryptography algorithms and which is equal to:

$$t_{Sec} = t_{sec(enc)} + t_{sec(dec)}, \qquad (7)$$

where $t_{sec(enc)}$ and $t_{sec(dec)}$ – the time spent on data encryption and decryption accordingly (sec).

According to [26] the total power consumption for data encryption and decryption TP_{Sec} (mW) is calculated as:

$$TP_{Sec} = P_{sec(enc)} \cdot t_{sec(enc)} + P_{sec(dec)} \cdot t_{sec(dec)} \qquad (8)$$

Where $P_{sec(enc)}$ and $P_{sec(dec)}$ – the power consumption for data encryption and decryption accordingly (mW).

Thus, variables $P_{sec(enc)}$ and $P_{sec(dec)}$ are defined using Equations:

$$P_{sec(enc)} = V_{enc} \cdot I_{enc}, \qquad (9)$$

$$P_{sec(dec)} = V_{dec} \cdot I_{dec}. \qquad (10)$$

where V_{enc} and I_{enc} – the voltage and current for encryption process respectively; V_{dec} and I_{dec} – the voltage and current for decryption process respectively.

Then the IoT device life cycle diagram will look like the one shown in Fig. 2, and the Equation for the calculation is as follows:

$$LT = \frac{E}{P_{IoT}(t_{c_i})}. \qquad (11)$$

Then, based on Eq. (1)–(10) we will calculate and assess the lifetime of the IoT device.

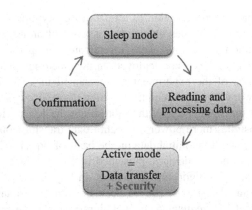

Fig. 2. IoT device operation cycle

6 Calculation of the IoT Device Lifetime Using the Proposed Solution

To calculate the lifetime the platform Arduino Mega 2560 have been chosen as an IoT device. This choice is due to the fact that this device has improved performance and is compatible with all expansion cards designed for platforms Uno or Duemilanove. In terms of battery life and battery capacity, the Arduino Mega can be powered both via USB connection and from an external power source. In our case, the research is focused on power consumption of the external battery connected to this device and, accordingly, on what capacity of the batteries should be used depending on the type of tasks.

The initial data for calculating the battery consumption during data transmission are presented in Table 1.

Table 1. Input data for transmission

Parameter, measurement unit (designation)	Value
The initial energy of the node, kJ (E)	20
Packet size, bytes (D_{packet})	50
The duration of one cycle of the IoT device, s (t_c)	0.5...2
CPU frequency, MHz (f_{proc})	16
Average number of elementary operations in one team (O)	3
Total number of commands in the data processing algorithm (K)	5000
Power consumption in transmission mode, mW (P_T)	52
Power consumption in active mode, mW (P_A)	20
Power consumption in sleep mode, mW (P_S)	0.03
Time in sleep mode, ms (t_{out})	8
Channel bit rate, kbps (C)	250
The size of the service fields of the packet, bytes (D_{field})	17

To calculate the battery consumption as a result of encrypting and decrypting the transmitted data, we took the values obtained as a result of an testbed experiment in which several encryption algorithms were investigated, in particular: AES, TEA, XTEA, RSA, Hight, Klien, they were deployed on the electronic platform of Arduino Mega 2560. The results of the experiment are presented in Table 2 and reflect the dependence of the power consumption on the duration of one cycle of the IoT device t_c.

Results show that the smaller cycle, the more often the IoT device transmits data and, accordingly, consumes more power for secure data transmission.

Then, in accordance with the initial data on the basis of Eq. (1)–(10) we obtain the following values of lifetime IoT devices, presented in Table 3. Summary for obtained results for each cryptography algorithms is present on Fig. 3.

Based on the obtained results it can be seen that under the same data transfer conditions (packet size, CPU frequency, channel bit rate, size of the packet and etc.) the lifetime of the IoT device directly depends on the encryption algorithm chosen. Thus, the most efficient algorithm when changing the duration of one cycle of the IoT device from 0.5 s to 2 s was the AES symmetric algorithm with the values of lifetime under the given conditions of 473 and 118 days, respectively.

Table 2. Total power consumption with using cryptography algorithms

Duration of one cycle, sec	Power consumption, mW					
	AES	TEA	XTEA	Hight	RSA	Klein
2.0	0.278	0.360	0.305	0.610	0.960	0.325
1.5	0.371	0.480	0.410	0.810	1.270	0.433
1.0	0.557	0.720	0.610	1.210	1.912	0.650
0.5	1.114	1.440	1.220	2.420	3.830	1.300

If we compare algorithms XTEA and TEA, then XTEA is considered more secure [26] but, like many other encryption algorithms, has its own vulnerabilities. However, this does not prevent these algorithms from implementing them on the devices with limited resources. During analysis of the IoT device lifetime using these algorithms for securing data, it is worth to note that under conditions of the same key length the modified XTEA compared to the original TEA algorithm is leading in all indicators; even though XTEA is based on operations with a 64-bit block and has 32 complete cycles, each with two rounds of the Feistel Network that means a total of 64 rounds of the network, and causes the performance degradation.

Table 3. Results of calculation of lifetime IoT devices (days)

Duration of one cycle, sec	Power consumption, mW					
	AES	TEA	XTEA	Hight	RSA	Klein
2.0	473	367	432	216	137	406
1.5	355	275	321	162	103	304
1.0	236	183	216	109	69	203
0.5	118	91	108	54	34	101

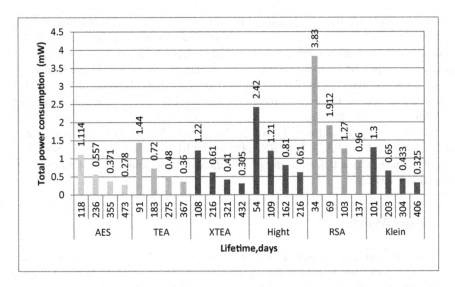

Fig. 3. Results of investigation lifetime IoT device with providing security techniques

Although the lightweight algorithm HIGHT was specifically designed to provide a low-resource implementation that is suitable for many such IoT and computing devices as USN or RFID tags, the final lifetime result was lower than the AES, XTEA, TEA and KLEIN algorithms (with key length of 128 bits) and ranged from 54 to 216 days.

Lightweight block cipher algorithm KLEIN showed quite good results from 101 to 406 days for the experiment conditions. This algorithm is mainly intended for sensor networks, however, it often finds its application in IoT systems. Due to the simple structure with an involutive S-block, the KLEIN algorithm has hardware efficiency, and the possibility of using of different key length characterizes it as a flexible algorithm with an average level of security.

The RSA algorithm demonstrated the worst result in terms of device lifetime, however, such results are dictated by the long key size of this asymmetric algorithm and, accordingly, its high cryptographic strength.

7 Conclusions

Nowadays, the IoT systems are increasingly penetrating into all spheres of human activity. At the same time, one of the most limiting factor in the implementation of the IoT concept is the energy efficiency and using of low-resourced devices. Therefore, the primary task is to ensure a long battery life of the device since in any wireless IoT device the battery life is critical parameter. To achieve long battery life for IoT devices it is necessary to determine the device's power consumption in active mode, idle mode, standby mode and in sleep mode. To this end, many studies have already been carried out to simulate various working conditions in order to estimate the energy consumption for each of the scenarios.

Along with this issue, the ensuring of the data security during transmission which is achieved by using various encryption algorithms is also quite relevant. Moreover, the choice of one or another cryptography algorithm depends not only on the degree of security, but also on its low-resource implementation. Therefore, in this paper, we solved an actual scientific problem related to estimating the lifetime of the end IoT device using the same data transfer parameters but using different encryption algorithms.

As a result of an experimental study on the Arduino Mega platform, a comparative analysis of the AES, TEA, XTEA, RSA, Hight and Klein algorithms was implemented that allows us to characterize the energy efficiency of these solutions for further use on the end IoT devices.

8   Future Work

The proposed approach focuses on estimating the lifetime of the end IoT device, which transmits only its own encrypted data to the main IoT node of the system and is not a transit node by itself. Therefore, future studies will be focused on the estimation of the lifetime of transit IoT devices taking into account structural (topology) and functional (data delivery efficiency, security, etc.) features of the IoT system construction.

References

1. Cisco Visual Networking Index: Forecast and Trends, 2017–2022. White Paper. Cisco (2019)
2. Top Strategic IoT Trends and Technologies Through 2023. Gartner report. Gartner (2019)
3. Wadud, Z., et al.: Lifetime maximization via hole alleviation in iot enabling heterogeneous wireless sensor networks. Sensors (Basel) 17(7), 1677 (2017)
4. Mallick, S., et al.: Performance appraisal of wireless energy harvesting in IoT. In: 3rd International Conference on Electrical Information and Communication Technology (EICT), IEEE, Khulna, Bangladesh, pp. 1–6 (2017)
5. Casals, L., Mir, B., Vidal, R., Gomez, C.: Modeling the energy performance of LoRaWAN. Sensors 17(10), 2364 (2017)
6. Carlsson, A., Kuzminykh, I., Franksson, R., Liljegren, A.: Measuring a LoRa network: performance, possibilities and limitations. In: Galinina, O., Andreev, S., Balandin, S., Koucheryavy, Y. (eds.) NEW2AN/ruSMART -2018. LNCS, vol. 11118, pp. 116–128. Springer, Cham (2018). https://doi.org/10.1007/978-3-030-01168-0_11
7. Fafoutis, X., Elsts, A., Vafeas, A., Oikonomou, G., Piechocki, R.: On Predicting the battery lifetime of IoT devices: experiences from the SPHERE deployments. In: RealWSN 2018 Proceedings of the 7th International Workshop on Real-World Embedded Wireless Systems and Networks, ACM, New York, Shenzhen, China, pp. 7–12 (2018)
8. Kim, T., Kim, S.H., Kim, D.: Distributed topology construction in ZigBee wireless networks. Wireless Pers. Commun. 103(3), 2213–2227 (2018)
9. Popov, O., Kuzminykh, I.: Analysis of methods for reducing topology in wireless sensor networks. In: 14th International Conference on Advanced Trends in Radioelectronics, Telecommunications and Computer Engineering (TCSET), IEEE, Lviv-Slavske, Ukraine, pp. 529–532 (2018)

10. Lemeshko, A.V., Evseeva, O.Y., Garkusha, S.V.: Research on tensor model of multipath routing in telecommunication network with support of service quality by great number of indices. Telecommun. Radio Eng. **73**(15), 1339–1360 (2014)
11. Sokolov, V., Carlsson, A., Kuzminykh, I.: Scheme for dynamic channel allocation with interference reduction in wireless sensor network. In: 4th International Scientific-Practical Conference Problems of Infocommunications, Science and Technology (PIC S&T), IEEE, Kharkov, Ukraine, pp. 564–568 (2017)
12. Kuzminykh, I., Snihurov, A., Carlsson, A.: Testing of communication range in ZigBee technology. In: 14th International Conference The Experience of Designing and Application of CAD Systems in Microelectronics (CADSM), IEEE, Lviv, Ukraine, pp. 133–136 (2017)
13. Morin, E., Maman, M., Guizzetti, R., Duda, A.: Comparison of the device lifetime in wireless networks for the internet of things. IEEE Access **5**, 7097–7114 (2017)
14. Ahmad, A., Swidan, A., Saifan, R.: Comparative analysis of different encryption techniques in mobile ad hoc networks (MANETS). Int. J. Comput. Networks Commun. (IJCNC) **8**(2), 89–101 (2016)
15. Panait, C., Dragomir, D.: Measuring the performance and energy consumption of AES in wireless sensor networks. In: 2015 Federated Conference on Computer Science and Information Systems (FedCSIS), IEEE, Lodz, Poland (2015), pp. 1261–1266 (2015)
16. Sahu, S.K., Kushwaha, A.: Performance analysis of symmetric encryption algorithms for mobile ad hoc network. Int. J. Emerg. Technol. Adv. Eng. **4**(6), 619–624 (2014)
17. Elminaam, D.S., Kader, H., Hadhoud, M.: Energy efficiency of encryption schemes for wireless devices. Int. J. Comput. Theory Eng. **1**, 302–309 (2009)
18. Yevdokymenko, M., Mayangani, M., Zalushniy, D., Sleiman, B.: Analysis of methods for assessing the reliability and security of infocommunication network. In: 4th International Scientific-Practical Conference Problems of Infocommunications, Science and Technology (PICST), IEEE, Kharkov, Ukraine, pp. 199–202 (2017)
19. Yeremenko, O., Lemeshko, O., Persikov, A.: Secure routing in reliable networks: proactive and reactive approach. In: Shakhovska, N., Stepashko, V. (eds.) CSIT 2017. AISC, vol. 689, pp. 631–655. Springer, Cham (2018). https://doi.org/10.1007/978-3-319-70581-1_44
20. Hong, D., et al.: HIGHT: a new block cipher suitable for low-resource device. In: Goubin, L., Matsui, M. (eds.) CHES 2006. LNCS, vol. 4249, pp. 46–59. Springer, Heidelberg (2006). https://doi.org/10.1007/11894063_4
21. Gong, Z., Nikova, S., Law, Y.W.: KLEIN: a new family of lightweight block ciphers. In: Juels, A., Paar, C. (eds.) RFIDSec 2011. LNCS, vol. 7055, pp. 1–18. Springer, Heidelberg (2012). https://doi.org/10.1007/978-3-642-25286-0_1
22. Buchanan, W.J., Li, S., Asif, R.: Lightweight cryptography methods. J. Cyber Secur. Technol. **1**(3–4), 187–201 (2017)
23. Goyal, T.K., Sahula, V.: Lightweight security algorithm for low power IoT devices. In: 2016 International Conference on Advances in Computing, Communications and Informatics (ICACCI), IEEE, Jaipur, India, pp. 1725–1729 (2016)
24. Deshpande, K., Singh, P.: Performance evaluation of cryptographic ciphers on IoT devices. In: International Conference on Recent Trends in Computational Engineering and Technologies (ICTRCET 2018), Bangalore, India, pp. 1–6 (2018)
25. Yevdokymenko, M., Elsayed, M., Onwuakpa, P.: Ethical hacking and penetration testing using Raspberry Pi. In: 4th International Scientific-Practical Conference Problems of Infocommunications, Science and Technology (PICST), IEEE, Kharkov, Ukraine, pp. 179–181 (2017)
26. Recommendation Y400/Y.2060: Overview of the Internet of things. ITU-T (2012)

Compression Methods for Microclimate Data Based on Linear Approximation of Sensor Data

Olli Väänänen[1]([✉])[iD] and Timo Hämäläinen[2][iD]

[1] Industrial Engineering, School of Technology,
JAMK University of Applied Sciences, Jyväskylä, Finland
olli.vaananen@jamk.fi
[2] Faculty of Information Technology, University of Jyväskylä,
Jyväskylä, Finland
timo.t.hamalainen@jyu.fi

Abstract. Edge computing is currently one of the main research topics in the field of Internet of Things. Edge computing requires lightweight and computationally simple algorithms for sensor data analytics. Sensing edge devices are often battery powered and have a wireless connection. In designing edge devices the energy efficiency needs to be taken into account. Pre-processing the data locally in the edge device reduces the amount of data and thus decreases the energy consumption of wireless data transmission. Sensor data compression algorithms presented in this paper are mainly based on data linearity. Microclimate data is near linear in short time window and thus simple linear approximation based compression algorithms can achieve rather good compression ratios with low computational complexity. Using these kind of simple compression algorithms can significantly improve the battery and thus the edge device lifetime. In this paper linear approximation based compression algorithms are tested to compress microclimate data.

Keywords: Edge computing · Internet of Things · Compression algorithm

1 Introduction

Edge computing has been one of the most significant research topics in the field of Internet of Things during these years. The edge computing means that part of the data analysis is carried out in so-called edge devices. The edge devices are devices located on the edge of the network. Wireless sensor nodes are one example of typical edge devices. The edge devices are often computationally constrained and light devices [1].

Edge computing is not going to substitute the cloud computing but it is more like a supplement concept in the IoT field. Most of the data analysis has been carried out in the cloud and this will be probably the case in the future as well. As the amount of the data from the sensing devices is increasing all the time and these sensing devices are often battery or energy harvesting powered, the energy efficiency of those so called edge devices has become very important. It is known that transmitting data wirelessly from the edge device is the most energy-consuming task in these devices. It is more

O. Galinina et al. (Eds.): NEW2AN 2019/ruSMART 2019, LNCS 11660, pp. 28–40, 2019.
https://doi.org/10.1007/978-3-030-30859-9_3

energy efficient to conduct some light data analysis or pre-processing locally and thus reduce the amount of data needed to send to the cloud. One possible pre-processing task for the sensor data is to filter clearly erroneous data and to compress the sensor data. The edge computing approach can also help to solve privacy and security issues concerning IoT data and offer minimized latency and improve the quality of service (QoS) [2].

There are different sensor data compression methods available. The suitability and efficiency of different methods depend on the data characteristics. Different methods differ in computational complexity, which is an important aspect in edge computing. This paper presents basic and light compression algorithms based on data linearity. Many environmental values are near linear in small time window. These compression algorithms' compression efficiency is tested for microclimate datasets. Datasets are temperature, air pressure and wind speed measurements from the Finnish Meteorological Institute's open data service.

The microclimate data is often nearly linear in short time window. For example, temperature normally changes slowly and if the measurement sampling rate is fast enough, the consecutive measurements cannot deviate much from each other [3]. Air pressure normally also changes slowly. Only approaching low pressure such as a thunderstorm, the air pressure can drop quickly [4]. Wind speed is slightly different because it can stay zero rather long periods. The wind speed also varies quite quickly and it is also quite abrupt in nature [3]. In this paper, the wind speed dataset is averaged data and thus represents more linear type data.

The microclimate data is very important for example in different agricultural applications. Agricultural applications for example for crop protection and to maximize crop production [5, 6] have been presented; however, microclimate measurements are important also in urban environment [7].

2   Lightweight Compression Methods for Sensor Data

In constrained edge devices, it is crucial to optimize resource usage. This means to optimize computational capacity, energy consumption and bandwidth usage [8]. These devices are often connected to the internet via wireless connection. Wireless transmitting is known to be the most energy-consuming task in these devices, thus it is in many cases more energy efficient to carry out data pre-processing and lightweight data analytics locally and thus reduce the amount of data needed to send via wireless link. A very simple method for reducing the amount of data is to compress the data. The other method is simply to reduce the sampling frequency of the sensor [3]. The drawback here is that information is lost between sampling points. Sampling a sensor is quite low energy operation compared to the energy consumption in radio transmission [3]. By using an effective and low computational complexity compression algorithm it is possible to keep radio transmitting rate low and thus keep the energy consumption on a low level, yet at the same time keep the accuracy of the higher sampling rate.

Typically, a simple edge device is a sensor node measuring some environmental magnitudes. Typical environmental magnitudes are for example temperature, humidity, air pressure and lightness. The measured values are then sent to the cloud and in the cloud, the data is combined with other data (for example open data) and together used for decision processing.

2.1 Lossy Methods and Lossless Methods

Sensor data compression methods are divided in lossy and lossless methods. Many different algorithms are presented for sensor data compression [9, 10]. The suitability of the compression algorithm is dependent on the sensor data characteristics. For example, many environmental magnitudes are nearly linear in short time scale, and thus some compression algorithms are more suitable for this kind of data. Some other type of data may require different types of compression algorithms.

If the reconstruction error accepted is more than zero, it is possible to use lossy compression algorithm. Compression ratio is dependent on accepted reconstruction error. Thus, the lossy compression algorithm will lead to loss of the information [11]. The advantages of lossy compression algorithms are the effective reduction of the data and in many cases, the computational simplicity. The compression and reduction of the data is done by eliminating some of the original information [11]. The accepted level of reconstruction error is very application dependent. In general, the lossy compression algorithms have higher a compression ratio together with lower computational complexity than lossless algorithms [12].

Many lossy algorithms have some latency and thus are not suitable for real-time applications. There are also lossy zero-latency compression algorithms. These compression methods are based on predictive filters (e.g. Kalman filter), which predict the data values from previous samples. In this method, the same filter is used in both sides of the network (sensor node and the user node where the data is analyzed further), thus the same estimation is used in both sides, and the new data is sent only if the value differs from the predicted value more than the tolerance level [8].

Lossless algorithms are able to reconstruct the original data without an error. The lossless methods perform two steps: the statistical model is first generated and then the second step uses this statistical model to map the input data to the bit sequences. In these bit sequences, the frequently occurred data generates a shorter output than infrequently occurred data. The two main encoding algorithms used are Huffman coding and arithmetic coding. The Huffman coding is computationally simpler and faster; however, it gives poor results in compression. Arithmetic coding is more efficient in compression but more complex. In many cases, the lossless algorithms are not suitable because the compression ratio is poor and computational complexity is higher than in lossy algorithms [13].

2.2 Lossy Compression Algorithms Based on Linear Approximation

Lossy data compression algorithms analyzed in this paper are based mainly on piecewise linear approximation. Piecewise linear approximation based compression algorithms are based on the fact that many environmental phenomena are near linear in short time window [3]. These kinds of phenomena are for example temperature, humidity, air pressure and wind speed.

A simple linear compression model is based on a regression line, which is calculated on the minimum of the first three measured values [13]. Least-squares regression line is used to approximation of discrete data [14]. In the least-squares regression line, the linear model is set to fit a set of data points. The least-squares method minimizes the sum of squares of the deviation between the data points and the fitting line thus gives a best fit to the data points. This is called a linear regression. A linear function $y = ax + b$ has two free parameters, a and b [14]. The general sum of squares of the deviation is [14]:

$$S = \sum_{k=1}^{N} [y_k - (ax_k + b)]^2 \tag{1}$$

Minimizing this equation and solving for a and b give [14]:

$$a = \frac{\sum_{k=1}^{N} x_k \sum_{k=1}^{N} y_k - N \sum_{k=1}^{N} x_k y_k}{\left(\sum_{k=1}^{N} x_k\right)^2 - N \sum_{k=1}^{N} x_k^2} \tag{2}$$

$$b = \frac{\sum_{k=1}^{N} x_k \sum_{k=1}^{N} x_k y_k - \sum_{k=1}^{N} x_k^2 \sum_{k=1}^{N} y_k}{\left(\sum_{k=1}^{N} x_k\right)^2 - N \sum_{k=1}^{N} x_k^2} \tag{3}$$

The parameters a and b give the best line fit to the N data points. To use these formulas it is needed to sum x_k, x_k^2, y_k, $x_k y_k$, and square the sum of x_k [14]. If the regression line is calculated from the first three measurements, then N is 3.

If the data is nearly linear, this regression line gives the prediction for the following measured data points with a certain error bound e. When the measured data point falls out of the error bound $\pm e$, then the new regression line is calculated. Hence, the data will be presented in piecewise linear segments.

There are several different versions of this kind of algorithm presented in literature. The algorithm is named here as Linear Regression based Temporal Compression (LRbTC). The algorithm is as follows:

1. Get the next three measured values and calculate the regression line to fit those three values.
2. Store (send) regression line point at time moment of the first measured value used to calculate regression line.
3. Get next measured value and compare it to the regression line.
4. If the difference is under the error bound e, then go to 3. Else, continue onto the next step.
5. Store (send) the regression line point when the measured value was last time under the error bound and go to 1.

Figure 1 shows an example of this linear regression based compression for sensor data. Original temperature data is marked in blue circle. The regression line is calculated from the first three measured values (20, 20.3 and 20.1). Then following measured values are compared to the regression line value on that time moment. Regression line continues until the difference between measured value and regression line exceed the error bound *e*, which is in this example set to 0.5. Regression line is the green line in Fig. 1. At time moment 11 the difference between regression line and measured value exceeds 0.5, thus the first regression line is set to end at time moment 10. From time 1 to 10, the compressed data includes only the starting point of the regression line and the end point of that line. The next regression line is calculated from the measured values in time moments 11 to 13. At time moment 15, the difference exceeds the error bound and thus the new line is calculated from the values at time moments 15–17. In 20, the difference exceeds again the error bound. The first 19 measured values (time moments 1 to 19) are compressed to 6 values (three regression lines).

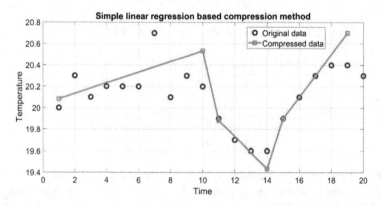

Fig. 1. Linear Regression based Temporal Compression (LRbTC) algorithm example.

In the example in Fig. 1, the error bound was set to 0.5. Thus, the measured value and regression line value should not exceed 0.5. This can anyway happen in time moments that are used to calculate the regression line.

The modified version of the LRbTC (M-LRbTC) algorithm corrects the problem if the difference between regression line and the data values used to calculate this regression line exceed the error bound *e*. The modified version of the algorithm is as follows:

1. Get the next three measured values and calculate the regression line to fit those three values.
2. Compare the regression line and three values used to calculate the line.
3. If the difference is greater than error bound *e*, then store (send) the first two data points and get the next two measurement values and calculate new regression line and go to 2, else continue onto the next step.
4. Get the next measured value and compare it to the regression line.
5. If the difference is under the error bound *e*, then go to 4. Else, continue onto the next step.
6. Store (send) the last regression line point when the measured value was under the error bound and go to 1.

Lightweight temporal compression (LTC) was introduced in [3]. It is simple and very efficient compression algorithm for microclimate type data in a small enough time window. LTC's effectivity to compress data depends on the data characteristics. For linear type environmental data, it can obtain up to 20-to-1 compression ratio [3]. Compression ratio is also dependent on error bound used. It is recommended to use the sensor manufacturer's specified accuracy value as the error bound in LTC algorithm [3]. For example if the sensor used is a temperature sensor with 0.5 degrees accuracy, it is reasonable to use 0.5 as the error bound.

The LTC algorithm is explained in detail in [3, 11, 15, 16] and a modified version in [17]. The linear model starts with the first data value as a starting point. The lower line and upper line (limit lines) are drawn from the starting point to the next measured value $\pm e$ as seen in Fig. 2a. The limit lines are tightened from the following values when error bound extreme or extremes are inside the previous limit lines as in Fig. 2b and c. The measured data is discarded from the linear model if the measurement cannot fit inside upper line and lower line determined by the previous data with the error bound $\pm e$. Then the new linear model starts using as a starting point the middle point of the upper line and lower line in last time moment included in the linear segment. This procedure of the algorithm can be seen in Fig. 2d.

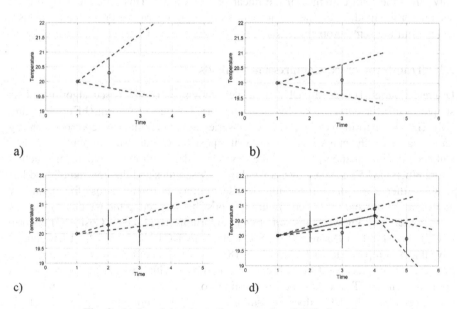

Fig. 2. Lightweight temporal compression (LTC) algorithm.

The reconstruction error never exceeds the error bound e in the LTC algorithm. The LTC algorithm has low computational complexity and thus it is suitable for constrained edge devices such as sensor nodes [17]. In Fig. 3, the LTC is compared to previously presented linear regression based algorithm.

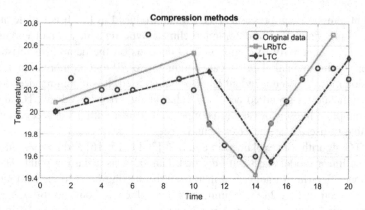

Fig. 3. LTC compared to the basic linear regression based algorithms.

The disadvantage of the LTC is that it is not well suited for real-time applications [8] and its suitability in general is very application dependent. LTC uses linear interpolation to represent the original signal, and the linear interpolation model is known only when the both extremes of the linear part is known. This introduces significant latency for the model [8]. The linear regression based algorithms presented previously suffer from the same problem.

2.3   Transform Based Compression Methods

Discrete Fourier Transform (DFT) is a well-known transform based algorithm. It is simple to use for compression by using the Fast Fourier Transform (FFT) algorithm [18]. The FFT algorithm expresses the time-series signal in frequency representation. By removing the coefficients with less energy, it is possible to reduce the amount of data and still keep the information to rebuild the time series data with reasonable reconstruction error. When the FFT is taken over a window of N samples and the first sample and last sample differ a lot, the information of discontinuity is spread across the frequency spectrum. To prevent this discontinuity it is possible to overlap the windows [18].

Another well-known transform based algorithm is Discrete Cosine Transform (DCT) and Modified Discrete Cosine Transform (MDCT) [12, 18, 19]. It has several advantages compared to FFT algorithm [18]. The DCT coefficients are real numbers; thus there is no need to deal with complex numbers. This saves memory and is less complex. The DCT also has the information concentrated to the few low-frequency components and the DCT does not suffer the edge discontinuity problem like FFT. The DCT is a well known and widely used compression algorithm for example in image compression and for time series type sensor data.

3 Testing the Algorithms with Real Microclimate Data

The linear approximation based compression algorithms are tested for microclimate type data and compared to the DCT algorithm. The datasets tested here are gathered from the Finnish Meteorological Institute's open data service [20]. Finnish Meteorological Institute has about 400 observation stations in Finland. Not all the stations have the same measured variables. For this research, Salla Naruska station's data from year 2018 in 10 min time sampling rate was chosen. The variables chosen were temperature, air pressure and wind speed. Salla Naruska measurement station is located in eastern Lapland and known as one of the coldest places in Europe. The exact situation of the station is: latitude 67.16226, longitude 29.17766 in decimal degrees. The temperature is in Celsius degrees (°C), air pressure in hectopascals (hPa) and wind speed in meters per second (m/s). The wind speed is measured in 10 min average. All variables are measured with one decimal resolution.

One year measurements in 10 min time interval mean 51,961 measurements for each variable. Some data was missing; however, the missing points were linearly interpolated. In air pressure data in total 102 points were missing, in temperature data 101 points were missing and in wind speed data 1,077 points were missing. For comparison, also the same data in one-hour measurement interval was used. This one-hour interval data for the whole year 2018 includes 8,761 measurements points for each magnitude. The missing values were also linearly interpolated.

The compression algorithms chosen were simple linear regression based approximation algorithm (LRbTC), modified linear regression based algorithms (M-LRbTC) and lightweight temporal compression (LTC). Basic discrete cosine transform (DCT) was used for comparison.

The algorithms were tested with MATLAB simulation. LRbTC, M-LRbTC and LTC algorithms were programmed in MATLAB by using mainly functions *polyfit* and *polyval*. *Polyfit* function was used for linear regression in LRbTC, and M-LRbTC and to create upper and lower lines in LTC instead of Eqs. 2 and 3 [21].

Discrete cosine transform (DCT) was tested by using the MATLAB built-in function *dct*. In this example, the DCT was used with window of five measured values to calculate DCT. It was then tested with different threshold values to cancel the smallest coefficient values. After rebuilding the signal, the maximum difference (variation) from the original values was calculated.

Algorithms were compared to each other by compression ratio versus reconstruction error. The compression ratio (*CR*) was calculated by:

$$CR = \frac{original\ data}{compressed\ data} \tag{5}$$

Thus, the bigger the *CR* value is, the more efficient the compression algorithm is. The *CR* varies significantly according the error bound *e* used.

Temperature data was tested first. In the total 51,961 measured values the highest temperature was +30.4 °C and the lowest temperature −33.8 °C. With 10-min measurement interval, the temperature data is mostly near linear; however, in some extremes the temperature changes quite a lot between consecutive measurements.

LRbTC algorithm showed very big reconstruction errors for tested temperature data, which is because when measuring the regression line, the values used to calculate may differ from the regression line more than the error bound e used. The data used is with 10-min interval and in extremes, the values may differ significantly from measurement moment to the next moment. Higher measurement sampling rate would help the situation.

The modified version of the basic linear regression based algorithm (M-LRbTC) works as it is intended. It was tested with different quantity of measurement values to calculate the regression line. In Fig. 4, the red line is used with three values to calculate the regression line; cyan line is with four values and magenta with five values.

Figure 4 illustrates the results for M-LRbTC, LTC and DCT algorithms for temperature data. With typical error bound $e = 0.5$ °C, the M-LRbTC algorithm can achieve 3.9–4.8 compression ratio. LTC is significantly more effective with $CR = 9.5$. DCT was tested with five values time window to calculate DCT. DCT compression ratio is significantly lower compared to the other tested algorithms. DCT suffers from the small window used and it can achieve higher compression ratios with bigger window used. Small time window was chosen to be more realistic for sensor data stream.

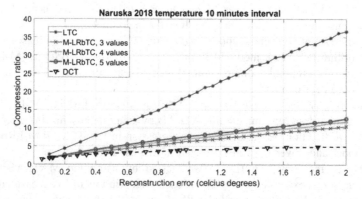

Fig. 4. Linear approximation based compression algorithms and DCT for temperature data. (Color figure online)

Figure 5 shows the results for the air pressure data. Air pressure values varied between 976.4 hPa–1056.4 hPa. The variation is nearly linear in short time window, however, the data includes few clear errors. Three times the air pressure value changes from measurement to the next more than it is normally possible. The biggest difference in consecutive measurements is 12.7 hPa (in 10 min), which is clearly an error. Normally the air pressure can change up to 5 hPa/h and only in some very quickly progressing low pressure it can be more than 5 hPa/h [4]. The quick changes in air pressure data can be due to clear measurement error or for example due to sensor calibration. The results for the air pressure data are similar to the temperature data, except the compression ratios are much higher. This indicates that the air pressure data is behaving very linearly with the 10-min measurement interval. The LTC algorithm can achieve high compression ratios.

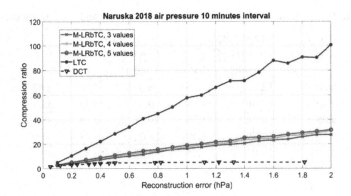

Fig. 5. Linear approximation based compression algorithms and DCT for air pressure data.

In Fig. 6 are the results of the wind speed measurements. Wind speed is measured in 10-min average values. Wind speed is a different characteristic compared to the temperature and air pressure data. Wind speed can remain quite a long period in 0 m/s. Wind speed can also change quickly and quite significantly; however, here the 10-min average measurement averages the results significantly. The compression ratios for wind speed data are on the same level as for the temperature data.

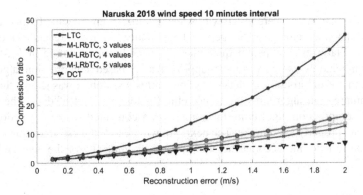

Fig. 6. Linear approximation based compression algorithms and DCT for wind speed data.

In every comparison, it can be seen that LTC is the most effective compression method. M-LRbTC also works well and it is a very simple algorithm and easy to apply. Table 1 illustrates a comparison of the compression algorithms between two different datasets with error bound e set to 0.5 for each quantity. The datasets are the same 10-min interval sets as used previously and also with 1 h measurement interval. It can be seen in Table 1 that all compression algorithms are significantly more effective for

10-minute sampling rate data. This is because with 10-min sampling rate, the data behaves more linearly.

Table 1. Comparison of the compression ratios for 10 min and 1 h interval datasets.

Compression algorithm	Temperature ($e = 0.5$ °C)		Air pressure ($e = 0.5$ hPa)		Wind speed ($e = 0.5$ m/s)	
	10 min	1 h	10 min	1 h	10 min	1 h
M-LRbTC, 3 values	3.9	1.85	8.94	3.05	2.62	1.88
M-LRbTC, 4 values	4.46	1.95	9.94	3.45	3.01	2.04
M-LRbTC, 5 values	4.78	1.86	10.75	3.79	3.18	1.97
LTC	9.49	3.19	28.22	8.28	5.09	3.03
DCT	3.07	1.75	4.63	2.72	2.6	1.8

The disadvantage in these linear approximation based algorithms is the latency. These methods are not directly suitable for real-time applications. LRbTC based methods are possible to modify to work better for almost real-time operations: After calculating the new regression line, the first point of the line and line coefficients can be sent. The receiver can use that information until the new point and line are received. Thus, the latency is in maximum when the new regression line is calculated, and it depends on how many point data is used for calculating the regression line.

4 Conclusions and Future Work

Compression algorithms were tested with some real measurement data. In this case, the environmental microclimate data such as temperature, air pressure and wind speed were used. Many environmental quantities are near linear in nature at least if the observation window is short. Linear approximation based compression algorithms benefit from this environmental data behavior. In this research, it was shown that these simple compression algorithms are rather efficient for this kind of data. The performance of compression algorithms for compression compared to reconstruction error was the main property to compare. The next step will be to test these algorithms in edge devices and to take into account the computational complexity of the algorithms.

References

1. Väänänen, O., Hämäläinen, T.: Requirements for energy efficient edge computing: a survey. In: Galinina, O., Andreev, S., Balandin, S., Koucheryavy, Y. (eds.) NEW2AN/ruSMART - 2018. LNCS, vol. 11118, pp. 3–15. Springer, Cham (2018). https://doi.org/10.1007/978-3-030-01168-0_1
2. Alrowaily, M., Lu, Z.: Secure edge computing in iot systems: review and case studies. In: 2018 IEEE/ACM Symposium on Edge Computing (SEC), Seattle, WA, pp. 440–444 (2018). https://doi.org/10.1109/SEC.2018.00060

3. Schoellhammer, T., Osterwein, E., Greenstein, B., et al.: Lightweight temporal compression of microclimate datasets. In: Proceedings of the 29th Annual IEEE International Conference on Local Computer Networks IEEE Computer Society, pp. 516–524 (2004)
4. Finnish Meteorological Institute. https://ilmatieteenlaitos.fi/ilmanpaine
5. Norbu, J., Pobkrut, T., Siyang, S., Khunarak, C., Namgyel, T., Kerdcharoen, T.: Wireless sensor networks for microclimate monitoring in edamame farm. In: 2018 10th International Conference on Knowledge and Smart Technology (KST), Chiang Mai, pp. 200–205 (2018) https://doi.org/10.1109/KST.2018.8426200
6. Muhammad, A.R., Setyawati, O., Setyawan R.A., Basuki, A.: WSN based microclimate monitoring system on porang plantation. In: 2018 Electrical Power, Electronics, Communications, Controls and Informatics Seminar (EECCIS), Batu, East Java, Indonesia, pp. 142–145 (2018). https://doi.org/10.1109/EECCIS.2018.8692849
7. Rathore, P., Rao, A.S., Rajasegarar, S., Vanz, E., Gubbi, J., Palaniswami, M.: Real-time urban microclimate analysis using internet of things. IEEE Internet Things J 5(2), 500–511 (2018). https://doi.org/10.1109/JIOT.2017.2731875
8. Giorgi, G.: A combined approach for real-time data compression in wireless body sensor networks. IEEE Sensors J. 17(18), 6129–6135 (2017)
9. Bose, T., Bandyopadhyay, S., Kumar, S., Bhattacharyya, A., Pal, A.: Signal characteristics on sensor data compression in IoT - an investigation. In: 2016 IEEE International Conference on Sensing, Communication and Networking (SECON Workshops), London, pp. 1–6 (2016)
10. Ying, Y.B.: An energy-efficient compression algorithm for spatial data in wireless sensor networks. In: 2016 18th International Conference on Advanced Communication Technology (ICACT), Pyeongchang, pp. 161–164 (2016). https://doi.org/10.1109/ICACT.2016.7423312
11. Fallah, S.A., Arioua, M., El Oualkadi, A., El Asri, J.: On the performance of piecewise linear approximation techniques in WSNs. In: 2018 International Conference on Advanced Communication Technologies and Networking (CommNet), Marrakech, pp. 1–6 (2018)
12. Alsalaet, J.K., Ali, A.A.: Data compression in wireless sensors network using MDCT and embedded harmonic coding. ISA Trans. 56, 261–267 (2015). https://doi.org/10.1016/j.isatra.2014.11.023. ISSN: 0019-0578
13. Aggarwal, C.C.: Managing and Mining Sensor Data. Springer, Boston (2013). https://doi.org/10.1007/978-1-4614-6309-2
14. Lopez, R.J.: Advanced Engineering Mathematics. Addison-Wesley, Boston (2001). ISBN: 0-201-38073-0
15. Sharma, R.: A data compression application for wireless sensor networks using LTC algorithm. In: 2015 IEEE International Conference on Electro/Information Technology (EIT), Dekalb, IL, pp. 598–604 (2015) https://doi.org/10.1109/EIT.2015.7293435
16. Azar, J., Makhoul, A., Darazi, R., Demerjian, J. and Couturier, R.: On the performance of resource-aware compression techniques for vital signs data in wireless body sensor networks. In: 2018 IEEE Middle East and North Africa Communications Conference (MENACOMM), Jounieh, pp. 1–6 (2018) https://doi.org/10.1109/MENACOMM.2018.8371032
17. Parker, D., Stojanovic, M. and Yu, C.: Exploiting temporal and spatial correlation in wireless sensor networks. In: 2013 Asilomar Conference on Signals, Systems and Computers, Pacific Grove, CA, pp. 442–446 (2013). https://doi.org/10.1109/ACSSC.2013.6810315
18. Zordan, D., Martinez, B., Vilajosana, I., Rossi, M.: On the performance of lossy compression schemes for energy constrained sensor networking. ACM Trans. Sensor Network 11(1), 34 (2014). http://dx.doi.org.ezproxy.jyu.fi/10.1145/2629660. Article 15
19. Tan, L.: Digital Signal Processing: Fundamentals and Applications. Academic Press, Elsevier, United States of America (2008). ISBN: 978-0-12-374090-8

20. Finnish Meteorological Institute's open data–service. https://en.ilmatieteenlaitos.fi/open-data

21. Matlab polyfit function documentation. https://se.mathworks.com/help/matlab/ref/polyfit.html

An Open Multimodal Mobility Platform Based on Distributed Ledger Technology

Robin Lamberti[1] ⓘ, Christian Fries[1] ⓘ, Markus Lücking[1] ⓘ,
Raphael Manke[1] ⓘ, Niclas Kannengießer[2] ⓘ,
Benjamin Sturm[2(✉)] ⓘ, Mikhail M. Komarov[3] ⓘ, Wilhelm Stork[4] ⓘ,
and Ali Sunyaev[2] ⓘ

[1] Embedded System and Sensors, FZI Research Center for Information
Technology, Karlsruhe, Germany
{lamberti,fries,luecking,manke}@fzi.de
[2] Institute of Applied Informatics and Formal Description Methods,
Karlsruhe Institute of Technology, Karlsruhe, Germany
{niclas.kannengiesser,benjamin.sturm,sunyaev}@kit.edu
[3] National Research University Higher School of Economics, Moscow, Russia
mkomarov@hse.ru
[4] Institute for Information Processing Technologies,
Karlsruhe Institute of Technology, Karlsruhe, Germany
stork@kit.edu

Abstract. The current challenges of many mobility solutions are based on an extremely fragmented booking system with complex service layers. A cross-company and user-friendly exchange of information and offers from different mobility providers is often not possible. Against this background, Distributed Ledger Technology (DLT) has the potential to revolutionize the existing mobility sector and enable completely new business models. Thus, we present a distributed mobility platform, which is valuable for a variety of mobility services. In contrast to conventional platform approaches, the data management of our infrastructure is distributed, transparent, and cost-efficient. By prototypically implementing the concept, we can demonstrate its technical feasibility and at the same time demonstrate that the introduction of our distributed mobility concept will benefit both the supply and demand sides of public transportation.

Keywords: Blockchain · Distributed Ledger Technology ·
Mobility as a Service · Multimodal transportation

1 Introduction

Urbanization is a global phenomenon [1] that challenges urban transportation systems in cities and their surrounding vicinities and raises demand for flexible, service-oriented mobility concepts [2–4]. These mobility challenges include, for instance, limited parking space, flexibility of transportation offerings, restrictions for CO_2 emissions, and increasing operational costs [5]. One approach to address many of these mobility challenges without vast expansions of existing infrastructures is to develop more

© Springer Nature Switzerland AG 2019
O. Galinina et al. (Eds.): NEW2AN 2019/ruSMART 2019, LNCS 11660, pp. 41–52, 2019.
https://doi.org/10.1007/978-3-030-30859-9_4

effective public transportation concepts by combining multiple transportation modes (e.g., bicycle, bus, train) [6, 7]. Such multimodal public transportation has many benefits over private transportation like cars, as it is usually cheaper (e.g., no fixed costs, utilization of scale effects), ecofriendly, and more reliable [8]. Correspondingly, the demand for public transportation is rising and plenty of mobility providers are introducing a broad spectrum of new mobility offers (e.g., rental bicycles and e-scooters) [9]. However, multimodal transportation often requires people to switch between systems of different mobility provider, to buy multiple tickets, to follow the respective payment processes (e.g., cash, credit card, PayPal), or even to install the mobility providers' applications [9]. People have to adapt to such circumstances, which produces high effort and can impede the use of mobility services.

On the other side, creating a seamless, multimodal mobility platform that consist of numerous service providers is a complex challenge as well, particularly in terms of accurate and fair distribution of costs and profits among all involved parties. For example, mobility providers may offer tickets that cover multiple mobility services of other providers. In such multimodal travel chains, each mobility provider receives a fraction of the overall ticket cost, regardless of the actual distance covered by the particular customer. Furthermore, smaller mobility providers become dependent on the larger competitors making it difficult to market their services or operate profitably [10].

An open and publicly usable mobility platform is required to enable mobility providers of any size to offer their services, to standardize usage and payment across providers, to resolve dependencies between service, and to create fair market conditions, and to facilitate cooperation between mobility providers in terms of payoffs. Based on published challenges in the field of public transportation, we pose multiple requirements on effective multimodal mobility platforms (cf. Table 1).

In order to overcome the challenges in multimodal public transportation, in this paper we propose an open mobility platform based on Distributed Ledger Technology (DLT) that facilitates intermodal mobility by enabling a reliable and immutable tracking of bookings of public transportations, while automating contracting between all involved parties.[1] To achieve efficient and confidential service provision over a publicly accessible distributed ledger we present an alternative approach for multi-channel messaging, which we call Random Access Authenticated Messaging (RAAM). The mobility platform allows for accurate, fast, and free payments using MIOTA (i.e., the common denominator of IOTA's cryptocurrency), which is based on IOTA [11].

2 Background

2.1 Distributed Ledger Technology

DLT enables the trustworthy operation of append-only databases, so-called ledgers, that are distributed among separated and potentially untrustworthy storage devices (so-called nodes). Each node maintains a local replication of the data stored on the ledger.

---

[1] The documentation and program code of the implemented prototype platform are available at https://github.com/cii-aifb/iota-mobility-platform.

Table 1. Requirements for the mobility platform

Name	Description	Exemplary references
Functional requirements		
Register vehicle	Vehicles should be able to register at the mobility platform	[12–14]
Vehicle check-in	If a vehicle is free it should automatically publish their location to be usable for people	[15]
Vehicle booking	People should be able to book vehicles in order to use them	[12]
Payment	The mobility platform should allow for payments for the mobility service	[15, 16]
Openness	The mobility platform should allow any mobility provider or private person to participate	[13]
Nonfunctional requirements		
Authentication	Orders and payments should only be done by authorized persons	[14, 15, 17]
Availability	The mobility platform should be available to at least 99.999%	[13, 14, 17]
Confidentiality	No personal data of users or of vehicles should be disclosed unauthorized	[14, 18, 19]
Cost	The mobility platform should not increase cost for mobility services compared to their current prices	[12, 13, 16]
Non-repudiation	All stored transactions (e.g., vehicle booking, payments) should be unambiguous and assigned to their respective issuer	[18]
Reliable payment	Both parties will be immediately payed according to the conditions of an individual agreement between the vehicle and the user	[12]
Scalability	The mobility platform should automatically scale with an increasing or decreasing amount of transactions	[12, 13]
Secure payment	No financial losses should occur after the payment was made	[12, 13, 17]
Throughput	The mobility platform should allow for at least 1.67 million transactions per second (14 million bookings per day per service provider [20], each for a mean distance of 10.3 km [21] and a payment transaction per meter)	[12, 13]

If new data should be added to the ledger all nodes need to synchronize. To reach an agreement on the data stored on each replication of the ledger a consensus mechanism is employed. Currently, the most prominent DLT concept is blockchain, which follows the idea of cryptographically linked transaction containers, so-called blocks, that include data formatted similar to financial transactions. The blockchain first solved the double-spending problem without the need for a trusted third party, which was prevalent to past attempts of cryptocurrencies. Double-spending describes the process of transferring the same digital coins to different users. A major concern regarding blockchain is its scalability, a challenge that is addressed by more recent DLT concepts, namely block-based directed acyclic graphs (blockDAGs) and transaction-based DAGs (TDAGs) [22, 23]. In both concepts, the individual data items do not form a straight chain but are linked to preceding or succeeding items.

2.2 IOTA

IOTA is a DLT design with high scalability that uses a TDAG, the so-called *Tangle*, as its underlying data structure [11]. In contrast to a blockchain, which has an inherent block creation rate because all participants must agree on a particular chain and discard forks, the Tangle allows different branches of the TDAG to eventually merge. This results in a much higher throughput and theoretically infinite scalability to fulfill high requirements on throughput. Another distinctive feature of IOTA is that transactions do not involve any fees, unlike in Bitcoin or Ethereum.

To transfer assets (e.g., MIOTA) or simple messages (e.g., using zero-value transactions) IOTA nodes issue transactions to a certain receiver address. Transactions issued by nodes constitute the Tangle's set of vertices. The Tangle's set of directed edges between its vertices is based on the approvals of transactions: when a new transaction arrives, it must approve its predecessor's validity [11]. Each transaction is part of a transaction bundle, whereas a bundle comprises one or more transactions. Before a node can issue a transaction bundle, the node must approve two other transaction bundles (i.e., confirm their validity), which will become the direct predecessors of the new transaction bundle. The nodes thereby contribute to the security of the Tangle when issuing transactions. In the course of the transaction approval, it is assumed that the nodes check if the approved transactions are not conflicting. If a node finds a transaction conflicting with the Tangle history (e.g., double spending), the node will not approve the transaction. In IOTA, nodes do not have to reach consensus within a certain period on which transactions are included into the distributed ledger in what order like in Bitcoin or Ethereum. All transactions are stored in the Tangle. However, in the case of conflicting transactions, the nodes need to decide which transactions should be approved. Therefore, the nodes adhere to the following rule: a node runs the tip (a still not approved transaction) selection algorithm multiple times and assumes which of the two transactions is more likely to be indirectly approved by the selected tip [11].

Masked Authenticated Messaging. A prevalent challenge in public DLT designs such as IOTA is confidentiality of stored data. IOTA already integrated an approach to overcome this challenge by introducing messaging channels called Masked Authenticated Messaging (MAM1) [24]. MAM1 channels include cryptographically linked and chronologically ordered sets of transaction bundles stored on the Tangle, where the ownership of the channel is secured by the owner's seed. Users can only add new messages to a MAM1 channel if they know the owner's seed which generates the key pairs required to sign the transactions to be added to the channel [24]. An issue regarding MAM1 pertains to the authentication of messages because the singing key for a message always depends on the preceding channel message.

Micropayments. IOTA supports micropayments where a stream of small sums of MIOTA is efficiently transferred between two IOTA addresses. Such micropayments are realized by flash channels, which are mainly based on off-chain communication between contracting parties. A flash channel is opened by the contracting parties as they create a multi-signature address on the Tangle. Both parties transfer an equal amount of coins to the new multi-signature address to incentivize both parties to not close the flash channel until all micropayments are done. Both parties then locally create a transaction bundle that will distribute the coins transferred to the multi-signature address according to the micropayments made in the flash channel. Each micropayment updates this transaction bundle and is signed by both parties. As both parties already transferred a certain amount of coins to the multi-signature address, the party receiving a payment also receives a certain amount of her own deposit of the multi-signature address and a certain amount of coins of the contracting party's deposit. Since only a final bundle of transactions is appended to the Tangle while all individual transactions are processed off-chain, multiple transactions can be finalized within a short period.

3   Multimodal Mobility Platform

One of the biggest challenges when it comes to creating multimodal transportation services involving multiple service providers is an accurate and fair billing system that is able to invoice the use of different means of transport on a fine granular basis. To address this challenge, the proposed mobility platform creates a permissionless marketplace for transportation service offerings that runs on the IOTA protocol and does not rely on central intermediaries. The mobility platform is open for any service provider or means of transportation and charges service users on a highly accurate per-use basis (e.g. per meter driven) using the MIOTA cryptocurrency. The basic concept of the mobility platform is illustrated in Fig. 1. Vehicles publish their mobility offers and vehicle information publicly in the IOTA Tangle, where they can be queried by users. Once a user found a suitable offer, the settlement of the remaining details for the trip (e.g., destination and price) and the pay-per-use micropayments during the trip are handled off-chain between the user and the vehicle.

3.1   Mobility Platform Use Case

The following use case describes a typical application of the multimodal mobility platform from a user's perspective, in which a person P wishes to travel from location A to location B. To do this, she can use a smartphone app or web interface (cf. Fig. 2) to query all mobility offers available at location A, or more specifically, all offers published on the IOTA address that the mobility platform associates with location A. In this use case, it is assumed that at location A at least one vehicle – a car – has been registered. The registration process essentially only requires sending a check-in message for a vehicle to a location's IOTA address, a process that creates little to no organizational overhead for mobility providers.

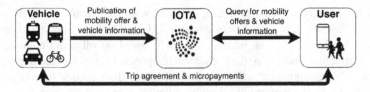

Fig. 1. Multimodal mobility platform concept

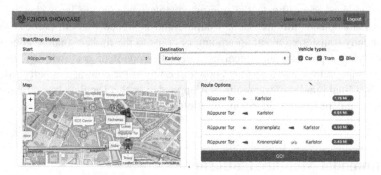

Fig. 2. Screenshot of the graphical user interface of the implemented mobility platform

Once P has chosen one of the available vehicles, in this instance the aforementioned car, she uses the smartphone app to directly connect via Bluetooth and sends her intended destination B to the car. The car then calculates the maximum travel fare to the destination, which is accepted by P. Next, an IOTA micropayment channel is opened between P and the car. After P transfers the first microtransaction, the car door unlocks, and P can start her trip. During the trip, P continues to transfer payments in short intervals over the micropayment channel to the vehicle (e.g., for each meter traveled or elapsed second). Upon arrival at B, the micropayment channel is closed, and the car automatically checks into the stop location nearest to its current location, ready to be booked by another mobility platform user.

3.2 Multi-channel Architecture

The architecture of the multimodal mobility platform features multiple messaging channels as depicted in Fig. 3. The architecture comprises four different channel classes (i.e., vehicle, vehicle info, trip, and micropayment) that are implemented using MAM1, RAAM, or IOTA flash channels. The different channels are linked through unidirectional or bidirectional references. In the following, an overview of the different channel classes and relations between them is provided.

Every vehicle that is available on the mobility platform is represented by exactly one **vehicle channel**. The main function of the channel is to link all information related to the particular vehicle and to document the history of provided mobility offerings. The *channelID* (see Sect. 3.3) serves as the unique identity for the vehicle. All other vehicle meta-information is stored in a second channel, the **vehicle info channel**, which is referenced by the first message in a vehicle channel. All following messages in the vehicle channel are so-called *stop welcome messages*. Each stop welcome message represents a mobility offer (i.e., the vehicles use for one specific trip), that allows to identify the physical location where the vehicle is/was checked-in before the beginning of a trip (i.e., reference to a check-in message by its hash) and whether the vehicle is available for booking or on a trip (i.e., by referencing the respective **trip channel** for the service transaction by its *channelID*). By referencing both a check-in message and the corresponding trip channel, stop welcome messages provide an additional verification that both belong to the same vehicle. This ensures that only the owner of the vehicle channel can create a valid check-in message for the respective vehicle. Stop welcome messages can optionally be encrypted with a password to allow the vehicle to conceal its trip history. Over time, the number of stop welcome messages stored in the vehicle channel will continue to grow, particularly fast in case in case of highly frequented vehicles (e.g., trams or trains). Vehicle channels are therefore based on RAAM that allows to directly access an individual message via its index in $O(1)$ instead of $O(n)$ of an MAM1 channel (see Sect. 3.3).

The second channel class is the **vehicle info channel**, which provides the meta-information for a particular vehicle, like the vehicle type, maximum speed, comfort level, or its schedule (e.g., for trains and trams). This information is stored in the form of *meta-info messages* and updated or removed by overwriting the current state of individual data fields by attaching a new meta-info message with the corresponding information to the channel. This overwrite mechanism requires traversing the channel's messages to retrieve the most current information state. The vehicle info channel is implemented using MAM1.

Trip channels are implemented based on RAAM with a capacity of two messages, whereas only the first message is currently in use. This message, the so-called *departed message*, is sent when the vehicle departs from its current stop place (i.e., the last position where the vehicle was checked-in the mobility platform). The departed message indicates the current state of the trip. Each (potential) stop place is given a **stop address** (81-character Tangle address) that uniquely identifies a location in the physical world (e.g., bus stop) and to which all vehicles in its near proximity are associated. By publishing a check-in message to a stop address, vehicles indicate that they are available at this stop to start a trip. These *check-in messages* reference *stop welcome messages* to prove identity and may references a trip channel to show that the vehicle is not available anymore. Check-in messages carry additional information, like the vehicle's payment address and price information.

The architecture's final channel class is the **micropayment channel** that is based on the IOTA flash channel concept. The channel's purpose is to provide granular pay-per-use mode for vehicle use. IOTA flash channels provide a trustworthy and efficient way for transferring MIOTA off-chain between two parties (see Sect. 2.2). Throughout the duration of the trip, every time a billing unit defined by the vehicle (e.g., distance or time) is used up, the corresponding monetary value is transferred from the user to the vehicle via the micropayment channel.

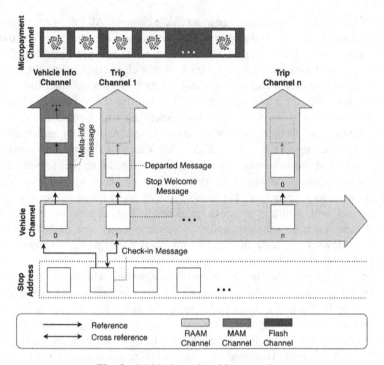

Fig. 3. Multi-channel architecture view

3.3 Random Access Authenticated Messaging

MAM1 comes with a complexity of $O(n)$ since each channel transaction needs to be verified from the first message to the message of interest, which results in an infeasible overhead for read operations as the number of channel messages increases over time. To overcome this challenge, we developed RAAM. Unlike MAM1, RAAM does not have to fetch all messages in a messaging channel from the beginning to ensure that two messages have been created by the same author, which results in a complexity of $O(1)$ instead of $O(n)$. RAAM allows direct access to a particular message in the channel, reducing the complexity for read operations to $O(1)$. In RAAM, each channel message is kept in one transaction bundle and is stored across the respective payload field of one or more transactions of this bundle. To assure integrity of the channel messages all messages are signed using the Winternitz signing scheme [25] based on the Kerl (Keccak) hash function [26]. Only the owner of the private key can create a signature for the message payload that can be verified by the respective public key. For signing different security levels are provided which differ in their signature length. The Winternitz signing scheme [25] is a one-time-signature scheme, which is why each key pair is used only once. Otherwise, attackers could be able to forge correctly signed messages with the same key pair. To authenticate message authors, we employ Merkle Trees [25] to cryptographically proof that someone who holds the private key for a message also holds all public keys used in the RAAM channel. Therefore, all key pairs for the channel messages are created beforehand using a certain seed (cf. Sect. 3.4). The root of the Merkle Tree is the *channelID* of the RAAM channel. Only someone who owns all public keys can create this Merkle root. Thus, if a message is signed with a private key and the corresponding public key was used to create the Merkle Tree, only the Merkle Tree's root is needed to verify the message author's ownership of the RAAM channel.

Message Address Generation. To find channel messages, a message index $msgIdx \subseteq \mathbb{N}_0$ is generated for each channel message. A RAAM channel can either have exactly one or no message at each $msgIdx$ and each message has a unique address $msgAdrs$. The $msgAdrs$ is deterministically calculated by numerically adding the $msgIdx$ to the $channelID$ and subsequently hashing their sum s. Therefore, RAAM channel messages can be read and written in an arbitrary order. The only information needed to read a certain message is the $channelID$ and the $msgIdx$. Messages can be protected from unpermissioned read attempts using a message password. If a message password $msgPass$ is set, it is used for the encryption of the message and for the calculation of $msgAdr = hash_{KECCAK}(s + msgPass)$, where + equals the concatenation of s and $msgPass$. Hashing of the addresses comes with the advantage that it is infeasible to guess an address, if the $channelID$ is not known.

Encryption. To grant read permissions for RAAM messages, RAAM messages can be encrypted using a one-time-pad encryption mechanism, where a password can only be used once. For making this process convenient in usage, RAAM ensures that the password for a message depends on the index of the message. The default password is the index of the message added to the $channelID$. This means that everybody who knows the $channelID$ can read the message. This way, the channel is public. However, it is possible to restrict the readability of messages. In order to do so, a channel password can be set that is applied to all messages. It is also possible to use different passwords for each message. This enables fine granular access restriction.

Branching. RAAM allows flexible and fast access of arbitrary messages. However, that comes with a trade-off. Because the $channelID$ must be created before a message can be published, all key pairs must be created as well. That means that a RAAM channel can only hold a finite number of messages, depending on the height of the Merkle Tree. While the key pairs can always be recreated by saving the seed of the channel, creating key pairs takes time, as it involves $2^n - 1$ hash values with a Merkle Tree's height n. The RAAM protocol allows for connecting RAAM channels through branching, which enables to extent RAAM channels despite their fixed message count. For branching a new $channelID$ is included in a RAAM message (e.g., in the last message available in the RAAM channel). The ownership of the new channel is authenticated because it was referenced by its $channelID$ in the old channel.

3.4 Seed Management

One important challenge that had to be addressed by our prototype is seed management. In the proposed architecture, any vehicle has to be able to create its vehicle channel and all its subchannels (e.g., vehicle info channel and trip channels) using only one seed. Accordingly, the seed for all subchannels must be generated from this one seed. However, because a seed is used to sign n (RAAM) or even infinite (MAM1) messages, it is impossible to use the same seed for two channels, as singularity of the resulting keys could no longer be guaranteed. The problem gets even more complex because of the approach IOTA implements for generating keys from a seed. Due to the one-time-nature of the applied Winternitz signing scheme [25], a signature reveals information that allows to successfully forge valid signatures. That is why a seed of a subchannel cannot be created by using some form of *hash(seed + index)*. Otherwise the signature from key with the index *i* generated from the master seed would reveal information allowing to forge a signature generated by message *i* using the so created sub-seed. Generating sub-seeds therefore require a mapping function that is more sophisticated than IOTA's default hash function Keccak. The hash function should be one-way to ensure the unfeasibility of deducing the root seed form a revealed sub-seed and should be collision-free. To realize an effective solution for the seed management problem, we implemented a function *explode(seed, index)*. *Explode* produces a string of trytes that is longer than the given *seed + index* (ternary sum) and can be fed into Keccak, which in turn provides the one-way-property and collision resistance. To this end, *explode* basically uses *seed + index* (ternary sum) as a trits input and splits it into three equally long parts. The individual parts are circularly permuted three times and then sequentially passed to the hash function. The function's output is a 243 trit string (i.e., 81 trytes), which conforms with the required length of seeds on IOTA. The resulted string can be used to create subchannels whose keys do not collide with the ones from the vehicle channel.

4 Discussion and Conclusion

The presented multimodal mobility platform provides a foundation for innovative mobility services and applications by enabling unified and interoperable booking and payment of mobility offerings. By prototypically implementing the platform based on the functional requirements (cf. Table 1), we were able to demonstrate its technical feasibility and at the same time demonstrate that the introduction of our decentralized mobility concept will benefit both the supply and demand sides. Extant studies on the use of DLT in the field of Mobility as a Service (MaaS) predominantly employ the DLT concept blockchain. The throughput of solutions based on blockchains is restricted to the individual block creation interval (e.g., [23]), which leads to poor scalability. Thus, the use of an asynchronous and alternative DLT concept such as TDAG is more promising in the mobility service context, because they have the potential to better scale than blockchains. Thus, the presented mobility platform fulfills the performance requirements (cf. Table 1) such as throughput and scalability to a higher extent than extant prototypes based on blockchain (e.g., [13, 17]). Additionally, our implementation is more cost-efficient compared to blockchain solutions related to mobility and DLT requiring transaction fees (e.g., [12]), while it is still open for a broad range of users to register and book vehicles.

However, our prototypical implementation comes with several limitations. First, snapshots are an essential part of the IOTA protocol to reduce the amount of data stored on each node. During a snapshot pruning of the transaction history is executed, which is why the complete history of trips with a particular vehicle can only be retrieved from permanodes in the long-term. Furthermore, the presented architecture could only be tested in a simulation because the adoption of DLT is still in an early stage.

Future work should consider the downsides of the current implementation for the seed and key management in RAAM. Due to the fact that the key generation is repeated after each re-start of the local client a vehicle operates, key generation is not performant. Since MAM2 [27] has been presented with very similar features like RAAM, a comprehensive comparison of both approaches is of interest for the improvement of both protocols. MAM2 is very similar to our channeling approach but has been finished after the development of the proposed RAAM. Although MAM2 [27] offers even more features than RAAM, we presented RAAM as an easy to use and lightweight protocol to setup messaging channels in IOTA with a complexity of $O(1)$. Finally, from a design science perspective, further evaluation of the platform implementation in a real-world setting may yield a deeper understanding of the underlying concept's feasibility and applicability and would enable future research on generalizable knowledge for the design of DLT-based mobility platforms (e.g., [28]).

Acknowledgements. Part of this work has been carried out in the scope of the project COOLedger which is funded by the bilateral funding program "Helmholtz-RSF Joint Research Groups". In Germany research is funded by the Helmholtz Association of German Research Centres (Project No. HRSF-0081), in Russia research is supported by the Russian Science Foundation (Project No. 19-41-06301).

References

1. O'Neill, B.C., Ren, X., Jiang, L., Dalton, M.: The effect of urbanization on energy use in India and China in the iPETS model. Energy Econ. **34**, 339–345 (2012)
2. Murray, A.T., Davis, R., Stimson, R.J., Ferreira, L.: Public Transportation Access. Transp. Res. Part D Transp. Environ. **3**, 319–328 (1998)
3. Saif, M.A., Zefreh, M.M., Torok, A.: Public transport accessibility: a literature review. Period. Polytech. Transp. Eng. **47**, 36–43 (2018)
4. Benlian, A., Kettinger, W.J., Sunyaev, A., Winkler, T.J., Guest Editors: Special section: the transformative value of cloud computing: a decoupling, platformization, and recombination theoretical framework. J. Manage. Inf. Syst. **35**, 719–739 (2018)
5. Martins, V.W.B., Anholon, R., Quelhas, O.L.G.: Sustainable transportation methods. In: Leal Filho, W. (ed.) Encyclopedia of Sustainability in Higher Education, pp. 1–7. Springer, Cham (2019). https://doi.org/10.1007/978-3-319-63951-2
6. Strasser, M., Albayrak, S.: Smart city reference model: interconnectivity for on-demand user to service authentication. Int. J. Serv. Oper. Manage. **3**, 17 (2016)
7. Haahtela, T., Viitamo, E.: Searching for the potential of MaaS in commuting – comparison of survey and focus group methods and results. In: Proceedings of the International Conference on Mobility as a Service (2017)
8. Oostendorp, R., Krajzewicz, D., Gebhardt, L., Heinrichs, D.: Intermodal mobility in cities and its contribution to accessibility. Appl. Mobilities **4**, 183–199 (2019)
9. Strasser, M., Weiner, N., Albayrak, S.: The potential of interconnected service marketplaces for future mobility. Comput. Electr. Eng. **45**, 169–181 (2015)
10. Catapult Transport Systems: Blockchain Disruption in Transport - Are You Decentralized Yet? (2018). https://s3-eu-west-1.amazonaws.com/media.ts.catapult/wp-content/uploads/2018/06/06105742/Blockchain-Disruption-in-Transport-Concept-Paper.pdf. Accessed 15 Jan 2019

11. Popov, S.: The Tangle (2018). https://assets.ctfassets.net/r1dr6vzfxhev/2t4uxvsIqk0EUau6g2sw0g/45eae33637ca92f85dd9f4a3a218e1ec/iota1_4_3.pdf. Accessed 11 June 2018

12. Pustisek, M., Kos, A., Sedlar, U.: Blockchain based autonomous selection of electric vehicle charging station. In: 2016 International Conference on Identification, Information and Knowledge in the Internet of Things (IIKI), IEEE, Beijing, pp. 217–222 (2016)

13. Naser, F.: Review: the potential use of blockchain technology in railway applications: an introduction of a mobility and speech recognition prototype. In: 2018 IEEE International Conference on Big Data, Seattle, WA, USA, pp. 4516–4524 (2018)

14. Yahiatene, Y., Rachedi, A.: Towards a blockchain and software-defined vehicular networks approaches to secure vehicular social network. In: 2018 IEEE Conference on Standards for Communications and Networking (CSCN), IEEE, Paris, pp. 1–7 (2018)

15. Brousmiche, K.L., Heno, T., Poulain, C., Dalmieres, A., Hamida, E.B.: Digitizing, securing and sharing vehicles life-cycle over a consortium blockchain: lessons learned. In: 2018 9th IFIP International Conference on New Technologies, Mobility and Security (NTMS), pp. 1–5 (2018)

16. Xiong, Z., Feng, S., Niyato, D., Wang, P., Han, Z.: Optimal pricing-based edge computing resource management in mobile blockchain. In: 2018 IEEE International Conference on Communications (ICC), IEEE, Kansas City, MO, pp. 1–6 (2018)

17. Lopez, D., Farooq, B.: A blockchain framework for smart mobility. In: 2018 IEEE International Smart Cities Conference (ISC2), IEEE, Kansas City, MO, USA, pp. 1–7 (2018)

18. Xu, C., Liu, H., Li, P., Wang, P.: A remote attestation security model based on privacy-preserving blockchain for V2X. IEEE Access 6, 67809–67818 (2018)

19. Dehling, T., Sunyaev, A.: Secure provision of patient-centered health information technology services in public networks—leveraging security and privacy features provided by the German nationwide health information technology infrastructure. Electron. Markets 24, 89–99 (2014)

20. Uber: Statistics Facts & Figures as of December 2018 (2019). https://www.uber.com/en-GB/newsroom/company-info/. Accessed 15 Apr 2019

21. SherpaShare: Uber trips are becoming longer and faster, but are they more profitable? In: SherpaShare Blog (2016). https://web.archive.org/web/20181208125258/http://www.sherpashareblog.com/tag/uber-trip-distance/. Accessed 15 Feb 2019

22. Kannengießer, N., Lins, S., Dehling, T., Sunyaev, A.: Mind the Gap: Trade-Offs between Distributed Ledger Technology Characteristics (2019). arXiv:190600861 [cs]

23. Kannengießer, N., Lins, S., Dehling, T., Sunyaev, A.: What does not fit can be made to fit! Trade-offs in distributed ledger technology designs. In: 52nd Hawaii International Conference on System Sciences (2019)

24. Handy, P.: Introducing masked authenticated messaging. In: IOTA News (2017). https://blog.iota.org/introducing-masked-authenticated-messaging-e55c1822d50e. Accessed 14 Mar 2019

25. Buchmann, J., Dahmen, E., Szydlom, M.: Hash-based digital signature schemes. In: Bernstein, D.J., Buchmann, J., Dahmen, E. (eds.) Post-Quantum Cryptography, pp. 35–93. Springer, Heidelberg (2009). https://doi.org/10.1007/978-3-540-88702-7_3

26. Bertoni, G., Daemen, J., Peeters, M., Van Assche, G., Van Keer, R.: Keccak implementation overview (2012)

27. Martinez, T.: MAM2 (2019). https://web.archive.org/save/https://github.com/iotaledger/entangled/blob/develop/mam/spec.pdf. Accessed 12 Apr 2019

28. Sturm, B., Sunyaev, A.: Design principles for systematic search systems: a holistic synthesis of a rigorous multi-cycle design science research journey. Bus. Inf. Syst. Eng. 61, 91–111 (2019)

Semantic Interoperability in IoT: A Systematic Mapping

Saymon Castro de Souza[1,2](✉) [iD] and José Gonçalves Pereira Filho[2](✉) [iD]

[1] Federal Institute of Espírito Santo, Vitória, ES, Brazil
saymon@ifes.edu.br
[2] Federal University of Espírito Santo, Vitória, ES, Brazil
zegonc@inf.ufes.br

Abstract. There has been a growing interest in adopting ontologies to address semantic interoperability in the domain of Internet of Things (IoT). Recent research claims that the lack of semantic interoperability support in IoT infrastructures hinders the potential of IoT for offering large scale value-added services. This article explores this fundamental issue through a systematic mapping of literature. The mapping intends to answer a set of research questions and identifies research gaps and trends that will guide future research in IoT semantic interoperability modeling.

Keywords: Internet of Things · Ontology ·
Semantic interoperability · Systematic mapping

1 Introduction

The initiative of bringing intelligence on everyday objects and connect them to the Internet, has launched a new phase of the Internet development, in which the main promise is the realization of the Internet of Things (IoT). In this vision of Future Internet, real-world objects - i.e, things of the physical world, such as tools, home appliances, personal and clothing items - are embedded with processing, storage, sensing, and wireless communication capabilities, integrated to the Internet, and seen by the applications as autonomous entities with proactive behavior, knowledge about the surrounding context, as well as collaborate with each other to reach a common goal. The basic idea is to bring the state of things that form the physical world into the applications, making these applications aware, in real-time, of the changes observed in the physical world and, consequently, allowing to promote faster adaptations and responses to these changes.

The research in Internet of Things has been supported by collaborative efforts of academia, industry and standards organizations, in various communities. The ongoing initiative towards the construction of the Internet of Things is

---

The authors are grateful for the financial support by FAPES (grant No. 21/2018 - Universal).

O. Galinina et al. (Eds.): NEW2AN 2019/ruSMART 2019, LNCS 11660, pp. 53–64, 2019.
https://doi.org/10.1007/978-3-030-30859-9_5

sustained by continued progress in hardware platforms, which accurately allow the sensing, processing and transmission of intelligent data, as well as the creation of software infrastructures to access, storage, manipulate and sharing of the data obtained from the devices [2]. However, the Internet of Things is far from reaching the full convergence of IT services and technologies. Technological challenges regarding IoT hardware heterogeneity, data security, connectivity and communication models, the lack of standards, the processing and analysis of huge volumes of data in real-time, among others, constitute examples of obstacles to the expansion of the Internet of Things on a global scale [40].

In recent years, academia and IoT industry, started driving research efforts on a fundamental issue: *IoT semantic interoperability*. According to the IERC Report "IoT Semantic Interoperability: Research Challenges, Best Practices, Recommendations and Next Steps" [28], semantic interoperability is the next step in consolidating the Internet of Things, and it is already happening. According to [39], the IoT's promise in terms of interoperability has not yet been delivered and the lack of semantic interoperability support hinders the potential of IoT for developing and offering large-scale value-added services.

A promising approach to deal with scenarios where interoperability is a fundamental issue is the use of *Ontologies* to enrich the information model. Ontologies [29] have been widely used by Software Engineering and Information Systems communities to enrich conceptual models and to provide semantic expressiveness to information, thus promoting the interchange between applications, services and systems at different levels of abstraction. Likewise, in IoT domain, there has also been growing interest in adopting ontologies as a means to address semantic interoperability issues. In fact, in the last decade, many formalisms of knowledge representation were developed and also adopted in Internet of Things projects [4, 8, 19, 21, 25, 28, 32, 36].

This article presents a systematic mapping of the literature to analyze initiatives that adopt ontologies for IoT semantic enrichment. Six research questions were formulated to investigate four main aspects of the selected studies: (i) the adoption of high-level ontologies; (ii) the representation languages used to construct the ontologies; (iii) the methodologies adopted in the ontology design process and (iv) the characteristics of these ontologies. These questions aimed to characterize and identify potential gaps and research trends that are not fully explored in the works presented in the scope of this mapping.

The paper is structured as follows: Sect. 2 introduces the dimensions of interoperability and the theoretical basis of ontologies. It also clarifies some terminological aspects. Section 3 describes the research method and the protocol used in this systematic mapping. Section 4 gives a summary of the data collected from the studies in the light of the research questions and discusses important points of data analysis, which may be useful for further research investigations in IoT. Section 5 concludes the paper, presenting the final considerations and future research perspectives.

2   Background

Interoperability, in a broader perspective, can be defined as the ability to operate together or, more specifically in the context of computational systems, the ability to exchange data and services between applications or application components [42]. According to [22], there are several dimensions of integration that contribute to characterize the notion of interoperability. This integration can happen in different levels, namely: hardware level, platform level, syntactic level and semantic level. The hardware level involves the integration of different computers, computer networks, etc. At this level network protocols are used so that two or more networks can communicate. The platform level deals with operating systems, database systems, and so on. The syntactic level refers to the way data models are written. Finally, the semantic level encompasses the intended meaning of the concepts in the data schema.

In the field of Wireless Sensor Networks - one of the classical Internet of Things' communication infrastructures - efforts have been made to build standards to mitigate problems related to interoperability. The Sensor Web Enablement (SWE) [7], for example, is an Open Geospatial Consortium (OGC) initiative that defined data and Web service encoding standards to access and store sensor data. These standards, such as SensorML [6] and O&M [10], provide syntactic interoperability [33]; however, an additional layer is needed to deal with semantic compatibility [24]. Semantic Web Technologies have also been proposed as a means to enable interoperability between sensors and detection systems, and can be used in isolation or as a way to expand SWE standards in the form of Semantic Sensor Web [33].

According to [39], IoT current research efforts aim at connecting heterogeneous devices to a common platform and, once connected, the devices are able to exchange data, but the semantics of the data and the characteristics of those devices are not described in a machine-readable form. This set of factors hinders the potential of IoT in order to allow the development of large-scale value-added services.

As aforementioned, a promising approach in scenarios where interoperability is a fundamental requirement is the use of Ontologies (Gruber, 1995) (Calhau and de Almeida Falbo, 2010). The term *Ontology* has its origin in philosophy and is used in this field to refer to a discipline and also as a domain independent category system [17]. A widely accepted definition in computing science community describes ontology as *"a formal specification of a shared conceptualization"* [5].

There are different ways of classifying ontologies, one of them according to their level of generality [15]: high-level ontologies, domain ontologies, task ontologies and application ontologies. High-level ontologies or foundation ontologies describe very general concepts, regardless of a particular problem or domain, such as an object, event, or action. Domain ontologies describe the conceptualization related to a generic domain, e.g. law or biology. Task ontologies describe the conceptualization related to a generic task, such as sales or service. Application ontologies describe concepts that are dependent on a domain and a specific task.

High-level ontologies are useful in designing lower-level ontologies (e.g., domain and application ontologies) because they provide important ontological distinctions. Among the well-known foundation ontologies, it is possible to cite DOLCE [14], SUMO [30] and UFO [17].

Another important way of classifying ontologies is the one proposed by [18], also discussed in [11], in which an approach that distinguishes reference ontologies and operational ontologies is defended. Reference ontology is developed with the objective of making the best possible description of the domain, being a special kind of conceptual model, an engineering artifact with the additional requirement of representing a consensus model within a community. On the other hand, once users have already agreed to a common conceptualization, operational versions of the reference ontology can be created. Unlike reference ontologies, operational ontologies are designed with the focus on guaranteeing desirable computational properties.

Important aspects should be considered in relation to the adoption of languages for the representation of reference and operational ontologies. In [16], for example, a discussion is made about ontological-level languages and epistemological-level languages. The author argues that ontological-level languages are more appropriate for clear, precise and unambiguous representation of knowledge about the domain but tend not to be computationally treatable since this type of language is more appropriately used to represent reference ontologies. Provided that guaranteeing computational properties according to the needs of the application is desired - a typical scenario of operational ontologies - it is more suitable to use languages at a logical/epistemological level, which are designed with that objective in mind.

3   Research Method

The method of research applied to this systematic review was based on [26,27] and inspired by other works [29,38]. The mapping aimed to answer the following research questions, considering the context of IoT initiatives:

- (RQ1) How broad has the adoption of ontologies in IoT initiatives been over the years?
- (RQ2) Which studies use or propose the use of high-level ontologies? What are these ontologies?
- (RQ3) What are the forms of representation used in the construction of ontologies in the context of IoT?
- (RQ4) What are the types of ontologies (considering their level of generalization) that have been employed? What representation languages were used in each level?
- (RQ5) Do the studies follow a methodological approach to develop the ontologies? What are these methodologies?
- (RQ6) How many studies reuse existing ontologies in their approaches? What are these ontologies?

These questions aim to investigate: (i) the interest, influences and tendencies of the scientific community, over time, regarding the use of ontologies to deal with the interoperability problem in IoT initiatives; (ii) whether and how high-level ontologies are being used to promote semantic interoperability in the context of IoT, considering that they describe very general concepts, regardless of particular domain; (iii) if the representation of an ontology can influence the quality of the produced model, according to the computational or expressivity requirements; (iv) if the adoption of a systematic process for the construction of ontologies tends to generate models with better quality [35]; (v) if reusing ontologies has the benefits of increasing semantic interoperability between systems and can reduce application development time [20].

Sources. Automatic searches were used to collect the studies. The search was applied in five representative electronic databases, frequently used in systematic reviews. The sources are: IEEE Xplore (http://ieeexplore.ieee.org), ACM Digital Library (http://dl.acm.org), Scopus (http://www.scopus.com), Science Direct (http://www.sciencedirect.com), Compendex (http://www.engineeringvillage2.org).

Search String. For the search string, two sets of terms were used (Table 1):

Table 1. Search string.

Part 1 - Search Terms	Part 2 - Search Terms
(("iot" OR "internet of things" OR "sensor" OR "wsn"OR "wireless sensor network"OR "sensor network"OR "web of things")	AND ("ontology" OR "reference model"OR "semantic model" OR "information model")

3.1 Selected Works

The search process was conducted in September 2017 covering the five mentioned electronic databases. The studies published until that date have been the focus of our recovery process. A total of 2006 records were retrieved: 91 from ACM, 957 from Compendex, 443 from IEEE Xplorer, 107 from Science Direct, and 408 from Scopus.

The entire process of searching and selecting papers was repeated in November 2018 to update our studies and trends. Disregarding all articles found in the first search process, a total of 472 records were retrieved: 21 from ACM, 204 from Compendex, 78 from IEEE Xplorer, 53 from Science Direct, and 116 from Scopus. In the first step of the selection process we eliminated duplicate studies (143 articles). In the second stage, the inclusion and exclusion criteria were applied considering only the title and the summary, resulting in a reduction of

66.22%. In the third step, considering the complete reading of the studies and applying the inclusion (IC) and exclusion (EC) criteria, 32 articles were selected.

Table 2 summarizes the stages and their results, showing a progressive reduction in the number of studies throughout the selection process (from 2478 to 103 studies, with a reduction rate of about 95.84%).

All references obtained in the consultation process were made available in the Mendeley Dataset[1].

Table 2. Results of selection process (per stage).

Stage	Criteria	Analyzed content	Initial N° of studies	Final N° of studies	Rate of reduction
		September 2017			
1st	Eliminating duplications, and wrongly retrieved studies by the engines	Title and abstract	2006	1238	38.28%
2nd	IC1, EC1, EC2, EC3, and EC4	Title and abstract	1238	226	81.74%
3rd	IC1, EC2, EC3, EC4, EC5 and EC6	Whole text	226	70	69.02%
		November 2018			
1st	Eliminating duplications, and wrongly retrieved studies by the engines	Title and abstract	472	329	30.30%
2nd	IC1, EC1, EC2, EC3, and EC4	Title and abstract	329	121	66.22%
3rd	IC1, EC2, EC3, EC4, EC5 and EC6	Whole text	121	32	85.84%

4   Synthesis of Data and Discussion

We start the discussion analyzing the distribution of relevant studies over the past 12 years, from 2007 to 2018. The results clearly indicate a tendency towards the growth of IoT solutions that employ models for syntactic and semantic descriptions. In 2007, [1] first raised the concern about data interoperability in heterogeneous sensor networks. The authors then proposed an architecture and a primitive syntactic description of data, adopting XML as a data exchange language. In this same year, the Open Geospatial Consortium (OGC) presented the Observations and Measures model O&M [10] and the Sensor Model Language sensorML [6], a syntactic description pattern for sensor networks, which

---

[1] http://dx.doi.org/10.17632/pghwxhnrmd.3.

reflected the kind of approach used to address the interoperability challenges of that time. These works inaugurate what became more evident later in 2010, when we observed the first publications that effectively addressed the issue of semantic interoperability using ontologies as the basis for conceptual modeling in the IoT domain. In fact, from 2010 on there has been a growing research interest toward semantics, since the syntactic description imposed many limitations to reach effective data interoperability [24].

The W3C Semantic Sensor Network Incubator Group (W3C-XG) then began working to integrate and align data-related interoperability research on sensor networks with Semantic Web technologies. The result of this work group was the development of the Semantic Sensor Network, SSN ontology [8]. Since this work, many research studies have been carried out to extend SSN, with the intention of achieving varied application domains. Considering the protocol adopted in this review, 50 studies have somehow reused the SSN ontology, demonstrating the importance of this ontology and the standardization efforts in the IoT community.

The extensions of SSN ontology has taken three directions [41]: (i) extend SSN to meet a specific application; (ii) combine SSN with another ontology to solve a real problem; and (iii) add new concepts to SSN in a way that makes it more general to use, allowing its use to be leveraged in different domains.

In contrast to the aforementioned directions, the work by [3] proposed a simplification of existing concepts in SSN ontology in order to enhance its use in real-world applications, especially with regard to performance requirements. In the same direction, SSN-XG started rethinking the SSN ontology based on the lessons learned in recent years. This effort has resulted in the Sensor, Observation, Sample and Actuator ontology (SOSA) [23]. SOSA provides a formal, light, and general-purpose specification for modeling the interaction between the entities involved in observation, actuation, and sampling in IoT scenarios.

In summary, we can observe that the use of ontology-based models to manage semantic interoperability issues in IoT has generated great interest from researchers. The analysis of the results clearly reveal a tendency of growth throughout the studied years, and demonstrates an incipient movement in the direction of the adoption of lightweight ontologies.

Another important issue in this systematic review regards the use of foundation ontologies in IoT. It was observed that only 15 studies explicitly mentioned the use of foundation ontologies, in contrast to the 88 studies that did not mention use of these high-level ontologies. However, this type of ontology plays an important role in establishing conceptual bases to approach fundamental concepts such as time, event, role, property, and so on. Such distinctions provided by foundation ontologies are used to improve the quality of conceptual models [18]. In addition, while the IoT community seems to start advocating the use of lightweight ontologies and "easiness of use" in order to favor wide adoption and reuse, as in the aforementioned study [3], aspects related to the conception of domain ontologies, as advocated by [43], support the use of foundation ontologies in order to clearly and precisely reproduce the description of domain elements. Therefore, the trade-off between ontological basis, expressiveness, and computational treatability, in the conception of ontologies for IoT, emerged as an important and fundamental challenge to be overcome.

Concerning the knowledge representation languages used for building ontologies in IoT, the results of this mapping has shown a predominance of the use of Semantic Web technologies, especially OWL (55%), RDF (16%), and XML (2%). The analysis revealed a tendency toward the use of machine interpretable languages, in detriment to other approaches of higher level representation (19%) such as graphs, taxonomies, and conceptual maps. It is believed that this trend is related to the strong influence of the standards proposed by the OGC Sensor Web Enablement (SWE) group [7,33], which encouraged the use of standards of coding and services aligned with semantic web technologies to access sensor data on the Internet. SSN-XG, the W3C group responsible for the development of the SSN ontology was one of the works that were influenced by OCG's SWE group contributions [8]. As previously noted, [23] recently presented a complete revision of the SSN ontology, giving rise to the lightweight ontology SOSA. These works have a strong impact on the scientific community, and both advocate the use of machine interpretable languages aligned with Semantic Web fundamentals to more efficiently deal with semantic interoperability issues. In fact, the languages adopted in the aforementioned works (OWL and RDF) represent 71% of the works selected in this review.

Particularly, the wide adoption of OWL (55%) by the IoT community evidences, so far, the choice for languages of epistemological level. However, although interesting to facilitate automatic logical reasoning, this type of language is not suitable for real-world representation in the modeling stage [16]. The use of OWL, as well as UML and the category of knowledge representation, may result in the design of ambiguous models. On the contrary, ontological level languages restrict the meaning of the elements of the domain promoting the creation of unambiguous models that faithfully represent the elements of the real world. However, as this mapping has shown, the use of ontological level languages is still extremely timid in IoT, which, in turn, presents itself as an interesting research gap.

Regardless of the level of ontology, this mapping also revealed that the use of methodological approaches to build ontologies are still too timid and little explored in IoT, even given the difficulties for the construction and maintenance of ontologies. The use of methodological frameworks, advocated by Ontology Engineering discipline, could give IoT system developers important benefits, such as structuring the process; reducing its complexity to manageable tasks; clarifying the responsibilities of the participants in the process; increasing their traceability; and enabling systematic quality assurance procedures [9,35].

We analyzed which studies made explicit reference to the use of a methodology for the development of ontologies. It was possible to conclude that most of the studies (94) did not explicitly mention the use of methodologies in the ontology development process. Among the eight studies that adopted methodologies, we have the following outcome: Ontology Development 101 [31]; NeOn [37]; Methontology [12]; and Linked Open Terms. Recently, [13] emphasized the importance of methodological approaches in the development of ontologies in IoT. The authors also emphasized the limited reuse of existing ontologies, as we have observed in this mapping. In the same direction, [20] highlighted the deficiency of methodological approaches in IoT. As mentioned in the ontological-level languages discussion, this situation indicates a need for new research efforts to address the research gap. Particularly concerning the reuse of ontologies, this mapping has shown that most of the analyzed works (59%) did not explicitly indicate the reuse of ontologies. According to [20], the reuse of ontologies has the benefits of increasing the semantic interoperability between systems, since it allows the sharing of knowledge of the domain using a common vocabulary. This could reduce application development time and, additionally, potentially improve the quality of reused ontologies because they are continuously reviewed and evaluated by several parties through reuse [34]. Also, [20] proposed the development of an ontology catalog system and a methodology to encourage reuse. Among the works that reported reuse of existing ontologies, most of them (50) adopted the SSN ontology, reinforcing the importance of this ontology to state-of-the-art IoT domain modeling.

5 Conclusions

This work presented a systematic mapping of IoT literature, aiming to carry out a comprehensive review of scientific studies that propose ontologies as a semantic interoperability solution in IoT. Semantic interoperability is a concern highlighted by IoT research communities, industrial consortia and standards organizations with the intent of creating the next generations of interconnected IoT systems.

The results of this mapping show a growing interest towards semantic interoperability. The analysis and observations made during the execution of this work show aspects that can guide future research. In particular, we highlight the minor usage of Ontology Engineering approaches to deal with the different aspects related to the adoption of ontologies and to promote interoperability in large and integrated solutions. More importantly, the concept of interoperability is still restricted to a single dimension and should be discussed more broadly by the IoT community, considering different dimensions of interoperability, as presented in [22].

As future work, we intend to deepen our analysis on these different dimensions that characterize interoperability and also on the use of high-level ontologies in IoT initiatives. In particular, we intend to explore new approaches to semantic annotation of data and investigate how methodologies, languages, foundation ontologies, lightweight ontologies and reuse of ontologies influence semantic data annotation solutions. This choice is influenced directly by the results of this mapping, which showed that: (i) most studies do not use foundation ontologies in their models; (ii) there is a predilection for epistemological level languages; (iii) semantic web technologies continue to be widely adopted, regardless of the level of generality of the proposed ontology; and, (iv) despite the relevance and influence of some specific ontology-based standards there is a need for additional research efforts to promote the reuse of existing ontologies in IoT solutions.

References

1. Aberer, K., Hauswirth, M., Salehi, A.: Infrastructure for data processing in large-scale interconnected sensor networks. In: 2007 International Conference on Mobile Data Management, pp. 198–205 (2007). https://doi.org/10.1109/MDM.2007.36
2. Barnaghi, P., Wang, W., Henson, C., Taylor, K.: Semantics for the Internet of Things: early progress and back to the future. Int. J. Semant. Web Inf. Syst. (IJSWIS) 8(1), 1–21 (2012)
3. Bermudez-Edo, M., Elsaleh, T., Barnaghi, P., Taylor, K.: IoT-lite: a lightweight semantic model for the Internet of Things. In: 2016 International IEEE Conferences on Ubiquitous Intelligence Computing, Advanced and Trusted Computing, Scalable Computing and Communications, Cloud and Big Data Computing, Internet of People, and Smart World Congress (UIC/ATC/ScalCom/CBDCom/IoP/SmartWorld), pp. 90–97 (2016). https://doi.org/10.1109/UIC-ATC-ScalCom-CBDCom-IoP-SmartWorld.2016.0035
4. Bonino, D., Corno, F., De Russis, L.: A semantics-rich information technology architecture for smart buildings. Buildings 4, 880–910 (2014). https://doi.org/10.3390/buildings4040880. http://www.mdpi.com/2075-5309/4/4/880/
5. Borst, W.N.: Construction of Engineering Ontologies for Knowledge Sharing and Reuse. Ph.D. thesis, University of Twente, Netherlands (1997)
6. Botts, M., Robin, A.: OpenGIS Sensor Model Language (SensorML) Implementation Specification; OpenGIS Implementation Specification OGC 07-000. Open Geospatical Consortium Inc. (2007)
7. Botts, M., Percivall, G., Reed, C., Davidson, J.: OGC® sensor web enablement: overview and high level architecture. In: Nittel, S., Labrinidis, A., Stefanidis, A. (eds.) GSN 2006. LNCS, vol. 4540, pp. 175–190. Springer, Heidelberg (2006). https://doi.org/10.1007/978-3-540-79996-2_10
8. Compton, M., et al.: The SSN ontology of the W3C semantic sensor network incubator group. J. Web Semant. 17, 25–32 (2012). https://doi.org/10.1016/j.websem.2012.05.003
9. Corcho, O., Fernández-López, M., Gómez-Pérez, A.: Methodologies, tools and languages for building ontologies. Where is their meeting point? Data Knowl. Eng. 46(1), 41–64 (2003)
10. Cox, S.: Observations and Measurements-Part 1-Observation schema (OpenGIS Implementation Standard OGC 07-022r1). Open Geospatial Consortium Inc., Technical report 8 (2007)
11. Falbo, R.d.A., Guizzardi, G., Gangemi, A., Presutti, V.: Ontology patterns: clarifying concepts and terminology. In: Proceedings of the 4th Workshop on Ontology and Semantic Web Patterns (2013)
12. Fernández-López, M., Gómez-Pérez, A., Juristo, N.: METHONTOLOGY: from ontological art towards ontological engineering. In: Proceedings of the Ontological Engineering AAAI-97 Spring Symposium Series. American Asociation for Artificial Intelligence (1997). http://oa.upm.es/5484/

13. Fruhwirth, T., Kastner, W., Krammer, L.: A methodology for creating reusable ontologies. In: Proceedings - 2018 IEEE Industrial Cyber-Physical Systems, ICPS 2018, pp. 65–70. Saint Petersburg, Russia (2018). http://dx.doi.org/10.1109/ICPHYS.2018.8387639

14. Gangemi, A., Guarino, N., Masolo, C., Oltramari, A., Schneider, L.: Sweetening ontologies with DOLCE. In: Gómez-Pérez, A., Benjamins, V.R. (eds.) EKAW 2002. LNCS (LNAI), vol. 2473, pp. 166–181. Springer, Heidelberg (2002). https://doi.org/10.1007/3-540-45810-7_18

15. Guarino, N.: Formal Ontology in Information Systems: Proceedings of the First International Conference (FOIS'98), 6–8 June, Trento, Italy, vol. 46. IOS press (1998)

16. Guarino, N.: The ontological level: revisiting 30 years of knowledge representation. In: Borgida, A.T., Chaudhri, V.K., Giorgini, P., Yu, E.S. (eds.) Conceptual Modeling: Foundations and Applications. LNCS, vol. 5600, pp. 52–67. Springer, Heidelberg (2009). https://doi.org/10.1007/978-3-642-02463-4_4

17. Guizzardi, G.: Ontological Foundations for Structural Conceptual Models. CTIT, Centre for Telematics and Information Technology (2005)

18. Guizzardi, G.: On ontology, ontologies, conceptualizations, modeling languages, and (meta) models. Front. Artif. Intell. Appl. 155, 18 (2007)

19. Gyrard, A., Bonnet, C., Boudaoud, K., Serrano, M.: Assisting IoT projects and developers in designing interoperable semantic web of things applications. In: 2015 IEEE International Conference on Data Science and Data Intensive Systems (DSDIS), pp. 659–666. IEEE (2015)

20. Gyrard, A., Zimmermann, A., Sheth, A.: Building IoT based applications for smart cities how can ontology catalogs help. IEEE Internet Things J. (2018). http://dx.doi.org/10.1109/JIOT.2018.2854278

21. Hasan, S., Curry, E.: Approximate semantic matching of events for the Internet of Things. ACM Trans. Internet Technol. (TOIT) 14(1) (2014). Article 2. https://doi.org/10.1145/2633684. http://www.edwardcurry.org/publications/hasan_TOIT_2014.pdf

22. Izza, S.: Integration of industrial information systems: from syntactic to semantic integration approaches. Enterp. Inf. Syst. 3(1), 1–57 (2009)

23. Janowicz, K., Haller, A., Cox, S.J.D., Le Phuoc, D., Lefrancois, M.: SOSA: a lightweight ontology for sensors, observations, samples, and actuators. J. Web Semant. (2018). http://dx.doi.org/10.1016/j.websem.2018.06.003

24. Janowicz, K., Schade, S., Bröring, A., Keßler, C., Maué, P., Stasch, C.: Semantic enablement for spatial data infrastructures. Trans. GIS 14(2), 111–129 (2010)

25. Jayaraman, P.P., Calbimonte, J.P., Quoc, H.N.M.: The schema editor of OpenIoT for semantic sensor networks. In: CEUR Workshop Proceedings, Bethlehem, PA, United States, vol. 1488, pp. 25–30 (2015)

26. Kitchenham, B.: Procedures for performing systematic reviews. Keele UK Keele Univ. 33(2004), 1–26 (2004)

27. Kitchenham, B.A., Budgen, D., Brereton, O.P.: Using mapping studies as the basis for further research-a participant-observer case study. Inf. Softw. Technol. 53(6), 638–651 (2011)

28. Serrano, M., Barnaghi, P., Carrez, F., Cousin, P., Vermesan, O., Friess, P.: Internet of Things IoT semantic interoperability: Research Challenges, Best Practices, Recommendations and Next Steps (2015)

29. Nardi, J.C., de Almeida Falbo, R., Almeida, J.P.A.: A panorama of the semantic EAI initiatives and the adoption of ontologies by these initiatives. In: van Sinderen, M., Oude Luttighuis, P., Folmer, E., Bosems, S. (eds.) IWEI 2013. LNBIP, vol. 144, pp. 198–211. Springer, Heidelberg (2013). https://doi.org/10.1007/978-3-642-36796-0_17

30. Niles, I., Pease, A.: Towards a standard upper ontology. In: Proceedings of the International Conference on Formal Ontology in Information Systems-Volume 2001, pp. 2–9. ACM (2001)

31. Noy, N.F., Mcguinness, D.L.: Ontology Development 101: A Guide to Creating Your First Ontology. Technical report (2001)

32. Qu, C., Liu, F., Tao, M.: Ontologies for the transactions on IoT. Int. J. Distrib. Sens. Netw. **2015** (2015). https://doi.org/10.1155/2015/934541

33. Sheth, A., Henson, C., Sahoo, S.S.: Semantic sensor web. IEEE Internet Comput. **12**(4), 78–83 (2008). https://doi.org/10.1109/MIC.2008.87. http://doi.ieeecomputersociety.org/10.1109/MIC.2008.87

34. Simperl, E.: Reusing ontologies on the semantic web: a feasibility study. Data Knowl. Eng. **68**(10), 905–925 (2009). https://doi.org/10.1016/j.datak.2009.02.002. http://www.sciencedirect.com/science/article/pii/S0169023X0900007X

35. Simperl, E.P.B., Tempich, C.: Ontology engineering: a reality check. In: Meersman, R., Tari, Z. (eds.) OTM 2006. LNCS, vol. 4275, pp. 836–854. Springer, Heidelberg (2006). https://doi.org/10.1007/11914853_51

36. Song, Z., Cárdenas, A.A., Masuoka, R.: Semantic middleware for the internet of things. In: 2010 Internet of Things, IoT 2010 (2010). https://doi.org/10.1109/IOT.2010.5678448

37. Suárez-Figueroa, M.C., Gómez-Pérez, A., Fernández-López, M.: The NeOn methodology for ontology engineering. In: Suárez-Figueroa, M.C., Gómez-Pérez, A., Motta, E., Gangemi, A. (eds.) Ontology Engineering in a Networked World, pp. 9–34. Springer, Heidelberg (2012). https://doi.org/10.1007/978-3-642-24794-1_2

38. Teixeira, S., Agrizzi, B.A., Filho, J.G.P., Rossetto, S., de Lima Baldam, R.: Modeling and automatic code generation for wireless sensor network applications using model-driven or business process approaches: a systematic mapping study. J. Syst. Softw. **132**, 50–71 (2017). https://doi.org/10.1016/j.jss.2017.06.024

39. Thuluva, A.S., Anicic, D., Rudolph, S.: IoT semantic interoperability with device description shapes. In: Gangemi, A., et al. (eds.) ESWC 2018. LNCS, vol. 11155, pp. 409–422. Springer, Cham (2018). https://doi.org/10.1007/978-3-319-98192-5_56

40. Wang, F., Hu, L., Zhou, J., Zhao, K.: A survey from the perspective of evolutionary process in the Internet of Things. Int. J. Distrib. Sens. Netw. **2015**, 1–9 (2015). https://doi.org/10.1155/2015/462752. http://www.hindawi.com/journals/ijdsn/2015/462752/

41. Wang, X., Zhang, X., Li, M.: A survey on semantic sensor web: sensor ontology, mapping and query. Int. J. u- e- Serv. Sci. Technol. **8**(10), 325–342 (2015). https://doi.org/10.14257/ijunesst.2015.8.10.32

42. Wegner, P.: Interoperability. In: ACM Computing Surveys. Citeseer (1996)

43. Zamborlini, V., Gonçalves, B., Guizzardi, G.: Codification and application of a well-founded heart-ECG ontology. In: Proceedings of the 3rd Workshop on Ontologies and Metamodels in Software and Data Engineering, Campinas, Brazil. Citeseer (2008)

Malware Squid: A Novel IoT Malware Traffic Analysis Framework Using Convolutional Neural Network and Binary Visualisation

Robert Shire[1], Stavros Shiaeles[1(✉)], Keltoum Bendiab[1],
Bogdan Ghita[1], and Nicholas Kolokotronis[2]

[1] Centre for Security, Communications and Network Research,
University of Plymouth, Plymouth PL48AA, UK
sshiaeles@ieee.org
[2] Department of Informatics and Telecommunications,
University of Peloponnese, 22131 Tripolis, Greece
nkolok@uop.gr

Abstract. Internet of Things devices have seen a rapid growth and popularity in recent years with many more ordinary devices gaining network capability and becoming part of the ever growing IoT network. With this exponential growth and the limitation of resources, it is becoming increasingly harder to protect against security threats such as malware due to its evolving faster than the defence mechanisms can handle with. The traditional security systems are not able to detect unknown malware as they use signature-based methods. In this paper, we aim to address this issue by introducing a novel IoT malware traffic analysis approach using neural network and binary visualisation. The prime motivation of the proposed approach is to faster detect and classify new malware (zero-day malware). The experiment results show that our method can satisfy the accuracy requirement of practical application.

Keywords: Traffic analysis · Neural network · Binary visualization ·
Network anomaly detection · Intrusion detection system

1 Introduction

The explosive development of the concept of Internet of Things (IoT) is accompanied by an unprecedented revolution in the physical and cyber world. Smart, always-connected devices provide real-time contextual information with low overhead to optimize processes and improve how companies and individuals interact, work, and live. An increased number of businesses, homes and public areas are now starting to use these devices. The number of interconnected devices in use worldwide now exceeds 17 billion, number that is expected to grow to 10 billion by 2020 and 22 billion by 2025, according to a recent report [1].

On one side, the IoT devices offer extended features and functionality; on the other side, their security level is still low, with well-known weaknesses and vulnerabilities, such as easily guessable passwords and insecure default settings [2]. This gives cybercriminals the opportunity to easily exploit these vulnerabilities and create

O. Galinina et al. (Eds.): NEW2AN 2019/ruSMART 2019, LNCS 11660, pp. 65–76, 2019.
https://doi.org/10.1007/978-3-030-30859-9_6

backdoors into a typical organisation infrastructure. To ensure their protection against potential vulnerabilities, these devices need to be updated and patched regularly; however, given they are not perceived as critical IT infrastructure, it is likely that they are less likely to be upgraded. Further, given their hardware, some IoT devices may not be patchable and the only option is to replace them entirely when they become vulnerable [3]. Beyond the convenience or simplicity of patching, insecure Internet-connected IoT devices represent a security risk. According to a recent report of Symantec, IoT devices will increasingly represent an exploitation target; Symantec already found a 600% increase in overall IoT attacks in 2017 [4].

Botnets are the most common type of malware when an IoT device is compromised [5], either standalone or aggregated to become part of a botnet, capable of launching devastating DDOS (Distributed Denial of service) attacks. Given its uncommon architecture, once a botnet infects an IoT device, it can be very hard to detect the malware. Most conventional antimalware tools rely on a syntactic signature for their detection methods [6], where the signature of a file is compared to a list of known malicious ones. Thus, these systems all require a database with every known malware signature contained within it. This is a very time-consuming process and requires already analysing the malware or its instruction sequence [7]. Moreover, the signature generation involves manual intervention and requires strict code analysis [6, 7], this pushes for enhanced, automated analysis. In this paper, we present a novel IoT malware traffic analysis method that addresses this issue by using a TensorFlow convolutional neural network paired with a binary visualization technique. The main contribution of this proposal is an automated malware traffic analysis method that combines binary visualisation of IoT traffic with the TensorFlow learning model. The combination is ideal for faster analysis of real-time traffic data compared to other approaches and makes it more appropriate to detect and analyse unknown zero-day malware. The proposal utilizes sockets to monitor devices network traffic, the Binvis binary data visualisation technique to convert the binary content of packets into 2D images, and the TensorFlow machine learning method to analyse the produced images. The objective of this analysis is to identify malware in the recent packets, based on the assumption that malware traffic tends to have a more clustered appearance of its patterns on the produced images whereas classic traffic presents more consistent and static. Obviously, both sides had anomalies and expectations.

The overall structure of the paper is organised as follows: Sect. 2 describes the prior works done in malware traffic analysis and classification. In Sect. 3, we present the methodology of the proposed method using neural network TensorFlow and binary visualization. Section 4 presents experiment results and analysis as well as a comparison with other methods. Finally, Sect. 5 provides concluding remarks and future work.

2 Related Works

Detection of malware and its associated traffic is still a persistent challenge for the security community. Research in this area is always needed to keep one step ahead of the hackers. However, IoT devices are upcoming new technology, especially inside a home environment, so anti-malware tools and associated research have been minimal compared to normal technologies [2]. Most attempts to detect or prevent malware traffic are performed by firewalls and intrusion prevention systems, for those in a home environment there is not much security other than the regular patches [8].

Several approaches have been proposed in the literature to detect or mitigate malware traffic. Signature-based detection techniques are the most commonly used, however, they are unable to detect unknown malware traffic for which there exists no signature and involve manual interventions [6–8]. Machine learning is one of the most efficient techniques that have been employed to overcome this issue. Over the years, many machine learning approaches have been proposed for malware traffic analysis and classification. In [9], authors introduced the deep learning method of DBN (Deep Belief Networks) to the intrusion detection domain. In the proposed approach, authors used the DBN for malware traffic classification. Following the same direction, recent work in [10] proposed a malware traffic identification method using a sparse autoencoder. In this work, authors proposed a novel classifier model by combining the power of the Non-symmetric Deep Auto-Encoder (NDAE) (deep-learning), and the accuracy and speed of Random Forest (RF) (shallow learning), leading to high accuracy in malware detection. However, they both used a hand-designed flow features dataset as input data. On the other hand, Convolutional Neural Networks (CNN) and recurrent neural networks (RNN) are also used in many studies to perform malware traffic classification tasks based on spatial and temporal features. For example, authors in [11] transformed the network traffic features into a sequence of characters and then used RNNs to learn their temporal features. The RNN was then applied to detect malware traffic. While in this study the RNNs are used alone and learned a single type of traffic feature, authors in [12], authors used CNN to learn the spatial features of network traffic and achieved malware traffic classification using an image classification method. The proposed method needed no hand-designed features but directly took raw traffic as input data of the classifier, and the classifier then can learn features automatically. The CNN is then used to perform image classification of the images that were created from traffic sample PCAP files. This method has proven the efficiency of malware traffic classification using representation learning approach, being very successful in identifying classic traffic and even malware. However, it does not focus on unknown malware traffic. Thus, if the neural network is not already trained on the type of attack traffic, it will not be able to classify the traffic, or it will falsely categorise it. Moreover, potentially missing Zero-day exploits traffic of viruses lets the work down as such threats are possibly the most serious ones to a network.

The study in [13] that also covered IoT intrusion detection used a different detection factor to help with the traffic analysis, more specifically the data associated with the CPU and memory usage of the IoT device. This is based on the observation that the CPU and memory usage tend to increase when a malware component is detected on the device. Although the CPU and memory features were effective, they require a lot of set up time and reconstruction of a testing network, making the method rather difficult to implement.

In [14], the authors built a similar malware detection tool that focused on malware executables as opposed to traffic. This work also had analysis of binary visualisations through a neural network. The proposed approach uses binary visualisation to convert a binary file data into an image, and self-organizing incremental neural networks (SOINN) for the analysis and detection of malicious payloads. The limitations of this work stemmed from the limited availability of samples, leading to restricting neural network training options.

3   The Proposed Method

The proposed IoT malware traffic analysis method consists of three main steps, as shown in Fig. 1. The first step is the network traffic collection, through either directly sniffing the network or using files containing pre-captured network traffic that can be replayed through tcpreplay for the sniffer to collect again. The second step is the binary visualisation phase, which takes the collected traffic stored in ASCII (American Standard Code for Information Interchange) and convert it into a 2D image. In the final step, the binary image is then processed by the TensorFlow module, which analyses it against its training modules.

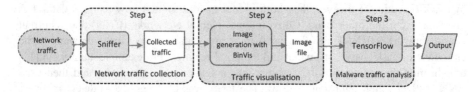

Fig. 1. Overview of the proposed method.

3.1   Network Traffic Collection

Packet capturing is the most commonly used scheme to accomplish the goal of network data collection [15]. Typically, packet-based collection mechanisms use sniffers to implement network data collection through centralized management such as Wireshark, nmap, Airodump, and TCPdump. A sniffer is regarded as a convenient and efficient tool to detect traffic and capture packets [15]. In our approach, we proposed a network traffic collection method using a Python-based tool [16] which ensures two major tasks are accomplished: collection and storage.

1. *Traffic collection*: When the sniffer is loaded into memory, it can collect all packets that are either traversing the network or are replayed. The used Sniffer utilizes Python sockets at a low-level networking interface to collect packets. It is worth noting that the proposed approach is also applicable for the case of very sporadic IoT traffic as it creates profile of what is normal traffic and compare it with abnormal.
2. *Traffic storage*: Received data is passed out to a file that contains the data from the payload in the packet, this data is turned to hexadecimal so that Binvis can plot it into a 2D image in the second step.

As shown in Fig. 2, the dataset used two main collection methods, one that the sniffer collected of both normal traffic and malware traffic and that used traffic from pcap files. All the collected files came from real-world network environments rather than be artificially generated data. If traffic samples have a large size, only parts of the PCAP file is used. Similarly, traffic samples that are too small will only be used on a slower tcpreplay speed. The collected pcap-based files are added to the dataset and replayed through the module tcpreplay using various speeds.

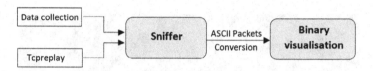

Fig. 2. Overview of the collection process.

3.2   Traffic Visualisation

In this work, we use a visual representation algorithm of the traffic collected that is based on Binvis [17]. This binary data visualization tool converts the contents of a binary file to another domain that can be visually represented (typically a two-dimensional space) [14]. Binvis represents the different ASCII values by using red, green and blue colour classes as shown in Table 1, while black (0x00) and white (0xFF) classes are used to represent null and (non-breaking) spaces.

Table 1. Binvis colour divisions

Colour	Division
Blue	if the ASCII character is printable
Green	if the character is control
Red	if the character is extended ASCII
Black	0x00
White	0xFF

To convert a binary file into a 2D image, its data are seen as a byte string, where each byte value is compared against the ASCII table and is attributed to a colour according to the division it belongs, as outlined in Table 1. In our approach, the binary file is made from network packets collected by the sniffer, these are then converted into a string of hexadecimal characters which is later used to create the image of the traffic, using a clustering algorithm. The final output of Binvis is an image that represents the features of network traffic. The Hilbert space-filling curve clustering algorithm is used in Figs. 3 and 4. This algorithm surmounts other curves in preserving the locality between objects in multi-dimensional spaces [14, 18], thus creating a much more appropriate imprint of the image. This helps the machine learning neural network analysis the image for anomalies in normal traffic.

Fig. 3. Binary visualisation of backdoor pcap

Fig. 4. Binary visualization of a normal traffic pcap

3.3 Malware Traffic Analysis

In this step, TensorFlow is used to analyze the produced images against its in-depth training. TensorFlow is a machine learning system that operates at large scale and in heterogeneous environments [19], providing full flexibility for implementing any type of model architecture [20]. Moreover, TensorFlow is an effective machine learning algorithm to analyse images and classify them accordingly; it is easy to retrain and learns quickly from updates to the neural network [19]. Its outstanding artificial intelligence feature is its excellent image recognition ability, which is specifically why it is being utilised within this application. The TensorFlow AI could easily detect differences between the images, including differences that the human eye could not detect [20].

The TensorFlow module utilizes a CNN which works like a classic neural network but has an extra layer at the beginning called the convolution. The binary output from Binvis is broken up into a number of tiles and, while the machine learning aims to predict what each tile is, the AI then aims to determine the combination of tiles that the picture is based on. This allows TensorFlow to parallelize operations and detect the object regardless of where it is located in the image [21].

The machine learning process is separated into two stages. The first stage is the training phase, where the MobileNet module is employed for the retraining element [21]. MobileNet is a neural network that is very small and efficient, chosen for its lightweight element. It is designed specifically to be mindful of the resources it takes up on a device or application [21]. In the second stage, the image files are tested against the samples of the database to perform classification.

4 Experimental Result

In this section, we present the performance analysis results of the prototype that was implemented based on the methodology presented in the previous section. Accuracy (A), precision (P), recall (R) and f1 value (F1) metrics were used to evaluate the overall performance of the proposed malware traffic classification approach.

$$A = \frac{TP + TN}{TP + FP + TN + FN} \quad P = \frac{TP}{TP + FP} \quad R = \frac{TP}{TP + FN} \quad F_1 = 2 \times \frac{P \times R}{P + R}$$

Where TP is the number of instances correctly classified as good traffic, TN is the number of instances correctly classified as bad traffic, FP is the number of instances incorrectly classified as good traffic, and FN is the number of instances incorrectly classified as bad traffic.

4.1 Experiment Setup

The simulation experiments were performed on a virtual machine built on VM workstation, running Ubuntu 18.0.4. The ISO was not updated during this time to keep prevent technology incompatibility. The dataset that was used in testing, contained a set of 100 pcap samples, collected from external repositories. It is composed of a mixture of 30 normal and 70 malware traffic samples. Samples were classified into the unknown section if the pcap collected was unnamed or Wireshark testing came back inclusive. Table 2 summarizes the percentage of malicious traffic samples of the whole data set.

Table 2. Malicious traffic sample percentage according to type of malware

Malware type	Trojan	DDoS	Botnets	Other		
				OS scan	Keylogger	Backdoors
Percentage	25%	16%	19%	8%	6%	10%

In the training stage, the TensorFlow algorithm was trained by 500 iterations of the data set in a static environment. Knowing that the more training, the better the accuracy of traffic analysis and classification, however incorrect training samples will result in a flawed neural network that can only produce inaccurate results. The minimum training requirements are 30 images for each section, which was 30 images for normal traffic and 30 for malware traffic. Since the TensorFlow was unable to detect whether traffic was good or bad at this stage, the samples used in this stage had to be labelled as being good or malware traffic. For the testing process, the set up was a home scenario, with the thermometer was chosen for the malware host because it is one of the most common home IoT devices and in recent years has been responsible for some of the most devastating attacks [23]. Therefore, it was fitting to use it as the testing scenario. Collected sample files were replayed using tcpreplay on the same network interface card to homogenize the network behavior exhibited by the datasets.

72   R. Shire et al.

4.2   Experiments Results and Analysis

Several tests were carried out to determine the accuracy of the proposed classifier after the addition of more samples, in each test more samples were added to the training data for the machine learning to be retrained on. Figure 5 shows the average accuracy of traffic in the different tests. Four set tests were completed, where the fourth test is the final test with the most training samples that were collected being used. It is apparent from the four tests that the overall accuracy rate of normal traffic stayed consistent throughout, only varying from 78% to 90% because most normal traffic was found to have very similar characteristics throughout the data stream. This makes the classifi-cation of this traffic very easy since the training data would almost always be able to match the sample pcap files against the current traffic being tested. However, even from the start of testing the good traffic had a high accuracy rate, which was surprising given the number of samples in the training at that point. This is the result of test 1 which has a high false positive rate. Knowing that, during early testing, if the algorithm did not recognise a binary visualisation then it would classify it as good traffic, leading to the 60%–40% split for malware traffic (*see* Fig. 5).

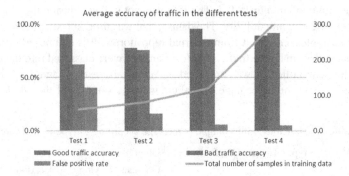

Fig. 5. Average accuracy of malware and good traffic throughout the 4 sets of tests with comparison to number of training data used.

As shown in Fig. 6, the issue with the high false positive rate was slowly being phased out, with the addition of more samples to the neural network throughout the tests, dropping from 40% to 5% by the final set of tests. The addition of more training data mainly contained more specific types of malware samples, in the first and second tests, one pcap of a Trojan was used to train the neural network, while no backdoor attacks were used. This made it near impossible for the algorithm to detect more of these types of traffic without more in-depth training. Malware traffic accuracy varied massively across all the tests, starting at a low accuracy (60%) but by the final test ended up reaching a good accuracy of 91%. Figure 6 shows the final stage of testing results. The stage has been broken down to show the accuracy of the individual types of traffic, DDOS traffic had clearly the best accuracy rate by the end test which even though it made up only 16% of the overall traffic it had a clear pattern where its malware samples were mainly covered in green pixels. This indicates the use of the control character being used over the staple amount, which the machine learning has clearly learned this malicious trait over four sets of tests and uses it to classify DDOS attacks. The algorithm had a much higher probability of showing false positives than false negatives, thus false negative data was not added to the graph results. In summary, the amount of training was the variable that had the most effect on the accuracy of the neural networks.

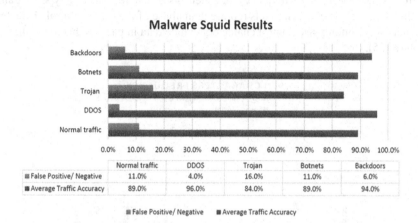

Fig. 6. Final test individual average accuracy for each malware type.

Table 3 shows that the proposed approach achieved an accuracy of 91.32%, which meets accuracy of practical use. It has got a high precision 91.67% and recall (91.03%), which shows the ability of our approach in classifying bad and good traffic.

Table 3. Results for the last test

	Accuracy (A)	Precision (P)	Recall (R)	F1 value (F1)
Test 4	91.32%	91.67%	91.03%	91.35%

4.3 ASCII Characters Frequency Throughout Traffic Results and Analysis

As can be seen in Fig. 7, the different types of malware have distinctive features to differentiate them. Whereas normal traffic can be spotted by their more even distribution of ASCII characters or colours across an image, most of the malware samples follow the same pattern of having more predominance of black (Null Bytes) or white areas (Spaces) in their samples, however, the DDOS is an exception with its extremely high frequency of Control characters. Malware samples do not follow the same pattern as normal traffic, the large volume of null and white spaces might indicate that code was present in the traffic stream. Null bytes which are normally used in coding to mark the end of the string or its termination point [24], could indicate the use of traffic containing a back-door attack or similar. Null bytes are also the main factor in injection exploitation techniques used to bypass security filters, the null bytes are added to user-supplied data to manipulate application behaviour that called a null byte injection attack [25]. Null bytes are also commonly not contained within the default ASCII web request [26], an indication of potential botnet usage that is targeting web servers with no intention to establish a legit connection. DDOS attacks also had an interesting pattern that did not match up with the rest of the other malware, as displayed in Fig. 7, the images had a very high frequency of green pixels, more than any type of traffic recorded. High levels of green pixels represent an abuse of the use in control characters [14], attacks commonly use control characters to hide data in packets that are malicious in nature [25].

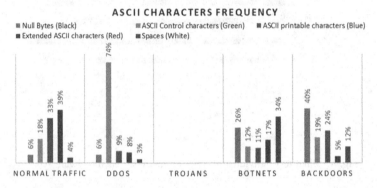

Fig. 7. Average ASCII character frequency between malware and normal traffic PCAPS.

4.4   Comparison

It is not easy to conduct a fair comparison among various malware classification approaches due to the differences between the datasets of traffic used, image visitation tools used and target environments. Thus, our comparison will be based on some significant features. Table 4 overviews a general comparison between our approach and the well-known IDS (Intrusion Detection System) Snort [28] and Suricata [29].

Table 4. Comparison with other methods

Features	Malware squid	Snort IDS	Suricata
Low false alarm rate	Yes	No	Yes
Lightweight	Yes	Yes	No
Protocol independent	Yes	No	No
Raw traffic input	Yes	Yes	Yes

The results from the experiments show that malware squid has a low false alarm rate, only beaten by Suricata due to its more modern nature and detection methods. Whereas snort has a high false alarm rate due to problems with extracting malware footprints from traffic, the means of which its Snort rule set runs off [12]. Malware squid is also a lightweight program for one that utilises an AI, this is due to the MobileNet algorithm being as minimalistic as possible, which is also similar to the older Snort [29], however Suricata Is not lightweight due to its increased memory consumption used in multithreading [30]. Both methods use a set rule set to detect malicious traffic, if traffic matches these sets it will trigger an alarm [12], Malware squid uses image classification so has no knowledge of rule sets making it protocol independent. Finally, all three approaches can take raw traffic input into their datasets [27], this seems to be a staple in IDS detection technologies.

5   Conclusion

This paper proposed a novel IoT malware traffic analysis method, leveraging multi-level artificial intelligence that uses a combination of neural network paired with a binary visualization. The method can be used to protect IoT devices on gateway level bypassing the limitations associated with the IoT environment. From our initial experimental results, the method seems promising and being able to detect unknown malware. Moreover, the method learns from the misclassifications and improve its efficiency. Future work would involve the use of more samples for training and testing and utilising GPU for binary visualization and CNN classification, and testing the propose approach for encrypted traffic as well.

Acknowledgement. This project has received funding from the European Union's Horizon 2020 research and innovation programme under grant agreement no. 786698. This work reflects authors' view and Agency is not responsible for any use that may be made of the information it contains.

References

1. IOT Analytics. https://iot-analytics.com/state-of-the-iot-update-q1-q2-2018-number-of-iot-devices-now-7b/. Accessed 02 Apr 2019
2. Anthony, O., John, O., Siman, E.: Intrusion detection in Internet of Things (IoT). Int. J. Adv. Res. Comput. 9(1) (2018)
3. Schneier on Security. https://www.schneier.com/blog/archives/2018/06/e-mail_vulnerab.html. Accessed 02 Apr 2019
4. Symantec. https://www.symantec.com/content/dam/symantec/docs/reports/istr-23-2018-en.pdf. Accessed 02 Apr 2019
5. McAfee. https://securingtomorrow.mcafee.com/consumer/mobile-and-iot-security/top-trending-iot-malware-attacks-of-2018/. Accessed 10 Mar 2019
6. Gandotra, E., Bansal, D., Sofat, S.: Malware analysis and classification: a survey. J. Inf. Secur. 5(02), 56–64 (2014)
7. Santos, I., Nieves, J., Bringas, P.G.: Semi-supervised learning for unknown malware detection. In: Abraham, A., Corchado, J.M., González, S.R., De Paz Santana, J.F. (eds.) International Symposium on Distributed Computing and Artificial Intelligence, pp. 415–422. Springer, Heidelberg (2011). https://doi.org/10.1007/978-3-642-19934-9_53

8. Garcia-Teodoro, P., Diaz-Verdejo, J., Maciá-Fernández, G., Vázquez, E.: Anomaly-based network intrusion detection: techniques, systems and challenges. Comput. Secur. **28**(1–2), 18–28 (2009)
9. Gao, N., Gao, L., Gao, Q., Wang, H.: An intrusion detection model based on deep belief networks. In: 2014 Second International Conference on Advanced Cloud and Big Data, IEEE, Huangshan, China, pp. 247–252 (2014)
10. Shone, N., Ngoc, T.N., Phai, V.D., Shi, Q.: A deep learning approach to network intrusion detection. IEEE Trans. Emerg. Top. Comput. Intell. **2**(1), 41–50 (2018)
11. Torres, P., Catania, C., Garcia, S., Garino, C.G.: An analysis of recurrent neural networks for botnet detection behavior. In: 2016 IEEE biennial congress of Argentina (ARGENCON), IEEE, pp. 1–6 (2016)
12. Wang, W., Zhu, M., Zeng, X., Ye, X., Sheng, Y.: Malware traffic classification using convolutional neural network for representation learning. In: 2017 International Conference on Information Networking (ICOIN), IEEE, Da Nang, Vietnam, pp. 712–717 (2017)
13. Bezerra, V.H., da Costa, V.G.T., Martins, R.A., Junior, S.B., Miani, R.S., Zarpelao, B.B.: Providing IoT host-based datasets for intrusion detection research. In: SBSeg 2018, SBC, pp. 15–28 (2018)
14. Baptista, I., Shiaeles, S., Kolokotronis, N.: A Novel Malware Detection System Based On Machine Learning and Binary Visualization. arXiv preprint arXiv:1904.00859 (2019)
15. Zhou, D., Yan, Z., Fu, Y., Yao, Z.: A survey on network data collection. J. Network Comput. Appl. **116**, 9–23 (2018)
16. Python. Python.org, https://docs.python.org/3/library/socket.html. Accessed 03 Jan 2019
17. binvis.io. http://binvis.io/#/. Accessed 12 Mar 2019
18. Jagadish, H.V.: Analysis of the Hilbert curve for representing two-dimensional space. Inf. Process. Lett. **62**(1), 17–22 (1997)
19. Abadi, M., et al.: TensorFlow: a system for large-scale machine learning. In: 12th {USENIX} Symposium on Operating Systems Design and Implementation ({OSDI} 16), pp. 265–283 (2016)
20. Géron, A.: Hands-On Machine Learning with Scikit-Learn and TensorFlow: Concepts, Tools, and Techniques to Build Intelligent Systems. O'Reilly Media, Inc. (2017)
21. MobileNet. https://ai.googleblog.com/2017/06/mobilenets-open-source-models-for.html. Accessed 23 Feb 2019
22. Abdellatif. A.: Image Classification using Deep Neural Networks—A beginner friendly approach using TensorFlow. https://medium.com/@tifa2up/image-classification-using-deep-neural-networks-a-beginner-friendly-approach-using-tensorflow-94b0a090ccd4. Accessed 23 Feb 2019
23. McAfee. https://securingtomorrow.mcafee.com/consumer/consumer-threat-notices/casinos-high-roller-database-iot-thermometer/. Accessed 15 May 2019
24. Huseby, S.H.: Common security problems in the code of dynamic web applications. Web Application Security Consortium (2005). www.webappsec.org
25. Afianian, A., Niksefat, S., Sadeghiyan, B., Baptiste, D.: Malware Dynamic Analysis Evasion Techniques: A Survey. arXiv preprint arXiv:1811.01190 (2018)
26. Büschkes, R., Laskov, P.: Detection of intrusions and malware and vulnerability assessment. In: Proceedings of Third International Conference DIMVA, pp. 13–14, July 2006
27. Snort-IDS. https://www.snort.org/. Accessed 10 Mar 2019
28. Suricata. https://suricata-ids.org/. Accessed 10 Mar 2019
29. Roesch, M.: Lightweight intrusion detection for networks. In: Proceedings of LISA, vol. 99 (2005)
30. Shah, S.A.R., Issac, B.: Performance comparison of intrusion detection systems and application of machine learning to Snort system. Future Gener. Comput. Syst. **80**, 157–170 (2018)

Context- and Situation Prediction for the MyAQI Urban Air Quality Monitoring System

Daniel Schürholz[1(✉)], Arkady Zaslavsky[1], and Sylvain Kubler[2]

[1] Deakin University, Melbourne, Australia
{daniel.schurholz,arkady.zaslavsky}@deakin.edu.au
[2] Université de Lorraine, Nancy, France
s.kubler@univ-lorraine.fr

Abstract. Predicting the time and place where concentrations of pollutants will be the highest is critical for air quality monitoring- and early-
warning systems in urban areas. Much of the research effort in this area
is focused only on improving air pollution prediction algorithms, disregarding valuable environmental- and user-based context. In this paper
we apply context-aware computing concepts in the MyAQI system, to
develop an integral air quality monitoring and prediction application,
that shifts the focus towards the individual needs of each end-user, without neglecting the benefits of the latest air pollution forecasting algorithms. We design and describe a novel context and situation reasoning
model, that considers external environmental context, along with user
based attributes, to feed into the prediction model. We demonstrate the
adaptability and customizability of the design and the accuracy of the
prediction technique in the implementation of the responsive MyAQI web
application. We test the implementation with different user profiles and
show the results of the system's adaptation. We demonstrate the prediction model's accuracy, when using extended context for 4 air quality
monitoring stations in the Melbourne Region in Victoria, Australia.

Keywords: Air quality · Context-aware computing ·
Internet of Things · Visualisation · Environmental monitoring

1 Introduction

Throughout the last years, even decades, there has been a steady rise of air pollution in major cities around the world. This has brought many health compli-
cations to citizens and even increased the mortality rate in urban areas. Already
in 2010 for example, a loss of 25 million healthy years and more than 1.2 million
premature deaths in China were attributed to outdoor air pollution [22]. A very
thorough study [4] done by the Global Burden of Diseases study published in
2017 showed that 4.2 million deaths were attributed to the influence of air pollution in 2015, from which 1.3 million happened in China and 1.2 million in India.

© Springer Nature Switzerland AG 2019
O. Galinina et al. (Eds.): NEW2AN 2019/ruSMART 2019, LNCS 11660, pp. 77–90, 2019.
https://doi.org/10.1007/978-3-030-30859-9_7

As a result of these terrible effects, the need for accurate monitoring and reasoning about environmental phenomena and creating effective measures to mitigate the damage caused by air pollution is clear. A way to improve the understanding of how air pollution behaves throughout time is by applying prediction solutions.

Much of the effort done to predict air quality levels has been aimed at improving the machine learning algorithms used for the forecasts as well as understanding the statistical correlation between the different input parameters [2,20]. However, too little has been done to make the prediction algorithms aware of the context in which the end-user operates (e.g., depending on the end-user's location, identity, activity), which is referred to as context-aware computing in the literature [15]. Some studies have applied context-aware computing on Air Quality (AQ) monitoring and prediction systems, as in [3,21], but there is still room for improvement. First, pollutant sources can be considered as contextual information (e.g., surrounding air pollution incidents such as bushfires, traffic volumes). Second, low-frequency high peaks of airborne pollutant concentration could be predicted, by knowing the source of pollution. Third, provide a customised and adaptive view of the real-time situations, through a accessible web context-aware application.

The objective of this paper is to research and propose an enhanced context-aware AQ prediction model, and evaluate it by a system implementation for the The Melbourne city urban area, in Victoria, Australia, that suffers of poor AQ caused by seasonal bushfires and high traffic levels. Figure 1 gives a glimpse of the structure of this study, to prove the benefits of context-aware computing on AQ prediction. We first review in Sect. 2, some context-aware computing concepts, along with AQ monitoring definitions and prediction algorithms for outdoor AQ prediction. In Sect. 3 we present the theoretical and architectural details of the proposed context-aware AQ prediction system, which is called "MyAQI", standing for "My Air Quality Index". Section 4 explain the architecture and implementation of the system. Next, Sect. 5 presents a real-life scenario in Melbourne (Australia) that both shows (i) how MyAQI can be used for predicting AQ in urban areas; and (ii) how the My Air Quality Index (MyAQI) system's prediction model performs over a historical test dataset. Finally, in Sect. 6 we discuss the contributions of this study and the possible future work in the AQ prediction area.

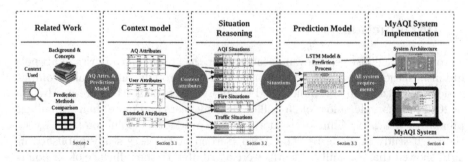

Fig. 1. Research structure for the MyAQI system context and situation prediction model.

2   Background

The core of context awareness is, obviously, the *context*. According to the widely acknowledged definition given in [1], context is "any information that can be used to characterize situation of an entity, where an entity is a person, place, or object that is considered relevant to the interaction between a user and an application, including the user and application themselves." The Context Space Theory (CST) is a method to design contextual models [14]. The approach taken in this method represents the context as a multidimensional space that results in a model that considers context attributes and situations defined to the specific use case.

In the case of this research, the field of application is air quality. The urge to monitor AQ levels is directly linked to the health risks that high levels of airborne pollutants or allergenic agents can have on humans [19]. There is big debate concerning which pollutants are more hazardous and a thorough list of pollutants and their impact on humans is offered in [19]. The selection of these attributes for different outdoor air pollution prediction algorithms and approaches can be seen in Table 1.

Table 1. Use of different air quality characteristics on Deep Learning Neural Network prediction algorithms.

Approach	PM$_{2.5}$	PM$_{10}$	NO$_x$	O$_3$	SO$_2$	CO	RH	TEMP	WIND	P	VIS	LUM	RP
[3]	✓	✓	✓	-	-	-	-	-	-	-	-	-	-
[20]	✓	✓	✓	✓	✓	✓	✓	✓	✓	-	-	-	-
[2]	-	✓	-	-	-	-	✓	✓	✓	✓	✓	-	-
[16]	✓	✓	✓	✓	✓	✓	✓	✓	✓	-	-	-	✓
[24]	✓	-	-	-	-	-	✓	✓	✓	-	-	✓	✓
[23]	✓	✓	✓	✓	✓	✓	✓	✓	✓	✓	-	-	-
[21]	✓	-	-	-	-	-	✓	✓	✓	-	-	-	-
[11]	✓	-	-	-	-	-	✓	✓	✓	-	✓	-	-
[13]	✓	-	-	-	-	-	✓	✓	✓	✓	-	✓	-
[8]	✓	-	-	-	-	-	-	-	✓	✓	-	-	-
Total	9	5	4	3	3	3	8	8	9	4	2	2	2

The goal of Table 1 is to examine which pollutants are being currently monitored and which ones are considered to be critical for the health of citizens; thus allowing us to select the proper group of variables for the MyAQI system monitoring and prediction features. These values are taken from air quality prediction techniques existing in the literature, which use different artificial intelligence algorithms (incl. machine learning) to estimate the possible levels for pollutants in the future. On this work we focus on methods that use Deep Learning (DL), specifically Deep Learning Neural Networks (DNNs), as their main prediction algorithm.

In recent years the focus of machine learning techniques has hugely shifted towards DNN algorithms, because of their complexity and accuracy for solving previously unattainable problems; and AQ forecasting is not an exception. Applying DL to the AQ problem was arguably started by B. Ong in [13] introducing a novel DL model for AQ prediction and in [24] a composite approach is presented to predict Particle Matter under μof diameter ($PM_{2.5}$) concentrations. Long Short-Term Memory Neural Networks (LSTM) is one of the most widely applied DNNs implementation in AQ problems, as well as in many other use cases, with high performance outcomes. Many approaches with high accuracy are presented for AQ prediction, such as in [2,8,11,16,20,23] where authors improve LSTM algorithms and combine them with input tuning and optimisation techniques. But little has been done to extend the context of the information used as input by the DNNs.

The data structured by a context model is aimed at improving context prediction of future context information, starting from low-level context prediction and ending with situation prediction [18]. Some studies for AQ prediction have applied these algorithms and have been designed considering context awareness. Examples of such a context-aware system, mixed with DNNs are shown in [3,21]. Authors include auxiliary data such as meteorological and aerosol optical depth data and geographic dependency between measuring stations, traffic information, points of interest, social media check-ins, building types in the surroundings, amongst other characteristics, improving the accuracy of the prediction by expanding the applied information.

3   Air Quality Context and Prediction Model

The MyAQI system is meant to be context-aware, thus, the context model used to describe its parts is of critical importance. As mentioned before we use CST to provide a structure to our context model. First we must define the context attributes, which are comprised by AQ attributes, extended extra attributes and user attributes.

3.1   Air Quality Attributes

As stated in the previous section we consider *Context Attributes* to be of critical importance for the functioning of the MyAQI system. Airborne pollutants play the main role in air quality and must be included. As mentioned before we refer to [19] and our comparison Table 1 to select the most relevant pollutants in relation to human health. These pollutants are **$PM_{2.5}$**, **Particle Matter under 10 μm of diameter, Nitrogen Dioxide (NO_2),Ozone (O_3),Sulphur Dioxide (SO_2)** and **Carbon Monoxide (CO)**. These 6 pollutants are broadly monitored in environmental systems, and are mapped to a human-readable scale using an **Air Quality Index (AQI)**, which are widely implemented by governments worldwide [5,6,19].

Meteorological Variables are crucial for understanding the behaviour of "already emitted" pollutants, as they affect their location, distribution and temporality. We consider **Temperature (TEMP)** as it affects the characteristics of gases, by making more or less airborne [9], **Relative Humidity (RH)** is also relevant as lower humidity enables pollutant particles to become more airborne [17], **Wind Speed (WSPEED)** and **Wind Direction (WDIR)** are clearly related to the dynamics of air pollutants, as they influence the present and future locations of a mass of pollutants.

3.2   Extended External Attributes

Air pollutants are measured once they are released into the air and move through the environment. The sources that emit them are usually not considered, although it can supply extra information to the spatio-temporal behaviour of pollutants. One such sources is **Traffic volume**; that is, the amount of motor vehicles driving a certain segment of road or road crossing over a certain period of time. Vehicle emissions are considered one of the primary sources of pollution in cities and contribute largely to high NO_2 and CO levels [5,6]. Other source of pollution, specially in countries such as Australia, where summers can reach high temperatures and low humidity, are **Fire incidents**. They contribute largely to the pollution in urban areas surrounded by dry vegetation areas. Bushfires, specifically, contribute largely to high $PM_{2.5}$ and PM_{10} levels [5,6].

3.3   User Attributes

Finally, any context-aware system has to consider relevant user features, that are significant in their interaction with the problem at hand. AQ monitoring is no exception, as air pollutants can affect users differently depending on certain characteristics. We consider the following user attributes: a **User Id** identifies a user to the system and separates information for a customised experience, **Geolocation** determines the spatial reference of a user's location, crucial for outdoor AQ monitoring, **Timestamps** give the specific time and date of interaction with the system to provide updated information and a novel **Pollutant sensitivity** scale, which represents the level of influence that a given pollutant has on the user. Each user has 6 pollutant sensitivity levels assigned, which are derived from answering a small questionnaire [12] at the system's profile section. Each level can take a value between 0 and 4. The values represent the following sensitivities: 0 - "neutral", 1 - "low", 2 - "moderate", 3 - "high" and 4 - "extremely high".

3.4   Situation Reasoning

After defining the context attributes, another major building block of a context-aware system is defining the *Situation Space*. For AQ monitoring the following situations and their characteristics were defined. Table 2 shows the situations for AQI levels for different user pollutant sensitivities, traffic volumes for severity levels and fire incidents for severity levels as well, and their corresponding triggering *Context States*. The traffic volumes are dependant of the road and crossing type where the data was collected, so percentages are used for defining the severity of a given measurement. Fire incidents are divided into urban and suburban (and/or countryside) scenarios, because of the size of incidents in each case and the obstacles that could stop the spread of smoke and ashes.

Table 2. Situation spaces definitions for AQ attributes, traffic volumes and fire incidents and their mapped *Context States*.

AQI categories	User Sensitivity Levels				
	0	1	2	3	4
Very Good	0 - 33	0 - 33	0 - 33	0 - 33	0 - 23
Good	34 - 66	34 - 66	34 - 66	34 - 54	24 - 44
Moderate	67 - 99	67 - 99	55 - 79	55 - 79	45 - 59
Poor	100 - 149	100 - 124	80 - 99	80 - 89	60 - 69
Very Poor	$\geqslant 150$	$\geqslant 125$	$\geqslant 100$	$\geqslant 90$	$\geqslant 70$

Traffic volume	Situation Id	Quantile	Value range
Very low	0	Q_1 (0%-20%)	$[0, q_1]$
Low	1	Q_2 (20%-40%)	$]q_1, q_2]$
Moderate	2	Q_3 (40%-60%)	$]q_2, q_3]$
High	3	Q_4 (60%-80%)	$]q_3, q_4]$
Extremely High	4	Q_5 (80%-100%)	$]q_4, +\infty[$

Fire severity	Situation Id	City range (kms)	Suburban range (kms)
No fire	0	$]20, +\infty[$	$]100, +\infty[$
Very low	1	$[16, 20]$	$[80, 100]$
Low	2	$[12, 16[$	$[60, 80[$
Moderate	3	$[8, 12[$	$[40, 60[$
High	4	$[4, 8[$	$[20, 40[$
Extremely High	5	$[0, 4[$	$[0, 20[$

3.5 Prediction Model

As previously mentioned, the reasoning module implements context-aware prediction for future AQ levels. The selected algorithm for this purpose is an LSTM. LSTMs are being widely used in forecasting solutions throughout different fields and were first introduced in [7]. They improve regular Recurrent Neural Networks (RNN) to keep information for long-term capabilities. In the MyAQI system the algorithm takes the measurements from AQ and meteorological variables for the previous 24 h as input, as depicted in Fig. 2. The notifications then sent to the user will prioritise those pollutants to which the user has more sensitivity towards.

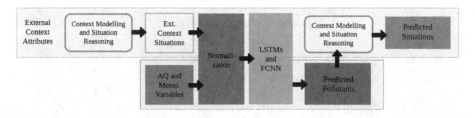

Fig. 2. The MyAQI system prediction algorithm structure, components and data flow.

4   System Architecture and Implementation

4.1   System Architecture

The MyAQI system architecture, depicted in Fig. 3, is divided into two major layers: *Backend* and *Frontend*. The backend layer comprises two other major layers: the *Data Layer* and the *Logic Layer*. The data layer retrieves the data required to fuel the context model attributes from external Application Programming Interfaces (API) and stored in relational databases, along with user-entered information. The logic layer is comprised by three modules: (i) the context modelling module, that maps the raw data into usable context attributes, (ii) the prediction algorithm module which executes data analysis, such as prediction, on the context attributes, to augment the known information, and (iii) the MyAQI API module which has two interfaces, a Representational State Transfer (RESTful) Hyper Text Transfer Protocol (HTTP) API for regular exchange of information with the frontend modules and a Web Sockets (WS) interface for push notification from the backend server to the user devices. Finally, the frontend layer has only one sub-layer, in charge of the context-aware visualisation of AQ data. The modules of the *Visualisation Layer* are: (i) API consumer, which maps incoming sever information into memory objects, which are then passed to the (ii) *Situation Reasoning* module, where the context states are mapped into real-life events, which are then visualised in the (iii) end-user device views.

4.2   Implementation

Previously the main functionalities, models and goals of the MyAQI system where introduced. In this subsection the implementation of the system is presented. Considering the system architecture presented in Sect. 4.1, the following step is to map each structural element to hardware equipment and it's functioning software.

The backend layer runs in a virtual server in the cloud for better availability. The data modules such as the user and the *PostgGIS* database are installed in this server. PostGIS is used for geographical queries, needed for the distance from users to fires incidents, for example. The logic modules are implemented in the *Python* programming language. Reasoning functionalities, such as

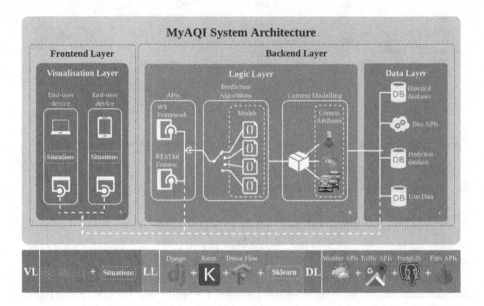

Fig. 3. MyAQI system layers and overall architecture.

prediction, are implemented with the machine learning frameworks *Keras* (with a *TensorFlow* backend). Finally, the API modules are implemented using *Django Rest Framework* for the RESTful API and *Django Channels*, plus the in-memory database *Redis*, for the WS interface.

The frontend layer runs in end-user devices, such as mobile phones, tablets and laptops. Thus, the chosen technology for its development is a *JavaScript* responsive web application framework, called *ReactJS*. All the API consumer, situation reasoning and context-aware visualisation logic is developed with this framework and presented via Hyper Text Mark-up Language (HTML) and Cascade Style Sheet (CSS) views. Figure 4 shows different views of the MyAQI web application rendered on different devices, to show the tools responsive nature.

5   Experiments and Results

The previous section introduced the architecture and implementation of the MyAQI system and highlighted its different building blocks. This section presents the experiments that were undertaken to specifically tackle those objectives. First, the experiments' setup and structure is explained, together with the required datasets, and finally, the results of the experimentation are shown.

5.1   Experiments

The AQ attributes required in the context model are presented in Sect. 3. As the data source for the experiments, the Australian Environmental Protection Agency (AU-EPA) in Victoria, Australia, provides a live data API that

Fig. 4. The MyAQI system rendered on different end-user devices, for more accessibility.

supplies both historical and current (updated hourly), and has almost all the needed context variables for many of the sensor stations distributed throughout Victoria, including meteorological data. For traffic information, data from one to four nearby vehicle crossing stations (close to the AQ monitoring stations) where selected to obtain the number of vehicles driving past the site every hour, measured by the *SCATS* system (developed by the government of New South Wales, Australia). Finally, the fire incidents are obtained from a historical dataset, provided by the Victoria government, in which it keeps track of every fire incident in its region since the year 1930. The prediction experiments use data from the previously described data sources, from January 2017 to December 2018.

5.2   Results

In this section the accuracy comparisons of the prediction algorithm are presented, comparing an LSTM that is run considering external context attributes versus one without the extended information. And, finally, the context-aware system views are presented, to complete the user-oriented nature of context-aware computing.

To evaluate AQ predictions we compare them against the ground truth and against the model without the extended environmental context. Table 3 shows the comparison of predictions' MAE, RMSE and precision values for the four stations, once with extended context and once without against the ground truth. The indicators are defined by the following equations:

$$MAE = \frac{1}{n} \sum_{i=1}^{n} |y_i - \hat{y}_i| \tag{1}$$

$$RMSE = \sqrt{\frac{1}{n} \sum_{i=1}^{n} (y_i - \hat{y}_i)^2} \tag{2}$$

$$precision = \frac{|\{true\ AQI\ situations\} \cap \{total\ AQI\ situations\}|}{|\{total\ AQI\ situations\}|} \tag{3}$$

where n is the number of measurements, y_i are the forecasted values and $\hat{y}_i$ are the ground truth values. Precision is a measure used for classification problems, and we apply it see how well the AQI situations corresponding to the ground truth values are kept after the forecast. For all stations except Alphington the improvement in prediction is clear when using the extended environmental context. The case with Alphington can be interpreted as a lack of correlation between extended context variables and the AQ in the area, probably coming from another pollution source. Precision is always improved in the other three stations, specially in Mooroolbark and Traralgon, which are influenced the most by seasonal bushfires.

Table 3. Comparison of MAE, RMSE and precision (Prec) for the prediction results from the LSTM model with and without extended context values, for all four AQ stations.

Station	Attributes	Performance indicators		
		MAE	RMSE	Prec
Traralgon AQ station	*+1hr PM$_{2.5}$ prediction*			
Without Extended Context	PM$_{2.5}$, PM$_{10}$, NO$_2$, SO$_2$, CO	1.678	2.411	0.916
With Extended Context	+ Fires	1.477	2.262	0.943
Mooroolbark AQ station	*+1hr PM$_{2.5}$ prediction*			
Without Extended Context	PM$_{2.5}$, PM$_{10}$	4.295	6.769	0.872
With Extended Context	+ Traffic, Fires	2.124	8.775	0.909
Alphington AQ station	*+1hr PM$_{2.5}$ prediction*			
Without Extended Context	PM$_{2.5}$, PM$_{10}$, NO$_2$, SO$_2$, CO	1.364	1.922	0.956
With Extended Context	+ Traffic, Fires	1.389	1.949	0.957
Melbourne CBD AQ station	*+1hr PM$_{2.5}$ prediction*			
Without Extended Context	PM$_{2.5}$	2.869	4.115	0.912
With Extended Context	+ Traffic	2.797	3.85	0.93

A snapshot of the prediction algorithm accuracy is presented in Fig. 5. The background colours correspond to the AU-EPA AQI categories and show that the prediction stays between the value range of each situations, excepting on some few steps, explaining the overall good precision values.

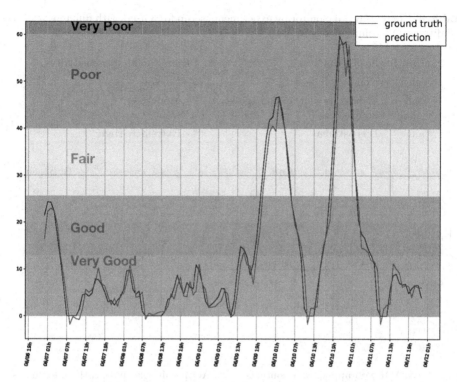

Fig. 5. Alphington AQ measuring station $PM_{2.5}$ levels prediction, considering $PM_{2.5}$, PM_{10}, NO_2, SO_2, CO and nearby traffic volumes. The background colours correspond to the AU-EPA AQI categories, proving that the prediction of AQ situations is accurate.

Another important contribution of the MyAQI system is the personalisation and context-aware views and notifications that the offered to users. Table 4 show three user profiles with different health conditions and corresponding pollutant sensitivities. Using this table's values the MyAQI was run and the notifications received by the user tested. Figure 6 presents the MyAQI web application context-aware air quality monitoring notifications for users with different sensitivity levels to main pollutants, changing the severity according to user health conditions. Finally, the notifications also highlight the source for the pollution (e.g., high traffic volumes) expanding the knowledge of the user over the AQ problem.

6   Conclusion

In this work we research and propose a context-aware model and system architecture for AQ monitoring and prediction systems and we prove its benefits by implementing the MyAQI system. It includes the proposed context- and situation model, the selected prediction algorithm, a thorough architectural design, the implementation, the coupling with selected data sources and the layout of

Table 4. Context-aware monitoring experiments setup for 3 users with different health conditions and pollutant sensitivities.

User Id	health condition	General pollutant sensitivity
alice	Completely healthy	0 - Neutral
bob	Unhealthy diet, casual smoker, no exercise.	2 - Moderate
ana	Has asthma	4 - Extremely High

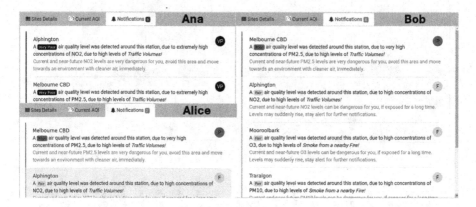

Fig. 6. MyAQI web application context-aware AQ monitoring notifications for users with different health conditions and pollutant sensitivity levels.

experiments to prove the expected performance of the system. A web application was developed to provide a user-friendly interface to allow user interaction with the monitoring and prediction functionalities. The application is a customisable tool that gives users a highly individualised experience and augments their understanding of the air pollution problem. All the data used in the system was obtained from trusted historical and current data sources and reflect real-life situations obtained from the Victoria EPA in Australia, allowing for an objective assessment of the research results. We prove that the use of extended environmental context sources on the AQ prediction problem improves prediction accuracy, in our specific case, for 3 out of 4 AQ stations.

Finally, the MyAQI system, implemented in this work, provides a proof-of-concept of a context-aware system for the use case of AQ monitoring and prediction in the real-world scenario of the Melbourne area in Victoria, Australia. It showed the benefits that can be drawn from using existing IoT data sources and extracting more information from them by relating them to real-life situations and phenomena.

Acknowledgment. This research was funded by the PERCCOM Erasmus Mundus Joint Masters Program of the European Union [10]. Part of this study has been carried out in the scope of the project bIoTope, which is co-funded by the European Commission under Horizon-2020 program, contract number H2020-ICT- 2015/688203-bIoTope. The research was also supported by Deakin University, Australia. Air pollution data in the city of Melbourne was freely obtained from Victoria EPA API (http://sciwebsvc. epa.vic.gov.au/aqapi/).

References

1. Abowd, G.D., Dey, A.K., Brown, P.J., Davies, N., Smith, M., Steggles, P.: Towards a better understanding of context and context-awareness. In: Gellersen, H.-W. (ed.) HUC 1999. LNCS, vol. 1707, pp. 304–307. Springer, Heidelberg (1999). https://doi.org/10.1007/3-540-48157-5_29

2. Athira, V., Geetha, P., Vinayakumar, R., Soman, K.P.: Deepairnet: applying recurrent networks for air quality prediction. Procedia Comput. Sci. **132**, 1394–1403 (2018)

3. Chen, L., Cai, Y., Ding, Y., Lv, M., Yuan, C., Chen, G.: Spatially fine-grained urban air quality estimation using ensemble semi-supervised learning and pruning. In: Proceedings of the 2016 ACM International Joint Conference on Pervasive and Ubiquitous Computing - UbiComp '16, pp. 1076–1087 (2016)

4. Cohen, A.J., et al.: Estimates and 25-year trends of the global burden of disease attributable to ambient air pollution: an analysis of data from the global burden of diseases study 2015. Lancet **389**(10082), 1907–1918 (2017)

5. EEA: Air quality in Europe - 2017 report. Technical report 13, European Environmental Agency (EEA) (2017)

6. EPA Victoria: Future air quality in victoria - final report future air quality in victoria - final report. Technical report. Environmental Protection Agency Victoria Australia, Melbourne (2013)

7. Hochreiter, S., Schmidhuber, J.: Long short-term memory. Neural Comput. **9**(8), 1735–1780 (1997)

8. Huang, C.J., Kuo, P.H.: A deep cnn-lstm model for particulate matter (pm2.5) forecasting in smart cities. Sensors **18**(7), 2220 (2018). Switzerland

9. Kalisa, E., Fadlallah, S., Amani, M., Nahayo, L., Habiyaremye, G.: Temperature and air pollution relationship during heatwaves in Birmingham, UK. Sustain. Cities Soc. **43**, 111–120 (2018)

10. Klimova, A., Porras, J., Andersson, K., Rondeau, E., Ahmed, S.: PERCCOM: A master program in pervasive computing and communications for sustainable development, April 2016

11. Li, X., et al.: Long short-term memory neural network for air pollutant concentration predictions: method development and evaluation. Environ. Pollut. **231**, 997–1004 (2017)

12. Nurgazy, M., Zaslavsky, A., Jayaraman, P., Kubler, S., Mitra, K., Saguna, S.: CAVisAP: Context-aware visualization of outdoor air pollution with IoT platforms. In: International Conference on High Performance Computing and Simulation (HPCS) (2019)

13. Ong, B.T., Sugiura, K., Zettsu, K.: Dynamically pre-trained deep recurrent neural networks using environmental monitoring data for predicting pm2.5. Neural Comput. Appl. **27**(6), 1553–1566 (2016)

14. Padovitz, A., Wai Loke, S., Zaslavsky, A.: Towards a theory of context. In: Second IEEE Annual Conference on Pervasive Computing and Communications (Workshops, PerCom), pp. 38–42 (2010)
15. Perera, C., Zaslavsky, A., Christen, P., Georgakopoulos, D.: Context aware computing for the internet of things: a survey. IEEE Commun. Surv. Tutor. **16**(1), 414–454 (2014)
16. Qi, Y., Li, Q., Karimian, H., Liu, D.: A hybrid model for spatiotemporal forecasting of pm2.5 based on graph convolutional neural network and long short-term memory. Sci. Total Environ. **664**, 1–10 (2019)
17. Qiu, H., Tak, I., Yu, S., Wang, X., Tian, L., Tse, L.A.: Season and humidity dependence of the effects of air pollution on COPD hospitalizations in Hong Kong. Atmos. Environ. **76**, 74–80 (2013)
18. Sigg, S., Gordon, D., Zengen, G., Beigl, M., Haseloff, S., David, K.: Investigation of context prediction accuracy for different context abstraction levels. IEEE Trans. Mob. Comput. **11**(6), 1047–1059 (2012)
19. USEPA: Technical assistance document for the reporting of daily air quality - the air quality index (AQI). Environmental Protection, pp. 1–28, May 2013
20. Wang, J., Song, G.: A deep spatial-temporal ensemble model for air quality prediction. Neurocomputing **314**, 198–206 (2018)
21. Wen, C., et al.: A novel spatiotemporal convolutional long short-term neural network for air pollution prediction. Sci. Total Environ. **654**, 1091–1099 (2019)
22. Yin, P., et al.: Particulate air pollution and mortality in 38 of china's largest cities: time series analysis. Bmj **667**, j667 (2017)
23. Zhou, Y., Chang, F.J., Chang, L.C., Kao, I.F., Wang, Y.S.: Explore a deep learning multi-output neural network for regional multi-step-ahead air quality forecasts. J. Clean. Prod. **209**, 134–145 (2019)
24. Zhu, S., et al.: PM2.5 forecasting using SVR with PSOGSA algorithm based on CEEMD, GRNN and GCA considering meteorological factors. Atmos. Environ. **183**, 20–32 (2018)

Data Mining Algorithms Parallelization in Logic Programming Framework for Execution in Cluster

Aleksey Malov[✉], Sergey Rodionov, and Andrey Shorov

Saint Petersburg Electrotechnical University "LETI", Saint Petersburg, Russia
alexeimal-2@yandex.ru, {sv-rodion,ashxz}@mail.ru

Abstract. This article describes an approach to parallelizing of data mining algorithms in logical programming framework, for distributed data processing in cluster. As an example Naive Bayes algorithm implementation in Prolog framework, its conversion into parallel type and execution on cluster with MPI system are described.

Keywords: Parallel algorithms · Distributed algorithms · Data mining · Distributed data mining · MapReduce · Logical programming

1 Introduction

Currently, there is a rapid growth of stored information volumes. The amount of information from different devices increases constantly. The set of such devices includes sensors, video cameras, mobile devices, etc. The set of such devices, united in global network, is called Internet of Things (IoT). The amount of information received from these devices will only increase with time. [1, 2] Currently, this kind of data is called Big data. They are characterized by: large volumes, different storage formats and a high rate of change.

Data mining of Big data is associated with the analysis of large amounts of data, which, accordingly, requires the use of significant computational resources to ensure an acceptable analysis time. Therefore, the current direction in the development and implementation of Data mining algorithms is their parallelization.

MapReduce (US patent No. 9612883) is the known method of performing data processing in a distributed and parallel processing environment.

The main disadvantages of this method are that the known method is suitable only for some tasks of Data mining. At the same time, there are Data mining algorithms, for example, Apriori, PageRank, etc., for which this method is not suitable, since the algorithms do not have a list homomorphism [3–5].

Also, this method does not show approaches to the implementation of Data mining algorithms in logical programming environment. Implementation and parallelization of Data mining algorithms in logical programming environment is separate part of Data mining algorithm realization. This separate part appears due to the significant differences between logical programming approach and the traditional imperative approach.

O. Galinina et al. (Eds.): NEW2AN 2019/ruSMART 2019, LNCS 11660, pp. 91–103, 2019.
https://doi.org/10.1007/978-3-030-30859-9_8

Therefore, this direction should be supported by special methods of implementing Data mining algorithms with the aim of their parallelization.

Imperative programming languages (Java, C/C++, Fortran and others.) are not suitable for parallel execution. They are designed for operation in accordance with the model of Turing machine. Its basic idea is to change the state of Turing machine in execution of each program statement. Thus, programs, written in imperative programming languages, presuppose the program state and its change during execution. The main problem of such programs in their parallel execution is requirement for simultaneous access to the program state from parallel branches. This gives rise to such problems as access synchronization, blocking, race and others. Solving these problems is a difficult task at stages of development and debugging.

Existing solutions for imperative languages for parallel execution (Ada, High Performance FORTRAN, High Performance C++ and others.) allow to parallelize only individual structures such as cycles.

Programs in logical programming languages can theoretically be parallelized automatically [6] due to the following features:

(1) In the process of logical inference, proving of a set of rules of a logic program occurs, the process of proving each rule could be performed independently and in parallel.
(2) The right side of each of the set of rules may contain many predicates. Their calculation can also be performed in parallel when proving the rules is performed in the process of logical inference.
(3) The unification process can also be performed in parallel by the inference engine.

2   Logical Programming Features and Related Works

Logical programming is a programming approach based on the proof of statements by the logical inference machine based on specified facts and inference rules. The most famous logical programming language is Prolog.

Facts in the Prolog language are unconditionally true statements. In terms of the Prolog language, they are predicates. A predicate consists of a name, or of a name and the sequence of arguments following it, which are enclosed in brackets. The combination of facts in Prolog is called the knowledge base of a logic program.

Among the advantages of logical programming, we can highlight the following properties of logical programming languages:

- programs are more easy for understand because they do not contain the details of the solution of the problem which are difficult to understand, but describe what the result of the decision is;
- the program looks like a description of what is the result of the decision; this feature makes it easier to check and make sure that the program actually implements what is required of it.

The disadvantages of logical programming include the fact that it is not intended for solving problems of any subject area. In particular, logical programming is not intended for solving problems that require large volume and high optimality of arithmetic calculations.

One of the advantages of logical programming languages in solving problems of data mining is the ability to present the results of work in a logical form, which is more understandable and informative, compared to the traditional presentation of results in numerical form [7]. These features can reduce the probability of user error. However, there are additional time costs in solving problems related to the functioning of logic programs, as well as the need to take additional measures to minimize them, which can be attributed to disadvantages.

When solving Data mining tasks, it is convenient to present and process various data arrays in a logical form, including the Data mining database, as well as the final knowledge model. The built-in functionality of the logic programming environment is suitable for these purposes. These capabilities allow creating universal patterns of decision for data mining algorithms [7–10].

One of the most popular logical programming languages is Prolog. The functionality built into the Prolog language for working with facts in the knowledge base of a logic program provides developers a lot of features. There exist features in Prolog allowing to process all the facts of a particular type and form a final result based on this. To solve the problems of Data mining, it is often necessary to traverse array in a particular order with aim to handle its elements. The Prolog language provides the ability to present and process arrays of data in the form of facts in the knowledge base of a logic program.

Research in the field of parallelization of data mining algorithms is performed for quite a long time. There are two main direction of research in the field of the parallelizing of data mining algorithms: parallel data mining (PDM) and distributed data mining (DDM) [11].

PDM is the parallelizing of data mining algorithms for systems shared memory, clusters of work stations with quick interaction. DDM is the parallelizing of data mining algorithms for systems consisting of nodes linked into a local network or distributed geographically and united through the global network.

The biggest part of the research in the field of the parallelization of data algorithms mining focused on particular data mining algorithms and optimization of parallel structure of algorithms for execution under particular conditions. Those conditions can be different [12]. For this reason the parallel algorithms developed under particular conditions, may not be efficient under other conditions.

There are a lot of examples of data mining algorithms for specific types of computing systems. Decision tree algorithms: Parallel Decision Tree [13] and others. Association algorithms: Common Candidate Partitioned Database [14], Partitioned Candidate Common Database [14], Asynchronous Parallel Mining [15] and others. Clustering algorithms: P-CLUSTER [16], Collaborative [17], Distributed Information Bottleneck [18] and others. All of these algorithms were implemented in imperative programming languages.

There are a lot of algorithms implementations in logic programming frameworks [8, 9]. However, existing decisions are intended not for parallel execution or intended for particular algorithm.

Thus, the most part of approaches resolve task of parallelization for particular Data Mining algorithm that requires significant effort to adopt such approach for new algorithm or create one new approach. The approach describing in this article allows converting sequential Data Mining algorithm, implemented in Logic Programming frame work, into different parallel versions.

3   Implementation of Data Mining Algorithm in Logic Programming Framework

3.1   Data Mining Algorithm Representation as a Function

Data mining algorithms at the input receive a set of input data, and the result of their work is the final output knowledge model. Thus, each Data mining algorithm represents a function, since it receives as input a set of input data and returns the result of applying the algorithm to the input data in the form of the final output knowledge model. Consequently, the Data mining algorithm can be represented as a predicate of a logic program, the input data for which is a set of input data D as facts of a knowledge base of a logic program, and the result of its satisfaction in the process of inference is a final output knowledge model M in the form of facts in the knowledge base of a logic program.

During the work, Data mining algorithms analyze the data set and build an intermediate knowledge model at each step. The made intermediate knowledge model is sent to the next stage. Thus, the Data mining algorithm can be represented by a functional expression in the form of a composition of functions [13]:

$$f = fn°fn - 1°\ldots°fi°\ldots°f1 \tag{1}$$

3.2   Data Mining Algorithm Representation in Logic Programing Framework

Each function fi from (1) can be implemented as a predicate. For each of such predicates input data is a data set (D) and an intermediate knowledge model in the form of knowledge base facts of a logic program. The result of the work is a new intermediate knowledge model (MP) in the form of facts in knowledge base of a logic program.

Thus, the Data mining algorithm can be represented as a predicate. Wherein this predicate is the left side of the rule, in the right side of which there are predicates representing the steps of the algorithm. Further, these predicates will be called distinct predicates. Hereinafter, the examples will use the notation of logic programming language SWI Prolog [http://www.swi-prolog.org].

```
algorithm_induce() :-
/*auxiliary predicates*/
    predicate_1()
/*auxiliary predicates*/.

predicate_1 ():-
/*auxiliary predicates*/

    predicate_2()
/*auxiliary predicates*/.
........
predicate_n-1 ():-
/*auxiliary predicates*/
    predicate_n()
/*auxiliary predicates*/.
```

$$(2)$$

where algorithm_induce() – a predicate implementing the Data mining algorithm, in the process of logical inference,

predicate_i - a predicate implementing the i-th step of the Data mining algorithm, during the process of logical inference and when satisfying this predicate, the input data set (D) of the Data mining algorithm and the intermediate knowledge model in the form of knowledge base of the logic program are used.

The predicate predicate_1, which implements the 1st step of the Data mining algorithm, uses only the initial data set in the form of facts of the knowledge base of the logic program.

The result of satisfying a predicate in the resolution process is an intermediate knowledge model in the form facts in the knowledge base of a logic program that is used in the next step, and the result of satisfying predicate_n is the final output knowledge model in the form of facts in the knowledge base of a logic program.

The set of initial data (D) of the Data mining algorithm in the knowledge base of a Prolog program can be represented by several types of predicates that define a list of attributes and a list of vectors. In the SWI Prolog language, the input data set (D) can be represented as a set of predicates in the knowledge base of a logical program, specifying a list of attributes (predicate attr_list, attr_values), a list of vectors (predicate db_vector) and a list of all possible classes of the target attribute (class_list):

A list of K attributes can be defined using the attr_list predicate in the following way

attr_list([attr1, ..., attrK]).

where attr<i> - the name of the i-th attribute.

A list of attribute values can be defined using the attr_values predicate in the following way

attr_values(attr_N, attr_N_val_list),

where attr_N – the name of some input attribute

attr_N_val_list – the list of possible values for this input attribute.

Below is an example for attribute 1, which can have five values:

attr_values(attr1, [2–5, 7]).

A list of L vectors can be represented by a set of L db_vector predicates in the knowledge base of a logic program. The db_vector predicate can be defined in the following way

db_vector(Id, class_N, attr_val_list),

where Id - unique numeric identifier for a specific predicate db_vector,

class_N - name of some target attribute,

attr_val_list - the list of attribute values for this occurrence of the target attribute class_N in the input data set (D) of the Data mining algorithm.

The value of the i-th attribute is located on the i-th place in the list of attr_val_list. Below is an example for vector 1:

db_vector(1, class1, [attr1=2, attr2=4, attr3=5, attr4=5]).

The list of classes can be defined using the class_list predicate. The predicate class_list is a list containing all possible target attributes.

class_list([class1, ..., classN]).

where class<i> - the name of the i-th class.

The structure of the intermediate (MP) and final output knowledge model (M) depends on the algorithm that implements Data mining. It is convenient to represent the intermediate (MP) and the final output knowledge model (M) as predicates that define lists. For example, a knowledge model in the form of decision trees is often found among the algorithms of Data mining. For decision trees, tree nodes and their corresponding sets can be represented as lists. In general, an intermediate knowledge model can be represented as a set of predicates in the knowledge base of a logic program, setting the parameters of the current step (step_context) and the set of predicates of rules (rule).

The parameters of the current step (step_context) may contain, for example, a list of attributes that have not yet been used to build a knowledge model, and, therefore, must be processed in the following steps. The set of predicates (rule) in the knowledge base of the logic program, define a part of the final knowledge model formed at the current moment.

3.3 Basic Features for the Implementation of Data Mining Algorithms in Logic Programing Framework

Data mining applications are mainly characterized by data parallelism. In this case, to implement parallel computing, it is necessary to add a data partitioning functional and functional for merge the results obtained in different working processes. Execution in parallel mode of each individual step, rather than the entire algorithm at once, allows, in contrast to MapReduce, to use this method to parallelize algorithms that do not have a list homomorphism.

For parallel computation of an algorithm, it is need to add functions that allow the application to be parallelized for a distributed computation system, for example, the Message Passing Interface (MPI). At the same time, a properly configured MPI system installed on a set of interconnected systems, each of which has one or several processors and memory, must have functionality for

- sending data from several working processes to the main process,
- receiving data in the main process from several working processes,
- determining the number of working process in the system,
- performing several working processes, including the main process
- storing facts in a logic program knowledge base,
- executing an inference process;

The functionality that needs to be implemented for parallel execution of the application includes:

- predicate for splitting the input data into parts for the work of a one from several working processes;
- a predicate for merging the results of several working processes into a single intermediate knowledge model in the form of facts in the knowledge base of a logic program;
- predicate of the main process for receiving data, created with the ability to receive data from several working processes;
- the predicate of the main process, which is the right part of the rule, while the left part consists of the predicates:
  - sending data to the several working processes,
  - receiving data in the main process,
  - recording the received facts in the knowledge base of the logic program of the main process
  - merging the results of several working processes into a single intermediate knowledge model;
- predicates of working processes, which is the right part of the rules, while the left part consists of the predicates for:
  - converting input data D from files or databases to facts in the knowledge base of a logic program
  - splitting the input data into parts for the work of one of the several working processes
  - receiving intermediate knowledge model for current step from the main process
  - distinct predicates, performing algorithm's step;
  - sending results to the main process.

Input data D from files or databases must be read by the main process into a set of list data structures containing attribute lists and vector lists. The configured MPI system should have a functional for reading from the input files or databases of a set of initial data specifying the task. Since the main process can be run on any machine into the system, then any machine in the system should have access to required files or databases.

4   Parallelization of Naive Bayes Algorithm in Logic Programming Framework

As examples, we will consider probability algorithm Naive Bayes [10].

>   for all attributes a
>   if a is not target attribute
>   for all vectors w
>   increment count of vectors for value of attribute a
>   equaled value of the attribute a of the vector w
>   end for all vectors;
>   end if
>   end for all attributes;
>   for all classes c
>   for all vectors w
>   increment count of vectors for class of vector w;
>   end for all vectors
>   end for all classes c

The Naive Bayes algorithm can be represented as composition of functions (1):
for_all_classes_cycle°for_all_attributes_cycle(D)
where

- for_all_classes_cycle: is function which adds to the final or partial model row with the count of classes entrances into the data base
- for_all_attributes_cycle: is function which executes iteration by attributes and executes iteration by all vectors or by the selected part of the vectors and calculates vectors count having the particular value of particular attribute and particular class

Below is the predicate bayes_model () in the form (2), in the right side of which the predicates cycle_attr () and cycle_class () are used. By the satisfaction of this predicates a final knowledge model for a Naive Bayes algorithm is formed during the process of logical inference.

>   bayes_model() :-
>   cycle_attr(),
>   cycle_class().

A knowledge model for a Naive Bayes algorithm may be represented as a set of predicates of two types:

result_model(class_N, count) - a predicate setting the number of occurrences (count) of some target attribute class_N in the input data. In this predicate class_N is the name of some target attribute, count is the number of occurrences of the target attribute class_N in the input data. This predicate is defined for each target attribute;

result_model(attr_N, attr_val, class_M, count) - a predicate setting the number of occurrences (count) of the target attribute class_M in the input data, it should be true that the input attribute attr_N has value attr_val.

The example of a predicate cycle_class illustrating the organization of a cycle by classes is below.

```
cycle_class() :-
    class_list(CLASS_LIST),
    forall((member(CLASS,CLASS_LIST),
        class_count(CLASS,COUNT)),
        assert(result_model(CLASS,COUNT))).
```

In the right side of the predicate cycle_class the predicate class_list (CLASS_LIST) is used. The predicate class_list in the process of logical inference sets the variable CLASS_LIST with a set consisting of all possible target attributes. The next predicate is forall. This predicate is embedded in the SWI Prolog framework. In the process of inference, it performs a loop on all classes. During the cycle the number of occurrences of each class in the vectors of the input data of the Naive Bayes algorithm is calculated using the class_count predicate described above. Then, on this basis, the predicate result_model (CLASS, COUNT) is formed and with the help of the assert operator this predicate is added to the knowledge database of the prolog program. Thus, a knowledge model is made for a Naive Bayes algorithm.

The implementation of predicate cycle_attr is missing for simplicity.

For parallel execution of the Naive Bayes algorithm, it is necessary to add functionality for splitting the input data and merging the results computed in several working processes.

Below is a variant of a parallel algorithm, which performs parallel processing on attributes, implemented with usage of the offered approach.

```
algorithm():-
    mpi_init, /*MPI initialization*/
    /* Predicates for preparing lists with input data */,
    mpi_comm_rank(R), /*Get the number of current process*/
    (R == 0->     /*If it's main process*/
        /*predicate mpi_main_process_predicate will be satisfied */
    mpi_main_process_predicate(), !.
    ;       /* If the process is one of several working processes */
        /*predicate mpi_work_process_predicate will be satisfied */
    mpi_work_process_predicate()
    ),
    mpi_finalize.
```

where mpi_main_process_predicate – predicate of the main process, according to offered approach;

mpi_work_process_predicate - a predicate of a working process that handles a part of input data intended for a particular working process.

In the process of logical inference, the process number in the MPI system is determined. If the process number is not the number of the main process, then the distinct predicates in the particular working processes will be satisfied. At the end of its work, the working process sends the computation results represented by a set of predicates in the knowledge base of the logic program to the main process. The predicates representing computation results are made in the process of satisfying distinct predicates in a particular working process. In order to get a part of the initial data for particular working process in each working process, the predicate of splitting the input data into parts for one of the several working processes is satisfied in each working process.

The predicate of the main process can be defined as

```
mpi_main_process_predicate ():-
        /*Predicates for send attributes list ATTR_LIST*/
        /* to working processes for computations*/
        mpi_merge_attr_parall().
```

where mpi_merge_attr_parall - a predicate for merging the results of several working. In the process of it satisfying, the computation results of each working process are read and the corresponding result_model predicates are written to the knowledge base of the logic program of the main process for each handling attribute.

After the finish of algorithm there is the final output knowledge model M in the form of a set of result_model predicates in the knowledge base of the logic program of the main process.

The predicate of the working process can be defined as

```
mpi_work_process_predicate():-
        /* Predicates for receive attributes list ATTR_LIST */
        /*from main process*/
        /* if list is empty, then the work of a working process will
    be finished */
        mpi_divide_attr_parall(CUR_ATTR_LIST,
    ATTR_LIST),
        bayes_model (CUR_ATTR_LIST),
        /*Predicates for send data to the main process */
```

where mpi_divide_attr_parall - predicate for splitting the input data into parts for the work of a one from several working processes; After satisfaction of this predicate, the variable CUR_ATTR_LIST stores the list of attributes intended for current working process. The splitting based on the number of the working process in the MPI system on which logic inference is now performed.

bayes_model - the predicate is based on the earlier described predicate bayes_model, but works only with the part of the attributes specified by the variable CUR_ATTR_LIST.

Thus, the implementation of parallelization by the attributes of the Naive Bayes algorithm on the basis of the offered approach was overviewed.

5   Performance Evaluation

There were performed several experiments for the implemented Naive Bayes algorithms. The experiments have been performed with various input data sets (Table 1). These data sets contain various numbers of attributes. Thus, we have used the data sets for which the Naive Bayes algorithm works with various loading.

Table 1. Experimental results

Input data set	Number of vectors	Number of attributes	Avg. number of classes
A1	200	200	100
A5	200	1000	100
A10	200	2000	100

The experiments have been done on a multicore computer the following configuration: CPU Intel Xeon (4 cores), 2.80 GHz, 4 Gb, Ubuntu 18.04 LTS, SWI Prolog 7.6.4. The parallel algorithms have been executed for the numbers of cores equal to 2 and 4, respectively. The experimental results are provided in Table 2.

Table 2. Experimental results (s)

Algorithm	Cores	A1	A5	A10
Sequential form of Naive Bayes algorithm	1	1,28	26,1	100
Parallel form of Naive Bayes algorithm	2	0,7	13,18	50,85
	4	0,4	7,4	28,68

The experiments show that parallel execution of the Naive Bayes algorithm for data sets with various parameters is different. The parallel form of algorithm performs faster with grow of the count of using CPUs. The coefficient of performance increasing is approximately equivalent to the count of using CPUs, but a little bit less. We can see such result because part of the time in the case of parallel execution consumes for data transfer between parallel processes. Also we can see next regularity: if the count of attributes grows, percent of time for data transfer between parallel processes will decrease. This occur because grow of the time required for computation increases when the count of attributes grows. By this reason when the count of the attributes of input data sets grows, computation time consumes the biggest part of entire task time.

6 Conclusion

There is a big amount of different kind of data in the word of Internet of Things. This kind of data also called Big data. Big data analyzing and Data mining is difficult task requiring big computational resources. For this reason parallelization of such task is actual question. Standard approaches like MapReduce are not suitable for all tasks of Data mining. This article describes an approach to implementation and parallelization of data mining algorithms in logic programming framework. Advantages and disadvantages of logical programming framework for implementing Data mining algorithms were observed. The approach requires representing an algorithm as a functional expression. Each function from the expression can be implemented in the form of logic programming predicate.

For parallel execution of the algorithm, according to approach it is necessary to add predicates for a data partitioning and predicates for merging of the results calculated in different working processes.

Using the proposed approach Naive Bayes algorithm was implemented. An experiment results showed the effectiveness of parallel version for different data sets.

There could be obtaining parallel forms of algorithm in logic programming framework by using the proposed approach. Parallel forms of algorithm are more effective than sequential when several CPUs are available.

Acknowledgments. This work was supported by the Ministry of Education and Science of the Russian Federation in the framework of the state order "Organization of Scientific Research", task #2.6113.2017/ВУ.

References

1. Santucci, G.: From internet to data to internet of things. In: Proceedings of the International Conference on Future Trends of the Internet (2009)
2. Tsai, C.-W., Lai, C.-F., Vasilakos, A.V.: Future internet of things: open issues and challenges. Wireless Netw. **20**(8), 2201–2217 (2014)
3. Gorlatch, S.: Extracting and implementing list homomorphisms in parallel program development. Sci. Comput. Program. **33**, 1–27 (1999)
4. Dean, J., Ghemawat, S.: MapReduce: simplified data processing on large clusters. In: Proceedings of the USENIX Symposium on Operating Systems Design & Implementation (OSDI), pp. 137–147 (2004)
5. Urbani, J., Kotoulas, S., Oren, E., van Harmelen, F.: Scalable distributed reasoning using MapReduce. In: Bernstein, A., Karger, D.R., et al. (eds.) ISWC 2009. LNCS, vol. 5823, pp. 634–649. Springer, Heidelberg (2009). https://doi.org/10.1007/978-3-642-04930-9_40
6. Gupta, G., et al.: Parallel execution of prolog programs: a survey. https://cliplab.org/papers/partut-toplas.pdf
7. Amanda, C., King, R.: Data mining the yeast genome in a lazy functional language. http://users.aber.ac.uk/afc/papers/ClareKingPADL.pdf
8. Giannotti, F., Manco, G., Wijsen, J.: Logical languages for data mining. http://informatique.umons.ac.be/ssi/jef/ch9.pdf

9. Mooney, R., et al.: Relational data mining with inductive logic programming for link discovery. http://www.cs.utexas.edu/~ml/papers/ld-nsf-wkshp-02.pdf
10. Malov, A., Rodionov, S., Kholod, I.: The realization of Naive Bayes algorithm in the logic programming framework PROLOG. In: Proceeding of 2016 IEEE North West Russia Section Young Researchers in Electrical and Electronic Engineering Conference, pp. 290–294. IEEE (2016)
11. Paul, S.: Parallel and distributed data mining. In: Funatsu, K. (ed.) New Fundamental Technologies in Data Mining, pp. 43–54 (2011)
12. Zaki, M.J., Ho, C.-T. (eds.): LSPDM 1999. LNCS (LNAI), vol. 1759. Springer, Heidelberg (2000). https://doi.org/10.1007/3-540-46502-2
13. Kufrin, R.: Decision trees on parallel processors. In: Geller, J., Kitano, H., Suttner, C. (eds.) Parallel Processing for Artificial Intelligence, vol. 3. Elsevier-Science (1997)
14. Zaki, M.J., Ogihara, M., Parthasarathy, S., Li, W.: Parallel data mining for association rules on shared-memory multi-processors. In: Supercomputing 1996 (1996)
15. Cheung, D., Hu, K., Xia, S.: Asynchronous parallel algorithm for mining association rules on shared-memory multi-processors. In: 10th ACM Symposium Parallel Algorithms and Architectures (1998)
16. Kashef, R.: Cooperative clustering model and its applications. Ph.D. thesis, University of Waterloo, Department of Electrical and Computer Engineering (2008)
17. Hammouda, K.M., Kamel, M.S.: Distributed collaborative web document clustering using cluster keyphrase summaries. Inf. Fusion 9(4), 465–480 (2008)
18. Deb, D., Angryk, R.A.: Distributed document clustering using word-clusters. In: IEEE Symposium Computational Intelligence and Data mining, CIDM 2007, pp. 376–383 (2007)
19. Kholod, I., Malov, A., Rodionov, S.: Data mining algorithms parallelizing in functional programming language for execution in cluster. In: Balandin, S., Andreev, S., Koucheryavy, Y. (eds.) ruSMART 2015. LNCS, vol. 9247, pp. 140–151. Springer, Cham (2015). https://doi.org/10.1007/978-3-319-23126-6_13

Application of an Autonomous Object Behavior Model to Classify the Cybersecurity State

Viktor V. Semenov[(⊠)] [ID], Ilya S. Lebedev [ID],
Mikhail E. Sukhoparov [ID], and Kseniya I. Salakhutdinova [ID]

SPIIRAS, 14-th Linia, VI, No. 39, St. Petersburg 199178, Russia
v.semenov@iias.spb.su

Abstract. This paper considers the issues of ensuring the cybersecurity of autonomous objects. Prerequisites that determine the application of additional independent methods for assessing the state of autonomous objects were identified. Side channels were described, which enable the monitoring of the state of individual objects. A transition graph was proposed to show the current state of the object based on data from side channels. The type of sound signals used to analyze and classify the state of information security was also shown. An experiment intended to accumulate statistical information on the various types of unmanned object maneuvers was conducted using two audio recorders. The data obtained was processed using two-layer feed-forward neural networks with sigmoid hidden neurons. The autonomous object behavior model can be used as an additional element to determine the state of cybersecurity. Using a segmented model, it was possible to improve the accuracy of determining the cybersecurity state. The proposed model enabled the identification of differences in the states of autonomous object cybersecurity with probabilities that were, on average, more than 0.8.

Keywords: Information security · Acoustic channel · Data processing · Neural networks

1 Introduction

The development of the Internet of Things, the industry of remote automated production and minimization of sensor dimensions are the main trends of current technologies aimed at the use of remote autonomous objects that are computing devices of wireless networks and unmanned vehicles.

The relative availability, the controlled area absence, the importance of various systems of technological processes for the critical infrastructure make such objects attractive targets for malicious actions by potential intruders, which can lead to disastrous consequences.

Increasing the frequency of using malicious software such as Stuxnet, Flame, Duqu, and others does not allow us to fully discuss the absence of compromise even at the level of controller firmware, without mentioning the built-in protection tools or

© Springer Nature Switzerland AG 2019
O. Galinina et al. (Eds.): NEW2AN 2019/ruSMART 2019, LNCS 11660, pp. 104–112, 2019.
https://doi.org/10.1007/978-3-030-30859-9_9

application software code used as basic elements for building typical systems [1, 2]. Therefore, information obtained through side channels [3] may be one of the additional independent elements of assessing the condition of autonomous agents. More than ten side channels were identified within the framework of the research, based on which one can monitor the conditions of individual autonomous objects, including the acoustic channel, electromagnetic radiation, the time channel, etc. [4, 5].

The data obtained via these channels can be used to conduct various attacks [6, 7] and to monitor and analyze the state of the software and hardware environments [8, 9].

2 Problem Statement

Ubiquitous autonomous computing devices necessitate the solution of challenging issues, associated with the detection of unauthorized access to the main units at the software level, the analysis and identification of anomalies in the technological cycles of unmanned vehicle operation, the detection of destructive information impact on programs and algorithms, and the detection of undeclared capabilities [10].

The impossibility of guaranteeing protection against unauthorized reprogramming and the introduction of malicious software code makes it necessary to develop external monitoring systems. Monitoring models and methods based on external behavioral characteristics that acoustic, electromagnetic, and mechanical side channels identify may be one of the approaches to the implementation of such systems. In the case under study, states C were determined by signals F, which were the set of amplitudes A_1, A_2..., A_n. The time-synchronized sequence of amplitude values $\{\{a_1(t_1),\ a_1(t_2),\ ...\ a_1(t_m)\},$ $\{a_2(t_1),\ a_2(t_2),\ ...\ a_2(t_m)\},\ \{a_n(t_1),\ a_n(t_2),\ ...\ a_n(t_m)\}$ will condition the values of indicators obtained as a result of the autonomous object performing a certain action.

$$F(t) = \begin{vmatrix} a_1(0) & a_1(1) & ... & a_1(m) \\ a_2(0) & a_2(1) & ... & a_2(m) \\ ... & ... & ... & ... \\ a_n(0) & a_n(1) & ... & a_n(m) \end{vmatrix} \qquad (1)$$

Sequences having the form (1) make it possible to reduce the solution to the classification problem, where the set of classes takes the values $C = \{C_0,\ C_1\}$, with C_0 being a safe state, where the identified action caused by the control command is performed, and C_1 being an unsafe state in which the identified action currently differs from the one caused by the control command.

3 The Proposed Approach

The possibilities of using acoustic and electromagnetic emanations of microcircuits, principles of analyzing the state of objects, applied during non-destructive testing were considered in [11, 12].

It is possible to use various behavioral characteristics of objects as a supplement to the side channel that enable the detection of emerging anomalies associated with delayed command execution, and the appearance of background noise and vibrations.

Such analysis requires a transition graph that shows the current state of the object: whether it performs any command or makes the appropriate maneuver. These actions cause certain external and internal changes that are fixed based on the characteristics of the side channels.

The sequence of state transitions is determined by an autonomous object behavior model. For example, an unmanned vehicle that moves forward can have only two states: the halted state S_0 and the state of forward motion S_1. The graph model describes the behavior of the object during various events, where the probabilities of a dependent variable of class C are determined from several values of the variable:

$$P(C| \; a_1, \; a_2, \; ..., \; a_n) \tag{2}$$

One approach to state identification is to use an artificial neural network. Commands are reiterated to learn such artificial neural network where each signal track is converted into a spectral sequence. Each change in the implementation of behavior strategy is determined by a set of spectral sequences that are used as a training sample.

Thus, it becomes possible to classify the state of the object by inputting the current spectral characteristics of signals received from the side channels while actions are performed.

An external monitoring system of the autonomous unmanned object can be implemented on the basis of sensors that record the signal in real time mode when the object performs various actions. Although signals are obtained by retrieving information retrieval from several points, an acoustic channel is taken as the main source of information in the conducted experiment.

A state model is formed to determine the object behavior, where a sequence of states $\{S\}$ is identified as: S_0 – halted state, S_1 – forward motion, S_2 – backward motion, S_3 – forward-left motion, S_4 – forward-right motion, S_5 – backward-left motion, S_6 – backward-right motion.

Figure 1 shows a state model where the probabilities for an object to be in one of the listed states $\{S\}$ are determined by expression (2). For simplicity we assume that each state of an unmanned object should have a certain time increment, determined by the duration of responses to the received commands, and that there is a certain time interval between the executions of two subsequent commands that is associated with the computational processes on the object control device.

This assumption makes it possible to carry out analysis on the basis of discrete states and use the mathematical apparatus on the basis of the neural network approach to classification.

A neural network is trained by repetitive actions to obtain spectral information in states S_0-S_6. Spectral data from sensors are accumulated in each state depending upon how the network is trained. The class of the current state C_i will be calculated by a set of characteristics that result in the greatest probability:

$$C = argmax_i \, p_i \tag{3}$$

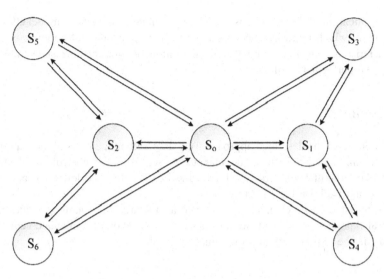

Fig. 1. An unmanned object state model

It was shown in [13, 14] that this approach is sufficient for obtaining classification results with an accuracy level of 0.7. At the same time, the application of the state model makes it possible to determine the states to/from which the transition of a system can be made.

For example, from state S_3 at the next discrete point in time, an unmanned object may get in state S_1 or S_0, from state S_1 it may pass into state S_0, S_3, S_4, and so on (see Fig. 2). The implementation of segmentation allows us not to consider all possible states, but to classify a limited set.

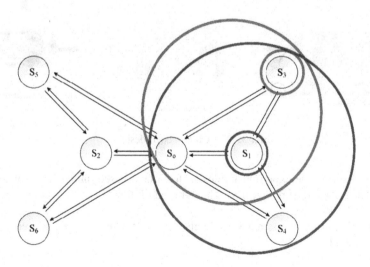

Fig. 2. An example of the state model segmentation by possible transitions from state S_1 (green oval) and S_3 (red oval). (Color figure online)

The application of a trained network and segmentation of the state model allows for classification, which takes into account not only the appearance of anomalous values of characteristics, but also eliminates consideration of unacceptable state transitions affecting the performance of actions.

4   Experiment

The capabilities of the described approach were assessed on the basis of the experiment where the unmanned vehicle made maneuvers set from the control panel straight forward (↑), straight backward (↓), forward-right (↗), forward-left (↖), backward-right (↘), and backward-left (↙) motion.

An external acoustic channel was selected as a source for obtaining values of the behavioral characteristics. Acoustic sensors were mounted on an unmanned vehicle and stationary in the experiment area for signal retrieval.

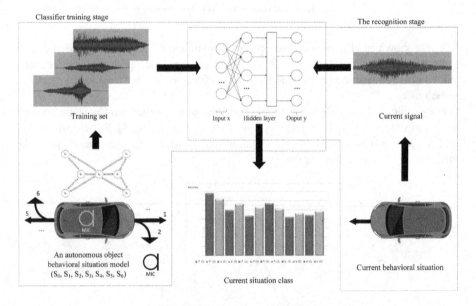

Fig. 3. The experimental design.

Figure 3 shows the experimental design for a land-based unmanned vehicle.

The data obtained were digitized. The duration of different maneuvers varied in the range from 2.5 to 3.5 s.

Figure 4 shows the external view of the signals for different maneuvers via two acoustic channels.

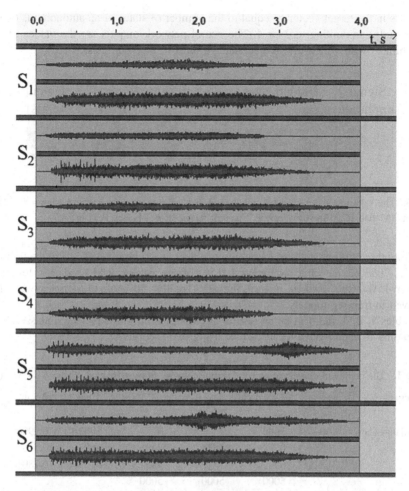

Fig. 4. The external view of the signals for two channels S_1–S_6. The acoustic sensor was located on the unmanned vehicle (top signal) and the acoustic sensor was statically mounted to the right of the experiment area (bottom signal).

For the purposes of analyzing the advantages and disadvantages of the method, the experimental conditions were reproduced exactly as those described in [15].

During signal sampling 5000 samples were taken evenly from each time-continuous signal containing amplitude values for two channels (from two acoustic sensors). The values obtained were processed using two-layer feed-forward neural networks with sigmoid hidden neurons (see Fig. 5). Software Neural Network Pattern Recognition application from DeepLearning Toolbox package of Matlab R2018b was used. Matlab was used because it is a simple and convenient tool for demonstrating the proposed approach. There were two input neurons; their number was determined by the measured parameters – signal amplitudes retrieved from two acoustic sensors, the number of hidden neurons was 300. The number of output neurons, i.e. the number of elements in the target vector, is equal to the number of states of an autonomous object under study, amounting to 3 or 4. The neural network outputs are the values of the probabilities with which the studied state belongs to a certain class.

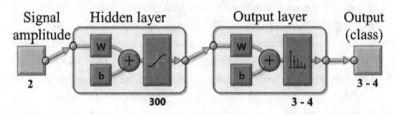

Fig. 5. The scheme of the used two-layer feed-forward neural networks with sigmoid hidden neurons (Matlab R2018b software); w – weight matrices, b – biases vectors.

From three to four states of autonomous object were classified. To form the training sample, 70% of the values were used, 15% of the values were used as a test, and another 15% were used as a verifying set. Totally, 30% of all samples were not involved in training models.

Tables 1, 2, 3 and 4 give the results of classification into segmented classes, corresponding to each maneuver in case of using two recorders.

Table 1. The results of classification by a neural network into segmented classes $S_3 - S_1 - S_0$.

Maneuvers		Actual class			Total	Accuracy
		S_0	S_1	S_3		
Calculated class	S_0	**5000**	2	0	5002	0.9996
	S_1	0	**3828**	849	4677	0.8185
	S_3	0	1170	**4151**	5321	0.7801
Total		5000	5000	5000		
Accuracy		1.0000	0.7656	0.8302		*0.8653*

Overall accuracy of the selected classifier for case 1 makes 12979/15000 = 0.8653.

Table 2. The results of classification by a neural network into segmented classes $S_1 - S_0 - S_3 - S_4$.

Maneuvers		Actual class				Total	Accuracy
		S_0	S_1	S_3	S_4		
Calculated class	S_0	**5000**	2	0	0	5002	0.9996
	S_1	0	**4153**	907	730	5790	0.7173
	S_3	0	183	**3489**	308	3980	0.8766
	S_4	0	662	604	**3962**	5228	0.7578
Total		5000	5000	5000	5000		
Accuracy		1.0000	0.8306	0.6978	0.7924		*0.8302*

Overall accuracy of the selected classifier for case 2 makes 16604/20000 = 0.8302.

Table 3. The results of classification by a neural network into segmented classes $S_6 - S_2 - S_0$.

Maneuvers		Actual class			Total	Accuracy
		S_0	S_2	S_6		
Calculated class	S_0	**5000**	2	2	5004	0.9992
	S_2	0	**3691**	619	4310	0.8564
	S_6	0	1307	**4379**	5686	0.7701
Total		5000	5000	5000		
Accuracy		1.0000	0.7382	0.8758		*0.8713*

Overall accuracy of the selected classifier for case 1 makes 13070/15000 = 0.8713.

Table 4. The results of classification by a neural network into segmented classes $S_2 - S_0 - S_5 - S_6$.

Maneuvers		Actual class				Total	Accuracy
		S_0	S_2	S_5	S_6		
Calculated class	S_0	**5000**	1	0	2	5003	0.9994
	S_2	0	**3719**	553	624	4896	0.7596
	S_5	0	412	**3617**	453	4482	0.8070
	S_6	0	868	830	**3921**	5619	0.6978
Total		5000	5000	5000	5000		
Accuracy		1.0000	0.7438	0.7234	0.7842		*0.8129*

Overall accuracy of the selected classifier for case 2 makes 16257/20000 = 0.8129.

As can be seen from Tables 1, 2, 3 and 4, the external monitoring system based on two recorders allows revealing differences in the maneuver parameters with a probability of more than 0.8, which significantly exceeds the performance when using a single recorder in [15].

5 Conclusion

The lack of permanent control over the object, the need for technical purposes of access to it using open networks necessitates the solution of problems related to identifying the internal and external processes. The use of internal monitoring tools does not fully guarantee the absence of a computing device compromise at the level of a software code, therefore, devices using independent channels based on the principles of measuring physical parameters and search for correlation of their changes with energy consumption, computation time, EMR, etc., become increasingly relevant.

The proposed approach can be applied to analyze the state of objects, in the case when the current state of the computational process is closed from the observer.

The accuracy of determining the cybersecurity state directly depends on the accuracy of the classification of the data processed by the system, which can be improved by the segmentation of the state model.

During segmentation the proposed model enabled to identify differences in the states of autonomous object cybersecurity with probabilities exceeding 0.8.

References

1. Heller, K., Svore, K., Keromytis, A., Stolfo, S.: One class support vector machines for detecting anomalous windows registry accesses. In: Proceedings Workshop Data Mining for Computer Security, Melbourne, FL, vol. 9, no. 108 (2003)
2. Chien, E., O' Murchu, L., Falliere, N.: W32. Duqu: the precursor to the next Stuxnet 2012. https://www.symantec.com/connect/w32_duqu_precursor_next_stuxnet. Accessed 5 July 2019
3. Semenov, V., Lebedev, I., Sukhoparov, M.: Identification of the state of individual elements of cyber-physical systems based on external behavioral characteristics. Appl. Inform. **13** (5/77), 72–83 (2018)
4. Hayashi, Y., Homma, N., Watanabe, T., Price, W., Radasky, W.: Introduction to the special section on electromagnetic information security. Proc. IEEE Trans. Electromagn. Compat. **55**(3), 539–546 (2013)
5. Han, Y., Christoudis, I., Diamantaras, K., Zonouz, S., Petropulu, A.: Side-channel-based code-execution monitoring systems: a survey. IEEE Signal Process. Mag. **36**(2), 22–35 (2019)
6. de Souza Faria, G., Kim, H.: Differential audio analysis: a new side-channel attack on PIN pads. Int. J. Inf. Secur. **18**(1), 73–84 (2019)
7. Gupta, H., Sural, S., Atluri, V., Vaidya, J.: A side-channel attack on smartphones: deciphering key taps using built-in microphones. J. Comput. Secur. **26**(2), 255–281 (2018)
8. Semenov, V., Lebedev, I., Sukhoparov, M.: Approach to classification of the information security state of elements for cyber-physical systems by applying side electromagnetic radiation. Sci. Tech. J. Inf. Technol. Mech. Opt. **18**(1/113), 98–105 (2018). https://doi.org/10.17586/2226-1494-2018-18-1-98-105
9. Semenov, V., Sukhoparov, M., Lebedev, I.: An approach to classification of the information security state of elements of cyber-physical systems using side electromagnetic radiation. In: Galinina, O., Andreev, S., Balandin, S., Koucheryavy, Y. (eds.) NEW2AN/ruSMART - 2018. LNCS, vol. 11118, pp. 289–298. Springer, Cham (2018). https://doi.org/10.1007/978-3-030-01168-0_27
10. Lebedev, I., et al.: The analysis of abnormal behavior of the system local segment on the basis of statistical data obtained from the network infrastructure monitoring. In: Galinina, O., Balandin, S., Koucheryavy, Y. (eds.) NEW2AN/ruSMART -2016. LNCS, vol. 9870, pp. 503–511. Springer, Cham (2016). https://doi.org/10.1007/978-3-319-46301-8_42
11. Han, Y., Etigowni, S., Liu, H., Zonouz, S., Petropulu, A.: Watch me, but don't touch me! Contactless control flow monitoring via electromagnetic emanations. In: Proceedings of the ACM Conference on Computer and Communications Security, pp. 1095–1108 (2017)
12. Genkin, D., Shamir, A., Tromer, E.: Acoustic cryptanalysis. J. Cryptol. **30**(2), 392–443 (2017)
13. Sukhoparov, M., Semenov, V., Lebedev, I.: Monitoring of cybersecurity elements of cyberphysical systems using artificial neural networks. Methods Tech. Means Ensuring Inf. Secur. **27**, 59–60 (2018)
14. Semenov, V., Lebedev, I.: Analysis of the state of cybersecurity of transport systems objects. Regional Informatics (RI-2018). In: Proceedings of the XVI St. Petersburg International Conference "Regional Informatics (RI-2018)", pp. 324–325 (2018)
15. Semenov, V., Lebedev, I.: Processing of signal information in problems of monitoring information security of unmanned autonomous objects. Sci. Tech. J. Inf. Technol. Mech. Opt. **19**(3/121), 492–498 (2019). https://doi.org/10.17586/2226-1494-2019-19-3-492-498

Decision Support Based on Human-Machine Collective Intelligence: Major Challenges

Alexander Smirnov and Andrew Ponomarev$^{(\boxtimes)}$ (iD)

St. Petersburg Institute for Informatics and Automation of the Russian
Academy of Sciences, 14th Line 39, St. Petersburg, Russian Federation
{smir,ponomarev}@iias.spb.su

Abstract. The paper discusses a novel class of decision support systems, based on an environment, leveraging human-machine collective intelligence. The distinctive feature of the proposed environment is support for natural self-organization processes in the community of participants. Most of the existing approaches for leveraging human expertise in a computing system rely on a pre-defined rigid workflow specification, and those very few systems that try to overcome this limitation sidestep current body of knowledge of self-organization in artificial and natural systems. The paper outlines the general vision of the proposed environment, identifies main challenges that has to be dealt with in order to develop such environment and describes ways to address them. Potential applications of such decision support environment are ubiquitous and influence virtually all areas of human activities, especially in complex domains: business management, environment problems, and government decisions.

Keywords: Collective intelligence · Artificial intelligence ·
Human-Machine systems · Human-in-the-Loop · Decision support

1 Introduction

Today, the rise of complexity of the problems we have to deal with, as well as the advances in the IT led to the emergence of new ways of organization of joint work of humans and machines: crowd computing, human computation, human-based optimization. In all the technologies mentioned above run-time workflow or algorithm is always fixed. However, for many problems occurring in highly dynamic domains it is not viable to arrange a well-defined workflow in advance, because the situation changes very fast. A natural technological answer to that is the creation of a new generation of human-machine systems, characterized by *adaptivity*. First steps in the direction of creating computational systems including humans and giving them more freedom to define possible actions and a course of solving a problem has already been done ("flash organizations" [1, 2], collective adaptive systems (CAS) [3], hybrid CAS [4] etc.).

It has to be noted, that decision support (in general) is one of that types of problems that really require flexible workflow, because decision-making very often is based on interactive and iterative exploration of the problem. Therefore, it is important to create methods (and a set of technologies implementing them) that would allow a collective of

© Springer Nature Switzerland AG 2019
O. Galinina et al. (Eds.): NEW2AN 2019/ruSMART 2019, LNCS 11660, pp. 113–124, 2019.
https://doi.org/10.1007/978-3-030-30859-9_10

people and software services with elements of AI to provide decision support defining the required activities in a flexible way.

The rest of the paper is structured as follows. In Sect. 2, we describe several areas related to the topic under research and enumerate most promising results that can be used in environments that support human-machine collective intelligence. Section 3 provides an overview of a novel concept of how human-machine collective intelligence environment for decision support. In Sect. 4, we formulate main challenges that has to be dealt with to develop such environment and blueprint ways to address them.

2   Related Work

In the very heart of the research there are two existing areas, specifically modern methods of programming joint human-computer effort (discussed under many names – from crowdsourcing to socio-cyberphysical systems), and self-organization. In this section, we briefly describe relevant results in each of the areas and underline the differences of the proposed approach.

2.1   "Programming" of Human-Machine Computing Processes

The overwhelming part of research in the field of human-machine computing (crowdsourcing, crowd-computing) systems position human as a certain type of "computing device" that can process requests of a certain type (for example, process images by identifying and labeling objects on them). Typical differentiation of duties in such systems is as follows. The system designer describes the necessary sequence of steps required for information processing, identifies steps that cannot be effectively performed using software and hardware only, and forms a set of solutions for performing these steps with a help of humans participating in such an environment (paying attention to motivation, quality assurance and other issues, caused by the specifics of the actual inclusion of human into a computing system). The function of a human participant, in turn, is reduced to performing a specific task, proposed by the system designer, interacting with the system in a manner strictly limited to the form of displaying the details of the task and entering the result. A human is a performer of some relatively simple function. This is partly due to the fact that the inclusion of a human is considered simply as an adaptation of the existing principles of the computing system construction. This is especially noticeable in the example of a service-oriented architecture (SOA), originally conceived for software systems, but formulated using sufficiently high-level concepts that allow different ways of implementing the services. For example, there are a number of SOA adaptations for representing services implemented by people (for example, WS-HumanTask, the works of D. Schall [5, 6]).

Based on the way the workflow is designed and encoded the existing approaches can be categorized [7] into three groups: (a) programming-level approaches; (b) parallel-computing approaches; and (c) process modeling approaches.

Programming level approaches focus on developing a set of libraries and language constructs allowing general-purpose application developers to instantiate and manage tasks to be performed on human-machine platforms (e.g., TurKit [8], CrowdDB [9] and AutoMan [10]).

Parallel computing approaches rely on the divide-and-conquer strategy that divides complex tasks into a set of subtasks solvable either by machines or humans (e.g., Turkomatic [11], Jabberwocky [12]).

The process modeling approaches focus on integrating human-provided services into workflow systems, allowing modeling and enactment of workflows comprising both machine and human-based activities (e.g., CrowdLang [13], CrowdSearcher [14] and CrowdComputer [7]).

Although such rigid division of roles (designer and participant of the system) and strict limitation of the participant's capabilities pays back in a wide range of tasks (usually simple ones like annotation, markup, etc.), the creative and organizational capabilities of a human in such systems are discarded. The first experimental systems where human participants could refine the workflow appeared in 2012 [15], but the problem is getting the closest attention of the research community only now. In particular, in the works of M. Bernstein, who studies the limitations of solutions based on the fixed flow of work [1] and the formation of dynamic organizations from members of the crowd community (the so-called flash organizations [2]).

This means that at present the challenge is to develop a set of theoretical and practical means to enable forming of such environments that would provide a wide range of opportunities to support self-organization – from describing fixed workflows that are well-proven in simple applications to flexible and adaptive self-organizing communities of participants for solving complex problems.

2.2 Self-organization

For the purpose of a brief overview, we can divide self-organization research into three streams: descriptive studies, formal studies and constructive studies.

Descriptive studies are aimed mostly on the analysis of how self-organization occurs in natural systems, what mechanisms are used. Due to the inter-disciplinary character of self-organization phenomenon descriptive studies may come from various scientific domains. The most important here is the research on human systems dynamics and self-organization in human collectives (e.g. [16]). And of that, the most relevant is research on how self-organization occurs by means of modern social media technologies, as it directly shows us how human collectives self-organize with a help of information and communication technologies. A prominent impulse for this kind of self-organization is an emergency situation. E.g., [17] discusses the phenomenon of creation of new social ties in the process of self-organization and problem solving by people affected by natural disasters. In particular, the paper focuses on methods that require shared information space (shared site of work and visible record of activity).

Formal studies are aimed mostly on the analysis of self-organization from the formal (mostly mathematical) point of view, developing new formal models of it: usually, based on differential equations, automata, or multi-agent paradigm. A prominent approach to formalize self-organization in human system has roots in game theory and describes humans as rational agents, which leads to self-organizing market models.

Constructive studies are aimed mostly on transferring principles of self-organization found in nature into artificial systems. This kind of research is mostly relevant in the paradigm of multi-agent systems, where various approaches to self-organization become a foundation of agent communication protocols. Self-organization mechanisms used in construction of computational (multi-agent) systems [18]: (a) mechanisms based on reinforcement learning; (b) cooperation-based mechanisms; (c) mechanisms based on the use of gradient fields; (d) market self-organization mechanisms; (e) mechanisms using the holonic system model.

3 Vision of the Decision Support Environment Based on Human-Machine Collective Intelligence

3.1 Positioning

The environment aims at supporting the process of making complex decisions and/or making decisions in complex problem domains. The complexity of making such decisions generally stems from problem uncertainty in many levels and the lack of relevant data at decision maker's disposal. Therefore, while in the upper level the methodology of decision-making stays quite definite (identification of the alternatives, identification of the criteria, evaluation of the alternatives etc.), the exact steps required to collect all the needed data, analyze it and present to the decision maker may be unclear. That is why decision support requires *ad hoc* planning of the low-level activities and should leverage self-organizing capabilities of the participants of the decision support process. Besides, currently most of the complex decisions are based not only on human intuition or expertise, but also on the problem-relevant data of various types and sources (starting from IoT-generated, to high-level Linked Data), processed in different ways. In other words, decision support is in fact human-machine activity, and the environment just offers a set mechanisms and tools to mitigate this activity.

The problem of supporting complex decisions touches the interests of the following roles. First, decision-makers, who need an access to the relevant expertise (not only in terms of domain knowledge, but also operational – what should be done to estimate absent parameters). Second, experts, who can provide this expertise, and, under some remuneration model may even be motivated to do so. Third, data and service providers, who are motivated to provide some problem-specific data and tools in exchange for payments.

The problem of making such decisions is ubiquitous and its impact spreads to virtually all areas of human activities, especially in complex domains: smart city, business management, environment problems, government and so on.

Successful solution would be an environment where participants of different nature (human and machine) could be able to communicate and decide on the particular steps of decision support process, perform these steps and exchange results, motivated by some external or internal mechanisms, making the whole environment profitable for all parties.

3.2   Users and Their Profiles

As it was already mentioned, there are three main categories of users: decision-makers, experts, service providers.

Decision-makers are responsible for the analysis of a situation and making a decision. In some cases, where the uncertainty associated with the situation is too high, the decision-maker requires some additional expertise that may be provided by participants of a human-machine collective intelligence environment. Bearing in mind, that using collective expertise is usually rather expensive and can be justified only for important problems, the decision-maker is usually a middle-to-top level manager in terms of typical business hierarchy. After the decision-maker posts the problem to the collective intelligence, he/she may oversee the process of solution and guide it in some way.

Experts possess problem-specific knowledge and may contribute into decision support process in several ways. First, they can propose procedures of obtaining relevant judgements, constructing in an *ad hoc* way elements of the whole workflow. This can be done not only in a direct manner, but also indirectly, by posting various incentives for other participants. Second, they can use their expertise by providing data as well as processing it to come to some problem-related conclusions. In general, an expert can be anyone – within or without the organization boundary, the difference is mostly in the incentives important for the particular expert.

Service providers design and maintain various software tools, services and datasets that can be used for decision support. Their goal is to receive remuneration for the use of these tools, that is why they are interested in making these services available for other participants of the environment. This is a direct evolution of the on demand service provisioning.

4   Major Challenges

To build an environment that meets the vision described in Sect. 3 several challenges have to be dealt with. This section enumerates these challenges and discusses possible ways to address them.

Meeting Collective Intelligence and Artificial Intelligence. Methods of collective intelligence (construed as methods for making people to work together to solve problems) and methods of artificial intelligence are two complementary (in some industries even competing) methods of decision support. Mostly, these approaches are considered as alternative (some tasks due to their nature turn out to be more "convenient" for artificial intelligence methods, and others – for collective), however, the scientists currently tend to speak about possibility of their joint usage and the potential that human-machine technologies have (*cf.* [19–21]).

A detailed analysis of existing (and potential) attempts to converge the approaches has revealed the following main options (Fig. 1):

(1) The use of artificial intelligence in collective intelligence systems:

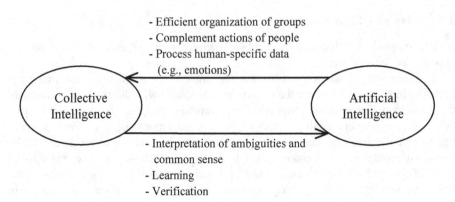

Fig. 1. Interrelations of collective intelligence and artificial intelligence.

a. The use of artificial intelligence methods for efficient and rational organization of people groups for joint problem solving. Existing solutions in this area include, for example, profiling of participants who perform tasks on crowdsourcing sites for the efficient work distribution, recommending tasks in crowdsourcing systems [22–24]. In a certain sense, this way of convergence can be understood as an application of artificial intelligence technologies at a meta level to the tasks of organizing the process of problem solving.

b. Application of problem-oriented methods of artificial intelligence to complement actions for information processing by people. This category, for example, includes optimization of human efforts in systems primarily focused on operations performed by people—partial automation (bots on Wikipedia). This also includes the use by a human (who is a part of a collective intelligence system) of some kind of artificial intelligence (at his/her discretion) and the interpretation of the result.

c. The application of artificial intelligence to the processing of specific human characteristics - recognition of the emotional state and other, in order to take it into account during the processing and integration of the results obtained by the community.

(2) The use of collective intelligence in artificial intelligence systems:

a. Ensuring interaction between a person (end user) and an artificial intelligence system (for example, to interpret user requests and translate these into a form "understandable" by a system of formal reasoning), the use of common sense, intuition, etc. for a more accurate specification of a task to be solved by an artificial intelligence system.

b. Learning artificial intelligence models while monitoring human activities [25], addressing a person through active learning protocols [26]. Learning the experience of a group of people, but not only subject experience (solving problems that arise in a particular subject area), but also meta-level experience – general social techniques (http://moralmachine.mit.edu).

c. Verification of the results of the work of artificial intelligence models for compliance with hardly formalized ethical principles and social norms.

From the point of view of orientation to a specific application area, the listed tasks can be divided into two levels: basic and problem-oriented. The basic level is associated with tasks, the solution of which is important regardless of the application area. This level includes tasks 1(a) and 2(a). The problem-oriented level, in turn, is associated with tasks, whose solving methods essentially depend on the application area of DSS (the remaining tasks listed above). We focus on the development of models and methods for solving basic problems.

There are four types of intelligent software services that can take part in the functioning of the environment (Fig. 2):

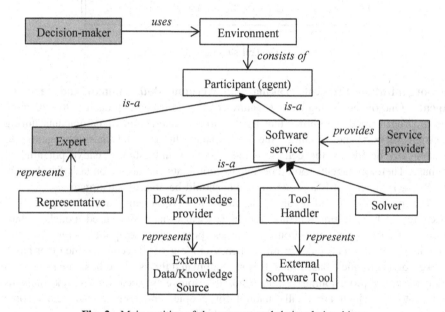

Fig. 2. Main entities of the concept and their relationships.

- Solver. A software code that can transform a task description in some way, enriching it with some derived knowledge.
- Data/Knowledge provider. Interface-wise almost similar to the previous type, however, only provides some problem-specific information.
- Tool handler. A utility agent that manages human access to some software tools (with GUI). In many cases, certain data processing routines required for decision-making can be implemented with some software (or, SaaS). It is not practical to re-implement it in a new way, however, granting an access to such tools might be useful for all the involved parties.
- Representative. Allowing expert to communicate with other services.

General structure of all types of software services contains several elements (see Fig. 3). Dashed part in the figure corresponds to communicative structure of the service, responsible for making agreements with other participants of the environment based on goals (given by the provider of the service) and competencies.

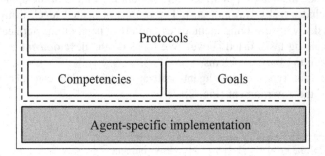

Fig. 3. General structure of an agent.

Self-organization Protocols Taking into Account Both Human and Machine Agents. One of the distinguishing features of the proposed approach is to overcome the preprogrammed workflows that rigidly govern interaction of participants during decision support and to allow the participants (human and machine agents) to dynamically decide on the details of the workflow unleashing creative potential of humans. Therefore, agents should be able to coordinate and decide on task distribution, roles etc., in other words a group of agents should be able to self-organize.

The protocols of self-organization in such environment have to respect both machine and human requirements. The latter means that widely used models of bio-inspired self-organization turn out to have less potential to be applied, as they mostly are taken from the analysis of primitive behaviors (e.g., of insects). On the other hand, market (or, economics) based models best of all match the assumed business model (on demand service provisioning). Another possible source are socio-inspired mechanisms and protocols which are totally natural for people, and there already exist some attempts to adapt them for artificial systems [30].

Particularly, the necessary activities are: (a) research on and formalization of the mechanisms of self-organization of human groups, (b) development of technological solutions based on the information technology to support the self-organization mechanisms, (c) transformation of the self-organization mechanisms into the form of protocols that can be used by software services (components) (socio-inspired self-organization).

The research efforts will further develop the models of the socio-inspired self-organization [30] earlier proposed by one of the authors and consistently apply these models for the organization of different kinds of interactions in the course of decision support in the human-machine collective intelligence environment.

Interoperability of Agents. To sustain various coordination processes, as well as information flow during decision-making there multilevel interoperability has to be provided inside the collaborative environment. This is especially acute in the case of mixed collectives, consisting of human and machine agents.

To implement any self-organization protocols, the participants of the system have to exchange several types of knowledge:

- Domain knowledge. What object and what relationships between objects are in the problem area.
- Task knowledge. Both goal description, and possible conceptualization of the active decision support task, e.g., mapping some concepts to alternatives, functions to criteria.
- Protocol knowledge. Terms of interaction, incentives, roles etc.

It is proposed to use ontologies as the main means ensuring the interoperability. The key role of the ontology model is in its ability to support semantic interoperability as the information represented by ontology can be interpreted both by humans and machines. Potentially, ontology-based information representation can provide the interoperability for all kinds of possible interactions (human-human, human-machine, machine-human). However, it is necessary to solve a number of fundamental tasks to fulfil this potential. These tasks are: (a) development of ontology model for representation and processing of data produced by the decision support processes, and (b) support of conciliated ontologies that capture different views on the same problem.

Taking into account the heterogeneity of the participants of the human-machine collective intelligence systems and the multidimensionality of the decision support activities, it is proposed to use multi-aspect ontologies. Such approach will enable to develop a model that can be applied to a broad spectrum of activities arising from the description and interaction of participants of the considered type of systems. Ontologies have proven themselves as a means resolving the problem of semantic interoperability, but applying ontologies to digital ecosystems is still a problem due to different terminologies and formalisms that the members of the ecosystems use. Three main groups of approaches to solving this problem can be distinguished: development of a universal common ontology, development of an ontology ecosystem, and development of a multi-aspect ontology.

The development of a universal common ontology is complicated by the amount of information it is supposed to represent. Besides, the diversity of service providers that make up the ecosystem, as well as its dynamic development, also complicates the problem of defining some common terminology and formalism (it is often reasonable to use different formalisms for solving different problems).

The development of an ontology ecosystem assumes existence of correspondences between the ontologies that make up the ontology ecosystem. With this approach, there are also significant difficulties associated with the ecosystem dynamics (the correspondences between ontologies must be constantly updated). One of the techniques to solving this problem may be ontology matching. However, at present, the methods supporting automatic ontology matching are relatively reliable only for specific narrow domains, and manual ontology matching requires considerable time and efforts. There are studies on enrichment of ontology facilities (e.g., extensions to ontologies in DAML+OIL for representation of the configuration problem [27] are developed; semantic annotations are introduced [28], etc.), but these studies still cannot solve the problem of integrating heterogeneous information and knowledge with different terminologies.

It is proposed to develop multi-aspect ontologies that enable to resolve the problems above. The multi-aspect ontologies will avoid the need for standardization of all services of a digital ecosystem through providing one aspect (some viewpoint on the domain) to services of one ecosystem community (services of one producer, services that jointly solve a certain task, etc.) for the service collaboration. A number of approaches and notations to development of multi-aspect ontologies exist. These approaches has to be analyzed, and based on this analysis an approach to development of multi-aspect ontologies for semantic interoperability in the developing digital ecosystems will be elaborated.

Soft Guidance in Collective Action. Though the execution process in the proposed environment is self-orchestrated and driven by negotiation protocols, human participants, however, will need intelligent assistance when communicating with other agents in the environment. The role of this assistance is to offer viable organization structures and incentive mechanisms based on current goals. An important aspect during the soft guidance is mapping actions defined by decision-making methodologies to human-computer collaboration scenarios. It means that the environment (or representative service) uses the existing knowledge on decision making to offer agents viable collaboration structures. In the context of classic prescriptive (recommended) decision making models (for instance, Simon's model), the activities delegated to the human-machine environment could be identification of criteria for decision support in the current situation, ranking and determining the criteria importance, identification and comparison of alternatives. All these activities often require a comprehensive analysis of different dimensions of the problem situation, taking into account the experience and, sometimes, the intuition of experts, which makes it advisable to use human-machine environments to carry out the mentioned activities. In the decision support theory, a large number of approaches (*cf.* [29]) to solve the decision-making problems has been developed. Such approaches can constitute a kind of initial pattern for organization of a decision support process that can either be reproduced exactly or be refined and modified according to the degree of definiteness of the problem situation and the decision maker desires. In particular, one of the common tasks is generation of a set of alternatives for the problem situation by the self-organized human-machine communities for the decision maker that makes choice.

5 Conclusion

The paper discusses a novel class of decision support systems, based on an environment, leveraging human-machine collective intelligence. The distinctive feature of the proposed environment is support for natural self-organization processes in the community of participants. Review of the current state-of-the-art in the area of programming human-machine effort has shown that most of the existing approaches rely on a pre-defined rigid workflow specification, and those very few systems that try to overcome this limitation sidestep current body of knowledge of self-organization in artificial and natural systems.

We have outlined the general vision of the proposed environment, identified main challenges that has to be dealt with to develop such environment and described ways to address them.

Potential applications of such decision support environment are ubiquitous and influence virtually all areas of human activities, especially in complex domains: smart city, business management, environment problems, government and so on.

Acknowledgements. The research is funded by the Russian Science Foundation (project # 19-11-00126).

References

1. Retelny, D., Bernstein, M.S., Valentine, M.A.: No workflow can ever be enough: how crowdsourcing workflows constrain complex work. In: Proceedings ACM Human-Computer Interact, vol. 1, no. 2, article 89 (2017)
2. Valentine, M.A., et al.: Flash organizations. In: 2017 CHI Conference on Human Factors in Computing Systems – CHI 2017, pp. 3523–3537. ACM Press, New York (2017)
3. Viroli, M., Audrito, G., Beal, J., Damiani F., Pianini D.: Engineering resilient collective adaptive systems by self-stabilisation. J. ACM Trans. Model. Comput. Simul. **28**(2) (2018). Article 16
4. Dustdar, S., Nastic, S., Scekic, O.: Smart Cities: The Internet of Things, People and Systems. Springer, Cham (2017). https://doi.org/10.1007/978-3-319-60030-7
5. Schall, D.: Service-Oriented Crowdsourcing: Architecture, Protocols and Algorithms. Springer, New York (2012). https://doi.org/10.1007/978-1-4614-5956-9
6. Schall, D.: Service oriented protocols for human computation. In: Michelucci, P. (ed.) Handbook of Human Computation, pp. 551–559. Springer, New York (2013). https://doi.org/10.1007/978-1-4614-8806-4_42
7. Tranquillini, S., Daniel, F., Kucherbaev, P., Casati, F.: Modeling, enacting, and integrating custom crowdsourcing processes. ACM Trans. Web **9**(2), 7:1–7:43 (2015)
8. Little, G.: Exploring iterative and parallel human computation processes. In: Extended Abstracts on Human Factors in Computing Systems, ser. CHI EA 2010, pp. 4309–4314. ACM (2010)
9. Franklin, M.J., Kossmann, D., Kraska, T., Ramesh, S., Xin, R.: CrowdDB: answering queries with crowdsourcing. In: Proceedings 2011 ACM SIGMOD International Conference on Management of Data, ser. SIGMOD 2011, pp. 61–72. ACM (2011)
10. Barowy, D.W., Curtsinger, C., Berger, E.D., McGregor, A.: AUTOMAN: a platform for integrating human-based and digital computation. SIGPLAN Not. **47**(10), 639–654 (2012)
11. Kulkarni, A.P., Can, M., Hartmann, B.: Turkomatic: automatic recursive task and workflow design for mechanical turk. In: CHI 2011 Extended Abstracts on Human Factors in Computing Systems, ser. CHI EA 2011, pp. 2053–2058. ACM (2011)
12. Ahmad, S., Battle, A., Malkani, Z., Kamvar, S.: The jabberwocky programming environment for structured social computing. In: Proceedings 24th Annual ACM Symposium on User Interface Software and Technology (UIST 2011), pp. 53–64. ACM (2011)

13. Minder, P., Bernstein, A.: *CrowdLang*: a programming language for the systematic exploration of human computation systems. In: Aberer, K., Flache, A., Jager, W., Liu, L., Tang, J., Guéret, C. (eds.) SocInfo 2012. LNCS, vol. 7710, pp. 124–137. Springer, Heidelberg (2012). https://doi.org/10.1007/978-3-642-35386-4_10

14. Bozzon, A., Brambilla, M., Ceri, S., Mauri, A., Volonterio, R.: Pattern-based specification of crowdsourcing applications. In: Casteleyn, S., Rossi, G., Winckler, M. (eds.) ICWE 2014. LNCS, vol. 8541, pp. 218–235. Springer, Cham (2014). https://doi.org/10.1007/978-3-319-08245-5_13

15. Kulkarni, A., Can, M., Hartmann, B.: Collaboratively crowdsourcing workflows with turkomatic. In: Proceedings of the ACM 2012 Conference on Computer Supported Cooperative Work, Seattle, Washington, USA (2012)

16. Dale, R., Fusaroli, R., Duran, N.D., Richardson, D.C.: The self-organization of human interaction. Psychol. Learn. Motiv. **59**, 43–95 (2013)

17. Kogan, M.: Digital traces of online self-organizing and problem solving in disaster. In: Proceedings of the 19th International Conference on Supporting Group Work - GROUP 2016, pp. 479–483. ACM Press (2016)

18. Gorodetskii, V.I.: Self-organization and multiagent systems: I. Models of multiagent self-organization. J. Comput. Syst. Sci. Int. **51**(2), 256–281 (2012)

19. Kamar, E.: Directions in hybrid intelligence: complementing ai systems with human intelligence. IJCAI Invited Talk: Early Career Spotlight Track (2016)

20. Nushi, B., Kamar, E., Horvitz, E., Kossmann, D.: On human intellect and machine failures: Troubleshooting integrative machine learning systems. In: 31st AAAI Conference on Artificial Intelligence, pp. 1017–1025 (2017)

21. Verhulst, S.G.: AI & Society **33**(2), 293–297 (2018)

22. Dai, P., Mausam, Weld, D.S.: Decision-theoretic control of crowd-sourced workflows. In: National Conference on Artificial Intelligence – AAAI (2010)

23. Dai, P., Mausam, Weld, D.S.: Artificial intelligence for artificial artificial intelligence. In: The 25th AAAI Conference on Artificial Intelligence, pp. 1153–1159 (2011)

24. Yuen, M-Ch., King, I., Leung, K.-S.: TaskRec: a task recommendation framework in crowdsourcing systems. Neural Process. Lett. **41**(2), 223–238 (2015)

25. Abbeel, P., Ng, A.: Apprenticeship learning via inverse reinforcement learning. In: 21st International Conference on Machine Learning (ICML) (2004)

26. Settles, B.: Active Learning Literature Survey. Computer Sciences Technical Report 1648. University of Wisconsin–Madison. http://pages.cs.wisc.edu/~bsettles/pub/settles.activelearning.pdf. Accessed 5 July 2019

27. Felfernig, A., et al.: Configuration knowledge representations for Semantic Web applications. Artif. Intell. Eng. Des. Anal. Manuf. **17**, 31–50 (2003)

28. Liao, Y., Lezoche, M., Panetto, H., Boudjlida, N.: Semantic annotations for semantic interoperability in a product lifecycle management context. Int. J. Prod. Res. **54**, 5534–5553 (2016)

29. Forsyth, D.R.: Decision making. In: Group Dynamics, 5th edn., pp. 317–349. Cengage Learning (2006)

30. Smirnov, A., Shilov, N.: Service-based socio-cyberphysical network modeling for guided self-organization. Procedia Comput. Sci. **64**, 290–297 (2015)

FaceWallGraph: Using Machine Learning for Profiling User Behaviour from Facebook Wall

Aimilia Panagiotou[1], Bogdan Ghita[2], Stavros Shiaeles[2(✉)],
and Keltoum Bendiab[2]

[1] Faculty of Pure and Applied Sciences, Open University of Cyprus,
2220 Nicosia, Cyprus
aimilia.panagiotou@st.ouc.ac.cy
[2] Centre for Security, Communications and Network Research,
University of Plymouth, Plymouth PL48AA, UK
{bogdan.ghita,gueltoum.bendiab}@plymouth.ac.uk,
sshiaeles@ieee.org

Abstract. Facebook represents the current de-facto choice for social media, changing the nature of social relationships. The increasing amount of personal information that runs through this platform publicly exposes user behaviour and social trends, allowing aggregation of data through conventional intelligence collection techniques such as OSINT (Open Source Intelligence). In this paper, we propose a new method to detect and diagnose variations in overall Facebook user psychology through Open Source Intelligence (OSINT) and machine learning techniques. We are aggregating the spectrum of user sentiments and views by using N-Games charts, which exhibit noticeable variations over time, validated through long term collection. We postulate that the proposed approach can be used by security organisations to understand and evaluate the user psychology, then use the information to predict insider threats or prevent insider attacks.

Keywords: Facebook · Social media · Information collection · OSINT ·
Machine learning · Web crawler

1 Introduction

Our preferred methods of communication over the last years have witnessed the highest speed of change in the whole human history [1]. In a short period of time, social media has become an integral part of our social life [1, 2]. Facebook, WhatsApp, Tencent QQ, Instagram, Twitter, Google, and LinkedIn are all examples of the rapid transfer of people interpersonal and social interactions from physical, human contact towards vast digital social communities. This transfer is happening on an unprecedented scale - Facebook alone currently has over 1 billion users and is expected to reach 1.69 billion registered accounts by 2020 [3], making it the most used social network worldwide. In light of the large amount of information provided by users about their personal life, such social networks become an increasingly significant public space that is used by

© Springer Nature Switzerland AG 2019
O. Galinina et al. (Eds.): NEW2AN 2019/ruSMART 2019, LNCS 11660, pp. 125–134, 2019.
https://doi.org/10.1007/978-3-030-30859-9_11

organisations for data gathering, from academic research to business intelligence [1, 3, 4]. In this context, psychologists state that Facebook contains valuable indicators of mental health [1], and indeed the social media profiles of participants in several psychological tests seem to show some indicative content [5] and analysis of available data on Facebook can be used to produce user patterns, help to profile users and anticipate their future behaviour [6].

Facebook provides easy access to information provided by friends and partners, including changes to their profile [6], additions of new contacts, called friends, and messages posted on their virtual wall. This makes user information public within one's circle of friends, allowing their list of friends to be publicly harvested or exploited [6], providing a relevant platform for capturing behavioural attributes relevant to an individual's thinking, mood, communication, activities, socialisation and psychological changes [7, 8]. The emotion and language used in the messages posted on the user virtual wall may indicate feelings of loneliness, self-hatred, depression, and anxiety that, in turn, characterise mental illness [8, 9]. Inferring such relationships could also be relevant for national security organizations when trying to pre-empt malicious activities such as hiring hitmen, grooming the targets of paedophiles, or stealing identities [1].

Therefore, collecting and mining information of millions of users who are digitally expressing their feelings is of substantive value within many areas [1]. Generally, collection and manipulation of such information is done by Open Source INTelligence (OSINT) techniques [10]. Such techniques aim to discover useful "intelligence that is produced from publicly available information collected, exploited, and disseminated to an appropriate audience for the purpose of addressing a specific intelligence requirement", according to [10]. It differs from traditional intelligence techniques since it collects data from public sources such as social media, blogs and web communities [4]. OSINT can access to important personal information or links through social networks without any limitations [2]. To our best knowledge, OSINT and Facebook have not been used in any research to show the variations in user psychology over time. Thus, in this paper, we aim to contribute in this relevant field, by introducing a new approach that relies on machine learning and OSINT to detect alternations in the psychological profile of a Facebook user. The main contribution of this paper is the proposal of a novel machine learning method that allows information filtering, followed by the psychological user profile features mined in N-games layouts to feed the location finder framework proposed in [4]. In addition, the paper introduces a practical implementation for collection and processing of information from Facebook profiles, which may be stored in a database for further analysis. This research could be developed with more specific goals to serve other organizations; for example, it can be used by security organizations to detect and prevent malicious activities. It can also be used by public health to identify at-risk individuals or detect depression, and frame directions on guiding valuable interventions.

The paper is organised as follows: Sect. 2 describes the prior work discussing social media and its impact on individual feelings. In Sect. 3, we present the methodology of the proposed method using machine learning and OSINT to construct the information profile of a user. Section 4 presents experimental results and Sect. 5 provides concluding remarks and future work.

2 Related Works

Over the recent years, there has been growing interest in using social media as a tool for analysing emotions as expressed by people as part of their online interaction. In the context of Twitter, [12] highlighted that people publicly post information about their feelings and even their treatment on social media. Work in [13] studied linguistic and emotional correlation for postnatal changes of new mothers and built a statistical model to predict extreme postnatal behavioural changes using prenatal observations. This work highlights the potential of Twitter as a source of signals about likelihood of current or future episodes of mental illness. In a similar context, [9] demonstrated the potential of using Twitter as a tool for measuring and predicting major depression in individuals. The work proposed a variety of social media measures such as language, emotion, style, and user engagement to characterise depressive behaviour; supervised learning was used to construct classifier trained to predict depression, yielding 70% classification accuracy.

The authors of [14] focused on examining the relationship between social network, loneliness, depression, anxiety and quality of life in community-dwelling older people living in Dublin. In the study, thousands of community-dwelling people aged 65 and over were interviewed using the GMS-AGECAT diagnostic approach [15] and information from social networks was processed using the Practitioner Assessment of Network Type (PANT) schedule developed by Wenger [16]. The study found that loneliness and social networks independently affect the mood and wellbeing of the elderly, concealing a significant rate of underlying depression.

With the increasing societal importance of online social networks, their influence on the human collective mood state has become a matter of considerable interest and several academic research studies were undertaken. A long-term experiment was conducted over two years as part of [5] to study the influence of social networks on people's feelings. Several tests were performed on Facebook users, which consisted of manipulation of information posted on 689,000 user home pages, including news feeds, the flow of comments, videos, pictures and web links posted by other people in their social network. The study found that the mix of information could influence the overall mood of the user both positively or negatively through a process coined by the authors as "emotional contagion". Similarly [17] included a series experiments performed on social contagion, using several statistical approaches to characterise interpersonal influence with respect to a range of psychological and health-related characteristics such as obesity, smoking, cooperation, and happiness. Through the analysis of social network data, authors have suggested that the above characteristics do traverse and influence social networks; the study found evidence regarding social contagion in longitudinally followed networks. Work in [18], focused on studying the general happiness of Twitter users by recording and measuring their individual tweets. 129 million tweets were collected over a 6month period, then a mathematical model was used to measure emotional variation of each user in time and how it is spread across links. The study found that general happiness, or subjective well-being (SWB), of these users is assortative across the Twitter social network. The study results imply that online social networks replicate social mechanisms that cause assortative mixing in a real social network. Thus, their ability to connect users with similar levels of SWB is an important factor in how positive and negative emotions are preserved and spread through human society.

Based on outlined research in this section, many experimental studies analysed the emotional variations of people over time and the influence of online social networks on these variations. In these studies, researchers have developed different approaches that can be used to measure user's emotions variation through social networks. All the performed experiments demonstrated the potential of using social media as a tool to study the mental status of users, but the majority of them were focused on the Twitter social network. In addition, most of them were experimentally conducted by using statistical approaches and in some cases with the physical presence of people under investigation. None of these approaches combines the simultaneous usage of a crawler for collecting information with a machine learning approach for filtering it and storing it in a database. Moreover, OSINT and Facebook were not applied by prior research to investigate the variations in user psychology. In this paper we propose a new approach that combines the above elements aiming to converge faster to more accurate results when compared to the studied approaches. The validation results indicate the proposed approach is effective and accurate.

3 The Proposed Method

This section presents the methodology used to investigate the psychological states of individual users over time based on the information available as part of the online social network.

As illustrated in Fig. 1, the first step consists in gathering relevant data from the Facebook social network and placing it in a database for further analysis. In this step, a web crawler has been developed and is used to collect the wall information from our Facebook friends. All this information were stored in a database for father analysis. The classification process is undertaken using the N-Gram technique. N-Grams are used for natural language processing to develop not just unigram models but also bigram and trigram models that will for developing features for our Learning Machine system. The Learning Machine system is where the data are further filter and the psychological profile of user is produced.

3.1 Information Collection from Facebook

On Facebook, people frequently express and share their negative and positive emotions [19], which are later seen by their friends via a "News Feed" page [20], which dynamically updates the activities of the Facebook friends of a user [20]. Authors in [19] confirmed that people use News Feed every day to stay in touch with friends and family, and to stay informed about the world around them. It is worth nothing that, in this study, information was initially collected from the content shared by friends of the research team, then it was expanded to friends of friends, with their consent. The data was gathered via a web crawler and stored in a database (see Fig. 1). In the initial stage of the data gathering process, a FacebookBot [21] was created for automatic collection

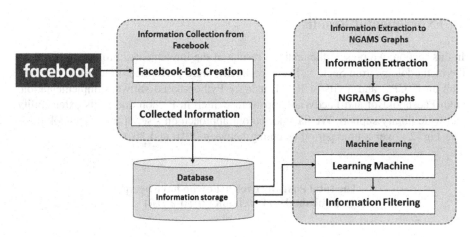

Fig. 1. Overview of the proposed method.

and storage of each user available information including posts (the content and the associated timestamp), personal details (name, Facebook URL, gender) and photos. The Facebook crawler uses Selenium [22] to automatically connect to Facebook and propagate to the news feed to scarp the HTML at that URL to gather, cache and extract the aforementioned information.

3.2   Information Extraction and Classification

The information collected from profiles of friends through Facebook Wall are sent to the machine learning components for pruning the posts from irrelevant content such as, URL links, Facebook user names and articles ("a", "an", "the"). Then, stored the result in the Database (see Fig. 1). The results stored in the database are used in the process of the N-grams graphs construction, where the information related to user psychology are mined using N-Grams layouts. The Python class NGrams [23] is used for the N-Grams graphs construction.

In this paper, we use English language; however, our approach can be updated to other languages. We assume that an emoticon within a post represents an emotion for the whole post and all the words of this post are related to this emotion. Thus, we split the psychological profiles in four classes: *Happy*, *Sad*, *Love*, and *Disappointment*.

- **Happy**: this class present word that design filing of happiness, such as "happy", ":-)", ":)", "=)", ":D".
- **Sad**: this class represent words that present emotion of sadness, such as "sad", the symbols '☹' ":-(", ":(", "=(", ";(".
- **Love**: this class consists of words that present feeling of love, such as "love", '<', '3', "<3".
- **Disappointment**: this class consists of words that present feeling of frustration, such as "disappointed" or "anger".

4 Experimental Results

In this section, we present the analysis results of the implemented prototype, based on the methodology from Sect. 3. The experiments were performed on a virtual machine with an i5 CPU and 8 GB of RAM, using a Python-based software implementation. Both FacebookBot and FaceWallGrap were written in Python due to its extensibility and wide library support. We have tested our approach on a set of real Facebook posts (see Fig. 2.) over a time period spanning between 2010 and 2016.

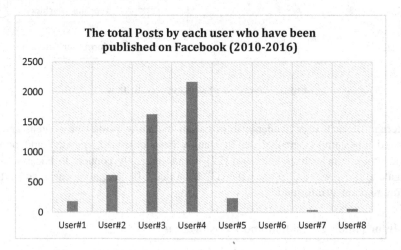

Fig. 2. The total Posts by each user who have been published on Facebook over the period (2010–2016)

The following graphs summarise the results of our Facebook experiment using our Facebook account wall.

4.1 The *Happy* N-Gram Graph

As illustrated in Fig. 3, the "Happy n-gram" shows the feeling of joy in time periods from a total result for all users. The words used to create this n-gram are words and symbols that belongs to the class "Happy". It seems that in this graph statistically in the past 2–3 years users have ups and downs in the subject of joy while in the past years they were stagnant.

4.2 The *Sad* N-Gram Graph

The "Sad n-gram" graph (Fig. 4) shows the feeling of regret in time periods from a total result for a user. The words used to create this n-gram are words and symbols that belongs to the class "Sad". Moreover, it seems that in this graph statistically the last two years users have very few and specific moments that are sad.

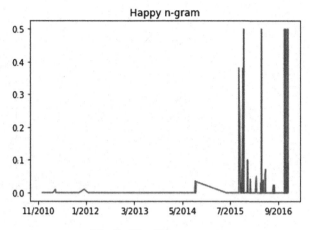

Fig. 3. The *Happy* n-gram

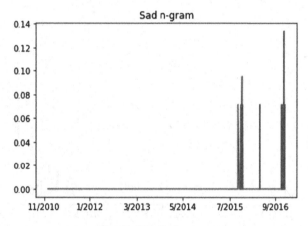

Fig. 4. The *Sad* n-gram

4.3   The *Love* N-Gram Graph

The Love n-gram graph (Fig. 5) shows the feeling of love in time periods from a total result for a user. The words used to create this n-gram are words, numbers and symbols that belong to the class "Love". Additionally, it seems that in this graph statistically the 2-year years users have increased moments who are in love or love their round.

4.4   The *Disappointed* N-Gram Graph

The *Disappointed* n-gram graph (Fig. 6) shows the feeling of frustration over time by a total result for all users. The words used to create this n-gram are from the *disappointed* class that reveals disappointment. In addition, it seems that in this graph statistically the last 3 years users have increased moments that are disappointed.

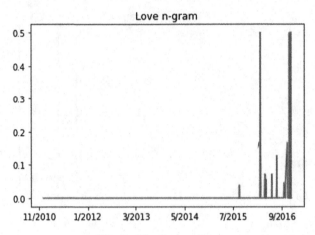

Fig. 5. The *Love* n-gram

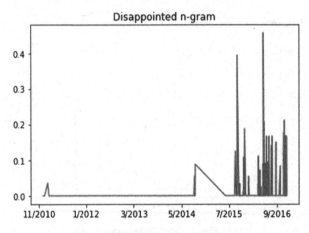

Fig. 6. The *Disappointed* n-gram

5 Applications

The experiment from Sect. 3 and the results presented in Sect. 4 are a preliminary analysis that aimed to determine whether predefined, major n-graphs can be used to discriminate between users overall psychological state and observe long-term trends. At this moment, the investigation is merely qualitative, aiming to highlight the technique. The graphs from Figs. 4, 5 and 6 indicate that long-term analysis does reveal a number of artefacts that can be further analysed for a more detailed or accurate conclusion.

As a baseline, the n-graphs can monitor one's overall state of mind and, should they be analysed accordingly, can correlate or reveal an underlying mental or psychological condition in order to prevent further deterioration. At a next level, the analysis can reveal the evolution of interaction and correlate the n-graphs with the context in order to highlight a possible negative attitude towards the other members of the society. As indicated by prior research, such a variation of perspective may align with more dangerous activities, from radicalisation to terrorism. The ability to determine such changes of perspective is not a new area of research, but the analysis of input data makes it rather challenging due to the volume and complexity of the context input data. In a long-term approach, observing any sudden changes in the n-graph characteristics can be used as an initial prompt to escalate the analysis and introduce more input data for analysis. Continuing this direction. one significant benefit of the proposed approach is the convenience of the data collection process. Given the openness of Facebook, the data already available for monitoring, as long as the data collection is following a GDPR-compliant procedure.

6 Conclusion

The purpose of this paper was to extend the system proposed in [4] by adding a suspicious user classification component based on psychological profiling. It can be concluded that social networks such as Facebook may be used, through n-gram extraction, for user profiling and it is also noteworthy that this procedure is complete transparent to the user. It can be also concluded that certain environmental and social context conditions are important factors and may affect emotions. Future extension of this work would be the integration with the proposed model in [4], the addition of more psychological profiles, retrieving information from other social networks such as Twitter, Instagram and finally process pictures uploaded from users and classify their emotion based on face detection. This combined framework could be proven beneficial for law enforcement and be able to prevent unwanted situations.

References

1. Omand, D., Bartlett, J., Miller, C.: Introducing social media intelligence (SOCMINT). Intell. Nat. Secur. **27**(6), 801–823 (2012)
2. Kanakaris, V., Tzovelekis, K., Bandekas, D.V.: Impact of AnonStalk (Anonymous Stalking) on users of Social Media: a case study. J. Eng. Sci. Technol. Rev. **11**(2) (2018)
3. Statista.   https://www.statista.com/statistics/490424/number-of-worldwide-facebook-users/. Accessed 12 Mar 2019
4. Pellet, H., Shiaeles, S., Stavrou, S.: Localising social network users and profiling their movement. Comput. Secur. **81**, 49–57 (2019)
5. Kramer, A.D., Guillory, J.E., Hancock, J.T.: Experimental evidence of massive-scale emotional contagion through social networks. Proc. Natl. Acad. Sci. **111**(24), 8788–8790 (2014)
6. Gritzalis, D., Kandias, M., Stavrou, V., Mitrou, L.: History of information: the case of privacy and security in social media. In: Proceedings of the History of Information Conference, pp. 283–310 (2014)

7. Muise, A., Christofides, E., Desmarais, S.: More information than you ever wanted: does Facebook bring out the green-eyed monster of jealousy? CyberPsychology Behav. **12**(4), 441–444 (2009)

8. Palumbo, C., Volpe, U., Matanov, A., Priebe, S., Giacco, D.: Social networks of patients with psychosis: a systematic review. BMC Res. Notes **8**(1), 560 (2015)

9. De Choudhury, M., Gamon, M., Counts, S., Horvitz, E.: Predicting depression via social media. In: Seventh International AAAI Conference on Weblogs and Social Media (2013)

10. Hribar, G., Podbregar, I., Ivanuša, T.: OSINT: a "grey zone"? Int. J. Intell. CounterIntelligence **27**(3), 529–549 (2014)

11. Williams, H.J., Blum, I.: Defining second generation open source intelligence (OSINT) for the defense enterprise. RAND Corporation Santa Monica United States (2018)

12. Sadilek, A., Kautz, H., Silenzio, V.: Modeling spread of disease from social interactions. In: Sixth International AAAI Conference on Weblogs and Social Media (2012)

13. Paul, M.J., Dredze, M.: You are what you tweet: analyzing twitter for public health. In: Fifth International AAAI Conference on Weblogs and Social Media (2011)

14. Golden, J., et al.: Loneliness, social support networks, mood and wellbeing in community-dwelling elderly. Int. J. Geriatr. Psychiatry J. Psychiatry Life Allied Sci. **24**(7), 694–700 (2009)

15. Copeland, J.R.M., Dewey, M.E., Wood, N., Searle, R., Davidson, I.A., McWilliam, C.: Range of mental illness among the elderly in the community: prevalence in Liverpool using the GMS-AGECAT package. Br. J. Psychiatry **150**(6), 815–823 (1987)

16. Wenger, G.C.: Support networks in old age: constructing a typology. In: Growing Old in the Twentieth Century, pp. 166–185 (1989)

17. Christakis, N., Fowler, J.: Social contagion theory: examining dynamic social networks and human behavior. Stat. Med. **32**(4), 556–577 (2012)

18. Bollen, J., Gonçalves, B., Ruan, G., Mao, H.: Happiness is assortative in online social networks. Artif. Life **17**(3), 237–251 (2011)

19. Bazarova, N.N., Choi, Y.H., Schwanda Sosik, V., Cosley, D., Whitlock, J.: Social sharing of emotions on Facebook: channel differences, satisfaction, and replies. In: Proceedings of the 18th ACM Conference on Computer Supported Cooperative Work & Social Computing, pp. 154–164. ACM (2015)

20. Facebook. https://www.facebook.com/facebookmedia/solutions/news-feed. Accessed 25 Apr 2019

21. Facebook for developers. https://developers.facebook.com/products/messenger/. Accessed 3 Mar 2019

22. Selenium with python. https://selenium-python.readthedocs.io/, Accessed 3 Apr 2019

23. Tutorial—Python NGram 3.3 documentation. https://pythonhosted.org/ngram/tutorial.html. Accessed 29 May 2019

Multi-agent Approach to Computational Resource Allocation in Edge Computing

Alexey Kovtunenko[1]([⊠]), Marat Timirov[2], and Azat Bilyalov[3]

[1] Ufa State Aviation Technical University, Karl Marx st. 12, Ufa, Russia
askovtunenko@ugatu.su
[2] Bashkir Production Association "Progress",
Kirovogradskaya st. 34, Ufa, Russia
TimirovM@progressufa.ru
[3] Bashkir State Medical University, Teatralnaya st. 2a, Ufa, Russia
azat.bilyalov@gmail.com

Abstract. The system of streaming data and complex events distributed processing is considered. A mathematical model for computational resource consumption and allocation in edge computing is offered. On the basis of mathematical model there was developed a multi-agent algorithm as a part of agent-based software architecture for distributed processing of streaming data, which had been developed by the authors previously. The efficiency of the developed algorithm has been investigated by means of simulation in AnyLogic. The resource allocation problem in edge computing was simulated and the algorithm demonstrated satisfying results both at static and dynamic regimes. Also, the perspectives of intellectual technologies using in resource allocation for edge computing are considered.

Keywords: Edge computing · Multi-agent algorithm ·
Computational resource allocation · Internet of things · Streaming data ·
Complex event processing

1 Introduction

Modern progress in computer engineering allows the coexistence of different data processing devices such as sensors, actuators and high-performance servers in the same communicative ecosystem. The most popular term to denote such systems is Internet of Things, and it is the best way to describe the gist of them - decentralizing, scalability and flexibility. Internet of things has the main difference from traditional internet of documents. It is aimed to operating with data streams (or data flows).

The implementation of the IoT concept requires the solution of many problems. Among them are the problems of security and reliability of networks, problems of network performance, as well as problems of resource management of network devices and services. Researchers have proposed a variety of different solutions to ensure

Supported by Ufa State Aviation Technical University.

O. Galinina et al. (Eds.): NEW2AN 2019/ruSMART 2019, LNCS 11660, pp. 135–146, 2019.
https://doi.org/10.1007/978-3-030-30859-9_12

reliability and security, in particular, Cloud Technologies, Blockchain, Software-Defined Networks (SDN) [1]. To increase the performance of D2D interactions, it is possible to use 5G communication technologies [2]. Since the market of IoT devices is largely represented by mobile devices, various virtualization technologies in the field of wearable devices have been developed [3]. Many researchers and developers are trying to solve the problem of computational resource allocation in both edge and cloud computing [4]. Let us consider several solutions.

For example, Thinnect [5] is an IoT edge network service provider which combines cloud and edge computing. Their service is complementary to cellular operators, offering high density, large scale, local coverage, utilizing cost-efficient technologies. Since their approach is more micro-controller based, for specific tasks we need specific devices. Nebula [6] is the project which goal is to be a Docker orchestration for IoT devices and distributed services (like CDN or edge computing). Users have the ability to update simultaneously tens of thousands of IoT devices all around the globe with a single API call, allowing users to treat IoT devices like another distributed Docker application. The project is at an early stage, but despite all of that, as an idea, it provides a lot of potentials. Mainflux [7] is highly secure, scalable, open-source IoT platform written in Go and deployed in Docker. It serves as software infrastructure and set of microservices for development of the Internet of Things Solutions and deployment of Intelligent products.

2 Problem Formulating

2.1 Understanding of Streaming Data in IoT

Data stream is a sequence of data elements that arrive one at a time, where each element can be viewed as a couple of the time stamp associated with the element, and the element payload, represented as a data structure. Many stream processing frameworks use data flow abstraction by structuring the application as a graph of operators [8].

In real world, phenomena can be considered from the perspective of either continuous or discrete-event models. Thus, transmitting data also can be divided into following categories.

- Streaming data (related to continuous model), when the efficiency of the system is fully depending on the timeliness of the new information delivery. Time stamps of all dataflow elements differ by the same value. As a rule, streaming data is the actual values of physical parameters (such as temperature, humidity, illumination etc.). Each transmitted value's usefulness is completely determined by its relevance, and, therefore, an untimely transmitted value can be lost without any harm to the system.
- Discrete events (the data about discrete phenomena), when efficiency directly depends on the data delivery fact, and therefore loss or corruption of even one element of dataflow (event) can significantly affect the system and make it work incorrectly. The time stamps of the dataflow elements point to event's occurrence time. As an event we can consider electric contact closure, micro switch tripping and more complex situations.

Dataflow processing operators may be of following types [9].

- Streaming data conversion operators. These operators perform functions such as counting, filtering, projection, and aggregation, where the processing of an input data stream by an element can result in the creation of subsequent streams that may differ from the original stream in terms of data structure, but not rate.
- Events generating operators. These operators transform the streaming data into discrete events by means of computational algorithms. When the algorithm detects in the incoming streaming data predetermined qualitative changes, it creates an outgoing dataflow element with the time stamp indicating the moment of occurrence of this change.
- Complex events processing operators. They aim to detect qualitative changes in the set of different event flows with the generation of new event flow.

Processing of streaming data regards the problem of data processing from a new point of view. Nowadays the efforts of researches are focused on the development of streaming data processing systems in terms of providing the following functions: processing continuous queries (CQ) based on some query language or algebra for data streams [10] and providing a Quality of Service (QoS) based on some data stream model [11].

Another aspect of scientific research in area of dataflow management systems is a complex event processing (CEP) systems: development of discrete-event models of different subject areas, formulating of logical rules for complex events generating.

As a rule, an IoT system is part of some global automated management and monitoring system. Traditionally, such systems implement a client-server hierarchical architecture, where at the bottom there are simple transducers (sensors and actuators), the nodes are data collection gateways, and at the top is a server (or several servers) that perform analytical functions: storing, complex processing of the data, generating of the control signals and also providing a special-policy-driven access to data. The client-server organization of the computing system simplifies the design and deployment, but on the other hand it has some flaws [12]. Among them is the lack of flexibility and scalability, as well as the low survivability of the software layer.

Nowadays engineers and developers prefer decentralized architectures of data processing and storing systems [13]. First of all, it refers to hardware of computing systems, when different analytical functions are physically deployed at different hosts. On the other hand it also may refer to software layer organizing, when different analytical functions are executed within different virtual containers or virtual environments [14].

Decentralization of data processing in IoT has spawned a new class of systems, where some tasks are performed directly on data collection gateways [15]. Thus, the concept of edge computing is realized. Edge computing provides fuller utilization of the hardware computing resources, increase the flexibility and survivability of the system.

Most of the considered systems represent a completed solution for event-based development, some of them even support distributed computations in common namespace, but they don't include a model of effectively computational resource allocation and don't allow dynamical management of computational resources.

2.2 Mathematical Model

The proposed mathematical model of processing of streaming data in IoT is represented by the following kit of sets and relations [16]:

-   attributes A – physical and abstract parameters which values can be got directly from sensors or calculated by means of computational algorithms from other parameters
-   computational operators P – computational procedures that allow to get new values of attributes
-   computing nodes W – computing devices on which all calculations are executed and all data are stored
-   a relation C that indicates which attributes are generated by what operator,

$$C = \{(p,a) : p \text{ calculates } a;\ a \in A,\ p \in P\} \tag{1}$$

-   a relation I that indicates which attributes are used in calculations in each operator, such that $C \cap I = \varnothing$,

$$I = \{(p,a) : p \text{ uses } a;\ a \in A,\ p \in P\} \tag{2}$$

-   a relation D that shows for each computational operator on what node it physically executes

$$D = \{(w,p) : w \text{ executes } p;\ w \in W,\ p \in P\} \tag{3}$$

It is traditionally considered that the computing resources of the network consist of the following components [17]: CPU performance, memory, network bandwidth performance.

   To formulate the problem of optimal computing resources allocation in the distributed system of streaming data processing, it is necessary to formulate a mathematical model for the resource consumption in computer networks in terms of numerical functions, as follows. The resource consumption of the processing problem:

-   $M : A \to \mathbf{N} \backslash \{0\}$, $M(a)$ – the amount of memory required to store the attribute a in bytes,
-   $V : P \to \mathbf{N} \backslash \{0\}$, $V(p)$ – the computational volume of operator p in the number of floating point operations,
-   $T : P \to \mathbf{R}^{+}$, $T(p)$ – the cycle period of the operator p in seconds.

   The node w resource reserve:

-   $S_{cpu}(w)$ flops – the maximum achievable performance;
-   $S_{mem}(w)$ bytes – the available RAM;
-   $S_{net}^{in}(w)$, $S_{net}^{out}(w)$ bytes/second – the maximum achievable network rate for the incoming and outgoing traffic, respectively.

3 Multi-agent Approach to Distributed Processing of the Streaming Data

3.1 Agent-Based Software Architecture for Distributed Data Processing Systems

In line to the proposed mathematical models the agent-based software architecture is developed [18]. It contains the description of the following components:

- the set of processing attributes implemented as in-memory database (IMDB), that provides distributed storing of streaming data and a low-latency access to them.
- the set of target agents, where each one is a software entity that implements some data processing operator;
- the set of service agents, which allow to manage creating, starting, stopping and destroying of target agents.
- agent container – a software environment that provides an access to the resources of each computing node and executes of target agents
- agent platform – a set of software components that supports the life cycle of target agents, interaction between them in the unified namespace, also the system event model
- XML-based declarative language that allows to develop specifications of distributed systems using text editors with the syntax highlighting as simple as using special XML-based visual diagrams editors.

If the target agent is not tied to any specific equipment of the computing node (to sensor or to actuator) within the resource management mechanism it can be moved between nodes without affecting the agent's task. It is offered to use multi-agent anytime algorithm to manage network computing resources.

3.2 Multi-agent Anytime Algorithm of Computational Resource Allocation

The offered algorithm of computational resource allocation in the distributed system of processing of streaming data allows to optimize resource utilization in edge computing systems. During operation, each agent performs self-profiling of its computational operations. If agent spends too many resources at the next cycle and consequently probability of lateness of data getting increase, the agent initiates a resource redistribution protocol [19]. It means that even if the resource reserve is random at each time, the probability of loosing of every value can be estimated and taken into account. The protocol assumes the following tasks.

- The detection of the computational resources shortage occurrence moment
- Monitoring of the current state of the nodes is carried out by the local node manager through continuous profiling of the test computing procedure. The test procedure execution time allows to evaluate indirectly the current reserve of computing resources. The resource intensity of the test procedure is chosen empirically and must be the same for all network nodes.

- Looking for the best node for the transposition of agent is carried out by the main manager. It uses information of nodes load collected earlier to build real-time resource model of the network. Actual resource model of the whole network helps to find the best node for transposition of that target agent, which initiated the resource allocation protocol.

Let $B : W \times R \to \mathbf{R}^+$ – function that shows $\forall w \in W$ amount of the involved resource. For each computational node w and every kind of resource it can be calculated as follows.

- amount of the involved CPU performance

$$B_{cpu}(w) = \sum_{\forall p \in D(w)} \frac{V(p)}{T(p)} \tag{4}$$

- amount of the involved memory

$$B_{mem}(w) = \sum_{\forall a \in (D \circ C)(w)} M(a) \tag{5}$$

- amount of the involved network bandwidth for incoming and outgoing traffic respectively

$$B_{net}^{in}(w) = \sum_{\forall p \in D(w)} \frac{\sum_{\forall a \in I(p) \cap A \setminus (D \circ C)(w)} M(a)}{T(p)} \tag{6}$$

$$B_{net}^{out}(w) = \sum_{\forall p \in D(w)} \frac{\sum_{\forall a \in I(p) \cap (D \circ C)(w)} M(a)}{T(p)} \tag{7}$$

The introduced notation makes it possible to formulate the problem of the optimal allocation of computational resources in the system of distributed processing of streaming data as the problem of combinatorial optimization. It is necessary to find such deployment relation $D*$, that gives maximum to predefined efficiency function F.

$$D* = \arg \max_{D \in 2^{W \times P}} F(D) \tag{8}$$

The following constraints apply.

$$\begin{cases} B_{CPU}(w) \leq S_{CPU}(w) \\ B_{mem}(w) \leq S_{mem}(w) \\ B_{net}^{in}(w) \leq S_{net}^{in}(w) \\ B_{net}^{out}(w) \leq S_{net}^{out}(w) \end{cases} \tag{9}$$

In real computer networks, node resources are spent not only to payload tasks, but also on tasks not related to the target data processing, such as, for example, system processes or services. At each time, they take a random amount of resources ξ, which, by virtue of the central limit theorem (CLT), can be declared as random variable with a normal distribution.

$$P(\xi) = \frac{1}{\sigma\sqrt{2\pi}} e^{\frac{(\xi-\mu)^2}{2\pi^2}} \tag{10}$$

The random variable ξ allows to take into account a lot of hard-formalizable factors that affect the available resources of nodes (temperature, system tasks, service or debugging tasks). It can be considered as the only source of uncertainty in IoT and the edge computing systems. For the streaming data processing system, it is especially necessary to ensure the low probability of a new value received later. The value α of the high limit of probability in this case determines the "criticality degree" of the data, and generally speaking, can be set for each computational operator, because it refers to the physical sense of attributes. And the upper α-quantile of the ξ distribution will determine the value of the "inviolable" resource reserve for each computing operator.

For some resources, it is impossible to determine rather accurately the reserve (for example, a CPU performance or a network resource). The values of computing procedures resource-intensiveness as well as their network loading cannot be measured directly. So, we offer the method of estimating indirectly such parameters. It is based on using measureable values: the amount of memory occupied and the time spent to perform calculations or data transmission – as well as a mathematical model of their interrelation. Thus, the generalized resource management algorithm for the distributed system of processing the streaming data consists of the following steps.

- The target agent stores the actual execution time of the calculations. The local manager estimates the execution time for the test procedure and sends it to the main manager. The main manager stores up-to-date information on the execution time of the test procedure on different nodes.
- If at some moment the time that some target agent spends on the operation "dangerously" gets close to the predefined period T (closer than a specified delta t), it sends to its local manager a transposition request.
- The local manager receives a transposition request and forwards it to the global manager.
- The global manager receives a transposition request and uses the available information about the current resource allocation to search for the suitable node. It sends to the local manager information about the node where the agent is to be moved, or about a failure if the target agent is already executing at the best node.
- The local manager sends information to the target agent, and also removes that agent from the control commands distribution list.
- The target agent suspends its computational operations, archives its state. The mechanisms of the platform move it to the node specified by the main manager.
- On the new node it resumes the execution from that point where it was stopped at the old node and sends a registration request to the new local manager.

– The local manager of the new node receives the request and registers a new target agent in its control commands distribution list.

The offered generalized algorithm makes it possible to ensure efficient resource allocation at all time of functioning for the distributed processing of the streaming data systems. It can be fine-tuned for different types of networks, that is especially important in the edge and cloud computing integrating.

4 Experimental Confirmation of the Proposed Algorithm

The efficiency of the offered algorithm was investigated by means of simulation using the AnyLogic software. The simulation model of network contains 10 computational nodes with the same reserve of CPU performance and memory. For each node its own system needs of CPU and memory can be set. The developed graphical interface of AnyLogic model allows to create a new target agent with any specified resource intensity and put it to any node. During the simulation each task is cyclically checked if the computational node has enough of available resources to perform the agent's computations. If the difference between the node's available resources and agent's needs becomes less than the specified threshold, the agent moves to the node which has the largest amount of the resource available. The algorithm allows to solve the resource allocation problem both at static and dynamic regimes.

4.1 An Example of a Static Resource Allocation

The static resource allocation is the assignment of computing nodes to the target agents at the initial stage of the system working under the assumption that the set of the target agents and the available amount of resources will not change during the working of the system. During the simulation, all target agents were created on one computing node, and the algorithm moved them in accordance with the established rules. The target agent that does not have enough resources of the current node is moved to the node that has the largest amount of the available resources, taking into account the internal consumption of resources by the nodes for system needs.

Since the simulated network model is homogeneous, the consumption of resources by the targeted agents is percentage set. The simulation results are shown in the Fig. 1. On average, every few seconds a new target agent was created on the first compute node.

It can be seen that when the agents exhausted the resources of this node, each new agent was moved by the algorithm to a free node (Fig. 2).

4.2 An Example of a Dynamic Resource Allocation

The dynamic allocation of resources means the automatic reconfiguration of the computing system in case of unpredictable changes on the computing nodes. In case of edge computing, these can be faults, diagnostic or debugging activities on a device that

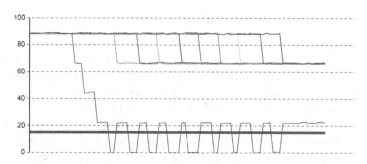

Fig. 1. Simulation of resource allocation on the static regime

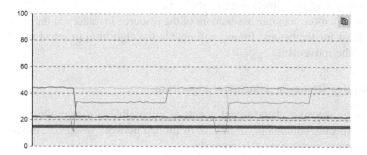

Fig. 2. Simulation of resource allocation on the dynamic regime

drastically reduce the amount of resources available for computation. In this case, the algorithm must ensure the target agents transposing to the most suitable nodes.

The dynamic modes of the computer network functioning were imitated as follows. On a randomly selected node, its own resource consumption doubled. The result of the simulation is shown in the figure. After the static distribution, the target agents are distributed over the compute nodes relatively evenly. Suddenly, a load jump occurs at node number 3. The algorithm moves one of the target agents to another node.

4.3   Further Research Prospects

The developed algorithm allows to manage the distribution of computational resources in the systems combining the edge and cloud computing. As it was mentioned above, it is possible to estimate the reserve and consumption of computing resources indirectly, based on the model [20]. Let us consider the possibility of multi-agent algorithms improving the resources distribution of heterogeneous computer networks through the use of intelligent resource models.

Due to the limited resources in organizing edge computing, resource-intensive training and forecasting procedures can be organized as cloud services.

4.4   Intellectual System for Assessing the Current State of the Node

The CPU performance and network bandwidth are resources directly related to the parameters of the computational operator itself, in particular, the period T. For the algorithm to work efficiently, it is necessary to estimate the current supply of resources in real time, and also to predict its changes. The main source of losing another value risk is the spontaneous reduction of available resources. Therefore, it is not necessary to determine accurately and quickly the moment of the change in the state of the node and, if possible, predict it. A two-level node state prediction system is proposed. On the first level, the current node state of the resources is estimated by indirect parameters. It is possible to use additional indicator procedures. On the second level, an autoprognostic model is built on the basis of convolutional neural networks, which are trained simultaneously with the functioning in real time.

The use of a more accurate assessment of the resource inventory at the node, as well as the forecast for the nearest future, will enable the algorithm to select more suitable options for the movement.

4.5   Adaptive Resource Allocation Algorithm

In case the transposing of the target agents is impossible (for example, due to interacting with specific equipment) or in case of limited network resources, the increase of the target agents cyclic period is possible.

To build such an adaptive period management system, it is necessary to estimate more accurately the probability of losing the next attribute value. This requires a more accurate statistical model of resource consumption, which would allow to take into account the peculiarities of resource allocation in the operational system of the node. In addition, the assessment of the spectral characteristics of the random variable xi in real-time can make it possible to identify and prevent possible sudden change because of resonance.

The extension of a multi-agent algorithm by the ability of period adapting will significantly increase its efficiency as applied to edge computing in the context of the shortage of performance and bandwidth of network interfaces.

4.6   Neural Network Prediction of Resource

While selecting a node for moving, the main manager should estimate as accurately as possible how long the calculations of the target agent will take at each node [21]. For this purpose, it is proposed to build a neural network approximation of the computation time of the target agent dependence from the execution time of the test procedure on the node. Basing on the data collection of the target agents self-profiling, it is possible to build a neural network predictive model for each target agent. The following structure of the network is offered: the input is represented by the execution time of test indicator procedures on the local manager, and the output is represented by the execution time of the target agent.

Thus, in the process of selecting an appropriate node, the main manager will be able to select the node to the target agent that meets the specified quality requirements of resource management best of all.

5 Conclusion

The main results of the work are presented below.

- The overview of state-of-the-art scientific research in the area of the computational resource management in IoT and edge computing.
- The mathematical model for computational resource consumption and allocation in edge computing.
- The formulation of the optimal resource allocation problem in IoT systems
- A multi-agent algorithm of the computational resource allocation in the distributed systems of the processing of streaming data
- AnyLogic simulation of the distributed system resource allocation by the offered algorithm on static and dynamic regimes.

References

1. Muthanna, A., et al.: Secure and reliable IoT networks using fog computing with software-defined networking and blockchain. J. Sens. Actuator Netw. **8**, 15 (2019)
2. Ateya, A., Muthanna, A., Koucheryavy, A.: 5G framework based on multi-level edge computing with D2D enabled communication. In: 2018 20th International Conference on Advanced Communication Technology (ICACT), pp. 507–512. IEEE, February 2018
3. Muthanna, A., Khakimov, A., Ateya, A.A., Paramonov, A., Koucheryavy, A.: Enabling M2M communication through MEC and SDN. In: Vishnevskiy, V.M., Kozyrev, D.V. (eds.) DCCN 2018. CCIS, vol. 919, pp. 95–105. Springer, Cham (2018). https://doi.org/10.1007/978-3-319-99447-5_9
4. Simic, M., Stojkov, M., Sladic, G., Milosavljević, B.: Edge computing system for large-scale distributed sensing systems. In: Proceedings of 8th International Conference on Information Society and Technology, ICIST 2018, vol. 1, pp 36–39. Society for Information Systems and Computer Networks, Belgrade (2018)
5. Thinnect. http://www.thinnect.com/. Accessed 8 May 2019
6. Nebula. https://nebula-orchestrator.github.io/. Accessed 8 May 2019
7. Mainflux. https://www.mainflux.com/. Accessed 8 May 2019
8. de Assuncao, M.D., Da Silva Veith, A., Buyya, R.: Distributed data stream processing and edge computing: a survey on resource elasticity and future directions. J. Netw. Comput. Appl. **103**, 1–17 (2018)
9. Mani Chandy, K.: Theory and implementation of a distributed event based platform. In: Proceedings of the 10th ACM International Conference on Distributed and Event-Based Systems, DEBS 2016, pp. 205–213. ACM New York (2016)
10. Soulé, R., et al.: A unified semantics for stream processing languages (extended). Technical Report 2010-924, New York University (2010)

11. Chakravarthy, S., Jiang, Q.: Stream Data Processing: A Quality of Service Perspective Modeling, Scheduling, Load Shedding, and Complex Event Processing. Springer, Boston (2009). https://doi.org/10.1007/978-0-387-71003-7
12. Uviase, O., Kotonya, G.: IoT architectural framework: connection and integration framework for IoT systems. In: First Workshop on Architectures, Languages and Paradigms for IoT, EPTCS 264, pp. 1–17 (2018)
13. Mocnej, J., Seah, W.K.G., Pekar, A., Zolotova, I.: Decentralised IoT architecture for efficient resources utilisation. IFAC-PapersOnLine 51(6), 168–173 (2018)
14. Inomata, A., et al.: Proposal and evaluation of a dynamic resource allocation method based on the load of VMs on IaaS, pp. 1–6 (2011)
15. Semyonova, D., Nagimov, T., Kovtunenko, A.: Distributed telemetric data collection in semi-natural modeling. In: Information Technologies for Intelligent Decision Making Support ITIDS 2015 Proceedings of the 3rd International Conference, Ufa (2015)
16. Kovtunenko, A.S., Valeev, S.S., Maslennikov, V.A.: The multi-agent platform for the distributed real-time data processing. Nat. Tech. Sci. 2(64) (2013)
17. Kovtunenko, A.S., Timirov, M.A., Valeev, S.S.: Resource management in the system of distributed processing of streaming data on the basis of a multi-agent approach. Nat. Tech. Sci. 10(124), 179–181 (2018)
18. Kovtunenko, A., Bilyalov, A., Valeev, S.: Distributed streaming data processing in IoT systems using multi-agent software architecture. In: Galinina, O., Andreev, S., Balandin, S., Koucheryavy, Y. (eds.) NEW2AN/ruSMART -2018. LNCS, vol. 11118, pp. 572–583. Springer, Cham (2018). https://doi.org/10.1007/978-3-030-01168-0_51
19. Nagimov, T., Semyonova, D., Kovtunenko, A.: Investigation of the computing resources management efficiency based on simulation. In: Information Technologies for Intelligent Decision Making Support ITIDS 2015, Proceedings of the 3rd International Conference, Ufa (2015)
20. Zhou, W., Fang, W., Li, Y., Yuan, B., Li, Y., Wang, T.: Markov approximation for task offloading and computation scaling in mobile edge computing. Mob. Inf. Syst. 2019, 12 (2019). Article ID 8172698
21. Chen, U., Wu, C., Lin, W.: Characteristic approximation for resources on computational grids. In: 2009 10th International Symposium on Pervasive Systems, Algorithms, and Networks, Kaohsiung, pp. 706–710 (2009)

The Use of Context-Dependent Modelling for the Construction of an Anti-fraud System in Transport

Yulia Shichkina$^{(\boxtimes)}$ (ID) and Alexander Koblov (ID)

Saint Petersburg Electrotechnical University "LETI",
ul. Professora Popova 5, St. Petersburg 197376, Russia
strange.y@mail.ru

Abstract. This article describes the results of applying a context-based approach to modelling when building a conditional system to counter fraud in transport. In this application, the exploration of the subject area into three contexts aimed at describing the characteristics of the driver's behaviour, his basic habits and the search for global anomalies based on historical data is considered The relationship and influence of context states is described by the system of the expert tree described. The description of the approach describes the advantages of the context-based approach when modelling analytical systems. At the end of the article, the results of testing the conditional system on synthetic data obtained from the collection of telematics data on the location and linear acceleration using a mobile device are described. It is shown that the proposed approach allows the detection of fraudulent actions of a telematics service user in more than 98% of cases with 16% of false positives.

Keywords: Context-based modelling · Driving behaviour · Driving style · GPS fraud

1 Introduction

Telematics is an interdisciplinary field that encompasses telecommunications, vehicular technologies, road transportation, road safety, electrical engineering (sensors, instrumentation, wireless communications, etc.), and computer science (multimedia, Internet, etc.) [1].

We would define telematics as a cross technology that integrates the technological areas of management, telecommunications and informatics. It is online technology – means of communication messages in the network information space, providing simultaneous exchange of information in real time and control objects at a distance.

Telematics data processing is currently required not only in transport systems, but also in other areas, such as medicine, smart homes, etc.

In the services of transport and logistics services based on the use of telematic devices there is also a vulnerability to fraudulent activities. For example, there are ways to substitute the readings of the sensors of a satellite navigation system, aimed at transmitting false data at frequencies of satellite equipment and magnetic field

© Springer Nature Switzerland AG 2019
O. Galinina et al. (Eds.): NEW2AN 2019/ruSMART 2019, LNCS 11660, pp. 147–156, 2019.
https://doi.org/10.1007/978-3-030-30859-9_13

generators that distort the readings of the sensors of acceleration and position [2]. These threats allow you to substitute the readings of telematics equipment sensors. This is a tangible barrier to the global implementation of transport and logistics services, because it entails financial risks for both users and owners of such systems [3].

In the market for telematics products, in addition to the substitution of sensor readings, there is still the possibility of transferring control of the vehicle to an unauthorized person, which is often a violation of the terms of the contract.

This article describes how to build an anti-fraudulent system in transport using characteristics of driving style, priorities of traffic routes and acceleration anomalies in historical data.

2 Related Works

The network of vehicles is a special network that uses advanced sensor technologies, network technologies, computing technologies for the purpose of collecting, analysing and managing the general condition of the roads and the flow of vehicles, monitoring traffic, providing various services to drivers, for example, determining their location, and finally, ensuring the safety and efficiency of road traffic [4].

The following areas are intensively developing in the network of vehicles

1. Prevent accidents on the roads. Information about potential collisions is delivered to the driver using sensors and information network. The quality of this actual information is largely dependent on the accuracy of data about time and space. This problem is solved in [5].
2. Providing traffic information. This is the most popular and rapidly developing field to improve traffic efficiency and avoid accidents through automatic driving, etc. [6].
3. Providing information about the current state of a road. Mostly these systems are designed for vehicles to provide contextual information about signs on the roads. But, there are other works in this area: the evaluation of traffic jams [8], the detection of potholes and road defects [7].
4. Driver identification is a new area of interest in the field of telematics of various vehicles, driving and insurance. Studies on human behaviour are becoming increasingly important, and a large number of theories appear on this topic [9].
5. Navigation on a curved road. If drivers know in advance about the bends on the roads and can adjust the speed accordingly, the vehicle's slipping or overturning can be less likely. Article [10] describes the anti-slip and anti-glare control of a vehicle on a curved road.
6. Information security. The integration of social networking concepts with the Internet of Vehicles (IoV) has led to a new paradigm of "Social Internet of Vehicles (SIoV)", which allows drivers to independently establish social relationships to improve traffic conditions. In most existing trust models, information about the reputation of each node is stored in other nodes with which it interacts. In [11], an RTM system was proposed in which each node stores its own information about the reputation of other nodes assessed by other nodes during past transactions.

In [12] researchers believe that applying social network principles to IoT can improve network navigation and speed up the process of finding objects and services. Establishing relationships between objects in transport networks is easier because of the smaller number of objects. In [13], the concept of "Social Internet of Things" SIoT extends to IoV and this leads to the new paradigm "Social Internet of Vehicles" (SIoV).

From this review there is a lot of research on the development of telematics vehicles and they are all very useful and interesting. However, among them, very little attention is paid to protecting systems from fraudulent threats.

The approaches proposed in this article may be applicable in conjunction with other systems and other research frameworks, partially reviewed in the overview above.

3 Formulation of the Problem

The basic data obtained from the telematics device are the position data (longitude and latitude), the speed of movement, and the acceleration readings. Due to the intensive development of V2x communications technology, the main source of such data can be the Basic Safety Message (BSM) data format, which is the American equivalent of the European data standard Cooperative Awareness Message (CAM). BSM and CAM messages are defined in SAE J2735 [18] and ETSI EN 302 637-2 [19] standards. For BSM and CAM messages in standards, a minimum message sending interval of 10 ms is set. With the use of such restrictions, the estimate of the amount of data received from sensors is slightly less than 1 TB per year from a single device. With the increase in the number of devices up to 1000 simultaneously connected, the amount of data received is more than 300 TB per month.

Such telematic data on the operation and use of automobiles is useful for many segments of the automotive industry. Insurance companies can benefit from using data to reduce their requirements and the risk of fraud and can also reduce the rates for customers who are willing to share data on vehicle use [20].

Based on this, the actual task is to determine the approach of identifying fraudulent actions when using telematics services, considering the need to analyse large amounts of data. The approach should be aimed at detecting the transfer of vehicle control to an unauthorized person and the fact of replacing telematics data through an external influence on the sensors of the telematics device. The next task is to evaluate the effectiveness of the approach on synthetic data. The solution to these problems is discussed below.

4 Solution Architecture

The materials of the Defense Advanced Research Projects Agency (DARPA) of the USA there is a description of three waves of development of artificial intelligence [14]. At the stage of development of the first wave, the descriptive approach was predominant. This approach used rules and procedures based on expert knowledge. During the evolution in the phase of the second wave, the use of cybernetic methods of data analysis. The main advantage of the approaches of the first wave is the possibility of explaining the obtained results, but these methods involve working only with pre-defined typed and formatted data. The second wave made it possible to work with data having different types, structure and form, but the possibility of restoring the entire sequence of decisions was lost. The main objective of the third wave of the approaches is to correct this situation. The context-based models, which constitute the third phase, combine the advantages of the previous phases.

To solve the problem of identifying fraudulent actions when using telematic equipment, we suggest using the following approach:

- to ensure the formation of the necessary assessment for unstructured user data;
- to develop an expert system for detect fraudulent behaviour. This will provide an opportunity to explain the final decision.

The context-based model will be called a description of a situation using contextual language tools [16]. The modelling process begins with an abstract description of the system (context definition). The context includes the definition of the subject of modelling, the purpose and point of view on the model.

The subject is the system itself. At the same time, it is necessary to establish exactly what is included in the system and what lies beyond its limits. The determination of the subject of the system will be significantly influenced by the position from which the system is considered and the purpose of modelling. These are questions that the constructed model should answer [15].

Context modeling is an approach to creating data analytics systems

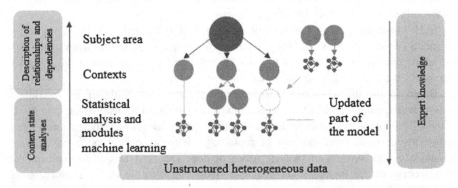

Fig. 1. Interconnection of expert knowledge and machine learning flexibility in context-based modelling.

The diagram presented in Fig. 1 shows the approach to the construction of context-based systems. According to this diagram, experts with extensive knowledge of the subject domain determine the contexts describing the state of the system, which reflects the subject domain, and formalize the dependencies between the factors influencing the state of the subject domain. After crushing the description of the subject area into basic components using the methods of statistical analysis and machine learning, the status and value of the state of the basic context are determined based on the analysis of unstructured heterogeneous data entering the analysis module.

Thus, a strong link is established between the expert understanding of the subject area and unstructured data located in the field of states on which the system describing the subject area is built. An expert description of dependencies allows you to get a deterministic solution with the possibility of modular updates without the need for a complete retraining of the data analytics model, and the statistical analysis and machine learning modules allow you to determine the state of contexts on unstructured large data sets.

This approach allows to conduct a study of each context separately and, at any time, change the degree of its influence on the final decision. This increases the determinism and reliability of decisions [15]. In addition, since the context describes a predetermined limited area, less data is necessary for analysing information than in the case when a similar situation would be considered as a whole [14].

The Fig. 2 shows a sketch of the architecture of the proposed solution.

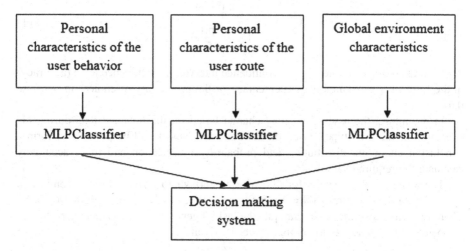

Fig. 2. Sketch of the implementation architecture of the approach to identifying fraudulent users of the telematics system.

This approach allows to achieve flexibility in data processing inherent in neural networks and the possibility of explaining the made decision about the fraudulent nature of user actions [14].

5 Personal Characteristics of User Behavior

The average characteristics of acceleration obtained at the beginning of the movement and during deceleration can be attributed to the personal characteristics of the user's behaviour. These characteristics are personalized for the user in the management of a specific vehicle. When using this parameter, we are guided by the assumption that during daily driving, the user develops certain preferences for choosing the effort of depressing the accelerator pedal and the brake. This assumption is indirectly confirmed in the article [17], which describes the study of the management style of calm and aggressive drivers.

In addition to using the average acceleration values, we propose the use of a parameter such as driving style, which is based on the number of accelerations and decelerations, defined as "aggressive", relative to the total number. It can be assumed that the value of this characteristic is in the range from 0 to 1, where 1 is safe driving, and 0 is aggressive.

This characteristic will make it possible to determine the probability of transferring control of the vehicle to an unauthorized person.

Thus, to obtain the most accurate recognition, it is required to determine a class other than the specified one. To determine the minimum amount of data (number of respondents) in this document is used corrected mathematical sample. The sample size (Vsample) is calculated using the following formula:

$$V_{sample} = \frac{\frac{z^2 p(1-p)}{e^2}}{1 + \left(\frac{z^2 p(1-p)}{e^2 N}\right)} \tag{1}$$

Where: N - total sample size; e - confidence interval; z - confidence level (accuracy) in the form of a z-evaluation: p - percentage of the probability of choice (in our case 0.5).

In the future, the assessment may change based on the obtained data during the execution of the prototype. Based on the obtained data, it will be possible to form a conclusion about the distribution and probability of selection and more accurately calculate the required sample size.

The availability of results for 100 trips is enough to provide a 10% confidence interval with 95% accuracy (N = 1,000,000, e = 0.1, z = 1.96, Vsample ≈ 96.0308). Therefore, after ensuring such a sample, the task of generating a class that characterizes the belonging of readings to another driver is trivial.

5.1   Personal Characteristics of the User Route

In addition to preferences in acceleration, most drivers adhere to a specific schedule and route of movement. Reason for this behavior is the constant work schedule of users, as well as personal habits. Based on this, the following characteristics will be the start time of the trip and its duration. For the stability of these characteristics, we are invited to add the frequency of use of the route and parking places by the user.

For the formation of these characteristics, need to solve the problem of clustering the user's travel history twice: the first time for temporal characteristics, and the second for spatial. After this need to check the obtained values for belonging to the existing clusters for each group separately. The result is the ratio of the number of values belonging to any of the clusters to the total number of obtained values.

5.2   Global Environment Characteristics

In addition, need to compare the obtained values with the parameters obtained considering of the environment. To determine fraudulent actions based on the substitution of location data or acceleration, we suggest localizing the location of abnormal accelerometer readings and vehicle speed.

In places with uneven road surface and also in front of crossroads, there will be jumps in the speed and acceleration readings relative to the rest state. After researching the readings of many users over a long period of time, it is possible to cluster the locations of such "anomalies".

Having performed a similar operation on the current data obtained during the movement, it is possible to correlate the location of the "anomalies" based on historical data accumulated in the telematics system. If there is a significant similarity in the location of "anomalies" to the user data, then it is possible to speak about the plausibility of the data obtained from the sensors.

However, when using equipment that provides for the substitution of readings of telematics device sensors it is possible to ensure the creation of anomalous zones in fictitious locations.

6   Experimental Results

As a test, was assessed synthetic data based on expert threshold values.

For the experiment, were prepared 30 sets of synthetic data, containing information on trips of various numbers (sets 1 to 10 contain 10 trips, from 11 to 20–100 trips, from 21 to 30–1000 trips). The data was generated by introducing random noise into the average travels obtained on ten trips from three different users, in the maximum deviation range.

For each set, three neural networks were trained for readings characterizing user behaviour, movement routes and proximity to global accelerometer readings. After training, readings were verified by generating sets that correspond to normal and fraudulent behaviour, exceeding a training set of 10, 100, and 1000 times for each behaviour model.

For the obtained results is used the following notation:

- 10 A 1000 - denotes the percentage of 1000 user trips identified as fraudulent, although they are not. This data is trained on a set of 10 trips (1);
- 10 B 1000 - denotes the percentage of 1000 fraudster trips identified as fraudulent, trained on a set of 10 trips (2).

Based on the obtained readings, it is possible to say that the proposed approach allows to determine fraudulent user actions when using telematics equipment, on average, in 94.78% of cases. At the same time, the number of false positives on legitimate user actions, even when conducting training after the first 10 trips, is no more than 20% of the total (see Fig. 3).

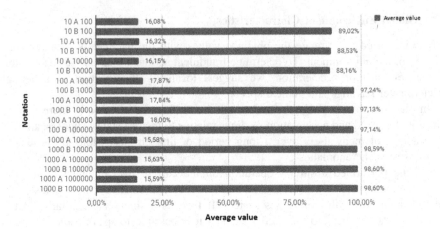

Fig. 3. Average value for considered parameters.

The diagram above clearly visualized that the magnitude of the error in determining user behaviour decreases with the largest training set and does not depend on the number of checks.

7 Conclusions

The use of context-based models should allow to increase the determinism of solutions that preserve the flexibility of neural networks. In addition, it should provide upgradeability without full retraining.

This article describes an approach to solving the problem of identifying fraudulent user telematics data, which consists of three main branches, implemented in parallel, with the aggregation of results for the final decision. These are branches: identification of the user based on the collected data on his behavior on the road, identification of the route based on the accumulated data on the routes of the user, identification of the environment. A detailed solution of these branches will be described in separate articles, since represents a significant amount of information and is a detail of the proposed approach.

The results obtained using context-based models in the field of fraud analysis on transport reveal fraud actions of telematics users in more than 98% of cases with 16% false positives, which is a good result for use in the insurance field.

Our further research will be conducted using training methods without additional combinations.

Funding Statement. The paper has been prepared within the scope of the state project "Initiative scientific project" of the main part of the state plan of the Ministry of Education and Science of Russian Federation (task № 2.6553.2017/8.9 BCH Basic Part) and was funded by RFBR and CITMA according to the research project № 18-57-34001.

References

1. Mayo-Wells, W.J.: The Origins of Space Telemetry, Technology and Culture (1963)
2. Telematics. https://en.wikipedia.org/wiki/Telematics. Accessed 1 Nov 2018
3. Ischenko, N.: Drivers cheat online taxi services in the fight for orders. Vedomosti 2016. August 02 In Russian. https://www.vedomosti.ru/business/articles/2016/08/02/651358-pogone-zakazami-voditeli-pitayutsya-vvodit-servisi-onlain-taksi-zabluzhdenie. Accessed 1 Nov 2018
4. Hu, L., Li, H., Xu, X., Li, J.: An intelligent vehicle monitoring system based on internet of things. In: Proceedings of the 7th International Conference on Computational Intelligence and Security (CIS 2011), vol. 59, pp. 231–233, December 2011
5. Chang, C.-Y., Xiang, Y., Shi, M.-L.: Development and status of vehicular ad hoc networks. J. Commun. **28**(11), 116–126 (2007)
6. Cheng, G., Guo, D.: The current status and development of vehicle networking. Mob. Commun. **17**, 23–26 (2011)
7. Muñoz-Organero, M., Ruiz-Blázquez, R.: Detecting different road infrastructural elements based on the stochastic characterization of speed patterns. J. Adv. Transp. Res. **2017**, 11 (2017). Article ID 3802807
8. Anbaroglu, B., Heydecker, B., Cheng, T.: Spatio-temporal clustering for non-recurrent traffic congestion detection on urban road networks. Transp. Res. Part C Emerg. Technol. **48**, 47–65 (2014)
9. Chowdhury, A., Chakravarty, T., Ghose, A., Banerjee, T., Balamuralidhar, P.: Investigations on driver unique identification from smartphone's GPS data alone. J. Adv. Transp. Res. **2018**, 11 (2018). Article ID 9702730
10. Lie, G., Yi-bing, Z., Lin-hui, L., Ping-shu, G., Xiao-hui, H.: Antisideslip and antirollover safety speed controller design for vehicle on curved road. Math. Prob. Eng. Res. **2014**, 12 (2014). Article ID 253176
11. Gai, F., Zhang, J., Zhu, P., Jiang, X.: Trust on the Ratee: a trust management system for social internet of vehicles. Wireless Commun. Mob. Comput. Res. **2017**, 11 (2017). Article ID 7089259
12. Atzori, L., Iera, A., Morabito, G.: SIoT: giving a social structure to the internet of things. IEEE Commun. Lett. **15**(11), 1193–1195 (2011)
13. Nitti, M., Girau, R., Floris, A., Atzori, L.: On adding the social dimension to the Internet of Vehicles: Friendship and middleware. In: Proceedings of the 2014 IEEE International Black Sea Conference on Communications and Networking, BlackSeaCom 2014, pp. 134–138, May 2014
14. A DARPA Perspective on Artificial Intelligence. Defense Advanced Research Projects Agency. https://www.darpa.mil/attachments/AIFull.pdf. Accessed 1 Nov 2018
15. Iosenkin, V.Y., Vykhovanets, V.S.: The use of contextual programming technology for solving large applied problems. In: Proceedings of the International Conference "Parallel Computing and Control Problems" (PACO 2001), vol. 4, pp. 121–139. Institute of Management Problems. V.A. Trapeznikova RAS (2001)
16. Iosenkin, V.Y., Vykhovanets, V.S.: The context model of the technological process of an enterprise. In: Proceedings of the II International Conference "System Identification and Control Problems" (SICPRO 2003), pp. 859–871 (2003)
17. Hong, J.-H., Margines, B., Dey, A.K.: A smartphone-based sensing platform to model aggressive driving behaviors. In: CHI 2014 Proceedings of the SIGCHI Conference on Human Factors in Computing Systems, Toronto, ON (2014)

18. Brian Cronin, Vehicle Based Data and Availability, United States Department of Transportation.   https://www.its.dot.gov/itspac/october2012/PDF/data_availability.pdf. Accessed 1 Nov 2018
19. ETSI EN 302 637–2 V1.3.1: Intelligent Transport Systems (ITS). Vehicular Communications. Basic Set of Applications. Part 2: Specification of Cooperative Awareness Basic Service.   https://www.etsi.org/deliver/etsi_en/302600_302699/30263702/01.03.01_30/en_30263702v010301v.pdf. Accessed 1 Nov 2018
20. Juliussen, E.: The future of automotive telematics. In: Business Briefing: Global Automotive Manufacturing & Technology (2003)

An Approach to the Analysis of the Vehicle Movement on the Organization Territory

Evgenia Novikova, Yana Bekeneva$^{(\boxtimes)}$, and Andrey Shorov

Department of Computer Science and Engineering,
Saint Petersburg State Electrotechnical University "LETI",
Professora Popova Street 5, Saint Petersburg, Russia
{evgeshka19,yana.barc,ashxz}@mail.ru

Abstract. Analysis of the vehicles' movement on the territory of the organization represents an area of considerable interest for both cyber-physical security and financial applications. In the paper an approach to the analysis of the vehicles routes on the territory of the organization is presented. The vehicle route is reconstructed on the basis of the data from different sources such as access control system, video surveillance and weight measuring devices. The approach presented consists of two stages: exploratory analysis of the data that allows constructing analysis models for detection deviations in the vehicle movement in the real time mode; and analysis of the sensor readings in the real time mode. We tested our data on both artificial data and real world data.

Keywords: Vehicle movement · Pattern detection · Anomaly detection · Clustering · Random forest classifier · Heterogeneous data

1 Introduction

The data describing the movement of the objects are one of the most widely spread type of the spatial data source. The amounts of them are constantly increasing due to the availability and, therefore, wide usage of the location-aware mobile devices. These data describe moving objects of any kind - pedestrians [1–3], vehicles [4], animals [5]. They can be collected by mobile phones, cars, proximity card readers, video cameras at public places, and checkpoints using RFID cards. The points can be obtained on regular basis or irregularly; they can be acquired by the moving object itself or by an external observing sensor. A wide range of applications may benefit from analysis of such data sets, also known as trajectories. The analysis of the objects' movement allows one to form object life patterns, discover constraints existing in the underlying environment, for example, rules or security policies, restricting access to the specified zones, or infrastructure available to the individuals, i.e. places of interests, ATMs, etc. [1]. Other important application of the trajectory mining is the generation of the features of the object life pattern to detect possible anomalies in observing data sets [3].

Systems for urban traffic regulation have been actively developed recently. They are able to predict the workload of various streets and provide drivers with options for less busy routes in order both to save drivers' time and reduce harmful vehicle emissions [6, 7]. Systems for tracking public transport and predicting the time of its

© Springer Nature Switzerland AG 2019
O. Galinina et al. (Eds.): NEW2AN 2019/ruSMART 2019, LNCS 11660, pp. 157–167, 2019.
https://doi.org/10.1007/978-3-030-30859-9_14

appearance at a particular stop have become very popular [8, 9]. These systems analyze the signals received from GPS sensors installed on vehicles. However, when analyzing movement of the commercial and industrial vehicles in order to identify different types of deviations in their behavior it is necessary to consider various factors simultaneously.

Production facilities are equipped with a large number of different devices for monitoring ongoing processes and measuring the parameters of these processes. Deviations in the vehicle behavior are not limited to the spatiotemporal deviations from a given route, the wrong weight of the vehicle is an example of the economical fraud. In addition, all movements of vehicles at production facilities are often strictly regulated, i.e. there are prescribing safety and security procedures. Therefore, the presence of a GPS tracker in the vehicle and the notification that the vehicle is in the specified area cannot indicate that all the actions required by the regulations are observed. For example, a vehicle may be in the weighing zone, but it does not pass the weighing procedure.

The rest of the paper is organized as follows. Section 2 discusses the related work in the analysis of the vehicle movement. In Sect. 3 we describe the proposed approach to the analysis of the vehicle movement. Section 4 presents case study used to evaluate approach, discusses results and defines directions of the future research. Conclusions sum up our contributions.

2 Related Works

The most of the research is devoted to monitoring vehicles moving on city streets [10] or highways [11]. In the most cases they address the problem of the real-time monitoring of traffic flows in order to identify and predict possible traffic jams [12], traffic accidents [13] and take timely measures to prevent them [14].

The most widely used data sources are data obtained from the various means of photo and video recording, as well as various radars that determine the vehicle speed and its direction [10, 15]. The tasks included in this research are the determination of the vehicles' number, their speed in a given area, which allows one to estimate the intensity of traffic flows and predict the occurrence of the traffic jam.

The works focusing on the car accident detection are targeted to recognize an accident that has been already registered by different monitoring tools; to determine the exact coordinates in order to send rescue services promptly to the place where the accident occurred; and to assess approximately the severity of the accident consequences [16, 17].

In the research devoted to the prevention of the road traffic incidents authors assess the vehicle speeds [18], their trajectories [19]. In [20] authors monitor the driver's state in order to identify the signs of fatigue [20].

Thus, it is possible to conclude that the existing research mostly focuses on the assessment of traffic flows or trajectories of individual vehicles on the separate parts of routes. The possible deviations are detected in the context of the traffic safety regulations.

There are systems for monitoring commercial vehicle movement [21, 22]. Such systems are based on signals received from GPS trackers or GSM stations and are not designed to analyze other data sources. In addition, this kind of system does not allow one to identify possible violations that are associated with the use of the commercial transport such as theft of products on cites, forgery of documents, etc.

The approach proposed in this paper is based on processing data from stationary monitoring tools used in enterprises, such as video surveillance system, access control systems, and measuring devices. The particular attention is made on data preprocessing step.

3 Description of the Approach

3.1 The Source Data

The territory of any organization can be presented as a set of zones, controlled by different devices, such as video cameras, proximity card readers of the access control system etc.

Let, zone z is controlled by a set of sensors S. Each sensor $s \in S$ generates an event e when the sensor registers the vehicle. In general case, event e can be presented as follows:

$$e = <timestamp, type, s_i, v, Attr_{s_i}>$$

where $timestamp$ is the timestamp of the event; $type$ stands for type of the event; s_i is a sensor that generated the event; v – vehicle that initiated the event; $Attr_{s_i}$ – a set of additional attributes, defined be the type of the s_i sensor. An attribute $a \in Attr$ may be either numerical (ratio or interval), nominal or ordinal.

Let us consider the following vehicle parameter measuring and monitoring devices: video cameras, proximity card readers and specialized weighing-machines.

An event generated by cameras is described as follows:

$$e^{cam} = <timestamp, cam\_id, v\_id, dir>,$$

where $timestamp$ is the timestamp of the event; cam_id is the camera identifier; v_id is the vehicle number recorded by camera and recognized by pattern recognition tools; attribute dir determines the direction of movement.

An event recorded using access control tools is described by such parameters as:

$$e^{acs} = <timestamp, acs\_id, card\_id, tp>,$$

where $timestamp$ is the timestamp of the event; acs_id is the identifier of the proximity card reader; $card\_id$ is the identifier of a proximity card assigned either to a specific vehicle or a specific driver; attribute tp determines the direction of movement when the pass is applied.

An event associated with vehicle measuring on specialized weighing-machine can be described using the following attributes:

$$e^{lib} = <timestamp, lib\_id, v\_id, wgh, time\_on, time\_off>$$

where *timestamp* is the timestamp of the weight measuring event, *lib_id* – is the weighing-machine identifier, *v_id* is the vehicle number recorded by camera and recognized by pattern recognition tools, *wgh* is the weight fixed; *time_on* is the time of arrival on the weighing-machine, *time_off* is the departure time from the weighing-machine.

The same event can be fixed both by one monitoring device or by several devices, simultaneously. Moreover, these devices may be of the same type or of different types. In this case, data from different sources can be used to cross validate the parameters of the vehicle when capturing an event and to merge several records into one [23, 24].

3.2   Data Preprocessing Step

The r_v route of the v vehicle moving inside some controlled zone z can be presented as a sequence of the events, produced by the different sensors: $r_v = \{e_0, e_1, \ldots e_n\}$. In each zone, the vehicle moves according to a specific route, performing a number of regulated actions. The sequence of the events may vary depending on the type of the vehicle and role designated to it. To analyze these data, we suggest converting the r_v sequence of events into the $\bar{r}_v$ vector of the fixed length. The following transformation procedure is proposed.

1. The selection of the events corresponding to one route. Firstly, we filter the events by the vehicle ID and then order the logs by the time of their occurrence. We also assume that the vehicle movement inside the controlled zone starts and ends with the events of the certain type generated by the sensor of the certain type. For example, the route may start with the entry through the checkpoint and the exit through it. Using these two events, we divide the data set into the separate routes.

2. Transformation of event sequence describing one route to a vector of finite length. Firstly, we group the selected events by sensor type and then sensor identifier. Then, for each type of the sensor we define a set of attributes that describe their operation, and for each attribute calculate numeric values characterizing events in the selected set. For example, for video cameras we calculate the number of vehicle registrations, grouped by the movement direction. For weighing machine, we also calculate the number of registrations and the mean weight of the vehicle, as well as mean duration of weighing process. Additionally, for each sensor regardless its type we calculate time between event of the given sensor and the event of the sensor that is previous in the route. This decision introduces an uncertainty in the calculation, given the fact that the first sensor in the route may be missing or sensor is present more than once in the route. In this case, such route will fall into a separate cluster consisting of routes, similar to it. The Fig. 1 gives a scheme of the process transforming the sequence of the events into a vector of the fixed length.

Let us consider the following example, the controlled zone is equipped by two cameras cam_0 and cam_1 monitoring the zone entrance/exit, one proximity card reader asc_0 and one weighing machine lib_0. Let assume for simplicity, that cameras and

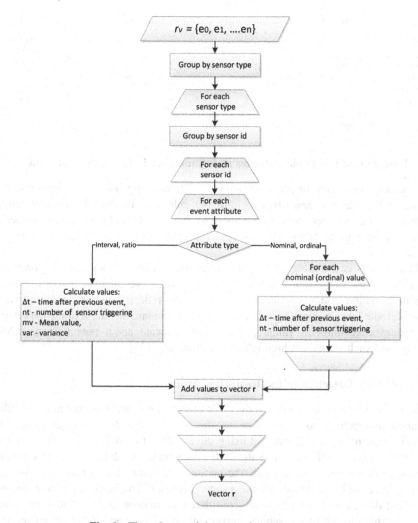

Fig. 1. The scheme of data transformation process

proximity card readers give us information on direction of the vehicle movement and weighing machine – the measured weight. Then the route of the vehicle is described by the following sequence of the events:

$$r_v = \{e_0^{cam_0} \xrightarrow{\Delta t_{01}} e_1^{acs_0} \xrightarrow{\Delta t_{12}} e_2^{lib_0} \xrightarrow{\Delta t_{23}} e_3^{acs_0} \xrightarrow{\Delta t_{34}} e_4^{cam_1}\}$$

is transformed to the following vector presented in the Table 1 in the row Resulting vector.

Table 1. The illustration of the data transformation process

Sensor type	Video camera				Access control system		Weighing machine
Sensor ID	Cam_0		Cam_1		Acs_0		Lib_0
Attribute nominal values	Entrance	Exit	Entrance	Exit	Entrance	Exit	
Resulting vector	0, 1	0, 0	0, 0	Δt_{34}, 1	Δt_{01}, 1	Δt_{23}, 1	Δt_{12}, 1, w, 0

3.3 Detection of the Typical Routes and Anomalies in the Historical Data

The vectors obtained on the previous step are clustered using *kmeans++* algorithm. It is similar to the classical *kmeans* algorithm but enhanced method for defining initial conditions for cluster centroids. In our approach the definition of the optimal number of clusters is done using silhouette analysis. The centroids of the clusters describe the typical routes.

The clusters constituting of one object can be considered as suspicious ones. To detect deviations in the routes belonging to one cluster we calculate z-score for each attribute deviation from cluster centroid. The Z-score reflects how far the current value of the attribute from its mean value considering the standard deviation of the parameter. The values exceeding the range [−2.58, 2.58] indicate about potential anomalous activity, and such routes are highlighted automatically [25].

3.4 Anomaly Detection in the Real Time Mode

After the definition of the typical routes it is possible to detect the maximum possible deviation for each sensor. It is done on the basis of the Z-score assessment. The anomaly determination is performed using the Random Forest classifier. The result of the algorithm is the table indicating the route type with a probability that the route has a given type of the route. Afterwards, the parameters of the current route are assessed using z-score with the values of the cluster centroid. The readings of the sensors exceeding the allowed deviations are considered anomalous and marked correspondingly. The monitoring of the vehicle movement on the territory of the organization is used using a map that shows the locations of the vehicles according to the sensors' logs. The vehicles exposing anomalous behavior are shown in red, vehicles with legitimate behavior are shown in green. The Fig. 2 shows an example of the map of some imaginary organization with 5 trucks moving on its territory. Four of them are drawn in red, and one is given in green.

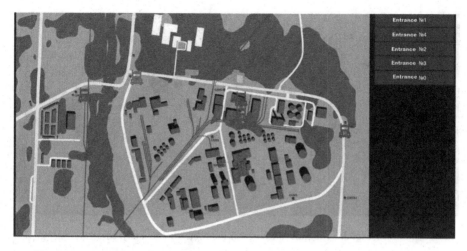

Fig. 2. The map with moving vehicle for real time monitoring (Color figure online).

4 Experiments and Approach Evaluation

To test the approach we developed a software prototype written in Python, the algorithms are implemented using scikit-learn library [26]. The approach was tested against real world data and synthetic data. As the real world data has no ground proof we developed the simulator according to the principles described in [27].

A typical scheme of a vehicle movement in a particular area of the organization is described as follows.

1. Vehicle arrival to the gate from the outside, the fact is recorded by a video camera.
2. The date and time of the gate opening is fixed.
3. Vehicle entry (fixed by a video camera).
4. The date and time of the gate closing is fixed.
5. Weighing the vehicle on specialized scales (recorded by video cameras, the result of measuring the weight is recorded and transmitted to the common data base).
6. Loading or unloading a vehicle.
7. The vehicle weighing.
8. Vehicle arrival to the gate from the inside (recorded by video cameras).
9. The opening of the gate (fixed).
10. Manual check out the vehicle.
11. Closing the gate (fixed).

In the existing scenario, the following violations can be implemented:

- violation of the events' sequence on the territory of the organization (except for weighing)
- violation of the time interval admissible between two monitoring devices adjacent in the vehicle route;
- theft: missing the weighing event or exceeding the period of time established by the regulations when transporting cargo from one object to another.

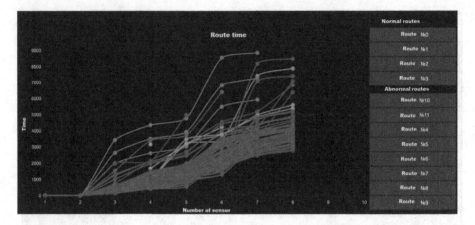

Fig. 3. The time attributes of the routes being analyzed

To test the approach, it was generated 102 routes of the vehicles with different sequence of the sensors' passage. There were 14 routes with anomalous sequence of sensors and time intervals between their readings, and 30 routes with deviations in time attributes and numeric attributes for weight measuring devices. Figure 3 presents temporal characteristics of the routes by showing time intervals between adjacent sensors. The sequence of the sensors on the line chart reflect the sequence of the order of the sensors in the route regardless their type. Figure 4 shows the number of the detected clusters, the stacked diagram reflects the number of the anomalous routes in each cluster. If the number of the routes in the cluster is less than 10% of the total number of the routes the routes in the cluster are considered anomalous.

According to the results of the z-scores assessment, it is possible to conclude about the type of the anomaly. If the z-score for the number of sensor readings for the given route is negative, it means that the sensor was skipped during the check-in, if the sensor z-score is positive, then the given sensor was passed through once more. If the value of the z-score for the time attributes exceeds 2, then this means that during the movement from the previous sensor to this, there was a delay of the vehicle movement.

The accuracy of this method strongly depends on the amount of the input data, because the number of routes having one type determines the accuracy of the centroid identification and anomaly detection.

This method can accurately identify the sensor on which the anomaly was reflected, which simplifies the further analysis of these anomalies, in order to identify the causes of the deviation. For example, if the sensor is skipped repeated several times on the same day, then it is possible to conclude about possible sensor malfunction. The fact that several routes of the vehicles on one day are characterized by a delay on the same sensor may indicate about appearance of an obstacle between it and the previous sensor.

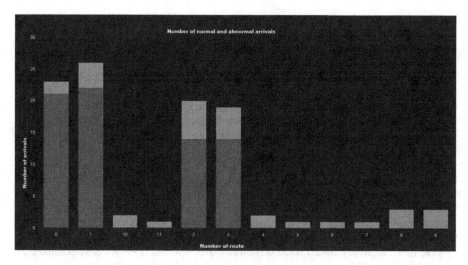

Fig. 4. The number of the detected clusters in the vehicle routes

5 Conclusion

In the paper we proposed the analytics approach to detection of the patterns and anomaly in vehicles' movement. The routes of the vehicle are reconstructed on the basis of the data from three types of data sources – access control system, video surveillance and weight measuring devices. We propose the approach how to combine these data in order to detect typical routes and reveal possible anomalies in the vehicle trajectories.

To illustrate our approach, we used 2 data sets – real world and synthetic ones. In the paper we discuss result obtained, and define future directions of work devoted to the enhancement of the prototype, elaboration of visualization techniques supporting the analysis of the raw data and implementation of the usability evaluation of the proposed visual analytical system.

Acknowledgements. This work was supported by the Ministry of Education and Science of the Russian Federation in the framework of the state order "Organization of Scientific Research", task #2.6113.2017/6.7.

References

1. Millonig, A., Maierbrugger, G.: Identifying unusual pedestrian movement behavior in public transport infrastructures. In: Proceedings of Movement Pattern Analysis Workshop (MPA2010), Zurich, pp. 106–110 (2010)
2. Versichele, M., Neutens, T., Delafontaine, M., van de Weghe, N.: The use of Bluetooth for analysing spatiotemporal dynamics of human movement at mass events: a case study of the Ghent Festivities. Appl. Geogr. **32**, 208–220 (2012)

3. Lerman, Y., Rofe, Y., Omer, I.: Using space syntax to model pedestrian movement in urban transportation planning. Geog. Anal. **46**(4), 392–410 (2014)
4. Guo, D., Liu, S., Jin, H.: A graph-based approach to vehicle trajectory analysis. J. Location Based Serv. **4**(3), 183–199 (2010)
5. Demšar, U., et al.: Analysis and visualisation of movement: an interdisciplinary review. Mov. Ecol. **3**(1), 5 (2015)
6. Brennand, C.A., da Cunha, F.D., Maia, G., Cerqueira, E., Loureiro, A.A., Villas, L.A.: FOX: a traffic management system of computer-based vehicles FOG. In: 2016 IEEE Symposium on Computers and Communication (ISCC), IEEE, pp. 982–987 (2016)
7. Nellore, K., Hancke, G.: A survey on urban traffic management system using wireless sensor networks. Sensors **16**(2), 157 (2016)
8. Bai, C., Peng, Z.R., Lu, Q.C., Sun, J.: Dynamic bus travel time prediction models on road with multiple bus routes. Comput. Intell. Neurosci. **2015**, 63 (2015)
9. Yu, B., Wang, H., Shan, W., Yao, B.: Prediction of bus travel time using random forests based on near neighbors. Comput. Aided Civ. Infrastruct. Eng. **33**(4), 333–350 (2018)
10. Petrov, V., Andreev, S., Gerla, M., Koucheryavy, Y.: Breaking the limits in urban video monitoring: massive crowd sourced surveillance over vehicles. IEEE Wireless Commun. **25**(5), 104–112 (2018)
11. Bekiaris-Liberis, N., Roncoli, C., Papageorgiou, M.: Highway traffic state estimation with mixed connected and conventional vehicles. IEEE Trans. Intell. Transp. Syst. **17**(12), 3484–3497 (2016)
12. Xu, B., Barkley, T., Lewis, A., MacFarlane, J., Pietrobon, D., Stroila, M.: Real-time detection and classification of traffic jams from probe data. In: Proceedings of the 24th ACM SIGSPATIAL International Conference on Advances in Geographic Information Systems, ACM, p. 79, October 2016
13. Ali, H.M., Alwan, Z.S.: Car accident detection and notification system using smartphone. LAP LAMBERT Academic Publishing, Saarbrucken (2017)
14. Kumar, J.M., Mahajan, R., Prabhu, D., Ghose, D.: Cost effective road accident prevention system. In: 2016 2nd International Conference on Contemporary Computing and Informatics (IC3I), IEEE, pp. 353–357, December 2016
15. Li, Y., Liu, W., Huang, Q.: Traffic anomaly detection based on image descriptor in videos. Multimedia Tools Appl. **75**(5), 2487–2505 (2016)
16. Fernandes, B., Alam, M., Gomes, V., Ferreira, J., Oliveira, A.: Automatic accident detection with multi-modal alert system implementation for ITS. Veh. Commun. **3**, 1–11 (2016)
17. Topinkatti, A., Yadav, D., Kushwaha, V.S., Kumari, A.: Car accident detection system using GPS and GSM. Int. J. Eng. Res. Gen. Sci. **3**(3), 1025–1033 (2015)
18. Goniewicz, K., Goniewicz, M., Pawłowski, W., Fiedor, P.: Road accident rates: strategies and programmes for improving road traffic safety. Eur. J. Trauma Emergency Surg. **42**(4), 433–438 (2016)
19. Tan, C., Zhou, N., Wang, F., Tang, K., Ji, Y.: Real-time prediction of vehicle trajectories for proactively identifying risky driving behaviors at high-speed intersections. Transp. Res. Rec. **2672**(38), 233–244 (2018)
20. Kaplan, S., Guvensan, M.A., Yavuz, A.G., Karalurt, Y.: Driver behavior analysis for safe driving: a survey. IEEE Trans. Intell. Transp. Syst. **16**(6), 3017–3032 (2015)
21. GPS Fleet Tracking. http://smpsolutions.in/route-alert/. Accessed 20 May 2019
22. Route Deviation Monitoring. https://unitedtracker.com/route-deviation-monitoring/. Accessed 20 May 2019

23. Ya, B., Lebedev, S., Kholod, I., Shorov, A., Novikova, E.: Method for transformation of data from heterogeneous monitoring devices for violations detection. In: 2017 XXI IEEE International Conference on Soft Computing and Measurements (SCM), St-Petersburg, 23–25 May 2018, pp. 753–756 (2017)
24. Bekeneva, Y.A., Kholod, I.I., Lebedev, S.I., Novikova, E.S., Shorov, A.V.: Violation detection in heterogeneous events streams. Procedia Comput. Sci. **150**, 381–388 (2019)
25. Caldas de Castro, M., Singer, B.: Controlling the false discovery rate: a new application to account for multiple and dependent test in local statistics of spatial association. Geogr. Anal. **38**, 180–208 (2006)
26. Pedregosa, F., et al.: Scikit-learn: machine learning in Python. JMLR **12**, 2825–2830 (2011)
27. Bekeneva, Y.A., Kholod, I.I., Lebedev, S.I., Novikova, E.S., Shorov, A.V.: Towards simulation of the processes related to transport movement within industrial objects. In: Proceedings of the IT&QM&IS – 2018, pp. 304–307 (2018)

Building Blocks of an Innovative Approach to Education in the Field of Cyber Operations in Smart Environment

Blaž Ivanc, Iztok Podbregar, and Polona Šprajc[(✉)]

Faculty of Organizational Sciences, University of Maribor,
Kidriceva Cesta 55a, 4000 Kranj, Slovenia
polona.sprajc@um.si

Abstract. Future cyber operations in an IoT-enabled smart environment will be much more demanding and complex. The purpose of this paper is to present the building blocks of an innovative approach to education in the field of cyber operations. Proposed model of an innovative approach to education in the field of cyber operations includes three building blocks with the set of key elements. The model is focused on the work process of cyber operations modelling, that will allow students with different levels of knowledge and specializations to work efficiently. The objective of the innovative approach to education in the field of cyber operations is to connect students with different knowledge in the field of cyber security and thus contribute to the development of highly secure IoT solutions and better security awareness. The main objective of the proposed model of an innovative approach to education in the field of cyber operations is to allow the students to develop a content-rich, in-depth knowledge in the field of cyber operations, especially in smart environment.

Keywords: Attack modelling · Cyber security · Fostering innovation · Internet of things · Secure infrastructure development

1 Introduction

Experts of different professions and specializations who have the necessary knowledge and experience are of key importance in the field of cyber operations. At the same time, knowledge in the field of cyber operations is quickly outdated so it is necessary that experts continuously follow global situation in the field.

As stated by Celik and co-authors (2016), cyber operations carried out in a cyberspace are a set of wide-ranging activities, which can be either offensive or defensive measures. As mentioned by Pipyrosa (2018), a group of experts defined a cyberattack as cyber operation, which can be offensive or defensive. On the other hand, cyber operations are defined as integration of cyber capabilities in order to achieve goals in a cyberspace. At the same time, the authors summarize that terms cyber operation and cyberattack involve a broad set of possible interpretations. As stated by Kim an Eom (2016), cyber operations are aimed at a target so as to achieve the desired effect. From this perspective, cyber operations are carried out against hostile cyber entities in order to attain previously defined effects. Within the framework of the

© Springer Nature Switzerland AG 2019
O. Galinina et al. (Eds.): NEW2AN 2019/ruSMART 2019, LNCS 11660, pp. 168–181, 2019.
https://doi.org/10.1007/978-3-030-30859-9_15

innovative approach to education in the field of cyber operations, it is necessary to consider approaches of authorized simulations of cyberattacks on IT systems in order to check the security of systems.

The purpose of the paper is to present model of an innovative approach to education in the field of cyber operations, which allows a group of students with different educational background and from different disciplines to work collectively. The proposed model will allow the students to develop a content-rich, in-depth knowledge in the field of cyber operations management.

2 Model of an Innovative Approach to Education in the Field of Cyber Operations

This section presents individual elements of the model of an innovative approach to education in the field of cyber operations. Figure 1 presents the model of an innovative approach with individual building blocks. Each building block consists of several elements. A key element of an individual building block is specified in the figure.

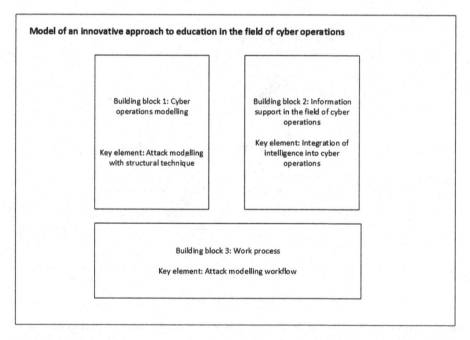

Fig. 1. Model of an innovative approach to education in the field of cyber operations with individual building blocks.

The objective of the proposed work process is to properly apply the attack tree model as a simplified methodology to recognize attackers' activities and objectives. This is a prerequisite for quality development of attack scenarios. Another objective of the presented work process is to enable students to better classify, combine and review the information within the tree structure.

2.1   Cyber Operations Modelling

Cyber operations modelling is one of the building blocks in the proposed model of an innovative approach to education in the field of cyber operations. This building block relies on the attack modelling method. Figure 2 shows the relationship between the terms used in order to provide a clear picture to the reader.

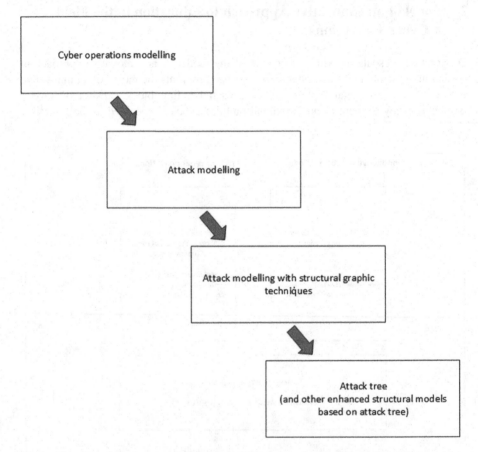

Fig. 2. The relationship between the terms used in order to provide a clear picture to the reader.

Attack modelling is an important method for identifying vulnerabilities. It is especially important for improving security awareness and preparing for unwanted scenarios. Unlike some other methods for presenting knowledge in the field of cybersecurity, attack models focus on the attacker's perspective. Attack modelling techniques can be divided into different categories. Chang, Jirutitijaroen and Ten (2010) distinguish between two approaches to attack modelling, namely, attack trees and stochastic models. Piètre-Cambacédès and Bouissou (2013) categorize formal graphic security models into static or structural and dynamic or behavioural ones. A typical technique that represents the first category is an attack tree whereas Petri nets are a typical representative of the dynamic models. Hong and Kim (2012) show hierarchical models for attack presentation and introduce the following techniques: attack tree, attack graph, Petri nets and Bayes networks. The main advantages of attack trees are the modular structure of the model, the relatively quick possibility of high-level model construction and intuitiveness.

As stated by Paul and Vignon-Davillier (2014), the attack trees allow for the overcoming of the human cognitive scalability issue by scaling down the problem. The method is thus used for the scaling down approach to solve a limited number of problems of high complexity.

After examining the approaches or proposed derivatives for attack modelling techniques, Kordy, Pietre-Cambacedes and Schweitzer (2014) list thirteen different aspects that serve as a basis for categorizing individual approaches. Within the framework of the building block of attack modelling, the selection of attack modelling technique will be based on classic structural model, that is, the attack tree model. Paul and Vignon-Davillier (2014) state the advantages of attack tree modelling, namely, the graphic presentation, which provides an excellent opportunity for collaboration between the participants constructing the model; possibility to quickly achieve tangible results; and the possibility to test a wide variety of different combinations using dedicated tools.

The so-called Enhanced Structural Model (ESM), proposed by Ivanc (2013, 2014) can be selected for the cyber operations modelling as a part of the innovative approach. The ESM eliminates certain weaknesses of the standard attack tree model, such as the difficult management of the tree structure when adding new nodes, management of the tree structure in terms of changes in attack targets, and version updates for a particular model.

To sum up, attack tree model is useful for the systematic presentation of possible attacks and for different approach to security analysis of the computer-network systems (Almasizadeh and Abdollahi, 2014). Attack tree model can be used to optimize the security function. A technique can provide a better support for decision-making (Dewri et al. 2012). We can further improve the usefulness of the attack tree technique with other enhanced versions.

2.2   Information Support in the Field of Cyber Operations

As stated by Kim and Eom (2016), the intelligence information on cyber targets is an important factor in achieving operational supremacy in cyber operations. When collecting data, it is vital to know the requirements and guidelines on the basis of which the data will be collected. When analysing the collected data, it is necessary to integrate and convert them into intelligence information. Common data sources are various databases, professional documentation, forums, dark web resources and, of course, social networks (Gibson, 2016). An analysis of social networks is also useful for better understanding of cyberattacks. Kumar and Carley (2016) analyse the effect of the sentiment on the existence and occurrence of cyberattacks.

Within the element of information support for cyber operations, it is especially important to rely on the method of Open Source Intelligence (OSINT). Open frameworks for communicating characteristics and severity of software vulnerabilities, such as CVSS (Common Vulnerability Scoring System), used by both software developers and security engineers (Mell, Scarfone and Romanosky 2007). The framework, such as CVSS, has an indispensable value in the field of security management, vulnerability assessments and research activities.

2.3   Work Process

The third building block of the proposed educational approach is the work process. It will logically link and guide the process itself through two previously presented building blocks and improve the situational awareness. Situational awareness about events in the cyberspace may be impaired due to inaccurate or incomplete vulnerability analyses, failure to adapt to new attacks, inability to convert raw data into useful information and poor coping with uncertainty (Jajodia et al. 2011). Attack trees are adequately simplified methodologies which facilitate the identification of the attackers' targets (Ten, Manimaran and Liu 2010). This is a prerequisite for the development of attack scenarios and consequently the production of analyses.

Figure 3 shows the items relevant inside work process building block. We can see, that it is necessary to previously know the users' security requirements of the developing system. It is then important to know the architectural design of the system. Next, we proceed with the attack modelling stage in accordance with the proposed workflow. This stage requires the knowledge of attack surface and attack techniques, tactics and procedures. The last stage involves the production of a set of security countermeasures based on different attack scenarios stemming from the prepared model.

The general course of the high-level attack modelling workflow as a main element of the work process building block can be presented in the following steps:

Step 1:   Selecting the purpose of cyber operations or a specific cyber operation.

Step 2:   Determining the main objective of cyber operations modelling.

Step 3:   Integrating the intelligence from information support building block.

Step 4:   Developing a cyber-operation scenario within the framework of all three building blocks of the proposed model.

Step 5:   Further use of developed content within the educational process

Due to limitations, the generic framework of the building block of work process is presented here. This high-level framework consists of five steps and it logically links the work through previously presented building blocks.

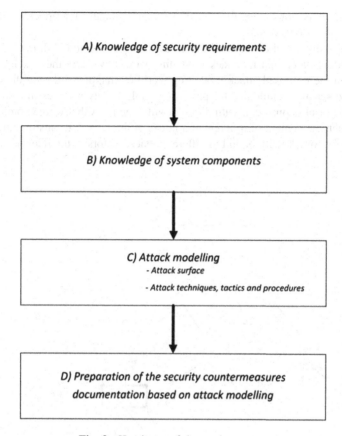

Fig. 3. Key items of the work process.

3 Illustration of an Example

The work and experience so far have shown that in terms of education and knowledge transfer, we mainly encounter two purposes of the attack modelling with structural techniques:

(a) The display of possible attacks on a specified target (primary purpose): It displays a possible range of attacks with the intention to identify weak points and to propose adequate security countermeasures.

(b) The display of the malware functioning: It displays the anatomy of the attack from the perspective of different sequential stages, carried out by the malware from the implementation of the infectious vector to the establishment of key malicious activities.

Based on the review of scientific literature, the activity in the field of higher education and the implementation of trainings for the employees in national security organizations, it is noted that the use of structural models for attack modelling is desirable for the primary purpose.

Below is a simplified example of the model that was created with the use of work process building block and its attack modelling workflow. Here the attack modelling that based on the proposed attack modelling workflow happened during the design of our observed set of technologies for providing public safety services in a smart environment. The discussions dealt with different vulnerabilities that were exploited in the past and match the content of different end nodes of the model. It must be said that the discussion was contextually bound to different attack vectors in the relevant individual nodes.

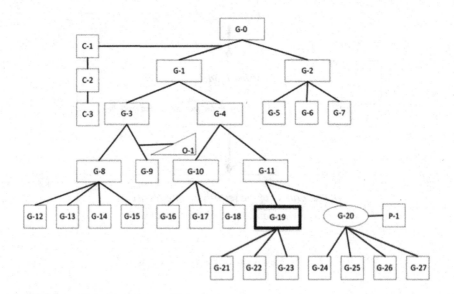

Legend

Node shape	Node type
	AND node
	OR node
	End-node / leave
	Conditional subordination node
	Housing node

Fig. 4. Partial attack model built on the proposed workflow.

Table 1. Description of the individual nodes from Fig. 4.

Node	Description
G-0	Attack on the observed system
G-1	WAN-based attacks
G-2	Local network attacks on end-user
G-3	Implementation attacks
G-4	Infrastructure attacks
G-5	Direct local access
G-6	Man-in-the-Middle implementations
G-7	Other unclassified implementations
G-8	SQL injections
G-9	Brute-force approaches
G-10	DoS/DDos attacks
G-11	Client side attacks
G-12	Tautologies
G-13	Piggy-backed queries
G-14	Stored procedures
G-15	Inference and alternate encodings
G-16	DNS amplification attack
G-17	ICMP flood attack
G-18	TCP SYN flood denial-of-service attack
G-19	Spear phishing
G-20	Watering-hole attack
G-21	Prompt user to enable macros
G-22	Drop malicious.vbs file
G-23	Trojan installation
G-24	Redirection of visitors to a malicious server
G-25	Fingerprinting script execution
G-26	Configuration information extraction
G-27	Installation attempt of a trojanized version of the software
P-1	Dropping the set of trojans
O-1	Dictionary attacks

Table 1 describes the individual nodes of the presented partial attack model, which directly concerns our observed system and possible attack paths. The presented partial model focuses on the subtree structure with the main goal G-1. It is developed up to level 5. The mentioned subtree structure thus forms its own segment in the model.

4 Evaluation of a Model of an Innovative Approach to Education in the Field of Cyber Operations

4.1 Evaluation Methodology

This subsection will present the evaluation methodology, including the explanation of the evaluation purpose and the selection of comparative processes. In general, the evaluation approach is divided into three sections:

Discussion Workshop with Security Experts
The purpose of the workshop was to check the usefulness of the cyber operations modelling based on the proposed work process as well as to identify possible directions of the practical use of a model of an innovative approach to education in the field of cyber operations by security experts in a specific environment. The purpose was also to identify tangible advantages of the proposed work process.

The discussion workshop was conducted with representatives of the General Police Directorate. The meeting lasted 90 min and involved experts from the Internal Investigation and Integrity Division, ICT Systems and Data Protection Division and Applications Development Division. First, the participants learnt about the basics of attack modelling and about the entire proposal of the work process of cyber operations modelling. Then, we started modelling a cyber-operation together, whose main goal was to "transfer personnel records". The participants constructed a model in accordance with the proposed attack modelling workflow through team work, mutual discussion and concepts of brainstorming.

The final model contained 4 levels with a total of 16 nodes. Four end nodes were present at the deepest, fourth level. Most intermediate nodes actually contained the OR operator, which means that the nodes, irrespective of model levels, mostly showed the alternative attacks.

The questions that arose during the work of the participants were the following:

- How do we solve an issue that has a different name, but is the same attack made by different analysts?
- Is there a connecting element or a rule for the same attacks that occur at dif-ferent places of the individual subtree structure within the model?

The opinions linked directly to the work of the group were the following:

- The development of the attack scenarios with cyber operations modelling is logical, so that security procedures can be further developed from results.
- The cyber operations modelling can be the basis for the development of a security checklist.
- Model of an innovative approach to education in the field of cyber operations offers the possibility to record ideas on how to improve certain segments of questionable security.

Taking into account the initial presentation and the meeting itself, it was concluded that the proposed work process is useful, especially for teamwork, because it focuses on the discourse. The process can be a good basis for the reports and definitions of security measures.

Working Meeting with Students

The purpose of the evaluation was to check how the proposed cyber operations modelling, as a part of the work process, contributes to the efficiency of the direct model construction by students at the higher education programme. Further, the purpose was to evaluate the direct advantages of the cyber operations modelling based on the proposed work process, such as the potential possibility to recognize a higher number of security countermeasures.

The working meeting with students was held in the form of a guest lecture and it lasted 90 min. The workshop was held within the course Business Application Development. The course is conducted in the third year of undergraduate study programme Business Informatics. The reader should also know that students enrolled in this course in the first year meet with subjects Business Informatics and Business Mathematics, in the second year with Business Process Modelling and Analysis, Business Information Systems, Software Development and in the third year with Information Systems Security, Operating Systems and Networks. Due to different levels of knowledge and experience of students, the workshop was an excellent opportunity to evaluate the attack modelling workflow.

Firstly, the concept of attack modelling was introduced to students, then they received a worksheet with a description of the situation. The situation gave students the role of the attacker with the intention that they design a structural attack model, specify the proposed security countermeasures and present the attack scenarios. Ad hoc, students were divided into two groups: the first group was modelling the offensive cyber operation with the so-called "intuitive" approach (Group A) and received a brief unstructured description of the attack modelling, which is a synthesis of the already presented descriptions of current approaches. The second group (Group B) received in a limited form a table with the attack modelling workflow as a key element of work process building block.

All students worked in pairs. During the working meeting, there were six final products made for the evaluation purpose, three in each group. None of the observing models contains a logical contradiction, as for example, the intermediate node further contains only one sub-node. Table 2 summarizes the characteristics of the models that were created during a working meeting with the students.

Table 2. Summarized characteristics of models created during a working meeting with students.

Attack model code	Number of levels	Total number of nodes	Number of AND nodes in the model	Number of proposed security countermeasures
A1	3	13	1	4
A2	3	9	0	0
A3	3	12	0	1
B1	4	21	2	4
B2	4	16	3	7
B3	4	14	2	6

Based on the comparison of the two groups, we can conclude that students are able to present more sophisticated models with the proposed work process and attack modelling workflow. T At the same time, they are more motivated to reasonably identify security countermeasures, which is the purpose of the training in the field of information security.

Structured Interviews with Higher Education Teachers

In order to evaluate predominantly subjective aspects, the interview method with key experts was chosen. The selected parameters, which are predominantly subjective in the evaluation of the work process of cyber operations modelling, are listed below with a brief explanation (Strahonja 2007):

- Replaceability – the opportunity and effort of using the specified method or concept in the place of other methods.
- Interoperability – the ability of one framework to interact with others.
- Adaptability – the opportunity for specialization and adaptation of concepts and methods for different domains.

Interviewees were recognized experts who meet the following criteria: they are university professors working in the field of cyber security and have adequate experience with the approaches used in cyberattack modelling. The interviews were carried out face-to-face and on average lasted an hour.

Interviewee 1 states that the work process in terms of replaceability of other methods seems appropriate to him and can substitute other methods. The reason lies in the fact that the work process is sufficiently universal and useful. In terms of the use of other specific methods, he believes that it can be complementary with other approaches, so that the cyber operations can be presented in more detail. In terms of adaptability, he evaluates the building blocks of the proposed model as accomplished. There is only a question of difficulty, which increases relatively with specific cyber operations. Interviewee 1 evaluates the proposed model of an innovative approach to education in the field of cyber operations as adequate and accomplished, since it can be followed by most of the potential users.

Interviewee 2 states that the proposed work process is suitable for delivering knowledge and it is useful in practice to work with students and produce cyber operations analyses. He evaluates the building blocks of the proposed model of an innovative approach to education in the field of cyber operations as clear and presentable. He also thinks that the model can be successfully used with other specific methods. In terms of adaptability, he emphasises its use for the wide display of differently conducted cyber operations. Interviewee 2 positively evaluates the proposed model and welcomes its implementation in the educational process. At the same time, he states that caution is needed when implementing the model, so that the students can understand it as well as possible.

4.2   Conclusions Based on the Evaluation

It is worth summarizing the segments of a proposed model of an innovative approach to education in the field of cyber operations to the reader, based on individual evaluation approaches (Fig. 5).

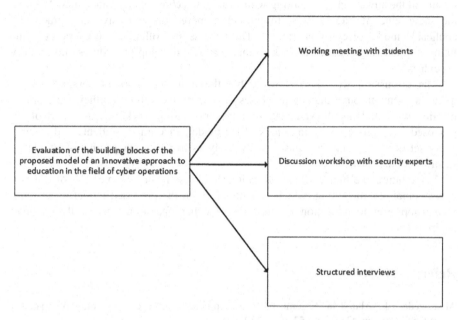

Fig. 5. Evaluation of the building blocks of the proposed model by the individual approach.

The discussion workshop evaluated the suitability of guiding field experts in cyber operations modelling with support of attack modelling workflow from work process building block. The working meeting with students evaluated the advantages of the cyber operations modelling building block which is an integral part of the proposed model. The interviews evaluated the suitability and the additional contribution of the entire model of an innovative approach to education in the field of cyber operations.

As illustrated by the evaluation, a model of an innovative approach to education in the field of cyber operations appropriately guides students through key processes of the cyber operations modelling. In terms of expert opinions, the findings from the discussion workshop with security experts are important. Based on the discussion workshop and participants' opinions, it can be observed that the work process as a building block of the proposed model adequately guides students through the key processes of cyber operations modelling.

It can be observed that the attack modelling workflow, as a part of the proposed work process, utilizes better guidance to the attack tree model construction. This increases the development of attack scenarios, better information classification and promotes the implementation of security countermeasures.

5   Conclusion

When modelling offensive cyber operations, it is important that the work process itself allows students who have different previous knowledge to work on the same structural model. Model of an innovative approach to education in the field of cyber operations retains all the advantages of working with structural techniques for attack modelling. At the same time, it directs users to present a more comprehensive and practically applicable model of cyber operations. The process also offers a good support to the future and existing security experts so that they can develop procedures and security checklists.

The discussion workshop showed that the proposed work process adequately guides students through the key processes of security modelling. Further, the working meeting with students showed that the attack modelling workflow, as a part of the proposed work process building block of the model, encourages students to develop a wider set of attack scenarios and security analyses from the perspective of setting up security countermeasures.

In the future, the subject of consideration should be the development of software for cyber operations modelling, which will be based on the proposed model of an innovative approach to education in the field of cyber operations, especially in smart environment.

References

Almasizadeh, J., Abdollahi Azgomi, M.: Mean privacy: a metric for security of computer systems. Comput. Commun. **52**, 47–59 (2014)

Celik, Z.B., et al.: Mapping sample scenarios to operational models. In: Military Communications Conference MILCOM (2016)

Chang, Y.H., Jirutitijaroen, P., Ten, C.W.: A simulation model of cyber threats for energy metering devices in a secondary distribution network. In: Proceedings of the 5th International Conference on Critical Infrastructure, pp. 1–7 (2010)

Dewri, R., Ray, I., Poolsappasit, N., Whitley, D.: Optimal security hardening on attack tree models of networks: a cost-benefit analysis. Int. J. Inf. Secur. **11**(3), 167–188 (2012)

Gibson, H.: Acquisition and preparation of data for OSINT investigations. In: Akhgar, B., Bayerl, P., Sampson, F. (eds.) Open Source Intelligence Investigation. ASTSA, pp. 69–93. Springer, Cham (2016). https://doi.org/10.1007/978-3-319-47671-1_6

Hong, J., Kim, D.S.: HARMs: Hierarchical Attack Representation Models for Network Security Analysis (2012)

Ivanc, B.: Modelling of information attacks on critical infrastructure by using an enhanced structural model. Master thesis (2013)

Ivanc, B.: Providing information privacy with attack modelling: from simple techniques to the enhanced structural model. In: Corporate Security: Open Dilemmas in the Modern Information Society, pp. 161–168 (2014)

Jajodia, S., Noel, S., Kalapa, P., Albanese, M.: Cauldron - mission-centric cyber situational awareness with defense in depth. In: Military Communications Conference (MILCOM 2011), pp. 1–6 (2011)

Kim, K.H., Eom, J.H.: Modeling of cyber target selection for effective acquisition of cyber weapon systems. Int. J. Secur. Appl. **10**(11), 293–302 (2016)

Kordy, B., Pietre-Cambacedes, L., Schweitzer, P.: DAG-based attack and defense modelling: Don't miss the forest for the attack trees. Comput. Sci. Rev.w **13–14**, 1–38 (2014)

Kumar, S., Carley, K.M.: Understanding DDoS cyber-attacks using social media analytics. In: Conference on Intelligence and Security Informatics (ISI), pp. 231–236 (2016)

Mell, P., Scarfone, K., Romanosky, S.: A Complete Guide to the Common Vulnerability Scoring System Version 2.0 (2007)

Paul, S., Vignon-Davillier, R.: Unifying traditional risk assessment approaches with attack trees. J. Inf. Secur. Appl. **19**(3), 165–181 (2014)

Piètre-Cambacédès, L., Bouissou, M.: Cross-fertilization between safety and security engineering. Reliab. Eng. Syst. Saf. **110**, 110–126 (2013)

Pipyrosa, K., Thraskiasb, C., Mitroua, L., Gritzalisa, D., Apostolopoulos, T.: A new strategy for improving cyber-attacks evaluation in the context of Tallinn manual. Comput. Secur. **74**, 371–383 (2018)

Strahonja, V.: The evaluation criteria of workflow metamodels. In: Proceedings of the International Conference on Information Technology Interfaces, pp. 553–558 (2007)

Ten, C.W., Manimaran, G., Liu, C.C.: Cybersecurity for critical infrastructures: attack and defense modeling. IEEE Trans. Syst. Man Cybern. Part A Syst. Hum. **40**(4), 853–865 (2010)

Next Generation Wired/Wireless Advanced Networks and Systems

Channel Switching Protocols Hinder the Transition to IP World: The Pentagon Story

Manfred Sneps-Sneppe[1(✉)], Dmitry Namiot[2], and Maris Alberts[3]

[1] Ventspils International Radioastronomy Centre,
Ventspils University of Applied Sciences, Ventspils, Latvia
manfreds.sneps@gmail.com
[2] Faculty of Computational Mathematics and Cybernetics,
Lomonosov Moscow State University, Moscow, Russia
dnamiot@gmail.com
[3] Institute of Mathematics and Computer Science, University of Latvia,
Riga, Latvia
alberts@latnet.lv

Abstract. In this paper, we target the strategy for telecommunications architectures during the transition to the IP-only models. The paper discusses the shifting from circuit switching to packet switching in telecommunications. Particularly, we analyze the coexistence of circuit switching and packet switching technologies in American military communications where each warfare object should have own IP address. The paper discusses the role of multifunction Soft Switches (MFSS). This Soft Switch plays the role of a media gateway between TDM channels and IP channels. As a case, we are passing through the transformation from SS7 signaling to internet protocol, ISDN-based government Defense Red Switch Network and, finally, the extremely ambitious cybersecurity issues and the cyber vulnerabilities of weapons found by Government Accountability Office. We conclude the growing cyber threats will provide a long-term channel-packet coexistence.

Keywords: IP protocol · Telecom · Packet switching

1 Introduction

In this paper, we discuss the government telecommunications strategy in the transition to the IP world using the Pentagon experience as a case.

In 2007, Pentagon published a fundamental program [1], in which we find three main points:

- GIG (Global Information Grid) as a unified solution,
- The network-centric war concept is accepted as a foundation,
- IP protocol accepted as the only means of communication between the transport layer and applications

It is illustrated in Fig. 1.

© Springer Nature Switzerland AG 2019
O. Galinina et al. (Eds.): NEW2AN 2019/ruSMART 2019, LNCS 11660, pp. 185–195, 2019.
https://doi.org/10.1007/978-3-030-30859-9_16

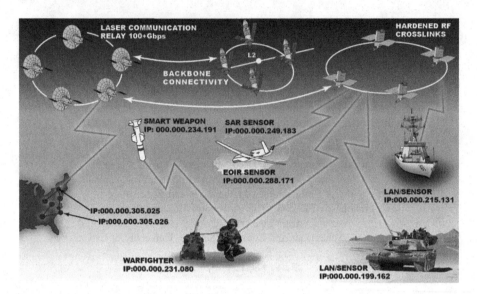

Fig. 1. GIG Communications Infrastructure. Here each warfare object has own IP address [4].

The third requirement (IP protocol only) is the most important one in this plan.

In this article, we use the novel unclassified open Defense Information Systems Agency (DISA) documents, particularly: Department of Defense (DoD) Information Enterprise Architecture [2, 3], DoD Unified Capabilities (UC) Requirements [4], 295-page description of the UC framework for the Army [5] and some others. In parallel, we apply to the analysis of DoD activities made by US Government Accountability Office experts [6]. Part of the materials of this work had published earlier in our papers [7–10] and the book on Pentagon telecommunications [11].

The rest of the paper is the following. Section 2 refers to U.S. "Joint Vision 2020" and Army Unified Capabilities. Section 3 considers Multifunctional Soft Switch as the turning point on IP-road and Sect. 4 – the transient DISN architecture. In Conclusion, Sect. 5, some cases are named, which testify about long-term channel-packet coexistence, particularly due to an increase of cyber threats.

2   On Requirements to Unified Capabilities

In general, Unified Capabilities are defined as the integration of telecommunication services independent of technology. So, Army Unified Capabilities tools are also targeting, for example, the integration of voice and data services delivered ubiquitously across the Army telecom infrastructure. The key goal is to provide increased mission effectiveness for DoD Components [12].

The following are the basic Voice Features and Capabilities:

- Call Forwarding (selective, on busy line, etc.)

- Multi-Level Precedence and Preemption (Interactions with call forwarding, No reply at the called station, etc.)
- Precedence Call Waiting (Busy with higher precedence call, busy with Equal precedence call, etc.)
- Call Transfer (at different precedence levels, at same precedence levels)
- Call Hold and Three-Way Calling.

The capabilities are provided through a collection of services, where a service is defined as 'a mechanism to enable access to a set of one or more capabilities'. The reference architecture is illustrated in Fig. 2.

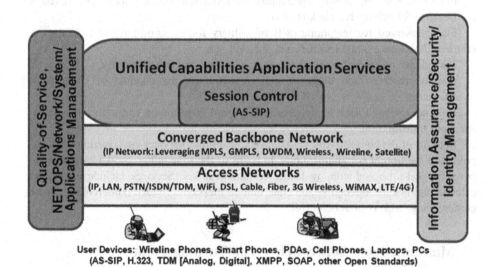

User Devices: Wireline Phones, Smart Phones, PDAs, Cell Phones, Laptops, PCs
(AS-SIP, H.323, TDM [Analog, Digital], XMPP, SOAP, other Open Standards)

Fig. 2. The UC reference architecture [12]

The UC architecture is based on the IP for the wide-area backbone network. The traffic engineering and QoS for the IP network are supported via MPLS (multiprotocol label switching protocol). In the same time, the system supports different access technologies. They may contain such technologies as IP, LAN, PSTN/ISDN/TDM, Wi-Fi, DSL, Cable, and Fiber, 3G Wireless, WiMAX, and LTE/4G Wireless. For example, the worldwide DISN backbone transport network will consist of both wireline (e.g. fiber) and wireless (e.g. satellite) network. Accordingly, some technologies (IP/Lan) will work with IP networks directly, somewhere (PSTN/ISDN/TDM) we will need intermediate gateways. We will discuss TDM networking with IP/MPLS network in Sect. 3.

Let us take attention to the AS-SIP protocol. The SIP, as a signaling protocol, does not support the ability to break into ongoing calls. It's critical to support Multi-Level Precedence and Preemption calls (e.g. emergency calls). It is supported in the so-called Assured Services SIP (AS-SIP) [13]. AS-SIP got many features for Unified Capabilities requirements [4], and had required support by up to 200 RFCs (the ordinary SIP uses only 11 other RFC standards).

The Multilevel Precedence and Preemption (MLPP) service is the key feature of AS-SIP. Calls have priority levels and can be ousted depending on this level. This capability lets high-ranking personnel reach critical organizations and personnel during network stress situations (e.g., a national emergency or degraded network performance).

RFC 4542 [15] describes six precedence classes, in descending order:

Executive Override (or Flash Override Override): used when declaring the existence of a state of war and cannot be preempted.

Flash Override: the calls by Presidential authorization (they cannot be preempted in the Defense Switching Network (DSN)).

Flash: reserved for telephone calls of military forces, continuity of federal government functions essential to national survival, etc.

Immediate: reserved for telephone calls devoted to national security.

Priority: reserved for telephone calls essential for the conduct of government operations.

Routine: reserved for telephone calls of official government communications with the rapid transmission.

In terms of the connectivity, the Unified Capabilities services are covering point-to-point and multipoint communications. In terms of the scheduling, the UC services are covering real-time and non- real-time communications. Services include email and calendaring, instant messaging and chat, unified messaging, video conferencing, voice conferencing, web conferencing [4].

3 Multifunctional SoftSwitch – The Turning Point on IP-Road

The DISN modernization to IP world includes the replacing of channel switches (electronic exchanges) by packet switches as well as adding local service controllers (LSC), wide area network SoftSwitches (WAN SS), and multifunction SoftSwitches (MFSS). Figure 3 shows a simple case: TDM-to-IP call switched over an IP backbone.

MFSS plays the role of a media gateway between TDM channels and IP channels and it is controlled via H.248 protocol. The Signaling Gateway (SG) provides communication between CCS7 (SS7) and SIP.

The Service Control Function (SCF) is AS-SIP based framework for creating service interactions for the above-mentioned LSCs, WAN SSs, and MFSSs. The task of SCF is to transfer (convert) existing traditional services (both wireline and wireless) into feature-rich advanced new services. SCF is cooperating with as many as 19 servers and supports plenty of protocols: SOAP, HTTP, LDAP, SQL, RADIUS, etc. (Fig. 4).

In general, the application servers include the following services: messaging (e.g. MS, MMS, email), conversation (e.g. Voice & Video, Web Conferencing, Collaboration & Whiteboard, Desktop Sharing), location and presence, special communications (e.g. E.911), etc. Note that some of them are obsolete (e.g. WAP).

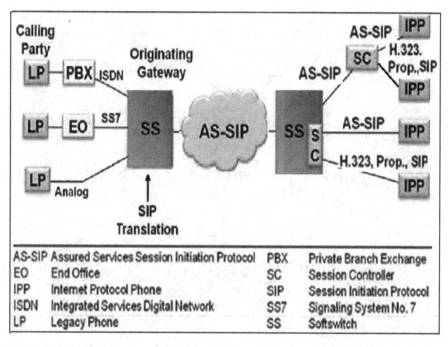

Fig. 3. TDM-to-IP call over an IP backbone [4]

Starting in 2011, CISCO - the largest contractor of the Pentagon - has installed 22 softswitches (both Wide Area Network SoftSwitch and MultiFunction SoftSwitch (MFSS) at military bases around the world [16]. First one MFSS has installed in Stuttgart (Germany). Unfortunately, up to now, there is no openly available information about any fully operating MFSS. The four GNSCs (Global Network Support Centers) are located in the US, Germany, and Bahrain [17].

4 On Transient DISN Architecture

To illustrate the current DISN architecture we refer to the certification of Avaya PBX by DISA Joint Interoperability Test Command in 2012 [18]. The SS7 network is the core communication system of DISN up to resent time (Fig. 6) connecting the channel mode MFS. That is, within the DISN network, the connections have established by means of SS7 signalling and, in the periphery, devices of any type are used.

The Fig. 5 illustrates the simplified DISN view. The key moment here is the usage of different protocols: 4-wire (4 W); classified LAN (ASLAN); ISDN BRI; Internet telephony (VoIP); Video-conferencing (VTC); proprietary protocols. Even though all new terminal equipment what appears is largely of IP type, SS7 network always keeps the central place. It means that the SS7 network itself does not prohibit the IP deployment. Not only does this not impede the transition to IP protocols, but actually, it is rather the opposite. This makes it easy to switch to packet switching step by step.

Fig. 4. The services and protocols in SCF [4]

The choice of SS7 as a basis for DISN is going back to program "Joint Vision 2010", approved in 1997. By this time, of course, there were many communication networks in use. Accordingly, this meant integrating multiple networks into DISNs. This integration inevitably limited the use of networks. In addition, such work, performed at rather low levels, consumed huge resources. The way out was the transition to open architecture and the reuse of architectural elements (As per DISA documents: "to build US military communications networks using the open architecture and commercial-off-the-shelf (COTS) products". The existing Bell Labs developments were chosen as such components: SS7 as the telephone signaling protocol (COTS) and the Advanced Intelligent Network as an open architecture. SS7 protocols finally were defined as ITU standards.

The details we found in one paper from Lockheed Martin Missiles & Space [19] – the well-known Defense contractor. Military communication systems have started to merge traditional circuit-switched voice with Internet and Asynchronous Transfer Mode (ATM) as the backbone networks. A critical issue is the interface of SS7 voice circuits with ATM.

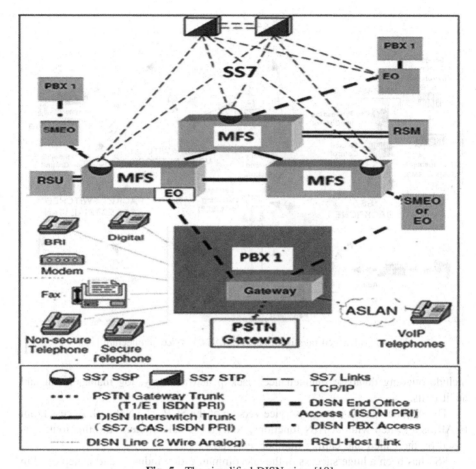

Fig. 5. The simplified DISN view [18]

The Advanced Intelligent Network (AIN) was originally designed as a critical tool to offer sophisticated services such as expert operator assistance and directory assistance. The functional structure of the SS7 makes it possible to create the AIN by putting together functional parts.

Figure 6 describes the AIN components that operate in the worldwide telecommunication network, as well as how they are deployed in SS7 backbone, the space Wide Area Network (WAN), circuit switched voice network and the packet switched terrestrial WAN. The AIN components include the Service Creation Environment (SCE), Service Management System (SMS), Service Control Point (SCP), Service Switching Point (SSP), Intelligent Peripheral (IP), Adjunct, and the Network Access Point (NAP).

The SCE provides design and implementation tools needed to assist in creating and customizing services in the SCP. The SMS is a database management system. It is used to manage the master database that controls the AIN warfighter services. These services

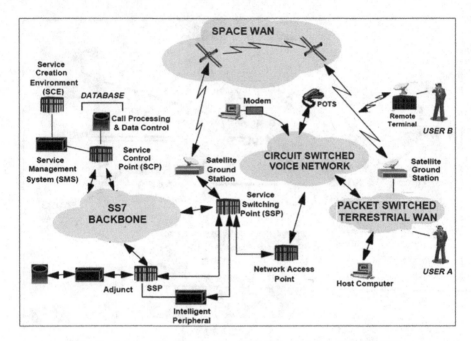

Fig. 6. Advanced intelligent network (AIN) service architecture [19]

include ongoing database maintenance, backup and recovery, log management, and audit trails.

The Adjunct here is just one-service version of the SCP. The Network Access Point (NAP) is a switch without AIN functions. NAP is the calls router. And this routing is based on the called and calling number received.

SS7 has been a huge success in the telecommunication industry and is deployed in all public telephone circuit switched networks by all carriers throughout the world. The key features of SS7 have found their way into other systems such as Global System for Mobile Communication (GSM), military communication, and even satellite signaling (see Fig. 6).

Now, more than 20 years later, according to Army regulation [14], DISN must be built on the basis of IP protocol as the only means of communication between the transport layer and applications. The Army regulator recognizes that up to now there is 'old' equipment on the network: Time-division multiplex equipment, Integrated services digital networking, channel switching Video telecommunication services.

The Army supports the moratorium on investment in legacy voice-switching equipment. It means that the Army will migrate as soon as practical to an almost-everything-over-Internet Protocol architecture, to include Unified Capabilities and collaboration, with an end state of end-to-end IP.

The DISN hybrid signaling design (Fig. 7) describes a local and a backbone model. At the local level (based on a multivendor assortment of SCs), SCs are located in secure enclaves. The backbone level (so-called Tier0 signaling), is a robust, homogeneous design based on current vendor-unique geographic cluster arrangements of Tier0 SS.

AS-SIP is used as the signaling method here. In the same time, during the transition period to AS-SIP there will be segments using H.323 signaling also as well ISDN signaling for the Defense Red Switch Network.

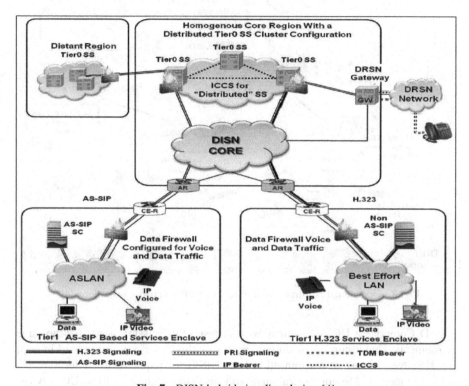

Fig. 7. DISN hybrid signaling design [4]

5   Conclusion. On Long-Term Channel-Packet Coexistence

Look for some cases who testify about long-term channel-packet coexistence.

ISDN-based government network DRSN. The Defense Red Switch Network (DRSN) uses 40 years old ISDN technology. It looks like some kind of birthmark in the IP environment. This is the most secure telecommunications network for the US military. At the same time, STE (Secure Terminal Equipment or so-called "Red Phone") connects to the network via the ISDN line and still works at a speed of 128 kbps. Due to cyber threats, how to force to refuse ISDN channels?

On U.S. Cyber Command Activities. These activities significantly slow down the transition to IP world. The DISA (more precisely the DISA Network Operation Center) is responsible for the whole backbone infrastructure and UC services. But the limitations of cyber-attack protection seriously restrict the development of services and changes plans for the DISN network – see, for example, Accreditation Boundary in Fig. 8.

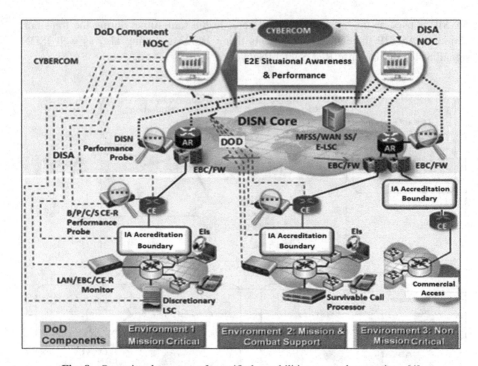

Fig. 8. Operational construct for unified capabilities network operations [4]

DoD Component UC infrastructures meet respective mission needs for the three environments. Under normal conditions, access to all UC services is required in each case. However, in case of disconnection from DISA, the requirements differ:

Environment 1: Mission critical environments. They require all basic UC services (including E.911 service). Examples include a combat support unit or operational flying wing;

Environment 2: Mission and Combat Support environments required limited voice-only services;

Environment 3: Non-Mission Critical Location. An example would be a small administrative function (e.g., recruiting office).

Environments 1 and 2 are under strong CYBERCOM supervision. Unfortunately, as cyber threats grow, grows the list of CYBERCOM cyber security rules complicating the transition to the IP world.

Summing up, the growing cyber threats will provide a long-term channel-packet coexistence. At least, it is highly likely.

References

1. U.S. Department of Defense. Global Information Grid. Architectural Vision, Version 1.0, June 2007

2. Vargas, A, et al.: Towards the development of the framework for inter sensing enterprise architecture. J. Intell. Manufact. **27**(1), 55–72 (2016)
3. Sneps-Sneppe, M., Seleznev, S., Kupriyanovsky, V.: DISN Network as the prototype of the network connection of civil defense NG 112. Int. J. Open Inf. Technol. **4**(5), 39–47 (2016)
4. U.S. Department of Defense Unified Capabilities Requirements (UCR 2013), August 2019. http://jitc.fhu.disa.mil/jitc_dri/pdfs/UCR_2013_Combined_signed.pdf
5. U.S. Army Unified Capabilities (UC) Reference Architecture (RA). Version 1.0, 11 October 2013
6. WEAPON SYSTEMS CYBERSECURITY: DOD Just Beginning to Grapple with Scale of Vulnerabilities. https://www.gao.gov/products/GAO-19-128, August 2019
7. Sneps-Sneppe, M.: On telecommunications evolution: pentagon case and some challenges. In: Proceedings of 2017 9th International Congress on Ultra Modern Telecommunications and Control Systems and Workshops (ICUMT), Munich, Germany, 6–8 November 2017
8. Sneps-Sneppe, M., Namiot, D.: On telecommunication architectures: from intelligent network to network functions virtualization. In: Proceedings of 8th International Congress on Ultra-Modern Telecommunications and Control Systems and Workshops (ICUMT2016), Brno, October 2016
9. Sneps-Sneppe, M., Sukhomlin, V., Namiot, D.: On cyber-security of information systems. In: 21st International Conference on Proceedings of Distributed Computer and Communication Networks DCCN 2018, Moscow, Russia, 17–21 September 2018
10. Sneps-Sneppe, M., Namiot, D.: Time to rethink the power of packet switching. In: Proceedings of the 23rd Conference of Open Innovations Association FRUCT. FRUCT Oy (2018)
11. Sneps-Sneppe, M.: Pentagon Telecommunications: Digital Transformation and Cyber Defense. (Telekommunikatsii Pentagona: tsifrovaya transformatsiya i kiberzashchita). Scientific and technical publishing house «Goryachaya liniya – Telekom» , Moscow (2017)
12. U.S. Department of Defense. Unified Capabilities Master Plan (UC MP), October 2011
13. U.S. Department of Defense. Assured Services (AS) Session Initiation Protocol (SIP) 2013 (AS-SIP 2013) Errata-1, July 2013
14. Army Regulation 25–13. Information Management. Army Telecommunications and Unified Capabilities. Headquarters Department of the Army. Washington, DC, 11 May 2017
15. Baker, F., Polk, J.: (Cisco Systems) "Implementing an Emergency Telecommunications Service (ETS) for Real-Time Services in the Internet Protocol Suite", RFC 4542, May 2006
16. Cisco Communication Strategy. Web. https://www.cisco.com/web/strategy/docs/gov/Cisco_LSC_Overview_Jan2011.pdf/. Accessed April 2019
17. Joint Interoperability Test Command. http://jitc.fhu.disa.mil/tssi/cert_pdfs/avaya_s8300d_v60_dtr1_jan13.pdf. Accessed August 2019
18. DARPA names Lockheed Martin to build intelligent network, 24 March 2005. http://www.militaryaerospace.com/articles/2005/03/darpanames-lockheed-martin-to-build-intelligent-network.html/. Accessed August 2019
19. Chao, W.W.: Emerging advanced intelligent network (AIN) for 21st century warfighters. In: IEEE Military Communications. Conference Proceedings MILCOM 1999 (Cat. No. 99CH36341), vol. 1. IEEE (1999)

Network Anomaly Detection in Wireless Sensor Networks: A Review

Rony Franca Leppänen[✉] and Timo Hämäläinen

Faculty of Information Technology, University of Jyväskylä, P.O. Box 35, (Agora),
40014 Jyväskylä, Finland
rony.e.m.leppanen@student.jyu.fi, timo.t.hamalainen@jyu.fi

Abstract. Wireless sensor networks function as one of the enablers
for the large-scale deployment of Internet of Things in various applications, including critical infrastructure. However, the open communications environment of wireless systems, immature technologies and
the inherent limitations of sensor nodes make wireless sensor networks
an attractive target to malicious activities. The main contributions of
this review include describing the true nature of wireless sensor networks through their characteristics and security threats as well as reflecting them to network anomaly detection by surveying recent studies in
the field. The potential and feasibility of graph-based deep learning for
detecting anomalies in these networks are also explored. Finally, some
remarks on modelling anomaly detection methods, using appropriate
datasets for validation purposes and interpreting complex machine learning models are given.

Keywords: Wireless sensor networks · Anomaly detection ·
Deep learning

1 Introduction

The availability of small, cheap and smart sensors has changed the nature of
wireless sensor networks (WSNs) and made them an attractive solution to many
real-time applications [1]. They are also an integral part of Internet of Things
acting as an interface between the physical and the digital world being one of
the most prominent technologies for the future. However, while sensor networks
share many similarities with other distributed systems, they experience a variety
of unique challenges and constraints which impact the design of a WSN. Dealing
with the constraints often leads to developing protocols and algorithms that
differ from their counterparts in other distributed systems. One of the main
challenges, among energy, design constraints and others, is security [3].

The open communications environment of wireless systems makes wireless
transmissions more vulnerable to malicious attacks than wired communications
where communicating nodes are physically connected through cables [4]. As more
physical devices, systems, facilities and critical infrastructure are connected to
Internet via wireless networks and the growth of cybercrime continues to acceler-
ate, the existence of such networks pose major security and privacy issues [5,6].
Protecting WSNs from cyberattacks and other factors that may impair the continuity of their secure and reliable operations is one of the most promising research
issues in WSNs [7].

© Springer Nature Switzerland AG 2019
O. Galinina et al. (Eds.): NEW2AN 2019/ruSMART 2019, LNCS 11660, pp. 196–207, 2019.
https://doi.org/10.1007/978-3-030-30859-9_17

2   Characteristics and Security Threats in WSNs

Wired and wireless networks both adopt the OSI[1] layered protocol architecture for transmitting and recovering encapsulated data packets. While the application, transport and network layers of wireless networks are in most cases identical to those of wired networks, the main differences lie in the physical transmission medium and how the devices access that medium to transmit data. Also, since different OSI layers rely on different protocols, every layer has its own security challenges and issues [4]. This is particularly true in the applications of Internet of Things where many systems are poorly designed with preference of functions over security and implemented using immature technologies including newly developed protocols for resource-scarce environments, among other issues [8,9]. What is more, the lack of physical protection, the resource restrictions and the wireless nature of communication make WSNs susceptible to not only internal but also to external attacks [5]. More information on how different types of common attacks against WSNs relate to the protocol stack can be found in [4,5,10].

WSNs present various different properties that change the nature of these communications networks [11]. Firstly, the measurements in different sensor nodes and the traffic between the nodes may exhibit many forms of interdependencies over different domains such as time and space. Secondly, the sensor nodes can be mobile or have self-organizing capabilities leading to dynamically changing network structure over time. Thirdly, these networks may experience unexpected events caused by environmental conditions or sophisticated cyber-criminals, among others. Moreover, similarly to other real-world systems, a relatively small event or change in one signal can have a significant impact to other signals due to multiplicative effects. As such, combined with the previous properties, the underlying distribution of the data generating process cannot be assumed as static, or to be known beforehand. In addition, node-specific attributes may be useful in detecting internal and external attacks. These factors already suggest that WSNs are inherently complex[2] and nonlinear which raises the question of how one should approach these systems when managing them. Especially in the field of security, the main idea must not be simplifying but rather understanding, incorporating and leveraging as much of the complex data that the system offers to detect all kinds of malicious activity, the previously unknown in particular. Therefore, the traditional protection systems based on predetermined rules, linear mechanisms[3] and statistical techniques[4], are simply not sufficient.

---

[1] In this article the 5-layer OSI-model is considered, as in [4,5].

[2] Complexity may not also be stable: the amount of complexity can vary according to changes in the system, the environment or transmitted data.

[3] Research across a range of disciplines is devoted to finding linear approximations of nonlinear phenomena, because they are easier to solve.

[4] Especially the assumption of identically and independently distributed (i.i.d.) variables and presuming the density distribution of data points a priori may not be realistic.

3   Network Anomaly Detection in WSNs

Anomaly detection is a well-established field with numerous applications in security, finance, healthcare, industry and many others. The problem of anomaly detection has had many definitions in the past often tailored for the specific application domain and has exhibited various names such as detecting outliers, anomalies, outbreaks, events, changes, fraud, faults, and so on. Even in some applications, outliers are called and treated as noise [12]. However, when considering security applications, anomaly detection is well-recognized as a branch of intrusion detection alongside signature-based detection, also commonly known as misuse detection. The problem with signature-based approach is that while it accurately detects known attacks, it is unable to detect any new attacks that emerge in the system, in contrast to anomaly-based approach. Still, there exists several issues with the anomaly-based approach such as false alarms may occur frequently in a non-stationary environment, or in other words, when the established normal behaviour of the system does not remain the same over time [13].

The power of deep learning based techniques, including their ability to learn representations and derive complex, nonlinear and hierarchical features with little or no prior knowledge from high-dimensional input data, has been demonstrated through breakthroughs in various applications such as computer vision, speech recognition, natural language processing and recommender systems [14,15]. The success has also led to the wide adoption of deep learning in many other fields as well, including anomaly-based intrusion detection [16]. However, even though the research community has explored machine learning based intrusion detection techniques for WSNs and recognized the potential of deep learning, remarks about some of the efforts made in the literature so far are in order.

In [17], a hybrid model was developed where raw data features were first captured and divided into subsets using spectral clustering, and then the subsets were fed into a subset-specific supervised 5-layer feed-forward neural network. Finally, the output of each sub-network was integrated for analyzing detection rates. One of the novelties of this work is that instead of random initialization, the sub-networks were pre-trained with normal traffic unsupervisedly using stacked autoencoders[5]. The authors further argued that the spectral clustering algorithm can extract similar features and gather more information from a dataset. The proposed method's superiority over basic artificial neural network, support vector machine, random forest and Bayesian tree models was validated using the 20-year-old KDD-CUP99 and NSL-KDD datasets as well as using a synthetic dataset created with the NS-2 network simulator. However, the hybrid model's performance depends highly on the spectral clustering algorithm whose parameters require careful empirical fine-tuning. Spectral clustering's inherent limitations such as scalability and computational efficiency also raise questions about its effectiveness. Moreover, changing the domain from graph-based spectral clustering to the sub-networks' sequential input may have an undesirable

---

[5] To overcome the limitations of traditional autoencoders, the study employs a sparse autoencoder and multiple denoising autoencoders for the sub-networks.

impact to the method's performance. Finally, no time-dependent dynamics were discussed, no take related to sensor hardware was given and more application-specific up-to-date as well as real-world datasets could have been used.

In a recent work [18], a model based on supervised 5-layer feed-forward neural network was developed for identifying malicious activities in industrial internet of things applications which is closely related to WSNs. In a similar vein to the previously presented work, the neural network was pre-trained with normal traffic in unsupervised manner using a standard deep autoencoder. The total time cost of the method was also discussed verifying that the proposed method is a practical solution for detecting intrusive activities in a real environment. To illustrate the proposed method's efficacy over several machine and deep learning based methods, two datasets were used, namely NSL-KDD and UNSW-NB15 created in 2015 [19]. While the achieved results with NSL-KDD are not directly comparable to the previously presented paper due to differences in data splitting, the proposed method achieves attractive results with simpler architecture. But, the method requires data pre-processing including (i) feature transformation since the model accepts only numerical features and (ii) feature normalization, where z-score was suggested to be used. The datasets used in testing are not also fully representative of industrial internet of things applications - they may apply to the traffic closer to cloud side but among edge devices, the nature of the traffic is inherently different. Also, using z-score as a normalization technique can be distorting since it makes the strong assumption of normally distributed data. What is more, no comments on interdependencies between observations over time were given and a more representative dataset could have been employed in testing.

Another recent work [20] proposed a plug-and-play lightweight network intrusion detection system. It is targeted for simple network devices such as routers and it is a first attempt in designing an online[6] anomaly detection system for computer networks. The method is based on an ensemble of small autoencoders - the authors argued and empirically verified that over the same feature space, they are more efficient and can be less noisier than a single autoencoder. More specifically, the method receives data packets in raw binary and after parsing, the meta information of a packet is forwarded to a feature extractor. Then, 115 traffic-related statistics are retrieved for describing the current state of the channel from which the packet came and for extracting a behavioural snapshot of the hosts and protocols which communicated the given packet. These statistics are then processed by a feature mapper where they are grouped into equally sized subsets using agglomerative hierarchical clustering with correlation-based similarity measure. The subsets are then fed as input to the respective autoencoders in the ensemble. To deal with temporal interdependencies, a damped window framework is leveraged where the weight of older values are exponentially decreased over time and selected statistics are incrementally updated.

---

[6] In practical terms, there must not be a queue of packets awaiting to be processed by the model. Technically the packet processing rate must be higher than the expected maximum packet arrival rate.

The proposed method was evaluated in terms of detection and runtime performance in a real IP camera video surveillance network and in a more heterogeneous and noisy real environment. The authors found that the developed online algorithm performed nearly as well as other offline algorithms and in some cases better. The algorithm was also efficient enough to run on a single core of a Raspberry PI. However, the used clustering algorithm makes strong assumptions about the data which can be critical since this phase is essential for the method's performance. The damped incremental statistics are also traditional in a sense that if the data exhibits nonlinearities, the measures do not work properly.

Many of the developed models in the past focus on detecting specific types of attacks that differ from experiment to experiment, including the environment, dataset, scale and so on [2]. The transition to more generalized[7] end-to-end semi- and unsupervised deep learning methods could extend the detection space to more elaborated and previously unknown attacks as well as increase the usability.

Many scientific fields study data with a complex underlying structure that is a non-Euclidean space, including social networks in computational social sciences, functional networks in brain imaging, regulatory networks in genetics, meshed surfaces in computer graphics and sensor networks in communications. However, while deep learning has recently proven to be powerful in a variety of problems, it has been most successful on data with an underlying Euclidean or grid-like structure, where the familiar properties of Euclidean space have applied. Currently, most of the methods applied to non-Euclidean domains require the conversion of non-Euclidean data to Euclidean form which results into information loss. The core idea in machine learning on graphs is to find a way to incorporate information about the network structure, or graph, into a machine learning model [21]. In the past years, the interest in machine and especially deep learning on graphs has exploded resulting in numerous attempts to apply these methods in a broad spectrum of problems that has triggered a new field in the area of data analysis. Through graphs it is possible to monitor and inspect the nonlinear interactions and dependencies between participants in the target system, its structural changes over different domains and participant-specific characteristics. Moreover, many problems are best expressed with a more complex structure such as graphs that have more capacity to encode complicated relationships in the data. One extension of geometric deep learning[8] problems is coping with signals defined over a dynamically changing structure where it is not possible to assume a fixed domain. This could prove to be useful in detecting abnormal activity in many types of networks, such as WSNs [22,23].

---

[7] A detection scheme is expected to have the capability of addressing a wide range of security issues.

[8] Geometric deep learning is an umbrella term for emerging techniques, such as deep learning on graphs and manifolds, attempting to generalize deep neural models to non-Euclidean domains. For a comprehensive introductory-level reference with descriptive graphics and further details, see [22].

For the first time, authors in [24] investigate the research problem of anomaly detection on attributed networks by designing a graph-based deep learning model. The developed model, called DOMINANT, is essentially a deep autoencoder with slight modifications. Different from convential autoencoders, DOMINANT leverages in the encoding and decoding components a graph convolutional network (GCN) in order to work with the graph-structured data directly and to deal with the inherent characteristics of attributed networks. GCN takes not only the plain network structure as an input but also the nodal attributes thus being able to handle data from the two information modalities. In the encoding part, the attributed network is fed to the GCN[9] that learns low-dimensional node embedding representations. Then, the topological structure and the node-specific attributes are reconstructed separately with corresponding decoding functions from the encoded latent representations. More specifically, the network structure is predicted by the means of a sigmoid-equipped link prediction layer and the nodal attributes are approximated with a graph convolutional layer. The reconstruction errors of nodes are finally used for detecting anomalous nodes on the given input.

The effectiveness of the proposed DOMINANT was tested in experiments where three real-world attributed network datasets, namely BlogCatalog, Flickr and ACM, were used. While the method outperformed other state-of-the-art methods, there are some caveats that require further discussion. In DOMINANT, the GCN is based on the first-order approximation of ChebNet, or 1stCheb-Net, introduced in [25]. While it bridges the gap between spectral-based[10] and spatial-based[11] methods and is considered as a strong baseline in the research community, its main drawback is that the computation cost increases exponentially while the number of 1stChebNet layers increase during batch training [26]. What is more, the GCN does not support edge features, is limited to undirected graphs and is not suitable for very large and densely connected graph datasets [25]. It would be interesting to verify whether employing such form of convolutional layers that are capable of dealing with directed graphs, such as in the MotifNet developed in [27], improves the overall performance of the proposed method. Another remedy for the convolution operation is to make use of lightweight-based dynamic convolutions as proposed in [28].

What is more, DOMINANT cannot be applied to dynamic graphs that evolve over time. For making it suitable for WSNs, the method requires modifications for extending it to dynamic environments with more flexibility and less computational load. A natural idea is to combine a graph neural network (GNN) with a recurrent mechanism where the typical approaches use the GNN as a

---

[9] The proposed method uses three hidden layers in the convolutional encoder, but the authors note that more layers can be stacked for building a deeper network.

[10] In general, spectral-based methods are computationally more expensive, assume a fixed input graph and are limited to work on undirected graphs whereas spatial-based methods can are more generalizable and can deal with directed graphs.

[11] Graph neural networks with attention mechanisms are an example of spatial-based approaches.

feature extractor and a recurrent neural network (RNN)[12] for sequence learning from the extracted features to capture the dynamism [30]. The authors in [30] follow this concept, but with a different approach: they propose carrying the dynamism directly to the GCN's network parameters[13] using the RNN[14] so that the model gets updated at every time step. In essence, their model oriented approach updates the GCN based on the dynamic evolution of the graphs without requiring historical information for nodes that did not appear before. This is an appealing property for the model to have because in detecting anomalies from dynamic networks, (i) there always exists the concern of how often and with what mechanism to update the model for adapting to non-stationary environment and (ii) flexibility is a rare commodity to have. The proposed method showed promising results in a variety of tasks, including node and edge classification as well as link prediction. However, the proposed method has the disadvantages of the chosen GCN as discussed before.

Only less than five years ago the shift from anomaly detection in static to time-evolving dynamic graphs was more widely recognized in the scientific community and a comprehensive survey was produced [31]. What is more, the literature on graph-based deep learning is still in its infancy, not to mention the related applications in network anomaly detection and especially in WSNs. There exists some recent studies on anomaly detection in dynamic networks [32–37], but since they are out of the scope of this article, they are left here only as a reference for interested readers.

4   Remarks on Developing Anomaly-Based Intrusion Detection Systems for WSNs

Modelling: In security, the commonly used definitions for categorizing different types of anomalies are not comprehensive enough for all anomalies that may occur in WSNs. When using fixed anomaly detection methods, the non-stationary nature of these networks can cause a model to classify normal data as anomalies and the same applies the other way round. An important aspect with updating models is to do it the minimum number of times possible due to the resource constrained environment. For saving resources, the previous model with old parameters can be used as a basis for constructing the new optimal model [29]. Deep learning techniques inherently have this property of incrementally updating the model for performance gains and the model oriented approach proposed in [30] could do so in dynamic, unsupervised and automated manner. In WSNs, the responsibility of updating the model can be delegated to a less-constrained device such as a router or other administrative device and then the new model parameters can be propagated to the model-equipped sensor nodes.

A network consists of a set of entities (nodes) and of interactions (links) among them and a network can be modelled as a graph [38]. The links between

---

[12] Typically enhanced with LSTM or GRU.

[13] The authors use the GCN introduced in [25] and used in DOMINANT.

[14] The authors apply GRU but LSTM could be tested as well.

the nodes can have a certain direction and both, the nodes and the links, can have features. When considering WSNs, the links represent the exchange of data packets between nodes and nodal features can be metadata in a more condensed format, traffic-related statistics or other information that is neither packet- or flow-based [41]. The advantage of graph-based methods is that all of this information can be naturally incorporated in them and as it was discussed before, WSNs are vulnerable to internal attacks that targets the protocol stack and to external attacks that are physical in nature. In other words, internal attacks leave protocol-related traces that can be captured from data packets and external attacks may represent themselves as alterations in physical properties of the devices. In graphs, these produce anomalies in nodes, links and in subgraphs inside the graph. Also, particularly in dynamic graphs, events and changes in the topological structure, nodes and links as well as related attributes may occur over time.

One issue with existing anomaly detection methods is the need for manipulating input data with feature extraction and dimensionality reduction methods, among others, which can be distorting from outliers' perspective. In data packets, all information related to the different protocols used in the system, are included: when dealing with traffic from interconnected sensors, the data itself could be transformed into more raw form, such as hexadecimal or binary, and fed to the anomaly detection method [39]. This approach is promising since working with raw data permits for more granular inspection and may reveal traces from all kinds of activities in the network, including previously unknown ones. Moreover, for improving detection performance, the approach can be combined with generalized device-related attributes which can be established after a careful investigation of how different types of attacks affect the devices and the communication between them.

Datasets: For the successful training and evaluation of anomaly-based network intrusion detection systems, labeled datasets are necessary. However, outdated datasets are still frequently used for benchmarking the developed models even though their flaws are well-known [40]. One of the reasons may be that there is no overall index of existing data sets and it is difficult to keep track on the latest development [41]. Systematic reviews and comparisons on currently available datasets are provided in [40] and [41], to name a few. The authors in [40] strongly advice for using the more modern and updated dataset UNSW-NB15 in forthcoming research related to anomaly-based network intrusion detection systems. Other recently published datasets that cover a wide range of attacks include CICIDS 2017[15] and TRAbID[16]. However, when these models are to be deployed in WSNs, a more representative dataset is still needed. One option, while still not optimal, could be the AWID-dataset[17]: it focuses on 802.11 networks, it is publicly available and is adopted by over 400 organizations from

---

[15] https://www.unb.ca/cic/datasets/ids-2017.html.

[16] https://secplab.ppgia.pucpr.br/?q=trabid.

[17] https://icsdweb.aegean.gr/awid/index.html.

universities, research labs and companies around the world. An ideal dataset would be 802.15-based which is still yet to be found. Finally, own datasets in real-world environments can be created.

Interpretability: Followed by their success, machine learning models are pervading to mission-critical applications where risks are higher. This leads to the question of reliability: can the user be sure that the model's functioning is based on a proper representation of the problem at hand. Therefore, efforts on understanding complex machine learning models are required. The authors in [42] provide an overview of techniques for interpreting and explaining complex machine learning models. While the focus is on deep neural network models, the techniques introduced can be applied to a general class of nonlinear machine models despite how they were trained or who trained them. However, it stays unclear how applicable the methods are for graph-based deep networks. The authors also remind that the tutorial seeks for functional understanding to characterize the behaviour of black-box models instead of trying to illustrate their inner workings - interpreting deep networks remains a young and emerging, while an essential, field of research. Also, as in many real-world problems, pervasive noise are commonplace and access to clean training data may not be realistic [43]. Thus, another important aspect of anomaly detection methods is their sensitivity and robustness to incomplete and noisy data. When validating the developed models, it is essential to assess how the model's performance is affected by unexpected disturbances.

Resources: Anomaly detection algorithms deployed in WSNs must have low computational complexity and require little memory space due to the inherent constraints of sensor nodes. Energy conservation is also a priority: the use of the wireless radio receiver consumes significantly more energy than any other component on the sensor node and in general, the cost of receiving is comparable to the cost of transmitting data. The algorithms developed thus have to pursue the objective of minimizing energy consumption while maximizing the lifetime of the network [13, 29].

5   Conclusion

Systems deployed in the real world, such as WSNs, exhibit a variety of characteristics that cannot be overlooked when designing mechanisms for ensuring the smooth and reliable functioning of these systems. Understanding the true nature of complexities that lie under the data gathered from real world applications may turn out to be crucial for security-related anomaly detection, among other fields. Especially when new advances and innovations are created at ever-growing speed in today's complex real-world systems, it is important to shift from simplified abstractions and conversions into making sense of different kind of systems as they are. While graph-based deep learning is one of the emerging and promising techniques used for understanding data produced by real world systems, the field is still in its infancy and requires further research. Additionally, the need for manipulating data must be revisited, other forms of data investigated and contributions to model understanding are required.

Thoughts and ideas presented in this article can be extended to other types of anomaly detection tasks, such as detecting faults using machine health monitoring systems, finding fraudulent behavior in credit card payments or systemically important organizations in financial networks. Moreover, networks with wired and wireless interconnections among heterogeneous devices could be strengthened with notions presented. Finally, anomaly detection presents only a piece of the information security puzzle: solutions and countermeasures for security issues must be viewed holistically as well as employed and maintained accordingly. After all, there is no such thing as a completely secure system.

References

1. Rawat, P., Singh, K.D., Chaouchi, H., Bonnin, J.M.: Wireless sensor networks: a survey on recent developments and potential synergies. J. Supercomput. **68**(1), 1–48 (2014)
2. Xie, M., Han, S., Tian, B., Parvin, S.: Anomaly detection in wireless sensor networks: a survey. J. Netw. Comput. Appl. **34**(1), 1302–1325 (2011)
3. Dargie, W., Poellabauer, C.: Fundamentals of Wireless Sensor Networks - Theory and Practice. Wiley Series on Wireless Communications and Mobile Computing (2010)
4. Zou, Y., Zhu, J., Wang, X., Hanzo, L.: A survey on wireless security: technical challenges, recent advances, and future trends. Proc. IEEE **104**(9), 1727–1765 (2016). https://doi.org/10.1109/JPROC.2016.2558521
5. Tomić, I., McCann, J.: A survey of potential security issues in existing wireless sensor network protocols. IEEE Internet Things J. **4**(6), 1910–1923 (2017)
6. Butun, I., Salvatore, D.M., Sankar, R.: A survey of intrusion detection systems in wireless sensor networks. IEEE Commun. Surv. Tutorials **16**(1), 266–282 (2014). https://doi.org/10.1109/SURV.2013.050113.00191
7. Kumar, D.P., Amgoth, T., Annavarapu, C.S.R.: Machine learning algorithms for wireless sensor networks: a survey. Inf. Fusion **49**(1), 1–25 (2019)
8. Sha, K., Wei, W., Yang, T.A., Wang, Z., Shi, W.: On security challenges and open issues in internet of things. Future Gener. Comput. Syst. **83**(1), 326–337 (2018)
9. CSA. https://downloads.cloudsecurityalliance.org/whitepapers/Security_Guidance_for_Early_Adopters_of_the_Internet_of_Things.pdf. Accessed 3 Mar 2019
10. Mosenia, A., Jha, N.K.: A comprehensive study of security of internet-of-things. IEEE Trans. Emerg. Top. Comput. **5**(4), 586–602 (2017)
11. Labatut, V., Ozgovde, A.: Topological measures for the analysis of wireless sensor networks. Procedia Comput. Sci. **10**, 397–404 (2012). https://doi.org/10.1016/j.procs.2012.06.052
12. Akoglu, L., Tong, H., Koutra, D.: Graph based anomaly detection and description: a survey. Data Min. Knowl. Disc. **29**(3), 626–688 (2015)
13. Rajasegarar, S., Leckie, C., Palaniswami, M.: Anomaly detection in wireless sensor networks. IEEE Wirel. Commun. **15**(4), 34–40 (2008)
14. Ball, J.E., Anderson, D.T., Chan, C.S.: A comprehensive survey of deep learning in remote sensing: theories, tools and challenges for the community. J. Appl. Remote Sens. **11**(4), 042609 (2017). https://doi.org/10.1117/1.JRS.11.042609

15. Liu, W., Wang, Z., Liu, X., Zeng, N., Liu, Y., Alsaadi, F.E.: A survey of deep neural network architectures and their applications. Neurocomputing **234**(1), 11–26 (2017)
16. Chalapathy, R., Chawla, S.: Deep Learning for Anomaly Detection: A Survey (2019). A preprint. https://arxiv.org/abs/1901.03407. Accessed 3 May 2019
17. Ma, T., Wang, F., Cheng, J., Yu, Y., Chen, X.: A hybrid spectral clustering and deep neural network ensemble algorithm for intrusion detection in sensor networks. Sensors (Basel) **16**(10), 1701 (2016). https://doi.org/10.3390/s16101701
18. Al-Hawawreh, M., Moustafa, N., Sitnikova, E.: Identification of malicious activities in industrial internet of things based on deep learning models. J. Inf. Secur. Appl. **41**(1), 1–11 (2018)
19. Moustafa, N., Slay, J.: UNSW-NB15: a comprehensive data set for network intrusion detection systems (UNSW-NB15 network data set). In: Military Communications and Information Systems Conference (MilCIS) (2015). https://doi.org/10.1109/MilCIS.2015.7348942
20. Mirsky, Y., Doitshman, T., Elovici, Y., Shabtai, A.: Kitsune: an ensemble of autoencoders for online network intrusion detection. In: Network and Distributed Systems Security Symposium (NDSS) (2018). https://doi.org/10.14722/ndss.2018.23211
21. Hamilton, W.L., Ying, R., Leskovec, J.: Representation Learning on Graphs: Methods and Applications (2018). https://arxiv.org/abs/1709.05584. Accessed 3 Mar 2019
22. Bronstein, M.M., Bruna, J., LeCun, Y., Szlam, A., Vandergheynst, P.: Geometric deep learning: going beyond Euclidean data. IEEE Signal Process. Mag. **34**(4), 18–42. IEEE (2017). https://doi.org/10.1109/MSP.2017.2693418
23. Xu, K., Wang, Z., Witbrock, M., Wu, L., Feng, Y., Sheinin, V.: Graph2Seq: Graph to Sequence Learning with Attention-based Neural Networks (2018). https://arxiv.org/abs/1804.00823. Accessed 3 May 2019
24. Ding, K., Li, J., Bhanushali, R., Liu, H.: Deep anomaly detection in attributed networks. In: SIAM International Conference on Data Mining (2019)
25. Kipf, T.N., Welling, M.: Semi-supervised classification with graph convolutional networks. In: Proceedings of the International Conference on Learning Representations (2017)
26. Wu, Z., Pan, S., Chen, F., Long, G., Zhang, C., Yu, P.S.: A Comprehensive Survey on Graph Neural Networks (2019). http://arxiv.org/abs/1901.00596. Accessed 4 May 2019
27. Monti, F., Otness, K., Bronstein, M.M.: MotifNet: a motif-based graph convolutional network for directed graphs. In: Data Science Workshop (DSW). IEEE (2018). https://doi.org/10.1109/DSW.2018.8439897
28. Wu, F., Fan, A., Baevski, A., Dauphin, Y.N., Auli, M.: Pay less attention with lightweight and dynamic convolutions (2019). https://arxiv.org/abs/1901.10430
29. O'Reilly, C., Gluhak, A., Imran, M.A., Rajasegarar, S.: Anomaly detection in wireless sensor networks in a non-stationary environment. IEEE Commun. Surv. Tutorials **16**(3), 1413–1432 (2014)
30. Pareja, A., et al.: EvolveGCN: Evolving Graph Convolutional Networks for Dynamic Graphs (2019). https://arxiv.org/abs/1902.10191. Accessed 7 May 2019
31. Ranshous, S., Shen, S., Koutra, D., Harenberg, S., Faloutsos, C., Samatova, N.F.: Anomaly detection in dynamic networks: a survey. WIREs Comput. Stat. **7**(3) (2015). https://doi.org/10.1002/wics.1347

32. Ding, K., Li, J., Liu, H.: Interactive anomaly detection in attributed networks. In: Proceedings of the Twelfth ACM International Conference on Web Search and Data Mining (WSDM 2019) (2019). https://doi.org/10.1145/3289600.3290964

33. Eswaran, D., Guha, S., Mishra, N.: SpotLogiht: detecting anomalies in streaming graphs. In: Proceedings of the 24th ACM SIGKDD International Conference on Knowledge Discovery & Data Mining (KDD 2018), pp. 1378–1386. ACM, New York (2018). https://doi.org/10.1145/3219819.3220040

34. Miz, V., Ricaud, B., Benzi, K., Vandergheynst, P.: Anomaly detection in the dynamics of web and social networks. (2019). https://arxiv.org/abs/1901.09688

35. Xue, L., Luo, M., Peng, Z., Li, J., Chen, Y., Liu, J.: Anomaly detection in time-evolving attributed networks. In: Database Systems for Advanced Applications (2019). https://doi.org/10.1007/978-3-030-18590-9_19

36. Yu, W., Cheng, W., Aggarwal, C.C., Chang, K., Chen, H., Wang, W.: NetWalk: a flexible deep embedding approach for anomaly detection in dynamic networks. In: In Proceedings of the 24th ACM SIGKDD International Conference on Knowledge Discovery & Data Mining (KDD 2018), pp. 2672–2681. ACM, New York (2018). https://doi.org/10.1145/3219819.3220024

37. Ahmad, S., Lavin, A., Purdy, S., Agha, Z.: Unsupervised real-time anomaly detection for streaming data. Neurocomputing **262**(1), 134–147 (2017)

38. Cimini, G., Squartini, T., Saracco, F., Garlaschelli, D., Gabrielli, A., Galdarelli, G.: The Statistical Physics of Real-World Networks (2018). https://arxiv.org/abs/1810.05095

39. Bodström, T., Hämäläinen, T.: A novel deep learning stack for APT detection. MDPI Appl. Sci. **9**(6), 1055 (2019)

40. Diverak, A., Parekh, M., Savla, V., Mishra, R.: Benchmarking datasets for anomaly-based network intrusion detection: KDD CUP 99 alternatives. In: 3rd International Conference on Computing, Communication and Security (ICCCS), pp. 1–8. IEEE (2018)

41. Ring, M., Wunderlich, S., Scheuring, D., Landes, D., Hotho, A.: A Survey of Network-based Intrusion Detection Data Sets (2019). https://arxiv.org/abs/1903.02460

42. Montavon, G., Samek, W., Müller, K.: Methods for interpreting and understanding deep neural networks. Digital Sig. Process. **73**(1), 1–15 (2018)

43. Zhou, C., Paffenroth, R.C.: Anomaly detection with robust deep autoencoders. In: Proceedings of the 23rd ACM SIGKDD International Conference on Knowledge Discovery and Data Mining (KDD 2017), pp. 665–674. ACM, New York (2018). https://doi.org/10.1145/3097983.3098052

Polarization Direction Finding Method of Interfering Radio Emission Sources

Alexey Simonov[1], Grigoriy Fokin[2](✉) (iD), Vladimir Sevidov[1], Mstislav Sivers[2], and Sergey Dvornikov[1]

[1] Military Academy of Communications Named After S.M. Budyonny,
3 Tikhoretskiy Prospekt, 194064 St. Petersburg, Russia
simonovalexey1971@gmail.com,
sevidovvlad1979@gmail.com, vasvdvornikov@gmail.com
[2] The Bonch-Bruevich St. Petersburg State University of Telecommunications,
22 Prospekt Bolshevikov, 193232 St. Petersburg, Russia
grihafokin@gmail.com, m.sivers@mail.ru

Abstract. Rapid growth of radio emission sources (RES) leads to the interference of signals from various transceivers and thus complicates its positioning based on direction finding (DF). Traditional DF use structural-statistical redundancy to resolve multi-signal and multipath situations, which makes impossible to implement it in compact direction finders. The aim of current work is to investigate possibilities of polarization DF accounting interference conditions. It is proposed to use frequency and phase differences of received rays in the interference mixture of signals in tri-orthogonal antenna system (TOAS). The solution to polarization DF task of RES during interference utilize methods of search for normals to planes that are envelopes from a family of curves, which use the resultant field, induced in TOAS. The novelty of presented solution is in the realization of the idea that in the case of RES interference with frequency or phase differences in TOAS, the resulting field vector forms three-dimensional shape, which contains spatial parameters of the rays. Contribution of current investigation validates the use of relatively simple DF techniques in a complex interference environment, both for several sources and separation of rays from one source.

Keywords: Direction finding · Radio emission sources · Positioning · Interference of radio emissions

1 Introduction

Positioning of radio emission sources (RES) is the topic of ongoing research both for indoor [1] and outdoor [2] scenarios. Positioning can be implemented from passive measurements of the arrival times, directions of arrival, or Doppler shifts of received electromagnetic waves. Positioning techniques had already got considerable attention in the past years and can be subdivided by primary measurements into Time Difference of Arrival (TDOA), Angle of Arrival (AOA) and Received Signal Strength Indication (RSSI) positioning. Every measurement processing technique based on TDOA AOA or RSSI has its advantages and shortcomings. Existing accuracy results in local

© Springer Nature Switzerland AG 2019
O. Galinina et al. (Eds.): NEW2AN 2019/ruSMART 2019, LNCS 11660, pp. 208–219, 2019.
https://doi.org/10.1007/978-3-030-30859-9_18

positioning systems are far from high precision satellite systems [3, 4] however, still remains in demand. Direction finding (DF) is one of the most popular techniques, because AOA often remains the only reliable information from RES [5–8].

There are many effective methods of DF, which are well studied and successfully implemented in practice [5–10]. Historically, the first were amplitude direction finders from frame rotary to wide-base stationary. With the development of electronic technology, in particular with the advent of highly accurate, highly stable reference generators, phase direction finders have become more common. Frequency direction finders are based on the use of the Doppler effect, which appears when the receiving antenna element is moving relative to the source of radio emission. In addition to classical direction finders, modern ones use correlation methods of direction finding, based on the use of multi-element antenna systems and multichannel receiving paths [11]. Attention should also be paid to self-structural methods with enhanced spatial resolution, based on the use of statistical properties signals and noise [12, 13].

The operation of the direction finders currently takes place in a complex signal-interference environment, due to fast increase in the number of RES. This leads to the case when several signals from various sources fall into the passband of the receiving device of the direction finder, leading to their interference. In radio communications in this situation, the problem of interference protection arises, when it is necessary to isolate the subscriber's useful signal and eliminate interfering signals from other radio emission sources. During direction finding process, all received radio emissions are useful, that is, they are obtained from objects of positioning thus the problem arises of the simultaneous direction finding of several interfering RES.

Typical direction finders are not designed for use in interference conditions in the presence of two or more RES and result in significant bearing error. However, there are methods [14], which provide an estimation of the azimuth and elevation angle for several RES with identical carrier frequencies. These methods usually utilize virtual spatial scanning according to the amplitude-phase distribution of the antenna array. The disadvantage of these methods is that they require structural or statistical redundancy, that is, the number of antenna elements and reception paths must exceed the number of RES, and increase in accuracy of direction finding needs time to accumulate statistical measurements and thus complicates its real-time application. Traditional approaches to direction finding are based on measurements of the amplitude, phase and frequency parameters of RES induced in antenna system (AS), and are based on the difference in the time of signal arrival from the RES at spatially distributed points. In addition, there are methods that use polarization-sensitive antenna elements and provide joint estimation of the polarization and bearing to RES [15–19]. These methods utilize polarization as an additional parameter, which allows to increase the accuracy of bearing estimation. Tri-orthogonal antenna system is the most popular [20–22] for polarization direction finding and utilize synthetic or asymmetrical vibrators, as well as loop antennas, as antenna elements. Such antenna systems are concentrated and have a single phase center, which simplifies the measurements processing.

The contribution of the presented article is a new solution to the problem of polarization DF of RES by a small-sized antenna system in the conditions of interference. Unlike other works in this subject area, presented solution is based on extracting spatial information from a three-dimensional figure, formed by the end of the resultant field vector in the presence of frequency and phase differences of interfering RES.

The material in the paper is organized in the following order. Polarization direction finding method of interfering sources is presented and evaluated Sect. 2. Simulation results are given in Sect. 3. Finally, we draw the conclusions in Sect. 4.

2 Polarization DF Method of Interfering Sources Analysis

2.1 Formulation of DF of Interfering Sources Problem

Method for determining the direction to a RES using polarization was described in [23, 24] and illustrated in Fig. 1a. The instantaneous spatial position of the electric field strength vector $\mathbf{E}$ at times t_i is determined from its projections E_{xi}, E_{yi}, E_{zi} (Fig. 1b), by measuring the electromotive forces (EMF) in the antenna elements AE_x, AE_y, AE_z.

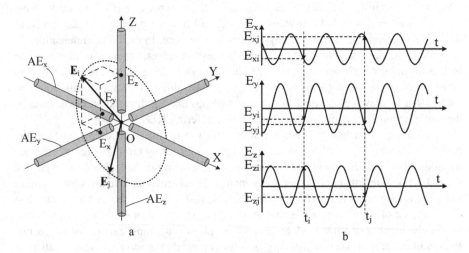

Fig. 1. Tri-orthogonal antenna system.

To determine the direction to RES that is, to find azimuth θ and elevation angle β, it is necessary to measure at least two values of field vector and at different points in time t_i and t_j (Fig. 1) to build positions perpendicular to these vectors Ω_i and Ω_j, passing through the center of AS, and at the intersection of these planes get a direction vector $\mathbf{L}$ to the source, as a result of the vector product of vectors $\mathbf{E}_i$ and $\mathbf{E}_j$ (Fig. 2).

The task of DF can be solved for multipath and multi-signal cases, which leads to interference of radio waves from various RES. Interfering RES will be combined in AS according to the principle of vector addition. The result of such combination and its effectiveness [23, 24] under these conditions are to be investigated further.

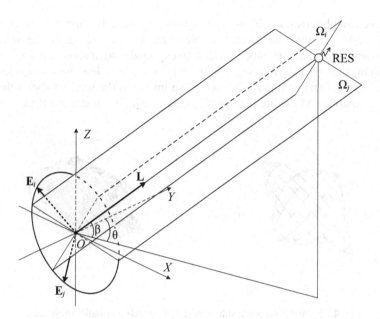

Fig. 2. Graphic representation of polarization direction finding.

2.2 Polarization Direction Finding Method

To test the possibility of polarizing DF method of interfering RES, computer simulations were carried out, the results of which showed that for RES operating at the same frequency ($f_1 = f_2$), the vector $\mathbf{E}_\Sigma$ obtained by addition of $\mathbf{E}_1$ and $\mathbf{E}_2$ is equivalent to the resulting radio emission with, in general, case, elliptical polarization (Fig. 3).

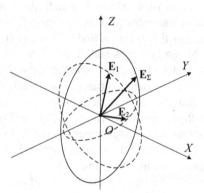

Fig. 3. The original radio emissions and the result of their interference when $f_1 = f_2$.

For the situation represented at Fig. 3, spatial parameters of interfering radio waves, that is, the azimuths θ_1, θ_2 and elevation angles β_1, β_2 are not derivable from resulting radio emission with $\theta_\Sigma = \theta_1 + \theta_2$ and $\beta_\Sigma = \beta_1 + \beta_2$, since summed items cannot be determined straightforwardly. It led to disappointing conclusion that it is impossible to simultaneously find direction from two interfering RES operating on the same frequency using polarization method [23, 24]. Unexpected insight about the way to solve the problem emerged after investigation of DF basics, which describe visual selectivity property of traditional amplitude DF, when the image on the receiver shows the possibility to separate and predict two (Fig. 4a), three (Fig. 4b) and more RES.

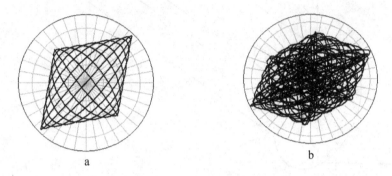

a   b

Fig. 4. Indicator screen result of receiving interfering radio emissions.

In [7, 8] it is shown that the envelope of the spiral line drawn on the screen under a two-signal effect is a parallelogram, with each side of it being the line that would have been on the screen in the absence of another signal. Thus, the position of sides of parallelogram corresponds to the azimuths of two RES. With three sources, the image has the form of a parallelepiped, presented in axonometric projection, and the azimuths of all three directional RES are determined from its three edges. The key condition for visual selectivity of the receiver-indicator is the difference in frequencies of RES.

It was found that the result of interference is radically changed if there is a difference in the frequencies of the radio emission sources ($f_1 \neq f_2$), provided that they fall into the bandwidth ΔF of the direction finder channels ($f_1 - f_2 = \Delta f < \Delta F$). The difference in frequencies Δf can be very small, if only it would lead to the appearance of beats in the direction finder channels. Computer simulation has shown that the presence of this feature, when frequency difference drastically change the interference pattern — the end of the resulting electric field intensity vector describes in space a complex volume figure with continuously changing shape and orientation (Fig. 5a). At first glance, the image seems to be uninformative, but if to find the correct angle (Fig. 5b), then this figure is projected into a parallelogram (Fig. 5c), the normals to the sides of which indicate directions to interfering RES.

The time of "drawing out" of each elliptical element of a spiral is determined by the period of the average frequency of two signals $f = (f_1 + f_2)/2$. The filling density of a volume figure or the rate of modification of the links of the spiral is determined by the difference frequency Δf. To obtain the flat sides of the shaped figure, it is necessary that the phase relations of the two signals change from 0° to 180°. It happens in time $\Delta t = 1/(2\Delta f)$. For the next half-period of the difference frequency, the cycle repeats. As already noted, the difference in frequencies with polarization direction finding can be

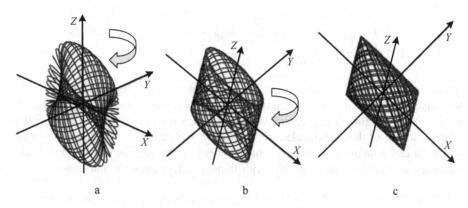

Fig. 5. The result of the interference of radio emissions when $f_1 \neq f_2$.

very small ($\Delta f \ll \Delta F$), its value only affects the display inertia: the smaller Δf, the longer the corresponding figure will be drawn.

Presented in Fig. 5 images, despite their high visibility, still cannot be considered three-dimensional. In fact, the figures are flat, so they require additional turns and projections to search for a good angle. In this situation, volume indicators, for example, holographic, should of course become more effective.

The visual resolution of interfering RES has a great visibility and, due to the heuristic features of human perception, can resolve the most complex situations. However, for automated processing it is necessary to obtain a formalized solution of the polarization DF problem under interference conditions. To this end, using the techniques described in [9, 10], we consider the physical nature of the described property and the conditions for its implementation, present a simplified derivation of the figure equation on indicator screen under the influence of two harmonic RES and compose a system of equations whose solution will give the required bearing on each RES.

2.3 Polarization Direction Finding Method of Interfering Sources

The interference stability property reveal itself only if all the channels of the polarization direction finder are identical in terms of transmission coefficients and linear with respect to both signals. Consider two sinusoidal oscillations from RES with different angular coordinates, comparable in amplitude and with frequencies ω_1 and ω_2, and $|\omega_2 - \omega_1| = \Omega < \Delta\omega_{\Pi}$, $\frac{|\omega_2 - \omega_1|}{\omega} < < 1$, where $\omega = \frac{\omega_1 + \omega_2}{2}$. Then the total voltage induced in the corresponding antenna elements will be equal to

$$u_{x\Sigma}(t) = u_{x1}(t) + u_{x2}(t); \quad u_{y\Sigma}(t) = u_{y1}(t) + u_{y2}(t); \quad u_{z\Sigma}(t) = u_{z1}(t) + u_{z2}(t);$$

$$u_{xi}(t) = U_{\|i} \cos \theta_i \sin \omega_i t + U_{\perp i} \sin \beta_i \sin \theta_i \sin(\omega_i t + \phi_i);$$

$$u_{yi}(t) = U_{\perp i} \sin \beta_i \cos \theta_i \sin(\omega_i t + \phi_i) - U_{\|i} \sin \theta_i \sin \omega_i t;$$

$$u_{zi}(t) = -U_{\perp i} \cos \beta_i \sin(\omega_i t + \phi_i);$$

$$U_{\|i} = U_{mi}\sqrt{\cos^2\gamma_i + r_i^2\sin^2\gamma_i}; \quad U_{\perp i} = U_{mi}\sqrt{\sin^2\gamma_i + r_i^2\cos^2\gamma_i};$$

$$\phi_i = \mathrm{arctg}\frac{2r_i}{(1-r_i^2)\sin 2\gamma_i}.$$

where $U_{mi} = l_d E_{mi}$ – maximum voltage amplitude; E_{mi} – electromagnetic wave intensity; r_i – ellipticity; γ_i – the angle of the major semi-axis of the polarization ellipse; θ_i, β_i – azimuth and elevation angle to RES respectively (i = 1,2); l_d – the effective length of the antenna element; i = 1, 2 – RES number. The voltages of two signals in each channel can be represented as beats in the form of oscillations of one frequency ω_1, but with time-varying (with frequency Ω) amplitude and phase:

$$\begin{cases} u_{x\Sigma}(t) = U_x(t)\sin(\omega_1 t + \Phi_x(t)); \\ u_{y\Sigma}(t) = U_y(t)\sin(\omega_1 t + \Phi_y(t)); \\ u_{z\Sigma}(t) = U_z(t)\sin(\omega_1 t + \Phi_z(t)); \end{cases} \qquad (1)$$

where

$$U_x(t) = \sqrt{U_{xm1}^2 + U_{xm2}^2 + 2U_{xm1}U_{xm2}\cos\Omega t};$$

$$U_y(t) = \sqrt{U_{ym1}^2 + U_{ym2}^2 + 2U_{ym1}U_{ym2}\cos\Omega t};$$

$$U_z(t) = \sqrt{U_{zm1}^2 + U_{zm2}^2 + 2U_{zm1}U_{zm2}\cos\Omega t};$$

$$\Phi_x(t) = \mathrm{arctg}\frac{U_{xm2}\sin\Omega t}{U_{xm1} + U_{xm2}\cos\Omega t};$$

$$\Phi_y(t) = \mathrm{arctg}\frac{U_{ym2}\sin\Omega t}{U_{ym1} + U_{ym2}\cos\Omega t};$$

$$\Phi_z(t) = \mathrm{arctg}\frac{U_{zm2}\sin\Omega t}{U_{zm1} + U_{zm2}\cos\Omega t}.$$

Determining the phase shift relative to the phase of the voltage in the antenna element AE_x, which we take as zero, we obtain the projections of the vector $\mathbf{E}_\Sigma$ in the three-dimensional indicator on the axes Ox, Oy and Oz:

$$\begin{aligned} x &= kU_x(t)\sin\omega_1 t; \\ y &= kU_y(t)\sin(\omega_1 t - \Delta\Phi_y(t)); \\ z &= kU_z(t)\sin(\omega_1 t - \Delta\Phi_z(t)); \end{aligned} \qquad (2)$$

where k – a proportionality factor corresponding to the amplitude coefficient of the receiving and indicator channels of the polarization direction finder; and

$$\Delta\Phi_y(t) = \Phi_y(t) - \Phi_x(t); \quad \Delta\Phi_z(t) = \Phi_z(t) - \Phi_x(t).$$

As a result, the equation of the spatial curve obtained on the three-dimensional indicator will have the following form

$$2\frac{x^2}{S_x^2(t)} + \frac{y^2}{S_y^2(t)} + \frac{z^2}{S_z^2(t)} - 2\frac{xy}{S_x(t)S_y(t)}\cos\Delta\Phi_y(t)$$

$$-2\frac{xz}{S_x(t)S_z(t)}\cos\Delta\Phi_z(t) - \sin^2\Delta\Phi_y(t) - \sin^2\Delta\Phi_z(t) = 0; \tag{3}$$

where $S_x(t) = kU_x(t)$; $S_y(t) = kU_y(t)$; $S_z(t) = kU_z(t)$.

The resulting expression at a fixed point in time describes the ellipsoid, the regulation of the parameter t causes a change in the position and size of its semi-axes. If (3) is considered as a function of time, then we obtain the equation of a spatial helix with a continuously changing time configuration and inclination of the links. Each link of the helix is a figure close to a spatial ellipse, the frequency of changing the shape and tilt of the major axis of which is equal $\Omega = \omega_2 - \omega_1$. The surfaces bounding the spatial helix will be planes that are generally sides of a parallelepiped, the normals to which indicate directions to radio emission sources. The proof of this, that is, the derivation of expressions for the sides of the parallelepiped, is based on finding the envelope surface of a family of spatial curves [25], and requires the joint solution of the systems of equations obtained from (3):

$$\begin{cases} x(y, z, \Omega t) = 0; & \dfrac{\partial x(y, z, \Omega t)}{\partial(\Omega t)} = 0; \\[2mm] y(x, z, \Omega t) = 0; & \dfrac{\partial y(x, z, \Omega t)}{\partial(\Omega t)} = 0; \\[2mm] z(x, y, \Omega t) = 0; & \dfrac{\partial z(x, y, \Omega t)}{\partial(\Omega t)} = 0. \end{cases} \tag{4}$$

The solutions of system (4) will be planes that can be described by linear equations

$$A_n x + B_n y + C_n z + D = 0. \tag{5}$$

where the index n = 1,2 corresponds to the first and second pair of parallel sides of the envelope surface of the spatial spiral (3). The orientation of obtained planes (5) contains information about the spatial parameters of the interfering radio emissions, the directions to which indicate the normals to the found planes, described by the expressions

$$N_n = (A_n, B_n, C_n). \tag{6}$$

As a result, we obtain expressions for calculating azimuths and elevation angles on the first and second RES

$$\theta_1 = \operatorname{arctg}\frac{A_1}{B_1}; \qquad \beta_1 = \operatorname{arctg}\frac{C_1}{\sqrt{A_1^2 + B_1^2 + C_1^2}};$$

$$\theta_2 = \operatorname{arctg}\frac{A_2}{B_2}; \qquad \beta_2 = \operatorname{arctg}\frac{C_2}{\sqrt{A_2^2 + B_2^2 + C_2^2}}. \tag{7}$$

Obviously, a larger area of the volume figure will correspond to more powerful radio emission, and therefore bearing to this source will be more stable and accurate.

3 Polarization DF Method of Interfering Sources Simulation

The condition of the difference in the frequencies of interfering radio emissions for the implementation of polarization direction finding in most cases means the possibility of resolving only the multi-signal situation when radio emissions come from different sources. The multipath situation, when two or more beams of the same frequency coming from the same source, having passed through different paths, seemed to be unsolvable by polarizing direction finders [23, 24].

However, further simulation studies have shown that such intractable multi-radiation will be observed only for monochromatic signals. If the signal occupies a certain frequency band, that is, it is modulated, the possibility of polarization direction finding of each beam exists. The reason for this phenomenon is the time shift between the interfering radio emission caused by the difference in the propagation path of each beam. As an example, we can take the multipath situation when two beams interfere with the phase-shifted keying (PSK) signal (Figs. 6a, 7a, 8a).

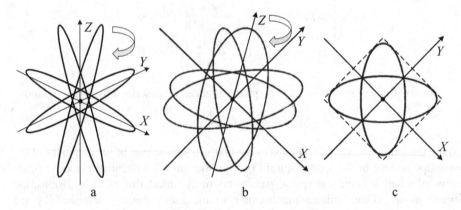

Fig. 6. The result of the interference of rays with PSK-4 signals.

As in Fig. 5a, the spatial resolution of the rays with PSK-4 (Fig. 6a), PSK-8 (Fig. 7a) and PSK-16 (Fig. 8a) seems impossible, but after selecting a proper angle (Figs. 6b, 7b, 8b) images become much more informative, and it becomes possible to approximate it by a parallelogram, the normals to the sides of which indicate directions to interfering rays (Figs. 6c, 7c, 8c).

With phase differences of radio emissions, in contrast to frequency, envelope (1) is no longer a smooth function of time, but becomes a step with discontinuity points at the moments of changing signal positions. However, the solution (5) of the system of Eqs. (4), remains the same, and the bearings are calculated according to (7).

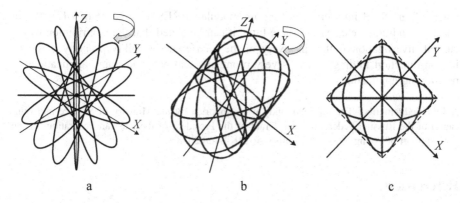

Fig. 7. The result of the interference of rays with PSK-8 signals.

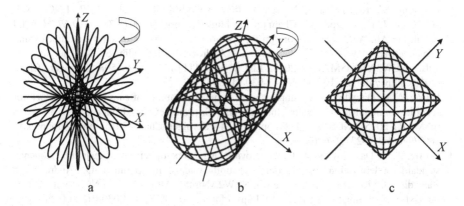

Fig. 8. The result of the interference of rays with PSK-16 signals.

It is noticeable that an increase in the modulation order leads to a "saturation of the image", that is, to an increase in the number of links in the spatial spiral inscribed in the approximating parallelogram. However, it should be noted that such a visual improvement of image quality, which should give an increase in the stability of polarization DF, is based on the accumulation of statistical data. Indeed, with the same manipulation speed, for RES with greater modulation order, more time is required for all possible signaling to appear.

4   Conclusion

The developed method validates the idea that in the case of RES interference with frequency or phase differences in TOAS, the resulting field vector forms three-dimensional shape, which contains spatial parameters of the rays. The directions of further research can address the influence of the nonlinearity and nonidentity of the polarization direction finder channels on the quality of DF, the possibility of polarization DF of interfering quadrature modulated RES, estimation of DF errors in various conditions, etc. At the same time, it can be stated that the contribution of the current investigation is the discovery of the interference stability property of polarization direction finding which is very promising and can find many applications in positioning applications.

Acknowledgements. The reported study was supported by the Ministry of Science and Education of the Russian Federation with Grant of the President of the Russian Federation for the state support of young Russian scientists № MK-3468.2018.9.

References

1. Sivers, M., Fokin, G., Dmitriev, P., Kireev, A., Volgushev, D., Hussein Ali, A.-O.A.: Wi-Fi based indoor positioning system using inertial measurements. In: Galinina, O., Andreev, S., Balandin, S., Koucheryavy, Y. (eds.) NEW2AN/ruSMART/NsCC -2017. LNCS, vol. 10531, pp. 734–744. Springer, Cham (2017). https://doi.org/10.1007/978-3-319-67380-6_69
2. Fokin, G., Ali, A.-O.A.H.: Algorithm for positioning in non-line-of-sight conditions using unmanned aerial vehicles. In: Galinina, O., Andreev, S., Balandin, S., Koucheryavy, Y. (eds.) NEW2AN/ruSMART -2018. LNCS, vol. 11118, pp. 496–508. Springer, Cham (2018). https://doi.org/10.1007/978-3-030-01168-0_44
3. Petrov, A.A., Davydov, V.V.: Improvement frequency stability of caesium atomic clock for satellite communication system. In: Balandin, S., Andreev, S., Koucheryavy, Y. (eds.) ruSMART 2015. LNCS, vol. 9247, pp. 739–744. Springer, Cham (2015). https://doi.org/10.1007/978-3-319-23126-6_68
4. Petrov, A.A., Davydov, V.V., Grebenikova, N.M.: Some directions of quantum frequency standard modernization for telecommunication systems. In: Galinina, O., Andreev, S., Balandin, S., Koucheryavy, Y. (eds.) NEW2AN/ruSMART -2018. LNCS, vol. 11118, pp. 641–648. Springer, Cham (2018). https://doi.org/10.1007/978-3-030-01168-0_58
5. Rembovsky, A., Ashikhmin, A., Kozmin, B.: Radiomonitoring – Problems, Methods, Tools, 4th edn. Goryachaya Liniya Telekom, Moscow (2018)
6. Lipatnikov, V., Tsarik, O.: Methods of Radiomonitoring. Theory and Practice. State Research Institute "National Development" Publishing, Saint Petersburg (2018)
7. Rembovsky, A., Ashikhmin, A., Kozmin, B.: Automated Monitoring Systems and Their Components. Goryachaya Liniya Telekom, Moscow (2017)
8. Poisel, R.: Electronic Warfare Target Location Methods. Artech House, Norwood (2005)
9. Mezin, V.: Automatic Direction Finders. Sovetskoe Radio, Moscow (1969)
10. Kukes, I., Starik, M.: Basics of Radio Direction Finding. Sovetskoe Radio, Moscow (1964)
11. Vinogradov, A., Rembovsky, A.: Problem issues and technical solutions for hardware and software implementation of modern correlation interferometers. Secur. Inf. Technol. **2**, 7–19 (2002)
12. Jonson, D., Graaf, S.: Improving the resolution of bearing in passive sonar arrays by eigenvalue analysis. IEEE Trans. V. ASSP **30**(4), 638–647 (1982). https://doi.org/10.1109/tassp.1982.1163915
13. Paulraj, A., Kailath, T.: Eigenstructure methods for direction of arrival estimation in the presence of unknown noise fields. IEEE Trans. V. ASSP **34**(1), 13–20 (1986). https://doi.org/10.1109/tassp.1986.1164776
14. Komarovich, V., Nikitchenko, V.: Methods of spatial processing of radio signals. Military Telecommunications Academy, Leningrad (1989)

15. Yuan, X.: Coherent sources direction finding and polarization estimation with various compositions of spatially spread polarized antenna arrays. Sig. Process. **102**, 265–281 (2014). https://doi.org/10.1016/j.sigpro.2014.03.041
16. Liu, Z.: DOA and polarization estimation via signal reconstruction with linear polarization-sensitive arrays. Chin. J. Aeronaut. **28**(6), 1718–1724 (2015). https://doi.org/10.1016/j.cja.2015.09.005
17. Zheng, G.: Two-dimensional DOA estimation for polarization sensitive array consisted of spatially spread crossed-dipole. IEEE Sens. J. **18**(12), 5014–5023 (2018). https://doi.org/10.1109/JSEN.2018.2820168
18. Chintagunta, S., Ponnusamy, P.: Integrated polarisation and diversity smoothing algorithm for DOD and DOA estimation of coherent targets. IET Sig. Process. **12**(4), 447–453 (2018). https://doi.org/10.1049/iet-spr.2017.0276
19. Wang, F., Xin, L., Mao, X.: Multidimensional parameter estimation for electromagnetic sources with diversely polarized sensor array — a review. In: 2017 8th IEEE Annual Information Technology, Electronics and Mobile Communication Conference (IEMCON), Vancouver, BC, 2017, pp. 343–347 (2017). https://doi.org/10.1109/iemcon.2017.8117133
20. Afraimovich, E.L., Chernukhov, V.V., Kobzar, V.A., Palamartchouk, K.S.: Determining polarization parameters and angles of arrival of HF radio signals using three mutually orthogonal antennas. Radio Sci. **34**(5), 1217–1225 (1999). https://doi.org/10.1029/1999RS900042
21. Wong, K.T.: Direction finding/polarization estimation-dipole and/or loop triad(s). IEEE Trans. Aerosp. Electron. Syst. **37**(2), 679–684 (2001). https://doi.org/10.1109/7.937478
22. Kitavi, D., Wong, K.T., Zou, M., Agrawal, K.: A lower bound of the estimation error of an emitter's direction-of-arrival/polarization, for a collocated triad of orthogonal dipoles/loops that fail randomly. IET Microwaves Antennas Propag. **11**(7), 961–970 (2017). https://doi.org/10.1049/iet-map.2016.0918
23. Simonov, A., Bogdanovsky, S., Volkov, R., Sevidov, V.: Method of polarization direction finding of radio signals. Patent for invention 2624449 (2017)
24. Simonov, A., Bogdanovsky, S., Teslevich, S.: Polarization method of direction finding of radio emission sources in space. High Technol. **17**(12), 40–43 (2016)
25. Korn, G., Korn, T.: Mathematics Handbook. Science, Moscow (1977)

Coexistence Management Approach for Densification of Randomly Deployed Low Power Nodes in TVWS Spectrum

Inam Ullah[1], Edward Mutafungwa[2], Muhammad Zeeshan Asghar[1(⊠)],
and Jyri Hämäläinen[2]

[1] Faculty of Information Technology, University of Jyväskylä, Jyväskylä, Finland
seeinam@gmail.com, muhammad.z.asghar@jyu.fi
[2] Department of Communication and Networking,
Aalto University School of Electrical Engineering, Espoo, Finland
{edward.mutafungwa,jyri.hamalainen}@aalto.fi

Abstract. Expansion of modern wireless communication technologies have led to spectrum scarcity due to inefficient spectrum utilization by conventional TV broadcasting technologies. While the transition of television from analogue to digital leaves an unused spectrum aka white space (WS). A need for a sophisticated coexistence mechanism observe to enable smooth operation of unlicensed secondary systems in TVWS sub-one-gigahertz (Sub-1GHz). To that end, this paper proposes a two-step algorithm to enables an efficient coexistence mechanism among cells in TVWS. This approach assumes a WS geo-location database (WSDB) as central entity allowing a new cell to operate in a given geographic area. When a new cell query is sent to WSDB, it ensures that the new cell will not overlap with already-deployed cells. This enables the cells to be in coexistence with neighbouring cells without interfering each other transmissions. Simulation results show that the proposed algorithm enables enhanced performance gains in terms of cell density and area capacity. The required cell density is notably increased in a given geographic area leading to improved area capacity while it also ensures the efficient TVWS utilization among cells.

Keywords: Coexistence management · TV white space ·
Spectrum sharing · Throughput · Area capacity · Cognitive radio

1 Introduction

In recent years, intensive research has been carried out on 5th generation (5G) cellular network, e.g., a network enabling extremely high data rates services per device (several Gigabits) per area (bps/km^2) with Ultra-reliability and Ultra-low latency [2]. Moreover, fast developments in communication devices along with avalanche of sophisticated data hungry applications increase the challenges to forthcoming 5G communication technologies. For example, mobile subscriptions

© Springer Nature Switzerland AG 2019
O. Galinina et al. (Eds.): NEW2AN 2019/ruSMART 2019, LNCS 11660, pp. 220–232, 2019.
https://doi.org/10.1007/978-3-030-30859-9_19

reached to 7.9 billion by now, while forecasted to be 1.5 billion 5G subscriptions by 2024 [1]) and data consumption per subscriber (e.g., 50 GB per month by 2024 [1]), new Internet-of-Things (IoT) service drivers (22 billion IoT connected device along with 4.1 billion of 5G cellular IoT connections [1]. These facts obliged the cellular operators to implement continuously evolve the network capacity and coverage.

Hence, keeping in view of the aforementioned evolution, 5G networks require developments on several fronts. Among them is the efficient utilization of the available limited radio resources [3] by enabling network densification through dense deployment of low power cognitive radios(10W or lower power) to complement legacy high-power macro sites [4]. In case of spectrum sharing, studies (e.g. [4]) signified a significant underutilization of spectrum and a general consensus support a flexible and managed spectrum sharing principles.

To that end, the migration from Analogue to Digital Terrestrial Television (DTT) leaves an unused frequency spectrum in Sub-1GHz range, for example, 790–862 MHz, known as Digital Dividend 1 (DD1) and 694–790 MHz known as Digital Dividend 2 (DD2) [5,6]. These unused frequency bands are also known as TV white space (TVWS), whose propagation characteristics make it able to travel more distantly, less interfered and penetrate easily into any obstruction such as hills, buildings, vegetation and so on [7].

These TVWS Sub-1GHz spectrum enable an opportunity to be utilized for secondary systems operation via sophisticated Dynamic Spectrum Access (DSA) schemes. This DSA approach ensures coexistence of primary and secondary system operation on TVWS with negligible or no interference to primary DTV transmissions as well as unlicensed secondary systems. To that end, several simulation and measurement trials have been conducted to analyse the cell deployment and reliability of TVWS Sub-1GHz [8–11]. The Federal Communications Commission (FCC) [12–14] and Ofcom [15], initiated the standardization process for secondary systems operation in TVWS bands.

This work is using a dense deployment of low power nodes (LPNs) operating in TVWS spectrum bands in a coexistence manner. We proposed a coexistence mechanism which enable the densification of LPN deployments through channel-splitting of the TVWS band as well as different target cell edge throughputs. The remainder of the paper is organized as follows. Section 2 provides an overview of the previous work and highlights the contribution of this study. Section 3 provides a descriptions of system and simulation framework. The discussion of the simulation results and conclusive remarks are presented in Sects. 4 and 5, respectively.

2 Previous Work and Our Contribution

Previously, several works have addressed the coexistence challenges to the wireless systems operating in TVWS spectrum bands. To that end, Sun et al. in [18] introduced a coexistence scheme among coexistent secondary users and primary users. Similarly, Villardi et al. in [19] employ the dynamic frequency selection algorithm among the neighbouring coexistent cognitive wireless networks in TVWS. In [20], Zhao et al. proposed a spectrum sensing scheme to minimize the interference among the coexisting access networks operating in TVWS. Moreover, in [21], Filin et al. evaluate the performance of secondary systems in IEEE 802.19.1 coexistence frameworks where, WSDB is a centralized entity allocating a frequency channel to a new WS radio system from the available spectrum pool. In [22], Wang et al. proposed a coexistence protocol for IEEE P802.19.1 system's discovery service which enables the operating networks to take coexistence decisions.

In [23], Bahrak et al. devised a centralized algorithm for coexistence management of heterogeneous networks namely *Fair Al-algorithm for coexistence decision making in TV whitespace* (FACT). In addition, Khalil et al. in [24] proposed a two-stage coexistence management algorithm that first seeks to achieve fairness in bandwidth allocation before splitting and allocating bandwidth into channel slots. In [25], Yuan et al. proposed a coexistence model to tackle the coexistence of two heterogeneous networks namely as IEEE 802.22 and IEEE 802.11af. In [26], Ameigeiras et al. proposed a coexistence algorithm for unplanned small cell networks that seeks to reduce network-wide reconfiguration by only reassigning channels to those nodes in the vicinity of newly activated small cells. In [27], Filin et al. implement the IEEE 802.19.1 based coexistence mechanism among different radio system operating on TVWS in the same area. In [28], the author examines the coexistence of LTE and Wi-Fi systems over unlicensed frequency spectrum in the 5 GHz range. In [29], Liu et al. studies the coexistence management between IEEE 802.11ah and IEEE 802.15.4g technologies which are designed for outdoor IoT applications.

This study proposes a coexistence algorithm for the case where LPN coexists with each other in the given geographic area. We assume that the LPN operate on frequency channels with bandwidth of 1.4 MHz, 3 MHz and 5 MHz. These channel bandwidths are complaint with LTE standards as the research study being done regarding LTE system operation on unlicensed bands [16, 17]. This combination of channels with certain target throughputs on cell edge enable several options to WSDB for granting operation permission to the LPN. The simulation results show that proposed algorithm enables notable performance gains in terms of maximum LPN density in a given geographic area as well as enable efficient utilization of TVWS spectrum. Unlike [18], this work mainly focused on to enable a coexistence among the secondary users operating in TVWS. Unlike [21, 26] and [19], our model further addresses the case where new LPNs are not permitted when WSDB is run out of frequency channels. In addition to effective use of available channels, our model enables the option for the utilization of multiple cell edge throughput requirements (with their corresponding cell ranges as explained in Table 1) to WSDB in case of running out of available channels. From [31], it is known that cell range and throughput possess inverse relation at fixed transmission power and cell load. Hence, the cell ranges mentioned in our model actually replicate the corresponding target cell edge throughputs. Unlike [20, 22], we assume a coordinated coexistence mechanism for the LPNs deployment in a given geographic area by employing WSDB as a central entity.

3 System Model

We consider a network in a geographic area of 1×1 km with X number of LPNs where each LPN w_x is randomly deployed such that $x = 1, \cdots, X$. Each LPN is required to setup a connection with WSDB, to determine its operational frequency channel f_y from a spectrum pool. The size of spectrum pool Y is such that $y = 1, \cdots, Y$. Moreover, LPN can be instructed to enables a certain target throughput T_z service on cell edge (with their corresponding cell range) for its operation. This throughput option T_z is provided by WSDB. This throughput pool Z available at WSDB such that $z = 1, \cdots, Z$. To that end, we assume a centralized coexistence management scheme, where each LPN can check the availability of channels with WSDB before initiating its operations. We split the available TVWS spectrum into multiple sub channels of fixed bandwidths to ensures the efficient and fair utilization of TVWS.

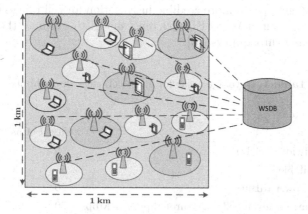

Fig. 1. LPN network model operating in TVWS spectrum

3.1 LPN Deployment and Operational Parameters

Two neighboring LPNs are said to be overlapping (means interfering with each other), if their corresponding serving cell areas (SCA) are overlapping with each other. These SCAs are cell ranges where the user equipment (UE) is enabled with network services. Each LPN can randomly query the WSDB to grant permission. This enables a LPN to discover its existing deployed neighbour LPNs. If new LPN interferes the SCA of existing deployed LPNs, then WSDB assigns next available channel and/or target throughput to a new LPN. The purpose of this allocation by WSDB is to enable a co-existence environment among the LPNs that minimizes the likelihood of interference in the network, caused by the newly deployed LPN.

We assume FCC rules for defining the maximum transmission power of LPN. In order to follow the FCC mandate of power spectral density (PSD) on given channel bandwidth, the portable devices (LPNs) can enable transmission at 40 mW on 6 MHz channel bandwidth when adjacent to TV channels. Otherwise, the LPN can enable transmission at 100 mW on non-adjacent channel [32]. Table 1 shows combination of frequency channels obtained by splitting the total bandwidth of 6 MHz into sub channels of 1×5 MHz, 2×3 MHz and 4×1.4 MHz for Case 1, 2 and 3, respectively.

We also developed a link budget to estimate the LPN cell ranges for the corresponding computed cell ranges with given channel bandwidth, transmission power and target cell edge throughput as shown in Table 1. We assumed that each LPN requires a single frequency channel for its operation. Moreover, LPNs with non-overlapping SCA can reuse the same frequency channel without interfering each other operations. We assume a network of large number of LPNs randomly distributed in a given geographic area as shown in Fig. 1. Similarly, a link budget gives a more credible justification for cell ranges of LPN. We employ the extension of WINNER+ channel model to estimate the cell ranges for corresponding throughputs [30].

Table 1. Coverage areas for given cell edge throughputs

System parameters		Case 1	Case 2	Case 3
Channel Bandwidth (MHz)		5	3	1.4
Channels Available		1	2	4
Transmission Power (dBm)		20	17	14
Cell Ranges (m) for given cell edge throughputs	1 Mbps	200	165	132
	3 Mbps	144	114	82
	5 Mbps	120	92	59

3.2 Simulation Algorithms

This section describes the proposed algorithms utilized in simulation campaign. We analyse the LPN density when WSDB have option of multiple channels availability and a single target cell edge throughput as modelled in Algorithm 1, while, Algorithm 2 analyses the impact of multiple channels and multiple throughput on LPN density.

Algorithm 1. WSDB comprises multiple channels and fixed cell edge throughput

---

1: **for** $\forall \, w_x$ **do**
2:     Generate locations in geographic area
3:     **for** x = 1 to X **do**
4:         Each w_x send query to WSDB for granting service permission. WSDB checks that new LPN w_x's SCA overlapping with existing LPNs.
5:         **if** overlapping doesn't exist **then**
6:             WSDB assigns first available channel f_m and proceed to step 3.
7:             **if** overlapping exist **then**
8:                 WSDB assigns next available channel to w_x from pool, which is different then the existing overlapping LPNs.
9:                 **if** no f_y available **then**
10:                     Note the index of w_x and proceed to step 3 for w_{x+1}.
11:                 **end if**
12:             **end if**
13:         **end if**
14:     **end for**
15: **end for**

Algorithm 2. WSDB comprises multiple channels and multiple cell edge throughputs

---

1: **for** $\forall \, w_x$ **do**
2:     Generate locations in geographic area.
3:     **for** x = 1 to X **do**
4:         Each w_x send query to WSDB for granting service permission. WSDB checks that new LPN w_x's SCA overlapping with existing LPNs.
5:         **if** overlapping exist **then**
6:             WSDB direct the w_x to raise the target throughput level available in pool, to shrink the coverage area.
7:             **if** no T_z option available **then**
8:                 WSDB assigns next available channel to new w_x from pool, which is different then the existing overlapping LPNs and proceed to step 3
9:                 **if** no f_y available **then**
10:                     Note the index of w_x and proceed to step 3 for w_{x+1}.
11:                 **end if**
12:             **end if**
13:         **end if**
14:     **end for**
15: **end for**

---

4   Performance Evaluation

4.1   Impact of Multiple Channels Availability

Let us first consider the simulation results generated for Algorithm 1. Figures 2, 3 and 4 illustrate the impact of multiple available channels by showing the cumulative number of LPN accepted by WSDB for the Case 1, Case 2, and Case 3, respectively.

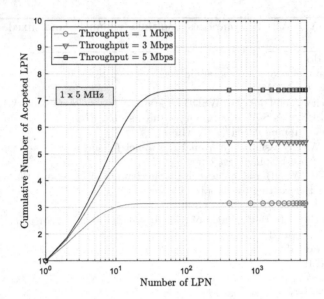

Fig. 2. LPN acceptance rate when WSDB available with 1 × 5 MHz channel and throughputs of 1 Mbps (O), 3 Mbps (▽) and 5 Mbps (□)

It is observed that the LPN density has been significantly increased, when WSDB ensures the availability of more frequency channels. It can be seen that the acceptance rate of LPNs, is significantly improved for the Case 3 as compared for the rest of cases. For example, in case of 5 Mbps target cell edge throughput, it is evident that with the increase of available frequency channels from 1 to 2 and to 4, the number of accepted LPNs increases i.e., 7, 39 and 250, respectively, which is due to the fact that, the WSDB ensures the availability of more frequency channels for LPN operation and thus, increases the likelihood of more LPNs to be accepted for the operation. The same behaviour is also observed for the rest of target throughput cases.

Moreover, the following Figs. 4 and 5 shows the cumulative number of accepted LPNs and probability of rejected LPNs by WSDB, respectively. Figure 4 shows that the acceptance rate of LPN increases rapidly, while increasing the target cell edge throughputs. For fixed number of available channels (i.e., 4 channels), LPN cell density with high throughput (and with corresponding small cell range) can be maximized. Similarly, it is evident from Fig. 5 that the probability of LPNs denial is higher for the case, when the target throughput decreases. It is due to the fact that the LPNs with small cell range have smaller likelihood to overlap with neighboring LPNs as compared to the case when the LPN having large cell ranges (and correspondingly low target throughputs).

We also analysed the LPN density performance, when WSDB employs multiple channels and multiple target cell edge throughput options as explained by Algorithm 2. It can be seen in Fig. 6 that on average, a WSDB can accept 135 LPNs for operation when WSDB is available with 3 target throughput options

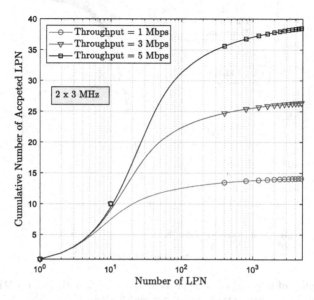

Fig. 3. LPN acceptance rate when WSDB available with 2 × 3 MHz channels and throughputs of 1 Mbps (O), 3 Mbps (▽) and 5 Mbps (□)

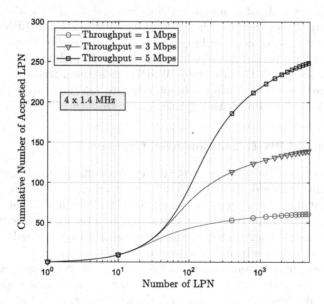

Fig. 4. LPN acceptance rate when WSDB available with 4 × 1.4 MHz channels and throughputs of 1 Mbps (O), 3 Mbps (▽) and 5 Mbps (□)

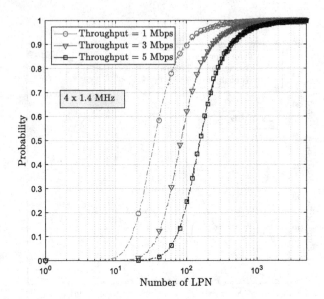

Fig. 5. Probability of rejected LPN when WSDB available with 4 × 1.4 MHz channels and throughputs of 1 Mbps (O), 3 Mbps (▽) and 5 Mbps (□)

in addition to 4 available channels. This shows a significant improvement gains in performance as compared to the case, when the WSDB is available with single target throughput of 1 Mbps and 4 available channels (i.e., only 35 LPNs have been accepted). This is because that WSDB can instruct a new LPN to reduce its cell range in case of cell overlapping and non availability of channels. We also observe reverse performance for the case when throughput is 5 Mbps (i.e., 153 LPNs are accepted). Because, here, WSDB can accommodate more LPNs of small coverage areas (i.e., 5 Mbps) as compared to multiple target throughput rates (which includes LPNs with larger coverage areas then 5 Mbps case).

4.2   Impact of LPN Density on Area Capacity

This section presents the obtained system improvements in terms of area capacity through our proposed algorithms as shown in Fig. 7. It is clearly evident in the results that the area capacity can be significantly enhanced, if the number of available frequency channels increase. It is shown that the area capacity for Case 3 (i.e., 4 available channels) achieves high area capacity as compared to the Case 1 (i.e., 1 available channel). It is due to the reason that multiple channel availability increases the acceptance rate of LPN in the given geographic area, which can be translated into area capacity. Furthermore, a high area capacity/km^2 can be achieved when the cell ranges of LPN are smaller. For example, in Case 3, it is observed that the area capacity of 59 m cell range (means 5 Mbps cell edge Throughput rate) is more as compared to 132 m cell range case. This area capacity improvements achieved via dense deployment of LPNs with smaller cell ranges.

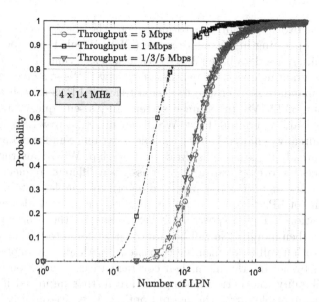

Fig. 6. Probability of rejected LPN when WSDB available with 4 × 1.4 MHz channels, fixed throughputs of 5 Mbps (O) and 1 Mbps (□) and multiple throughputs of 1/3/5 Mbps (▽)

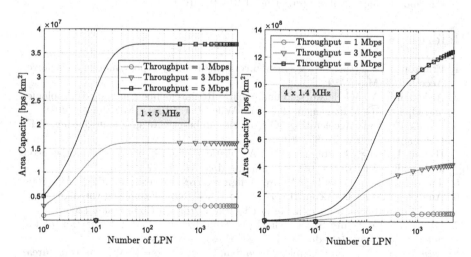

Fig. 7. Area capacity achieved when WSDB available with 1 × 5 MHz channels (Left: Case 1) and 4 × 1.4 MHz channels (Right: Case 3) and throughputs of 1 Mbps (O), 3 Mbps (▽) and 5 Mbps (□)

5 Conclusion

We proposed a two-step algorithm to implement the coexistence of low-power nodes (LPNs) operating in TV white space (TVWS) Sub-1GHz spectrum. We assumed a white space data base (WSDB) that acts as a centralized coordinator responsible for granting operation permission to new LPNs in a given geographic area. WSDB mandate is to ensure the coexistence of LPNs as well as the efficient utilization of TVWS. In assumed system, LPNs send query to WSDB for permission to enable operation. First, WSDB ensures that new LPN is not overlapping with already deployed-LPNs to enable coexistence and avoid interference towards the already-deployed cells. In case of overlap, WSDB instruct the new LPN to (reduces its coverage range) and/or assigns a different frequency channel. Simulation results show that our proposed algorithm enables notable improvements in terms of LPN density in a given geographic area. It also ensures the efficient utilization of TVWS spectrum. If WSDB ensures the availability of multiple channels, then more LPNs will be granted permission to initiate operation. Here, the condition will be of ensuring coexistence with already-deployed neighbouring LPNs by not interfering in their coverage areas. In addition to multiple channels availability, the performance gains are further improved if WSDB is comprised with multiple target throughput option. This also directly lead to the improved LPN acceptance rates. Finally paper also outlined the fact that, the more the LPNs density increase in a given geographic area, the higher will be the area capacity achieved.

Acknowledgement. This work supported by the Academy of Finland (grant no. 284634) and Business Finland (grant no. 1916/31/2017).

References

1. Ericsson AB. https://www.ericsson.com/en/mobility-report/mobility-visualizer. Accessed 20 May 2019
2. Al-Dulaimi, A.: 5G Networks: Fundamental Requirements, Enabling Technologies, and Operations Management, 2nd edn. Wiley-IEEE Press, New Jersey (2018)
3. Mitola, J., et al.: Accelerating 5G QoE via public-private spectrum sharing. IEEE Commun. Mag. **52**, 77–85 (2014)
4. Hwang, I., Song, B., Soliman, S.S.: A holistic view on hyper-dense heterogeneous and small cell networks. IEEE Commun. Mag. **51**, 20–27 (2013)
5. The digital dividend and the future of digital terrestrial television. DigiTAG (2009)
6. Analysys Mason, DotEcon, HOGAN and HARTSON: Exploiting the digital dividend - a European approach. Report for the European Commission (2009)
7. Oh, S.W.: TV White Space: The First Step Towards Better Utilization of Frequency Spectrum, 1st edn. Wiley-IEEE Press, New Jersey (2016)
8. Deshmukh, M., et al.: Wireless broadband network on TVWS for rural areas: an Indian perspective. In: 16th International Symposium on Wireless Personal Multimedia Communication (WPMC) (2013)
9. Holland, O., et al.: A series of trials in the UK aspart of the Ofcom TV white spaces pilot. In: 1st International Workshop on Cognitive Cellular Systems (CCS) (2014)

10. Kennedy, R., et al.: TV white spaces in Africa: trials and role in improving broadband access in Africa. In: AFRICON (2015)
11. Almantheri, H.M., et al.: TV white space (TVWS) trial in Oman: phase one (Technical). In: Sixth International Conference: on Digital Information, Networking, and Wireless Communication: (DINWC) (2018)
12. FCC Second report and order and memorandum opinion and order. ET Docket no. 08–260 (2008)
13. FCC, Second report and order and memorandum opinion and order in the matter of unlicensed operation in the television broadcast bands. ET Docket no. 10–174 (2010)
14. FCC, Third report and order and memorandum opinion and order in the matter of unlicensed operation in the television broadcast bands. ET Docket no. 12–36 (2012)
15. Ofcom Digital dividend: Cognitive access Consultation on licence-exempting cognitive devices using interleaved spectrum (2009)
16. Al-Dulaimi, A., et al.: 5G communications race: pursuit of more capacity triggers LTE in unlicensed band. IEEE Veh. Tech. Mag. 10(1), 43–51 (2015)
17. Zhiyi, Z., Fei, T., Jialing, L., Weimin, X.: Performance evaluation for coexistence of LTE and WiFi. In: International Conference on Computing, Networking and Communication, pp. 1–6 (2016)
18. Sun, C., et al.: Optimizing the coexistence performance of secondary-user networks under primary-user constraints for dynamic spectrum access. IEEE Trans. Veh. Tech. 61(8), 3665–3676 (2012)
19. Villardi, G.: Efficiency of dynamic frequency selection based coexistence mechanisms for TV white space enabled cognitive wireless access points. IEEE Wirel. Commun. 19(6), 69–75 (2012)
20. Zhao, B., et al.: Geo-location assisted spectrum sensing for cognitive coexistent heterogeneous networks. In: IEEE International Conference on Communication Workshops (ICC), pp. 347–351 (2013)
21. Filin, S., Baykas, T.: Performance evaluation of IEEE 802.19.1 coexistence system. In: IEEE International Conference on Communication, pp. 1–6, June 2011
22. Wang, J., et al.: Coexistence protocol design for autonomous decision-making systems in TV white space. In: Proceedings of IEEE WCNC, pp. 3249–3254 (2012)
23. Bahrak, B., et al.: Coexistence decision making for spectrum sharing among heterogeneous wireless systems. IEEE Trans. Wirel. Commun. 13, 1298–1307 (2014)
24. Khalil, K., et al.: Coexistence management for heterogeneous networks in white spaces. In: International Conference on Computing, networking and Communication, pp. 691–697 (2014)
25. Yuan, S., et al.: A selfishness-aware coexistence scheme for 802.22 and 802.11af networks. In: IEEE Conference on Wireless Communication and Networking, pp. 194–199 (2015)
26. Ameigeiras, P., et al.: Dynamic deployment of small cells in TV white spaces. IEEE Trans. Veh. Tech. 9, 4063–4073 (2015)
27. Filin, S., et al.: Implementation of TV white space coexistence system based on IEEE 802.19.1 standard. In: IEEE Conference on Standard for Communication and Networking, pp. 206–211 (2015)
28. Alhulayil, M., et al.: Coexistence mechanisms for LTE and Wi-Fi networks over unlicensed frequency bands. In: 11th International Symposium on Communication: Systems, Networks and Digital Signal Processing (CSNDSP) (2018)
29. Liu, Y., et al.: Coexistence of 802.11ah and 802.15.4g networks. In: IEEE Wireless Communication and Networking Conference (WCNC) (2018)

30. D5.3: WINNER+ Final channel models, CELTIC CP5-026 deliverable. In: Heino, p. (ed.) Wireless World Initiative New Radio-WINNER+ (2010)
31. Salo, J., et al.: Practical introduction to LTE radio planning. In: European Communications Engineering (2010)
32. FCC, Second report and order memorandum opinion and order about Small Entity Compliance Guide: Part 15 TV bands devices. DA 11–195, Washington, D.C. (2011)

Toward an Ultra-low Latency and Energy Efficient LoRaWAN

Mohammed Saleh Ali Muthanna[1], Ping Wang[2(✉)], Min Wei[2],
Abdelhamied A. Ateya[3,4], and Ammar Muthanna[3,5]

[1] School of Computer Science and Technology, Chongqing University of Posts
and Telecommunications, Chongqing, China
muthanna@mail.ru
[2] School of Automation, Chongqing University of Posts
and Telecommunications, Chongqing, China
{wangping,weimin}@cqupt.edu.cn
[3] St. Petersburg State University of Telecommunication,
22 Prospekt Bolshevikov, St. Petersburg, Russia
ammarexpress@gmail.com
[4] Electronics and Communications Engineering, Zagazig University,
Zagazig, Egypt
a_ashraf@zu.edu.eg@gmail.com
[5] Peoples' Friendship University of Russia (RUDN University),
6 Miklukho-Maklaya St, Moscow 117198, Russia

Abstract. LoRaWAN represents a promising Internet of things (IoT) technology that has recently gained high a market and academic interest. With the recent advances in sensory manufacturing, the number of connected devices increases dramatically and thus, a reliable IoT system able to provide the coverage for such system becomes a high demand. This paper proposed a reliable LoRaWAN architecture for ultra-dense IoT networks. The mobile edge computing is proposed to provide the computing and energy resources at the edge of the access LoRa network, while the software defined networking (SDN) is deployed at the core network to manage and control the whole network traffic. The novelty of the proposed system comes from the deploying of a distributed version of the SDN controller at the edge servers, to reduce the load on the core network. The system is simulated for various simulation scenarios and the results validate the proposed structure for ultra-dense and ultra-low latency IoT networks.

Keywords: LoRaWAN · MEC · SDN · Latency · Dense deployment

1 Introduction

The low-power wide-area network (LPWAN) is one of the most deployed Internet of things (IoT) networks that has attracted both market and academic sectors [1]. It provides many advantages over other IoT paradigms that include the energy efficiency and wide range coverage [2]. LoRa is one of the most efficient technologies that have been used for LPWAN [3]. Recently, LoRaWAN is used in wide range as an efficient

© Springer Nature Switzerland AG 2019
O. Galinina et al. (Eds.): NEW2AN 2019/ruSMART 2019, LNCS 11660, pp. 233–242, 2019.
https://doi.org/10.1007/978-3-030-30859-9_20

IoT network that can be deployed for many IoT applications and it is proper for most of the deploying environments. However, there are many design challenges and issues associated with the LoRaWAN that have not been solved yet [4]. One of the main challenges is the dense deployment of sensor devices [5]. By 2020, it is expected that the number of connected devices will be in billions order, which requires a reliable system that provide the coverage for such ultra-dense networks [6]. Thus, LoRaWAN should be adapted for dense operation, with the required performances. The massive number of deployed devices, results in enormous number of traffic that affect the whole network operation [7]. Another issue is the end-to-end latency associated with applications that run over the LoRaWAN networks [6].

To overcome these challenges and furthermore to achieve other important requirements such as reliability, scalability and system availability, LoRaWAN should deploy new communication paradigms such as the mobile edge computing (MEC) and the software defined networking (SDN) [9]. MEC is a recent communication paradigm that has been first introduced by the European Telecommunications Standards Institute (ETSI) in 2014 to break the centralization of cloud computing and moves from the centralized huge data centers to distributed small edge servers [10]. In another work, MEC moves the computing and energy resources from the remote cloud units to the edge of the access network; approximately one or two communication hops away from the end device, i.e. IoT device. MEC has been first introduced to support cellular networks, however deploying MEC for IoT networks can be used for integrating IoT networks with the cellular networks. It also can be used for connecting heterogeneous IoT technologies and exchanging data between them. Such example can be introduced by linking narrow-band IoT (NB-IoT) network with a LoRaWAN network through the MEC technology [11].

SDN is another communication paradigm that has been used recently for modern communication networks. It represents the way of separating data plane from the control plane, which starts a new era of softwarization networks [12]. The control plane represents the tire of devices that controls and manages the whole network. A single centralized SDN controller or distributed SDN controllers may be used for the control plane. Distributed OpenFlow switches that support the OpenFlow protocol interface are deployed in the data plane layer [9]. The deployment of SDN technology for IoT, and even other communication networks, achieves a high network flexibility and reliability [13].

The main contribution of this work is to provide a LoRaWAN architecture to support the expected dense deployment of IoT device. This can be achieved by the introduction of MEC and SDN technologies to the LoRaWAN. The organization of the paper will be as follows: In Sect. 2 the proposed MEC/SDN LoRaWAN scheme and the comprised methods are introduced. Section 3 provides the performance evaluation of the proposed system and methods. And the conclusion will be given in Sect. 4.

2 MEC/SDN Based LoRaWAN System

In this part, the proposed MEC/SDN based LoRaWAN system is introduced. As shown in Fig. 1, the proposed LoRaWAN architecture is given. Two main recent technologies are introduced to the traditional LoRaWAN system; MEC and SDN. The proposed LoRaWAN system can be viewed as four layers.

The first layer is the access network, which is represented by end devices.

The second layer includes MEC servers and LoRaWAN gateways. As introduced in Fig. 1, MEC servers are deployed at the edge of the access network. Each LoRaWAN gateway is connected to an edge cloud server via a high speed fiber cables. The introduction of MEC servers at the edge of the access network, one communication hop away from the end devices, provides the computing and energy resources near to end users.

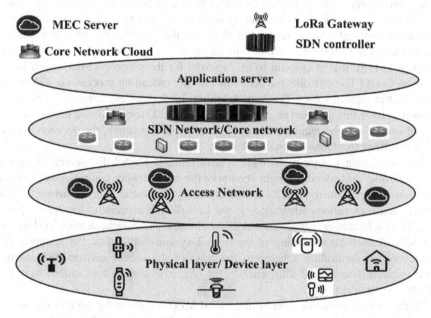

Fig. 1. Proposed MEC/SDN LoRaWAN system architecture.

The third layer is the core network, which deploys the SDN technology. This core network is built based on our proposed intelligent core network for ultra-low latency applications developed in [9]. A single centralized SDN controller is located at the core network that takes the labor of controlling and managing the distributed OpenFlow switches and edge cloud servers. This controller represents the control plane of the SDN network, while the distributed OpenFlow switches forms the data plane. The interaction and communication between both data plane and control plane is held over appropriate communication interface, e.g. OpenFlow protocol. Furthermore, the deployed OpenFlow switches are assumed to have limited computing resources, such that switches can perform some packet processing operations [14].

Distributed light versions of the SDN controller are set up over edge servers. The MEC server deploys a controller, which maps to the SDN controller deployed at the core network. A VM of the main controller located at the MEC server, moves the control operations to the edge of the access network and therefore, reduces the load on the core network and reduces the communication latency.

The fourth layer is for developing application server. It includes the database server for different applications.

One of the main features of the proposed system is the computation offloading. This advantage is added by the introduction of MEC system to the LoRaWAN. The property of computation offloading ads a lot to the IoT networks, since end-devices are always battery operated with limited computing resources. The offloading process enables the end-devices to transfer their computing tasks to the nearby MEC server and thus, reduces the power consumption of end devices and prolongs the devices' life time. Furthermore, the offloading of computing tasks to the nearby MEC server, one communication hop, reduces the round trip time and thus enables the ultra-low latency applications over LoRaWAN. We developed an energy and latency-aware offloading algorithm in [15], that is efficient to be deployed for the proposed MEC based LoRaWAN network. The offloading scheme decides the offloading process via a decision engine that implements the offloading algorithm. Each end-device, i.e. IoT device, is assumed to have this decision engine and also each MEC server. Based on the current level of energy and the required quality of service (QoS) latency, the decision engine decides whether to offload or process locally.

The other main feature is the edge control process. Each MEC server deploys a MEC controller that contains a light version of the core network controller and has the direct connection to the main SDN controller deployed at the core of the network. This property achieves various advantages to the LoRaWAN; especially, for dense operations. These advantages include the reduction of traffic and communication load on the core network and the increasing of the overall system throughput. Furthermore, the introduction of controlling scheme to the edge of the access network reduces the communication latency and thus, enables the ultra-low latency applications over the LoRaWAN networks.

Furthermore, the introduction of distributed MEC servers at the edge of the access network enables the implementation of artificial intelligence algorithms at the edge of the access network. This provides a great support for data compression and reduction and also, reduces the overall communication latency. The proposed system with the offloading methods introduced in [15] achieves the following:

1. Efficient bandwidth utilization,
2. Introduction of new services,
3. Reduction of end-to-end latency,
4. Efficient traffic management,
5. Reduction of blocking probability,
6. Higher energy efficiency,
7. Ultra-high reliability,

8. System scalability,
9. High availability, and
10. High system flexibility.

3 Performance Evaluation

3.1 Overview

In this part, the proposed system with offloading scheme introduced in [15] is simulated over a reliable environment for performance evaluation and system validation. In the first part, the simulation set up is hold, while the second part presents the simulations results and analysis.

3.2 Simulation Setup

The event-driven based simulator introduced in [15] and [16] is used for building and simulating the proposed system and the offloading scheme. The introduced system in Fig. 1 is built with distributed end-devices. With a LoRa gateway connected to a MEC sever. The simulation parameters used for performance evaluation of the proposed system are introduced in Table 1. The end-devices generate traffic based on an exponential distribution.

Table 1. Simulation parameters.

Parameter	Value
No. end devices	200/400/600
Simulation time	3000 s
Delay for one communication hop toward edge server/access network/LoRaWAN	1 ms
Delay for core network	2 ms
Packet size	32 Byte
Data traffic	3 packets
Frequency band of base station	868 MHz
End-device storage/RAM [IoT]	512 Mb
End-device processing/CPU	$\ni [0.1, 0.2]$ GHz
MEC serverstorage/RAM	2048 Mb
MEC serverstorage/HDD	5 Gb
MEC serverprocessing/CPU	$\ni [0.5, 2.5]$ GHz
SDN controller	ODL specifications
Switches	OpenFlow with the specifications introduced in [14]

Two main simulation scenarios are considered for performance evaluation. The first scenario is introduced to evaluate the system efficiency for dense operation and the scalability of the proposed system. For this scenario, the performance metrics used to evaluate the proposed system performance and the ability of scaling are the system reliability and the average latency. The second scenario is introduced to evaluate the performance of the offloading process and the efficiency of resources utilization. For both scenarios, three simulation cases are considered, in each case the network is operated with a certain number of end-devices. For case (1), the network is operated with 200 IoT devices, in the second case the number of deployed IoT devices is doubled and in case (3) the number of end devices is increased to 600.

3.3   Simulation Results

Figures 2, 3 and 4 provide the results for the first simulation scenario. Figure 2 illustrates the average number of received packets for each considered case. The value of received packet is recorded for two systems; the proposed system and the traditional LoRaWAN without the introduction of MEC and SDN. Results indicate that the proposed MEC/SDN LoRaWAN system achieves higher efficiency in terms of received packets for the three considered simulation cases. However, for case (3) that represents the dense deployment, as the number of deployed devices is increased, the proposed system achieves percentage improvement in performance, in terms of received packets, higher than that of the previous two cases. This indicates that the traditional LoRaWAN system fails for dense operation in comparison to the proposed MEC/SDN system. Since, reliability can be measure by the number of received packets, thus the proposed system achieves higher reliability efficiency.

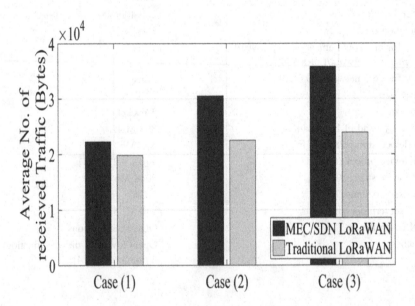

Fig. 2. Average number of received packets for the first simulation scenario.

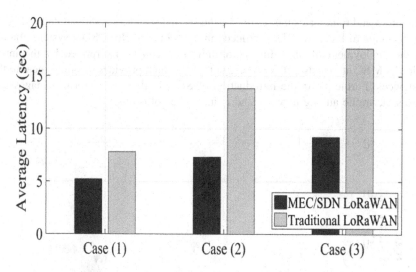

Fig. 3. Average latency for the first simulation scenario.

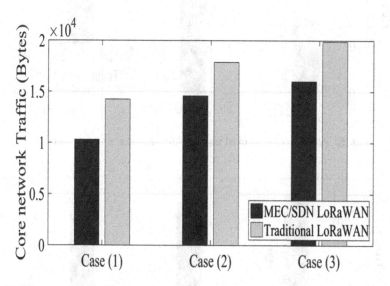

Fig. 4. The amount of traffic passed to the core network, for the first simulation scenario.

Figure 3 indicates the average latency of received traffic for the considered IoT applications; for both considered systems. Results indicate that the proposed MEC/SDN achieves higher latency efficiency than the traditional LoRaWAN system.

Furthermore, Fig. 4 illustrates the amount of traffic passed to the core network for the two considered systems in the three considered cases. Results indicate that the proposed system reduces the load on the core network, since the amount of traffic passed to the core network is reduced by a reasonable amount.

Figure 5 and 6 indicates the results for the second scenario. As illustrated in Fig. 5, the percentage of blocked tasks is reduced in the proposed MEC/SDN system; mainly with the deployment of offloading algorithm. This can be interpreted by the introduction of MEC at the edge of the access network, which provides resources near to the end-devices. Furthermore, the introduction of SDN in distributed form manages and controls the traffic among network and reduces the collisions.

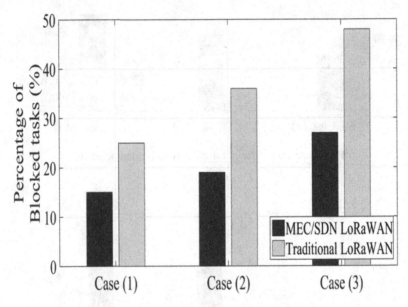

Fig. 5. Percentage of blocked tasks for the second simulation scenario.

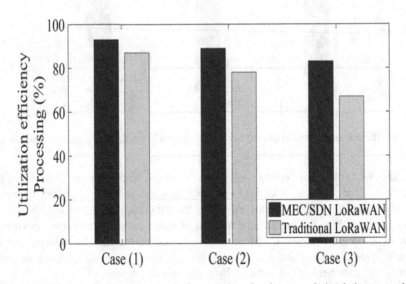

Fig. 6. Utilization efficiency, in terms of processing, for the second simulation scenario.

Figure 6 illustrate the system utilization in terms of processing for the considered two systems for the three considered simulation cases. As results indicate, the developed MEC/SDN LoRaWAN utilize the computing resources more efficient than the traditional LoRaWAN system.

4 Conclusion

This paper introduces a MEC/SDN LoRaWAN system structure to enable the dense deployment and ultra-low latency applications. The proposed system deploys MEC server connected to LoRa base station and a single centralized SDN controller located at the core network. Distributed versions of the SDN controller are located at each MEC server; mainly at the MEC controller. This reduces the load on the core network and moves the control scheme to the edge of the access network. The developed structure achieves higher system reliability, availability and flexibility. The proposed MEC/SDN system has been simulated over a reliable environment, for two simulation scenarios and three simulation cases per scenario, for performance evaluation. Results validate the system in terms of reliability, latency and resources utilization.

Acknowledgement. This paper is supported by The National Key Research and Development Program of China (2017YFE0123000) and the Chongqing S&T project (cstc2018jszx-cyztzxX0012).

References

1. Haxhibeqiri, J., Poorter, E.D., Moerman, I., Hoebeke, J.: A survey of LoRaWAN for IoT: from technology to application. Sensors **18**(11), 3995 (2018)
2. Sinha, R.S., Wei, Y., Hwang, S.H.: A survey on LPWA technology: LoRa and NB-IoT. IctExpress **3**(1), 14–21 (2017)
3. Finnegan, J., Brown, S.: A comparative survey of LPWA networking. arXiv preprint arXiv: 1802.04222 (2018)
4. Adelantado, F., Vilajosana, X., Peiro, P.T., Martinez, B., Segui, J.M., Watteyne, T.: Understanding the limits of LoRaWAN. IEEE Commun. Mag. **55**(9), 34–40 (2017)
5. Georgiou, O., Raza, U.: Low power wide area network analysis: can LoRa scale? IEEE Wirel. Commun. Lett. **6**(2), 162–165 (2017)
6. Ateya, A., Al-Bahri, M., Muthanna, A., Koucheryavy, A.: End-to-end system structure for latency sensitive applications of 5G. Электросвязь **6**, 56–61 (2018)
7. Sanchez-Iborra, R., Sanchez-Gomez, J., Ballesta-Viñas, J., Cano, M., Skarmeta, A.: Performance evaluation of LoRa considering scenario conditions. Sensors **18**(3), 772 (2018)
8. Ateya, A.A., Muthanna, A., Makolkina, M., Koucheryavy, A.: Study of 5G services standardization: specifications and requirements. In: 2018 10th International Congress on Ultra Modern Telecommunications and Control Systems and Workshops (ICUMT), pp. 1–6. IEEE, November 2018
9. Ateya, A.A., Muthanna, A., Gudkova, I., Abuarqoub, A., Vybornova, A., Koucheryavy, A.: Development of intelligent core network for tactile internet and future smart systems. J. Sens. Actuator Netw. **7**(1), 1 (2018)

10. Ateya, A., Vybornova, A., Kirichek, R., Koucheryavy, A.: Multilevel cloud based Tactile Internet system. In: IEEE-ICACT2017 International Conference, Korea, February 2017
11. Kim, D., Kim, S., Park, J.H.: Remote software update in trusted connection of long range IoT networking integrated with mobile edge cloud. IEEE Access **6**, 66831–66840 (2017)
12. Farris, I., Taleb, T., Khettab, Y., Song, J.: A survey on emerging SDN and NFV security mechanisms for IoT systems. IEEE Commun. Surv. Tutorials **21**(1), 812–837 (2018)
13. Salman, O., Elhajji, I., Chehab, A., Kayssi, A.: IoT survey: an SDN and fog computing perspective. Comput. Netw. **143**, 221–246 (2018)
14. Muthanna, A., et al.: Secure and reliable IoT networks using fog computing with software-defined networking and blockchain. J. Sens. Actuator Netw. **8**(1), 15 (2019)
15. Ateya, A.A., Muthanna, A., Vybornova, A., Darya, P., Koucheryavy, A.: Energy - aware offloading algorithm for multi-level cloud based 5G system. In: Galinina, O., Andreev, S., Balandin, S., Koucheryavy, Y. (eds.) NEW2AN/ruSMART -2018. LNCS, vol. 11118, pp. 355–370. Springer, Cham (2018). https://doi.org/10.1007/978-3-030-01168-0_33
16. Ateya, A.A., Vybornova, A., Samouylov, K., Koucheryavy, A.: System model for multi-level cloud based tactile internet system. In: Koucheryavy, Y., Mamatas, L., Matta, I., Ometov, A., Papadimitriou, P. (eds.) WWIC 2017. LNCS, vol. 10372, pp. 77–86. Springer, Cham (2017). https://doi.org/10.1007/978-3-319-61382-6_7

Novel AI-Based Scheme for Traffic Detection and Recognition in 5G Based Networks

Volkov Artem[1], Abdelhamied A. Ateya[1,2(✉)], Ammar Muthanna[1,3], and Andrey Koucheryavy[1]

[1] St. Petersburg State University of Telecommunication,
22 Prospekt Bolshevikov, St. Petersburg, Russia
Artem.n@5glab.ru, ammarexpress@gmail.com,
akouch@mail.ru
[2] Electronics and Communications Engineering, Zagazig University,
Zagazig, Egypt
a_ashraf@zu.edu.eg@gmail.com
[3] Peoples' Friendship University of Russia (RUDN University),
6 Miklukho-Maklaya St, Moscow 117198, Russia

Abstract. With the dramatic increase in the number of connected devices, the traffic generated by these devices puts high constraints on the design of fifth generation cellular systems (5G) and future networks. Furthermore, other requirements such as the mobility, reliability, scalability and quality of service (QoS) should be considered as well, while designing such networks. To achieve the announced requirements of the 5G systems and overcome the high traffic density problems, new technologies, such as the mobile edge computing (MEC) and software defined networking (SDN), and novel schemes, such as artificial intelligence (AI) algorithms and offloading algorithms, should be introduced. One main issue with the 5G networks is the heterogeneous traffic, since there are enormous number of applications and sub-networks. The main design challenge with the 5G network traffic is the recognition and classification of heterogeneous massive traffic, which cannot be performed by the current traditional methods. Instead, new reliable methods based on AI should be introduced. To this end, this work considers the problem of traffic recognition, controlling and management; mainly for ultra-dense 5G networks. In this paper, a novel AI algorithm is developed to detect and recognize the heterogeneous traffic at the core network. The algorithm is implemented at the control plane of the SDN network, located at the core network. The algorithm is based on the neural network. The system is simulated over a reliable environment for various considered cases and results are indicated.

Keywords: 5G/IMT-2020 · IoT · SDN · AI · RNN · Traffic recognition

1 Introduction

With the near release of 5G, it is expected that new enormous and cost efficient services will be supported [1]. Such services are expected to change the human life by introducing new paradigms such as the human to machine communications (H2M) [2].

© Springer Nature Switzerland AG 2019
O. Galinina et al. (Eds.): NEW2AN 2019/ruSMART 2019, LNCS 11660, pp. 243–255, 2019.
https://doi.org/10.1007/978-3-030-30859-9_21

Thus, creating an ecosystem for technical and business innovation for such systems become a great demand. There are many challenges associated with the design and realization of the 5G cellular system and announced use cases, these challenges are introduced in many literatures, such as [3, 4]. These challenges include the latency, reliability and availability. Some use cases of 5G systems requires an ultra-low end-to-end latency and ultra-high reliability and availability [5]. Tactile Internet is such example of these used cases, which require an end-to-end latency of 1 ms and a reliability of 99.99% [6]. Another main use case is the virtual reality applications that require a latency of 5 ms [7].

To enable such services, with the announced requirements, new technologies and infrastructure approaches should be involved. Another main challenge for 5G networks is the network traffic. By 2020, it is expected that the number of connected devices will be of order of billions, which puts high constraint on the design of networks that provides the coverage for such enormous number. This massive number of devices will generate a huge massive traffic that will pass over the network and load the core network [8]. For the current Internet, a significant part that always considered and monitored is the traffic generated by smart devices [9]. Thus, 5G networks should be provided with an infrastructure and novel technologies able to support and manage this enormous scalable number of traffic, with the required quality of services (QoS). One of the main novel technologies that will be deployed for 5G networks is the software defined networking (SDN). SDN enables the traffic management and control over the network separating data plane and control plane [10].

A significant number of the available Internet services and applications require the exact value of some network parameters such as latency, jitter, round trip time (RTT), and bandwidth. For such services and applications, SDN provides a vital solution that is able to extract and control such required network parameters. Furthermore, deploying SDN, especially for controlling network parameters, provides the way for implementing new services [11].

For the current networks, the QoS is measured and monitored using current existing differentiated services (DiffServ) technologies, as well as, some traffic engineering solutions [12]. A main example of these tools is the MPLS-TE, which uses the RSVP-TE protocol [13]. However, these solutions have several implementation disadvantages and cannot be used for a wide range of 5G/IMT-2020 services and applications. The main issues with these tools are the lack of dynamic control depending on the variability of the traffic type in the network, as well as the limited set of traffic classifiers.

The current Internet protocol (IP) systems deploys a field, in the packet IP header, dedicated with defining certain QoS requirements, which is referred to as the type of service (ToS) field [14]. The field of ToS can be used to indicate the requirement of high through put or request of a low latency route for the data packet associated with a certain application. Current IPv6 uses an alternate definition for the ToS fields, which is the differentiated services field (DS), which can be used for traffic classification. The DS field is defined by eight bits and contains two main sub-fields; differentiated services code point (DSCP) field and explicit congestion notification (ECN). The first six bits of the DS field define the DSCP sub-field and the last two bits define the ECN sub-field. Current existing mechanisms are efficient for ensuring QoS in applications and services such as telephony, video, television, etc. However, these mechanisms represents an efficient solutions when using for ensuring QoS of modern and upcoming applications such as Internet of things (IoT) applications, VR applications and most of expected 5G services. The main issue with these mechanisms, that makes them inefficient for 5G networks, is the lake of prober traffic engineering solution.

Besides the massive increase of traffic, another problem associated with the traffic, in 5G networks, is the traffic heterogeneity [15]. Such example is the IoT traffic that can go along with the traffic of the Video on-demand service, Web traffic and others. Furthermore, the unpredictability of 5G networks' traffic represents a challenge; mainly for the network operators.

Due to traffic heterogeneity, the computations and calculations of changes in traffic profile become more complex and thus, current existing static computational methods become improper and lose their relevance [16]. One way to ensure a rapid response of control systems to the heterogeneous traffic growth, including periodicity, and changes in the traffic profile, is to consider novel methods for computations and control processes associated with network traffic. In order to ensure rapid response, the communication network and, accordingly, the traffic switching and routing device should have the necessary level of abstraction. This necessary level of abstraction from physical processes can be realized by introducing network softwarization; including SDN and network function virtualization (NFV) paradigms [17].

For SDN/NFV based networks, it is required to ensure a traffic detection and classification with very high precision, to ensure ultra-reliable and ultra-low latency system. This process should be hold in an intelligent way, so that the pre-registered traffic flows, no recognition tasks should be deployed; such example is the VoIP and IPTV traffic. To this end, this paper provides an intelligent scheme for managing, controlling and classifying previously unregistered traffic over SDN/NFV based networks. IoT is considered as the main application in this work, since it is one of the main use cases of the 5G.

An intelligent comprehensive analysis of mathematical classification methods is carried out, considering the characteristics of the analyzed input data. These characteristics include the traffic flows and the initial requirements. The proposed methods are developed based on the neural network. The neural network architecture is chosen in accordance to the objectives and the characteristics of the input data set. The experimental evaluation of the proposed methods has been carried out in our SDN lab. A server monitoring and performing predictive analysis of traffic flow over SDN based networks is developed to be connected to the SDN controller. Experimental results validate the proposed system and confirm the feasibility of the proposed methods in Traffic detection and classifications for IoT networks; as a main use case of 5G.

2 Proposed SDN/NFV System Structure

Recently, there are many approaches that have been developed for traffic recognition in current existing communication networks [18]. These approaches are mainly developed based on the periodic capture of traffic and the analysis of its headers. Such methods have several disadvantages that include the delaying stream, the implementation of the analytical module requires additional hardware-software solutions and the high complexity of such methods and even the additional required hardware. To overcome these limitations, the proposed work aims to implement the analytical system at the service level of the proposed system. Figure 1 illustrates the proposed SDN based network structure.

The proposed SDN/NFV based structure ensures the system portability, the independence of the data transmission medium, and mostly the integration of data Plane. For an analytical system, all devices and flows are digital objects with a number of parameters and functionality represented by a set of methods. The level of abstraction of the proposed SDN/NFV based structure enables the implementation of an analytical system that works with traffic flows metadata. Thus, the proposed system doesn't introduce additional traffic delays, or make changes to its activity, e.g. changes in the distribution laws and intensity. Furthermore, the proposed system monitors the flow activity and provides a graphical indication of all routes among core network.

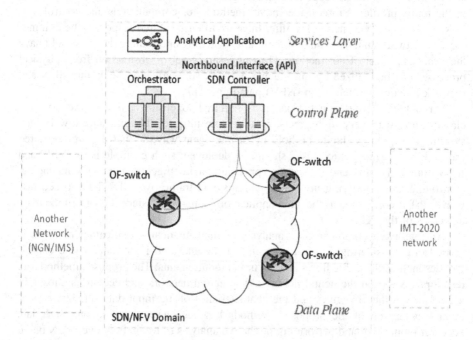

Fig. 1. Principle structure of the proposed scheme.

3   Input Data Set

The proposed SDN/NFV based system structure enables the implementation of the analytical system in the form of an application at the service level of the SDN/NFV network. The data received at the core network, via the northern interface of the SDN controller, represented the input data to the proposed analytical application. Since, one of the main objectives of this work is to examine the analytical activity of flows and to detect the flows at the data plane level, the analytical system has the ability to request flow tables data from all switches connected to the SDN control. The flow tables contain two main fields; the match field and the action filed. Analyzing data of both match and action fields, enables the development of a meta-model of flows. The general structure of the flow table is introduced in Fig. 2.

The circled data, in Fig. 2, is the data that should be used to form the meta-model for the network flow. These data includes two main counters; Byte Count and Packet Count. In addition to these counters, flow table contains another main parameter, which is the Time Stamp. Time Stamp enables the instantaneous calculation of ByteCount-delta and PacketCount-delta. One of the main features of these data is that, based on the "Byte Count" and "Packet Count" counters, it is impossible to accurately determine the exact packet length in the packet stream. Accordingly, based on these data it is impossible to accurately determine the length of each packet registered in the stream during a time interval ΔT.

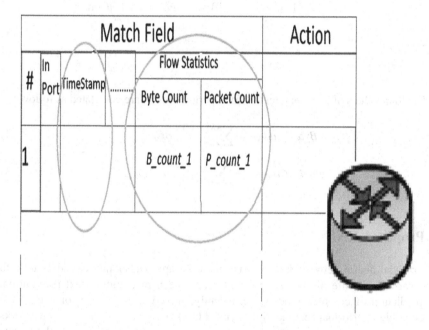

Fig. 2. Main fields of flow table.

However, for an arbitrary period of time ΔT, having samples of [Byte Count], [Packet Count], [Time Stamp] values, it is possible to create a data set with an established data structure, where each sample displays the instantaneous value [ByteCount-delta] and [PacketCount-delta]. The values of [Byte Count], [Packet Count] and [Time Stamp] are generated by instantaneous requests received by the SDN controller, via the RESTful application programming interface (API).

The data structure is formed on the basis of the $DataSet_{RQ}$ queries with "raw" data as in (1). Formula (2) is used to convert the data structure to the required $DataSet_{ML}$ format, while the instantaneous values are evaluated based on (3).

Let PacketCount-delta - PC_{delta},

ByteCount-delta - BC_{delta},

TimeStamp-deltas - TS = 1 [sec.] = const, Then:

$$DataSet_{RQ} = \begin{array}{cccc} [TimeStamp] & [ByteCount] & [PacketCount] \\ TimeStamp_{11} & ByteCount_{12} & PacketCount_{13} \\ TimeStamp_{21} & ByteCount_{22} & PacketCount_{23} \\ ... & ... & ... \\ TimeStamp_{N1} & ByteCount_{N2} & PacketCount_{N3} \end{array} \quad (1)$$

$$\begin{cases} BC\_delta_{N2} = ByteCount_{N2} - ByteCount_{(N-1)2}, if \ N \geq 1 \\ PC\_delta_{N2} = PacketCount_{N2} - PacketCount_{(N-1)2}, if \ N \geq 1 \end{cases} \quad (2)$$

$$DataSet_{ML} = \begin{array}{cccc} [TimeStamp] & [ByteCount] & [PacketCount] \\ TS & BC_{delta_{12}} & PC_{delta_{13}} \\ TS & BC_{delta_{22}} & PC_{delta_{23}} \\ ... & ... & ... \\ TS & BC_{delta_{N2}} & PC_{delta_{N3}} \end{array} \quad (3)$$

The total values of these parameters for a period of time are calculated as following:

$$ByteCount_{\Delta T} = \sum_{N=1}^{N=\Delta T/TS} BC_{delta_{N2}} \quad (4)$$

$$PacketCount_{\Delta T} = \sum_{N=1}^{N=\Delta T/TS} PC_{delta_{N2}} \quad (5)$$

4   Proposed RNN

The artificial neural network is one of the main AI approaches that is widely used to solve various real-life problems. Tasks such as speech recognition, text recognition, and prediction of complex models are, recently, solved with the help of neural networks; at high performance. One of the typical tasks introduced to the neural networks is the classification. There are many neural network based classification methods; one of the most efficient methods among these methods is the descriptions of objects using features [19]. Each object is characterized by a set of numeric or non-numeric features. However, for some types of data, open features do not provide classification accuracy, for example, the color of image points or the digital sound signal; this is because of these data contains hidden features. Another AI paradigm is the Deep Learning, which is a set of machine learning algorithms that attempt to simulate high-level abstractions in data and extract hidden features from input data.

Therefore, considering the object characteristics as traffic and its features as numerical - statistical series, a neural network with Deep Learning can be used for detection and classifications. A recurrent neural network (RNN) with Long Short-Term Memory (LSTM) is chosen as the neural network architecture as introduced in Fig. 3. The LSTM network is a universal network, such that with a sufficient number of network elements, it can perform any calculation that a conventional computer can do. This requires an appropriate matrix of weights that can be considered as a program.

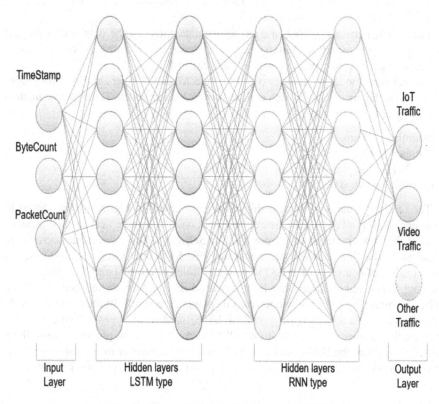

Fig. 3. The architecture of the proposed RNN.

LSTM part of the neural network enables the identification of patterns of influence of the considered samples, taking into account the correlations between samples' values. Such example is the periodicity in IoT traffic and the character self-similarity. Since, the selected neural network architecture implements the principle of data training, it is required to create training labeled data sets and save the state of the trained network. In order to train the neural network, the input $DataSet_{ML}$ is converted to a $DataSet_{MLtrain}$, by adding a new data column; each line of which contains the statistical sampling identifier. Accordingly, in order to recognize a wide range of traffic types, this training data set needs to be expanded by marking the appropriate statistical sample with a traffic label. Thus, the structure of the training $DataSet_{MLtrain}$, for IoT network, can be defined as following:

$$DataSet_{ML} = \begin{array}{cccc} [TypeOfTraffic] & [TimeStamp] & [ByteCount] & [PacketCount] \\ IoT & TS & BC_{delta_{12}} & PC_{delta_{13}} \\ IoT & TS & BC_{delta_{22}} & PC_{delta_{23}} \\ \ldots & \ldots & \ldots & \ldots \\ Video & TS & BC_{delta_{N2}} & PC_{delta_{N3}} \\ others & TS & BC_{delta_{(N+1)2}} & PC_{delta_{(N+1)3}} \end{array}$$

The LSTM network receives a fixed length data entry, so the data is divided into 200-line segments or 10 s. Activity tags are converted into unitary code. Data is divided into training and practice sets in the ratio of 8:2.

The network model contains two fully connected RNN layers and two fully connected LSTM layers; each of which contains seven hidden nodes. The hyperparameters of learning are as following:

- Optimizer: Adam,
- Number of epochs: 40,
- Number of samples per iteration: 1024, and
- Learning speed: 0.0025

The resulting neural network architecture is presented in Fig. 3. As introduced in the figure, the input layer of the neural network consists of three neurons; the input of which is the corresponding data from the DataSet$_{ML train}$. The three input neurons are connected to the first hidden layer, consisting of LSTM neurons. The second hidden layer consists of neurons of the LSTM type and it links both first and third layers. The third and fourth hidden layers are built based on the fully connected neurons of the RNN. The output neurons are two neurons displaying the results of the neural network. Accordingly, for a DataSet$_{ML train}$ with a large number of types of tagged traffic statistics, the number of output neurons will be increased. For the proposed RNN, there is a certain defined threshold, which is used for defining number of neurons associated with each hidden layer; mainly introduced for optimizing neural network training.

5   Network Model

For performance evaluation of the proposed system, a network model is introduced. This model represents a real model for a smart city network, since it comprises large heterogeneous amount of traffic. In order to generate IoT traffic, IoT traffic generators are used. These IoT traffic generators are used with the Internet of Things Data Management system (IoTDM), which is built according to the specifications of the international community of OneM2 M. These traffic generators have been developed as part of a previous research work, the results of which are introduced in details in [9].

The analytical system is referred to as the network application. A part of this network application is the module for detecting the IoT flows. This module has been developed using Python programming language as a WEB server operating on the MVC model. This server runs on the Northbound API controller and the orchestrator. The used SDN controller is OpenDaylight Beryllium SR4 and the OpenStack is used as the infrastructure orchestrator. The Data Plane layer is built on based of Mikrotik switches that support the OpenFlow protocol.

To train the data sets for the proposed neural network, the following assumptions are introduced.

1. Any traffic is turned off and the network is modified so that it could be accurately detect flows from smart city traffic in the switch flow table.
2. Next, application is given an exact flow number in the flow table and the application starts saving data.

A training data set for video on demand traffic has also been formed while software has been already used to generate video on demand traffic. After the trained data set is created, the developed neural network is activated and the state of the neural network is preserved after its successful training. At the next stage, the network application must get the ability to load the stored neural network to identify any flow in the network. To that end a testing data set is created, after specifying the flow number in the flow table. The end-to-end structure of the developed performance evaluation model is introduced in Fig. 4.

6 Performance Evaluation

For performance evaluation of the proposed RNN, the previous introduced network model is used. The proposed IoT traffic generator models the operation of 160 IoT devices. Each device generates an HTTP packet and transmits the values of three sensors in the packet body.

As a result, a $DataSet_{MLtrain}$ has been formed, which merges two labeled data sets, i.e. IoT and Video. Furthermore, the $DataSet_{MLtrain}$ is fed to the input of the proposed neural network. A scatter plot of the obtained $DataSet_{MLtrain}$ is introduced in Fig. 5. In Fig. 5, it is visually possible to identify several areas, i.e. clusters, of points' distribution, in the area where the corresponding values of the average value of the packet length in the stream prevails. For example, for the IoT flow, in most cases (dense area), the average packet length is approximately 400 bytes, and the parameter may vary. There is also a spike in the average packet length of 470 bytes.

During training, the parameter of accuracy is monitored, which is represented by the function inverse to the error function in recognition of activity. The other monitored parameter is the loss, which is the cross entropy determining a slightly near-predictable distribution to the true one. Figure 6 illustrates the training progress that indicates these parameters.

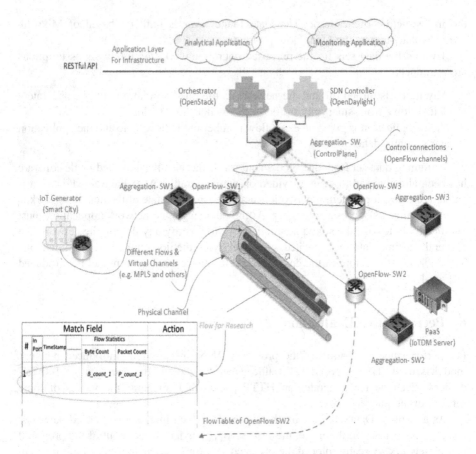

Fig. 4. End-to-end system structure of the considered network model.

Furthermore, Fig. 7 introduces the confusion matrix of the training neural network. Figure 6 mainly highlights the learning process of the network, in which it is clearly indicated that the network has successfully completed the learning process. As a result of the learning process of the neural network and testing its operation on test Data Sets, in the trained state, the developed neural network can identify the IoT flow with a probability of 99.97% and respectively the video stream. The confusion matrix indicates that the network fails identifying the Video flow; it is identified as an IoT flow, only one time. Furthermore, for IoT traffic it identifies all flows successfully. Thus, the proposed RNN architecture represents an optimal solution for traffic detection and classifications.

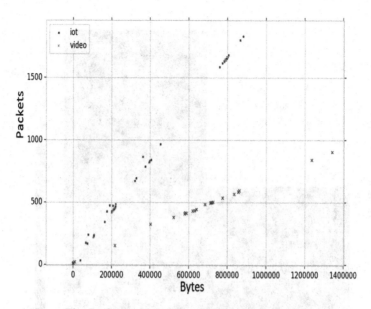

Fig. 5. Scatter plot of the obtained DataSet$_{MLtrain}$.

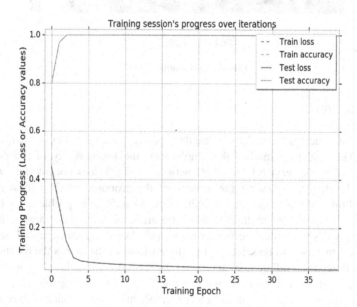

Fig. 6. Machine learning process.

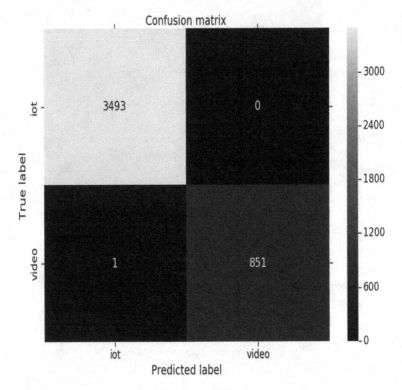

Fig. 7. Confusion matrix.

7   Conclusion

This paper introduces a method for traffic detection and classification of 5G/IMT-2020 networks. The analytical module that implements the functionality of the proposed system is built at the service level of 5G networks, which does not interfere with the data plane level; this represents the novelty of the proposed work. Ultimately, this approach allows solving most of the traffic flow identification problems at the data plane level. The proposed method is based on artificial intelligence algorithms. The paper presents RNN architecture for detecting and classifying traffic in 5G networks. The proposed method can be scaled in two dimensions; of the first is the parallel study of multiple flows from different switches and the second is the uniquely identifications of flows, e.g., various IoT services, not just one Smart City service and video stream. A network model based on IoT traffic and smart city has been introduced to provide an input data set for the proposed RNN. The experimental evaluation of the proposed system has been introduced, which indicates that the proposed methods efficiently detect and classify the heterogeneous traffic.

Acknowledgement. The publication has been prepared with the support of the "RUDN University Program 5-100".

References

1. Feasibility Study on New Services and Markets Technology Enablers, document 3GPP TR 22.891, ver. 14.2.0, September 2016
2. Ateya, A.A., Muthanna, A., Makolkina, M., Koucheryavy, A.: Study of 5G services standardization: specifications and requirements. In: 2018 10th International Congress on Ultra Modern Telecommunications and Control Systems and Workshops (ICUMT), pp. 1–6. IEEE, November 2018
3. 5G PPP Architecture Working Group white paper, "View on 5G Architecture," July 2016
4. Jiang, D., Liu, G.: An overview of 5G requirements. In: Xiang, W., Zheng, K., Shen, X. (eds.) 5G Mobile Communications, pp. 3–26. Springer, Cham (2017). https://doi.org/10.1007/978-3-319-34208-5_1
5. Tudzarov, A., Gelev, S.: Requirements for next generation business transformation and their implementation in 5G architecture. Int. J. Comput. Appl. **162**(2), 31–35 (2017)
6. Ateya, A. Vybornova, A., Kirichek, R., Koucheryavy, A.: Multilevel cloud based tactile internet system. In: IEEE-ICACT2017 International Conference, Korea, February 2017
7. Koucheryavy, A., Makolkina, M., Paramonov, A.: Applications of augmented reality traffic and quality requirements study and modeling. In: Vishnevskiy, V.M., Samouylov, K.E., Kozyrev, D.V. (eds.) DCCN 2016. CCIS, vol. 678, pp. 241–252. Springer, Cham (2016). https://doi.org/10.1007/978-3-319-51917-3_22
8. Ateya, A.A., Muthanna, A., Gudkova, I., Abuarqoub, A., Vybornova, A., Koucheryavy, A.: Development of intelligent core network for tactile internet and future smart systems. J. Sens. Actuator Netw. **7**(1), 1 (2018)
9. Volkov, A., Khakimov, A., Muthanna, A., Kirichek, R., Vladyko, A., Koucheryavy, A.: Interaction of the IoT traffic generated by a smart city segment with SDN core network. In: Koucheryavy, Y., Mamatas, L., Matta, I., Ometov, A., Papadimitriou, P. (eds.) WWIC 2017. LNCS, vol. 10372, pp. 115–126. Springer, Cham (2017). https://doi.org/10.1007/978-3-319-61382-6_10
10. Zaidi, Z., Friderikos, V., Yousaf, Z., Fletcher, S., Dohler, M., Aghvami, H.: Will SDN be part of 5G?. IEEE Commun. Surv. Tutorials **20**(4), 3220–3258 (2018)
11. Yousaf, F.Z., Bredel, M., Schaller, S., Schneider, F.: NFV and SDN-Key Technology Enablers for 5G Networks, arXiv preprint arXiv: 1806.07316 (2018)
12. BinSahaq, A., Sheltami, T., Salah, K.: A survey on autonomic provisioning and management of QoS in SDN networks. IEEE Access **7**, 2 (2019)
13. Attar, H., et al.: A mathematical model for managing the distribution of information flows for MPLS-TE networks under critical conditions. Commun. Netw. **10**(02), 31 (2018)
14. Haider, F., Chaudary, M.H., Naveed, M.S., Asif, M.: IPv6 QoS for multimedia applications: a performance analysis. In: Proceedings of the 2019 8th International Conference on Software and Computer Applications, pp. 501–504. ACM (2019)
15. An, J., Yang, K., Wu, J., Ye, N., Guo, S., Liao, Z.: Achieving sustainable ultra-dense heterogeneous networks for 5G. IEEE Commun. Mag. **55**(12), 84–90 (2017)
16. Qiu, T., Chen, N., Li, K., Qiao, D., Fu, Z.: Heterogeneous ad hoc networks: architectures, advances and challenges. Ad Hoc Netw. **55**, 143–152 (2017)
17. Haleplidis, E., Salim, J.H., Denazis, S., Koufopavlou, O.: Towards a network abstraction model for SDN. J. Netw. Syst. Manage. **23**(2), 309–327 (2015)
18. Velan, P., Čermák, M., Čeleda, P., Drašar, M.: A survey of methods for encrypted traffic classification and analysis. Int. J. Netw. Manage. **25**(5), 355–374 (2015)
19. Yang, Y., Luo, H., Xu, H., Wu, F.: Towards real-time traffic sign detection and classification. IEEE Trans. Intell. Transp. Syst. **17**(7), 2022–2031 (2015)

A Connectivity Game with Incomplete Information on Jammer's Location

Andrey Garnaev$^{(\boxtimes)}$ and Wade Trappe

WINLAB, Rutgers University, North Brunswick, NJ, USA
garnaev@yahoo.com, trappe@winlab.rutgers.edu

Abstract. In this paper we consider the problem of maintaining communication between a transmitter and a receiver in the presence of hostile interference. To maintain communication the transmitter must keep the SINR greater or equal to an SINR threshold, while the adversary aims to break the communication by making this SINR less than this threshold. In particular, we focus on investigating how incomplete information about the adversary's location can impact the rival's strategies. Namely, we assume that the transmitter does not know the adversary's location, but knows only an a priori distribution of possible adversary locations. The problem is formulated as a non-zero Bayesian game between the transmitter and the adversary in a Nash and Stackelberg framework for Rayleigh channel fading gains. Existence and uniqueness of both type of equilibria are proven and derived in closed form. We prove theoretically, and numerically illustrate, that the Stackelberg equilibrium strategy for the transmitter can be non-sensitive to the a priori information, while the Nash equilibrium strategy for the transmitter is always sensitive to such information. The condition when the Stackelberg transmitter equilibrium strategy is non-sensitive to a priori information is derived.

Keywords: Connectivity · Bayesian equilibrium

1 Introduction

Communication between a transmitter and a receiver under hostile interference is a well-studied problem in the wireless literature (see, for example, a survey [23]). Such problems are multi-objective problems since they deal with different agents (say, a transmitter and an adversary (jammer)), and each of these agents has its own objective. Game theory supplies concepts for analyzing and solving such multi-objective problems [17], and thus game theory is widely used to model jamming problems. Typically jamming problems are considered in two frameworks: to maintain communication reliability and communication connectivity. In problems where the transmitter aims to maintain communication reliability, the transmitter's payoff is a function of its throughput or SINR, and the transmitter intends to maximize such payoff [1, 2, 4, 5, 8–11, 14, 15, 22, 25]. Meanwhile,

© Springer Nature Switzerland AG 2019
O. Galinina et al. (Eds.): NEW2AN 2019/ruSMART 2019, LNCS 11660, pp. 256–268, 2019.
https://doi.org/10.1007/978-3-030-30859-9_22

in problems where the transmitter aims to maintain communication connectivity, the transmitter must keep its SINR greater or equal to a threshold value [6, 13, 16, 18, 19, 21, 24, 26].

In this paper, we consider a complimentary aspect of the connectivity problem. Namely, how incomplete information about the adversary's location can impact the rival strategies. To gain insight into this problem, we use the basic connectivity model suggested in [20, 21]. We generalize this model for the case where the transmitter does not know the exact adversary location, but knows only a priori distribution of the possible adversary locations. The problem is formulated as a non-zero Bayesian game between the transmitter and the adversary in Nash and Stackelberg framework for Rayleigh channel fading gains. Existence and uniqueness of both type of equilibria are proven, and are derived in closed form. For the convenience of the readers we give detailed derivation of the equilibrium strategies in closed form for the case of incomplete information, as well as complete information about the adversary's location. The organization of this paper is as follows: in Sect. 2, we present our basic connectivity communication model involving hostile interference. In Sect. 3, a Bayesian game with incomplete information about the adversary's location is investigated. In Sect. 4, a Bayesian Stackelberg game with incomplete information about the adversary's location is solved. Finally, in Sect. 5, discussion of the results is provided.

2 Basic Connectivity Model

Following [20, 21], we consider a wireless system in which one transmitter communicates directly to a single receiver. This system is affected by hostile interference from an adversary. Let G and H be the channel power gains from the transmitter to receiver, and interferer to receiver. In practice, the channel power gains depend on distances, fading and antenna characteristics. In general, the channel power gains are random variables (e.g., representing channel fading). Let p and q be the power levels used by the transmitter and interferer, respectively. Thus, the set of all non-negative numbers is the set of feasible strategies for the transmitter as well as for the adversary. The receiver also is affected by background noise power N. Thus, the SINR at receiver is given by

$$\mathrm{SINR}(p, q) = \frac{Hp}{N + Gq}. \tag{1}$$

We say that the communication from the transmitter to receiver is maintained if and only if the SINR at receiver is greater than or equal to a given SINR threshold ϵ, i.e., the following condition holds:

$$\mathrm{SINR}(p, q) \geq \epsilon.$$

This threshold ϵ depends on the system's requirements, such as bit rate and bit error rate (BER). Then, since G and H are random variables, the probability that the link between transmitter and receiver is maintained is given by

$$\pi_T(p, q) = I\!P\left(\mathrm{SINR}(p, q) \geq \epsilon\right). \tag{2}$$

Then, for the case of Rayleigh fading channels, i.e., G and H are exponential random variables with means $I\!E[G] = g$ and $I\!E[H] = h$, assuming accordingly to [20, 21] that background noise can be neglected, the probability that the link between transmitter and receiver is maintained (2) can be represented as follows:

$$\pi_T(p, q) = \frac{hp}{hp + \epsilon gq}, \tag{3}$$

while the probability that the link between transmitter and receiver is interrupted can be represented as follows:

$$\pi_A(p, q) = 1 - \pi_T(p, q) = \frac{\epsilon gq}{hp + \epsilon gq}. \tag{4}$$

3 Nash Game

In this section we consider the scenario where the transmitter does not know the adversary's exact location. The transmitter only knows that the adversary can be at distance d_i from the receiver with a priori known probability γ_i, for $i = 1, \ldots, n$, and $\sum_{i=1}^{n} \gamma_i = 1$. Such knowledge could have been obtained based on past observations of adversary's behavior. Equivalently, the transmitter knows that the gain of the adversary's channel to the destination is $g_i = g/d_i^2$ (with $g > 0$ some constant) with probability γ_i [27]. In the following, by *type-i adversary* we refer to an adversary employing strategy q_i and having fading channel gain g_i. Let the transmitter and adversary incur costs proportional to the power levels. Then, payoff of type-i adversary is given as follows:

$$u_{A,i}(p, q_i) = \pi_{A,i}(p, q_i) - C_A q_i, \tag{5}$$

where C_A is the jamming cost per a unit of power level and

$$\pi_{A,i}(p, q) = \frac{\epsilon g_i q}{hp + \epsilon g_i q}. \tag{6}$$

Let $q = (q_1, \ldots, q_n)$. Also, without loss of generality we can assume that adversary's types are arranged in decreasing order by destination to the receiver, i.e., $d_1 > d_2 > \ldots > d_n$, or, what is equivalent, in increasing order by fading channel gains, i.e., the following inequality holds:

$$0 = g_0 < g_1 < g_2 < \ldots < g_n. \tag{7}$$

The payoff for the transmitter is taken as the expected payoff, i.e.,

$$u_T(p, q) = \sum_{i=1}^{n} \gamma_i \pi_T(p, q_i) - C_T p, \tag{8}$$

where C_T is the transmission cost per a unit of power level.

We are looking for a Bayesian equilibrium [7]. Recall that (p, q) is a Bayesian equilibrium if and only if for any $(\tilde{p}, \tilde{q})$ the following inequalities hold:

$$u_T(\tilde{p}, q) \leq u_T(p, q) \text{ and } u_{A,i}(p, \tilde{q}_i) \leq u_{A,i}(p, q_i) \text{ for } i = 1, \ldots, n. \tag{9}$$

Denote this Bayesian game by $\Gamma^B_{T;\{A_i, \gamma_i\}}$.

To derive equilibrium strategies let us introduce the following auxiliary notation:

$$\xi_k \triangleq \sum_{i=k}^{n} \frac{\gamma_i}{\sqrt{g_i}} \left(\frac{1}{\sqrt{g_k}} - \frac{1}{\sqrt{g_i}} \right) \text{ for } k = 1, \ldots, n \tag{10}$$

and $\xi_0 \triangleq \infty$.

In the following proposition auxiliary properties of the sequence $\{\xi_k\}$ are given.

Proposition 1. (a) ξ_k is strictly decreasing from infinity for $k = 0$ to zero for $k = n$.

(b) There exists a unique $k_\star \in \{1, \ldots, n\}$ such that

$$\xi_{k_\star} \leq \frac{\epsilon C_T}{C_A h} < \xi_{k_\star - 1}. \tag{11}$$

Proof. Let $k < n$. Then, by (10), we have

$$\xi_k - \xi_{k+1} = \sum_{i=k}^{n} \frac{\gamma_i}{\sqrt{g_i}} \left(\frac{1}{\sqrt{g_k}} - \frac{1}{\sqrt{g_i}} \right) - \sum_{i=k+1}^{n} \frac{\gamma_i}{\sqrt{g_i}} \left(\frac{1}{\sqrt{g_k}} - \frac{1}{\sqrt{g_i}} \right)$$

$$= \left(\frac{1}{\sqrt{g_k}} - \frac{1}{\sqrt{g_{k+1}}} \right) \sum_{i=k+1}^{n} \frac{\gamma_i}{\sqrt{g_i}} > \text{(by (7))} > 0,$$

This and (7) yield (a). While (a) directly implies (b), and the result follows. ∎

Theorem 1. *The Bayesian game $\Gamma^B_{T;\{A_i, \gamma_i\}}$ has a unique Bayesian equilibrium, where*

$$p = \frac{\epsilon}{h C_A} \left(\frac{\sum\limits_{i=k_\star}^{n} (\gamma_i / \sqrt{g_i})}{\frac{C_T \epsilon}{C_A h} + \sum\limits_{i=k_\star}^{n} (\gamma_i / g_i)} \right)^2 \tag{12}$$

and

$$q_i = \frac{1}{h C_A \sqrt{g_i}} \frac{\sum\limits_{j=k_\star}^{n} (\gamma_j / \sqrt{g_j})}{\frac{C_T \epsilon}{C_A h} + \sum\limits_{j=k_\star}^{n} (\gamma_j / g_j)} \times \begin{cases} 1 - \dfrac{h}{\sqrt{g_i}} \dfrac{\sum\limits_{j=k_\star}^{n} (\gamma_j / \sqrt{g_j})}{\frac{C_T \epsilon}{C_A h} + \sum\limits_{j=k_\star}^{n} (\gamma_j / g_j)} & i \geq k_\star, \\ \\ 0, & i < k_\star \end{cases} \tag{13}$$

with $k_\star$ uniquely defined by (11).

Thus, not each jammer's type is active. Here we can observe a similarity with network protection games where an invader also might not intrude at each node [3,12].

Proof. By (9), $(p, \boldsymbol{q})$ is a Bayesian equilibrium if and only if each of these strategies is the best response strategy to the others, i.e.,

$$p = \mathrm{BR}_T(\boldsymbol{q}) = \underset{p \geq 0}{\mathrm{argmax}}\, u_T(p, \boldsymbol{q}), q_i = \mathrm{BR}_{A,i}(p) = \underset{q_i \geq 0}{\mathrm{argmax}}\, u_{A,i}(p, q_i), i = 1, \ldots, n.$$

(14)

Since $u_T(p, \boldsymbol{q})$ is concave in p, p is the best response strategy to $\boldsymbol{q}$ if and only if the following relation holds:

$$\frac{\partial u_T(p, \boldsymbol{q})}{\partial p} = \sum_{i=1}^{n} \gamma_i \frac{\epsilon h g_i q_i}{(hp + \epsilon g_i q_i)^2} - C_T \begin{cases} = 0, & p > 0, \\ \leq 0, & p = 0. \end{cases}$$

(15)

By (14), $\mathrm{BR}_{A,i}(p)$ is given as follows:

$$\mathrm{BR}_{A,i}(p) = \begin{cases} \sqrt{\dfrac{h}{\epsilon g_i C_A}p} - \dfrac{h}{\epsilon g_i}p, & p \leq \dfrac{\epsilon g_i}{C_A h}, \\ 0, & p > \dfrac{\epsilon g_i}{C_A h}. \end{cases}$$

(16)

By (7), for each p there exists a unique $k = k(p) \in \{1, \ldots, n\}$ such that

$$\frac{\epsilon g_{k-1}}{C_A h} < p \leq \frac{\epsilon g_k}{C_A h}.$$

(17)

Then, substituting (16) and (17) into (15) implies

$$\sum_{i=k}^{n} \gamma_i C_A q_i / p - C_T = C_A \sum_{i=k}^{n} \gamma_i \left(\sqrt{\frac{h}{\epsilon g_i C_A}} \frac{1}{\sqrt{p}} - \frac{h}{\epsilon g_i} \right) - C_T \begin{cases} = 0, & p > 0, \\ \leq 0, & p = 0. \end{cases}$$

(18)

Since $p > 0$, solving (18) by $\sqrt{p}$ we have that

$$\sqrt{p} = \frac{\sum_{i=k}^{n} \gamma_i \sqrt{\dfrac{hC_A}{\epsilon g_i}}}{C_T + C_A \sum_{i=k}^{n} \gamma_i \dfrac{h}{\epsilon g_i}}.$$

(19)

Substituting (19) into (17) yields

$$\sqrt{\frac{\epsilon g_{k-1}}{C_A h}} < \frac{\sum_{i=k}^{n} \gamma_i \sqrt{\dfrac{hC_A}{\epsilon g_i}}}{C_T + C_A \sum_{i=k}^{n} \gamma_i \dfrac{h}{\epsilon g_i}} \leq \sqrt{\frac{\epsilon g_k}{C_A h}}.$$

This is equivalent to

$$\sqrt{g_{k-1}} < \frac{\sum_{i=k}^{n} \gamma_i / \sqrt{g_i}}{\epsilon C_T / (C_A h) + \sum_{i=k}^{n} \gamma_i / g_i} \leq \sqrt{g_k}.$$

By (10), these inequalities are equivalent to

$$\xi_k = \sum_{i=k}^{n} \frac{\gamma_i}{\sqrt{g_i}} \left(\frac{1}{\sqrt{g_k}} - \frac{1}{\sqrt{g_i}} \right) \leq \frac{\epsilon C_T}{C_A h} < \sum_{i=k}^{n} \frac{\gamma_i}{\sqrt{g_i}} \left(\frac{1}{\sqrt{g_{k-1}}} - \frac{1}{\sqrt{g_i}} \right). \tag{20}$$

Note that

$$\sum_{i=k}^{n} \frac{\gamma_i}{\sqrt{g_i}} \left(\frac{1}{\sqrt{g_{k-1}}} - \frac{1}{\sqrt{g_i}} \right) = \frac{\gamma_{k-1}}{\sqrt{g_{k-1}}} \left(\frac{1}{\sqrt{g_{k-1}}} - \frac{1}{\sqrt{g_{k-1}}} \right)$$

$$+ \sum_{i=k}^{n} \frac{\gamma_i}{\sqrt{g_i}} \left(\frac{1}{\sqrt{g_{k-1}}} - \frac{1}{\sqrt{g_i}} \right) = \sum_{i=k-1}^{n} \frac{\gamma_i}{\sqrt{g_i}} \left(\frac{1}{\sqrt{g_{k-1}}} - \frac{1}{\sqrt{g_i}} \right) = \xi_{k-1}. \tag{21}$$

By (10) and (21), (20) is equivalent to $\xi_k \leq \epsilon C_T/(C_A h) < \xi_{k-1}$. Thus, $k = k_\star$ is uniquely given by (11). Substituting this $k = k_\star$ into (19) and solving it by p yields (12), while (16) implies (13), and the result follows. ∎

4 Stackelberg Game

In this section, we consider the scenario where the transmitter acts as the leader, but does not know the adversary's exact location, and while the adversary is the follower. The transmitter only knows that the adversary can be at a distance d_i from the receiver with a priori known probability γ_i, for $i = 1, \ldots, n$, and $\sum_{i=1}^{n} \gamma_i = 1$. Equivalently, the transmitter knows that the gain of the adversary's channel to the destination is $g_i = g/d_i^2$ (with $g > 0$ some constant) with probability γ_i [27]. In the following, we refer to the adversary employing strategy q_i and having fading channel gain g_i as a *type-i adversary*. Let $q = (q_1, \ldots, q_n)$. This is a Bayesian Stackelberg game. It can be solved in two steps by backward induction. *In the first step* of the two-level game, for a fixed p, assigned by the transmitter, each type-i adversary, where $i = 1, \ldots, n$, tries to maximize his payoff. Thus, a type-i adversary applies strategy $q_i = \mathrm{BR}_{A,i}(p)$, which is given in closed form by (16). *In the second step* of the two-level game, the payoff for the transmitter is taken as the expected payoff

$$F_T(p) = \sum_{i=1}^{n} \gamma_i u_T(p, \mathrm{BR}_{A,i}(p)). \tag{22}$$

The transmitter selects p to get a maximal payoff, i.e., to solve the following optimization problem:

$$p = \underset{p \geq 0}{\mathrm{argmax}}\, F_T(p). \tag{23}$$

Such $(p, q_i = \mathrm{BR}_{A,i}(p), i = 1, \ldots, n)$ is called a Bayesian Stackelberg equilibrium. Denote this Stackelberg game with incomplete information about the adversary's location by $\Gamma^S(T; \{A_i, \gamma_i\})$. To derive equilibrium strategies, let us introduce the following auxiliary notations:

$$\underline{\varphi}_k \triangleq \frac{\sqrt{g_{k-1}}}{\sum\limits_{i=k}^{n}(\gamma_i/\sqrt{g_i})} \quad \text{for } k = 1, \ldots, n, \text{ and } \underline{\varphi}_{n+1} \triangleq \infty, \tag{24}$$

$$\overline{\varphi}_k \triangleq \frac{\sqrt{g_k}}{\sum\limits_{i=k}^{n}(\gamma_i/\sqrt{g_i})} \quad \text{for } k = 1, \ldots, n, \tag{25}$$

In the following proposition, auxiliary properties of sequences $\{\underline{\varphi}_k\}$ and $\{\overline{\varphi}_k\}$ are given.

Proposition 2. (a) $\underline{\varphi}_k$ is strictly increasing on k from zero for $k = 0$ to infinity
for $k = n + 1$.
(b) $\overline{\varphi}_k$ is strictly increasing on k.
(c) For any $k \in \{1, \ldots, n\}$ the following inequalities hold:

$$\overline{\varphi}_{k-1} < \underline{\varphi}_k < \overline{\varphi}_k. \tag{26}$$

(d) There exists a unique $k_\# \in \{1, \ldots, n\}$ such that either

$$\underline{\varphi}_{k_\#} < \frac{hC_A}{2\epsilon C_T} < \overline{\varphi}_{k_\#} \tag{27}$$

or

$$\overline{\varphi}_{k_\#} \leq \frac{hC_A}{2\epsilon C_T} \leq \underline{\varphi}_{k_\#+1}. \tag{28}$$

Proof. Since $\sum\limits_{i=k}^{n}\gamma_i(1/\sqrt{g_i})$ is decreasing on k, then, by (7), (25) and (24) are increasing in k as products of two increasing function sequences, and (a) with (b) follow. By (7), (24) and (25), we have that

$$\underline{\varphi}_k < \overline{\varphi}_k. \tag{29}$$

By (24) and (25)

$$\underline{\varphi}_k - \overline{\varphi}_{k-1} = \frac{\gamma_{k-1}}{\sum\limits_{i=k}^{n}(\gamma_i/\sqrt{g_i}) \sum\limits_{i=k-1}^{n}(\gamma_i/\sqrt{g_i}} > 0. \tag{30}$$

Then, (29) and (30) yield (26), and (c) follows. (d) straightforward follows from (a)–(c). ∎

Theorem 2. The Stackelberg game $\Gamma^S(T; \{A_i, \gamma_i\})$ has a unique equilibrium $(p, q_i = BR_{A,i}(p), \ldots, BR_{A,n}(p), i = 1, \ldots, n)$ where

$$p = \begin{cases} \dfrac{C_A}{4\epsilon C_T^2}\left(\sum\limits_{i=k_\#}^{n}(\gamma_i/\sqrt{g_i})\right)^2, & \text{if (27) holds,} \\ \dfrac{\epsilon g_{k_\#}}{C_A h}, & \text{if (28) holds} \end{cases} \tag{31}$$

with $k_\#$ uniquely defined by (27) and (28), and

$$q_i = \begin{cases} \sqrt{\dfrac{h}{\epsilon g_i C_A}} p - \dfrac{h}{\epsilon g_i} p, & i \geq k_\#, \\ 0, & i < k_\#. \end{cases} \tag{32}$$

Proof: By (7) and (16), for each p there is $k = k(p)$ such that

$$F_T(p) = -C_T p + \begin{cases} \displaystyle\sum_{i=k}^{n} \gamma_i \sqrt{\dfrac{C_A h p}{\epsilon g_i}}, & \dfrac{\epsilon g_{k-1}}{C_A h} < p \leq \dfrac{\epsilon g_k}{C_A h}, \\ 0, & \dfrac{\epsilon g_n}{C_A h} < p. \end{cases} \tag{33}$$

Our goal is to find p where $F_T(p)$ achieves its maximum, i.e., to solve the problem (23). By (33), $F_T(p)$ is a continuous and piece-wise differentiable function and its derivative can have jumps only at a finite set points $\{\epsilon g_1/(C_A h) < \dots < \epsilon g_n/(C_A h)\}$. Let p be such that $F_T(p)$ achieves its maximum. Then, to derive such p, we have to consider separately two cases: (A) $\epsilon g_{k-1}/(C_A h) < p < \epsilon g_k/(C_A h)$ and (B) $p = \epsilon g_k/(C_A h)$.

(A) Let

$$\frac{\epsilon g_{k-1}}{C_A h} < p < \frac{\epsilon g_k}{C_A h}. \tag{34}$$

Then

$$\frac{dF_T(p)}{dp} = \frac{1}{2\sqrt{p}} \sum_{i=k}^{n} \gamma_i \sqrt{\frac{C_A h}{\epsilon g_i}} - C_T. \tag{35}$$

Thus,

$$\frac{dF_T(p)}{dp} = 0 \tag{36}$$

if and only if

$$\sqrt{p} = \frac{1}{2C_T} \sum_{i=k}^{n} \gamma_i \sqrt{\frac{C_A h}{\epsilon g_i}}. \tag{37}$$

Moreover,

$$\frac{dF_T(p)}{dp} \begin{cases} > 0, & \sqrt{p} < \dfrac{1}{2C_T} \displaystyle\sum_{i=k}^{n} \gamma_i \sqrt{\dfrac{C_A h}{\epsilon g_i}}, \\ = 0, & \sqrt{p} = \dfrac{1}{2C_T} \displaystyle\sum_{i=k}^{n} \gamma_i \sqrt{\dfrac{C_A h}{\epsilon g_i}}, \\ < 0, & \sqrt{p} > \dfrac{1}{2C_T} \displaystyle\sum_{i=k}^{n} \gamma_i \sqrt{\dfrac{C_A h}{\epsilon g_i}}. \end{cases} \tag{38}$$

Thus, (34) and (36) holds if and only if

$$\sqrt{\frac{\epsilon g_{k-1}}{C_A h}} < \frac{1}{2C_T} \sum_{i=k}^{n} \gamma_i \sqrt{\frac{C_A h}{\epsilon g_i}} < \sqrt{\frac{\epsilon g_k}{C_A h}} \tag{39}$$

Taking into account notations (24) and (25), the last inequalities are equivalent to

$$\underline{\varphi}_k = \frac{\sqrt{g_{k-1}}}{\sum\limits_{i=k}^{n} (\gamma_i/\sqrt{g_i})} < \frac{hC_A}{2C_T\epsilon} < \frac{\sqrt{g_k}}{\sum\limits_{i=k}^{n} (\gamma_i/\sqrt{g_i})} = \overline{\varphi}_k, \tag{40}$$

(B) Let

$$p = \frac{\epsilon g_k}{C_A h} \tag{41}$$

Then, for p to be point of maximum $F_T(p)$ has to be increasing for $p < \epsilon g_k/(C_A h)$ and decreasing for $p > \epsilon g_k/(C_A h)$, or, by (38), the following relation have to hold:

$$\lim_{p \uparrow \epsilon g_k/(C_A h)} \frac{dF_T(p)}{dp} \geq 0, \tag{42}$$

$$\lim_{p \downarrow \epsilon g_k/(C_A h)} \frac{dF_T(p)}{dp} \leq 0. \tag{43}$$

By (38), (42) is equivalent to

$$\sqrt{\frac{\epsilon g_k}{C_A h}} \leq \frac{1}{2C_T} \sum_{i=k}^{n} \gamma_i \sqrt{\frac{C_A h}{\epsilon g_i}} \tag{44}$$

while (43) is equivalent to

$$\sqrt{\frac{\epsilon g_k}{C_A h}} \geq \frac{1}{2C_T} \sum_{i=k+1}^{n} \gamma_i \sqrt{\frac{C_A h}{\epsilon g_i}}. \tag{45}$$

Taking into account notations (24) and (25), inequalities (44) and (45) can be presented in the following equivalent form:

$$\frac{\sqrt{g_k}}{\sum\limits_{i=k}^{n} (\gamma_i/\sqrt{g_i})} \leq \frac{hC_A}{2\epsilon C_T} \leq \frac{\sqrt{g_k}}{\sum\limits_{i=k+1}^{n} (\gamma_i/\sqrt{g_i})}. \tag{46}$$

By (40), (46) and Proposition 2, $k = k_\#$, and such $k_\#$ is defined uniquely. Solving (37) by p and taking into account (41) imply (31), and the result follows. ∎

5   Discussion of the Results

In this paper, we have proven that in a Bayesian connectivity game with incomplete information about the interfering adversary's location, and where the channels have Rayleigh fading gains, the Nash and Stackelberg equilibria exist and are unique. Moreover, Theorem 1 and 2 reveal an important difference between these Nash and Stackelberg equilibrium. By (12), the Nash transmitter equilibrium strategy generally is not monotonic with respect to a priori probabilities while (31) holds, yet the Stackelberg transmitter equilibrium strategy can be constant with respect to a priori probabilities while condition (28) holds. Moreover, the Stackelberg equilibrium strategy has essentially more switching points than the Nash equilibrium strategy, namely, $2n - 1$ switching points for the Stackelberg equilibrium strategy and and $n - 1$ switching points for Nash equilibrium strategy. We provide an illustration of the equilibrium strategies for the case $n = 2$. Here, $\gamma_1 = \gamma$ and $\gamma_2 = \overline{\gamma} = 1 - \gamma$. Then

$$\xi_1 = \frac{\overline{\gamma}}{\sqrt{g_2}}\left(\frac{1}{\sqrt{g_1}} - \frac{1}{\sqrt{g_2}}\right) \text{ and } \xi_2 = 0,$$

$$\underline{\varphi}_1 = 0, \underline{\varphi}_2 = \sqrt{g_1 g_2}/\overline{\gamma}, \ \overline{\varphi}_1 = \frac{\sqrt{g_1}}{\gamma/\sqrt{g_1} + \overline{\gamma}/\sqrt{g_2}} \text{ and } \overline{\varphi}_2 = \frac{g_2}{\overline{\gamma}}.$$

Thus, the Stackelberg equilibrium strategy has three switching points while the Nash equilibrium strategy has only one switching point. Moreover, by Theorem 1, the transmitter Nash equilibrium strategy is given as follows:

$$p = \frac{\epsilon}{hC_A} \times \begin{cases} \left(\dfrac{\overline{\gamma}/\sqrt{g_2}}{\dfrac{\epsilon C_T}{C_A h} + \overline{\gamma}/g_2}\right)^2, & \dfrac{\epsilon C_T}{C_A h} < \dfrac{\overline{\gamma}}{\sqrt{g_2}}\left(\dfrac{1}{\sqrt{g_1}} - \dfrac{1}{\sqrt{g_2}}\right), \\[6mm] \left(\dfrac{\gamma/\sqrt{g_1} + \overline{\gamma}/\sqrt{g_2}}{\dfrac{\epsilon C_T}{C_A h} + \gamma/g_1 + \overline{\gamma}/g_2}\right)^2, & \dfrac{\epsilon C_T}{C_A h} > \dfrac{\overline{\gamma}}{\sqrt{g_2}}\left(\dfrac{1}{\sqrt{g_1}} - \dfrac{1}{\sqrt{g_2}}\right), \end{cases}$$

while, by Theorem 2, the transmitter Stackelberg equilibrium strategy is given as follows:

$$p = \begin{cases} \dfrac{C_A\left(\gamma\sqrt{g_2} + \overline{\gamma}\sqrt{g_1}\right)^2}{4\epsilon C_T^2 g_1 g_2}, & \dfrac{C_A h}{\epsilon C_T} < \dfrac{\sqrt{g_1}}{\gamma/\sqrt{g_1} + \overline{\gamma}/\sqrt{g_2}}, \\[5mm] \dfrac{\epsilon g_1}{C_A h}, & \dfrac{\sqrt{g_1}}{\gamma/\sqrt{g_1} + \overline{\gamma}/\sqrt{g_2}} \leq \dfrac{C_A h}{\epsilon C_T} \leq \dfrac{\sqrt{g_1 g_2}}{\overline{\gamma}}, \\[5mm] \dfrac{C_A \overline{\gamma}^2}{4\epsilon C_T^2 g_2}, & \dfrac{\sqrt{g_1 g_2}}{\overline{\gamma}} < \dfrac{C_A h}{\epsilon C_T} < \dfrac{g_2}{\overline{\gamma}}, \\[5mm] \dfrac{\epsilon g_2}{C_A h}, & \dfrac{g_2}{\overline{\gamma}} \leq \dfrac{C_A h}{\epsilon C_T}. \end{cases}$$

Fig. 1(a)–(f) illustrate the dependence of the equilibrium strategies on a priori probability γ and the threshold value ϵ. The larger threshold value ϵ, then the harder it is for the transmitter to maintain communication. This is reflected by a decrease in the transmitter's payoff with respect to an increase in the threshold. Since $g_1 > g_2$, the type-1 adversary is located further from the receiver than type-2 adversary. That is why an increase in a priori probability γ, that type-1 adversary occurs, leading to an increase in the transmitter's payoff (Fig. 1(h) and (g)). The goal of our future work is to relax assumption on channel gains from Rayleigh fading to more general channel models.

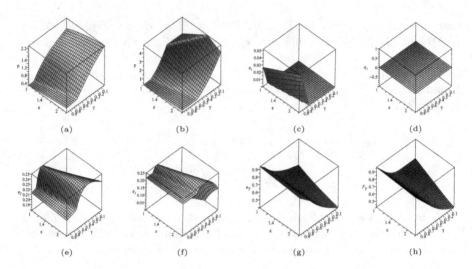

Fig. 1. (a) Nash transmitter strategy, (b) Stackelberg transmitter strategy strategy, (c) Nash type-1 adversary strategy, (d) Stackelberg type-1 adversary strategy, (e) Nash type-2 adversary strategy, (f) Stackelberg type-2 adversary strategy (g) payoff to the transmitter in Nash game, (h) payoff to the transmitter in Stackelberg game for $n = 2$, $g_1 = 0.3$, $g_2 = 3$, $h = 1$, $C_T = 0.1$, $C_A = 1$.

References

1. Abuzainab, N., Saad, W.: Dynamic connectivity game for adversarial internet of battlefield things systems. IEEE Internet Things J. **5**, 378–390 (2018)
2. Amariucai, G.T., Wei, S., Kannan, R.: Minimax and maxmin solutions to gaussian jamming in block-fading channels under long term power constraints. In: 41st Annual Conference on Information Sciences and Systems (CISS), pp. 312–317 (2007)
3. Baston, V.J., Garnaev, A.Y.: A search game with a protector. Naval Res. Logistics **47**, 85–96 (2000)
4. Cumanan, K., Ding, Z., Xu, M., Poor, H.V.: Secure multicast communications with private jammers. In: 17th IEEE International Workshop on Signal Processing Advances in Wireless Communications (SPAWC), Edinburgh, UK (2016)
5. Dabcevic, K., Betancourt, A., Marcenaro, L., Regazzoni, C.S.: Intelligent cognitive radio jamming - a game-theoretical approach. EURASIP J. Adv. Signal Process. **2014**, 171 (2014)
6. El-Bardan, R., Brahma, S., Varshney, P.K.: Power control with jammer location uncertainty: a game theoretic perspective. In: 48th Annual Conference on Information Sciences and Systems (CISS), pp. 1–6 (2014)
7. Fudenberg, D., Tirole, J.: Game Theory. MIT Press, Boston (1991)
8. Gao, Y., Xiao, Y., Wu, M., Xiao, M., Shao, J.: Game theory-based anti-jamming strategies for frequency hopping wireless communications. IEEE Trans. Wireless Commun. **17**, 5314–5326 (2018)

9. Garnaev, A., Hayel, Y., Altman, E., Avrachenkov, K.: Jamming game in a dynamic slotted ALOHA network. In: Jain, R., Kannan, R. (eds.) GameNets 2011. LNICST, vol. 75, pp. 429–443. Springer, Heidelberg (2012). https://doi.org/10.1007/978-3-642-30373-9_30

10. Garnaev, A., Liu, Y., Trappe, W.: Anti-jamming strategy versus a low-power jamming attack when intelligence of adversary's attack type is unknown. IEEE Trans. Signal Inf. Process. Netw. **2**, 49–56 (2016)

11. Garnaev, A., Trappe, W.: To eavesdrop or jam, that is the question. In: Mellouk, A., Sherif, M.H., Li, J., Bellavista, P. (eds.) ADHOCNETS 2013. LNICSSITE, vol. 129, pp. 146–161. Springer, Cham (2014). https://doi.org/10.1007/978-3-319-04105-6_10

12. Garnaev, A., Trappe, W.: A bandwidth monitoring strategy under uncertainty of the adversary's activity. IEEE Trans. Inf. Forensics Secur. **11**, 837–849 (2016)

13. Garnaev, A., Trappe, W.: Stability of communication link connectivity against hostile interference. In: 2017 IEEE Global Conference on Signal and Information Processing (GlobalSIP), pp. 136–140 (2017)

14. Garnaev, A., Trappe, W.: The rival might be not smart: revising a CDMA jamming game. In: IEEE Wireless Communications and Networking Conference (WCNC) (2018)

15. Garnaev, A., Trappe, W., Petropulu, A.: Equilibrium strategies for an OFDM network that might be under a jamming attack. In: 51st Annual Conference on Information Systems and Sciences (CISS), pp. 1–6 (2017)

16. Hajimirsaadeghi, M., Mandayam, N.B.: A dynamic colonel Blotto game model for spectrum sharing in wireless networks. In: 55th Annual Allerton Conference on Communication, Control, and Computing (Allerton), pp. 287–294 (2017)

17. Han, Z., Niyato, D., Saad, W., Basar, T., Hjrungnes, A.: Game Theory in Wireless and Communication Networks: Theory, Models, and Applications. Cambridge University Press, Cambridge (2012)

18. Liu, Y., Garnaev, A., Trappe, W.: Maintaining throughput network connectivity in ad hoc networks. In: 41th IEEE International Conference on Acoustics, Speech and Signal Processing (ICASSP), pp. 6380–6384 (2016)

19. Namvar, N., Saad, W., Bahadori, N., Kelleys, B.: Jamming in the internet of things: a game-theoretic perspective. In: IEEE Global Communications Conference (GLOBECOM) (2016)

20. Nguyen, G.D., Kompella, S., Kam, C., Wieselthier, J.E., Ephremides, A.: Wireless link connectivity under hostile interference: nash and stackelberg equilibria. In: 14th International Symposium on Modeling and Optimization in Mobile, Ad Hoc, and Wireless Networks (WiOpt) (2016)

21. Nguyen, G.D., Kompella, S., Kam, C., Wieselthier, J.E., Ephremides, A.: Impact of hostile interference on wireless link connectivity. IEEE Trans. Control Netw. Syst. **5**, 1445–1456 (2018)

22. Song, T., Stark, W.E., Li, T., Tugnait, J.K.: Optimal multiband transmission under hostile jamming. IEEE Trans. Commun. **64**, 4013–4027 (2016)

23. Vadlamania, S., Eksioglub, B., Medala, H., Nandia, A.: Jamming attacks on wireless networks: a taxonomic survey. Int. J. Prod. Econ. **172**, 76–94 (2016)

24. Wu, Y., Wang, B., Liu, K.J.R.: Optimal power allocation strategy against jamming attacks using the colonel Blotto game. In: IEEE Global Telecommunications Conference (GLOBECOM), pp. 1–5 (2009)

25. Yang, D., Xue, G., Zhang, J., Richa, A., Fang, X.: Coping with a smart jammer in wireless networks: a Stackelberg game approach. IEEE Trans. Wireless Commun. **12**, 4038–4047 (2013)

26. Yu, Y., Sankar, R.: A cooperative game-theoretic approach to cellular network hand-off. In: Wireless Telecommunications Symposium (WTS), pp. 312–317 (2010)
27. Zhu, Q., Saad, W., Han, Z., Poor, H.V., Basar, T.: Eavesdropping and jamming in next-generation wireless networks: a game-theoretic approach. In: Military Communications Conference (MILCOM), pp. 119–124 (2011)

Ray-Based Modeling of Unlicensed-Band mmWave Propagation Inside a City Bus

Aleksei Ponomarenko-Timofeev[1], Aleksandr Ometov[1], and Olga Galinina[1,2(✉)]

[1] Tampere University, Tampere, Finland
{aleksei.ponomarenko-timofeev,aleksandr.ometov,olga.galinina}@tuni.fi
[2] Peoples' Friendship University of Russia (RUDN University), Moscow, Russia

Abstract. In the wake of recent hardware developments, augmented, mixed, and virtual reality applications – grouped under an umbrella term of eXtended reality (XR) – are believed to have a transformative effect on customer experience. Among many XR use cases, of particular interest are crowded commuting scenarios, in which passengers are involved in in-bus/in-train entertainment, e.g., high-quality video or 3D hologram streaming and AR/VR gaming. In the case of a city bus, the number of commuting users during the busy hours may exceed forty, and, hence, could pose far higher traffic demands than the existing microwave technologies can support. Consequently, the carrier candidate for XR hardware should be sought in the millimeter-wave (mmWave) spectrum; however, the use of mmWave cellular frequencies may appear impractical due to the severe attenuation or blockage by the modern metal coating of the glass. As a result, intra-vehicle deployment of unlicensed mmWave access points becomes the most promising solution for bandwidth-hungry XR devices. In this paper, we present the calibrated results of shooting-and-bouncing ray simulation at 60 GHz for the bus interior. We analyze the delay and angular spread, estimate the parameters of the Saleh-Valenzuela channel model, and draw important practical conclusions regarding the intra-vehicle propagation at 60 GHz.

Keywords: mmWave · SBR · Channel model · Wearables · Intra-vehicular

1 Introduction

The rising consumer adoption of smart wearable devices is fueling the rapid growth of the wearable technology market, the revenues of which are expected to reach 30 billion by 2023 according to a CCS Insight forecast [1]. While a significant share of the heterogeneous wearables market is made up of gadgets with rather modest communication requirements (e.g., watches, wristbands, hearables, smart clothing, etc.), bandwidth-hungry devices dealing with video processing (head-mounted displays, smart glasses, and wearable cameras) are gaining increasing popularity as virtual, mixed, and augmented reality (AR/MR/VR) become mainstream.

© Springer Nature Switzerland AG 2019
O. Galinina et al. (Eds.): NEW2AN 2019/ruSMART 2019, LNCS 11660, pp. 269–281, 2019.
https://doi.org/10.1007/978-3-030-30859-9_23

Until recently, VR remained only a niche concept, its mass adoption severely hampered by the wired connections between the headset and external hardware, but this changed in 2018 by introducing a whole generation of untethered headsets. Although AR and MR devices have not been limited by the wires, their constraints come from lower hardware capabilities that may impair the user experience; fortunately, the processing capacity can be enhanced today by, e.g., computational offloading to the network edge [2,3].

In 2019, multiple hardware suppliers are already offering and shipping high-end wireless VR devices. The list of available products includes Oculus Quest (Project Santa Cruz), Lenovo Mirage Solo, HTC Vive Focus, DPVR M2 PRO, GenBasic Quad HD, and Xiaomi Mi VR. Some vendors have also supplemented older versions of headsets with wireless adapters, such as Intel's Wireless gigabit, TPCAST Wireless Adapter, and the HTC VIVE Wireless Adapter for same-brand devices, as well as the DisplayLink Wireless VR adapter. Just as important, the more affordable price (starting at $200 for some untethered headsets) could potentially serve as the main catalyst driving VR technology to the tipping point for the mass adoption [4].

In the wake of recent hardware developments, AR, MR, and VR applications – grouped under an umbrella term of *eXtended reality* (XR) [5,6] – are believed to have a transformative effect on customer experience and, in general, the way people interact with real and digital worlds. In that regard, successful operation of XR devices imposes stringent communication requirements that include extremely high throughput to the processing unit [7] and ultra-low over-the-air latency as the total latency of the rendered image should be less than 15 ms to avoid motion sickness [8].

Beyond controversy, the growing density of demanding XR devices will result in increased traffic congestion, which would be impossible to resolve with the existing capacity of microwave networks. To ameliorate the situation, we may let the wireless XR devices take advantage of the high volume of spectrum available in millimeter-wave (mmWave) bands. The best candidate in the unlicensed mmWave frequency range is the 60 GHz band, where the channel access may be controlled by the IEEE 802.11ad (2012) or brand-new IEEE 802.11ay (expected to be ratified in 2020) protocols [9,10].

Among several 802.11ad/ay use cases the IEEE 802.11 TGay Group adopted [11], of particular interest are crowded commuting scenarios, in which passengers are involved in in-bus/in-train entertainment, e.g., high-quality video or 3D hologram streaming and AR/VR gaming. To be able to predict the achievable performance of mmWave networks for these settings reliably, researchers need to develop a better understanding of the properties of the signal propagation in similar environments. The existing studies on intra-vehicular radio propagation are somewhat limited and address mainly the microwave spectrum: For example, the results of 2.4 GHz channel measurements and simulations inside a personal vehicle may be found in [12–14], while in [15–17], the authors build a propagation map for the airplane using ray-tracing simulation at 2.4 and 5.25 GHz. For the 28 GHz frequency, a study on the multiple-input-multiple-output (MIMO) performance in a subway tunnel is presented in [18].

In this paper, we make a first and important step toward the understanding of the structure of the 60 Hz channel for the intra-vehicle environment using an example of a city bus. We rely on the results of Wireless Insite, which is adopted by the community as a reliable ray-tracing simulator and based on the shooting-and-bouncing ray (SBR) method, and we also calibrate the obtained results with the outcomes of our measurement campaign in [19]. Our main contribution lies in the analysis of the variation of delay and angular spreads in the radio channel and in estimating the parameters of the Saleh-Valenzuela channel model, which can be used in subsequent higher-level studies for synthesizing the channel impulse responses (CIRs).

The rest of the paper is organized as follows. Section 2 describes the scenario of interest and the calibration procedure. In Sect. 3, we provide selected numerical results related to the delay and angular spreads in our system. Section 4 presents the method of estimating parameters of the Saleh-Valenzuela statistical channel model, while our main conclusions are drawn in the last section.

2 Simulation Scenario and Calibration with Measurements

In this work, we rely on simulation results obtained by the community-approved commercial tool, Wireless Insite [20], which exploits SBR method. Here, we provide a detailed description of our intra-vehicle, city bus scenario represented by a 3D model, and also verify the choice of selected simulation parameters by calibrating with the outcomes of our measurement campaign in [19].

The 3D model of the bus is designed to match the bus interior and its dimensions (2.5 by 2.3 m in cross-section and 12.5 m length). The materials of individual model components and their electromagnetic properties are also adjusted according to the reference environment of the measurement campaign. Figure 1 depicts the wireframe of our 3D model that replicates the actual bus interior and consists of four color-coded materials: particularly, window glass sections are colored green, metal parts (handles and railings) are black, interior body (plastic) is colored orange, and seats (velvet) are shown in blue. The corresponding material parameters, relative dielectric permittivity ϵ, conductivity σ, and thickness d, are given in Table 1.

Table 1. Material properties

Material	Conductivity, σ [S/m]	Permittivity, ϵ	Thickness, d [mm]
Glass (Window)	0.5674	6.27	7
Plastic (Body)	0.0586	1.50	3
Metal (Rails)	Perfect conductor		
Velvet (Seats)	Perfect absorber		

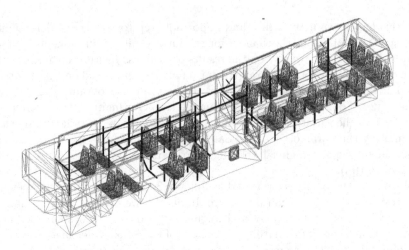

Fig. 1. Color-coded wireframe of the bus model: blue color corresponds to seat coating (velvet), green describes windows (glass), and red and black cover the bus body (plastic) and rails (metal). (Color figure online).

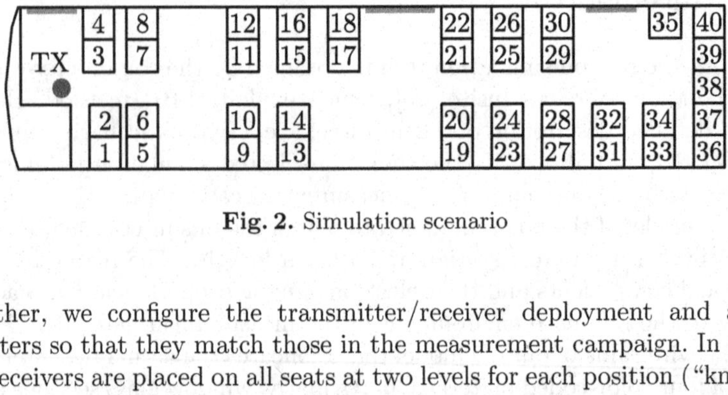

Fig. 2. Simulation scenario

Further, we configure the transmitter/receiver deployment and antenna parameters so that they match those in the measurement campaign. In our scenario, receivers are placed on all seats at two levels for each position ("knee" and "headset") that are suggested in the reference paper and represent a wearable headset and a handheld device. The top-down view of our scenario, illustrating the locations of the receivers and the transmitter, is shown in Fig. 2. The core parameters are given in Table 2.

We calibrate our simulation environment with the measurement data in [19] by comparing the logarithmic path loss models that describe both sets of the path loss values. Using the SBR simulation data, we first estimate the coefficients of the distance-dependent path loss function

$$L(d) = \alpha + 10\beta \log_{10}(d) + \chi(0, \sigma), \tag{1}$$

where α is the reference path loss value at the distance of 1 m, β is the path loss exponent, which demonstrates how the signal attenuation evolves with the distance, and $\chi(0, \sigma)$ is a Gaussian random variable with standard deviation σ that corresponds to the channel fading.

Table 2. Receiver/transmitter properties [19]

Parameter	Transmitter	Receiver
Antenna type	Omnidirectional	
Half Power Beamwidth (HPBW)	40°	
Max. gain	0 [dBi]	
Polarization	Vertical	
Input power	0 [dBm]	—
Transmission line loss	0 [dBi]	—
Noise figure	—	3 [dBi]
Waveform	Sinusiod	
Center frequency	60 GHz	
Height	2.1 [m]	[0.7,1.7] [m]

The comparison of the two path loss models is illustrated in Fig. 3. The coefficients obtained by regression are the following: $\alpha_{sim} = 88.66$ dB, $\alpha_{meas} = 85.28$ dB, $\beta_{sim} = 1.06$, $\beta_{meas} = 1.74$, $\sigma_{sim} = 2.54$, and $\sigma_{meas} = 6.93$. For our data set, α differs from that of the reference path loss model only by 3 dB; however, standard deviation σ_{meas} and exponent β_{meas} exceed the respective values obtained through the simulation, which can be explained by the fact that not every small detail is included in the model to optimize the simulation time.

In addition, we note that the parameters of the X3D (extensible 3D) engine performing the SBR simulation are configured to record three reflections, one diffraction, and transmission per path. In total, the engine records up to 250 propagation paths with a noise floor of −140 dBm. Ray spacing for the geometry engine is set to 0.2°. This choice of the engine parameters allows us to maintain the required precision while keeping the simulation times under eight hours for Intel Core 2 Quad CPU (model Q9500), 8 GB of RAM, nVidia GTX 550 Ti GPU, and 200 GB HDD (WD Blue at 5400 RPM).

3   RMS Delay and Angular Spread

In this section, we study the angular and delay spreads obtained from the simulation data. This analysis may help to estimate the coherence bandwidth and also explore, whether one may potentially mitigate the inter-symbol interference (ISI) by adjusting an antenna array pattern.

3.1   Delay Spread Analysis

One of the key metrics for wireless channels is the coherence bandwidth that allows estimating the frequencies, over which the channel has a flat frequency response. The information on the expected channel coherence bandwidth can

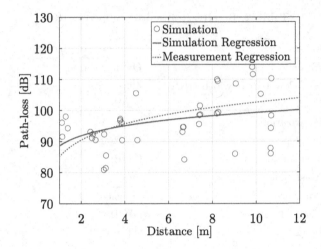

Fig. 3. Calibration results: comparison of the simulation path loss data (blue color) and the reference logarithmic path loss model that is based on measurements (dotted red curve) (Color figure online).

be used in designing hardware for the communication systems. Specifically, it may help the engineer in estimating the impact of ISI and the complexity of an equalizer. The coherence bandwidth can be calculated as an inverse of root-mean-square (RMS) delay spread σ_τ, which is defined as

$$\sigma_\tau = \sqrt{\frac{\sum\limits_{i=0}^{N} p_i \cdot (\tau_i - \hat{\tau})^2}{\sum\limits_{i=0}^{N} p_i}}, \tag{2}$$

where N is the total number of propagation paths, p_i is power delivered over path i, τ_i is the arrival time of path i, and $\hat{\tau}$ is the earliest arrival time in the CIR.

We calculate the delay spread for both positions (higher "headset" and lower "knee") of the receivers. While the difference between the two positions is relatively small in terms of the distance to the receiver, no line-of-sight (LOS) path

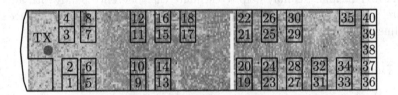

Fig. 4. Received power heat-map

exists in the case of the "knee" level. To obtain the data set required for analyzing the delay (and also angular) spread, we cover the entire plane by receivers at the "waist" (pocket) level, which would not be possible to accomplish during a measurement campaign.

For "waist"-level positions, we observe that the delay spread decreases with the distance between the transmitter and receiver, which is caused primarily by the fact that at low distances, the radiation pattern of the antenna impedes the communication over LOS links and results in multiple NLOS components. Also, since the root mean square (RMS) delay spread is a weighted variance of the arrival time (with the component power being the weight), even if a weak LOS component is present, it will have the weight comparable to those of the NLOS components.

In Fig. 5, we observe that at short ranges, the delay spread is higher due to the antenna patterns of both receiver and transmitter; irregularities in the environment explain a small rise at a range of 7–9 m. Moreover, the positions at "knee" level have lower delay spread, which may have been caused by the following two factors. First, the range of paths, over which signal may propagate, is limited since reflections can mainly occur either from the windows or the floor. Another important factor is the presence of a perfect absorber (velvet seats), which also reduces the number of propagation paths and ultimately leads to the narrower range of the path length and, hence, a decrease in the delay spread.

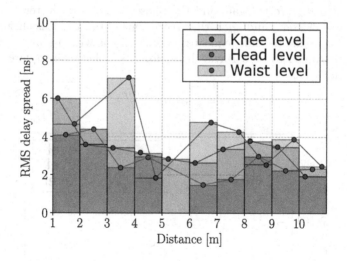

Fig. 5. Delay spread vs. distance

3.2 Angular Spread Analysis

Further, we study the root mean square (RMS) angular spread, which may serve in estimating the coherence distance of a channel that, in turn, helps to determine, whether specific hardware can exploit spatial diversity.

The histogram of RMS angular spread in Fig. 6 shows that at short ranges, the angular spread is smaller for the "headset" level. Coupled with the previous result, it may indicate that filtering undesired MPCs by tuning the receive antenna pattern might be unfeasible. For the "knee" level, however, the angular spread mostly maintains relatively consistent values. For all locations at the "waist" level, the angular spread increases with the distance since the number of interactions (reflection, diffraction) increases. However, due to the blockage by the seats (even at short ranges), the change in the angular spread is not as pronounced as for the "head" level.

4   Saleh-Valenzuela Statistical Model

In this section, we analyze the CIRs produced by the simulation and estimate the parameters of Saleh-Valenzuela clustering model [21]. The Saleh-Valenzuela model is a statistical multipath model, where the received rays arrive in clusters so that the clusters and the rays arrivals form a Poisson process and amplitudes decay exponentially. The model generates artificial CIRs corresponding to specific environment type and defines the CIR through the Dirac delta function as follows

$$h(t) = \sum_{j=0}^{\infty} \sum_{k=0}^{\infty} \beta_{jk} e^{i\theta_{jk}} \delta(t - T_j - \tau_{jk}), \tag{3}$$

where β_{jk} and θ_{jk} are the gain and the phase of ray k in cluster j, T_j is the arrival time of cluster j and τ_{jk} is the arrival time of ray k in this cluster. While the phase θ is a uniform independent random variable within a range of $(0, 2\pi)$, gain β_{jk} is assumed to be a Rayleigh random variable. Mean square values of β_{jk} can be found as

$$\overline{\beta^2(T_j, \tau_{jk})} = \overline{\beta^2(0,0)} e^{-T_j/\Gamma} e^{-\tau_{jk}/\gamma}, \tag{4}$$

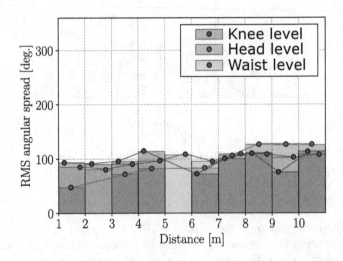

Fig. 6. Angular spread vs. distance

where $\overline{\beta^2(0,0)}$ is the first ray power in the first cluster [22], Γ and γ are the power decay constants for clusters and rays, correspondingly. Inter-arrival times for clusters and rays T_j and τ_{jk} are defined by independent exponential probability density functions [22] as

$$p(T_j|T_{j-1}) = \Lambda \exp\big[-\Lambda(T_j - T_{j-1})\big], \tag{5}$$

$$p(\tau_{jk}|\tau_{j,k-1}) = \lambda \exp\big[-\lambda(\tau_{jk} - \tau_{j,k-1})\big], \tag{6}$$

where Λ and λ are the cluster and ray arrival rates. In Fig. 7, we schematically illustrate parameters $\lambda, \Lambda, \gamma, \Gamma, \beta(0,0)$ of the model, binding them to the physical aspect.

Following the methodology in [22], we use visual observation to analyze the CIRs and identify clusters. Selected examples of CIRs with identified clusters for several receiver locations are shown in Figs. 8, 9, 10, and 11 (see Fig. 2 for reference locations).

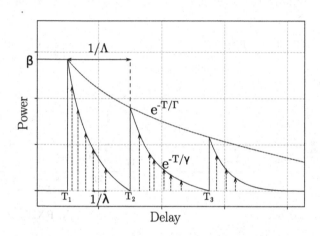

Fig. 7. Example of a synthetic CIR produced by the Saleh-Valenzuela model

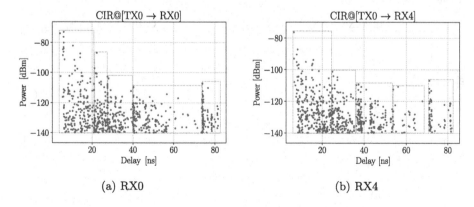

(a) RX0   (b) RX4

Fig. 8. Channel impulse response with identified clusters for receiver locations "0" and "4"

We begin with Fig. 8 illustrating the power delay profile (PDP), measured at two seats located closer to the transmitter, and corresponding identified ray clusters. At short ranges, we notice that the amplitude of the last clusters in the shown CIRs exceeds that of the preceding one, which corresponds to a path that experiences two reflections (from the windshield and the rear window) at nearly 90° angle and caused by the antenna patterns with a narrow mainlobe.

In Fig. 9, we observe more sparse CIR due to the increasing distance between the receiver and the transmitter. Components that are reflected from both windshield and rear window disappear in Fig. 9(b) as the corresponding paths become blocked at this position and further. Interestingly, in Fig. 10, the number of clusters decreases, which is due to the elements of the bus interior, which block or attenuate some of the clusters that are present in previous figures. Further, in Fig. 11, we observe that clusters merge, and the components are located within the first two clusters, which is caused by the fact that all propagation paths with more substantial delays are attenuated (based on the delay values, the average path length is around 13 m, which corresponds to the length of the bus).

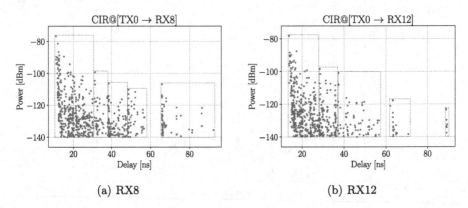

(a) RX8   (b) RX12

Fig. 9. Channel impulse response with identified clusters for receiver locations "8" and "12"

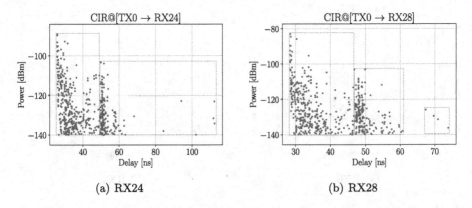

(a) RX24   (b) RX28

Fig. 10. Channel impulse response with identified clusters for locations "24" and "28"

Finally, based on these and other identified clusters, we can estimate the parameters of the Saleh-Valenzuela model as it has been done, e.g., in [22] or other similar literature on channel modeling. To calculate the ray arrival rate λ, we consider only dominant MPC components and omit nearby low-powered components. The resulting parameters of the statistical cluster model for our data set are $\Lambda = 0.29$, $\lambda = 3.45$, $\gamma = 0.67$, and $\Gamma = 0.85$. For convenience, all parameters are collected in Table 3.

Table 3. Parameters of the Saleh-Valenzuela model for our intra-vehicle scenario (city bus)

Cluster arrival rate, Λ	Ray arrival rate, λ	Cluster power decay, Γ	Ray power decay, γ
0.29 ns^{-1}	3.45 ns^{-1}	0.85 ns	0.67 ns

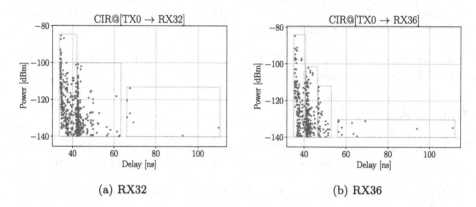

(a) RX32   (b) RX36

Fig. 11. Channel impulse response with identified clusters for locations "32" and "36"

5   Conclusions and Discussion

In this paper, we study the intra-vehicular propagation at 60 GHz using the ray-based simulation of a scenario that mimics a realistic interior of a typical city bus. We construct a highly detailed 3D model of the bus environment, calibrate the simulation scenario with our prior results obtained during a measurement campaign, and adjust the parameters, striking a balance between the simulation time and appropriate accuracy. Using the obtained simulation data, we extract CIRs and analyze the delay and angular spreads, which might prove useful for hardware engineers when estimating the required complexity of equalizers at the receiver based on the operational bandwidth.

We place the receivers at three different levels: the "knee" and "headset" levels on every seat, as well as the "waist" level covering the entire bus space. The delay spread at the "knee" level appears to be lower than that at the "headset" level, which translates to the narrower coherence bandwidth for "headsets". Hence, such aspects as ISI will be more pronounced at the "headset" level than for the "knee" locations. Moreover, the delay spread decreases with the growing distance due to the antenna directivity pattern; a sudden increase at 7 to 10 m is caused by the presence of the second exit in the rear part of the bus. Interestingly, the presence of exit and railings form a notable irregularity in the environment as we demonstrate on the power heat-map. Further, for the higher angular spread, it may be possible to isolate the ISI by configuring antenna pattern at receiver, which, however, might become infeasible for the "headset" level, if the antenna arrays in the respective hardware are simplistic.

Finally, we analyze a clustering structure of the power delay profile and derive statistical characteristics of the channel using the Saleh-Valenzuela cluster model. The obtained statistical multipath model can be directly implemented in link- and system-level studies to reproduce the structure of the 60 GHz channel in large intra-vehicle scenarios.

Acknowledgment. The publication has been prepared with the support of the "RUDN University Program 5-100".

References

1. CCSInsight, Optimistic outlook for wearables, March 2019. https://www.ccsinsight.com/press/company-news/optimistic-outlook-for-wearables/. Accessed Jul 2019
2. Satyanarayanan, M.: The emergence of edge computing. Computer **50**(1), 30–39 (2017)
3. Ometov, A., Kozyrev, D., Rykov, V., Andreev, S., Gaidamaka, Y., Koucheryavy, Y.: Reliability-centric analysis of offloaded computation in cooperative wearable applications. Wirel. Commun. Mobile Comput. **2017**, 15 (2017)
4. Aniwaa Pte. Ltd., The best all-in-one VR headsers of 2019, June 2019. https://www.aniwaa.com/best-of/vr-ar/best-standalone-vr-headset/
5. Forbes Media LLC, Preparing enterprises for the extended reality explosion, May 2019. https://www.forbes.com/sites/joemckendrick/2019/05/21/preparing-enterprises-for-the-extended-reality-explosion/#570d20fd24d9
6. Qualcomm Technologies Inc., The mobile future of eXtended Reality (XR), October 2018. https://www.qualcomm.com/media/documents/files/the-mobile-future-of-extended-reality-xr.pdf
7. Abari, O., Bharadia, D., Duffield, A., Katabi, D.: Cutting the cord in virtual reality. In: Proceedings of the 15th ACM Workshop on Hot Topics in Networks, pp. 162–168. ACM (2016)
8. Doppler, K., Torkildson, E., Bouwen, J.: On wireless networks for the era of mixed reality. In: Proceedings of European Conference on Networks and Communications (EuCNC), pp. 1–5. IEEE (2017)

9. IEEE 802.11 Working group, Wireless LAN Medium Access Control (MAC) and Physical Layer (PHY) Specifications. Amendment 3: enhancements for very high throughput in the 60 GHz band (2012)
10. Ghasempour, Y., da Silva, C.R., Cordeiro, C., Knightly, E.W.: IEEE 802.11 ay: next-generation 60 GHz communication for 100 Gb/s Wi-Fi. IEEE Commun. Mag. **55**(12), 186–192 (2017)
11. IEEE 802.11 Working group, IEEE 802.11 TGay use cases (2015)
12. Rao, T.R., Balachander, D., Sathish, P., Tiwari, N.: Intra-vehicular RF propagation measurements at UHF for wireless sensor networks. In: Proceedings of International Conference on Recent Advances in Computing and Software Systems, pp. 214–218. IEEE (2012)
13. Costa, C.A., et al.: Damper-to-damper path loss characterization for intra-vehicular wireless sensor networks. In: Proceedings of 47th European Microwave Conference (EuMC), pp. 1341–1344. IEEE (2017)
14. Bas, C.U., Ergen, S.C.: Ultra-wideband channel model for intra-vehicular wireless sensor networks beneath the chassis: from statistical model to simulations. IEEE Trans. Veh. Technol. **62**(1), 14–25 (2012)
15. Chetcuti, K., Debono, C.J., Farrugia, R.A., Bruillot, S.: Wireless propagation modelling inside a business jet. In: Proceedings of IEEE EUROCON 2009, pp. 1644–1649. IEEE (2009)
16. Debono, C.J., Farrugia, R.A., Chetcuti, K.: Modelling of the wireless propagation characteristics inside aircraft. In: Aerospace Technologies Advancements. IntechOpen (2010)
17. DeHaan, R.: Ray-tracing techniques and the simulation of large-scale in-flight wi-fi signal propagation. Ph.D. thesis, Baylor University (2018)
18. Jiang, Y., Zheng, G., Yin, X., Saleem, A., Ai, B.: Performance study of millimetre-wave MIMO channel in subway tunnel using directional antennas. IET Microwaves Antennas Propag. **12**(5), 833–839 (2017)
19. Semkin, V., Ponomarenko-Timofeev, A., Karttunen, A., Galinina, O., Andreev, S., Koucheryavy, Y.: Path loss characterization for intra-vehicle wearable deployments at 60 GHz. In: 2019 13th European Conference on Antennas and Propagation (EuCAP), pp. 1–4. IEEE
20. Mededović, P., Veletić, M., Blagojević, Ž.: Wireless insite software verification via analysis and comparison of simulation and measurement results. In: Proceedings of the 35th International Convention MIPRO, pp. 776–781. IEEE (2012)
21. Saleh, A.A., Valenzuela, R.: A statistical model for indoor multipath propagation. IEEE J. Sel. Areas Commun. **5**(2), 128–137 (1987)
22. Spencer, Q.H., Jeffs, B.D., Jensen, M.A., Swindlehurst, A.L.: Modeling the statistical time and angle of arrival characteristics of an indoor multipath channel. IEEE J. Sel. Areas Commun. **18**(3), 347–360 (2000)

Maximizing Achievable Data Rate in Unlicensed mmWave Networks with Mobile Clients

Nadezhda Chukhno[1](✉), Olga Chukhno[1], Sergey Shorgin[3],
Konstantin Samouylov[1,3], Olga Galinina[1,2], and Yuliya Gaidamaka[1,3]

[1] Peoples' Friendship University of Russia (RUDN University), Moscow, Russia
`nvchukhno@gmail.com`, `olgachukhno95@gmail.com`
`{gaydamaka-yuv,samuylov-ke}@rudn.ru`
[2] Tampere University, Tampere, Finland
`olga.galinina@tuni.fi`
[3] Federal Research Center "Computer Science and Control" of the Russian Academy
of Sciences, Moscow, Russia
`sshorgin@ipiran.ru`

Abstract. In millimeter-wave (mmWave) networks, where faster signal
attenuation is compensated by the use of highly directional antennas, the
effects of high mobility may seriously harm the link quality and, hence,
the overall system performance. In this paper, we study the channel
access in unlicensed mmWave networks with mobile clients, with particular emphasis on initial beamforming training and beam refinement
protocol as per IEEE 802.11ad/ay standard. We explicitly model beamforming procedures and corresponding overhead for directional mmWave
antennas and provide a method for maximizing the average data rate over
the variable length of the 802.11ad/ay beacon interval in different mobility scenarios. We illustrate the impact of the client speed and mobility
patterns by examples of three variations of the discrete random walk
mobility model.

Keywords: mmWave · Beamforming · 802.11ad/ay · Channel access ·
Mobility · Random walk

1 Introduction

Millimeter-wave (mmWave) has become one of the most discussed wireless technologies of today, being recently included in the top ten trends that will drive
innovation [1]. Even thought mmWave research has a long history, only recent
advances in microelectronics and, more importantly, the rapidly growing popularity of bandwidth-hungry applications, such as virtual, augmented, and mixed
reality (AR/VR/MR) have brought the technology closer to the mass adoption.

A promising mmWave carrier candidate to meet the growing consumer data
demand can be found in the 60 GHz band, which is identified by FCC as a

© Springer Nature Switzerland AG 2019
O. Galinina et al. (Eds.): NEW2AN 2019/ruSMART 2019, LNCS 11660, pp. 282–294, 2019.
https://doi.org/10.1007/978-3-030-30859-9_24

part of the unlicensed spectrum. At 60 GHz, two pioneering standards, Wireless HD and ECMA-387, published in 2008, and the first IEEE standard, the IEEE 802:15.3c-2009, approved in 2009, promised to provide real gigabit throughputs but did not become widely accepted due to a number of technical issues.

The first 60 GHz protocol in IEEE 802.11 family (Wi-Fi), 802.11ad, published in 2012, has already brought many commercial hardware products to the market. Its backward-compatible successor, the IEEE 802.11ay standard [2], has passed the Draft 3.1 stage and currently is approaching its final IEEE ratification, which is expected in March 2020[1]. The target use cases studied by the IEEE 802.11 TGay Group [3] include, among others, AR/VR headsets and other high-end wearables, where the access point and client devices could be highly mobile.

In the case of IEEE 802.11ad/ay protocols, where faster mmWave signal attenuation is compensated by the use of highly directional antennas, the effects of high mobility may seriously harm the link quality and the overall system performance. Fortunately, the respective IEEE specifications leave many protocol parameters flexible and implementation-dependent to be later tuned depending on the degree of mobility, as concrete use cases would require.

As the IEEE 802.11ad specification has been openly accessible since 2012, multiple research papers address evaluation, optimization, and other various aspects of the 802.11ad protocol [4–7], including a comparison of in-field experiments and simulation-based studies [8]. The latest 802.11ay, although not yet publicly available, has been described in detail in seminal papers [2,9,10] by Intel (one of the IEEE 802.11ay contributors), where [9] focuses on the beam refinement protocol (BRP) and asymmetric beamforming training, and [10] addresses fundamental elements of the single-carrier PHY of IEEE 802.11ay (the new frame format, modulation and coding schemes (MCSs), new beamforming training field, etc.).

Despite the surge of interest to IEEE 802.11ad/ay, the effect of mobility has received limited attention of the research community. For vehicular applications of 802.11ad, there have been proposed several learning-based sweeping solutions, including non-supervised online learning algorithms for beam pair selection and refinement [11], sparsity-aware beamforming design for radars [12], and multipath fingerprinting algorithm for the beam alignment [13]. The effects of mobility in mmWave have also been addressed in [14,15] in the context of blockers mobility, however, without any protocol linked.

In this paper, we study the channel access of IEEE 802.11ad/ay, with particular emphasis on initial beamforming training and beam refinement protocol. We explicitly model beamforming procedures and corresponding overhead for highly directional antennas and maximize the average data rate over the variable length of the 802.11ad/ay beacon interval. Our main contribution lies in providing a method for maximizing the average data rate in various mobility scenarios. The impact of the client speed and mobility patterns is illustrated by

---

[1] QUALCOMM has already announced new 802.11ay chipsets, QCA6431 and QCA6421, intended for mobile use; however, the corresponding hardware specifications have not been published as of July 2019.

selected numerical results when using examples of three diverse variations of the discrete random walk mobility model.

The rest of the paper is organized as follows. Section 2 introduces the system model and the main assumptions on system geometry, protocol settings, and client mobility. In Sect. 3, we propose a method for estimating the achievable data rate, define the objective function, and formulate an optimization problem. Finally, Sect. 4 presents selected numerical results and main conclusions.

2   System Model and Assumptions

In this section, we introduce the main system assumptions on geometry and link abstraction, as well as describe the considered client mobility patterns.

2.1   System Geometry and Link Abstraction

We consider a stationary IEEE 802.11ad/ay access point (AP), located at the origin A, and a dynamic device hereinafter termed client or user equipment (UE), which moves with the speed v in a selected direction (see Fig. 1). The UE mobility pattern is addressed below, and its initial location B is drawn from the uniform distribution in a circle of radius R around point A, where R defines the service area of the AP.

We assume that both devices can transmit directionally with the same antenna beam pattern, which is symmetrical w.r.t. the antenna boresight and characterized by the half power beam width (HPBW) θ. The resulting transmit/receive antenna gain may be approximated by the following function of the deviation α from the antenna boresight [16]:

$$G_{\text{tx/rx}} = D_0 \rho(\alpha), \tag{1}$$

where $D_0 = \frac{2}{1-\cos\frac{\theta}{2}}$ is the maximum antenna gain corresponding to $\alpha = 0$ and estimated as the ratio between the area of a sphere and the area of a cone antenna pattern, $\rho(\alpha)$ is the multiplier, which scales the antenna gain according to the deviation from the boresight, $\rho(\alpha) \in [0, 1]$ and $\rho(0) = 1$.

We note that $\rho(\alpha)$ and more precise value of D_0 can be estimated by using the data obtained in the course of measurement campaigns, in computational electromagnetics modeling tools, or via numerical evaluation of antenna arrays of a given structure, which is provided, e.g., by the corresponding MATLAB toolbox (that is, Sensor Array Analyzer). Here, we employ an approximation of the numerical function $\rho(\alpha)$ proposed in [16]. In particular,

$$\rho(\alpha) = \max\left(1 - \frac{\alpha}{\theta}, 0\right), \tag{2}$$

where θ is the HPBW and α is a current angular deviation of the transmit/receive direction from the antenna boresight. We let the AP (or UE) have M (N) predefined antenna beam patterns, i.e., for the AP, each m-th beam, $1 \leq m \leq M$, covers roughly a sector limited by angles in the interval $[2\pi\frac{m-1}{M}, 2\pi\frac{m}{M}]$. Then, we may assume HPBW of $\frac{2\pi}{M}$ and $\frac{2\pi}{N}$ for the AP and UE, correspondingly.

Further, we focus on the downlink transmission and assume that the line-of-sight connection between the devices exists at all times and, hence, the path loss can be estimated by the Friis equation. Consequently, we calculate the received power as

$$P_{\text{rx}} = P_{\text{tx}} G_{\text{tx}} G_{\text{rx}} \left(\frac{\lambda}{4\pi d}\right)^2, \tag{3}$$

where $G_{\text{tx/rx}}$ are transmit/receive antenna gains, λ is the wavelength, d is the current distance between the transmitter (AP) and receiver (UE) located at points A and B, correspondingly.

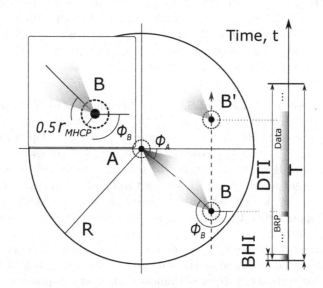

Fig. 1. Schematic illustration of the considered dynamic scenario: the client moves according to the rectilinear motion (gray dashed line) from point B to point B'. Initial beamforming (sector-level sweep for all UEs in the system) is performed during the time interval "BHI", and more refined alignment is achieved at the end of interval "BRP" (beam refinement protocol for this particular UE). After the data transmission "Data", the beam misalignment reaches the maximum, and the devices should initiate a new beamforming procedure.

2.2 Channel Access Abstraction

Beacon Interval Structure. If the UE moves on the plane while maintaining a highly-directional connection with the AP, the mutual alignment of the two antenna beams changes over time, and the devices need to update the beam directions by using, e.g., standard beamforming procedures. We assume that the time is divided into beacon intervals (BI) of length T that consist of (i) beacon header interval (BHI), where the devices perform initial beamforming (that is, sector-level sweep, SLS) and align their wider transmit beams, and (ii) the data transmission interval (DTI), incorporating service periods (SP) of different connected clients, while each SP contains beam refinement protocol (BRP) and the data transmission[2]. Figure 1 schematically illustrates two phases, while Fig. 2 depicts a more detailed structure of the considered protocol. We note that since our system relies on highly-directional communication and implies the use of narrow beams, we may assume that after the beamforming procedure antenna beams are perfectly aligned.

If both devices were stationary, then after beamforming and during the data transmission of an arbitrary length, the antenna gains would reach its maximum. However, if either of the devices moves, then the received power and, hence, the achievable rate, may substantially drop due to the decrease in receive/transmit gain, which should be redressed at the next beamforming opportunity, e.g., in the following BI. The duration T of BIs (i.e., between the beginning of the adjacent BHIs) is an essential parameter affecting the overall performance of mmWave systems as, e.g., throughput, energy efficiency, data transmission delay, etc.

Sector-Level Sweep Phase, SLS. At the beginning of each BI, in BHI, the AP performs SLS by sweeping through M_{SLS} transmit directions. We assume that the corresponding HPBW is given by $\theta_{\text{SLS,AP}} = \frac{2\pi}{M_{\text{SLS}}}$. Upon reception of the AP signal, the UE attempts to access the channel in the following interval (A-BFT, according to IEEE specifications) by selecting one of several time-divided transmission intervals, where the UE sweeps through N_{SLS} transmit beams with $\theta_{\text{SLS,UE}} = \frac{2\pi}{N_{\text{SLS}}}$. In A-BFT interval, if more than one clients select the same transmission opportunity, the signals collide, and devices cannot establish a connection in the current BI[3].

After SLS, both AP and UE know their best transmit beams, which are selected, in general case, based on the strongest received signal strength (RSSI). We remind that reception is (quasi-)omnidirectional at this phase[4].

The protocol limits the duration of the BHI, but we intentionally leave it flexible to investigate the corresponding system trade-offs. Hence, we assume that T_{SLS} can be derived according to the formula:

$$T_{\text{SLS}} = (M_{\text{SLS}} + n_a N_{\text{SLS}}) \tau_{\text{SLS}}, \tag{4}$$

---

[2] Here, we focus only on the scheduled operation, although DTI may as well contain contention-based access periods (CBAPs).

[3] In fact, one of the clients may succeed. We leave the evaluation of initial random access collisions out of the scope of this paper. The detailed description of the protocol may be found in [17].

[4] In reality, the antenna pattern is not omnidirectional. Moreover, devices may use several directional antennas, and all of them participate in sector sweep. For the sake of clarity, we omit consideration of more complex procedures with multiple antennas but note that the respective modifications can be easily incorporated into our system.

where τ_{SLS} is the duration of each sector sweep packet and n_a is the total number of transmission attempts in A-BFT. We assume that n_a is sufficiently large to minimize the channel collisions and neglect their effect in our model.

Beam Refinement Phase, BRP. After the SLS phase, the AP and the UE have established preliminary connection relying on relatively wide beams. The following service period (in DTI) allocated explicitly for the tagged UE, may start with beam refinement to improve the resulting instantaneous data rate.

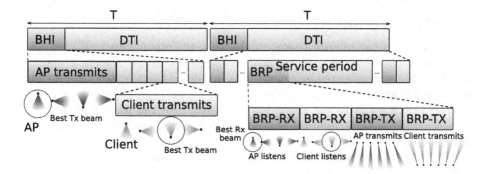

Fig. 2. Beacon interval structure.

Specifically, BRP may include both receive and transmit antenna training (as shown in Fig. 2) for the AP and the UE. We assume that the number of beams at this phase is defined by $M = bM_{\text{SLS}}$ ($N = bN_{\text{SLS}}$), where N_{SLS} is the number of the transmitter beams during SLS and $b > 1$ is a scaling multiplier, which we hereinafter term a beam refinement factor. Therefore, during the transmit antenna training, both devices sweep through exactly b narrower beams (within the initial transmit sector), while for receive training, all M or N directions should be covered. Thus, if we denote the duration of one BRP packet as τ_{BRP}, the total time spent on beam refinement is defined by

$$T_{\text{BRP}} = (2b + N + M)\,\tau_{\text{BRP}}. \tag{5}$$

Data Transmission. The DTI may contain SPs scheduled during BHI or contention-based access periods (CBAPs). The use of CBAPs proved much less effective [18] and, therefore, in this work, we assume that only SPs are used for the data transmission. Moreover, we focus on one link, abstracting away the presence of other clients. Therefore, for simplicity, we assume that the scheduler follows round-robin policy, and our tagged client obtains and fully occupies, on average, 1/U of the total time resource, where U is the average number of clients per AP.

The duration of the time when devices are exchanging data traffic may be then found from the following equality:

$$T = T_{\text{SLS}} + U\left(T_{\text{BRP}} + T_{\text{DT}}\right) + T_0, \tag{6}$$

where T_{DT} is the duration of data transmission for one UE and T_0 is the total signaling overhead independent of the number of beams. We note that in (6), $T_{\text{SLS}} + UT_{\text{BRP}}$ depends on the beam settings, i.e., the number (and the width) of beams M, M_{BRP} and N, N_{BRP}, while T_{DT} is a variable that is to be properly selected to maximize the data rate as detailed in Sect. 3.

2.3   Mobility Model

We assume that the receiver moves according to the random walk (RW) model and, hence, its path consists of a sequence of steps of length l taken in a random direction that is represented by angle γ (γ is calculated counterclockwise starting at the abscissa).

The length l can be expressed via the speed v of the moving client, i.e., $l = vT$, where T is a duration of one mobility step. Thus, the increments of coordinates

$$\Delta X_B = l \cdot \cos\gamma \quad \text{and} \quad \Delta Y_B = l \cdot \sin\gamma$$

determine new coordinates $(X_B + \Delta X_B, Y_B + \Delta Y_B)$ of the UE.

In our simulation setup, we guarantee the minimum distance between the receiver and the transmitter, introducing the minimum distance r_{MHCP} by analogy with Matern Hard Core Point process type-1 (MHCP-1) so that

$$d = \sqrt{(X_B - X_A)^2 + (Y_B - Y_A)^2} \geq r_{\text{MHCP}}.$$

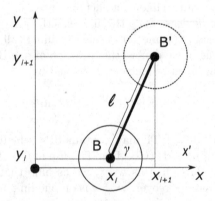

Fig. 3. Illustration of one step in the considered mobility models.

We consider three RW models: (i) rectilinear motion (no change in direction and, hence, in γ), (ii) random increment in direction γ that follows the normal distribution $N(0, \delta^2)$, and (iii) random walk on a grid, i.e., with the equiprobably distributed directions. Below we address these three models in detail.

Random Walk Model with the Constant Direction of Motion. Uniform rectilinear motion is a special case of RW with the constant distance l and constant direction of motion $\gamma = \frac{\pi}{2}$ as illustrated in Fig. 1. In this case, increments of coordinates equal $\Delta X_B = 0$, $\Delta Y_B = l$.

Random Walk with the Normally Distributed Direction. Here, the distance l per cycle is constant, and the deviation of the direction of movement from the abscissa is a normally distributed random variable $\gamma \sim N(0, \delta^2)$. We assume that the standard deviation δ is given by $\delta = \pi/2$ and, therefore, approximately 95% of γ fall into the interval $[-\pi, \pi]$.

Random Walk on a Grid. In this model, the distance l is constant, and the angle γ is a random variable, which at each step equiprobably accepts one of the following values: $\gamma \in \left\{0, \frac{\pi}{2}, \pi, \frac{3\pi}{2}\right\}$.

3 Optimization Problem

In this section, we formulate an optimization problem that maximizes the achievable data rate for the given beamwidth and client mobility model.

As the data is transmitted only during the SP excluding BRP, i.e., during the time T_{DT}, the average achievable data rate is calculated by subtracting the corresponding overhead on SLS and BRP from the instantaneous rate. The instantaneous data rate is estimated by Shannon's limit constrained by the choice of modulation coding scheme (MCS), i.e.,

$$D = W \log_2 \left(1 + \min\left(\frac{P_{\text{rx}}}{P_{\text{noise}}}, \text{SNR}_{\text{max}}\right)\right), \tag{7}$$

where SNR_{max} corresponds to the SNR value, at which the maximum MCS is selected, P_{rx} incorporates both transmit and receive antenna gains after the BRP phase (that is, w.r.t. the antenna beam misalignment, if any), $P_{\text{noise}} = W \cdot N_0 \cdot \text{NF}$, NF is the noise factor, and the interference is assumed to be negligible.

Consequently, if T_{DT} is significantly large, then during this time interval, the antenna alignment changes as the client moves forward and, hence, the instantaneous data rate gradually degrades. Alternatively, if the interval T_{DT} is short, and the configuration of the antenna beams does not noticeably change due to the client mobility, then the instantaneous data rate remains much higher; however, due to overhead $T_{\text{SLS}} + UT_{\text{BRP}}$, the actual average data rate during the BI decreases. This trade-off dictates the necessity of choosing an optimal value for the length T_{DT} (or, equivalently, T) that would maximize the average data rate, and results in an optimization problem below.

For our system with clients moving according to the RW model with the constant speed, the optimization problem can be formulated as

$$D(T) \xrightarrow[T]{} \max \quad \text{or, equivalently, find} \quad T^* = \arg \max D(T)$$
$$\text{subject to} \quad T > T_{\text{SLS}} + UT_{BRP}. \tag{8}$$

While the client moves, we measure the instantaneous data rate within the DTI at discrete time instants and average the rate over the actual duration of data transmission as

$$D_{\mathrm{DT}} = \frac{1}{J} \sum_{j=0}^{J} D_j, \tag{9}$$

where J is the number of time slots within one interval of data transmission. The resulting data rate per client is, therefore, defined by

$$D(T) = \frac{D_{\mathrm{DT}} T_{\mathrm{DT}}}{T}, \tag{10}$$

where the multiplier $\frac{T_{\mathrm{DT}}}{T}$ allows us to subtract the beamforming overhead and the time scheduled for other AP clients. The sought optimal BI duration T can be obtained numerically, which we demonstrate in the following section.

4   Numerical Results and Conclusions

Table 1. Simulation parameters.

Parameter	Value	Description
λ	0.005 m	Wavelength
f_c	60 GHz	Carrier frequency
R	40 m	Service area radius
P_{tx}	10 dBm	Transmit power
N_0	−174 dBm/Hz	Noise power per 1 Hz
NF	6 dB	Noise figure
W	2.16 GHz	Bandwidth
n_a	40	Number of slots in A-BFT
$\mathrm{SNR}_{\max}$	20 dB	SNR corresponding to choosing MCS19 (rate 13/16)
M	4, 8, 16	Number of AP transmit beams
N	4, 8, 16	Number of client transmit beams
τ_{SLS}	16 µs	Duration of one packet in SLS
τ_{BRP}	0.7 µs	Duration of one packet in BRP
v	var	Client speed

In this section, we present selected results of our numerical optimization. In general, we assume numerology of IEEE 802.11ay in its basic configuration (one channel, one spatial stream), the core simulation parameters are given in Table 1.

We begin with analyzing the dependence of the average data rate on the BI length for three considered mobility models and different values of the client speed (pedestrian 1.1 m/s, bicycle 5 m/s, and vehicle/drone 25 m/s). The rectilinear motion (movement along a straight line) is typical for cars on a highway or drones in the air. Another realistic pattern, with a pronounced drift, can be modeled by the normal distribution of the movement direction. Finally, RW on a grid may describe a chaotic movement of customers, for example, during a crowded large-scale open-air event. For the rectilinear motion, we set $\gamma = \pi/2$, while for the second mobility model $\gamma \sim N(0, \pi/2)$, and for the RW on a grid, γ is randomly and equiprobably selected, $\gamma \in \{0, \pi/2, \pi, 3\pi/2\}$.

As shown in Fig. 4, in the case of the rectilinear motion and mobility with normally distributed directions at high speed ($v = 25$ m/s), the data rate starts decreasing already after approximately 100 ms, while for the lower speeds, $v = 1.1$ m/s and 5 m/s, the optimal BI interval length increases up to 500 ms and 250 ms, respectively. In all cases, the rate curves rise abruptly at the beginning of the abscissa (where T remains small), since the overhead occupies noticeably long time w.r.t. the total BI duration. When the time allocated for the data transmission increases, the data rate quickly reaches the maximum and then declines much slower due to the beam misalignment.

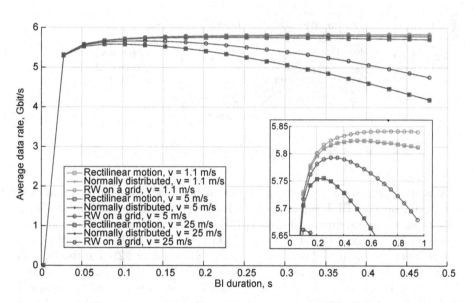

Fig. 4. The average data rate vs. the BI length for different mobility patterns and speeds. The HPBW $\theta = 45°$.

Moreover, for the rectilinear motion and mobility pattern with normally distributed directions, the data rate decreases visibly faster than in the case of the RW on a grid. This effect may be explained by the fact that clients moving on a grid stay within the same region longer than if they rapidly moved away along a concrete direction.

Further, we continue by varying the beamwidth, $\theta = 11°, 22°, 45°$ (see Fig. 5). For the lower speeds, the narrower beams guarantee better data rate due to higher antenna gains; however, for the higher speeds, there exists a trade-off between the increasing gains and even faster beam misalignment of narrow beams (see the group of curves for $\theta = 11°$ that). In general, the optimal BI duration becomes longer if we decrease θ due to the large overhead. Importantly, even for these settings, our system continues to function well since we assume 802.11ay BRP timings that are much more effective than those of 802.11ad.

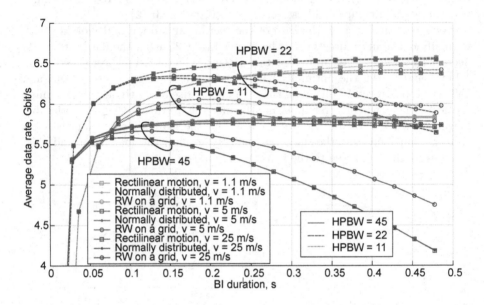

Fig. 5. The average data rate for different HPBW ($\theta = 11°, 22°, 45°$).

Finally, in Fig. 6, we provide an important practical trade-off that may help in selecting optimal BI duration in scenarios with different client mobility. Naturally, when the speed grows, the optimal BI duration decreases and for narrower beams remains larger in most cases (as confirmed by results in Fig. 6). However, at the lower speeds ($0 - 5\,\text{m/s}$), the optimal BI for $\theta = 22°$ due to the high antenna gain is significantly longer than that of for $\theta = 11°$. As the speed grows, the difference between the optimal BI values for narrow and wider beams is diminished and may fluctuate around approximately the same value of the BI duration, which should be further investigated in more details.

In summary, we may conclude that in order to determine the optimal BI duration in highly mobile scenarios, it is necessary to take into account a combination of multiple factors, such as mobility pattern, client's speed, and the structure of the wireless protocol. A promising future direction could be found, e.g., in defining practically feasible (or even adaptive) heuristics through analytical optimization of a simplified system, more detailed analysis of periodicity of BRPs within one SP, effects of heterogeneous time and space data demand, and mobility of blockers and APs.

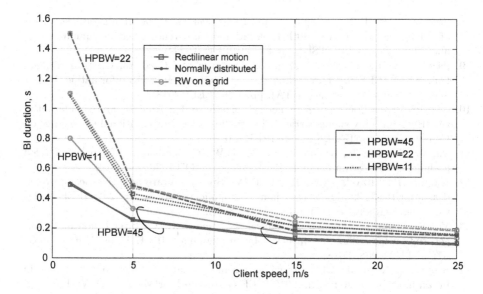

Fig. 6. The optimal BI vs. the client speed for three mobility models.

Acknowledgment. The publication has been prepared with the support of the "RUDN University Program 5-100" (recipients Nadezhda Chukhno, Olga Chukhno, Konstantin Samouylov). The reported study was funded by RFBR, project numbers 17-07-00845 and 18-07-00576 (recipients Yuliya Gaidamaka, Sergey Shorgin). This work has been developed within the framework of the COST Action CA15104, Inclusive Radio Communication Networks for 5G and beyond (IRACON).

References

1. Jones, N.: The top 10 wireless technologies and trends that will drive innovation (2019)
2. Ghasempour, Y., da Silva, C.R., Cordeiro, C., Knightly, E.W.: IEEE 802.11ay: next-generation 60 GHz communication for 100 Gb/s Wi-Fi. IEEE Commun. Mag. **55**(12), 186–192 (2017)
3. IEEE 802.11 Working group, IEEE 802.11 TGay use cases (2015)
4. Cordeiro, C., Akhmetov, D., Park, M.: IEEE 802.11ad: Introduction and performance evaluation of the first multi-Gbps Wi-Fi technology. In: Proceedings of the 2010 ACM International Workshop on mmWave Communications: From Circuits to Networks, pp. 3–8. ACM (2010)
5. Kim, J., Molisch, A.F.: Enabling Gigabit services for IEEE 802.11ad-capable high-speed train networks. In: 2013 IEEE Radio and Wireless Symposium, pp. 145–147. IEEE (2013)
6. Verma, L., Fakharzadeh, M., Choi, S.: Wi-Fi on steroids: 802.11ac and 802.11ad. IEEE Wirel. Commun. **20**(6), 30–35 (2013)
7. Nitsche, T., Cordeiro, C., Flores, A.B., Knightly, E.W., Perahia, E., Widmer, J.: IEEE 802.11ad: directional 60 GHz communication for multi-Gigabit-per-second Wi-Fi. IEEE Commun. Mag. **52**(12), 132–141 (2014)

8. Zeman, K., et al.: Emerging 5G applications over mmWave: Hands-on assessment of WiGig radios. In: 2017 40th International Conference on Telecommunications and Signal Processing (TSP), pp. 86–90. IEEE (2017)

9. da Silva, C.R., Kosloff, J., Chen, C., Lomayev, A., Cordeiro, C.: Beamforming training for IEEE 802.11ay millimeter wave systems. In: 2018 Information Theory and Applications Workshop (ITA), pp. 1–9. IEEE (2018)

10. da Silva, C.R., Lomayev, A., Chen, C., Cordeiro, C.: Analysis and simulation of the IEEE 802.11ay single-carrier PHY. In: 2018 IEEE International Conference on Communications (ICC), pp. 1–6. IEEE (2018)

11. Va, V., Shimizu, T., Bansal, G., Heath, R.W.: Online learning for position-aided millimeter wave beam training. IEEE Access 7, 30507–30526 (2019)

12. Kumari, P., Eltayeb, M.E., Heath, R.W.: Sparsity-aware adaptive beamforming design for IEEE 802.11ad-based joint communication-radar. In: 2018 IEEE Radar Conference (RadarConf18), pp. 0923–0928. IEEE (2018)

13. Va, V., Choi, J., Shimizu, T., Bansal, G., Heath, R.W.: Inverse multipath fingerprinting for millimeter wave V2I beam alignment. IEEE Trans. Veh. Technol. 67(5), 4042–4058 (2018)

14. Samuylov, A. et al.: Characterizing spatial correlation of blockage statistics in urban mmWave systems. In: 2016 IEEE Globecom Workshops (GC Wkshps), pp. 1–7. IEEE (2016)

15. Gapeyenko, M., et al.: On the temporal effects of mobile blockers in urban millimeter-wave cellular scenarios. IEEE Trans. Veh. Technol. 66(11), 10124–10138 (2017)

16. Chukhno, O., Chukhno, N., Galinina, O., Gaidamaka, Y., Andreev, S., Samouylov, K.: Analyzing effects of directional deafness on mmWave channel access in unlicensed bands. In: accepted to IEEE GLOBECOM 2019 (2019)

17. IEEE 802.11 Working group, Wireless LAN Medium Access Control (MAC) and Physical Layer (PHY) specifications. Amendment 3: enhancements for very high throughput in the 60 GHz band (2012)

18. Galinina, O., Pyattaev, A., Johnsson, K., Turlikov, A., Andreev, S., Koucheryavy, Y.: Assessing system-level energy efficiency of mmwave-based wearable networks. IEEE J. Sel. Areas Commun. 34(4), 923–937 (2016)

Runtime Minimization of the Threshold Distributed Computation Protocol in the Case of Participants Failures

Alexandra Afanasyeva$^{(\boxtimes)}$ (iD), Ivan Evstafiev (iD), and Andrey Turlikov (iD)

State University of Aerospace Instrumentation,
B. Morskaya 67, 190000 Saint-Petersburg, Russia
{alra,ev.ivan,turlikov}@k36.org

Abstract. Threshold (k, n)-schemes are applied to ensure fault tolerance in many protocols of distributed computing, where k is a minimum amount of nodes and $n \geq k$ is a total number of nodes in the system. When $n = k$ in such a system, if one of the nodes fails during the execution of this protocol, it is replaced with a new one and the procedure is restarted, while all the time and traffic spent during the previous stages will be lost. If it is initially involved more than the minimum threshold of nodes $(n > k)$, then in case of any node fails, the protocol will not be restarted. In such a case, an optimization problem arises, which consists in runtime minimizing of the protocol with given constraints on the threshold k and the failure characteristics. In this paper, using the example of the threshold calculation protocol of the digital digest, such an optimization problem will be formulated and solved.

Keywords: Threshold protocols · Transport encoding ·
k-th order statistic

1 Introduction

The widespread use of cloud systems and computing clusters requires the development of distributed computing protocols. In the framework of the standard problem statement of the distributed computing, the main problem is divided into subtasks that can be calculated by different nodes. In this case, the failure of any system component leads to the inability to complete the computations. Any compute node becomes a point of vulnerability. In other words, one technical failure is enough to disrupt the operation of the entire system, leading to an irreversible violation of the integrity or availability of the processed data. Such systems utilize redundancy and replication to realize the goal of high reliability. Distributed computation architectures (DCA) systems utilize highly parallel computing resources to dynamical networks; the computing resources of DCA are built by potentially faulty and untrusted components.

The author Turlikov is supported by research project RFBR No. 17-07-00142.

O. Galinina et al. (Eds.): NEW2AN 2019/ruSMART 2019, LNCS 11660, pp. 295–304, 2019.
https://doi.org/10.1007/978-3-030-30859-9_25

Distributed threshold algorithms allow an excessive group of participants to perform calculations together even in the presence of failures [3]. Threshold algorithms increase robustness and flexibility at the cost of reduced computational efficiency. Reliability is a critical issue for component-based distributed computing systems, some distributed software allows the existence of large numbers of potentially faulty components on an open network. Faults are inevitable in this large-scale, complex, distributed components setting, which may include a lot of untrustworthy parts. How to provide highly reliable component-based distributed systems is a challenging problem and a critical research. The main goal of our work is to analyze the trade-off between reliability, flexibility and the cost in the terms of the runtime and traffic overhead. The main contributions of this paper are summarized in the following.

- (i) Paper introduces a new problem statement in redundancy distributed computing systems to analyze the runtime of the protocol based on its features.
- (ii) We formulate and solve the optimization problem. In order to make appropriate cost and reliability trade-offs.
- (iii) A formal theoretical analysis based on probability theory and experiments are designed to optimize parameters of the selected threshold protocol.

The rest of this paper is organized as follows. The background and related works are introduced in Sect. 2. In Sect. 3, a selected threshold protocol for distributed digital digest generation is described and its complexity analysis is proposed. Theoretical analysis and experiment results on proposed approach are given in Sect. 4. The conclusion of the paper is shown in Sect. 5.

2   Background and Related Works

The most popular application of redundancy in distributed computing systems is to increase robustness and reliability of complex systems, so the most part of work is devoted to the analysis of reliability and development of algorithms for failure detection of individual nodes [1,3].

Another application of distributed calculation was proposed in [2]. It has been observed and understood for some time now that, for machine resilience at least, there is a specific relationship between robustness and randomness, without which errors are amplified through the systems. The original motivation to investigate the provision of distributed redundancy came about from the authors' work on the analysis of communities that emerge in large scale Service-Oriented Architectures.

Our approach is similar to the solution proposed in [9] for the problem of transport coding in the transmission of messages in packet-switched networks. A detailed description and analysis of the effectiveness of this message transmission method are given in [7,8]. Let's briefly introduce the idea of transport coding. Suppose one wants to send a message consisting of k packets over the network. We will consider packages as symbols of some alphabet. Let the code $G(n, k)$ be given over this alphabet. Packages are encoded using the specified code before sending the message. Instead of k original packets n encoded ones $(n > k)$ are sent. To receive a message, you must, in order for at least k packets to reach the destination, the message will be considered accepted. The speed of the code $R = k/n$ is fixed during transmission from source to receiver. By analogy with transport encoding in the framework of this work the problem of introducing redundancy in the computing nodes is considered to ensure fault tolerance of the system as a whole in the presence of failures.

A lot of similar problems exist in cryptography, as there are distributed protocols and applications such as blockchain [4–6], and the threshold (k, n)-schemes commonly use for robustness and security reasons. The paper investigates the efficiency of distributed (k, n)-threshold computing on the example of the problem of distributed computing of digital digest.

3 Protocol Description

In this section the description of the distributed protocol of digital digest generation is proposed. It works with any cyclic group G_1 of order q with generator g. The detailed description of the Protocol is given in algorithm 1.

Algorithm 1. (k, n)-threshold distributed digital digest generation

1: **Initialization phase:**
2:      Each user generates a polynomial $f_i(x)$ such as $deg(f_i(x)) = t$ in a random way over Z_q.
3:      Each user i sends its part $f_i(ID_j)$ to another user with ID_j for $j = 1...n$.
4:      Each user calculates its own share in a following

$$x_i = \sum_{j=1}^{n} f_j(ID_i).$$

5:      Each user sends to all $y_i = x_i \cdot g$ over G_1.
6:      Each user interpolates $y = x \cdot g = \sum_{i=1}^{n} x_i g$ over G_1.
7: **Digital digest distributed generation:**
8:      Parameters k, a distributed generation.
9:      Share k and a like x.
10:      Calculate $r = k \cdot g$.
11:      Calculation $s_i = k_i m + x_i r + a_i$ and message exchanging.
12:      Interpolation $s = km + xr$.

As a result the digital digest is a pair (r, s) for given message m.

For further analysis of the Protocol runtime, it is necessary to assess its complexity. The Protocol includes two stages: distributed parameter generation and digital digest generation. The first stage can be attributed to the pre-calculations and not taken into account when estimating the complexity. The second stage is distributed digest generation. For each message (m), the Protocol participants must work together to produce a value of r and s sequentially. For each value at least k active participants should exchange messages with each other. After the exchange, the value is interpolated by all participants independently. Therefore, the main time of the Protocol will be allocated for network communications and can be evaluated as a function of the order $O(n^2)$.

4   Optimization Problem and System Model

The main feature of the distributed protocols is the need for multilateral distributed computing, requiring participants to several stages of interaction, while part of the steps includes the exchange of data with each other. In the case when the node during the execution of the Protocol ceases to transmit data, you must start the procedure from the beginning after adding a new active participant, and all the time and traffic spent in the previous stages will be lost. The computation will be guaranteed to be success if number of node unlimited, but depending on the number of failures and the moment of failure, the time for this and the amount of data transferred may vary significantly. Therefore, it is possible to modify the procedure of distributed calculations so that the procedure involved more than the minimum threshold nodes, then in case of failure of one of the parties to the Protocol, the Protocol setting is restarted. With this modification of the calculation procedure an optimization problem arises, which consists in minimizing the Protocol runtime under the specified restrictions on the threshold k and the probability of failure p. For this task it is necessary to find the optimal number of active participants(n) at the beginning of the procedure. Formally

$$T_{k,p}(n) \to \min.$$

The solution of this problem can be obtained by modeling or by using probabilistic calculations. For the theoretical solution it is necessary to describe the system in terms of mathematical statistic. We will consider the threshold computation procedure as a random process involving n compute nodes.

A number of assumptions are made:

Assumption 1 *The lifetime (X) of each node during the execution of the Protocol is independent of the others. The time in proposed model is measured in abstract timeslots equal to one step of the Protocol.*

Assumption 2 *For the calculation Protocol to complete successfully, it is necessary that at least k participants work longer than the runtime of the Protocol without failures.*

Assumption 3 *As you can see from the description of the Protocol, the operating time will consist of two iterations of data exchange with all and, accordingly, can be estimated as $Tr = O(n^2)$.*

Assumption 4 *The lifetime (X) of each node during the execution of the Protocol is distributed exponentially with λ intensity.*

Under these assumptions the failure probability p is a function of λ, so the optimization problem can be reformulate in different terms:

$$T_{k,\lambda}(n) \to \min.$$

Such a system can be considered as a series of multiple rounds, each of which is interrupted either upon reaching the threshold (successfully completed calculations), or if the network has less than k nodes.

The temporary chart of the protocol is shown in Fig. 1.

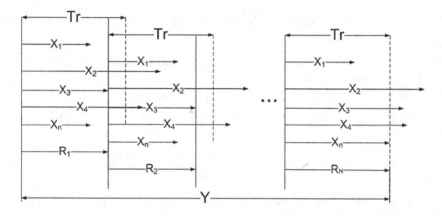

Fig. 1. Temporary chart of the protocol

Total runtime is denoted by random variable Y. Then the objective function $T_{k,\lambda}(n)$ is equal to $E[Y]$.

5   Optimization Problem Solving

The running time consists of a random number of rounds (N_{round}), each of which last a random time (R_i) and the last successful iteration over time Tr. The number of rounds is a random value from a geometric distribution, with

$$p = Pr\{failure\} = Pr\{X_{(r)} < Tr\},$$

where $X_{(r)}$ – is r-th order statistic of exponential distribution with parameter λ, and $r = n - k + 1$. This means that in a variational series

$$x_1 < x_2 < ... | \underbrace{x_r < ... < x_n}_{k},$$

at least one of last k values less than Tr. Total runtime can be considered as a random variable Y:

$$Y = \sum_{i=1}^{N_{round}-1} R_i + Tr, \tag{1}$$

where R_i are a i.i.d. for all $i \in \{\overline{1, N_{round}}\}$ and each R_i can be considered as $X_{(r)}$. The expectation of a random variable Y allows to obtain the objective function of the optimization problem $T_{k,\lambda}(n)$. The expectation of this sum we can estimate by using the Wald's equation [11] and present it in a following way

$$T_{k,\lambda}(n) = E\left[Y\right] = (E\left[N_{round}\right] - 1)E\left[X_{(r)}|X_{(r)} < Tr\right] + Tr. \qquad (2)$$

We first independently estimate two mathematical expectations. The first one is the expectation of random variable N_{round} with geometric distribution.

$$E\left[N_{round}\right] = \sum_{i=1}^{\infty} ip^{i-1}(1 - p) = \frac{1}{1 - p}. \qquad (3)$$

Here p is the probability that less than k nodes will remain operational for one round. So, we can find p as

$$p = Pr\left\{X_{(r)} < Tr\right\} = F_{X_{(r)}}(Tr) = \sum_{i=r}^{n} C_n^i F_X(Tr)^i (1 - F_X(Tr))^{n-i}$$

$$= \sum_{i=r}^{n} C_n^i (1 - e^{-\lambda Tr})^i (e^{-\lambda Tr})^{n-i}, \qquad (4)$$

where $F_X(\cdot)$ is the CDF of random variable X, and $F_{X_{(r)}}(\cdot)$ is the CDF of r-th order statistic of X.

The second one is the conditional expectation of one round runtime which is the expectation of r-th order statistic less than Tr.

$$E\left[X_{(r)}|X_{(r)} < Tr\right] = \int_0^{Tr} \frac{xf_{X_{(r)}}(x)}{Pr\left\{X_{(r)} < Tr\right\}} dx$$

$$= nC_{n-1}^{r-1} \int_0^{Tr} \frac{xF(x)^{r-1}(1 - F(x))^{n-r}f(x)}{p} dx$$

$$= nC_{n-1}^{r-1} \int_0^{Tr} \frac{x(1 - e^{-\lambda x})^{r-1} e^{-\lambda x(n-r)} \lambda e^{-\lambda x}}{p} dx$$

$$= \frac{\lambda nC_{n-1}^{r-1}}{p} \int_0^{Tr} x(1 - e^{-\lambda x})^{r-1} e^{-\lambda x(n-r+1)} dx. \qquad (5)$$

After substituting formulas 3 and 5 into 2, we obtain an expression for 1 in a following form:

$$T_{k,\lambda}(n) = \frac{\lambda nC_{n-1}^{r-1}}{p} \int_0^{Tr} x(1 - e^{-\lambda x})^{r-1} e^{-\lambda x(n-r+1)} dx \left(\frac{1}{1-p} - 1\right) + Tr$$

$$= \frac{\lambda nC_{n-1}^{r-1}}{1-p} \int_0^{Tr} x(1 - e^{-\lambda x})^{r-1} e^{-\lambda x(n-r+1)} dx + Tr. \qquad (6)$$

Now if the values k and λ are fixed the minimum of $T_{k,\lambda}(n)$ can be found.

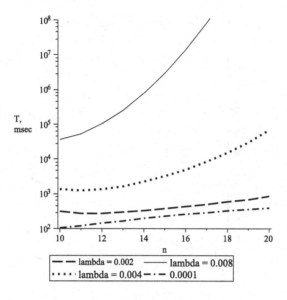

Fig. 2. Runtime dependency on n for $k = 10$

For example, there was received the results for fixed $k = 10$, and increasing λ, they are presented in the following Fig. 2.

According to the graphs shown in the Fig. 2, it can be concluded that the efficiency of introducing redundancy into the computing procedure depends on the failure intensity (λ) of individual nodes. With increasing intensity, the average runtime $(T_{k,\lambda})$ increases, but the redundancy begins to bring benefits. With a fixed dependence of the threshold function $Tr = O(n^2)$ and $k = 10$, the greatest gain is achieved at $\lambda = 0.004$. It means that average uptime is slightly (2 times) higher than Tr without redundancy. In the case $\lambda = 0.0001$ and $\lambda = 0.008$ (i.e. average uptime is significantly higher than Tr), redundancy is useless in terms of runtime.

For example, there were received the results for fixed $\lambda = 0.001$, and increasing k from 5 to 15, according to the definition of threshold schemes $n \geq k$, they are presented in the following Fig. 3.

The calculation results show, that optimal value n increases with k but there exists such optimum. For small values of k the redundancy gain is insignificant, in case of node failure it is easier to restart the procedure. With the growth of k, the choice of the optimal value of n gives an increasingly significant gain, since each restart of the procedure entails large overhead.

An accurate runtime estimate allows us to find the optimum of the function $T_{k,\lambda}(n)$ by numerical methods and can be applied to small values k and n. But the conditional expectation of one round runtime $E\left[X_{(r)} \mid X_{(r)} < Tr\right]$ can be substituted to the unconditional one. This substitution allows us to obtain an

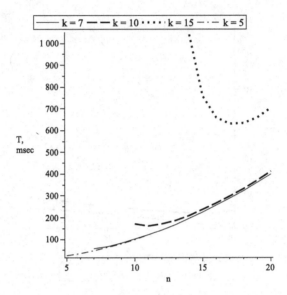

Fig. 3. Runtime dependency on n for $\lambda = 0.001$ and $n \geq k$

upper bound for the value of $T_{k,\lambda}(n)$, and the dependence on n gets a simpler form. This allows us to further solve the optimization problem using standard approaches: differentiation, search for extremum of the function.

$$E\left[X_{(r)}\right] = \int_0^\infty x f_{X_{(r)}}(x)dx = \sum_{i=k}^n \frac{1}{\lambda i}. \tag{7}$$

After substituting formulas 7 into 6, we obtain the upper bound for 1 in a following form:

$$T_{k,\lambda}(n) \leq \left(\frac{1}{\sum_{i=0}^{n-k} C_n^i (1 - e^{-\lambda Tr})^i (e^{-\lambda Tr})^{n-i}} - 1\right) \sum_{i=k}^n \frac{1}{\lambda i} + Tr. \tag{8}$$

The accuracy of the obtained upper bound can be estimated by using the numerical values of the function $T_{k,\lambda}(n)$ obtained in the previous examples. Comparison of accurate estimate with the upper bound allows understanding the possibility of using a simplified formula to obtain theoretical solutions. An example of such comparison for $k = 10$ and different values λ are shown in the Fig. 4.

The graphs shown in the Fig. 4 show that the used of the unconditional mathematical expectation insignificantly overestimates the absolute values of the estimate. Similar results were obtained in a series of experiments on different input data, which suggests that the use of the upper bound of $T_{k,\lambda}$ to solve the optimization problem will not lead to a significant decrease in accuracy.

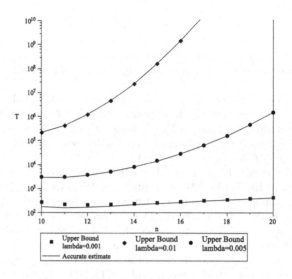

Fig. 4. Comparison of the accurate estimate and the upper bound

6 Conclusion

The paper deals with the problem of adding redundancy by using threshold (k, n)-scheme in the distributed computing procedure in order to improve fault tolerance in the presence of failures on individual nodes. For the considered protocols, an optimization problem was formulated to minimize the runtime of distributed computing protocol in failure conditions. As a result of solving the optimization problem, it was found the conditions of the redundancy efficiency. The following conclusion can be made if the average uptime of one node significantly exceeds the threshold (Tr) of the Protocol in (k, k)-scheme, the introduction of redundancy into the computing procedure has often not led to substantial gains. The effectiveness of the use of redundancy was shown in the cases where the average uptime is slightly (up to 2–3 times) higher than Tr of (k, k)-scheme. In such a case the optimum value of $n > k$ can be found for distributed threshold (k, n)-scheme. From the moment when one node uptime becomes less or equal to Tr of (k, k)-scheme adding redundancy again loses its effect. It means that the optimum value of n found by optimization problem solution is equal to k, i.e. (k, k)-scheme has the minimal runtime. All the above results were obtained for the case when the value of the runtime threshold (Tr) is a function of n^2. The proposed solution can be generalized for other types of Tr and n dependence and for other types of time delay distribution. The only requirement for its applicability is to satisfy Assumption 1 and 2.

References

1. Wang, H., Wang, Y.: Designing fault tolerance strategy by iterative redundancy for component-based distributed computing systems. Math. Probl. Eng. **2014**, 11 (2014)
2. Randles, M., Lamb, D., Odat, E., Taleb-Bendiab, A.: Distributed redundancy and robustness in complex systems. J. Comput. Syst. Sci. **77**, 293–304 (2011)
3. Brun, Y., Edwards, G., Bang, J.Y., Medvidovic, N.: Smart redundancy for distributed computation. In: 31st International Conference on Distributed Computing Systems, pp. 665–676, Minneapolis (2011)
4. Farley, N., Fitzpatrick, R., Jones, D.: BADGER - Blockchain Auditable Distributed (RSA) key GEneRation. In: IACR Cryptology ePrint Archive (2019)
5. Di Pietro, R., Mancini, L., Zanin, G.: Efficient and adaptive threshold signatures for ad hoc networks. Electron. Notes Theor. Comput. Sci. **171**, 93–105 (2007)
6. Wang, B., Cai, C., Zhou, Q.: A rational threshold signature model and protocol based on different permissions. J. Appl. Math. **2014**, 9 (2014)
7. Krouk, E., Semenov, S.: Transmission of a message during limited time with the help of transport coding. In: Proceedings of ICETE 2005 – International Conference on E-business and Telecommunication Networks, pp. 88–93, Reading (2005)
8. Krouk, E., Semenov, S.: Delivery of a message during limited time with the help of transport coding. In: Fifth IEEE Workshop on Signal Processing Advances in Wireless Communications, pp. 1–5. Lisbon (2004)
9. Kabatiansky, G., Krouk, E., Semenov, S.: Error Correcting Coding and Security for Data Networks: Analysis of the Superchannel Concept. Wiley, New Jersey (2005)
10. David, H., Nagaraja, H.N.: Order Statistics, 3rd edn. WileyInterscience, New York (2003)
11. Mitzenmacher, M., Upfal, E.: Probability and Computing. Cambridge University Press, Cambridge (2005)

Preemptive Priority Queuing System with Randomized Push-Out Mechanism and Negative Customers

Polina Shorenko, Oleg Zayats⓪, Alexander Ilyashenko$^{(\boxtimes)}$ ⓪, and Vladimir Muliukha ⓪

Peter the Great St. Petersburg Polytechnic University, St. Petersburg, Russia
shorenko.po@edu.spbstu.ru, zay.oleg@gmail.com, ilyashenko.alex@gmail.com, vladimir@mail.neva.ru

Abstract. A single-server priority queuing system with limited buffer size, Poisson arrivals, an exponentially distributed service time is considered. The primary customers take preemptive priority over secondary customers. We also consider a *randomized push-out mechanism*. It allows pushing secondary customers out of the system to free up space that could be taken by primary customers. Studied a new model where in addition to mentioned above two kinds of regular arriving customers, there are negative arrivals. A negative arrival has the effect of removing a customer from the buffer. The type of customer to be removed is determined in accordance with the following kill strategy. If at the moment of the occurrence of the next negative customer, both types of positive customers were presented in the system, then the primary customer is getting removed with a given probability. If there is only one type of customers in the system, then the customer of the existing type is deleted. Finally, if the system does not contain any positive customers at all, then a negative customer does not affect it. It is shown that such a queuing system can be investigated using the technique developed earlier by the authors for similar systems without negative customers. Using the method of generating functions, loss probabilities for both types of positive customers are obtained. The dependence of these loss probabilities on the basic parameters of the model (such as the probability of pushing out and the probability of crowding out a positive customer by a negative one) is investigated.

Keywords: Queuing systems · Randomized push-out mechanism · Negative customers · Preemptive priority · Poisson arrival

1 Introduction

Real data flows in modern computer networks are very complex in structure. In practice, packet traffic with a sufficient degree of accuracy can be considered a superposition of several homogeneous data streams. Separate traffic components can vary in many ways, such as the speed of transmission over communication channels, bandwidth used by virtual connections, processing algorithms in switching nodes, available buffer memory, methods of information protecting, economic characteristics of traffic and other characteristics similar to those listed.

© Springer Nature Switzerland AG 2019
O. Galinina et al. (Eds.): NEW2AN 2019/ruSMART 2019, LNCS 11660, pp. 305–317, 2019.
https://doi.org/10.1007/978-3-030-30859-9_26

Network technologies used in engineering practice are extremely diverse, and they are far ahead in their development of the level of their theoretical substantiation, as well as the development of adequate mathematical models. The stochastic nature of the functioning of computer networks and the structure of virtual connections formed in this case predetermines the expediency of using methods and models of queuing theory for their analysis. The methods of this theory make it possible to solve numerous and diverse problems of calculating the characteristics of the functioning of computer networks.

It is well-known that the main analytical method for the study of telematic devices is based on their consideration as specific queuing systems [1]. By telematics, we mean the section of computer science covering the field of telecommunications. For the early stage of the theoretical analysis of network interactions, the use of the simplest single-stream queuing system models was typical. Unfortunately, at present, such models do not properly reflect the main features of modern computer networks. The simplest models allow you to get only the most general qualitative understanding of the processes of network interaction, which is often insufficient for modern computer engineers.

As explained above, only multi-stream models of service systems can adequately describe the complex traffic transmission processes in real computer networks. This not only increases the level of detail, accuracy, and reliability of the analysis results but also allows solving fundamentally new tasks, the most important of which are management tasks. One of the relatively simple and the same time fairly accurately describing real network interactions, a class of queuing system models forms priority queuing systems with a probabilistic push-out mechanism. Such queuing systems were first introduced into science fifteen years ago in paper [2]. In specified work, a two-stream Markov model with non-preemptive priority and a randomized push-out mechanism was first studied. Subsequently, the authors of this work studied in detail a similar system but with preemptive priority [3, 4]. In combination with a probabilistic push-out mechanism, some other practically interesting options for setting priorities, including alternating priority and randomized priority, were analyzed in considered task [5, 6].

In all the previously mentioned papers, only the primary customers were taken into account, that is, those customers that for the first time entered the system. Meanwhile, in real conditions there is a technical possibility to check whether the customer reached the receiving device, and, if it did not, send it again. It is well-known that taking into account such retrial customers can drastically change the probabilistic characteristics of a queuing system. Accounting for repeated calls in systems with probabilistic pushing is a rather complicated task. The retrial queuing system, subject to the presence of preemptive priority was studied in the works [7, 8]. The paper [9] considers a similar task when non-preemptive priority is in effect. An interesting task was stated and solved in the article [10]. At the entrance of the retrial priority queuing system, there is only one stream of primary customers, which is treated as a high-priority. The flow of secondary customers was considered as low-priority (two types of priorities were dismantled: preemptive and non-preemptive). As it turned out, in such a system with a sufficiently high intensity of the primary flow, the phenomenon of "explosion" of the flow of secondary requirements is observed, the intensity of which may increase several thousand times.

This paper will consider a queuing system similar to the system [3, 4], but in the presence of a stream of so-called "negative" customers. Service systems with negative customers are widely used in computer, communication and manufacturing systems. The concept of negative customers was introduced by E. Gelenbe in the early nineties. Since then, many works have been published on this topic [11, 12]. We consider the simplest classical variant of the behavior of negative customers. We assume that when another negative customer arrives in the system, it captures one "positive" (that is, high-priority or low-priority) customer and leaves the system with it immediately. The probability of selecting a high-priority customer in this capture is given by the condition of the problem. For example, if a virus infection in a computer network leads to a system failure, in such a network, the appearance of a virus can be viewed as a negative customer admission, and performing all the usual calculations using processor represents positive customer.

A good example of the possibility of using such models of queuing systems is the problem of controlling a remote robotic object, the results of which, together with telemetry data and video stream, are transmitted using global networks [7, 15, 16]. In the considered example, a reasonable choice of the push-out probability makes it possible to balance such system performance indicators as the loss probability of control packets and the quality of video surveillance data for various network environment conditions.

2 Queuing System with Negative Customers

In this article, we consider the queuing system from [3–10] with another one incoming flow addition, with negative customers [13–15]. All parameters have the same meaning. Intensity of high priority positive customer flow is denoted as λ_1. Intensity of low priority positive customer flow is λ_2, service rate is μ and it's the same for all types of positive customers. For a randomized push-out mechanism, we use pushing-out probability α.

Negative customers flow has Poisson arrival rate and its intensity denoted as λ_0. Negative customers work as in [9, 10]. When a negative customer comes to the queuing system with positive customers it removes one positive customer and both are leaving the system. If it comes when the system is empty then it leaves without any action. If the system has only high-priority positive customers or only low-priority positive customers then a single negative customer removes one positive customer with probability equal to one. In another case it removes with probability β high-priority positive customer and with probability $1 - \beta$ low-priority positive customer.

In work [14] two strategies of pushing out positive customers by negative were considered. The first one is RCH (Remove the customer at the head) and the second one is RCE (Remove the customer at the end). In the first case, negative customer pushes out customer from the start of a buffer and in the second case from the end of a buffer. In this article, we used RCE strategy. When the buffer is empty it'll push out customer from service channel.

The scheme of this queuing system is presented in Fig. 1. $a_1(\tau)$, $a_2(\tau)$ denote distributions of time intervals between incoming positive customers. $b_1(\tau), b_2(\tau)$ denote distributions of service times for positive customers from both incoming flows. $a_0(\tau)$ is used for time intervals between negative customers coming to the system. k is a size of a buffer in the model and $\alpha \in [0, 1]$, $\beta \in [0, 1]$ are probabilities of pushing-out.

The phase space of the system introduced as pair of numbers:

$$\Omega = \left\{ (i,j) : i = \overline{0,k}; j = \overline{0,k-i} \right\} \tag{1}$$

Steady-state probabilities of queuing system states are

$$P_{i,j} = p\{N_1 = i, N_2 = j\}, \left(i = \overline{0,k}; j = \overline{0,k-i}\right), \tag{2}$$

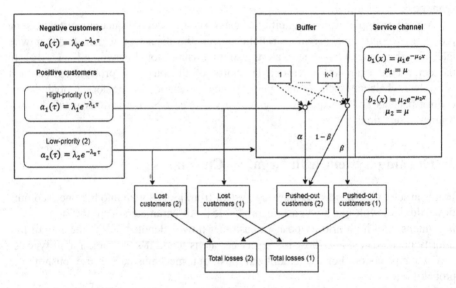

Fig. 1. The scheme of queuing system $\overrightarrow{M_2}/M/1/k/f_2^1$ with negative customer flow

where N_i is an amount of l-th type customers in queuing system storage and service channel. Using phase space (1), the state graph of the considered system will look like a graph from [6] with minor changes because of using negative customers. This graph is presented in Fig. 2. Using state graph from Fig. 2 and probabilities (2) system of balance equations can be obtained by using standard rules from [3–10] with agreement that $p_{i,j} \equiv 0$ ($i < 0$ or $j < 0$ or $i + j > k$).

Theorem 1. The system of balance equations for the queuing system $\overrightarrow{M_2}/M/1/k/f_2^1$ with negative customers is

$$- \left[\lambda_1\left(1 - \delta_{i+j,k}\right) + \alpha\lambda_1\left(1 - \delta_{i+j}\right)\delta_{i+j,k} + (1 - \alpha)\lambda_1\delta_{i+j,k}\delta_{i,0}\right.$$
$$+ \lambda_2\left(1 - \delta_{i+j,k}\right) + \left(\mu + \beta\lambda_0\right)\left(1 - \delta_{i,0}\delta_{j,0}\right)\Big]P_{i,j} + (\mu + \beta\lambda_0)P_{i+1,j}\left(1 - \delta_{j,0}\right)$$
$$+ (\mu + \lambda_0)P_{i+1,j}j\delta_{j,0} + (\mu + \lambda_0)P_{i,j+1}\delta_{i,0} + \lambda_1 P_{i-1,j} + \lambda_2 P_{i,j-1} \tag{3}$$
$$+ (1 - \beta)\lambda_0\left(1 - \delta_{i,0}\right)P_{i,j+1} + \alpha\lambda_1 P_{i-1,j+1}\delta_{i+j,k}$$
$$+ (1 - \alpha)\lambda_1 P_{i-1,j+1}\delta_{i+j,k}\delta_{i,1} = 0 \left(i = \overline{0,k}, j = \overline{0,k-i}\right).$$

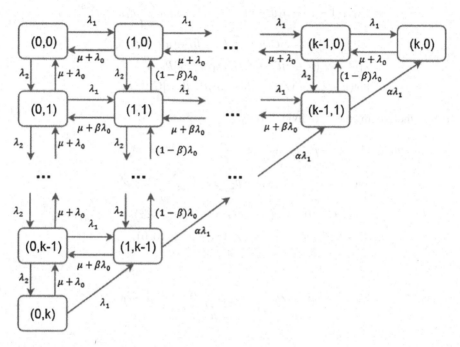

Fig. 2. The state graph for queuing system $\overrightarrow{M_2}/M/1/k/f_2^1$ with negative customers

3   Generating Functions Method Application

Using generating functions method lets introduce a generating function using probabilities (2) as

$$G(u, v) = \sum_{i=0}^{k}\sum_{j=0}^{k-i} P_{i,j}u^i v^j \tag{4}$$

Normalization condition will be

$$G(1, 1) = \sum_{i=0}^{k}\sum_{j=0}^{k-i} P_{i,j} = 1. \tag{5}$$

Then by summarizing Eqs. (3) following theorem can be introduced.

Theorem 2. The generating function for queuing system $\overrightarrow{M_2}/M/1/k/f_2^1$ with negative customers and probabilities (2) is

$$
\begin{aligned}
&[uv(\lambda_1(1-u)+\lambda_2(1-v)+(u-\lambda_0))-v(u+\beta\lambda_0)\\
&\quad -u(\lambda_0(1-\beta))]G(u,v)\\
&= [\lambda_1(v(1-u)+\alpha(u-v))+\lambda_2 v(1-v)]u\sum_{i=0}^{k}P_{i,k-i}u^i v^{k-i}\\
&\quad -\lambda_0(1-\beta)(u-v)G(u,0)+(\mu+\beta\lambda_0)(u-v)G(0,v)\\
&\quad +\lambda_1(1-\alpha)(u-v)P_{0,k}uv^k-\alpha\lambda_1(u-v)P_{k,o}u^{k+1}\\
&\quad +[uv(\mu+\lambda_0)-v\lambda_0(1-\beta)-u(\mu+\beta\lambda_o)]G(0,0).
\end{aligned}
\tag{6}
$$

Here is more convenient to divide (6) by μ

$$
\begin{aligned}
&[uv(\rho_1(1-u)+\rho_2(1-v)+\gamma_3)-v\gamma_1-u\gamma_2]G(u,v)\\
&= [\rho_1(v(1-u)+\alpha(u-v))+\rho_2 v(1-v)]_u\sum_{i=0}^{k}P_{i,k-i}u^i v^{k-i}\\
&\quad -\gamma_2(u-v)G(u,0)+\gamma_1(u-v)G(0,v)+\rho_1(1-\alpha)(u-v)P_{0,k}uv^k\\
&\quad -\alpha\rho_1(u-v)P_{k,o}u^{k+1}+[uv\gamma_3-v\gamma_2-u\gamma_1]G(0,0),
\end{aligned}
\tag{7}
$$

where

$$
\gamma_1 = \frac{\mu+\beta\lambda_0}{\mu}, \gamma_2 = \frac{\lambda_0(1+\beta)}{\mu}, \gamma_3 = \frac{\mu+\lambda_0}{\mu}
\tag{8}
$$

and

$$
\rho_i = \frac{\lambda_i}{\mu}, (i = \overline{0,2}), \rho = \sum_{i=0}^{2}\rho_i
\tag{9}
$$

Substituting u = v into (7), we obtain the generating function for the total amount of customers in our model

$$
G_\Sigma(u) = \frac{1-\eta}{1-\eta^{k+1}}\frac{\left(1-(u\eta)^{k+1}\right)}{1-u\eta},
\tag{10}
$$

where

$$
\eta = \frac{\rho_1+\rho_2}{\gamma_3}.
\tag{11}
$$

So, the probabilities that model has n customers in a buffer will be

$$r_n = P\{N = n\} = \frac{1 - \eta}{1 - \eta^{k+1}} \eta^n, \ (n = \overline{0, k}). \tag{12}$$

It is the well-known truncated geometric distribution.

Using technique from [6, 7] Eq. (7) can be expanded into the series in powers of u and v. Then probabilities of all model states can be expressed through two sets of probabilities: "diagonal" $P_{k-i,i}$ and "boundary" $P_{k-i,0}$:

$$
\begin{aligned}
P_{k-j,l} = {} & \alpha P_{k,0} \left(\theta_{j+1,l}^{(+)} - \theta_{j,l-1}^{(+)} \right) + \frac{\gamma_2}{\rho_1} \sum_{i=0}^{j-l-1} (\theta_{j-i,l+1}^{(+)} - \theta_{j-i-1,l}^{(-)}) P_{k-i,0} \\
& + \sum_{i=0}^{l} \theta_{j-i+1,l-i}^{(+)} P_{k-i,i} + \sum_{i=l+1}^{\frac{i+l}{2}} \theta_{j-i+1,-l+i}^{(-)} P_{k-i,i} \\
& - \alpha \sum_{i=0}^{l} \theta_{j-i+1,l-i}^{(+)} P_{k-i,i} - \alpha \sum_{i=l+2}^{\frac{i+l+1}{2}} \theta_{j-i+1,-l-1+i}^{(-)} P_{k-i,i} \\
& + ((\alpha - 1) - \tilde{\varepsilon}) \sum_{i=0}^{l} \theta_{j-i,-l+i}^{(-)} P_{k-i,i} + \tilde{\varepsilon} \sum_{i=0}^{l-1} \theta_{j-i,l-i-1}^{(+)} P_{k-i,i} \\
& + \tilde{\varepsilon} \sum_{i=l}^{\frac{i+l-2}{2}} \theta_{j-i,-l+i+1}^{(-)} P_{k-i,i} + (\alpha - 1) P_{0,k} \delta_{j,k} \delta_{l,j-1},
\end{aligned}
\tag{13}
$$

where

$$
\begin{aligned}
\theta_{j,l}^{(+)} &= \left(\frac{\gamma_1}{\rho_1} \right)^{\frac{i-1}{2}} \sum_{m=l}^{\frac{i-l+1}{2}} U_{j-1}^{(2m-l)}(\vartheta) \frac{\varepsilon^m \omega^{m-l} (-1)^l}{m!(m-l)!}, \\
\theta_{j,l}^{(-)} &= \left(\frac{\gamma_1}{\rho_1} \right)^{\frac{i-1}{2}} \sum_{m=0}^{\frac{i-l-1}{2}} U_{j-1}^{(2m+l)}(\vartheta) \frac{\varepsilon^m \omega^{m+l} (-1)^l}{m!(m+l)!}, \\
\end{aligned}
\tag{14}
$$

$$
\vartheta = \frac{\tilde{\vartheta}}{2\sqrt{\frac{\gamma_1}{\rho_1}}}, \varepsilon = \frac{\tilde{\varepsilon}}{2\sqrt{\frac{\gamma_1}{\rho_1}}}, \omega = \frac{\tilde{\omega}}{2\sqrt{\frac{\gamma_1}{\rho_1}}}, \tilde{\vartheta} = \frac{1+\rho}{\rho_1}, \tilde{\varepsilon} = \frac{\rho_2}{\rho_1}, \tilde{\omega} = \frac{\gamma_2}{\rho_1}.
$$

As a result, all probabilities can be calculated using only $2k + 1$ unknown probabilities.

It remains to build a shortened system of linear equations for probabilities which was mentioned above. The set of equations for "diagonal" probabilities can be obtained using the technique from [6, 7] by summation all diagonal probabilities of the state graph. The set of equations for "boundary" probabilities can be obtained using the method from [5, 6] by consideration limit $v \to 0$ of generating function (6):

$$(1 + \rho - \gamma_3)P_{0,0} - \gamma_3 P_{1,0} - \gamma_3 P_{0,1} = 0, (i = 0),$$

$$(1 + \rho)P_{i,0} - \rho_1 P_{i-1,0} - \gamma_3 P_{i+1,0} - \gamma_2 P_{i,1} = 0, (i = \overline{1, k-1}), \qquad (15)$$

$$(1 + \rho_0)P_{k,0} - \rho_1 P_{k-1,0} - \alpha \rho_1 P_{k-1,1} = 0, (i = k).$$

4 Computational Results

In this part, we calculate customer loss probability for both types of customers in considered queuing system depending on probabilistic parameters α and β using equations above. The following theorem will provide a formula for calculating numerical results.

Theorem 3. Loss probability for both types of positive customers in $\overrightarrow{M_2}/M/1/k/f_2^1$ queuing system with negative customers are:

$$P_{loss}^{(1)} = p_0 + (1 - \alpha) \sum_{i=1}^{k-1} p_i + \frac{\lambda_0}{\lambda_1} \sum_{i=1}^{k} P_{i,0} + \frac{\lambda_0}{\lambda_1} \beta \sum_{i=1}^{k-1} \sum_{j=1}^{k-i} P_{i,j},$$

$$P_{loss}^{(2)} = r_k + \frac{\lambda_1}{\lambda_2} \alpha \sum_{i=1}^{k-1} p_i + \frac{\lambda_1}{\lambda_2} p_k + \frac{\lambda_0}{\lambda_2} \sum_{i=1}^{k} P_{0,i} + \frac{\lambda_0}{\lambda_2} (1 - \beta) \sum_{i=1}^{k-1} \sum_{j=1}^{k-i} P_{i,j}, \qquad (16)$$

where $p_i = P_{k-i,i}$ are probabilities of states when system is full and
r_k – probabilities of a total amount of customers in queuing system.

4.1 Randomized Pushing-Out Parameter Dependency

Computational results for dependencies of high-priority and low-priority customers loss probability from α are presented below. For comparison decided to take same intensities as in [3–10] when $\rho_1 = 1.2, \rho_2 = 0.2$ and $\rho_1 = 0.2, \rho_2 = 0.9$. The dependency of loss probability for both types of positive customers with relative intensities $\rho_1 = 0.2, \rho_2 = 0.9$ is shown in Fig. 3. Here we can see linear behavior of dependency for both customers.

The probability of loss increases with increasing ρ_0 for both customers and also increases for high-priority customers with increasing β and decreases for low-priority under the same conditions. The dependency of loss probability for both types of positive customers with intensities $\rho_1 = 1.2, \rho_2 = 0.2$ is shown in Fig. 4. Here we can see the "closing" effect for low-type customers which was also observed in [6]. Also, we observe that the dependency of loss probability becomes closer to linear in case of increasing ρ_0.

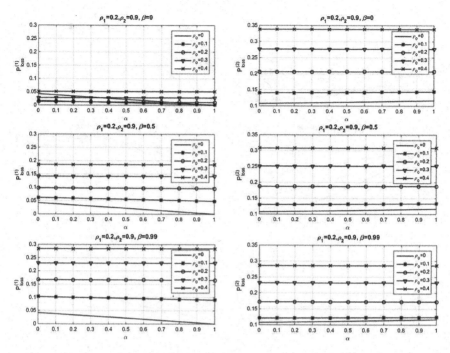

Fig. 3. Loss probability dependency of high-priority packets (left column) and low-priority packets (right column) on α and ρ_0 with intensities $\rho_1 = 0.2, \rho_2 = 0.9$ ($\beta = [0, 0.5, 0.99]$)

4.2   Areas of Linear Behavior Effect

In [6, 7] we introduced the effect of linear behavior for loss probabilities in such systems when loss probability dependency from parameter α for specific intensities is closed to linear. Areas of linear behavior high-priority packets (left column) and low-priority packets (right column) for levels $[0, 0.01, 0.05, 0.1, 0.25]$ are investigated in Fig. 5 as it was done in [6]. These areas were well studied in [6, 7]. For this model more interesting to investigate how the form of calculated areas changes for different intensities of negative customer flow. For high-priority customers, we can see that areas get smaller in the case of increasing ρ_0 and β. Also, we observe that the maximum deviation doesn't exceed level 0.25. For low-priority customers, we see that areas get smaller in case of increasing ρ_0 and the maximum deviation also doesn't exceed level 0.25. Although, in the case of increasing β we can see that maximum deviation grows and can exceed level 0.25 in the case of $\beta = 0.99$. It happens because in the case of $\beta = 0.5$ low-priority customers still cannot get into the system, so the probability of loss here approximately equals 1 and the maximum deviation here is too small. But when we increase β, low-priority customers have more chances to get into the system and probability of loss decreases. As a result, the maximum deviation grows. In common results in Fig. 5 correspond with previous results on Figs. 3 and 4 where the dependency of loss becomes closer to linear.

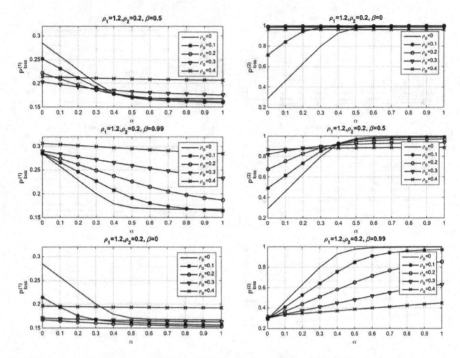

Fig. 4. Loss probability dependency of high-priority packets (left column) and low-priority packets (right column) on α and ρ_0 with intensities $\rho_1 = 1.2, \rho_2 = 0.2$ ($\beta = [0, 0.5, 0.99]$)

4.3   Areas of Closing Effect

One more effect was introduced in [6, 7] as an effect of model closing for low-priority positive customers. The model could be closed for the low-priority positive customer using only parameter α. On Fig. 6 and 7 are presented areas of closing for low-priority customers for levels $[0, 0.5, 0.9, 0.95, 0.99, 0.999, 0.9999]$ when $\alpha = 0.5$ and $\alpha = 0.9$. Each area corresponds to introduced above interval. In particular, the last area (white) corresponds to values of probability of loss which exceeds 0.9999. We can see that areas get smaller in the case of increasing β. Such behavior corresponds to the meaning of this parameter. When β increases, the amount of high-priority customers decreases in the system, so low-priority customers have more chances to get into the system and be served. Also, we observe that areas of closing slightly decrease in the case of increasing ρ_0.

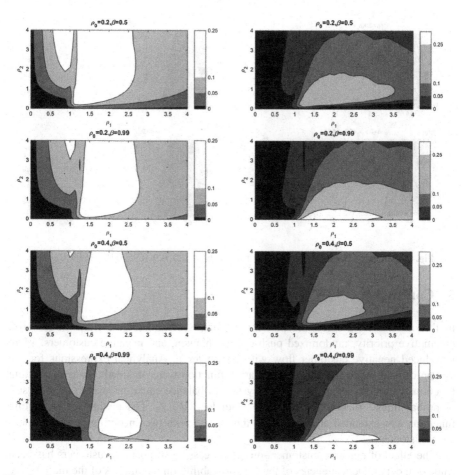

Fig. 5. Areas of linear behavior high-priority packets (left column) and low-priority packets (right column) for levels $[0, 0.01, 0.05, 0.1, 0.25]$

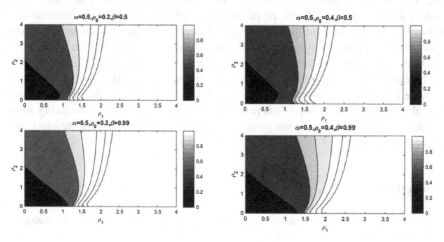

Fig. 6. Areas of closing with $\alpha = 0.5$ for levels $[0, 0.5, 0.9, 0.95, 0.99, 0.999, 0.9999]$

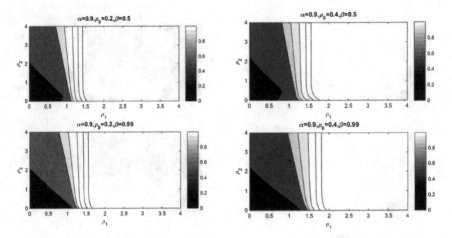

Fig. 7. Areas of closing with $\alpha = 0.9$ for levels $[0, 0.5, 0.9, 0.95, 0.99, 0.999, 0.9999]$

5   Conclusion

In this article, we studied the priority queuing system with the finite buffer size, preemptive priority, randomized push-out mechanism, and negative customers. Were introduced negative customer flow with parameter β. Analytical expressions for the system of balance equations and generating function were obtained. Using generating functions method was built shortened system of balance equations using only two sets of probabilities: "diagonal" $P_{k-i,i}$ and "boundary" $P_{k-i,0}$. The method of generating functions was extended and applied to the system with two incoming flows and negative customers.

The effect of negative customers on the processing of positive customers has been studied in detail. Dependencies of the loss probability on parameters of the model were found. The phenomenon of "closing" the system, as well as the effect of "linear behavior" of loss probabilities, are studied. These results allow to rationally select pushing-out parameter. It allows, on the one hand, effectively protect the information of high-priority traffic, and, on the other hand, drastically reduce the amount of computation, thereby increasing system performance. The service model studied in this article makes it possible to strictly quantitatively describe such features of the real network environment with hardware and software failures, etc. Also, using the studied model we were able to consider network packets transfer failures in used for remote robotics control satellite channel and to improve remote control quality by preventing model "closing" effect and improved the model characteristics calculation time by using areas of linear behavior.

References

1. Vishnevsky, V.M.: Theoretical bases of designing computer networks. Technosphere, Moscow (2003)

2. Avrachenkov, K.E., Vilchevsky, N.O., Shevlyakov, G.L.: Priority queueing with finite buffer size and randomized push-out mechanism. Perform. Eval. **61**(1), 1–16 (2005)
3. Ilyashenko, A., Zayats, O., Muliukha, V., Laboshin, L.: Further investigations of the priority queuing system with preemptive priority and randomized push-out mechanism. In: Balandin, S., Andreev, S., Koucheryavy, Y. (eds.) NEW2AN 2014. LNCS, vol. 8638, pp. 433–443. Springer, Cham (2014). https://doi.org/10.1007/978-3-319-10353-2_38
4. Muliukha, V., Ilyashenko, A., Zayats, O., Zaborovsky, V.: Preemptive queueing system with randomized push-out mechanism. Commun. Nonlinear Sci. Numer. Simul. **21**(1/3), 147–158 (2015)
5. Ilyashenko, A., Zayats, O., Muliukha, V., Lukashin, A.: Alternating priorities queueing system with randomized push-out mechanism. In: Balandin, S., Andreev, S., Koucheryavy, Y. (eds.) ruSMART 2015. LNCS, vol. 9247, pp. 436–445. Springer, Cham (2015). https://doi.org/10.1007/978-3-319-23126-6_38
6. Ilyashenko, A., Zayats, O., Muliukha, V.: Randomized priorities in queuing system with randomized push-out mechanism. In: Galinina, O., Balandin, S., Koucheryavy, Y. (eds.) NEW2AN/ruSMART -2016. LNCS, vol. 9870, pp. 230–237. Springer, Cham (2016). https://doi.org/10.1007/978-3-319-46301-8_19
7. Zayats, O., Korenevskaya, M., Ilyashenko, A., Lukashin, A.: Retrial queueing systems in series of space experiments "Kontur". Procedia Comput. Sci. **103**, 562–568 (2017)
8. Ilyashenko, A., Zayats, O., Korenevskaya, M., Muliukha, V.: A retrial queueing system with preemptive priority and randomized push-out mechanism. In: Galinina, O., Andreev, S., Balandin, S., Koucheryavy, Y. (eds.) NEW2AN/ruSMART/NsCC -2017. LNCS, vol. 10531, pp. 432–440. Springer, Cham (2017). https://doi.org/10.1007/978-3-319-67380-6_39
9. Korenevskaya, M., Zayats, O., Ilyashenko, A., Muliukha, V.: Retrial queueing system with randomized push-out mechanism and non-preemptive priority. Procedia Comput. Sci. **150**, 716–725 (2019)
10. Korenevskaya, M., Zayats, O., Ilyashenko, A., Muliukha, V.: The phenomenon of secondary flow explosion in retrial priority queueing system with randomized push-out mechanism. In: Galinina, O., Andreev, S., Balandin, S., Koucheryavy, Y. (eds.) NEW2AN/ruSMART - 2018. LNCS, vol. 11118, pp. 236–246. Springer, Cham (2018). https://doi.org/10.1007/978-3-030-01168-0_22
11. Do, T.V.: An initiative for a classified bibliography on G-networks. Perform. Eval. **68**(3), 385–394 (2011)
12. Do, T.V.: Bibliography on G-networks, negative customers and applications. Math. Comput. Model. **53**(2), 205–212 (2011)
13. Gelenbe, E.: Product form networks with negative and positive customers. J. Appl. Probab. **28**(3), 656–663 (1991)
14. Gelenbe, E., Glynn, P., Siegman, K.: Queues with positive and negative arrivals. J. Appl. Probab. **28**(3), 245–250 (1991)
15. Zaborovsky, V., Muliukha, V., Ilyashenko, A.: Cyber-physical approach in a series of space experiments "Kontur". In: Balandin, S., Andreev, S., Koucheryavy, Y. (eds.) ruSMART 2015. LNCS, vol. 9247, pp. 745–758. Springer, Cham (2015). https://doi.org/10.1007/978-3-319-23126-6_69
16. Zaborovsky, V., Guk, M., Muliukha, V., Ilyashenko, A.: Cyber-physical approach to the network-centric robot control problems. In: Balandin, S., Andreev, S., Koucheryavy, Y. (eds.) NEW2AN 2014. LNCS, vol. 8638, pp. 619–629. Springer, Cham (2014). https://doi.org/10.1007/978-3-319-10353-2_57

Development of Analytical Framework for Evaluation of LTE-LAA Probabilistic Metrics

Maksym V. Korshykov[1]([✉]) [iD], Anastasia V. Daraseliya[1] [iD],
and Eduard S. Sopin[1,2] [iD]

[1] Department of Applied Probability and Informatics, Peoples' Friendship University
of Russia (RUDN University), 6 Miklukho-Maklaya Street,
Moscow 117198, Russian Federation
1032144224@pfur.ru
[2] Institute of Informatics Problems, FRC CSC RAS,
Vavilova 44-2, Moscow 119333, Russia

Abstract. Nowadays, the number of mobile subscribers has increased significantly. It is predicted that the number of subscribers will reach 9.5 billion by 2020. With such rapid growth, telecommunications networks suffer from several problems, such as lack of bandwidth, the occurrence of collisions and interference. To cope with the problems, it was proposed to fill the lack of the licensed spectrum with an unlicensed frequency bands. However with this approach, it is necessary to provide fair coexistence in the unlicensed band with existing users. The use of LTE - LAA technology is designed to solve this problem. The paper builds an analytical model of the LAA work scheme using a discrete time Markov process, which provides a probabilistic estimate of the collision and data transmission probabilities. This work also outlines all the principles and specifications necessary to describe the functioning of the model. A description of the LBT mechanism for the unlicensed spectrum is given. After analytically obtained expressions for collision and transmission probabilities, numerical results are presented.

Keywords: LTE · LAA · Analytical model · Queuing system

1 Introduction

This paper analyses Licensed-Assisted-Access (LAA) technology. At the moment, the availability of frequencies for traffic transmission is the most important aspect, therefore LAA is widely considered as a solution to this problem. The main idea of the technology implies that it is possible to use license and

The publication has been prepared with the support of the "RUDN University Program 5-100" (mathematical model development). The reported study was funded by RFBR, project numbers 17-07-00845 and 19-07-00933 (numerical analysis).

O. Galinina et al. (Eds.): NEW2AN 2019/ruSMART 2019, LNCS 11660, pp. 318–328, 2019.
https://doi.org/10.1007/978-3-030-30859-9_27

unlicensed carrier aggregation for different frequencies utilisation in the data transmission process [1]. The technology uses unlicensed spectrum free capacity to create a connection with LTE in the transmission process, and decrease load on the licensed spectrum [2]. This paper is devoted to the problem of estimating the probability of collisions and the probability of successful data transfer in case of using the LTE-LAA technology.

2 Standards and Specifications

2.1 Description of LAA Operation Using LTE

The LAA is described in 3GPP as an improvement of the LTE, with extension to the unlicensed spectrum, which was standardised in 3rd Generation Partnership Project (3GPP) release 13 [3]. In LAA, licensed carriers are aggregated with unlicensed, in order to improve the throughput of a user downstream signal, offering seamless mobility support. In order to coexist with Wi-Fi in an unlicensed spectrum, improvements must include communication protocols based on Listen Before Talk (LBT) mechanisms, intermittent data transmission on a carrier with a limited maximum transmission frequency. Also, dynamic frequency selection in order to avoid certain frequencies will be an important aspect of communication. Multicore data transmitting through multiple unlicensed channels should also be used for this kind of improvement.

Modified LTE broadcast used to operate in the unlicensed spectrum. Such an improvement is required for fair coexistence between LTE and Wi-Fi, as well as between operators operating in LTE. To ensure the coexistence of this kind, it is necessary to apply such technologies as LBT, dynamic frequency selection (DFS), carrier selection, etc.

Aggregation in LAA is performed using primary and secondary cells. Primary cells operating in the licensed frequency range are used to ensure the transmission of critical information and guaranteed quality of service (QoS). Secondary cells, which in turn operate in the unlicensed frequency range, will increase the data transfer rate due to aggregation with the primary cells.

2.2 Carrier Aggregation

For the first time carrier aggregation appeared in the 3GPP 10th release [4]. It allowed operators to use their frequencies more efficiently. In the 3GPP 12th release [5], aggregation appeared, with support for bandwidths up to 1000 MHz and up to 5 carriers. Up to 32 carriers were supported in 3GPP release 13 [3]. The use of carriers in an unlicensed spectrum in conjunction with the LAA is assumed. After improvements, it is expected that LTE terminals will cope with frequencies up to 640 MHz, most of which will be located in the unlicensed spectrum, thereby greatly increasing the data transfer rate.

3 Related Works

There are a number of papers that analyse licensed and unlicensed spectrum coexistence using LAA technology [6–8]. Here, we consider several important works. Processor sharing was widely investigated by research community and applied for the analysis of wire and wireless communication systems and networks. This approach was proposed as a decision for spectrum sharing. Technology suggests to rent free bands between different operators [9,10].

Papers [11]–[12] address the fair coexistence process as a capturing part of the unlicensed spectrum by LTE taking into account existing users of the unlicensed spectrum. Nowadays, there are two main approaches LWA and LTE-LAA [13]. In this regard, the development of the concept of fair coexistence, especially for LTE-LAA, is of great importance. Initially, it was assumed that coexistence with LAA would be based on the technology of multiple access/collision avoidance with carrier control (CSMA/CA) [14], but this approach revealed a number of shortcomings that reduce network efficiency. A DCF Wi-Fi scheme was proposed. Subsequent work addressed such optimisation issues as the size of the contention window (CW) for LAA nodes [15].

The development of the contention window theme resulted in the fact that the technology of adaptive CW was developed, which made it possible to optimise channel occupancy [16]. As a result, a description of the two modes was suggested. The first is LBT without a random backoff, which uses a fixed CW, and LBT with a binary exponential CW, to coexist on the network with other devices already present in paper [17]. Later, the 3GPP evolution of the coexistence of Wi-Fi and LTE-LAA LBT [18] was made and standardised.

An important work is [19], on which we mainly relied in the development of the paper. This work establishes an accurate analytical model that allows to analytically evaluate the LTE-LAA coexistence mechanism, as described in the standardised technical specification (TS) 36.213 of 3GPP release 13 and 14 [18] of LTE-LAA LBT and outlines the effects of the proposed coexistence mechanism and its parameters by extensive analysis.

In this paper, we enhance the Markov chain approach from [19], which in turn resulted in a significant simplification of the analytical expressions for transmission and collisions probabilities. Moreover model takes into account number of repeated retransmissions K may be set at any value $K \geq 1$, in contrast to work [19], where only $K = 1$ is possible.

4   System Model

As already mentioned, to ensure the fair coexistence of LAA - LTE, it is necessary to apply LBT technology. Using this mechanism, the equipment performs a clear channel estimate (CCA), which allows to evaluate the channel availability before transmission. Although LBT is not a requirement in all regions, this mechanism is the basic principle of ensuring fair coexistence in an unlicensed spectrum within a global decision.

Now we will describe the principle of operation of LBT in more detail. It is believed that the LAA technology will use the CoLBT mechanism (LBT based on channel observation) [20]. If the communication channel is idle, the eNB switches from CCA (pure channel estimate) to ECCA (extended pure channel

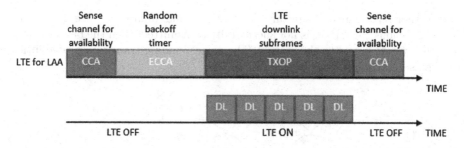

Fig. 1. LBT mechanism of LAA

estimate) by starting the back-off timer with a random value selection. After the CAA is an idle channel, the observation time is divided into slots with a specific duration, which can be a constant value (Fig. 1).

After the specified interval, during which the channel was empty, the value of the back-off timer is determined by one. Then TXOP (transmission opportunity) appears, but it is available only at the beginning of the time slot, when the value of the back-off timer reaches one. If, as a result of the observation, the channel was busy, then the eNB pauses the back-off timer and continues to listen to the channel. The timer starts only when the channel is free.

It is important to note that the eNB has the ability to estimate the channel based on the probability of a collision, which occurs when several devices on the network transmit in one time slot. In such a situation, an increase in CW occurs. The probability of collisions also depends on the number of devices in the unlicensed range connected to the network. Thus, an increase in CW, taking into account the probability of collisions, ensures the fair coexistence of LAA - LTE using LBT mechanisms [18].

5   Analysis of Transmission and Collision Metrics

The behaviour of a single user can be described using a discrete time Markov process $X_n, n \geq 0$. Values of X_n are from $\{0, 1, ..., M, M+1, ..., M+K\}$. $X_n = m$ denotes stage of transmission (number of unsuccessful transmissions of current packet) means that the size of CW is uniformly distributed in $[0, 2^m W - 1]$. In the initial state 0, where the backoff counter is decreased to 0, the system may move to a new state 1, with a probability γ, or return to state 0, with a probability $1 - \gamma$.

Here γ is the probability of a collision. With the transition to the new state m, the CW size changes to another $W = 2^m W - 1$, where $m = \overline{0, M}$, and W is the minimum value of CW. Parameter M determined as the maximum number of unsuccessful transmissions during which it is possible to increase CW. In state m, the pocket can be transferred successfully with the probability of $1 - \gamma$ and change state to 0. The collision occurs with the probability γ and system state change to $m+1$. At the state M the maximum CW size is reached and further retransmissions occur with the same maximum CW size. If after K retransmissions the packet is not successfully transmitted, then the packet is lost and the system resets to the state 0. The corresponding transition probabilities diagrams depicted on Fig. 2.

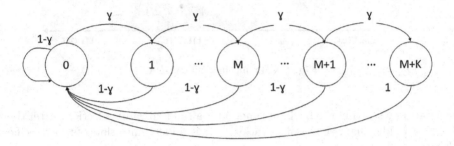

Fig. 2. Transitions intensity diagram

The transition probabilities matrix Q of $\{X_n\}$ is easily derived from Fig. 2.

$$
Q = \begin{pmatrix}
1-\gamma & \gamma & 0 & & 0 \\
1-\gamma & 0 & \gamma & \cdots & 0 \\
1-\gamma & 0 & 0 & & 0 \\
\vdots & & & \ddots & \vdots \\
1-\gamma & 0 & 0 & \cdots & \gamma \\
1 & 0 & 0 & & 0
\end{pmatrix} \tag{1}
$$

Let q_m be the stationary probability of $\{X_n\}$: $q_m = \lim_{n\to\infty} P\{X_n = m\}$. The system of equilibrium equations is given below:

$$
q_0 = \sum_{m=0}^{M+K-1} (1-\gamma)\, q_m + q_{M+K} \tag{2}
$$

$$
q_m = \gamma q_{m-1} \qquad m = 1, \dots, M+K \tag{3}
$$

With the equilibrium Eqs. (2) and (3), the expressions for the stationary probabilities are obtained.

$$
\sum_{m=0}^{M+K} q_m = \sum_{m=0}^{M+K} \gamma^m q_0 = q_0 \frac{1-\gamma^{M+K+1}}{1-\gamma}; \tag{4}
$$

Finally, we use the normalisation condition to express q_0 and explicitly write down probabilities in analytical way.

$$
1 = q_0 \frac{1-\gamma^{M+K+1}}{1-\gamma} \Rightarrow q_0 = \frac{1-\gamma}{1-\gamma^{M+K+1}}; \tag{5}
$$

The interrelations of collision probability γ and transmission probability π_{tr} determined by the system of two Eqs. (6) and (7).

$$\pi_{tr} = \left(\sum_{m=0}^{M+K} q_m b_m \right)^{-1} \tag{6}$$

$$\gamma = 1 - (1 - \pi_{tr})^{n-1} \tag{7}$$

The first Eq. (6) for a π_{tr} is obtained as follows. In any state of $\{X_n\}$ the system spends one time slot for transmission. Thus, the transmission probability is one time slot divided by the average number of time slots in any state. This means, that in denominator we need to sum b_m the average number of time slots that packet spends in state m, multiplied by q_m, the probability, that packet is in the m state.

The equation number (7) describes the probability of collisions γ. This can happen if at least one of $n - 1$ users transmit data in the system in the same timeslot b_m. In this equation, $(1 - \pi_{tr})^{n-1}$ - is the probability, that no one is transmitting data, except the current user. Then $1 - (1 - \pi_{tr})^{n-1}$ means, that at least one of $n - 1$ user is transmitting data.

The average number of time slots b_m that the system spends in the state m is given in (8).

$$b_m = \sum_{i=0}^{2^m W - 1} i \frac{1}{2^m W} = \frac{1}{2^m W} \frac{2^m W(2^m W - 1)}{2} = \frac{2^m W - 1}{2} \tag{8}$$

Then denominator of π_{tr} is expressed as follows.

$$\begin{aligned}
\sum_{m=0}^{M+K} q_m b_m &= \sum_{m=0}^{M+K} \gamma^m q_0 \frac{2^m W - 1}{2} = \frac{q_0}{2} \sum_{m=0}^{M+K} \gamma^m (2^m W - 1) \\
&= \frac{q_0}{2} \sum_{m=0}^{M+K} \gamma^m 2^m W - \frac{q_0}{2} \sum_{m=0}^{M+K} \gamma^m \\
&= \frac{q_0}{2} \left(W \frac{1 - (2\gamma)^{M+K+1}}{1 - 2\gamma} - \frac{1 - \gamma^{M+K+1}}{1 - \gamma} \right);
\end{aligned} \tag{9}$$

Finally, system (6)–(7) can be written in the following form.

$$\begin{cases} \pi_{tr} = \frac{q_0}{2} \left(W \frac{1 - (2\gamma)^{M+K+1}}{1 - 2\gamma} - \frac{1 - \gamma^{M+K+1}}{1 - \gamma} \right)^{-1} \\ \gamma = 1 - (1 - \pi_{tr})^{n-1} \end{cases} \tag{10}$$

6   Numerical Results

To obtain numerical results, a program implementing the Newton method with quadratic convergence was written in the Python programming language. The numerical method finds a solution of a system of two equations. The following parameters were taken as initial data: $W = 16$, $M = 2, 4, 6$. Probabilities were calculated for the following value of $K = 0, 4$. Below are graphs of the results.

Numerical estimation of transmission probabilities is shown on Figs. 3 and 4. It is easy to notice, that with increasing parameter K, dependence on M decreases. Behaviour with different M becomes similar. It can be said, that transmission probability is significantly decreasing in some initial interval. Starting with the number of users $n = 20$ probability starts to fall less.

Estimation of collision probability is given on Figs. 5 and 6. There is a similar behaviour kind, when parameter K is increasing. Curves tend to the same values.

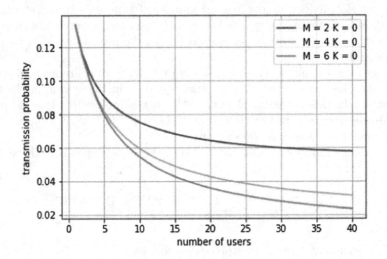

Fig. 3. Transmission probabilities for LAA configurations

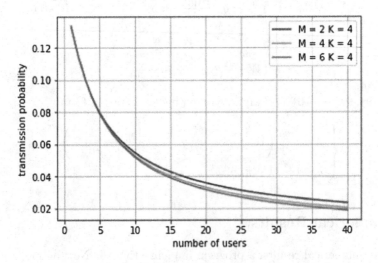

Fig. 4. Transmission probabilities for LAA configurations

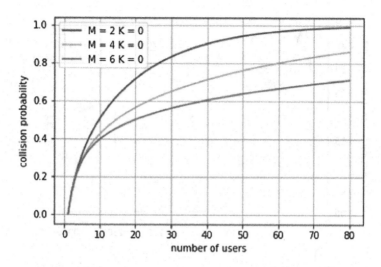

Fig. 5. Collision probabilities for LAA configurations

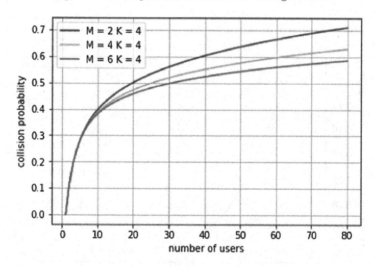

Fig. 6. Collision probabilities for LAA configurations

This can be explained by the fact, that with larger number of retransmissions K, on maximum CW size there are more possibilities for package transmission. Numerically considered, the matter can be summed up thus: growing number of K retransmissions allows system to keep an acceptable level of transmission probability, and helps us to reduce collisions.

Here is an important result. On Figs. 7 and 8 there is an estimation of total successful transmission with dependence on number of users $\pi_S = n(1 - \gamma)\pi_{tr}$. Let's call this metric as saturation point. On pictures you can find maximum for each configuration. This means, that with this estimation it's possible to obtain an optimal number of users for different parameters values.

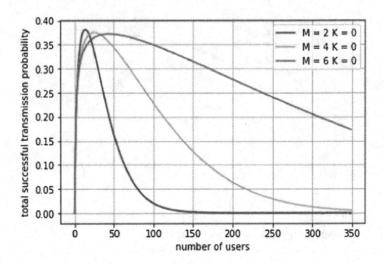

Fig. 7. Maximum total transmission probability for LAA configurations

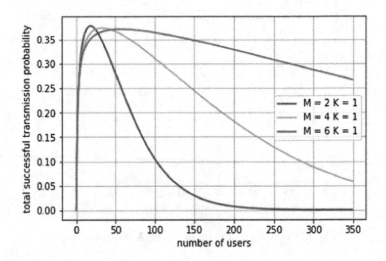

Fig. 8. Maximum total transmission probability for LAA configurations

It can be observed that as K increases, the acceptable transmission probability grows for larger number of users. Maximum total transmission probability is depicted for every configuration an has it's own saturation point. On the left hand of any saturation point there are not so many users as it could be. This means, that collisions occurs not often and there are more resources could be used. On the right hand it's absolutely opposite picture. There are too many users require service. This means, that collisions occur much rapidly and quality of service decreases. Saturation point could be used for selection of characteristics or calibrations and further network deployment.

7 Conclusion

In conclusion, it would be important to note that all necessary specifications and operating principles of LTE-LAA were taken into account. The paper proposed a simple discrete time Markov Process that describe the system behaviour which using the LBT mechanism. This discrete time Markov Process allows to calculate the probability of transmission or collisions. Then a system of two equations was obtained, which allows us to determine the probability of transmission and the probability of a collision in the network. Based on the obtained results in finding of local maximum with help of probabilistic estimation it's possible to choose the optimal parameters to search the largest number of users in various configurations. Formulation and solution of the optimisation problem is a priority for further research.

References

1. TR36.889, v1.0.1, Study on Licensed-Assisted Access to Unlicensed Spectrum
2. 3GPP TD RP-141664: Study on Licensed-Assisted Access using LTE
3. 3GPP TSG CT Meeting 69, Release 13, RP-151569, Phoenix, Arizona, USA, 14–15 September 2015
4. Overview of 3GPP Release 10 V0.2.1 (2014–06)
5. Overview of 3GPP Release 12 V0.2.0 (2015–09)
6. Markova, E., et al.: Analytical models for schedule-based license assisted access (LAA) LTE systems. In: Galinina, O., Andreev, S., Balandin, S., Koucheryavy, Y. (eds.) NEW2AN/ruSMART -2018. LNCS, vol. 11118, pp. 210–223. Springer, Cham (2018). https://doi.org/10.1007/978-3-030-01168-0_20
7. Xiao, J., Zheng, J., Chu, L., Ren, Q.: Performance modeling of LAA LBT with random backoff and a variable contention window. In: 10th International Conference on Wireless Communications and Signal Processing (WCSP) (2018). https://doi.org/10.1109/WCSP.2018.8555559
8. Chen, Q., Yu, G., Ding, Z.: Enhanced LAA for unlicensed LTE deployment based on TXOP contention. IEEE Trans. Commun. **67**(1), 417–429 (2019). https://doi.org/10.1109/TCOMM.2018.2868694
9. Samouylov, K., Naumov, V., Sopin, E., Gudkova, I., Shorgin, S.: Sojourn time analysis for processor sharing loss system with unreliable server. In: Wittevrongel, S., Phung-Duc, T. (eds.) ASMTA 2016. LNCS, vol. 9845, pp. 284–297. Springer, Cham (2016). https://doi.org/10.1007/978-3-319-43904-4_20
10. Samouylov, K., Sopin, E., Gudkova, I.: Sojourn time analysis for processor sharing loss queuing system with service interruptions and MAP arrivals. In: Vishnevskiy, V.M., Samouylov, K.E., Kozyrev, D.V. (eds.) DCCN 2016. CCIS, vol. 678, pp. 406–417. Springer, Cham (2016). https://doi.org/10.1007/978-3-319-51917-3_36
11. Xu, S., Li, Y., Gao, Y., Liu, Y., Gačanin, H.: Opportunistic coexistence of LTE and WiFi for future 5G system: experimental performance evaluationand analysis. IEEE Access **6**, 8725–8741 (2018). https://doi.org/10.1109/ACCESS.2017.2787783

12. Ericsson: LTE License Assisted Access. https://web.archive.org/web/20161014143936/, http://www.ericsson.com/res/thecompany/docs/press/media-kits/ericssonlicense-assisted-access-laa-january-2015.pdf. Accessed May 2018
13. Alternative LTE Solutions in Unlicensed Spectrum: Overview of LWA, LTE-LAA and Beyond. Intel Corporation, White paper (2016)
14. Mukherjee, A., et al.: System architecture and coexistence evaluation of licensed-assisted access LTE with IEEE 802.11. In: Proceedings IEEE International Conference Communication Workshop (ICCW), June 2015, pp. 2350–2355. https://doi.org/10.1109/ICCW.2015.7247532
15. Song, Y., Sung, K.W., Han, Y.: Coexistence of Wi-Fi and cellular with listen-before-talk in unlicensed spectrum. IEEE Commun. Lett. **20**(1), 161–164 (2016). https://doi.org/10.1109/LCOMM.2015.2504509
16. Mushunuri, V., Panigrahi, B., Rath, H.K., Simha, A.: Fair and efficient listen before talk (LBT) technique for LTE licensed assisted access (LAA) networks. In: Proceedings IEEE 31st International Conference Advanced Information Networking and Applications (AINA), Taipei, Taiwan, March 2017, pp. 39–45. https://doi.org/10.1109/AINA.2017.135
17. Feasibility Study on Licensed-Assisted Access to Unlicensed Spectrum, Release 13, document TR 36.889, 3GPP, June 2015
18. Evolved Universal Terrestrial Radio Access (E-UTRA) Physical Layer Procedures, document TS 36.213 V13.6.0, 3GPP, June 2017
19. Bitar, N., Al Kalaa, M.O., Seidman, S.J., Refai, H.H.: On the coexistence of LTE-LAA in the unlicensed band: modeling and performance analysis. IEEE Access **6**, 52668–52681 (2018). https://doi.org/10.1109/ACCESS.2018.2870757. Digital Object Identifier
20. Ali, R., Shahin, N., Musaddiq, A., Kim, B.-S., Kim, S.W.: Fair and efficient channel observation-based listen-before talk (CoLBT) for LAA-WiFi coexistence in unlicensed LTE. In: Tenth International Conference on Ubiquitous and Future Networks (ICUFN) (2018). https://doi.org/10.1109/ICUFN.2018.8436776

Beamforming Signal Processing Performance Analysis for Massive MIMO Systems

Irina Stepanets[1] and Grigoriy Fokin[2(✉)]

[1] Deutsche Telekom, Technische Planung und Rollout,
Landgrabenweg 151, 53227 Bonn, Germany
irina.stepanets@telekom.de
[2] The Bonch-Bruevich St. Petersburg State University of Telecommunications,
22 Prospekt Bolshevikov, 193232 St. Petersburg, Russia
grihafokin@gmail.com

Abstract. ITU-R requirements related to technical performance for IMT-2020 radio interfaces claims average spectral efficiency (SE) up to 9 bit/s/Hz for indoor hotspot, 7.8 bit/s/Hz for dense urban and 3.3 bit/s/Hz for rural environment. The purpose of this work is to analyse the performance of beamforming signal processing techniques for various massive MIMO configurations in terms of SE and reveal scenarios, which meet ITU-R requirements. Analysis is performed for single user (SU) and multi-user (MU) cases with 14, 50, 100 users, various number of antenna ports (8, 32, 64 and 128) and three beamforming techniques: matched filter (MF), minimum mean square error (MMSE) and zero-forcing (ZF). Simulation results compare SE as a function of signal-to-noise ratio (SNR) for various beamforming techniques and conclude that at higher SNR MMSE acts as ZF and at low SNR MMSE acts as MF for the case, when the number of users is much lower than the number of antenna ports. MU case analysis reveal, that ITU-R average SE requirements hold only in rural environment for high SNR values and 128 antenna ports.

Keywords: ITU-R requirements · 5G · Massive MIMO · Beamforming · MF · ZF · MMSE · Spectral efficiency

1 Introduction

The growing demands for higher data rates dictates more stringent requirements towards the next 5th generation wireless communication radio networks (5G). ITU-R requirements related to technical performance for IMT-2020 radio interfaces claims that 5G technology should guarantee peak spectral efficiency (SE) up to 30 bit/s/Hz in downlink (DL), 15 bit/s/Hz in uplink (UL), and the average user spectral efficiency up to 9 bit/s/Hz for indoor hotspot, 7.8 bit/s/Hz for dense urban and 3.3 bit/s/Hz for rural environment in DL [1]. These requirements give rise the application of novel transceiver technique like massive multiple-input multiple-output (MIMO), which utilizes a large number of antenna ports and enable to leverage the spatial diversity maintained by beamforming signal processing techniques.

© Springer Nature Switzerland AG 2019
O. Galinina et al. (Eds.): NEW2AN 2019/ruSMART 2019, LNCS 11660, pp. 329–341, 2019.
https://doi.org/10.1007/978-3-030-30859-9_28

In this paper beamforming techniques classification, analysis and simulation is carried out for three signal processing algorithms: matched filter (MF), minimum mean square error (MMSE) and zero-forcing (ZF). These techniques are evaluated in terms of average spectral efficiency (SE) for various massive MIMO configurations with 8, 32, 64 and 128 antenna ports number, which are currently offered by vendors to mobile providers [2, 3], in order to analyze its ability to meet SE requirements related to technical performance for IMT-2020 radio interfaces.

The material in the paper is organized in the following order. Beamforming signal processing techniques are classified and analyzed in Sect. 2. Simulation scenarios and results are given in Sect. 3. Finally, we draw the conclusions in Sect. 4.

2 Beamforming Signal Processing Techniques Analysis

2.1 Beamforming Signal Processing Techniques Classification

Valuable property of massive MIMO systems is its ability to produce and steer the adaptive antenna lobe, which is termed beamforming. There are two important aspects for enabling required beamforming gains for 5G requirements. One is proper antenna design and another is appropriate signal processing technique. According to [4] the last one is implemented in the radio frequency (RF) processing chain as depicted in Fig. 1.

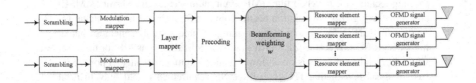

Fig. 1. Beamforming placement in a signal processing chain.

Antenna design has an effect on the physical parameters of beamforming techniques and consequently defines the form of the adaptive beam and its pointing accuracy toward the user equipment (UE) of interest as well as null steering in the direction of interfering UE. These aspects were discussed and classified in [5].

Beamforming signal processing techniques enables transmission and reception in required directions by applying proper antenna port phases and amplitudes by so called *weighting* approach. The complex weights w_k are instantaneously calculated by an adaptive algorithm, in order to direct the maximum antenna radiation pattern toward the desired UE and to steer null antenna pattern toward interfering UEs. Corresponding weights w_k can be described by [6]:

$$w_k = p_k e^{j\varphi_k}, \qquad (1)$$

where p_k is a gain magnitude and φ_k is a phase shift of kth RF antenna port.

Transmit beamformer signal processing technique at BS estimates weights w in order to transmit signal $\hat{X}_{BS}$ via proper antenna lobe in the direction of desired UE and to place nulls of the antenna diagram into direction of interfering UEs, which mathematically can be represented by

$$x_k = \sum_{k=1}^{M} w_{Txk}\hat{X}_{BS} \tag{2}$$

where x_k is an output beamformed signal from the transmit BS antenna (Figs. 2, 3 and 4), M is a number of antenna elements at BS and w_{Txk} – corresponding BS antenna weight.

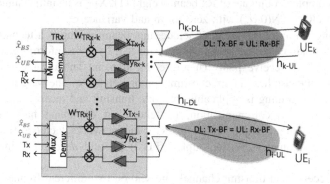

Fig. 2. Transmit and receive beamforming in TDD.

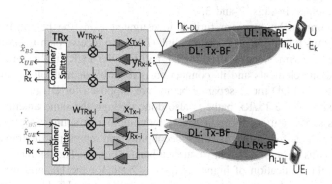

Fig. 3. Transmit and receive beamforming in FDD.

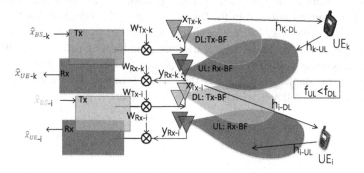

Fig. 4. Transmit and receive beamforming in 5G NR.

Receive beamformer signal processing technique at BS has a reverse task, namely, to estimate UE signal of interest $\hat{X}_{EU}$, minimizing received interference [7]:

$$y_k = \sum_{k=1}^{M} w_{Rxk}^* h_k \hat{X}_{EU} + w_{Rx}^* n, \tag{3}$$

where y_k is a received signal at BS, M is a number of antenna ports at BS and h_k is k^{th} channel coefficient, which represents a channel in terms of attenuation between UE and BS; w_{Rxk}^* is complex conjugate of Rx beam weight (1); $\hat{X}_{UE}$ is an initial signal from UE to BS, n – AWGN $CN(0,\sigma_n)$ with zero mean and variance σ_n.

Comparing (2) and (3) we can conclude, that beamforming is a technique, which corresponds the signal to the propagation channel **h** by a beamforming weight **w** in transmit as well in receive path. Therefore, for transmit and receive chains beamforming technique task lies in correct estimation of weights **w**. The weighting approach for transmit beamforming is applicable for both, transmit and receive chains [7].

For mobile radio communication, the choice of duplexing has influence on beamforming technique implementation. 5G radio networks includes time-division-duplex (TDD), frequency-division-duplex (FDD) [8] and *decoupling* in new radio (NR) [9].

In FDD receive and transmit channels are sent on two different frequency carriers, whereas in TDD mode both Rx and Tx channels are collocated on the same frequency carrier with separation through time slots. Structures of beamforming in TDD and FDD mode are presented in Figs. 2 and 3.

Beamforming techniques are further investigated from the perspective of base station (BS), thus transmit path is in downlink (DL), and receive path is in uplink (UL). In TDD mode the same beam is used for DL and UL, the Tx/Rx paths are laid through the same antenna elements and its connection is done by de/multiplexor at the transceiver (Fig. 2). In FDD mode separate beams are supposed for DL and UL, however the transmission of the Tx/Rx paths is also done through the same antenna elements, but the paths separation occurs respectively by combiner or splitter (Fig. 3) [10]. The last duplexing mode for 5G NR is called decoupling and represents the transmission of DL and UL paths through different frequency bands (Fig. 4). The distinctive feature of this mode is the dedication of higher frequency band for the DL and the lower frequency band for UL, because of lower power capability at the UE [11], so that UE can benefit from the farther propagation on lower frequencies and can use more efficiently its energy; the transmission of Tx/Rx paths is provided by different antenna arrays.

Note, the common notation in the Figs. 2, 3 and 4. The signal of interest (SOI) to UE_k is marked by index k. Interfering signals are marked by index i. Beam paths are presented as follows: DL Tx-BF means transmit beamforming path, UL Rx-BF means receive beamforming path. The weighting approach for beamforming technique in the above introduced duplex modes holds the same. However, there are three different signal processing techniques which are analyzed further.

2.2   Beamforming Signal Processing Techniques Analysis

Beamforming signal processing techniques for antenna ports weight estimation are investigated for the three algorithms: matched filter (MF), zero-forcing (ZF) and minimum mean square error (MMSE). We will investigate these algorithms, firstly, from optimization point of view and then analyse its weighting solution.

Spectral efficiency (SE) metric for various massive MIMO configurations is known from information theory, called Shannon ergodic capacity C, and is defined as:

$$C = \log_2\left(1 + \frac{P_{\tilde{s}}|w^H h|^2}{\sigma_n^2 \|w\|^2}\right) \tag{4}$$

where $P_{\tilde{s}}$ is a power of a signal after beamforming processing, σ_n^2 is a noise power.

Matched Filter. This transmission technique is also known as maximum-ratio-combiner (MRC). The main idea of this algorithm is to maximize the signal-to-noise ratio (SNR) by adaptation of the weight **w** to the transmission channel **h**. So the optimization problem being solved by MF, can be generally (without separation into Tx and Rx paths) formulated as estimation of weight **w**, that maximizes SNR:

$$\text{argmax}_w \frac{|E[s^H \tilde{s}]|}{E[\|wn\|_2^2]} \tag{5}$$

where SNR is presented as rate between the expectation of signal power to expectation of noise power. Here and further s is signal before MF processing, $\tilde{s}$ is signal after MF processing for both transmit and receive cases.

From (4) we can conclude, that MF algorithm is optimal if the channel is orthogonal, or, in other words, if corresponding coefficients h_k are independent. The input signals for MF precoder are weighted by **w** and summed up, so that the signal after MF is [12]:

$$s_{MF} = \sum_{k=1}^{M} w_k^*(h_k \tilde{s} + n_k) = \sum_{k=1}^{M} w_k^* h_k \tilde{s} + \sum_{k=1}^{M} w_k^* n_k \tag{6}$$

where the first term describes the signal energy and the second term is noise energy, M is antenna element number of massive MIMO system at the BS; then SNR is:

$$\text{SNR}_{MF} = \frac{\left|\sum_{k=1}^{M} w_k^* h_k\right|^2}{\left|\sum_{k=1}^{M} w_k^* n_k\right|^2}. \tag{7}$$

From the Cauchy–Schwarz inequality:

$$\left|\sum_{k=1}^{M} w_k^* h_k\right|^2 \le \left|\sum_{k=1}^{M} w_k^*\right|^2 \left|\sum_{k=1}^{M} h_k\right|^2 \tag{8}$$

it follows, that the equality in (8) can be reached only if:

$$w_k^* = h_k \tag{9}$$

The consequence from (8) and (9) is that the maximum SNR (7) can be reached only if the weight takes a value of the channel fading $(w_k^*)^* = h_k^*$ or

$$w_k = h_k^* \tag{10}$$

Expression (10) represents a key principle of MF and is valid for transmit as well as receive beamforming. Rewriting (10) using matrix $\mathbf{H}$ for all antenna ports, we get:

$$\mathbf{w}_{MF-Rx} = \mathbf{w}_{MF-Tx} = \mathbf{H}^H \tag{11}$$

where the superscript ^{H} is the Hermitian operator (conjugate transpose) and the bold type of $\mathbf{w}$ means a vector.

The conclusions above show that MF processing concentrates the energy as much as possible towards the SOI, but cannot cope with interferences [6]. On the other hand, ZF is able to exclude the interferences, but cannot manage the energy towards the SOI. Next ZF processing is discussed for transmit and receive beamforming.

Zero-Forcing. ZF signal processing algorithm solves the optimization problem in a similar way for transmit and receive paths minimizing the expectation of mean-squared-error (MSE) between the initial signal $\mathbf{s}$ before ZF and the signal $\tilde{\mathbf{s}}$ after ZF algorithm under the constraint of complete interference cancellation. However, ZF has different approaches for transmit and receive paths [7]:

$$\mathrm{argmin}_w E\left[\|\mathbf{s} - \tilde{\mathbf{s}}\|_2^2\right] = \begin{cases} \text{for Tx: } w = \mathrm{argmin}_w E\left[\|\mathbf{ws}\|_2^2\right] \\ \text{for Rx: } w = \mathrm{argmin}_w E\left[\|\mathbf{wn}\|_2^2\right] \end{cases}$$

$$\text{s.t.} \mathbf{w}_{ZF-Rx} \mathbf{H} \mathbf{w}_{ZF-Tx} = 1 \tag{12}$$

Signal after ZF processing is defined by:

$$s_{ZF} = w_{ZF}^* h_k \tilde{s}_k + \sum\nolimits_{i \neq k}^{M} w_{ZF}^* h_i \tilde{s}_i + \sum\nolimits_{k=1}^{M} w_{ZF}^* n \tag{13}$$

where the interfering signal $\tilde{s}_i$ is suppressed by the weight w_{ZF}^* due to their orthogonality to each other. For transmit path MSE algorithm process s in order to send it in a direction of interest suppressing the interference impacts. For receive path MSE is applied on s from the noise perspective, in order to extract only an estimated signal of interest $\tilde{s}$ and to suppress interference sources.

The weighting algorithm of ZF beamformer is based on a *channel matrix inversion*. Utilizing massive MIMO systems, suppose that the number of antenna elements at BS M is larger than the number of users K in the cell, M > K [13]. Then the channel matrix cannot be square, because it has dimension N × M in DL (transmit beam-forming) and dimension M × N in UL (receive beamforming), so its inverse does not exist. Instead, pseudo-inverse is used in such cases. Notice that for UL the number of receivers means the number of antenna ports at the BS while the number of transmitters is the number of the users in a cell, and vice versa for DL, the number of transmitters means the number of BS antenna ports, while the number of receivers is the number of users in a cell. Then channel matrix pseudo-inverse can be calculated for transmit and receive ZF beamforming. Solving (12) follows by taking its derivation and equating to zero:

$$\frac{d\|s - \tilde{s}\|_2^2}{d\tilde{s}} = 0 \tag{14}$$

In case of ZF beamforming transmit and receive paths have to be observed separately. To estimate beamformer weights, lets substitute (2) for Tx paths and (3) for Rx paths in (14). For transmit beamforming (14) yields:

$$\|x_k - H\hat{x}_{BS}\|^2 = (x_k - H\hat{x}_{BS})^H (x_k - H\hat{x}_{BS}) = x_k^H x_k - \hat{x}_{BS}^H H^H x_k - x_k^H H\hat{x}_{BS} + \hat{x}_{BS}^H H^H H x_k \tag{15}$$

Taking a derivation from (15) by transmitted signal $d\hat{x}_{BS}$ and equating to 0 gives:

$$-2H^H x_k + 2H^H H\hat{x}_{BS} = 0$$

$$H^H H\hat{x}_{BS} = H^H x_k \tag{16}$$

So ZF transmitted signal is precoded by:

$$x_k = H^H \left(H^H H\right)^{-1} \hat{x}_{BS} \tag{17}$$

Repeating the same procedure for receive path with (3), we get the derivation of MSE by the estimated signal $d\hat{x}_{UE}$:

$$\frac{d\|\hat{x}_{UE}H - y_k\|^2}{d\hat{x}_{EU}} = -2H^H y_k + 2H^H H\hat{x}_{UE} = 0 \tag{18}$$

So ZF estimated received signal is obtained by:

$$\hat{x}_{UE} = \left(H^H H\right)^{-1} H^H y_k \tag{19}$$

If the number of transmitters is larger than the number of receivers $M > N$, which means the above mentioned DL case, and the M rows of channel matrix H are linearly independent [14], then the weighting w_{ZF-Tx} of ZF transmit beamformer is:

$$\mathbf{w}_{\text{ZF-Tx}} = \mathbf{H}^{\text{H}}\left(\mathbf{H}^{\text{H}}\mathbf{H}\right)^{-1} \tag{20}$$

Likewise, if the number of receivers is larger than the number of transmitters $N > M$, which means the above mentioned UL case, and the N columns of channel matrix H are linearly independent, then the weighting $\mathbf{w}_{\text{ZF-Rx}}$ of ZF receive beamformer is:

$$\mathbf{w}_{\text{ZF-Rx}} = \left(\mathbf{H}^{\text{H}}\mathbf{H}\right)^{-1}\mathbf{H}^{\text{H}} \tag{21}$$

Returning now to (12), we note that the noise weighted by $\mathbf{w}_{\text{ZF}}$ can be increased if the term $\mathbf{H}^{\text{H}}\mathbf{H}$ is getting small, this negative effect of ZF is called *noise-enhancement*.

MMSE-Beamforming. The purpose of MMSE is also to minimize the MSE of symbols like ZF does, but the difference is that MMSE does not usually eliminate ISI completely, but minimizes the total power of the noise and ISI components in the output [15]. The optimization problem solved by MMSE can be expressed as:

$$\operatorname{argmin}_{\mathbf{w}}\mathbf{E}\left[\|\mathbf{s} - \tilde{\mathbf{s}}\|_2^2\right] \tag{22}$$

In contrast to ZF, MMSE approach works as follows [16]:

$$\mathbf{w}_{\text{MMSE-Tx}} = \mathbf{H}\left(\mathbf{H}^{\text{H}}\mathbf{H} + \sigma_\eta^2\mathbf{I}\right)^{-1} \tag{23}$$

$$\mathbf{w}_{\text{MMSE-Rx}} = \left(\mathbf{H}^{\text{H}}\mathbf{H} + \sigma_\eta^2\mathbf{I}\right)^{-1}\mathbf{H}^{\text{H}}$$

Obtained weights for observed beamforming signal processing techniques, terms **w** in Eqs. (2) and (3), are summarized in Table 1.

Table 1. Summary of weights for beamforming signal processing techniques.

BF Tech.	Weight matrix for receive BF (UL)	Weight matrix for transmit BF (DL)
ZF	$\mathbf{w}_{\text{ZF-Rx}} = \left(\mathbf{H}^{\text{H}}\mathbf{H}\right)^{-1}\mathbf{H}^{\text{H}}$	$\mathbf{w}_{\text{ZF-Tx}} = \mathbf{H}^{\text{H}}\left(\mathbf{H}\mathbf{H}^{\text{H}}\right)^{-1}$
MF	$\mathbf{w}_{\text{MF-Rx}} = \mathbf{H}^{\text{H}}$	$\mathbf{w}_{\text{MF-Tx}} = \mathbf{H}^{\text{H}}$
MMSE	$\mathbf{w}_{\text{MMSE-Rx}} = \left(\mathbf{H}^{\text{H}}\mathbf{H} + \sigma_\eta^2\mathbf{I}\right)^{-1}\mathbf{H}^{\text{H}}$	$\mathbf{w}_{\text{MMSE-Tx}} = \mathbf{H}\left(\sigma_\eta^2\mathbf{I} + \mathbf{H}\mathbf{H}^{\text{H}}\right)^{-1}$

In the next section lets simulate summarized beamforming signal processing techniques in terms of spectral efficiency (4) for various massive MIMO configurations.

3   Beamforming Signal Processing Techniques Simulation

3.1   Simulation Scenario

Simulation model concludes following approach in brief. Firstly, the number of antennas, users, Monte Carlo realizations were initialized, and the range of SNR was set. Then based on these channel matrix was initialized by random standard normally distributed parameters. After that the weights for each beamforming method were calculated. Finally, based on the Shannon capacity theorem and considering the beamforming weights in the SNR part of the theorem, the channel rates were estimated.

Initial conditions for selected simulations scenarios were chosen regarding to massive MIMO implementations in mobile telecommunication [2, 3, 17]. Leading antenna vendors offer to increase data rate by applying massive MIMO systems with 8, 32, 64 or 128 antenna ports and these configurations are already presented for the 5G market [2]. The simulation scenarios were set by combination of different number of users and various antenna port numbers at massive MIMO systems as well as with different beamforming signal processing techniques presented in Table 1 and were carried out for both, single-user (SU) and multi-user (MU) cases.

SU case includes a transmission from BS with multiple antenna porta to a single UE, which also has multiple antenna ports. In the MU case BS massive MIMO antenna system transmits simultaneously several flows in separate beams to different users with only one antenna port at each UE. Both SU and MU cases utilize same time and frequency resource. For SU simulation case one user was considered, and for MU simulation case 14, 50 and 100 users in a cell were considered. In each simulation scenario spectral efficiency depending on SNR was calculated for three beamforming signal processing techniques: MMSE, ZF, MF. The following assumptions were taken during the simulations: presence of perfect CSI, spectral efficiency was estimated by ergodic capacity, as it was supposed to deal with a fast fading channel. MU simulation scenarios accounted for the interference in a cell [17].

3.2   Simulation Results

Simulation results of beamforming signal processing techniques for massive MIMO configurations with 8, 32, 64 and 128 antenna ports are illustrated in Figs. 5, 6, 7 and 8. In Fig. 5 results of the SU and MU mode for 14 users are presented for MMSE, ZF and MF beamforming techniques with 8, 32, 64 or 128 antenna ports. In Figs. 6 and 7 corresponding MU mode results are presented for 50 and 100 users respectively.

From Fig. 5 we can make the following conclusion for the case, when the number of users is much lower than the number of antenna ports: at higher SNR the MMSE acts as ZF and at low SNR MMSE acts as MF.

From Figs. 6 and 7 we can make the following conclusion for the case, when the number of users is close to the number of antenna ports: at high and low SNR values MMSE still provides better spectral efficiency compared with ZF and MF. Thus we will focus in following conclusions on the MMSE results.

Simulation results of beamforming signal processing techniques for massive MIMO configurations for both SU and MU mode revealed, that it cannot achieve IMT-

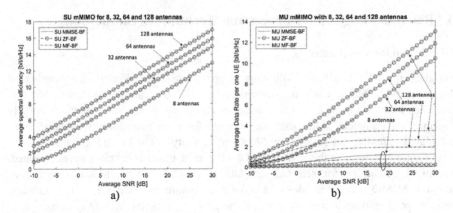

Fig. 5. Average spectral efficiency for (a) SU and (b) MU with 14 UEs with 8, 32, 64 and 128 antenna ports.

2020 *peak spectral efficiency* of 30 bit/s/Hz in DL [1], but can sustain a half required peak value 15 bit/s/Hz using 128 antennas for SU case with high SNR > 25 dB (Fig. 5).

To analyze ability of beamforming technique with M = 8, 32, 64 and 128 antenna ports to meet ITU-R IMT-2020 *average spectral efficiency* requirements for 5G, lets observe and compare MU mode results for 100 UE in Fig. 7 with average spectral efficiency requirements for various test environments, summarized in Table 2 [1].

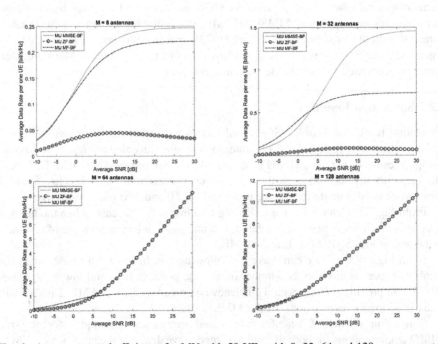

Fig. 6. Average spectral efficiency for MU with 50 UEs with 8, 32, 64 and 128 antenna ports.

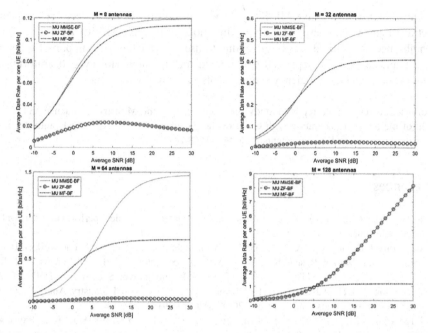

Fig. 7. Average spectral efficiency for MU with 100 UEs with 8, 32, 64 and 128 antenna ports.

Table 2. Average spectral efficiency [1].

Test environment	Downlink (bit/s/Hz)	Uplink (bit/s/Hz)
Indoor hotspot	9	6,75
Dense urban	7,8	5,4
Rural	3,3	1,6

From Figs. 6 and 7 and Table 2 we can make the following conclusion for DL:
(a) Indoor Hotspot requirements do not to hold even for high SNR values with M ≤ 128;
(b) Dense Urban requirements hold only for high SNR > 25 dB with M ≥ 128;
(c) Rural requirements hold only for medium-high SNR > 15 dB with M ≥ 128.

4   Conclusion

Simulation results of beamforming signal processing techniques for massive MIMO configurations with 8, 32, 64 and 128 antenna ports revealed, that it cannot achieve IMT-2020 peak spectral efficiency of 30 bit/s/Hz in DL [1], but can sustain a half required peak value 15 bit/s/Hz using 128 antennas for SU case with high SNR = 25 dB. Average spectral efficiency requirements per user for MU case hold only for medium-high SNR > 15 dB with M ≥ 128 in rural environment.

Thus, the use of massive MIMO systems in beamforming mode decrease the peak spectral efficiency compared to multiplexing mode of massive MIMO. Reason for this is an absence of spatial diversity effect due to use of available antenna ports for beam steering and not for the separated data stream transmission. However, beamforming meets average spectral efficiency requirements in rural environment.

Acknowledgements. The reported study was supported by the Ministry of Science and Education of the Russian Federation with Grant of the President of the Russian Federation for the state support of young Russian scientists № MK-3468.2018.9.

References

1. Report ITU-R M.2410-0: Minimum requirements related to technical performance for IMT-2020 radio interface(s) (2017)
2. Ericsson white paper GFMC-18:000530: Advanced antenna systems for 5G networks (2018)
3. Transparency Market Research: Massive MIMO Technology Market (Antennas - 8T8R, 16T16R & 32T32R, 64T64R, and 128T & 128R and above; Spectrum - TDD, FDD; Technology - LTE Advanced, LTE Advanced Pro, 5G) - Global Industry Analysis, Size, Share, Growth, Trends, and Forecast, 2018-2026 (2019)
4. ETSI TS 136 211 V15.2.0: Evolved Universal Terrestrial Radio Access (E-UTRA); Physical channels and modulation (2018)
5. Stepanets, I., Fokin, G., Müller, A.: Beamforming techniques performance evaluation for 5G massive MIMO systems. In: CEUR Workshop Proceedings, vol. 2348, pp. 57–68 (2019)
6. Brown, T., Kyritsi, P., De Carvalho, E.: Practical Guide to MIMO Radio Channel: with MATLAB Examples. Wiley, Hoboken (2012)
7. Joham, M., Utschick, W., Nossek, J.A.: Linear transmit processing in MIMO communications systems. IEEE Trans. Signal Process. **53**(8), 2700–2712 (2005). https://doi.org/10.1109/TSP.2005.850331
8. R1-164018: Analog/digital/hybrid beamforming for massive MIMO. In: 3GPP TSG RAN WG1 #85 (2016)
9. Liu, X., Li, R., Luo, K., Jiang, T.: Downlink and uplink decoupling in heterogeneous networks for 5G and beyond. J. Commun. Inf. Netw. **3**(2), 1–13 (2018). https://doi.org/10.1007/s41650-018-0023-4
10. Kim, H. et al.: A 28 GHz CMOS direct conversion transceiver with packaged antenna arrays for 5G cellular system. In: 2017 IEEE Radio Frequency Integrated Circuits Symposium (RFIC), Honolulu, HI, pp. 69–72 (2017). https://doi.org/10.1109/rfic.2017.7969019
11. Fokin, G., Volgushev, D., Kireev, A., Bulanov, D., Lavrukhin, V.: Designing the MIMO SDR-based LPD transceiver for long-range robot control applications. In: 2014 6th International Congress on Ultra-Modern Telecommunications and Control Systems and Workshops (ICUMT), St. Petersburg, pp. 456–461 (2014). https://doi.org/10.1109/icumt.2014.7002144
12. Gelgor, A., Pavlenko, I., Fokin, G., Gorlov, A., Popov, E., Lavrukhin, V.: LTE base stations localization. In: Balandin, S., Andreev, S., Koucheryavy, Y. (eds.) NEW2AN 2014. LNCS, vol. 8638, pp. 191–204. Springer, Cham (2014). https://doi.org/10.1007/978-3-319-10353-2_17
13. Marzetta, T., Larsson, E.G., Yang, H., Ngo, H.Q.: Fundamentals of Massive MIMO. Cambridge University Press, Cambridge (2016). https://doi.org/10.1017/CBO9781316799895

14. Stüber, G.L.: Principles of Mobile Communication. Springer, Cham (2017). https://doi.org/10.1007/978-3-319-55615-4
15. Malik, D., Batra, D.: Comparison of various detection algorithms in a MIMO wireless communication receiver. Int. J. Electron. Comput. Sci. Eng. 1(3), 1678–1685 (2012)
16. Lee, H., Sohn, I., Kim, D., Lee, K.B.: Generalized MMSE beamforming for downlink MIMO systems. In: 2011 IEEE International Conference on Communications (ICC), Kyoto, pp. 1–6 (2011). https://doi.org/10.1109/icc.2011.5962843
17. Björnson, E., Bengtsson, M., Ottersten, B.: Optimal multiuser transmit beamforming: a difficult problem with a simple solution structure. IEEE Signal Process. Mag. 31(4), 142–148 (2014). https://doi.org/10.1109/MSP.2014.2312183

Autonomous UAV Landing on a Moving Vessel: Localization Challenges and Implementation Framework

Carlos Castillo[1], Alexander Pyattaev[2], Jose Villa[3], Pavel Masek[4],
Dmitri Moltchanov[3], and Aleksandr Ometov[3(✉)]

[1] Nokia Bell Labs, Oulu, Finland
[2] Peoples' Friendship University of Russia (RUDN University), Moscow, Russia
[3] Tampere University, Tampere, Finland
aleksandr.ometov@tuni.fi
[4] Brno University of Technology, Brno, Czech Republic

Abstract. The number of Unmanned Aerial Vehicle (UAV) applications is growing tremendously. The most critical ones are operations in use cases such as natural disasters, and search and rescue activities. Many of these operations are performed on water scenarios. A standalone niche covering autonomous UAV operation is thus becoming increasingly important. One of the crucial parts of mentioned operations is a technology capable to land an autonomous UAV on a moving surface vessel. This approach could not be entirely possible without precise UAV positioning. However, conventional strategies that rely on satellite localization may not always be reliable, due to scenario specifics. Therefore, the development of an independent precise landing technology is essential. In this paper, we developed the localization and landing system based on Gauss-Newton's method, which allows to achieve the required localization accuracy.

Keywords: UAV · Positioning · Automatic landing · Simulation

1 Introduction

Unmanned Aerial Vehicle (UAV) operation is a topic that has been under careful research community attention for more than a decade [1,2]. While its use has been a spreader, more applications have been found for the UAVs in life-rescuing and natural disaster scenarios [3,4]. These involve border surveillance to rescue people in the water [5], where the performance of the UAVs has to be as perfect as possible [6].

Work of the last author is supported by Nokia Foundation under a personal grant. The publication has been prepared with the support of the "RUDN University Program 5-100". The described research was supported by the National Sustainability Program under grant LO1401. For the research, the infrastructure of the SIX Center was used.

O. Galinina et al. (Eds.): NEW2AN 2019/ruSMART 2019, LNCS 11660, pp. 342–354, 2019.
https://doi.org/10.1007/978-3-030-30859-9_29

However, new limitations appear with the new drone-based applications together with the need to overcome the UAV operation challenges. Commonly, UAVs are controlled by the operator having direct sight to the UAV [7–9]. In this case, the performance degradation may bring a mission failure when video transmission experience network delays, or if the UAV is not in the direct Line-of-sight (LOS). In order to prevent it, modern UAVs are equipped with Global Navigation Satellite System (GNSS) receivers that can estimate the position of the UAV and can trigger the UAV to "return home" mode in the emergency scenario, i.e., to return to the original position from which the UAV was launched or to a preprogrammed location. However, the widely used Global Positioning System (GPS) trackers still face an error of approximately 1–10 m[1]. GPS modules with less positioning error are also present on the market (or could be mitigated by utilizing multiple receivers [10]).

Autonomous and Collaborative Offshore Robotics (aColor) project[2] started in January 2018 in Tampere University of Technology, Finland. The core of this project is to achieve a shared intelligence between different offshore vehicles (Unmanned Surface Vehicle (USV), UAVs, and Autonomous Underwater Vehicle (AUV)), as well as the situational awareness of these subsystems. The ultimate aColor's goal is to build an autonomous and cooperative multicomponent robotic system, as well as to demonstrate it in a challenging open environment.

The aColor UAV will be used for water surveillance, tracking disruptions in the water surface, and also as a communications relay [11] if the distance between the vessel and the shore station is greater than the distance capability of a direct radio link. However, one essential key is missing, to be a fully autonomous system is needed a landing system for the UAV that can be performed on top of a moving surface without human aid. The aim of this paper is to introduce a safe method to land the UAV without any form of vision on a moving surface vessel not relying on GNSS. This goal is based on the desired outcome to transform the whole aColor concept into a fully autonomous system. The correct performance of the UAV has to be independent of weather conditions, as very dense fog, rain, snow or wind, which are especially common for next to the shore and maritime operation. These cases describe the drawback of having a ground operator, because of the lack of visual aid, the operator cannot anticipate the behavior of the UAV and prevent a crash.

This paper proposes to rely on antenna-anchors located on the vessel to perform the landing. The localization would be based on the real time processing of the received signal strength in order to approximate the distance from the antennas. Nonetheless, the reduction of the GNSS location error is required in the selected scenario. The main concept is shown in Fig. 1. This architecture is composed of four anchors forming a surface in which center the drone is supposed

---

[1] See "GPS Accuracy" by United States Air Force, 2017: https://www.gps.gov/systems/gps/performance/accuracy/.

[2] See "Autonomous and Collaborative Offshore Robotics (aCOLOR)" by Kari T. Koskinen, 2017: https://techfinland100.fi/mita-rahoitamme/tutkimus/tulevaisuuden-tekijat/autonomous-and-collaborative-offshore-robotics-acolor/.

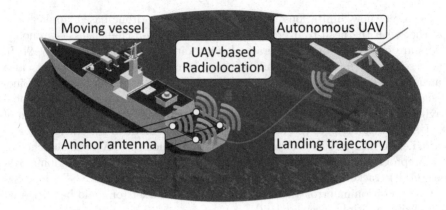

Fig. 1. Concept of the vessel-UAV landing.

to land, implementing conventional IEEE 802.11n technology (due to its broad market adoption [12]) and an autonomous UAV.

This paper is organized as follows. In Sect. 2, the background information, and positioning strategy are briefed. The proposed solution is described in Sect. 3. Simulator and related numerical results are presented in Sect. 4. Finally, the last section concludes the paper.

2   Background Information

This section provides an overview of different autonomous landing approaches. In conventional aircraft, the landing is the most critical operation since passengers might get injured if the landing is not performed correctly. Today passenger aircraft is not yet unmanned, and a crew is controlling the flight on board and from the ground station. Over the years, significant improvements have been developed in this matter to aid the pilots and the ground crew, creating an approach to a more autonomous landing system [13].

In UAVs, the autonomous landing capability has been studied and tested in more details, UAV presents an advantage due to the lack of direct risks. These techniques usually employ GNSS positioning and own UAV's autopilot to land on a steady base station [14]. However, in this approach, positioning error brought by the GNSS might be expected. In order to mitigate the landing accuracy error from GNSS, a visual analysis technique is commonly applied [15]. This hybrid landing system consists of taking the UAV to specific GNSS coordinates and, then, the UAV will recognize the visual pattern given in the landing point using its cameras and signal processing.

However, all these previous solutions have limitations in case of moving platforms. In this paper, the landing points will have continually changing the coordinates compared to ones provided at the beginning of the mission. To overcome this situation, the UAV requires a stable communication link to the landing platform to receive the new coordinates, or another technique has to be employed.

Visual aid provided by some given pattern in the landing platform can continue to be used to minimize the location error in the surface target position. In the scenario of interest, the moving platform is located on the water surface, thus, creating a risk of damaging or losing the UAV. In order to land an UAV on a moving surface vessel, the use it's equipment to capture and retrieve should be applied.

Most approaches to autonomous UAV landing methods assume that the locations of the UAV and landing platform are known. The default localization system commonly used by many devices is the GNSS [16]. Due to these technologies, it is possible to locate any device that has a receiver in any part of the planet having LoS to the satellites. Moreover, GNSS positioning accuracy gets worse if multi-path propagation is present due to buildings, trees, bridges, etc., surrounding the receiver. Satellite-based localization requires the presence of four or more satellites. These satellites follow an orbit that it might imply that it is possible to have a "dead-zone" for some localization technologies. If the desired positioning error is in the magnitude of a few centimeters, then, satellite positioning technologies are not the best choice. Even though these technologies can reduce that error, the expense of a receiver increments the cost of the project greatly. For this purpose, it is necessary to find a technique that provides a better outdoors accuracy, and the cost is not very significant.

In contrast, high accuracy is more desired in indoor scenarios than outdoors because an error of few meters might mean that the object is on another floor. The most common approach for indoor positioning is the use of low range radio signals [17]. A transmitter acts as an anchor and the device that is meant to be located will measure the received signal estimating the distance to the anchor. If the indoor plan is known, the position estimation has an error of 0.5 m measured in [18]. Works [19] and [20] have shown the possibility and related challenges to use not only RFID signals but IEEE 802.11 signals for indoor localization.

3 Localization Based on Modified Gauss-Newton's Method

In this paper, the indoor positioning-based method is proposed as an optimal solution for outdoor positioning. In the proposed scenario, the idea implies that the layout is not known beforehand and could dynamically be changed due to the mobility of the vessel and the effects of waves. Therefore, a method to calculate the position of an object in an unknown and potentially moving space is the problem statement. In this paper, a modification of the Gauss-Newton method for non-linear models is proposed to be used for localization. This method is iterative, meaning, the calculated intersection will be a better approximation to the root of the function than the original guess. This is why it is not known beforehand how many iterations are required to find the root of the function.

The calculation of the position of the UAV can be accomplished using the knowledge of Friis distances (applying Free Space Path Loss (FSPL)) from the anchor point to the UAV location. The position coordinates are non-linear, creating an extra difficulty to the UAV location calculation. One of the most studied methods to calculate non-linear regression by least squares is the use of the Gauss-Newton algorithm. This solution allows solving non-linear least square problems [21]. The following mathematical development modifying Gauss-Newton's method is shown to prove that the system is robust. The goal is to allow modeling a system by a non-linear function $y = f(x, a_1, a_2, \dots)$ composed by a set of parameters $a = [a_1, a_2, \dots]^T$ able to minimize the residual error between the actual location and the calculated position

$$\epsilon(a) = \sum_{i=1}^{N} r_i^2 = \sum_{i=1}^{N} [y_i - f(x_i, a)]^2 = \sum_{i=1}^{N} [y_i - f_i(a)]^2 \to min, \qquad (1)$$

where a is the set of parameters that will define the system, $\epsilon(a)$ is the total residual error depending on the parameter set a, r_i is the residual error in every iteration, N is the number of data points $(x_i, y_i), (i = 1, \dots, N)$, y_i is the real location of the UAV, and $f(x_i, a), (i = 1, \dots, N)$ is the location calculated in every iteration. It has been defined $f_i(a)$ as $f(x_i, a)$.

The residual error is given by the cumulative sum of every iteration of the system. The minimization of this residual error is the ultimate goal. The residual error expressed in vector form to represent the general approach is as follows

$$r = \begin{pmatrix} r_1 \\ \dots \\ r_N \end{pmatrix} = \begin{pmatrix} y_1 - f_1(a) \\ \dots \\ y_N - f_N(a) \end{pmatrix} = \begin{pmatrix} y_1 \\ \dots \\ y_N \end{pmatrix} - \begin{pmatrix} f_1(a) \\ \dots \\ f_N(a) \end{pmatrix} = y - f(a), \qquad (2)$$

where $y = [y_1, \dots, y_N]^T$, $f(a) = [f_1(a), \dots, f_N(a)]^T$.

Therefore, the general equation in vector form is

$$\epsilon(a) = \sum_{i=1}^{N} r_i^2 = r^T r = ||r^2|| = ||y - f(a)^2||, \qquad (3)$$

where the total residual error $\epsilon(a)$ is the absolute value obtained in the difference between the real and the calculated location of the UAV.

Once the method to find the residual error is obtained, it is necessary to find a that minimizes $\epsilon(a)$. In order to do so, the equation where the gradient of the vector is equal to zero is used

$$\frac{\delta}{\delta a_j} \epsilon(a) = \frac{\delta}{\delta a_j} \sum_{i=1}^{N} [y_i - f_i(a)]^2 = -2 \sum_{i=1}^{N} [y_i - f_i(a)] \frac{\delta f_i(a)}{\delta a_j} = -2 \sum_{i=1}^{N} [y_i - f_i(a)] J_{ij} = 0, \qquad (4)$$
$$(j = 1, \dots, M),$$

where J is the Jacobian matrix defined as

$$J_{ij} = \frac{\delta f_i(a)}{\delta a_j}, (i = 1, \dots, N, j = 1, \dots, M), \qquad (5)$$

where the Jacobian is the first derivative with respect to the parameters required to be minimized (a in this case), $i = 1, \dots, N$ defines the number of iterations, and $j = 1, \dots, M$ defines the number of parameters a.

However, it might happen that Eq. (5) does not have a closed solution for a. Finding the optimal parameters $a = [a_1, \ldots, a_M]$ that will minimize the residual error $\epsilon(a)$ when the previous equations do not have a solution, can be done with the use of iteration $a_{n+1} = a_n + \Delta a$, where it is required to find $\Delta a = a_{n+1} - a_n = [\Delta a_1, \ldots, \Delta a_M]^T$. If Taylor expansion is considered $f_i(a_{n+1})$ at a_n, then the following equation is found (in vector form)

$$f(a_{n+1}) \approx (a_n) + J\Delta a. \tag{6}$$

After substituting (6) in (4), we arrive at

$$\sum_{i=1}^{N} J_{ij} \sum_{k=1}^{M} J_{ik} \Delta a_k = \sum_{i=1}^{N} J_{ij}[y_i - f_i(a_n)], (j = 1, \ldots, M). \tag{7}$$

Adapting Eq. (7) to matrix form and solving for Δa

$$\Delta a = a_{n+1} - a_n = (J^T J)^{-1} J^T (y - f(a_n)) = J^-(y - f(a_n)),$$
$$a_{n+1} = a_n + \Delta a = a_n - J^-(f(a_n) - y), \tag{8}$$

where $J^- = (J^T J)^{-1} J^T$ is the pseudoinverse, and is obtained the iteration

$$a_{n+1} = a_n + \Delta a = a_n - J^-(f(a_n) - y). \tag{9}$$

Finally, it is possible to see that the solution to the modeling problem is (by Newton's method for solving the multivariate non-linear equations) is

$$f'(a) = f(a) - y = 0. \tag{10}$$

The situations may appear that the system may not converge at any iteration, this is why a parameter $0 < \gamma < 1$ is introduced to reduce the step size of the iteration. This parameter γ has to be calculated in order to maximize the performance of the system. This performance measures if the system has found the solution, and how long it has taken to find the solution.

The following would be further used to estimate the UAV position

$$a_{n+1} = a_n + \gamma \Delta a, \tag{11}$$

where a_n in the first iteration is the initial guessed position, γ is the step parameter to smooth the step Δa obtaining a value a_{n+1} closer to the actual position of the UAV. The residual error is minimized at every iteration. In this paper, the initial guess position is analyzed and provided an optimal solution as well as the parameter γ.

4   Selected Numerical Results

This section provides the description of the scenario and the simulation with respect to the antenna spacing and elaborates on the modified Gauss-Newton's algorithm to calculate the UAV coordinates.

In order to estimate the location of the UAV, the following elements are needed: (i) Antennas acting as anchor points; (ii) The distance between anchors and UAV; (iii) Gauss-Newton's method estimation; (iv) Residual of the actual target location and estimated location. The anchors are antennas forming a surface in which center the UAV will land. The anchor-antennas are isotropic radiating in every direction as initially the location of the UAV is unknown. The antennas acting as the anchors form a flat surface. The center will be equidistant to every antenna, and the antennas will have the same distance between each other for the simplest case.

In order to find the distance between every antenna and the UAV, Friis distance formula was used. First, the Euclidean-Distance was calculated between every antenna and the UAV. Then, FSPL was obtained using as parameters the calculated distance and a frequency of 2.4 GHz. From the antenna specification, it is known that it has an error that follows a normal distribution of 1 dB, this error is also added to the path losses. Once the path loss is known, we proceed to calculate the Friis distance. For the simulation, it is necessary to randomize the initial location of the UAV. The UAV will appear in a volume of 1000 cubic meters (10 m in each of the three coordinates). A vector of 5000 different UAV positions will be created in order to be able to check the performance of the system. Once the actual position of the UAV is obtained, as well as the knowledge on how to calculate the Friis distance to each antenna, the values will be submitted to customized Gauss-Newton's algorithm in order to estimate the UAV location. This algorithm is a variation of Gauss-Newton's method explained previously. It has been implemented manually to suit our purposes, no external library with this algorithm has been used. Here, the initial coordinate guess should be provided, i.e., the coordinates of the anchor-antennas are situated and the original Friss distance between the UAV actual position and the anchor.

The Gauss-Newton parameter γ defines if the system converges. It allows specifying the step to take if the coordinate has not yet been found on every iteration. The error tolerance is used as the margin error between the calculated position using Newton's method and the measured position between the anchors and the UAV. If the error tolerance is not fulfilled, then a variation in the guessed position of the UAV will be applied depending on the direction of the vector connecting the anchor and the UAV. Providing the initial position guess, the Friis-distance from every anchor to the real UAV position and the anchors' coordinates are input parameters. After the allowed error of the calculated and real distances is met, the new position is returned as well as the number of iterations required to find it. Once the UAV location has been found, the landing procedure should be executed. The landing surface might be moving, which makes the requirement of fast calculation of a trajectory from the current position to the landing point a must. To perform a soft landing, an algorithm should be able to be executed fast enough to estimate the new position taking into account bad weather conditions. The best solution that fulfills all the criteria is to have a parabolic trajectory between the UAV and the anchor-antennas. In this approach, the center of the anchors will act as one of the points of the parabola and the calculated position of the UAV will act as the vertex of the parabola $y = a(x - h)^2 + k$, where a is to be found, $[h, k]$ is the vertex coordinates, in this case the UAV coordinates, and $[y, x]$ are the coordinates of the center of the anchors.

In this paper, the error tolerance with value 0.1 m was used aiming to achieve the smallest error when calculating the position of the UAV. In order for the system to be able to converge, the error between the calculated distance and the measured distance must be smaller or equal to the error tolerance. At the same time, the spacing between the anchors will be taken into consideration, which influences the spatial diversity of the antenna, and together with the path loss, the error performance of the system will vary. During the first calculation, the idea was to represent the maximum positioning error, from the calculated position to the actual location of the UAV, that was calculated by the simulation tool. Every case has been run 1000 times with different UAV location every time. The maximum positioning error was defined as it is the most representative value that will affect the proposed system directly. On every simulation step, the number of iterations needed for convergence was estimated when calculating the UAV coordinates. Convergence rate can be improved by allowing to have a higher positioning error between the calculated position and the actual location of the UAV. This might mean that a trade-off between convergence and positioning error should be considered.

4.1   Number and Spacing Between Anchors

The Newton parameter γ is set to 0.4. Together with the number of anchors, the spacing between the anchors was evaluated in order to get the optimal environment solution. In this experiment, the performance of the system using a different set of antennas and spacing between them is shown. For Fig. 2, the mean of errors at every distance to the center were calculated and then the overall mean was also obtained. The mean error is lower than 20 cm in every case, asserting that the proposed system is reliable. In case of *three antennas*, a significant improvement in the mean error when the spacing between the antennas is 50 cm compared to two antennas can be observed. With *four antennas*, there is an improvement compared to three antennas, but it is not that significant as the previous improvement. Since the goal is to have an accurate landing as possible, thus, the positioning error should be minimal when the UAV is close to the center of the anchors. The system that best fulfills this criterion is the one with four anchors. Therefore, it is proposed that the system should utilize *four anchors* with a distance of 1 m between them.

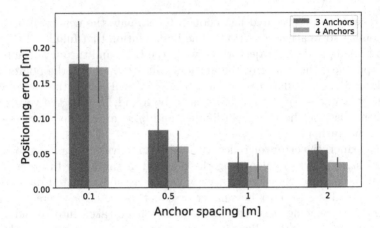

Fig. 2. Mean positioning error.

4.2 Influence of Newton's Method Parameter

Since error tolerance and path-loss are not variables, but the convergence must be improved, the Gauss-Newton's method parameter, γ, in Eq. (11) should improve reaching a solution with a minimized residual error. Antenna spacing is further set to 1 m. Figure 3 shows the influence of the parameter γ on positioning error and the number of iterations to achieve convergence. Parameter γ determines what will be the next movement of the drone in every iteration. If γ is too big, the calculated position of the drone will be far from the actual location, and if γ is too small, then the system will take longer to calculate the drone position with an error distance of 10 cm, causing the system to not converge (meaning that the position was not calculated). It is evident that the best overall performance is achieved when it is chosen γ value between 0.3 and 0.5. In all the simulations

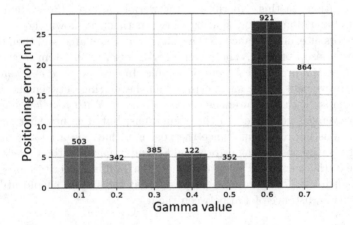

Fig. 3. System performance due to Newton's γ.

γ of 0.4 was used to maximize the performance of the system. The relationship between the number of iterations to achieve convergence of the system and the positioning error is also provided in this figure on top of every bar representing the positioning error for the analyzed γ value. With lower positioning error the system converges steadily. However, when γ takes values 0.3 and 0.4 a similar positioning accuracy can be achieved, except the convergence of the system is 30% faster for the case when γ is 0.4. A final analysis is that the needed number of iterations to converge is not directly related to the positioning error, but to the chosen γ value.

4.3 System Performance Depending on the Error Tolerance

During the evaluation, we faced low system convergence (the number of times the UAV was found, satisfying the permitted error calculation) of around 60% in most of the cases. It could be explained as a trade-off for having a low positioning error. According to the results given in Table 1, the performance of the system related to the mean and the maximum values shows that a chosen error tolerance of 10 cm improves the accuracy but a number of iterations to achieve convergence is very high. On the other hand, the accuracy of the system for error tolerance of 30 cm degrades in less than 10 cm in the mean calculation and around half a meter in the maximum accuracy error (with less number of iterations to reach convergence). Even though the overall performance of the system is similar for both cases (error tolerance has values of 10 or 30 cm), the convergence is faster for all cases when error tolerance is 30 cm. This analysis does not affect the previously obtained system performance results if an error tolerance of 10 cm was used. However, it was decided to study both cases for implementation of the system in a real-life scenario.

4.4 Unmanned Aerial Vehicle Landing

The results of applying the optimized parameters from Sects. 4.2 and 4.3 for two different landing trajectories are detailed in this subsection.

This work aims to propose a solution to create an accurate landing system for the aColor project that will utilize two different UAVs being fixed-wing (landing according to parabola) and multi-rotor (landing vertically) ones. The velocity is shown as a number of points that the UAV has to overcome in order to move from its initial position to the center of anchors. The performance of the system can be improved by controlling the UAV velocity. If the drone travels too fast to the center of the anchors, the system might not converge in time for all the steps that the drone takes. Two examples represented in Figs. 4 and 5, are the simulation proofs that the system can be utilized in real life. When this solution gets applied to the real scenario, a series of tests will need to be done to maximize the performance of the system. It is possible to see that when the UAV is far away from the anchor points the calculated position does not coincide with the position of the UAV, but as soon as the Euclidean distance gets reduced the calculated position matches the actual position of the UAV, being almost perfect

Table 1. Comparison between two error tolerance values.

Error tolerance [m]	Init position	Spacing [m]	Mean [m]	Max [m]	Convergence [Num iterations]
0.1	[0,0,0]	0.1	0.32	8.95	498
		1	0.273	8.98	501
	[5,5,5]	0.1	0.31	16.8	914
		1	0.04	4.13	254
	[10,10,10]	0.1	0.28	16.25	683
		1	0.05	4.5	525
0.3	[0,0,0]	0.1	0.25	8.65	251
		1	0.31	8.99	232
	[5,5,5]	0.1	0.3	17.5	487
		1	0.1	4.87	186
	[10,10,10]	0.1	0.3	16.73	330
		1	0.11	4.51	101

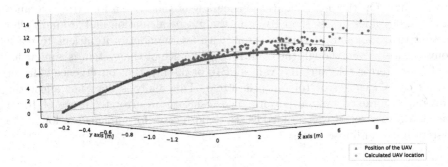

Fig. 4. Example fixed-wing UAV landing trajectory.

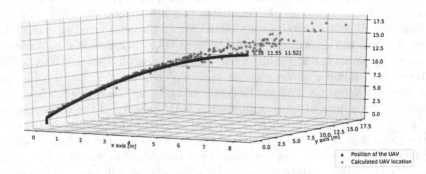

Fig. 5. Example multi-rotor UAV landing trajectory.

in the last meters to the landing point. The simulation of the system has proven to be almost perfect for the critical points of the different descent methods for both types of UAVs.

5   Conclusion and Future Work

The number of UAV applications is growing daily. A crucial niche of UAV development is related to the automated UAVs where positioning plays a significant role, especially during landing on moving objects, e.g., vehicles or vessels. In this work, a system for automatic landing support was developed. The paper focuses on the relationship between positioning errors and system configurations, aiming and keeping the landing surface as small as possible. The moving platform may change its coordinates over time, thus, a pre-decided location cannot be reliable. Moreover, a stable communication link between moving objects and UAV cannot be constantly assumed as there might be some disturbances in the radio link due to environmental and radio factors. A modified Newton-Gauss's method was selected to enable the localization of the UAV. The system designed in this paper could be proposed as a possible solution to achieve a fully automated UAV because of the accomplishment of a landing system with a high positioning accuracy. Currently, W.I.N.T.E.R. and aColor teams are in the final phase of developing a full-scale prototype aiming at testing the developed system in the real-life case.

References

1. Trotta, A., Di Felice, M., Montori, F., Chowdhury, K.R., Bononi, L.: Joint coverage, connectivity, and charging strategies for distributed UAV networks. IEEE Trans. Rob. **34**(4), 883–900 (2018)
2. Orsino, A., et al.: Effects of heterogeneous mobility on D2D-and drone-assisted mission-critical MTC in 5G. IEEE Commun. Mag. **55**(2), 79–87 (2017)
3. Merwaday, A., Guvenc, I.: UAV assisted heterogeneous networks for public safety communications. In: Proceedings of IEEE Wireless Communications and Networking Conference Workshops (WCNCW), pp. 329–334. IEEE (2015)
4. Erdelj, M., Król, M., Natalizio, E.: Wireless sensor networks and multi-UAV systems for natural disaster management. Comput. Netw. **124**, 72–86 (2017)
5. Erdelj, M., Natalizio, E.: Drones, smartphones and sensors to face natural disasters. In: Proceedings of 4th ACM Workshop on Micro Aerial Vehicle Networks, Systems, and Applications, pp. 75–86. ACM (2018)
6. Gupta, S.G., Ghonge, M.M., Jawandhiya, P.: Review of unmanned aircraft system (UAS). Int. J. Adv. Res. Comput. Eng. Technol. (IJARCET) **2**(4), 1646–1658 (2013)
7. Kong, W., Zhou, D., Zhang, D., Zhang, J.: Vision-based autonomous landing system for unmanned aerial vehicle: a survey. In: Proceedings of International Conference on Multisensor Fusion and Information Integration for Intelligent Systems (MFI), pp. 1–8. IEEE (2014)
8. Pokorny, J., et al.: Concept design and performance evaluation of UAV-based backhaul link with antenna steering. J. Commun. Netw. **20**(5), 473–483 (2018)

9. Pyattaev, A., Solomitckii, D., Ometov, A.: 3D folded loop UAV antenna design. In: Chowdhury, K.R., Di Felice, M., Matta, I., Sheng, B. (eds.) WWIC 2018. LNCS, vol. 10866, pp. 269–281. Springer, Cham (2018). https://doi.org/10.1007/978-3-030-02931-9_22

10. Gowda, M., Manweiler, J., Dhekne, A., Choudhury, R.R., Weisz, J.D.: Tracking drone orientation with multiple GPS receivers. In: Proceedings of 22nd Annual International Conference on Mobile Computing and Networking, pp. 280–293. ACM (2016)

11. Kovalchukov, R., et al.: Analyzing effects of directionality and random heights in drone-based mmWave communication. IEEE Trans. Veh. Technol. **67**(10), 10064–10069 (2018)

12. Vattapparamban, E., Güvenç, İ., Yurekli, A.İ., Akkaya, K., Uluağaç, S.: Drones for Smart Cities: issues in cybersecurity, privacy, and public safety. In: Proceedings of International Wireless Communications and Mobile Computing Conference (IWCMC), pp. 216–221. IEEE (2016)

13. Gharapurkar, A.A., Jahromi, A.F., Bhat, R.B., Xie, W.: Semi-active control of aircraft landing gear system using H-infinity control approach. In: Proceedings of International Conference on Connected Vehicles and Expo (ICCVE), pp. 679–686, December 2013

14. Clarke, R.: Autonomous multi-rotor aerial vehicle with landing and charging system. US Patent 2017/0139409 A1, May 2017

15. Chae, H., Park, J., Song, H., Kim, Y., Jeong, H.: The IoT based automate landing system of a drone for the round-the-clock surveillance solution. In: Proceedings of IEEE International Conference on Advanced Intelligent Mechatronics (AIM), pp. 1575–1580, July 2015

16. Lohan, E.S., et al.: 5G positioning: security and privacy aspects. In: Liyanage, M., Ahmad, I., Abro, A.B., Gurtov, A., Ylianttila, M. (eds.) A Comprehensive Guide to 5G Security, pp. 281–320. Wiley, Hoboken (2018)

17. Liu, H., Darabi, H., Banerjee, P., Liu, J.: Survey of wireless indoor positioning techniques and systems. IEEE Trans. Syst. Man Cybern. Part C (Appl. Rev.) **37**(6), 1067–1080 (2007)

18. Lohan, E.S., Talvitie, J., Figueiredo e Silva, P., Nurminen, H., Ali-Löytty, S., Piché, S.: Received signal strength models for WLAN and BLE-based indoor positioning in multi-floor buildings. In: Proceedings of International Conference on Location and GNSS (ICL-GNSS), pp. 1–6, June 2015

19. Lohan, E.S., Koski, K., Talvitie, J., Ukkonen, L.: WLAN and RFID propagation channels for hybrid indoor positioning. In: Proceedings of International Conference on Localization and GNSS 2014 (ICL-GNSS 2014), pp. 1–6, June 2014

20. Basiri, A., et al.: Indoor location based services challenges, requirements and usability of current solutions. Comput. Sci. Rev. **24**, 1–12 (2017)

21. Yan, J., Tiberius, C., Bellusci, G., Janssen, G.: Feasibility of Gauss-Newton method for indoor positioning. In: Proceedings of IEEE/ION Position, Location and Navigation Symposium, pp. 660–670, May 2008

Features of Multi-target Detection Algorithm for Automotive FMCW Radar

Vladimir D. Kuptsov$^{(\boxtimes)}$ ⓘ, Sergei I. Ivanov ⓘ,
Alexander A. Fedotov ⓘ, and Vladimir L. Badenko ⓘ

Peter the Great St. Petersburg Polytechnic University, St-Petersburg, Russia
vdkuptsov@yandex.ru

Abstract. An algorithm is considered for estimating speeds and distances to targets in the multi-target mode as applied to the linear frequency modulation continuous wave (FMCW) car radar of a millimeter-wave range. The algorithm is based on an estimate of the fast Fourier transform (FFT) amplitude spectrum of the received microwave signals. The reasons for the occurrence of false targets when using frequency modulation continuous wave radars and the probability of their occurrence are identified. Using a computer experiment in the LabVIEW environment, the detection efficiency in additive white noise was investigated and the probability of detection was calculated for various ratios signal to the noise at the radar receiver input.

Keywords: FMCW radar · Multi-target detection · Time-frequency analysis · Ghost target · Detection probability

1 Introduction

Accurate determination of the location and speed of moving objects is necessary in many technical applications, including the vehicular communications. To determine the location and radial speed of objects, various principles can be used: ultrasonic, optical lidar, use of video cameras and microwave [1–10]. The first three listed principles have a significant drawback - a strong dependence on weather conditions (precipitation, fog) and have a small detectable distance to the object (ultrasound and video camera). For this reason, microwave radars have found the greatest use in automotive telecommunications networks and motion control systems. Radar in automotive applications are used to automatically control the distance between moving vehicles, cross traffic alert and lane change assist, parking aid, obstacle, pedestrian and blind spots detection. In addition, radars are used by traffic police to ensure traffic safety using administrative measures.

Microwave car radars for the simultaneous measurement of speed and distance use the formation and emission of signals of a special form, which include: frequency-modulated continuous wave (FMCW), frequency shift keying (FSK) continuous wave. In [1], a waveform was proposed that became classic for FMCW radars. Such radars have high measurement accuracy, high resolution in distance and speed of targets and high performance. The disadvantage of such radars is the small value of the uniquely

© Springer Nature Switzerland AG 2019
O. Galinina et al. (Eds.): NEW2AN 2019/ruSMART 2019, LNCS 11660, pp. 355–364, 2019.
https://doi.org/10.1007/978-3-030-30859-9_30

measured target speed. In [2], for a radar with a sequence of 64 chirps with a duration of 1 ms each, sweep bandwidth of 150 MHz at a radio frequency of 24 GHz, the maximum detectable speed of only 22.5 km/h is determined, which is not enough for use in vehicular communications applications.

In [3–6], the problem of unambiguously determining high target velocities is solved by using a signal consisting of multiple chirp sequence segments with different chirp repetition intervals (including random). These methods are quite difficult to implement and are accompanied by a decrease in the maximum measured range. In [2, 7] to extend the range of target speed estimation, a signal in the form of a Chirp Sequence Waveform is used with a shift of the initial frequency in the adjacent chirps of the frame.

We have shown [10] that the conclusions of the authors of [2, 7] are incorrect, including when exposed to noise on the receiving radar system. In [10], it was also shown that the use of phase methods to determine the target's moving speed significantly reduces the noise immunity of the radar and, therefore, the minimum measured value of the target scattering area. From this point of view, methods based on processing the amplitude spectrum of the fast Fourier transform (FFT) of the received radio signal are more preferable. In [11–13] various waveforms of the generated signals are proposed with subsequent processing methods for estimating the speed and distance in the multi-target mode.

In this paper, we consider an algorithm for estimating velocities and distances to targets based on an estimate of the amplitude spectrum of the received microwave signals. The additive white Gaussian noise was used as a model to describe the effect of clutter in real radar system. The causes of the occurrence of false targets are determined and the detection efficiency in the additive white Gaussian noise in the multi-target mode is investigated. These issues are of practical interest and were not considered in [11–13].

2 Model Derivation and Parameter Estimation Algorithm

Figure 1 shows the chirp sequence waveform of the radar transmitter's probe signal. The choice of the chirp sequence waveform is determined by the simplicity of the technical implementation and the wide choice of millimeter-wave FMCW radar for automotive and industrial applications serially supplied to the market by various companies [14]. One frame of a transmitter signal with a duration of $T_\Sigma = T_0 + T_1 + T_2$ contains a sequence of one pair of periodically repeating signals with linear frequency modulation. In the first interval, the frequency deviation is 0, i.e. the signal is emitted with a constant frequency equal to f_0. The duration of the chirp T_k, the deviation of the frequency of the Δf_k, is determined by the variable k ($k = 0, 1$). The initial frequency of the signal f_0.

The waveform of the radar receiver's input signal (echo-signal waveform) is delayed with respect to the transmitter's sounding signal by a time equal to τ. For practically important radar applications, the delay τ can be calculated using the formula

$$\tau(t) = 2R(t)/c = 2R(0)/c - 2V_Rt/c = \tau_0 - 2V_Rt/c, \quad t \in [0, T_\Sigma],$$

where c – is the speed of light, $R(0)$ – is the initial coordinate of the target, and the radial component of the velocity is related to the Doppler shift of the f_D frequency as follows $2V_R/c = f_D/f_0$.

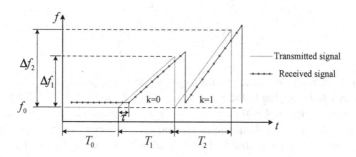

Fig. 1. Proposed waveform

In Fig. 1, a dashed line represents the echo waveform. At the radar receiver, the input analog microwave signal is converted into an intermediate frequency (IF) signal and then converted into digital form.

For further construction of the mathematical model of the considered radar signals and the synthesis of the algorithm for their processing, we use the results of [10]. The signal $SI(t_L, k)$ at the output of the mixer in the in-phase channel I of intermediate frequency of the radar receiver can be written as

$$SI(t_L, k) = U_M \cos\left\{ \begin{array}{l} 2\pi\left[\dfrac{\Delta f_k}{T_k}\tau_0 + f_D + \Delta F(k)\right]t_L \\ + 2\pi[f_D T(k) + f_0\tau_0] + \Delta\varphi(k) \end{array} \right\} \cdot \text{rect}\left[\dfrac{t_L - \tau}{T_k} - \dfrac{1}{2}\right] \\ + n(t_L, k) \tag{1}$$

Formula (1) describes the mathematical model of the IF signal of the odd k chirp in a waveform sequence under the conditions of the additive noise $n(t_L, k)$. The rect(x) function is a symmetric function of a rectangular window. The functions $\Delta F(k)$ and $\Delta\varphi(k)$ clarify the mathematical model of the intermediate frequency radar signal, their form is given in [10]. Time t_L is "fast" (local time) [10].

The additive noise $n(t_L, k)$ is described by a white Gaussian noise model with zero mean and dispersion (power in the receiver frequency band) σ^2.

The mathematical model of the intermediate frequency signal of the odd k chirp $SQ(t_L, k)$ in the quadrature channel Q is described by a similar expression (1) replacing the function $\cos(x)$ with $\sin(x)$. The noise correlation in quadrature channels depends on the ratio of the levels of external and internal noise. The model of radar signals allows you to develop a strategy for the simultaneous assessment of target distance R and its speed V_R in multi-target mode.

According to (1), the beat frequencies f_{R1} and f_{R2}, i.e. the frequency of the signal at the mixer output in the I and Q channels of the intermediate frequency of the radar receiver for chirps, respectively, with the numbers $k = 0$ and $k = 1$ are determined by the expressions

$$f_{R1} = \frac{\Delta f_1}{T_1} \tau_0 - \frac{2V_R}{c} f_0,$$ (2)

$$f_{R2} = \frac{\Delta f_2}{T_2} \tau_0 - \frac{2V_R}{c} \left(f_0 - \frac{3\Delta f_2}{2} \right)$$ (3)

The most difficult task in the multi-target mode is the problem of identifying the beat frequency f_R belonging to the same target at $t \in T_1$ and $t \in T_2$, i.e. for a sequence of chirps with numbers $k = 0$ and $k = 1$. From relations (2) and (3) in the first approximation for the beat frequencies f_{R1} and f_{R2} belonging to the same target, the following condition must be fulfilled (the condition of "pairing") for the functional $\Xi(f_{R1}, f_{R2})$

$$\Xi(f_{R1}, f_{R2}) = f_{R1} \frac{T_1}{\Delta f_1} - f_{R2} \frac{T_2}{\Delta f_2} - f_D \left(\frac{T_1}{\Delta f_1} - \frac{T_2}{\Delta f_2} \right) \to \min$$ (4)

The processing algorithm for the received signal includes a one-dimensional fast complex Fourier transform with a Hamming window, an estimate of the frequencies of the maxima of the amplitude spectrum, a calculation of the distances and speeds of the targets based on the frequencies of the maxima and the use of the algorithm for pairing the targets and the speed of the targets. On the interval where the frequency of the emitted signal is constant (T_0), the speed of all targets is determined from the maxima of the amplitude spectrum. The resolution of determining the frequency is determined by the duration of the corresponding interval. Thus, the speed resolution δV_R and range are respectively equal to

$$\delta V_R = \frac{c}{2T_0 f_0} ; \ \delta R = \frac{c}{2\Delta f}$$

Each Doppler target frequency corresponds to a 2D matrix of beat frequencies for T_1 and T_2 gaps. The choice of the minimum matrix values according to the algorithm (4) determines the Doppler frequency (speed) and two beat frequencies corresponding to the range R to the target.

The results of a computer experiment to estimate the parameters of R and V_R in LabVIEW environment based on the developed algorithm are given in Sect. 4.

3   The Genesis of False Targets in Multi-target Mode

The analysis of the target parameter estimation algorithm in the multi-target mode, described in Sect. 2, shows the possibility of false targets with virtual (not corresponding to the actual) values of the radial velocity V_R and the distance R. The reason for the occurrence of false targets is the situation when the beat frequencies f_{R1} and f_{R2}, corresponding to two different speeds V_{R1} and V_{R2} of two distinguishable targets, coincide with the beat frequencies f_{R1} and f_{R2} of the third target, which has a speed V_{R3}. A false target has a speed corresponding to one of the real targets, but the distance is erroneously determined. The maximum possible number of false targets n_F with a total number of observable targets N with distinct speeds is proportional to the number of placements without repetitions of N elements of 2 each.

$$n_F = N \cdot A_2^N = N \cdot (N-1) \cdot (N-2)$$

The value of n_F rapidly increases with increasing N. Thus, when the number of observed targets is $N = 9$ with distinct velocities V_R, the maximum possible number of false targets is 504, which can be comparable with the number of resolvable samples in the distance N_R, determined by the formula.

$$N_R = \frac{R_{MAX}}{\delta R}$$

To exclude false targets, complex multi-segment waveform types [2, 3] or methods of tracking the trajectory of targets with subsequent filtering algorithms are used. The multiple-input multiple-output (MIMO) design of radars, which is mainly used to increase the angular resolution by increasing the aperture of the antenna array, will also reduce the probability of false targets. The using of MIMO in the proposed multi-target algorithm requires further study.

4   Results of a Computer Experiment

The standard mode of car radar operation is the simultaneous assessment of parameters (distance and speed) of a significant number of targets, which corresponds to the actual situation on the road. As noted earlier, in radar data processing algorithms there is a technical problem of pairing measured distance values with speed. In this process, ghosts can occur - false targets, the occurrence of which is explained by the discrepancy between the distance-speed pair and the true values of the target. Using a computer experiment in LabVIEW, we investigated the efficiency of multi-target detection in additive white Gaussian noise and calculated the probability of detection for various ratios of the signal to noise at the input of the radar receiver. Simulation parameters are shown in Table 1.

The choice of relatively extended modulation period ($T_{0,1,2} = 7$ ms) is due to the high-resolution requirements for target distances and speeds. The speed resolution for

Table 1. Simulation parameters

	Constant frequency	First chirp	Second chirp
Modulation period, ms	7 (T_0)	7 (T_1)	7 (T_2)
Sampling frequency, MHz	1,2	1,2	1,2
FFT samples	8400	8400	8400
Carrier frequency f_0, GHz	76,5	76,5	76,5
Sweep bandwidth, MHz	0	350 (Δf_1)	500 (Δf_2)

the parameters indicated in Table 1 is 1 km/h, for the distance - 0.42 m at the first chirp and 0.3 m at the second.

Computational experiments are implemented in the LabVIEW design environment. The block diagram contains a modulator unit and a radar intermediate frequency signal processing unit. The modulator unit sets the target parameters (distance, speed) and the technical characteristics of the radar (sampling frequency, modulation period $T_{0,1,2}$ of waveform segments, carrier frequency and sweep bandwidth). Based on these parameters, a discrete complex intermediate frequency signal is formed in the modulator unit, consisting of samples of in-phase (I) and quadrature signals (Q), to which is added an additive white Gaussian noise with standard deviation σ. Thus, the modulator unit generates an intermediate frequency signal at the output of the radar mixer. The processing unit performs a complex Fourier transform with the Hamming window of the complex intermediate frequency signal, converts the spectrum indices to bring the first Nyquist zone to zero frequency, which is necessary to determine positive and negative speed values, calculates the amplitude and phase spectrum of the intermediate frequency signal, determines the peak frequencies of the amplitude spectrum on all three waveform segments, forms a functional $\Xi(f_{R1}, f_{R2})$ in accordance with the expression (4) for the speeds and distances of all targets. As a result, each speed corresponds to a 2D distance matrix of size ($N_{target} \times N_{target}$), in which cells with the minimum value of the functional $\Xi(f_{R1}, f_{R2})$ are searched. The minimum correspond to the speed from the constant frequency segment and to the distances from the first and second chirps, which coincide with the accuracy of the radar resolution in distance. Thus, pairing of speeds and target distances in the multi-target mode is carried out.

To compare the degree of influence of additive noise on the amplitude and phase spectrum of the FFT, the standard deviations of 50 samples of the maximum amplitude spectrum of the target and phase values at the maximum of the amplitude spectrum under Gaussian noise with a signal-to-noise ratio of −18 dB were measured in the radar. The noise was determined by integrating the entire FFT frequency band, the signal amplitude is 1 V, the root-mean-square deviation (standard deviation - SD) of the Gaussian noise is $\sigma = 4$. The simulation results are summarized in Table 2.

The normalized SD, defined as the ratio of the root-mean-square deviation (standard deviation) to the maximum level of the amplitude spectrum and phase at the maximum in the absence of noise, for phase noise is more than three times greater than for amplitude noise. Thus, the use of the amplitude spectrum (without the phase spectrum) in the radar processing algorithm leads to an improvement in the sensitivity

Table 2. Simulation results

	Maximum amplitude spectrum value	Phase value at amplitude spectrum maximum
SD	330,27 V	0,0711 rad
Normalized SD	0,07374	0,1754

of the radar to low signal levels. The waveform of the intermediate frequency, amplitude and phase spectrum of the first chirp for 4 targets in the absence of noise a) and at signal-to-noise ratio (SNR) = −18 dB b) is shown in Fig. 2.

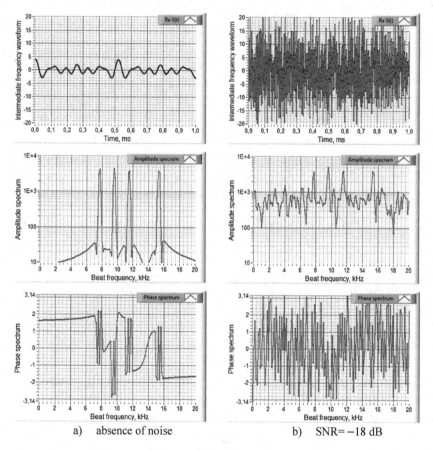

a)   absence of noise   b)   SNR=−18 dB

Fig. 2. Intermediate frequency waveform on the T1 interval, its amplitude and phase spectrum

The signal-to-noise ratio (SNR) is determined as the ratio of the signal power at the beat frequency to the Gaussian noise dispersion σ^2. The effect of additive noise most significantly affects the shape of the phase spectrum.

The simulation of the work of the automotive radar in the multi-target mode for the 4 targets under the influence of Gaussian noise was carried out according to the following procedure. At a given level of SNR, 100 computational experiments were carried out to determine the parameters (speed and range) of 4 targets. If at least one of the targets was determined in the experiment incorrectly, the result was recorded as negative. If the parameters of all 4 targets are determined without errors, the result was recorded as positive. The target parameters set in the experiments are shown in Fig. 3. As a result, we obtained the dependence of the probability of the velocity and distance true estimation (%) on the signal-to-noise ratio (dB) shown in Fig. 4.

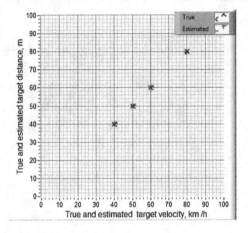

Fig. 3. Parameters of targets in computational experiments

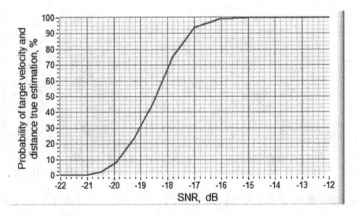

Fig. 4. The dependency of probability of targets velocity and distance true estimation, % from signal-to-noise ratio.

The most dramatic slump in the probability of true detection of target parameters is observed when the signal-to-noise ratio decreases from −17 dB to −20 dB. As shown in [10], the operability of phase algorithms [2–7] under the influence of noise on the useful signal is determined by the conditions of SNR > 10 dB.

5 Conclusion

Our studies have shown the advantage of signal processing algorithms based on an amplitude spectrum estimate as compared to phase methods for determining the radial velocities V_R and the distances R to targets in the automotive FMCW radar of the millimeter range. A computer experiment in the LabVIEW environment showed the efficiency of the proposed algorithm when noise is applied to a useful signal when fulfilling the conditions $SNR > -15$ dB. A specific feature of signal processing algorithms based on an amplitude spectrum estimate is the occurrence of false targets when estimating the radial velocities V_R and distances R. The probability of false targets occurring increases dramatically with an increase in the total number of accompanied targets in the monitoring process.

Acknowledgements. The research was supported by Ministry of Science and Higher Education of the Russian Federation (project unique ID RFMEFI58418X0035).

References

1. Stove, A.G.: Linear FMCW radar techniques. IEE Proc. Radar Signal Process. **139**(5), 343–350 (1992)
2. Kronauge, M., Rohling, H.: New chirp sequence radar waveform. IEEE Trans. Aerosp. Electron. Syst. **50**, 2870–2877 (2014)
3. Wojtkiewicz, A., Misiurewicz, J., Nalecz, M., et al.: Two-dimensional signal processing in FMCW radars. In: Proceedings of the XXth National Conference on Circuit Theory Electronic Networks, Kolobrzeg, Poland, pp. 626–635 (1997)
4. Kronauge, M., Schroeder, C., Rohling, H.: Radar target detection and Doppler ambiguity resolution. In: Proceedings of the International Radar Symposium, Vilnius, Lithuania, pp. 126–129 (2010)
5. Thurn, K., Shmakov, D., Li, G., et al.: Concept and implementation of a PLL-controlled interlaced chirp sequence radar for optimized range-Doppler measurements. IEEE Trans. Aerosp. Electron. Syst. **64**, 3280–3289 (2016)
6. Wang, Y., Xiao, Z., Wu, L., Xu, J., Zhao, H.: Jittered chirp sequence waveform in combination with CS-based unambiguous Doppler processing for automotive frequency-modulated continuous wave radar. IET Radar Sonar Navig. **11**(12), 1877–1885 (2017)
7. Rohling, H., Kronauge, M.: New radar waveform based on a chirp sequence. In: International Radar Conference, pp. 1–4. (2014)
8. Zanina, M.A., Belov, A.A., Volvenko, S.V.: Estimation of accuracy of algorithm for measuring radiofrequency pulse parameters. In: Proceedings of the IEEE International Conference on Electrical Engineering and Photonics (EExPolytech), St. Petersburg, Russia, pp. 98–102 (2018)

9. Privalov, V.E., Shemanin, V.G.: Lidar system for monitoring radioactive contamination of atmospheric air. J. Opt. Technol. (A Transl. Opticheskii Zhurnal) **84**(5), 289–293 (2017)
10. Ivanov, S.I., Kuptsov, V.D., Fedotov, A.A.: The signal processing algorithm of automotive FMCW radars with an extended range of speed estimation. J. Phys. Conf. Ser. **1236**(1), 012081 (2019)
11. Zhou, H.Y., Cao, P.F., Chen, S.J.: A novel waveform design for multi-target detection in automotive FMCW radar. In: Proceedings of the Radar Conference (RadarConf), Philadelphia, PA, USA, pp. 1–5 (2016)
12. Son, Y., Heo, S.W.: A novel multi-target detection algorithm for automotive FMCW radar. In: Proceedings of the International Conference on Electronics, Information, and Communication, Honolulu, HI, USA, pp. 1–3 (2018)
13. Duan, Z., Wu, Y., Li, M., Wang, W., Liu, Y., Yang, S.: A novel FMCW waveform for multi-target detection and the corresponding algorithm. In: Proceedings of the IEEE 5th International Symposium on Electromagnetic Compatibility, Beijing, China, pp. 1–4 (2017)
14. Texas Instruments Inc. http://www.ti.com/lit/wp/spyy006/spyy006.pdf. Accessed 25 May 2019

Cell State Prediction Through Distributed Estimation of Transmit Power

Muhammad Zeeshan Asghar[1]([⊠]), Farhan Azhar[2], Muhammad Nauman[2],
Nouman Ali[2], Muaz Maqbool[2], Muhammad Saqib Ilyas[2],
and Mirza Mubasher Baig[2]

[1] University of Jyväskylä, Jyväskylä, Finland
muhammad.z.asghar@jyu.fi
[2] National University of Computer and Emerging Sciences, Lahore, Pakistan
{1154051,1154310,1154053,1154160}@lhr.nu.edu.pk,
{saqib.ilyas,mubasher.baig}@nu.edu.pk

Abstract. Determining the state of each cell, for instance, cell outages, in a densely deployed cellular network is a difficult problem. Several prior studies have used minimization of drive test (MDT) reports to detect cell outages. In this paper, we propose a two step process. First, using the MDT reports, we estimate the serving base station's transmit power for each user. Second, we learn summary statistics of estimated transmit power for various networks states and use these to classify the network state on test data. Our approach is able to achieve an accuracy of 96% on an NS-3 simulation dataset. Decision tree, random forest and SVM classifiers were able to achieve a classification accuracy of 72.3%, 76.52% and 77.48%, respectively.

Keywords: 5G cellular networks · Cell outage detection ·
Machine learning

1 Introduction

Traditionally, the deployment, operational optimization and troubleshooting have been extensive manual tasks. For large scale networks, this manual effort becomes intractable. Significant recent research focus has been directed towards Self Organizing Networks (SONs), which aim to reduce human operator involvement in the running of a network [3,9].

Three main tasks in the realm of SON are self-configuration, self-healing and self-optimization [3,9,11]. Within self-healing, an important sub-task is to automatically detect faulty or failed cells. Cell failures can result from a variety of reasons, such as component failures and mis-configuration [17,19,22]. Traditional techniques to detect cell failures include manual scanning of alarms from network management systems and manual drive tests. With network sizes growing, not only geographically, but also in density, this process is way too slow and ad hoc to achieve competitive network operations. Thus, automation of cell outage detection has become a necessity.

© Springer Nature Switzerland AG 2019
O. Galinina et al. (Eds.): NEW2AN 2019/ruSMART 2019, LNCS 11660, pp. 365–376, 2019.
https://doi.org/10.1007/978-3-030-30859-9_31

Some prior work has used insights from the way cellular networks operate to detect cell outages. For instance, [19] used neighbor cell list (NCL) reports to construct a visibility graph. The authors observed that topology changes in this graph indicate cell outages. In [17], a weighted combination of the distribution of channel quality indicator (CQI), time-correlation of CQI differential and registration request frequency was used to detect cell outages. Incoming handover request statistics were used to detect cell outages in [8,24]. A hidden Markov model based cell outage detection scheme was proposed in [2]. A framework for self healing for LTE networks is presented in [6]. In [7] the use of contextual information for self healing was proposed in order to pre-emptively handle expected future faults. An inter-cell key performance indicator (KPI) based approach for detecting sleeping cells is proposed in [4]. Another approach that uses correlation between cells to detect degraded cells is proposed in [5].

Recently, many researchers have taken a machine learning approach to cell outage detection. The authors in [18] applied clustering to cell outage detection while Bayesian networks were tried in [16]. Onireti proposed k nearest neighbors and local outlier factor based classifiers for heterogeneous cellular network in [21]. Zoha et al. applied local outlier factor and SVM based classification algorithms to the cell outage detection problem in [25]. In [10], Chernov et al. compared k nearest neighbor, self organizing map and several probabilistic data structures for sleeping cell detection in LTE networks. Gurbani et al. applied Chi-Square test and Gaussian Mixture Models trained on LTE network log data to detect cell outages in [14]. Wang et al. proposed an RBF neural network based approach to detect cell outages in [23]. Mulvey et al. applied a recurrent neural network to the sleeping cell detection in [20].

In this work, similar to works such as [21], we use Minimization of Drive Test (MDT) reports, which are transmitted by each active user in a network, periodically [1]. Each report consists of current channel and carrier characteristics. Each user transmits their location along with signal strength and quality for four best reference signals received from neighboring base transceiver stations (BTSs).

We make the simplifying assumption that the best strength signal comes from the geographically nearest BTS, the second best strength signal form the second nearest BTS and so on. Under this assumption, the received signal strength depends solely on two factors: the transmit power and the distance between the user and a BTS. Given BTS locations and a channel propagation model, from each MDT report, we can estimate the transmit power being used by the four nearest BTSs. In the present work, we only consider the BTS nearest to each user and the highest strength signal to estimate the transmit power being used by the BTS nearest to each user.

In case of normal network operation, our estimated transmit power should be somewhat similar for all users. In case of a cell outage, our estimated transmit power for the users in the vicinity of the failed BTS should be different from others. Consider a toy example, illustrated in Fig. 1. User A's nearest BTS is BTS1. Under normal conditions, user A would receive the best strength signal from BTS1[1]. If we adjust the path loss for the distance between user A and BTS1 to the best received signal strength, we get an estimated transmit power. Most users in the network would arrive at similar estimate transmit power levels. However, if BTS1 fails, user A's best received signal strength now comes from BTS2, which would be significantly weaker than before. If we adjust the path loss for the distance of user A from its nearest BTS, i.e., BTS1, we would arrive at a significantly lower estimated transmit power. Users that are not in the vicinity of the failed BTS would estimate a higher transmit power. We hypothesize that this method can be used to detect the network's operational state.

Fig. 1. A user in the proximity of two BTSs, BTS1 and BTS2. Normally, user A's best received signal strength should depend on distance from BTS1. However, if BTS1 fails, the best received signal strength would no longer be a function of geographical distance from the nearest BTS, i.e., BTS1.

We develop an algorithm based on the above technique. Our approach is able to achieve an accuracy of 96% on an NS-3 simulation dataset. Decision tree, random forest and SVM classifiers were able to achieve accuracies of 72.3%, 76.52% and 77.48%, respectively.

The rest of the paper is structured as follows. In Sect. 2, we provide necessary background and problem formulation. In Sect. 3, we describe our proposed classifier. In Sect. 4, we describe our simulation setup. The results obtained through our proposed classifier on the simulation traces are discussed in Sect. 5. We, then, conclude in Sect. 6.

---

[1] Under obstructions or irregular terrains, this would not hold for many users. Nevertheless, several users may still be located such that this condition holds.

2   Background and Problem Formulation

In wireless communication, a signal travels through air from the transmitter to the receiver. If the signal is transmitted at a power level t (measured in dB), the strength of the signal received by the receiver r (also in dB) is typically lower than t. The difference in transmit and receive power levels, also known as power loss (l), is due to various factors such as attenuation, fading and noise.

Power loss depends on various factors such as communication signal frequency, distance between the transmitter and the receiver and the terrain. Researchers have developed various models that allow estimation of the channel power loss given certain input parameters. Examples include free space path loss model [13] and Hata model [15]. The two most important parameters in path loss estimation are the distance between the communicating entities and the frequency of the radio signal. To emphasize this dependence, we will denote path loss as $l(d, f)$, where f is the frequency of the signal and d is the distance between the transmitter and the receiver.

A cellular network consists of a number of BTSs dispersed in a geographical area, such that each BTS provides coverage to users in its vicinity. Each BTS transmits radio signals that allow users in its vicinity to communicate using the network. One of these signals is called a reference signal - each BTS has a separate one. Every user is likely to receive the signals transmitted by several nearby BTSs, however weak some of these signals may be.

While each BTS uses several channels at different frequencies, these frequencies do not differ greatly. Furthermore, the path loss is a function of the log of the frequency. Thus, for a given distance between transmitter and receiver, the path loss is almost the same for all channels within a given operator's network. Thus, we consider path loss to be purely a function of the distance between the transmitter and receiver, denoted $l(d)$.

Consider a cellular network consisting of m BTSs b_1, b_2, ..., b_m located at Cartesian coordinates (x_1, y_1), (x_2, y_2) ..., (x_m, y_m), serving n users. Every active user periodically transmits an MDT report to a network entity. This MDT report consists of the user's current location, signal strength and quality of the four best reference signals the user is able to receive. In this paper, we consider a slotted time operation whereby at discrete intervals, each user transmits an MDT report.

We assume availability of training data in the form of MDT reports while operating the network in various discrete known states where in each state all BTSs are transmitting at the same power level. For instance, we may operate all BTSs transmitting for some time at 43 dBm and call this state 0 of the network. Then, for some time we may operate all BTSs transmitting at 40 dBm and call this state 1 of the network. Then, for some time we may operate all BTSs transmitting at 46 dBm and call this state 2 of the network. We label all the received MDT reports with the state label of the network at that time.

An MDT report m_i^j received from the jth user for the ith time interval consists of the following fields:

- User location (x_i^j, y_i^j): The current x and y coordinate of the user element (UE).
- RSRP1-4 $(r_i^j[1], r_i^j[2], r_i^j[3], r_i^j[4])$: The power of the four strongest reference signals received from nearby BTSs, measured in dB. $r_i^j[1]$ is the strongest and $r_i^j[4]$ is the weakest.
- RSRQ1-4 $(q_i^j[1], q_i^j[2], q_i^j[3], q_i^j[4])$: The signal quality of the above four signals, also measured in dB. The signal power does not necessarily indicate the quality of the signal. The signal quality may be thought of as its ability to carry out intelligible high-speed communication. Factors such as interference and signal to noise ratio (SNR) affect signal quality.
- Label (s_i^j): We assume that we have labelled training data. Label s_i^j represents the network state as perceived by user j in the ith interval. The label could have different integer values representing different transmit power levels, while one network state represents a failure of one of the BTSs.

3   Proposed Solution

Consider MDT report m_i^j, emanating from user j in interval i. We calculate the Euclidean distance of the corresponding user from each of the BTSs as:

$$d(u_i^j, b_k) = \sqrt{\left(x_k - x_i^j\right)^2 + \left(y_k - y_i^j\right)^2} \tag{1}$$

The user is then considered associated to the base station b_o where:

$$o = \arg\min_k d(u_i^j, b_k) \tag{2}$$

The corresponding distance from the nearest BTS is picked as: $d_i^j = d(u_i^j, b_o)$. Then, the path loss $l(d_i^j)$ is calculated corresponding to the distance d_i^j. In the present work, we use the free space path loss model, but any of the more sophisticated models could be used easily.

Since the power levels are in dB:

$$r_i^j[1] = t_i^j - l(d_i^j) \tag{3}$$

Since the left hand side in the above equation is known from the MDT reports, and the path loss is given by a path loss model, the transmit power can be estimated as:

$$t_i^j = r_i^j[1] + l(d_i^j) \tag{4}$$

The estimated transmit power for all MDT reports in a given interval can be averaged to give:

$$t_i = \frac{\sum_{i=0}^n t_i^j}{n} \tag{5}$$

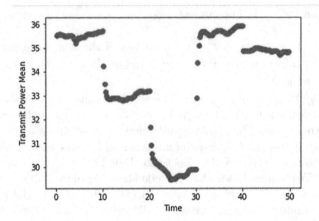

Fig. 2. Estimate transmit power vs simulation time

We conducted a simulation study, which is formally described in Sect. 4. From the simulation trace data, we have plotted the mean estimated transmit power t_i for the duration of the simulation in Fig. 2. We setup the simulation such that network was in state 0 for the first 10 s, state 1 for the next 1 s, state 2 for the next 10 s and state 0 for the next 10 s. During the last ten seconds, one of the BTSs was forecefully failed, so some MDT reports were also labeled 3. State 3 represents the network state whereby one of the BTSs has failed.

From Fig. 2, it appears that the average estimated transmit power is normally distributed around a different mean for each class. We may safely assume that the actual transmitted power is additively affected by a Gaussian variable with certain mean and standard deviation. Furthermore, the additive Gaussian distributions are independent for each class (different operating states vis a vis transmit power level), but have the same standard deviation. Under these conditions, we may calculate mean estimate transmit power for each class using training data and classify a test MDT report to the network state that has the closest average transmit power to the estimated transmit power for the test instance [12].

4   Experiment Setup

A simulation study has been performed in NS-3 whereby 105 mobile users were uniformly spread around 7 base stations. The base stations were spread regularly on the grid as shown in Fig. 3, where each circle represents a base station. The users were mobile and spread uniformly across the grid. The periodic (once every 0.2 ms) MDT report generated by each of the users were recorded for about 50 s. The data has been hand labeled so that each MDT report has been assigned a label from the set {0, 1, 2, 3}. Labels 0, 1 and 2 represent normal network conditions with distinct transmit power levels. An MDT report with label 3 indicates that the corresponding user was previously being served by a BTS that has failed. Only one BTS failure is simulated. The main parameters used in the system simulation are summarized in Table 1.

Table 1. Simulation parameters

Parameter	Value
Cell Layout	Hexagonal grid
Number of cells	57 cells (19 eNodeBs, 3 sectors per eNodeB)
Number of sectors for each base station	3
Inter-site Distance	500 m
eNodeB default Tx Power	46 dBm
Number of UEs in the scenario	105
UE Distribution	Uniform
Traffic Type	Downlink Full Buffer
Simulation lengths	150 s
Cell Outage Tx Power	−50 dBm

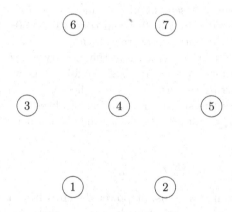

Fig. 3. Network topology for the simulation

5 Results

We partitioned the simulation data into training and test split, where 80% of the data was taken as training data. Due to sparsity of the class 3 instances, we partitioned the data such that a random 80% of each class' instances were in the training set and the rest were in the test set.

From all the t_i values from the training set that have label k, we pick the maximum and minimum values as t_{max}^k and t_{min}^k, respectively. Here, $k \in \{0, 1, 2\}$. Then, for classifying a test MDT report, we selected the thresholds shown in Table 2.

Table 2. Classification rules for each class

Predicted class label	Threshold on estimated transmit power
0	$t_i^j > \frac{t_{min}^0 + t_{max}^1}{2}$
1	$\frac{t_{min}^0 + t_{max}^1}{2} \leq t_i^j < \frac{t_{min}^1 + t_{max}^2}{2}$
2	$t_i^j \leq \frac{t_{min}^1 + t_{max}^2}{2}$

The simple rules given in Table 2 form our classifier for cell state prediction. However, this simple nearest distance classifier did not prove to be effective for discriminating label 3, i.e., cell outage, from the remaining classes. The reason is that with only one BTS failing, several users that are not in the vicinity of the failed BTS are not affected by the failure. Such users would continue to perceive the network in one of states 0, 1 or 2. The few users that are affected by the failure estimate a different transmit power level. If a minority of users is affected by a failure, it skews the average estimated transmit power for network state 3 only slightly from that of a "healthy" network state. Thus, the average estimated transmit power level for class 3 differs only slightly from one of the other classes. A threshold-based classifier would not perform well.

This can be observed from Fig. 2. From time 40 s to 50 s, most of the BTSs are operating in state 0, but the average estimated transmit power is pulled down slightly by the few users that are in the vicinity of the failed cell.

To classify class 3 instances, we devised the following two additional rules. If either of the rules is true, the instance is classified as class 3.

Rule 1: Calculate the standard deviation of all estimated transmit power values as $\sigma = \text{stdev}(t_i^j)$. Instance m_i^j is classified as class 3, if:

$$t_i^j < \frac{t_{i-1} + t_{i-2}}{2} - \sigma \tag{6}$$

That is, if the estimated transmit power deviates from the mean estimated transmit power for the last two intervals by more than the standard deviation of estimated transmit power, an anomaly is flagged.

Rule 2: Label instance m_i^j as class 3, if:

$$r_i^j[1] < \frac{r_{i-1}^j[1] + r_{i-2}^j[1]}{2}. \tag{7}$$

That is, if RSRP 1 has dropped below the mean RSRP 1 value observed by the same user over the last two intervals, an outage is flagged. This is based on the hypothesis that a user near a failed BTS would notice a sudden drop in the best strength reference signal that they receive. For instance, in Fig. 1, if BTS1 fails, user A's RSRP1 value will drop significantly, because the best known reference signal now traverses a much greater distance, thereby suffering much greater path loss.

Based on this classifier, we were able to achieve the results given in Table 3. The average accuracy for the classifier was 96%. The confusion matrix for the classifier is given in Table 4.

Table 3. Accuracy, precision and recall for our cell state predictor

Average accuracy	Average F1 score	Average precision	Average recall
96.09%	90.33%	87.62%	94.48%

Table 4. Confusion matrix for our classifier

		Predicted Label			
		0	1	2	3
	0	14536	105	105	334
True	1	210	5040	0	0
Label	2	0	210	5040	0
	3	59	0	0	506

For comparison purposes, we trained three standard machine learning classifiers on the same feature set. We performed k-fold validation on these classifiers with k=5.

For the decision tree classifier, the average classification accuracy achieved was 72.3%. The confusion matrix for one of the folds for decision tree classifier on our dataset is given in Table 5.

Table 5. Confusion matrix for decision tree classifier

		Predicted Label			
		0	1	2	3
	0	2469	359	147	20
True	1	359	546	129	16
Label	2	161	134	710	45
	3	17	13	36	47

For the random forest classifier, the average classification accuracy achieved was 76.52%. The confusion matrix for one of the folds for decision tree classifier on our dataset is given in Table 6.

For the SVM classifier, the average classification accuracy achieved was 77.48%. The confusion matrix for one of the folds for decision tree classifier on our dataset is given in Table 7.

Table 6. Confusion matrix for random forest classifier

		Predicted Label			
		0	1	2	3
	0	2562	268	160	5
True	1	336	577	129	8
Label	2	145	93	794	18
	3	17	19	39	38

Table 7. Confusion matrix for SVM classifier

		Predicted Label			
		0	1	2	3
	0	2618	216	161	0
True	1	385	534	131	0
Label	2	130	79	836	5
	3	9	21	54	29

6   Conclusion

We used MDT reports from a simulation run to classify a cellular network's state in terms of its current transmit power and cell outage. We estimated the transmit power from the MDT reports and applied thresholding to classify the healthy network's state. We developed two simple rules to classify the single cell outage state. Our classifier accuracy is 96%. Using the same feature set, the standard SVM classifier with 5-fold validation achieved an accuracy of 77.48%.

References

1. 3GPP: Universal Mobile Telecommunications System (UMTS); LTE; Universal Terrestrial Radio Access (UTRA) and Evolved Universal Terrestrial Radio Access (E-UTRA); Radio Measurement Collection for Minimization of Drive Tests (MDT); Overall Description; Stage 2. Technical Specification (TS) 37.320, 3rd Generation Partnership Project (3GPP) (04 2011), version 10.1.0 Release 10
2. Alias, M., Saxena, N., Roy, A.: Efficient cell outage detection in 5G HetNets using hidden Markov model. IEEE Commun. Lett. **20**(3), 562–565 (2016)
3. Aliu, O.G., Imran, A., Imran, M.A., Evans, B.: A survey of self organisation in future cellular networks. IEEE Commun. Surv. Tutor. **15**(1), 336–361 (2012)
4. Asghar, M., Nieminen, P., Hämäläinen, S., Ristaniemi, T., Imran, M.A., Hämäläinen, T.: Cell degradation detection based on an inter-cell approach. Int. J. Dig. Content Technol. Appl. **11** (2017)

5. Asghar, M.Z., Fehlmann, R., Ristaniemi, T.: Correlation-based cell degradation detection for operational fault detection in cellular wireless base-stations. In: Pesch, D., Timm-Giel, A., Calvo, R.A., Wenning, B.-L., Pentikousis, K. (eds.) MONAMI 2013. LNICST, vol. 125, pp. 83–93. Springer, Cham (2013). https://doi.org/10.1007/978-3-319-04277-0_7

6. Asghar, M.Z., Hämäläinen, S., Ristaniemi, T.: Self-healing framework for LTE networks. In: 2012 IEEE 17th International Workshop on Computer Aided Modeling and Design of Communication Links and Networks (CAMAD), pp. 159–161. IEEE (2012)

7. Asghar, M.Z., Nieminen, P., Hämäläinen, S., Ristaniemi, T., Imran, M.A., Hämäläinen, T.: Towards proactive context-aware self-healing for 5G networks. Comput. Netw. 128, 5–13 (2017). https://doi.org/10.1016/j.comnet.2017.04.053. http://www.sciencedirect.com/science/article/pii/S1389128617301895, survivability Strategies for Emerging Wireless Networks

8. de-la Bandera, I., Barco, R., Munoz, P., Serrano, I.: Cell outage detection based on handover statistics. IEEE Commun. Lett. 19(7), 1189–1192 (2015)

9. Van den Berg, J., et al.: Self-organisation in future mobile communication networks. In: Proceedings of ICT-Mobile Summit 2008, Stockholm, Sweden, 2008 (2008)

10. Chernov, S., Cochez, M., Ristaniemi, T.: Anomaly detection algorithms for the sleeping cell detection in LTE networks. In: 2015 IEEE 81st Vehicular Technology Conference (VTC Spring), pp. 1–5. IEEE (2015)

11. Combes, R., Altman, Z., Altman, E.: Self-organization in wireless networks: a flow-level perspective. In: 2012 Proceedings IEEE INFOCOM, pp. 2946–2950. IEEE (2012)

12. Duda, R.O., Hart, P.E., Stork, D.G.: Pattern Classification and Scene Analysis, vol. 3. Wiley, New York (1973)

13. Garg, V., Wilkes, J.E.: Wireless and Personal Communications Systems, 1st edn. Prentice Hall, Upper Saddle River (1996)

14. Gurbani, V.K., Kushnir, D., Mendiratta, V., Phadke, C., Falk, E., State, R.: Detecting and predicting outages in mobile networks with log data. In: 2017 IEEE International Conference on Communications (ICC), pp. 1–7. IEEE (2017)

15. Hata, M.: Empirical formula for propagation loss in land mobile radio services. IEEE Trans. Veh. Technol. 29(3), 317–325 (1980)

16. Khanafer, R.M., et al.: Automated diagnosis for UMTS networks using bayesian network approach. IEEE Trans. Veh. Technol. 57(4), 2451–2461 (2008)

17. Liao, Q., Wiczanowski, M., Stańczak, S.: Toward cell outage detection with composite hypothesis testing. In: 2012 IEEE International Conference on Communications (ICC), pp. 4883–4887. IEEE (2012)

18. Ma, Y., Peng, M., Xue, W., Ji, X.: A dynamic affinity propagation clustering algorithm for cell outage detection in self-healing networks. In: 2013 IEEE Wireless Communications and Networking Conference (WCNC), pp. 2266–2270. IEEE (2013)

19. Mueller, C.M., Kaschub, M., Blankenhorn, C., Wanke, S.: A cell outage detection algorithm using neighbor cell list reports. In: Hummel, K.A., Sterbenz, J.P.G. (eds.) IWSOS 2008. LNCS, vol. 5343, pp. 218–229. Springer, Heidelberg (2008). https://doi.org/10.1007/978-3-540-92157-8_19

20. Mulvey, D., Foh, C.H., Imran, M.A., Tafazolli, R.: Cell coverage degradation detection using deep learning techniques. In: 2018 International Conference on Information and Communication Technology Convergence (ICTC), pp. 441–447. IEEE (2018)

21. Onireti, O., et al.: A cell outage management framework for dense heterogeneous networks. IEEE Trans. Veh. Technol. **65**(4), 2097–2113 (2015)
22. Wang, W., Zhang, J., Zhang, Q.: Cooperative cell outage detection in self-organizing femtocell networks. In: 2013 Proceedings IEEE INFOCOM, pp. 782–790. IEEE (2013)
23. Wang, Y., Long, P., Liu, N., Pan, Z., You, X.: A cooperative outage detection approach using an improved RBF neural network with genetic ABC algorithm. In: 2018 10th International Conference on Wireless Communications and Signal Processing (WCSP), pp. 1–6. IEEE (2018)
24. Zhang, T., Feng, L., Yu, P., Guo, S., Li, W., Qiu, X.: A handover statistics based approach for cell outage detection in self-organized heterogeneous networks. In: 2017 IFIP/IEEE Symposium on Integrated Network and Service Management (IM), pp. 628–631. IEEE (2017)
25. Zoha, A., Saeed, A., Imran, A., Imran, M.A., Abu-Dayya, A.: A learning-based approach for autonomous outage detection and coverage optimization. Trans. Emerg. Telecommun. Technol. **27**(3), 439–450 (2016)

Performance Study of 5G Downlink Cell

Aymen I. Zreikat$^{(\boxtimes)}$ and Suat Mercan

College of Engineering and Technology,
American University of the Middle East, Egaila, Kuwait
{aymen.zreikat,suat.mercan}@aum.edu.kw

Abstract. 5G new radio millimeter wave (5G NR mmWave) is the upcoming technology with a new interface which is developed to be an extension to the existing 4G technology. The main target for 5G is to have a wide range of services with the high data rate, high coverage, reduced delay, reduced cost, high system capacity and multiple connectivity for users everywhere. In this paper, a performance study of 5G cell in the downlink is considered. The main idea of 5G is to provide high performance regarding throughput and spectral efficiency in the dense urban area which is not possible to be provided by a Wi-Fi network. Based on OFDM modulation, the 5G cell is divided into three virtual zones to study the performance of 5G in the inner zone compared to outer zones for licensed and unlicensed spectrum alike. Different performance indicators are considered in the analysis such as; loss probability, delay, throughput as well as aggregate average bit rate in different zones. The provided numerical results show that 5G performance is always better in the most inner zones (i.e. pico) compared to the outer zones (i.e. micro and macro) zones, consequently, the overall cell performance is also improved. Besides, 5G performance is compared with LTE performance under the same simulation parameters to show that 5G always provide a better performance, especially in the most inner zone.

Keywords: Performance study · MOSEL-2 · Downlink · Virtual zones · 5G

1 Introduction

Fifth-generation wireless (5G) is a promising technology that has the potential to enable a myriad of applications in various areas such as health, smart city, augmented reality, etc. 3GPP Release 15 is the first full set standard developed by the focus group [1]. Two characteristics of 5G distinguishing from the previous generation are high bandwidth capacity (>10 Gbps) and short delay (<5 ms). A high capacity communication channel is expected to foster the development of new applications. Offloading computing task from the mobile device to edge computing facility requires such infrastructure. Autonomous driving, vehicular networks, and smart city are some examples which are very sensitive to delay and are a good fit for this upcoming model. These two features are achieved by using higher frequency millimeter wave (30 GHz, mmWave) band which hinders long-range communications (small coverage area) in addition to high loss rate and sensitivity to blockage. This calls for the development of new techniques to improve the range. Various approaches and solutions in physical and network layer have been proposed to address the challenges posed by 5G. MIMO

© Springer Nature Switzerland AG 2019
O. Galinina et al. (Eds.): NEW2AN 2019/ruSMART 2019, LNCS 11660, pp. 377–389, 2019.
https://doi.org/10.1007/978-3-030-30859-9_32

(Multi Input Multi Output) is one of the prominent solutions. It is basically an array of antennas each of which is directed at a different angle and is sending beams at that direction. Moreover, smaller cell size is used to improve spectral efficiency which will cause many handover situations between the cells in mobile environments. In the physical layer, there are mainly two operating modes, orthogonal frequency division multiplexing (OFDM) and single carrier (SC) [2]. OFDM is a better fit for high-performance applications, SC, on the other hand, is more suitable for low power applications. 4G LTE means fourth generation long term evolution. LTE is a kind of 4G that provides fastest connection to mobile Internet experience-10 times faster than 3G. Compared to 5G, LTE has different versions that in the process of development in order to improve the performance of LTE such as; LTE-A, LTE-Pro, and LTE-U. The infrastructure of these technologies will be the base for the new 5G technology. Therefore, the operators nowadays will keep using the LTE technology with new improvements until the new 5G technology is completely evolving by 2020. On the other hand, and due to the increasing number of customers, mobile operators start thinking about the so-called 5G-Wi-Fi debate on whether 5G technology will replace Wi-Fi or the two technologies will be merged to fulfill the requirements of the users anywhere any time. It is expected that Wi-Fi technology will continue evolving even with the existence of 5G technology. Therefore, to achieve this objective, Wi-Fi will coexist in the future with 5G and become a part of many 5G use cases [3, 4].

2   Related Work

The fifth generation of mobile networks (5G) is the new generation of mobile that will replace the fourth generation (4G) to provide services with higher data rates, minimum delay, and maximum spectrum efficiency compared with 4G. Therefore, performance study and analysis of such systems is considered as an important issue and should be deliberated in the literature. Many researchers have studied this in Literature [5–9]. According to [5], the enhancement of spectral efficiency is the main challenge facing mobile operators. Therefore, in this paper, 5G radio access framework is presented using inter-block and intra-block nonorthogonality. It is shown that this framework is able to unify a large number of link and hence increase the spectral efficiency. In [6], the cooperative location estimation algorithms in 5G cellular networks is studied by performance evaluation based on time of arrival (TOA) triangulation method to obtain sensitive location between the mobile terminal and the cell. The given numerical results in this paper show that this algorithm can help in location determination. The paper in [7] presented the higher order sectorization scenario that can be used in future 5G networks operating in the millimeter wave band. The provided simulation results for the data throughput and coverage probability show that higher order sectorization in millimeter wave band will provide capacity increase and data speed for next-generation networks. The performance of a single cell and multi-cell downlink channel over MIMO-OFDM of 5G mmWave propagation channel is studied in [8]. The main idea was to optimize the cyclic prefix parameter by studying the statistics of the power delay profile of the channel. For single cell scenario, the performance was better compared to the multi-cell scenario due to the increase in the number of users in the cell. Because wireless backhaul (WB) channel in heterogonous networks is an important parameter to be considered and studied in the research, the performance of the backhaul channel model in the urban area is studied in [9]. The backhaul scheme is evaluated via random and hotspot user equipment distribution. The given numerical results show that the frequency and the distribution have a major effect on the performance. For the sake of comparison, the same parameters in [10] are used in this analysis and some results are compared in the numerical results section to show how the suggested model did improve the performance of 5G cell in the downlink compared to LTE model.

3 Modeling

3.1 5G Cell Model Description

The main objective of 5G is to provide high throughput in dense area (i.e. Pico cells) and to provide connectivity for the end user in different environments and with control to heterogeneous network along with the availability of high frequency to support internet-of-things (IOT) and internet-of-buddies (IOB). For this reason, 5G model is introduced in Fig. 1 based on OFDM new wireless based standard which has been already used by the 4G technology as well as the Wi-Fi technology. However, 5G OFDM uses a large number of parallel, narrow-band subcarriers instead of a single wide-band carrier to transport information with a wider range of frequency (i.e. 6 GHz to 100 GHz). Based on this concept, three zones are assumed in the analysis; zone 1 represents the indoor/Pico applications where it was covered by Wi-Fi network with limited spectrum efficiency, zone 2 represents the Micro or urban area and zone 3 represents the Marco or suburban area.

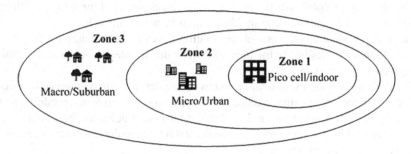

Fig. 1. 5G OFMD cell model

3.2 Modeling Assumptions

1. A single 5G cell is assumed with a maximum capacity given for the whole cell. The capacity in the cell depends on the modulation schemes of 5G and the maximum number of users in the cell.

2. Based on [5], a cyclic prefix-based OFDM (CP-OFDM) waveform is assumed which provides better spectrum separation than LTE to meet the diverse applications expected by 5G.
3. A Poisson inter-arrival time λ is assumed in the cell. The traffic load is calculated based on the area of the zones ($\alpha_i * \lambda$, $i = 1...$ zone i), and the density of users in the cell.
4. The admission control algorithm is based on a threshold value of a minimum bit rate to be satisfied for the user to be admitted to the system.
5. Three types of services are assumed in the analysis:
   (a) Web browsing-over-5G (WBo5G): the possibility to browse, locate and download web pages from the internet with gigabyte plus speed. Most of the applications on the internet are web browsing, therefore, 60% portion of the services is given to this service.
   (b) Video Streaming-over-5G (STRo5G): which includes real-time video streaming, video on demand (VoD) as well as virtual and augmented reality video. To provide multimedia services with a minimum delay to the end user, this type of service is given a 30% portion of the total services in the analysis.
   (c) Voice-over-5G (Vo5G): also called voice over new radio which provides customers with high definition (HD) voice communications for smartphones. This type of service is not frequently used. It is used only for emergency calls. Therefore, it is given the rest of the 10% portion of the total available services.

3.3   Performance Measures

The Continuous Time Markov Chain (CTMC) solver of MOSEL-2 language [11] produce the following performance parameters of the analysis:

1. The blocking probability is the probability to reject the call if the bit rate of the call goes below the minimum threshold bit rate assumed in the cell.
2. The loss probability is the probability to drop the call after admission when the bit rate of the user goes below the threshold value due to interference reason or due to degradation in the channel quality.
3. The utilization is the mean total bit rate with respect to the maximum cell capacity.
4. The cell delay is the time needed by the message to arrive successfully to the destination with respect to the time the first bit is sent out from the source.
5. The average bitrate of the cell is the mean total bit rate in the cell with respect to the average session duration.
6. The aggregate bit rate of the cell is the mean total bit rate to all services in the cell.
7. The aggregate bit rate per service is calculated as the mean number of jobs in each service multiplied by the average bit rate of the zone where the service is required.

3.4   The Model Solution by MOSEL-2 Language

Using MOSEL-2 language [11], the proposed model is solved numerically. MOSEL-2 stands for modeling specification and evaluation language version 2 which is developed by the operating system department in Erlangen University by the group of

MOSEL members to help in performance evaluation of complex systems. By MOSEL-2 language, Fig. 2 describes the modeling and evaluation process done by MOSEL-2 step by step.

In step 1, the system description is written by MOSEL-2 in a simple and easy to learn language similar to C language.

In step 2 and step 3, MOSEL-2 program ("filename.mos") translates the description of the system into the required tool: i.e. SPNP or Time NET.

In step 4 and step 5, the model is solved by using the numerical analysis or by using discrete-event-simulation. This can be done using some options given to each case.

In step 6, if the analysis in step 4 or step 5 was done successfully then two files are generated automatically; one result file (filename.res) which contains the numerical results and one graphical file ("filename.igl") which contains the generated figures. Intermediate Graphical Language (IGL) is a package installed with MOSEL-2 to construct the numerical results and prepare them in an interactive and easy to use environment. Table 1 shows the simulation parameters that are assumed in the analysis.

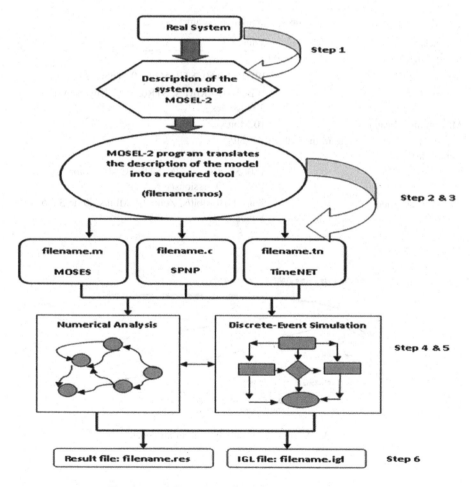

Fig. 2. Modeling and evaluation process in MOSEL-2

4 Numerical Results and Discussion

After solving the model numerically using MOSEL-2 language, the obtained results are used to prepare the figures (Figs. 3, 4, 5, 6, 7, 8, 9, 10, 11, 12, 13, 14, 15, 16) using the intermediate graphical language package called IGL which is already installed with MOSEL-2 package. There are two groups of results in this section. The first group of results (Figs. 3, 4, 5, 6, 7, 8, 9, 10, 11, 12) is to show 5G performance and the second group of results (Figs. 13, 14, 15, 16) is to compare 5G performance with LTE performance under the same simulation parameters and conditions. Figure 3 shows the aggregate average bit rate in all three zones. It can be noticed from Fig. 3 that the aggregate average bit rate in Zone 1 has the highest bit rate. It is expected as most of the calls are served by the inner zone (i.e. zone1). This behavior reflects the main objective of 5G networks to improve the performance of dense area networks. Figure 3 shows the aggregate average bit rate in Zone 1 for different types of services (i.e. WBo5G, STRo5G, and Vo5G).

Table 1. Simulation parameters

Parameter	Value
Number of zones/cell 3	2
Service ratio	WBo5G-0.60, STRo5G-0.30, Vo5G-0.10
Session size	WBo5G-10 Kbytes, STRo5G-50 Kbytes, Vo5G-300 Kbytes
Maximum cell capacity	10 Mbit/s
Assumed minimum bit rate in the cell (threshold)	1 Mbit/s
Service rate	WBo5G-8 sessions/s, STRo5G-40 sessions/s, Vo5G-240 sessions/s
Max. bit rate/zone	Zone 1-10 Mbit/s, Zone 2-5 Mbit/s, Zone 3-2.5 Mbit/s

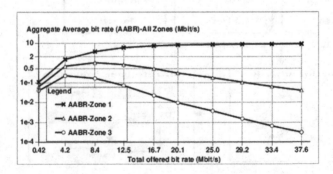

Fig. 3. Aggregate average bit rate in all zones

Looking at Fig. 4, one can notice that at the lower rate all Vo5G services will be firstly served and given higher priority over other services. When the load increases, the model starts serving services that need high bandwidth such as real-time video streaming and downloading big size documents from the internet. Therefore, the aggregate bit rate will be higher for these services. On the other hand, in the outer zones, zone 2 is shown in Fig. 5 and zone 3 is shown in Fig. 6. Better performance is shown for Vo5G as the bandwidth in the outer zones is lower than the inner zone, therefore, the outer zones will not be able to serve requests that require higher bandwidth. Only the best effort data will have the chance to be served in the outer zones. Consequently, the aggregate average bit rate for WBo5G and STRo5G will be lower at the outer zones compared to Vo5G. The mentioned behavior is not very clear in Zone 2 as when the load is low (i.e. less than 13 Mbit/s) some STR05G services can be served by Zone 2 to some extent. However, when the load increases, only Vo5G requests can be served. In Fig. 6, the aggregate average bit rate is given in Zone 3. Here, most of the low bandwidth requests are served, therefore, the aggregate average bit rate for Vo5G will be higher at different traffic loads.

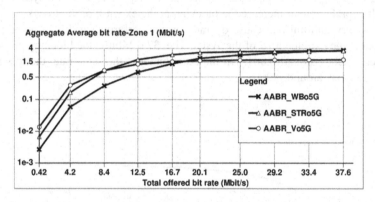

Fig. 4. Aggregate average bit rate for all services in zone 1

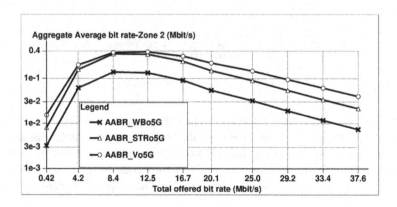

Fig. 5. Aggregate average bit rate for all services in zone 2

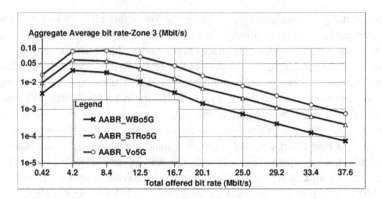

Fig. 6. Aggregate average bit rate for all services in zone 3

Cell blocking probability and loss probability are given in Figs. 7 and 8. The main objective of the model is to reduce the level of loss probability in the cell compared to blocking probability as the loss probability has a major effect on the system performance. It is preferable to increase the blocking at the expense of the loss as increasing the loss is very harmful and cause degrade on the performance. It can be seen from Fig. 7 that at the extreme load of the system (i.e. 37.6 Mbit/s) the blocking probability of the cell reaches around 0.001. On the other hand, at the same rate, the loss probability in Fig. 8 is around 0.0001.

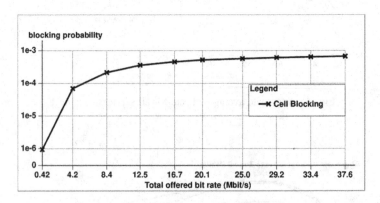

Fig. 7. Cell blocking probability

Cell delay is given in Fig. 9. It is known that 5G networks are aiming at a threshold value of delay of less than 5 ms. Therefore, the ideal values of the delay are shown in Fig. 9 for our suggested model at different loads. It is shown that the delay results are very high at small loads, starts to decrease at higher loads until it reaches to around 0.10 ms. Cell throughput is shown in Fig. 10. The throughput is increased when the

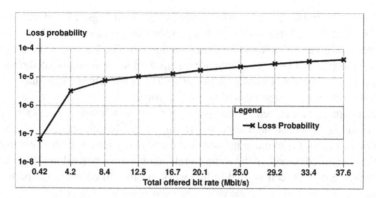

Fig. 8. Cell loss probability

load is increased. This means that at higher loads with maximum needed bandwidth, the model is able to serve the maximum number of users in the cell, hence the cell performance will be improved.

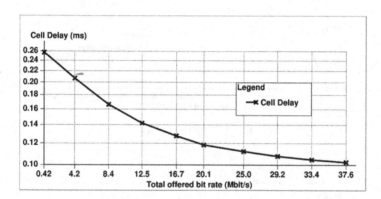

Fig. 9. Cell delay

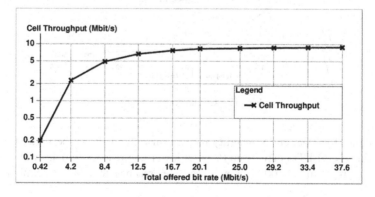

Fig. 10. Cell throughput

The prove for improved cell performance is shown in Figs. 11 and 12. In Fig. 11 the utilization of the cell reaches 0.99. On the other hand, it is shown in Fig. 12 that the average bit rate of the cell is high (2.6 Mbit/s) at lower traffic loads, reaches small value (1.0 Mbit/s) at higher loads. This situation again explains the ability of the model on improving the cell performance in dense areas (i.e. pico zone) more than the outer zones. Therefore, the average bit will be high at lower loads as at lower loads most of the requests are served at cell border (i.e. outer zones) when the load increases, we become close to the dense area zone where higher bandwidth ranges are needed for most services, hence, the average bit rate of the cell is decreased. In this research work and for the sake of comparison, the same simulation parameters used in [10] to evaluate the performance of LTE cell is used in this work in order to compare the performance of both technologies. In Fig. 13 the cell utilization is shown. One can notice that LTE cell utilization starts from 2% at a lower rate and reaches to below 70% at a higher rate. On the other hand, 5G cell utilization starts from 3% and reaches 99% at a higher rate.

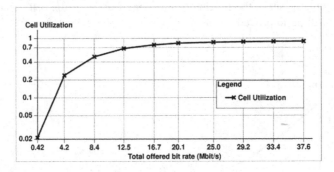

Fig. 11. Cell utilization

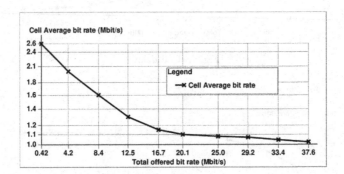

Fig. 12. Cell average bit rate

The figure for blocking probability is given in Fig. 14. It can be noticed that cell blocking in case of LTE starts from 0.001 at lower rate reaches to 0.45, while 5G cell blocking starts from 0.000001 at a lower rate, reaches to 0.001 at a higher rate.

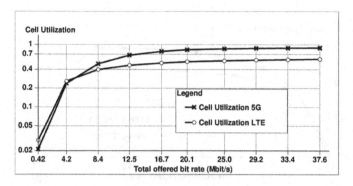

Fig. 13. Cell utilization comparison

A significant difference in the cell delay between the two technologies is also shown in Fig. 15 with better performance for 5G technology which is always below 1 ms at different traffic loads. The last figure is Fig. 16. It is shown in this figure that the aggregate average bit rate is improved in zone 1 which represents the dense area (i.e.

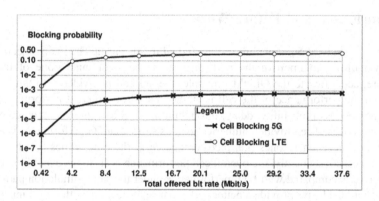

Fig. 14. Cell blocking probability comparison

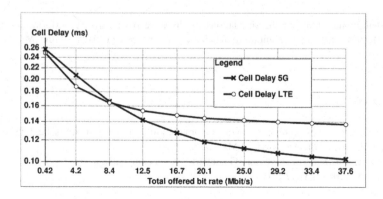

Fig. 15. Cell delay comparison

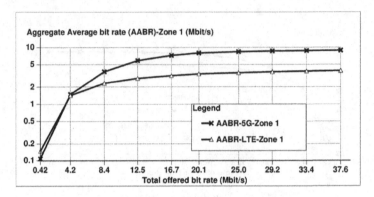

Fig. 16. Aggregate average bit rate-zone 1 comparison

Pico zone) in both technologies, however, the results show that 5G network works better in the most inner zone of the cell regarding the aggregate average bit rate and hence the cell performance will be improved.

5   Conclusions and Future Work

Performance study of 5G downlink cell is studied in this paper. The main objective of a 5G new radio is to improve the performance of the most inner zone of the cell. Therefore, according to the OFDM principle, the 5G cell is divided into three virtual zones and the performance of the cell is studied based on different performance indicators such as loss probability, delay, throughput as well as aggregate average bit rate in different zones. The provided numerical results show that 5G performance is always better in the most inner zones (i.e. pico) compared to the outer zones (i.e. micro and macro) zones, consequently, the overall cell performance is also improved. Besides, 5G performance is compared with LTE performance under the same simulation parameters to show that 5G always provide a better performance, especially in the most inner zone. As future work, coexistence between Wi-Fi and 5G from one side and coexistence between 5G and LTE from the other side can be considered.

Acknowledgment. Some of the simulations results have been partially performed using the Phoenix High-Performance Computing facility at American University of the Middle East, Kuwait.

References

1. 3gpp release 15- 5G. https://www.qualcomm.com/media/documents/files/the-3gpp-release-15-5g-nr-design.pdf. Accessed 29 May 2019
2. Gupta, A., Jha, R.K.: A survey of 5G network: architecture and emerging technologies. IEEE Access **3**, 1206–1232 (2015)

3. Will 5G replace Wi-Fi. https://www.sdxcentral.com/5g/definitions/will-5g-wifi/. Accessed 6 May 2019
4. Foundation to 5G. https://www.qualcomm.com/invention/5g. Accessed 29 May 2019
5. Tong, W., Ma, J., Zhu, P.: Enabling technologies for 5G air-interface with emphasis on spectral efficiency in the presence of very large number of links. In: 21st Asia-Pacific Conference on Communications (APCC), Kyoto, Japan, pp. 184–187 (2015)
6. Namdar, M., Basgumus, A., Guney, A.: Performance analysis of the TOA cooperative localization estimation algorithm for 5G cellular networks. In: 26th Signal Processing and Communications Applications Conference (SIU), Izmir, Turkey, pp. 1–4 (2018)
7. Al-Falahy, N., Alani, O.Y.: The impact of higher order sectorisation on the performance of millimetre wave 5G network. In: 10th International Conference on Next Generation Mobile Applications, Security and Technologies (NGMAST), Cardiff, UK, pp. 1–5 (2016)
8. El Hassan, M., El Falou, A., Langlais, C.: Performance assessment of linear precoding for multi-user massive MIMO systems on a realistic 5G mmWave channel. In: 2018 IEEE Middle East and North Africa Communications Conference (MENACOMM), Jounieh, Lebanon, pp. 1–5 (2018)
9. Shi, R., Ai, B., He, D., Guan, K., Wang, N., Zhao, Y.: Channel analysis and performance evaluation of wireless backhaul at 5G frequency bands. In: 2018 IEEE International Symposium on Antennas and Propagation & USNC/URSI National Radio Science Meeting, Boston, MA, USA, pp. 2001–2002 (2018)
10. Zreikat, A.I.: Performance evaluation of downlink LTE-advanced CELL by MOSEL-2 language. In: 29th European Conference on Modelling and Simulation, Technical University of Sofia, Albena (Varna), Bulgaria, pp. 662–668 (2015)
11. MOSEL Home Page. https://www4.cs.fau.de/Projects/MOSEL/. Accessed 15 July 2018

Downlink Power Allocation
in Delta-OMA (D-OMA) 6G Networks

Jerzy Martyna[✉]

Faculty of Mathematics and Computer Science, Institute of Computer Science,
Jagiellonian University, ul. Prof. S. Lojasiewicza 6, 30-348 Cracow, Poland
`jerzy.martyna@uj.edu.pl`

Abstract. The authors in [9] introduced basic concept of delta-orthogonal multiple access (D-OMA) scheme. This scheme for massive wireless
connectivity in wireless communications allows the implementation of 6G
cellular networks. The practical implementation of 6G cellular networks
requires solving many problems, including downlink power allocation of
all multiple access points (APs). In this paper, an optimal downlink
power allocation model is developed. To overcome the substantial computational complexity of the power allocation optimisation approach,
a heuristic algorithm based on the Stackelberg game was proposed. The
theoretical results were confirmed by simulation.

Keywords: Beyond 5G (B5G)/6G wireless ·
Massive wireless connectivity ·
Orthogonal and non-orthogonal multiple access ·
Delta-orthogonal multiple access · Power allocation · Stackelberg game

1 Introduction

The 5G networks currently being implemented in many countries transform in
a fundamental way current industries, create new industries and impact societies and completely change the method of communication between people and
devices, as well as the communication of the devices themselves. Thanks to the
5G network, single delivery platforms are created that provide a lot of services
starting from Enhanced Mobile Broadband communication starting from links
with Gbps bandwidth, automated driving, massive machine type communication, Industry 4.0, etc. These technologies have contributed to the creation of two
important tools: softwarerisation and virtualisation of network functionalities.
These tools allowed software resources to be provided wherever they are required.
In turn, at the physical layer meeting, the stringent requirements of data rate
and latency resulted in the introduction of millimeter-wave (mmW) communications and delivery of massive multiple-input/multiple-output (MIMO) links and
ultra dense deployment of radio access points.

Despite these unquestionable achievements associated with the 5G network,
each new decade is characterized by the introduction of new generation of wireless communication systems. New generations of wireless networks are emerging

© Springer Nature Switzerland AG 2019
O. Galinina et al. (Eds.): NEW2AN 2019/ruSMART 2019, LNCS 11660, pp. 390–401, 2019.
https://doi.org/10.1007/978-3-030-30859-9_33

at the confluence of two major paths, namely: when they provide both new communication technologies and new social trends. And we are now dealing with such facts. As noticed by Strinati *et al.* [1] the debate "6G or not 6G" has already started. There are ongoing discussions by academicians, researchers and industry practitioners who are to provide a definition and identification of relevant key enabling technologies, which is confirmed by publications on the so-called "beyond 5G" (B5G) or sixth generation (6G) [2–4].

Based on the findings already made, one can give the basic features that the 6G network will have [5]. These are:

(1) *All networks connected:* Each wireless device should be connected to one or more wireless access networks to be served by multiple access points (APs) or base stations (BS). They form a radio network in the cloud. Examples of the services of such a network are systems for autonomous driving, smart network applications, virtual reality, etc. Conventional wireless networks will not be able to provide such complex and demanding services that will be required for future applications.

(2) *Minimising energy at every level of the network:* Energy consumption in radio devices and transceivers will have to be minimised at millimeter and nanometer frequencies. Providing a very dense number of access points and edge servers in a wireless network will enable implementation of new concepts of energy saving, its energy consumption and cooperation between power nodes.

(3) *Effective use of the spectrum after its departure:* The 5G system extends the network frequency range 4G (0.6–6 GHz) for some higher frequency bands (30–300 GHz). 6G systems will have to be developed in new bands for wireless access also including coexistence unlicensed spectrum.

6G cellular networks have been the subject of numerous articles. Among others things, based on artificial intelligence, as well as satellite roaming and even direct communication through implants, proposed for this generation of cellular systems in the work of Rommel *et al.* [7]. Hossain and Al-Eryani in the paper [8] gave the concept of large scale implementation of *non-orthogonal multiple accesses* (NOMA) in machine type communication. Applying this technique allows for unified collaboration of all nearby users who use it so that it can be done at the same time. The same authors in the paper by Al-Eryani *et al.* [9] presented a new access method for 6G mobile networks, referred to as D-OMA (Delta-OMA).

The main goal of the paper is to introduce a new method of power allocation in a downlink in the cellular network 6G based on the D-OMA access method. It presents the optimal power allocation in the downlink in this network. And in addition, heuristic downlink power allocation algorithm based on the Stackelberg game is presented. This allows suboptimal values to be obtained with considerably lower computational complexity.

The remainder of the paper is organised as follows. The next chapter presents methods of access in 5G and 6G cellular networks. Optimal power allocation in the downlink in the 6G network is presented in Sect. 3. Section 4 shows the heuristic algorithm downlink power allocation in the 6G network. The results of simulation tests are given in Sect. 5. Summary conclusions of the work are included in Sect. 6.

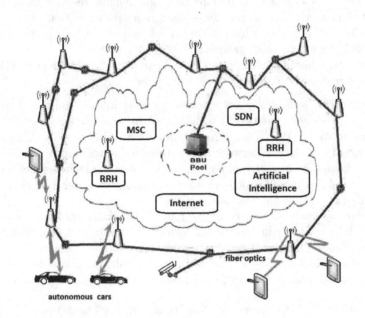

Fig. 1. Cell-less architecture of 6G network.

2   Access Methods in Future Cellular Networks

Due to its nature, the 6G network will lose its cellular structure. Great concentration transceiver devices in the centres of big cities realise multi-point transmissions. This will enable the simultaneous service of hundreds of active devices. The cellular network will become a cell-free system with a dispersed structure (see Fig. 1). This will result in the unificcation of the spectrum [7]. The main elements of such a network will be *massive multiple-input multiple-output* (massive MIMO) systems, that will act as an AP (*access point*) for all active devices. They can be treated as remote heads radio (RRHs) that are used in the C-RAN system (*cloud radio access network*) [10]. The components of such a system are: a mobile switching centre (Mobile Switching Centre, MSC), a *pool of baseband units* (BBU Pool) and a *remote radio head* (RRH), which is connected to the network of *access points* (APs). A characteristic feature of such a system is that any active device can be operated by more than one remote radio head. Otherwise, used in such an AP network they will support all active devices which are in their radio range. This description is supplemented by the possibility of coordinating connections between them, management of interferences, etc., so that the bandwidth is as large as possible.

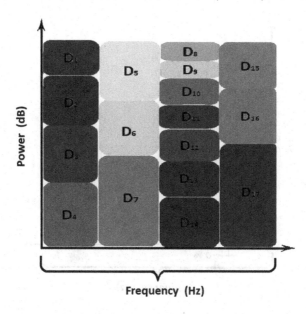

Fig. 2. The scheme of NOMA to serve multiple wireless devices simultaneously within the same sub-band.

The multi-access technology already used in 5G mobile networks as well as the next ones is the NOMA method [11–14]. The main idea of the NOMA method lies simultaneous servicing of many users in the same subbands in the same spectrum (time, frequency and space) at the expense of the resulting interference between them and the deteriorating link quality along with the increase in transmission power (see Fig. 2). The effect of this is to increase the transmission performance of links down and up in wireless networks. Signals of many users are imposed on themselves, although their basic the course can also be described using OFDM modulation or so-called DFT-spread OFDM. The NOMA method has been extended using the MIMO (multiple-input multiple-output) scheme, which allowed a number of clusters to be created covering the same type of users [15].

The new D-OMA access method is only intended for use on 6G [9] networks. It allows the creation of different NOMA clusters, with adjacent frequency bands overlapping with the assumed value δ the percentage of their maximum value of the allocated subband (i.e., $\delta = 0$ corresponds to the area for the NOMA method) (see Fig. 3). In the D-OMA method, the size of the NOMA cluster can be reduced, while maintaining the values of the parameters of the NOMA access method. This allows a significant increase in the number of end devices in the same frequency range. Also, the required energy consumption of these devices can be significantly reduced compared to the NOMA method while maintaining the same performance requirements.

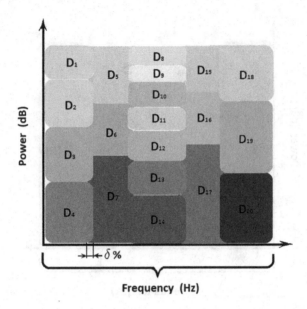

Fig. 3. A concept of the D-OMA scheme [9].

3   Optimal Power Allocation in a Downlink in a Single Cluster with the D-OMA Access Method

It is assumed that the 6G network is composed of a single cluster with the D-OMA method and only the one transmission channel is used. The cluster has M users. It is assumed that the signal for the first active device AD (*active device*) is decoded first, then the signal for second AD, etc. Then the AD can decode the desired signal, treating other users' signals as interference, using the SIC (Successful Mobile Cancellation) technique. The last active device, the M-th, can decode the desired signal after cancelling all INUI signals (*inter-NOMA-user interface*) [13].

The transmission intensity for m-th user inside the cluster using the D-OMA method is as follows [9]:

$$R_m = B \log_2 \left(1 + \frac{\sum_{k=1}^{K} P_{m,k} \mid h_{m,k} \mid^2}{\sum_{k=1}^{K} \Lambda_k \mid h_{m,k} \mid^2 + \delta I_{ICI} + N_m} \right) \tag{1}$$

and

$$\Lambda_k = \sum_{j=\beta+1}^{M} P_{j,k}, \quad \mathcal{M} \le M \tag{2}$$

where K is the number of APs, β is the degree of ordering rank of the m-th device with respect to all APs, M is the number of all ADs within the cluster, B is the bandwidth of the channel transmission, $P_{m,k}$ is the transmission power of

k-th AP for m-th of the device, I_{ICI} (*interstation interference*) gives the value of internal interference in the D-OMA method, N_m is the power of AWGN on the input m-th device AD, $h_{m,k}$ is the channel gain between AP and m-th device.

To perform a decoding operation using the SIC technique, the conditions for power allocation in the downlink must be met:

$$\sum_{k=1}^{K} P_{m,k} \mid h_{m,k} \mid^2 - \sum_{k=1}^{K} \Lambda_k \mid h_{m,k} \mid^2 - \delta I_{ICI} - N_m \geq \Theta,$$

$$\forall m = 1, \ldots, M - 1 \tag{3}$$

where Θ means the minimum difference between the power of an un-encoded signal INUI plus noise power [13].

The downlink optimal allocation power in a single cluster with the D-OMA method can be achieved by maximising the flow intensity for m this user for a given bandwidth B in the transmission channel, namely:

$$\max \sum_{m=1}^{M} \log_2 \left(1 + \frac{\sum_{k=1}^{K} P_{m,k} \mid h_{m,k} \mid^2}{\sum_{k=1}^{K} \Lambda_k \mid h_{m,k} \mid^2 + \delta I_{ICI} + N_m} \right) \tag{4}$$

subject to:

$$C1: \ \sum_{m=1}^{M} \sum_{k=1}^{K} P_{m,k} \leq P_c \tag{5}$$

$$C2: \ \log_2 \left(1 + \frac{\sum_{k=1}^{K} P_{m,k} \mid h_{m,k} \mid^2}{\sum_{k=1}^{K} \Lambda_k \mid h_{m,k} \mid^2 + \delta I_{ICI} + N_m} \right) \geq R_m,$$

$$\forall m = 1, \ldots, M \tag{6}$$

$$C3: \ \sum_{k=1}^{K} P_{m,k} \mid h_{m+1,k} \mid^2 - \sum_{k=1}^{K} \Lambda_k \mid h_{m+1,k} \mid^2 - 1 \geq 0,$$

$$\forall m = 1, \ldots, M - 1 \tag{7}$$

where the $C1$ condition represents a limitation for the energy budget; the $C2$ condition gives the limitation for the m-th user's transmission intensity. Last condition, $C3$, is a limitation due to the SIC technique.

If the parameter vector of this model is convex, then it meets the conditions of Karush-Kuhn-Tucker (KTT) [16]. Then you can specify the Lagrange function:

$$L(P, \alpha, \overline{\beta}, \overline{\gamma}) =$$

$$\alpha \left(P_c - \sum_{m=1}^{M} \sum_{k=1}^{K} P_{m,k} \right) + \sum_{m=1}^{M} \sum_{k=1}^{K} \beta_m \left((P_{m,k} \mid h_{m,k} \mid^2 \right.$$

$$\left. - \delta I_{ICI} - N_m) \times (2^{\frac{R_m}{B}} - 1) \right) + \sum_{m=1}^{M-1} \sum_{k=1}^{K} \gamma_m$$

$$\left(P_{m,k} \mid h_{m+1,k} \mid^2 - \sum_{k=1}^{K} \Lambda_k \mid h_{m+1,k} \mid^2 - P_T \right) \tag{8}$$

where $\alpha, \overline{\beta}, \overline{\gamma}$ are Lagrange multipliers for $\forall m = 1, 2, \ldots, M$; P_c is the maximum transmission power per c-th cluster, P_T is the total transmission power.

Taking derivatives of Eq. (8) with respect to P_m, α, β_m, γ_m and then equate them to zero, an optimal solution is obtained. In an m-th user cluster, there are $m - 1$ Langrange multipliers. Thus, there are 2^{m-1} combinations of Lagrange multipliers for which it must be checked whether the KTT conditions are satisfied [16]. However, the solution found is computationally complex. Therefore, heuristic solution is proposed here.

4   A Heuristic Power Allocation Algorithm in the Downlink for the D-OMA Access Method

The downlink power allocation algorithm below is based on the two-step Stackelberg game. It is assumed that the algorithm concerns a single cluster in which leaders of particular groups are located users. It is assumed that in the 6G system these leaders are RRH devices associated with a given BBU Pool. In the first stage, the leaders compete with each other, maximising their bandwidth. In the second stage individual users subject to a given leader compete with each other, they also maximise their bandwidth. It is assumed that the group leader has full information about his users and about all L channels and their parameters.

It has been assumed that the number of all channels for the group is equal to L. The channels of a given group form a set of $\mathcal{A}$ and the sub-channels moved out to set $\mathcal{B}$. Let F_j be the total power allocated to the channel $j, j \in \mathcal{A}$, and G_j is the interference power in this j channel. The moving of channels must respect to the total power constraints $P_{total} - \sum_{j \in \mathcal{B}} F_j$ and F_j. The adopted power changes are equal to $\Delta = F_j - G_j$, but they can be both positive and negative. It has been taken into account that in the case of the D-OMA method the power value in odd channels is changed by $\Delta \cdot \delta\%$. The goal of the game of leaders of each group is to achieve the Nash equilibrium. Users of the group, on the other hand, use the power shared by group leaders. This is accomplished through the use of the iterative water-filling (WF) algorithm algorithm [17,18].

The pseudocode of the downlink power allocation algorithm for the D-OMA access method is shown in Fig. 4.

Algorithm 1 Downlink power allocation in D-OMA cluster

1: **procedure** DOWNLINK PA WITHIN GROUP USERS(k)
2: **Initialisation**
3:     $\mathcal{A} = \{j \mid j = 1, 2, \ldots, L\}; \mathcal{B} = \emptyset;$
4:     $P = P_{total}; \quad j \leftarrow 0;$
5: **Label 1:**
6:     Using WF algorithm calculate for the set $\mathcal{A}$ with P the water level w;
7:     Reduce the downlink power by Δ;
8:     **if** j is an odd number **then**
9:         move channel j to set $\mathcal{B}$
10:    **end if**
11: **Label 2:**
12:    **while** $j \leq L$ or $\mathcal{A} \neq \emptyset$ **do**
13:        $j \leftarrow j + 1;$
14:        **if** $F_j > G_j$ **then**
15:            move channel j to set $\mathcal{B}$
16:            and update the set $\mathcal{A}$
17:        **end if**
18:        Increase the downlink power value by $\Delta = F_j - G_j$;
19:        **if** j is an odd number **then**
20:            increase the downlink power value by $\Delta \cdot \delta\%$
21:        **end if**
22:        Update the water level w;
23:        **if** $F_j > G_j$ **then**
24:            goto 2
25:        **else** goto 1
26:        **end if**
27:    **end while**
28: **end procedure**

29: **procedure** DOWNLINK PA FOR ALL GROUPS($k : group\ number$)
30:    **while** Stackelberg equilibrium is not found **do**
31:        Finding a Nash equilibrium by group leaders;
32:        Downlink PA within group users k
33:    **end while**
34: **end procedure**

Fig. 4. Pseudocode of the downlink power allocation algorithm for the cluster with D-OMA access method.

5   Simulation Results

This section will present the results of simulation tests for the 6G system with the D-OMA access method.

In the studies conducted, the single-cluster system with the D-OMA access method was taken into account, comparing its performance parameters to the same system with the NOMA access method. The main parameters of the simulated system are presented in Table 1. The simulation assumes that all users uniformly

Table 1. Simulation parameters.

Parameter	Value
System effective bandwidth, B	20 MHz
Bandwidth of one resource block	160 kHz
Number of frequency blocks	24
Number of user terminals per cell	parameterised from 5 to 50
Base station transmission power	24 dBm
Max. transmission power of user terminal	50 dBm
Distance-dependent path loss	$128 + 37 \log_{10}(d)$ dB - d kilometers
Lognormal shadowing	6 dB
Instantaneous fading	six-path Rayleigh
Receivernoise density	-160 dBm/Hz
Scheduling interval	1 ms

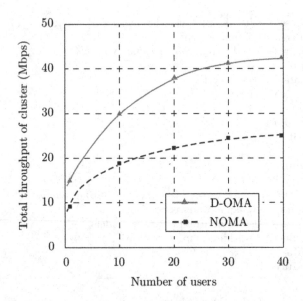

Fig. 5. Total throughput of cluster versus the number of users.

distributed from 10 to 500 m away from the BS. According to the NOMA access method the BS sends the number of messages equal to $\sum_{m=1}^{M} \sqrt{\alpha_m P} \cdot s_m$, where s_m is the message for the m-th user, P is the transmission power of BS, α_m is the power allocation coefficient, i.e. $\alpha_1 \geq \ldots \geq \alpha_M$ [19].

Figure 5 shows the total throughput of the cluster depending on the number of user for both access methods, namely NOMA access method using 16-QAM modulation technique and D-OMA. The D-OMA access method has a higher overall throughput, which is particularly evident in the case of a large number of users.

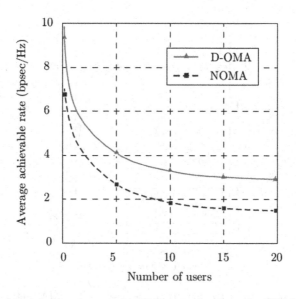

Fig. 6. Comparison of the average achievable rates between the NOMA and D-OMA systems with varying number of users.

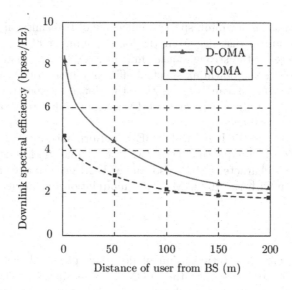

Fig. 7. Downlink spectral efficiency of transmission to individual users as a function of distance from the BS.

Figure 6 compares the average achievable rates to user between NOMA and D-OMA systems with varying number users. It was assumed that the average transmission power to noise for each user is equal to 10 dB. It can be observed that the average achievable rate of D-OMA is higher than that of NOMA system.

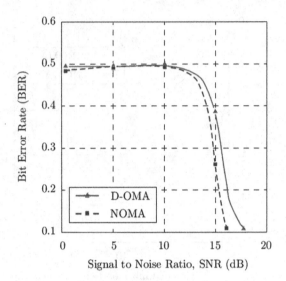

Fig. 8. Comparison of the BER performances between D-OMA method and NOMA using 16-QAM modulation technique.

Figure 7 shows the downlink spectral efficiency of transmission to individual users as a function of distance from BS for the number of users equal to 4. Since the BS decodes the strongest user first in D-OMA, the distance from the BS directly impacts the downlink spectral efficiency more for D-OMA than for NOMA. Nevertheless, it can be seen that the downlink spectral efficiency of transmission to individual user in D-OMA is comparable to that of NOMA, especially at long distances from BS.

Figure 8 shows the Bit Error Rate (BER) performance for NOMA using 16-QAM modulation technique and D-OMA. It is visible in this figure that the D-OMA scheme is characterized by also similar BER performance as in the case of NOMA system with the 16-QAM digital modulation technique.

6   Conclusion

The article presents the optimization of downlink power allocation in a 6G-based cellular network based on the D-OMA access method. The problem of such optimisation has been formulated, and a method of carrying it out has been proposed. The algorithm was given as a suboptimal solution based on the Stackelberg game. In the simulation tests, the obtained results were verified.

Further research must take into account the effectiveness of D-OMA, performance evaluation that assume practical coding and data modulations and the impact of residual interference in the SIC process. In the downlink, the SIC process has a significant impact on the achievable throughput performance and it has to be very carefully researched and defined.

References

1. Strinati, E.C., et al.: 6G: the next frontier (2019). https://arxiv.org/abs/1901. 03239
2. Li, R.: Towards a new Internet for the year 2030 and beyond. https://www.itu. int/en/ITU-T/studygroups/2017-2020/13/Documents/Internet2030%20.pdf
3. David, K., Berndt, H.: 6G vision and requirements: is there any need for beyond 5G? IEEE Veh. Technol. Mag. **13**(3), 72–80 (2018)
4. di Pietro, N., Merluzzi, M., Strinati, E.C., Barbarossa, S.: Resilient design of 5G mobile-edge computing over itermittent mmWave links. Submitted to IEEE Trans. Mobile Comput. https://arxiv.org/abs/1901.0189
5. IEEE 5G and Beyond Technology Roadmap White Paper (2018). https:// futurenetworks.ieee.org/images/files/pdf/ieee-5g-roadmap-white-paper.pdf
6. Chiani, M., Paolini, E., Callegati, F.: Open issues and beyond 5G. https://www. 5gitaly.eu/wp-content/uploads/2019/01/5G-Italy-White-eBook-Beyond-5G.pdf
7. Rommel, S., Raddo, T.R., Johannsen, U., Okonkwo, Ch., Monroy, I.T.: Beyond 5G - wireless data center connectivity. In: Proceedings of the SPIE 10945, Broadband Access Communication Technologies XIII, 109450M, 1 February 2019
8. Hossain, E., Al-Eryani, Y.: Large-scale NOMA: promises for massive machine-type communication. CoRR abs/1901.07106 (2019)
9. Al-Eryani, Y., Hossain, E.: Delta-OMA (D-OMA): a new method for massive multiple access in 6G. CoRR abs/1901.07100 (2019)
10. Chećko, A., Christiansen, H.I., Yan, Y., Scolari, L., Kardaras, G., Berger, M.S.: Cloud RAN for mobile networks - a technology overview. IEEE Commun. Surv. Tutor. **17**(1), 405–426 (2014)
11. Saito, Y., Kishiyama, Y., Benjebbour, A., Nakamura, T.: Non-orthogonal multiple access (NOMA) for cellular future radio access. In: IEEE 77th Vehicular Technology Conference (VTC Spring) (2013)
12. Tabassum, H., Ali, M.S., Hossain, E., Hossain, M.D.J., Kim, D.I.: Non-orthogonal multiple access (NOMA) in cellular uplink and downlink: challenges and enabling techniques. Cornell University (2016). https://arxiv.org/abs/1608.05783
13. Ali, M.S., Tabassum, H., Hossain, E.: Dynamic user clustering and power allocation in non-orthogonal multiple access (NOMA) systems. IEEE Access **4**, 6325–6343 (2016)
14. Ding, Z., Lei, X., Karagiannidis, G.K., Schober, R., Yuan, J., Bhargava, V.: A survey on non-orthogonal multiple access for 5G networks: research challenges and future trends (2017). https://arxiv.org/pdf/1706.05347.pdf
15. Ding, Z., Adachi, F., Pool, V.: The application of MIMO to non-orthogonal multiple access. IEEE Trans. Wirel Commun. **15**(1), 537–552 (2016)
16. Luenberger, D.G.: Linear and Nonlinear Programming. Addison-Wesley, Arlington (1984)
17. Hong, M., Garcia, A.: Averaged iterative water-filling algorithm: robustness and convergence. IEEE Trans. Signal Process **59**(5), 2448–2454 (2011)
18. Qi, Q., Minturn, A., Yang, Y.: An efficient water-filling algorithm for power allocation in OFDM-based cognitive radio systems. In: IEEE International Conference on CSE (2012). https://digitalcommons.unl.edu/cgi/viewcontent.cgi?article=1195& context=cseconfwork
19. Ding, Z., Yang, Z., Fan, P., Poor, H.V.: On the performance of non-orthogonal multiple access in 5G systems with randomly deployed users. IEEE Signal Process. Lett. **21**, 1501–1505 (2014)

Robust Estimation of VANET Performance-Based Robust Neural Networks Learning

Ali R. Abdellah[1,2,3(✉)], Ammar Muthanna[2,3,4],
and Andrey Koucheryavy[2,3]

[1] Electronics and Communications Engineering, Electrical Engineering
Department, Al-Azhar University, Cairo, Egypt
alirefaee@azhar.edu.eg
[2] The Bonch-Bruevich Saint Petersburg State University
of Telecommunications, Saint Petersburg, Russia
ammarexpress@gmail.com, akouch@mail.ru
[3] PJSC Rostelecom, Moscow, Russia
[4] Peoples' Friendship University of Russia, (RUDN University),
6 Miklukho-Maklaya St, Moscow 117198, Russia

Abstract. Vehicular ad hoc network (VANET) can manage live traffic and send emergency messages to the base station in any smart city and is emerging as a connectivity network. In VANET, every vehicle acts as a sensor node, which collects the surrounding information and sends information to the base station. VANET network is created when communication between cars with wireless transceiver is needed. Despite the fact that VANET and mobile ad hoc network (MANET) have some similarities; the dynamic nature of VANET has posed a challenge on routing protocols designing; VANET is composed of models based communication among vehicles and vehicle with a high mobility feature. Presently the artificial neural networks is often used in several fields. Neural networks are usually trained by conventional backpropagation learning algorithm that minimizes the training data mean square error (MSE). The goal of this paper is to investigate VANET performance in terms of packet loss rate and throughput using robust neural networks learning based on the robust M-Estimators performance function instead of the traditional MSE performance function. Robust M-estimators performance functions outperform the traditional MSE performance function in terms of RMSE and training speed as simulation results show.

Keywords: VANET · Robust neural networks · M-estimators ·
Robust statistics

1 Introduction

Vehicular Vehicular ad hoc network (VANET) is a special type of MANET that provides vehicle to vehicle and vehicle to roadside base stations communication with the aim to provide road safety and efficient transportation.

In general, VANET is created when vehicles need to transmit packets to each other through wireless channel. Hence, vehicles need to have wireless transceivers and

© Springer Nature Switzerland AG 2019
O. Galinina et al. (Eds.): NEW2AN 2019/ruSMART 2019, LNCS 11660, pp. 402–414, 2019.
https://doi.org/10.1007/978-3-030-30859-9_34

computerized modules that let the vehicles to act as network nodes. Many characteristics differentiate VANET from other types of ad hoc networks.

Due to the movement and speed of the nodes, VANET's topology is very dynamic compared to traditional MANET, and because of that, VANET's network is permanently partition, especially if vehicle density is low. Unlike the traditional MANET, VANET does not have any restrictions in term of energy and storage, since in VANET nodes are cars, not handheld devices.

Other characteristics that distinguish VANET from MANET are a geographical type of communication, mobility modelling and prediction, various communications environments, hard delay constraints and onboard sensors interaction [1]. Artificial Neural networks (ANNs) composed of large numbers of nonlinear computational elements (neurons) that operate in parallel and inspired by the way biological nervous systems works.

ANNs learn function is determined by the network structure, connections between elements, and the processing performed at computing elements or nodes. Once the network is configured for a specific problem, the network is ready to be trained, by adjusting the values of connections (weights) between elements.

Commonly, they are trained or adjusted so, that a particular input leads to a specific target output. ANNs are trained in order to perform complex functions in different fields, such as pattern recognition, function approximation, classification, and computer vision and communications systems.

Artificial neural networks can also be trained to solve problems that are difficult for conventional computers or human being [2]. Multilayer feed-forward neural networks are commonly trained by backpropagation learning algorithm that was developed for training multilayer neural networks with the objective of minimizing mean square error (MSE) between the actual and desired output.

In data modeling the use of MSE is known commonly as the least mean squares (LMS) method. The basic idea of LMS is to optimize the model fit with respect to the training data by minimizing the square of residuals. MSE is a quality measure of an estimator and it is considered as the preferred measure in many data modeling techniques.

Tradition and ease of computation account for the MSE popularity [3]. M-estimators are a wide class of estimators belonging to robust statistics [4] and generated from maximum likelihood estimators that are designed to be stable under minor noise perturbation and robust against outliers' presence in the data.

Many researchers propose M-estimators as performance function in order to robustify the NN learning process [2, 3, 8–11]. M-estimators attempt to reduce outlier effect by replacing the squared residual by another residual function.

In this paper robust trained neural networks is used; this type of NN will be trained using backpropagation learning algorithm that uses a family of robust statistics estimators called M-estimators as cost functions (performance functions) instead of the most famous traditional MSE – performance function to get the best performance.

Our main aim is to investigate VANET performance in terms of packet loss rate and throughput using robust neural network learning based on robust M-estimators performance functions instead of traditional MSE performance function in order to optimize the neural network learning.

This paper is organized as follows: Sect. 2 discusses VANET background and related works; Sect. 3 deals with VANET simulation using Matlab; Sect. 4 presents M-estimators and shows some common M-estimators; Sect. 5 discusses the M-estimators based back propagation learning algorithm; Sect. 6 shows our experimental result; Sect. 7 conclude the paper.

2 Background and Related Works

VANET provides communication between vehicles and between vehicles and roadside base stations with an aim of providing efficient and safe transportation. In VANET, the vehicle is considered an intelligent mobile node capable of communicating with its neighbors and other vehicles in the network. VANET presents many challenging aspects compared to MANET because of high mobility of nodes and fast topology changes in VANET.

Different routing protocols are developed after considering the major challenges involved in VANETs. This paper provides a survey of routing protocols for VANET. It covers application areas, challenges and security issues prevailing in VANETs [12]. VANET is a special type of MANET and integrates the capabilities of new generation wireless networks for vehicles.

VANET establishes a robust Ad-Hoc network between mobile vehicles and roadside units. It is class of MANET that establishes communication among nearby vehicles and adjacent fixed apparatus, usually described as roadside apparatus. VANET can achieve effective communication between moving node by using different ad-hoc networking tools such as Wi-Fi IEEE 802.11 b/g, WiMAX IEEE 802.10, Bluetooth, IRA, [13]. The main aim of VANET is to provide safety-related information and traffic management.

Safety and traffic management involves real-time information and directly affects people's lives on the road. VANET simplicity and security mechanism ensure greater efficiency. Safety is a prime attribute of Vehicular Ad Hoc Network (VANET) system. Many nodes in VANET are vehicles that are able to form self-organizing networks without prior knowledge of each other.

VANET with low-security level is more vulnerable to frequent attacks. VANETs are deployed in a wide range of application, such commercial establishments, consumers, entertainment, and it is very necessary to add security to these networks to prevent damage to life and property [14].

VANET is a special type of MANET where data is propagated through messages exchange between vehicles. In VANETs, restricted road topology requires a directional nature to the message flow. Also caused by the higher node speeds and unstable connectivity among the nodes, it becomes essential that data can be transmitted in the most efficient ways and with minimal delay. A VANET system architecture composed of different domains and many individual components is depicted in Fig. 1.

Figure 1 shows three distinct domains (in-vehicle, ad hoc and infrastructure), and individual components (application unit, on-board unit, and roadside unit) [15]. Vehicular Ad hoc Networks (VANETs) are group of wireless communication enabled vehicles.

RSU is considered as wireless LAN access point and can provide communications with infrastructure. Also it can have higher range of communication up to 1,000 m. Many researchers have made great efforts for improving of the VANET performance and for establishing a reliable communication. In addition, a lot of efforts was done to obtain the robust leaning algorithms, we summarized few of them and discussed below.

Li et al. in [5] proposed a new protocol (CSR) for VANETs in case of urban scenario. CSR uses information on the distribution of vehicles collected by the intersection of the infrastructure to help vehicles choose a road not only with progress to their destination, but also with better network connectivity.

Moreover, experimental results show that this protocol achieves lower end-to-end delay, higher throughput, and delivery rate, than another routing protocols. Authors in [6] he studied VANETs performance evaluation based on routing protocols with different scenarios. He focused and inspected several routing protocols: AODV, DSR and DSDV and have been done recommendations for all scenarios.

Based on various metric performance criteria, such as throughput, PDR, End delay or latency and network stability, have been done evaluation and were compared various routing protocols.

Authors in [7] addressed the performance evaluation between unicast protocols and multicast protocols for a vehicular environment, based based on Manhattan grid model for transmission between one node and multiple receivers.

Unlike in a multicast transmission in geocast routing, several receivers for the paper scenario are not located in a specific geographic region.

When a multicast routing geocast multiple receivers to the paper cases is not located in a particular geographic region. They studied the performance evaluated in term of packet delivery ratio, average end-to-end delay, throughput, and routing overhead. The results reveal a consistent performance for multicast protocols as the number of receiving nodes increases during the transmission.

Ellah et al. in [8] learned four different artificial neural network (ANN) training algorithms, which are conjugate gradient with Fletcher-Reeves updates, conjugate gradient with Polak - Ribiére updates, resilient backpropagation, and conjugate gradient with Powell - peal restart. He compared their performance based on Root Mean Square Error as a merit function and the training speed in seconds.

The studied neural networks had been trained by aforementioned backpropagation learning algorithms, that used the robust M estimators evaluation functions instead of MSE in order to obtain sustainable training in the presence of outliers. He noticed that Traincgf is the best algorithm in terms of RMSE, while the Traincgp is the best in terms of training speed.

El-Melegy et al. in [9] presented several methods to robustify neural network training algorithms. First, they exploited A family of statistical evaluations, known as M-estimators, to robustify the learning process of the backpropagation learning algorithm.

The trained NN performance using backpropagation learning algorithm, that uses M-estimators, was reviewed and evaluated for the task of function approximation and dynamical model identification. As these M-estimators don't have sufficient insensitivity to outliers, so they used the statistically another robust estimator of the least median of squares and developed a stochastic algorithm to minimize a related cost function.

Essai et al. in [10] addressed the problem of selecting a functional model to the data distorted by emissions using a multilayered feed forward neural network (MFNN). He proposed new functions based on M-estimators to replace the traditional activation functions.

The proposed activation function was evaluated on synthetic data contaminated with varying degrees of emissions, and compared them with existing neural network learning algorithms.

In [11] we studied the robust backpropagation learning algorithm study for feed forward Neural networks. He investigated of the robustness of artificial neural networks learning in presence of data corrupted with outliers. He used many of the methods to improve the robust learning process.

In addition, he implemented a new approach by selecting a new activation functions that are based.

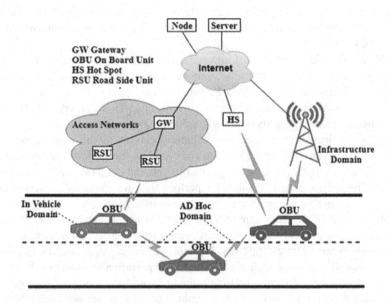

Fig. 1. VANET architecture

3 VANET Simulation

The performance of the VANET can be measured in real road environments, but costs, inaccurate results, and evaluation of the protocol in complex environments are negative factors. Therefore, simulation results are obtained similar to the actual environment [16].

VANET differs from MANET in that nodes must follow traffic laws and the movement patterns of nodes are very complex. In order to obtain good results from VANET simulations, close to real ad-hoc network communication It is most important to create a mobility model. A motion model is many rules that define the motion patterns of nodes in an ad-hoc network.

In VANET, motion models have constraints such as distance, building, vehicle, movement of vehicles, and inter-vehicle behavior. One way to create this real model is to create the actual pattern from the trace. However, since MANET is not yet fully exploitable, it is unrealistic to get an actual trace [17].

In this paper, the VANET in the Urban City simulation implemented as follow: We used Matlab for generating a realistic mobility model for VANETs. The AODV routing protocol have been implemented over the generated realistic mobility model to analyze & evaluate their behavior and performance. In order to generate a mobility map, firstly the road network needs to be created. It has 3 main modules: City Size, Nodes & RSU's.

Here one has to specify the city size, where the nodes will be travelling in random directions for the AODV implementation. Maximum the city size and large number nodes will be required for testing the work of the AODV. Similarly, more number of RSU's need to be installed. If the city size is larger than the simulation time extends automatically. So according to the work, we assumed the city size is 100 metres and both in respect to X-axis and Y-axis. In the next step, the urban city will be designed so the desired nodes could travel random directions on the dedicated paths. Before that, code for nodes number and mobility behaviour has to be coded.

The block size which act as junction and edges, which represent roadways or path through which the nodes will travel in random directions. The code is given below which shows all the properties required to configure urban city. The block size is taken 30 in the simulation. Figure 2 represents the urban city scenario.

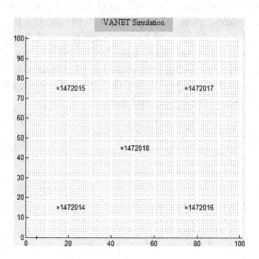

Fig. 2. Urban city scenario

We are creating the positioning the RSU's (Road Side Unit) in the urban city map for assistance to the nodes moving in random direction for communication with other nodes which are far from other nodes range.

RSU we can connect to the Internet using infrastructure network or using multihop for communication with each other directly.

Below given Fig. 3 shows the RSU's locations along with their ID numbers which is assigned to them by the network architecture or topology designer.

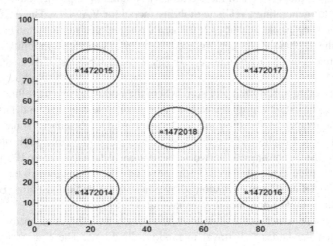

Fig. 3. RSU's in urban city map

For creating Node (vehicles), we built module, this module is responsible for defining number of nodes, flow of nodes that will specify the groups of nodes movements flow on the simulation and turning ratio that will define the probability of directions on each junction. Simulation module is used to visualize the configured topology, also specify the beginning, and end time of simulation. The source node in this model indicated by node number 20 and the destination node number 70. Figure 4 shows the nodes movement on the desired paths.

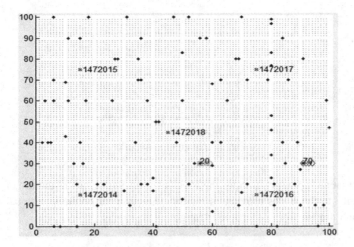

Fig. 4. Nodes movement on desired paths

4 M-Estimators

M-estimators have gained popularity in the neural network's community [2, 3, 8–10], Let determinate r_i as residual of the ith datum, the difference between the ith and its sitting value.

The mean different between standard least squares method and M-evaluators that's, first one optimize the training data by minimizing, second minimize the error by replacing the standard deviation using another function of the residuals, yielding

$$Min \sum\nolimits_i \rho(r_i) \tag{1}$$

Where $\rho(.)$ is a symmetric, positive definite function with a unique minimum at zero, and this was chosen for less increasing than square. Table 1 lists a few used M-estimators and their influence functions. M-estimators' influence functions can be illustrated graphically in Fig. 5a and b [18].

Table 1. Some used M-estimators

Type	$\rho(r)$	$\psi(r)$	$\omega(r)$
L2	$r^2/2$	r	1
L1	$\|r\|$	Sgn(r)	$\frac{1}{\|r\|}$
Fair	$c^2\left[\frac{\|r\|}{c} - \log\left(1 + \frac{\|r\|}{c}\right)\right]$	$\frac{r}{1+\|r\|/c}$	$\frac{1}{1+\|r\|/c}$
Huber $\begin{cases} if\,\|r\| \le k \\ if\,\|r\| \ge k \end{cases}$	$\begin{cases} r^2/2 \\ k(\|r\| - k/2) \end{cases}$	$\begin{cases} r \\ kSgn(r) \end{cases}$	$\begin{cases} 1 \\ k/\|r\| \end{cases}$
Cauchy	$\frac{c^2}{2}\log\left(1 + (r/c)^2\right)$	$\frac{r}{1+(r/c)^2}$	$\frac{1}{1+(r/c)^2}$
GM	$\frac{r^2/2}{1+r^2}$	$\frac{r}{(1+r^2)^2}$	$\frac{1}{(1+r^2)^2}$
LMLS	$\log\left(1 + \frac{1}{2}r^2\right)$	$\frac{r}{1+\frac{1}{2}r^2}$	$\frac{1}{1+\frac{1}{2}r^2}$

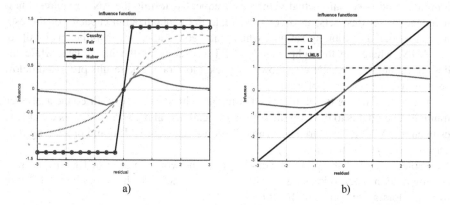

a)   b)

Fig. 5. M-estimators' influence functions for (a) Huber estimators, (b) LMLS estimators.

5 Backpropagation Learning Algorithm Based on M-Estimators

For implementing the traditional learning algorithm based on M-estimators concept, all we need to do is replacing the squared residuals r_i^2 by another function of the residuals, yielding

$$E = \sum_i \rho(r_i) \tag{2}$$

Where ρ is a symmetric positive definite function with a unique minimum at zero, and is selected so that it has less rising than the square [8–11].

6 Simulation Results

Comparison of all above-mentioned performance functions, we choose root mean square error of each model,

$$RMSE = \sqrt{\frac{\sum_{i=1}^{N}(t_i - y_i)^2}{N}} \tag{3}$$

Where the target t_i is the actual value of the function at x_i and y_i is the output of the network given x_i as its input.

In this paper, we used a multilayer feed-forward neural network structure with one hidden layer having 20 hidden neurons.

The network was trained with the backpropagation algorithm based on the M-estimators mentioned above. In our implementation, we use Matlab neural network toolbox with the following settings: network training function is Traincgf (conjugate gradient backpropagation with Fletcher-Reeves updates), the epochs maximum number to train is 5000 epochs, and the goal (Minimum Performance Value) is 0.001.

Input values need to be normalized in the range [−1, 1], which corresponds to the minimum and maximum actual values, Subsequently, testing the ANN requires a new independent set (test sets) in order to validate the generalization capacity of the estimation model. The training set data which had been studied was supplied as input for the ANN, where the throughput of VANET operating as the input data, while the desired output of the ANN was shown by packet loss rate. The results presented below are the average response of training.

The aim of this is to see the weights initial values and first of all connect of each training. The goal is to develop the robust artificial neural network, which is capable to estimate the VANET performance Table 2 shows values of RMSE and processing time for all performance functions.

Table 2 displays the performances of neural networks in term of RMSE and processing time in order to investigate the best performance, which use traditional MSE or M-estimators as performance function.

Table 2. Comparison between robust and traditional training NN performances

VANET Performance-based estimated packet loss		
Performance function	RMSE	Processing time
MSE	0.0620	2.1406
Cauchy	0.0471	2.7813
GM	0.0394	1.9531
Fair	0.0101	2.0625
L1	0.0129	2.0313
LMLS	0.0400	1.9063
Huber	0.0301	1.9531

It is clear from tabulated results that, Fair estimator has the best performance with RMSE value 0.0101, in comparison to other estimators, and the L1 estimator has RMSE value 0.0129, which is semi-equal to the fair estimator. In addition, the Cauchy, GM and Huber estimators have approximately semi-equal RMSE values in comparison to others. On the other hand, the traditional MSE performance function provides so poor performance with RMSE value 0.0620, in comparison with others.

It is clear from tabulated results that, the neural network trained using LMLS estimators has the shortest training time = 1.9063 s, and hence it the fastest training between its peers.

Looking at Figs. 6, 7, 8 and 9, all responses of neural networks trained using the traditional MSE performance function and robust M-estimators performance function in term of VANET throughput and packet loss rate.

As shown in Figs. 6, 7, 8 and 9 all responses are semi similar to actual model except the model predicted by traditional MSE performance function that slightly deviate from the actual model.

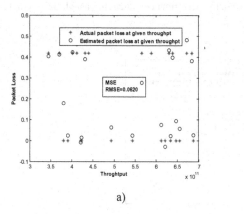

a)

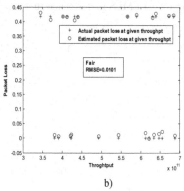

b)

Fig. 6. (a) Model predicted using traditional MSE performance function, (b) Model predicted using robust Fair performance function

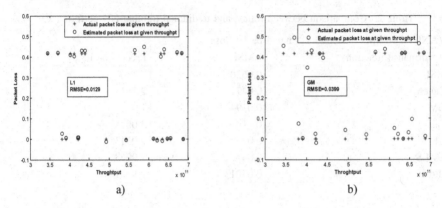

Fig. 7. (a) Model predicted using robust L1 performance function, (b) Model predicted using robust GM performance function.

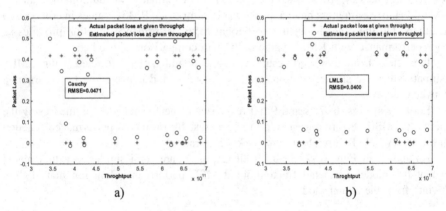

Fig. 8. (a) Model predicted using robust Cauchy performance function, (b) Model predicted using robust LMLS performance function

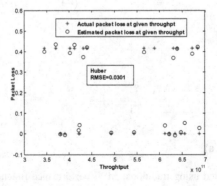

Fig. 9. Model predicted using robust Huber performance function

7 Conclusions

In this paper, ANN was proposed for investigating the VANET performance, in case of the throughput and packet loss rate as input and target for NN respectively.

The neural network performance was evaluated in case of RMSE and processing time. We introduced a group of robust statistics M estimators for training neural networks like robust performance functions. This family provides high reliability for robust NN training in several applications.

Based on the mentioned above result we noticed that the Fair estimator has the best performance in cases RMSE value and the LMLS estimator is the best in the term of speed of training. We recommend this estimators like a performance function for robust training neural network for comparison traditional function performance of MSE.

Acknowledgments. The publication has been prepared with the support of the "RUDN University Program 5-100".

References

1. Chadha, D.: Reena, vehicular ad hoc network (VANETs): a review. Int. J. Innov. Res. Comput. Commun. Eng. **3**(3), 2339–2346 (2015)
2. Zahra, M.M., Essai, M.H., Ellah, A.R.A.: Performance functions alternatives of MSE for neural networks learning. Int. J. Eng. Res. Technol. (IJERT) **3**(1), 967–970 (2014)
3. Zahra, M.M., Essai, M.H., Ellah, A.R.A.: Robust neural network classifier. Int. J. Eng. Dev. Res. (IJEDR) **1**(3), 326–331 (2013). ISSN 2321-9939
4. Huber, P.J.: Robust Statistics. Wiley, New York (1981)
5. Li, C., Wang, M., Zhu, L.: Connectivity-sensed routing protocol for vehicular ad hoc networks: analysis and design. Int. J. Distrib. Sens. Netw. **11**(8), 1–11 (2015). https://doi.org/10.1155/2015/649037
6. Rehman, M.U., Ahmed, S., Khan, S.U., Begum, S., Ahmed, S.H.: Performance and execution evaluation of VANETs routing protocols in different scenarios. EAI Endorsed Trans. Energy Web Inform. Technol. **5**(17), 1–5 (2018)
7. Hassan, A., Ahmed, M.H., Rahman, M.A.: Performance evaluation for multicast transmissions in VANET. In: 2011 24th Canadian Conference on Electrical and Computer Engineering(CCECE), Niagara Falls, pp. 001105–001108 (2011)
8. Ellah, A.R.A., Essai, M.H., Yahya, A.: Comparison of different backpropagation training algorithms using robust M-estimators performance functions. In: the IEEE 2015 Tenth International Conference on Computer Engineering and Systems (ICCES), 23–24 December, Cairo, Egypt, pp. 384–388 (2015)
9. El-Melegy, M.T., Essai, M.H., Ali, A.A.: Robust training of artificial feedforward neural networks. In: Hassanien, A.E., Abraham, A., Vasilakos, A.V., Pedrycz, W. (eds.) Foundations of Computational, Intelligence, Volume 1, vol. 201, pp. 217–242. Heidelberg, Springer (2009). https://doi.org/10.1007/978-3-642-01082-8_9
10. Essai, M.H., Ellah, A.R.A.: M-estimators based activation functions for robust neural network learning. In: the IEEE 10th International Computer Engineering Conference (ICENCO 2014), 29–30 December, Cairo, Egypt, pp. 76–81 (2014)
11. Ellah, A.R.A., Essai, M.H., Yahya, A.: Robust backpropagation learning algorithm study for feed forward neural networks. Thesis, Al-Azhar University, Faculty of Engineering (2016)

12. Vegni, A.M., Biagi, M., Cusani, R.: Smart vehicles, technologies and main applications in vehicular ad hoc networks. In: Vehicular Technologies - Deployment and Applications. INTECH Open Access Publisher (2013). https://doi.org/10.5772/55492

13. Bodhy Krishna, S.: Study of ad-hoc networks with reference to MANET, VANET. FANET. Int. J. Adv. Res. Comput. Sci. Softw. Eng. 7(7), 390–394 (2017). ISSN 2277-128X

14. Saggi, M.K., Sandhu, R.K.: A survey of vehicular ad hoc network on attacks and Security Threats in VANETs. In: International Conference on Research and Innovations in Engineering and Technology (ICRIET 2014), 19–20 December 2014

15. Bronsted, J., Kristensen, L.: Specification and performance evaluation of two zone dissemination protocols for vehicular ad-hoc networks. In: Proceedings of the 39th Annual Simulation Symposium (ANSS 2006). IEEE (2006)

16. Boban, M., Vinhoza, T.T.V.: Modeling and simulation of vehicular networks: towards realistic and efficient models. In: Xin W. (eds.) Mobile Ad-Hoc Networks: Applications, pp. 41–66. INTECH Open Access Publisher (2011) https://doi.org/10.5772/12846

17. Nam, J.: Implementation of VANET simulator using Matlab. J. Korea Inst. Inform. Commun. Eng. 20(6), 1171–1176 (2016)

18. Zhang, Z.: Parameter estimation techniques: a tutorial with application to conic fitting, October 1995

Multi-level Architecture for P2P Services in Mobile Networks

Rustam Pirmagomedov[1,2(✉)], Aram A. Ahmed[3], and Ruslan Glushakov[4]

[1] Peoples' Friendship University of Russia (RUDN University),
6 Miklukho-Maklaya St, Moscow 117198, Russian Federation
prya.spb@gmail.com
[2] Tampere University, Korkeakoulunkatu 10, 33720 Tampere, Finland
[3] The Bonch-Bruevich Saint-Petersburg State University of Telecommunications,
22 Prospekt Bolshevikov, St. Petersburg 193232, Russian Federation
[4] Saint-Petersburg State Pediatric Medical University,
Litovskaya St. 2, St. Petersburg 194100, Russian Federation

Abstract. Latency is an important metric of mobile applications performance. To reduce the latency, recently it was proposed to replace the standard centralized architecture of mobile applications by the mobile edge computing (MEC). Such an approach allows processing of users data closer to their location. Motivated by disaster response scenarios, in this paper we investigated the capabilities of MEC for the forwarding of first aid request as an illustrative example of P2P service discovery in an emergency situation. We proposed an analytical model of the system and executed performance evaluation using system level simulator. Our results show that the developed solution considerably reduces the request processing time. The proposed solution can be used not only for first aid but also for general purposes, e.g., searching various service providers in a certain location.

Keywords: Mobile networks · Edge computing · P2P · eHealth

1 Introduction

Recently, there has been broad interest in peer-to-peer (P2P) services on mobile networks. Such services consider users interaction in relative proximity, where one user is searching for services and another is ready to provide it (e.g., taxi applications). Currently, such services are enabled by centralized architecture, which completes all the requests on the central server (finding the service provider).

Such a centralized approach is often criticized for overheads since the required service provider can commonly be found locally while sending a request to the central server introduces additional overheads to request processing time and

The publication has been prepared with the support of the "RUDN University Program 5-100".

O. Galinina et al. (Eds.): NEW2AN 2019/ruSMART 2019, LNCS 11660, pp. 415–423, 2019.
https://doi.org/10.1007/978-3-030-30859-9_35

network load. These overheads considerably reduce the implementation of the centralized architecture in the case of delay-sensitive applications.

A considerable amount of recent research has been focused on enabling low-latency applications using mobile edge computing [1]. The improvements in processing time there were mostly achieved due to (i) offloading of computation hungry tasks from constrained user devices to the more powerful entities co-located with the base stations of cellular networks, and (ii) delegating a share of the server functions to these entities [2].

Less attention has been paid to the implementation of MEC for distributed P2P services in mobile networks. Therefore, in this paper, we attempted to implement MEC technology for reducing the request processing time in P2P service request processing. Particularly, we considered the scenario of Public Protection and Disaster Relief, where victims are distributed in a particular location and requesting first aid. Our metric of interest for this scenario is the request processing time.

For the considered scenario, we introduced a multi-level computing architecture for processing of P2P service requests and evaluated its performance in terms of delay. The proposed architecture is supported by an analytical model based on Queueing Theory, and a system level simulation. Our results demonstrate that multi-level architecture may considerably improve the request processing time.

2   Illustrative Use Case

Modern society is shaped by increasing urban population densities, as well as the threat of large-scale biological or ecological conditions. These factors can result in a higher number of emergency situations, often when there is widescale lack information or logistical resources. Medical response times in such circumstances can vary significantly due to vehicle traffic and physical obstacles. However, medical concepts as the "golden hour", which characterize the need for timely first aid remain crucial for many conditions (e.g., heart failure, drowning, the presence of foreign bodies in the airways, serious injury, severe bleeding). Understanding this need to response, an incident in a crowded location (e.g., concert, public transportation, shopping center) typically causes some level of hesitation or panic among bystanders due to lack of necessary skills. This delayed response can ultimately lead to the patient's death.

The situation develops somewhat better in the case of other catastrophic events (e.g., fires, traffic accidents, flooding, and terrorism). With effective mutual assistance, the adverse effects of these events can be reversed. These circumstances suggest that in order to achieve a certain level of public security, it is necessary to use resources focused on self-help and mutual assistance. Thus, as an illustrative scenario, we consider a mobile application which facilitates health-related assistance in the case of a disaster (e.g., flood, earthquake, storm). In this scenario, possible victims (patients) use the mobile application which gathers their geolocation. If a patient requests medical aid, the application notifies a qualified person in the same location (referred as a doctor in this paper), who can provide such help and also uses the application.

To illustrate the computational functions required by the application, let us assume that there are m doctors and n patients on the area of interests. The application processes every received request by (i) defining the distance between doctors and patients (matrix D), (ii) evaluating whether the qualification of each doctor is relevant to each patient request (matrix C), and notify the most relevant doctors about the requests.

$$
D = \begin{pmatrix} d_{11} & d_{12} & \cdots & d_{1n} \\ d_{21} & d_{22} & \cdots & d_{2n} \\ \vdots & \vdots & \ddots & \vdots \\ d_{m1} & d_{m2} & \cdots & d_{mn} \end{pmatrix}, C = \begin{pmatrix} c_{11} & c_{12} & \cdots & c_{1n} \\ c_{21} & c_{22} & \cdots & c_{2n} \\ \vdots & \vdots & \ddots & \vdots \\ c_{m1} & c_{m2} & \cdots & c_{mn} \end{pmatrix}
$$

The division of matrixes D and C results in a matrix of ratings $(\frac{D}{C})$, where the most suitable doctor is selected by the minimum value. After the doctor is selected, the application notifies him or her, indicating the location of the patient and a description of the problem that has arisen. If the request is accepted, doctor starting online consultation about required actions and moving towards the patient location.

$$
\frac{D}{C} = \begin{pmatrix} \frac{d_{11}}{c_{11}} & \frac{d_{12}}{c_{12}} & \cdots & \frac{d_{1n}}{c_{1n}} \\ \frac{d_{21}}{c_{21}} & \frac{d_{22}}{c_{22}} & \cdots & \frac{d_{2n}}{c_{2n}} \\ \vdots & \vdots & \ddots & \vdots \\ \frac{d_{m1}}{c_{m1}} & \frac{d_{m2}}{c_{m2}} & \cdots & \frac{d_{mn}}{c_{mn}} \end{pmatrix}
$$

It should be noted that we considered a simplified scenario which does not take into account the congestion of roads, which may be obtained from public services (e.g., using Google Maps API). The distance between the patient and the medical officer is taken as the shortest path between two points.

In further sections, we consider how the proposed system can be enabled in mobile networks.

3   Utilization of Multi-level Computing Systems for the Requests Processing

When implementing this medical application, it should be taken into account that an excessive number of patients, who simultaneously send signals for help, may overload the remote server with a vast number of requests and subsequent calculations [3]. To reduce the load on the server, Mobile Edge Computing (MEC) technology can be utilized [1,4]. In the considered scenario, MEC will include three levels of computing devices - Micro-cloud, Mini-cloud and Server (Cloud) [5]. The whole geographic area, which is under Server responsibility is

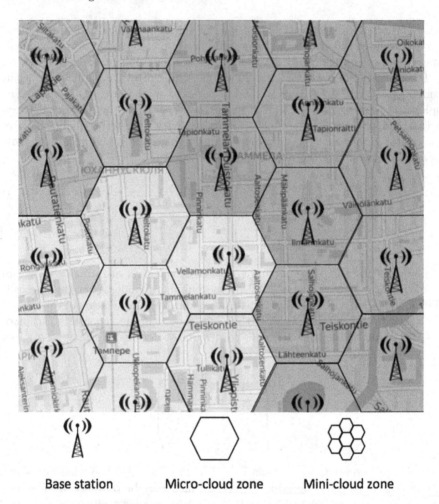

Base station   Micro-cloud zone   Mini-cloud zone

Fig. 1. The multi-level computing architecture deployment.

divided into zones of Mini-clouds, and the zone of each Mini-cloud is, in turn, divided into several Micro-cloud zones (Fig. 1).

Micro-cloud has relatively low computational capabilities and processes requests within the range of one base station. Mini-cloud is deployed at a cell's aggregation point and processes requests from the territory of several base stations, in case if the request cannot be satisfied at the Micro-cloud level (no doctor in the area). It is supposed that Micro-cloud, Mini-cloud, and Server contain tables with data on doctors that are located in their service range, while the Mini-cloud table contains the tables of the Micro-cloud included in it, and the Server contains information about all medical staff.

First, the request from the devices of patients (victims) comes to the Micro-cloud connected to the base station. If a medical officer is not found (available) in the service area of this Micro-cloud, then the request will be passed further along the hierarchy - for processing in the Mini-cloud [6]. Only if the request was not satisfied in the Mini-cloud area will the search will be performed on the Server (Cloud) (Fig. 1).

When a request is received by the node of a higher level, a segment of data already processed at the underlying Micro-cloud or Mini-cloud will be not processed again. Such an approach allows for more efficient use of computing resources. If the processing time in the nodes of the same level (Micro-cloud and Mini-clouds) will take the same amount of time, and the computational resources available on the nodes are proportional to the computational complexity of the task performed on each level (size of the database), the request processing time t will be the same for each computational entity.

Since the system allows more than one user's request simultaneously, a queue may occur. The mean delay introduced by waiting in a queue at Micro-cloud tq_{micro} can be estimated using M/D/1 queue model [7].

$$tq_{micro} = \frac{\lambda_{micro}}{2\mu_{micro}(\mu_{micro} - \lambda_{micro})} \tag{1}$$

Where: μ is a serving rate, for $\mu = 1/t$ (for all nodes of the same level μ is equal); λ_{micro} is a mean arrival rate of requests from users to the Micro-cloud (arrivals on Micro-cloud occur according to a Poisson process). If a request can be served at the Micro-cloud with the probability P_{micro}, then an arrival rate of requests to the Mini-cloud λ_{mini} is defined as:

$$\lambda_{mini} = (1 - P_{micro})\lambda_{micro}N_{micro} \tag{2}$$

where N_{micro} is a number of Micro-clouds in the zone of one Mini-cloud. Consequently, mean delay introduced by waiting in a queue at the Mini-cloud tq_{mini} can be expressed as:

$$tq_{mini} = \frac{(1 - P_{micro})\lambda_{micro}N_{micro}}{2\mu_{mini}(\mu_{mini} - (1 - P_{micro})\lambda_{micro}N_{micro})} \tag{3}$$

Finally, if a request can be served at the Mini-cloud with the probability P_{mini}, an arrival rate of requests to the Server (λ_{server}) is defined as:

$$\lambda_{server} = (1 - P_{mini})(1 - P_{micro})\lambda_{micro}N_{micro}N_{mini} \tag{4}$$

where N_{mini} is a number of Mini-cloud within the Server's (Cloud) service area. The mean delay introduced by waiting in a queue at the Server tq_s can be expressed as

$$tq_s = \frac{(1 - P_{mini})(1 - P_{micro})\lambda_{micro}N_{micro}N_{mini}}{2\mu_{server}(\mu_{server} - (1 - P_{mini})(1 - P_{micro})\lambda_{micro}N_{micro}N_{mini})} \tag{5}$$

Using the proposed model, the mean request serving time will be distributed as follows:

$$\begin{cases} tq_{micro} + tn_{micro} + t, & \text{with probability } P_{micro} \\ tq_{micro} + tq_{mini} + tn_{mini} + 2t, & \text{with probability } (1 - P_{micro})P_{mini} \\ tq_{micro} + tq_{mini} + tn_{server} + 3t, & \text{with probability } (1 - P_{micro})(1 - P_{mini}) \end{cases}$$

Where $tn_{micro}, tn_{mini}, tn_{server}$ are network delays for the Micro-cloud, Mini-cloud and the Server respectively.

4   Numerical Results

To evaluate the gain from using the multi-level architecture, we performed a simulation campaign in a WinterSIM simulation framework. During the experiment, we considered an area of interest divided into the zones of Micro-clouds. All Micro-clouds were associated with a certain Mini-Cloud (seven Micro-clouds per one Mini-Cloud). The mobility of doctors was defined by Random Direction Movement model (RDM). Users' requests generated in each Micro-cloud zone with a constant rate. Network delay was determined using Normal distribution. The set of simulation parameters is presented in Table 1.

Table 1. Simulation parameters

Parameter	Value
Mobility model	RDM
Duration of experiment (system time)	100 s
Size of the buffer at Micro-cloud	30 requests
Size of the buffer at Mini-cloud	120 requests
Size of the buffer at Server	960 requests
Mean network delay to Micro-cloud, tn_{micro}	3 ms
Mean network delay to Mini-cloud, tn_{mini}	12 ms
Mean network delay to Server, tn_s	90 ms
Request processing time, t	50 ms
Number of Mini-Clouds	8
Number of Micro-Clouds	56

Our metric of interests is the request processing time. During the experiment, we considered centralized architecture (all the requests processed on a single server) as a baseline scenario, and multi-level architecture as an alternative. The histogram of requests processing time obtained during the experiment is shown in Fig. 2.

The results indicate that multi-level architecture allows reducing the mean processing time if the free doctor is associated with the same Micro- or Mini-cloud with the user. However, if there is no doctor in the same Mini-cloud zone, then multi-level processing requires almost double the time when compared with centralized architecture. The mean processing time in multi-level architecture is slightly better than in the baseline scenario.

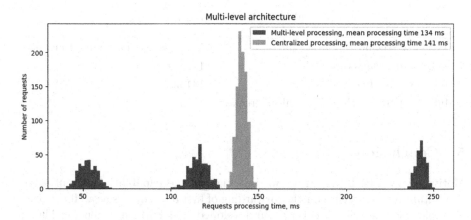

Fig. 2. Simulation results for centralized and multi-level architecture.

To improve the performance of the multi-level processing architecture, we considered an option with multiple requests. In this case, the request was simultaneously sent to the Micro-cloud, Mini-cloud, and Server (multicast send). The histogram of request processing time for this case is presented in Fig. 3.

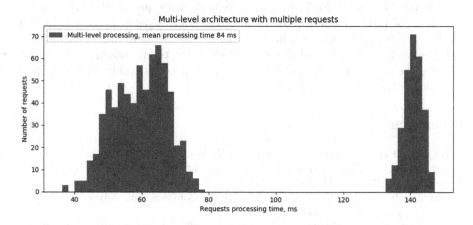

Fig. 3. Simulation results for multi-level architecture with multiple requests.

In the case of multiple requests, the experimental results demonstrate the obvious advantages over the classical cloud architecture, as well as multi-level architecture without multiple request option.

The mean request processing time for all processing schemes considered during the experiment are summarized in Table 2.

Table 2. Results summary

Processing scheme	Mean request processing time, ms
Centralized architecture	141 ms
Multi-level architecture	134 ms
Multi-level architecture with multiple requests	84 ms

5   Conclusion

Motivated by PPDR scenarios, we investigated the capabilities of multi-level computation architecture for the forwarding of first aid requests as an illustrative example of P2P service discovery. Our developed model utilizes Queueing Theory and employs system-level simulations.

Overall, our results demonstrate that standard multi-level architecture does not provide a significant improvement in request processing time. However, the performance of the standard multi-level architecture can be increased drastically if the requests sent simultaneously to all levels of the system. The proposed multi-level architecture with multiple requests enhances the performance of applications relying on P2P services, especially if these applications require low latency communication (e.g., real-time applications).

A natural extension for further improvement of the proposed solution is the utilization of D2D communication. The D2D option is particularly relevant for PPDR use case, where network infrastructure could be unavailable due to physical damage (e.g., in a case of disaster). In this case, the proposed model should consider the coexistence various data streams, as it considerably affects the performance of the communication [8].

References

1. Hu, Y.C., Patel, M., Sabella, D., Sprecher, N., Young, V.: Mobile edge computing-a key technology towards 5G. ETSI White Paper **11**(11), 1–16 (2015)
2. Beck, M. T., Werner, M., Feld, S., Schimper, S.: Mobile edge computing: a taxonomy. In: Proceedings of the Sixth International Conference on Advances in Future Internet, pp. 48–55, Citeseer (2014)
3. Pirmagomedov, R., Blinnikov, M., Glushakov, R., Muthanna, A., Kirichek, R., Koucheryavy, A.: Dynamic data packaging protocol for real-time medical applications of nanonetworks. In: Galinina, O., Andreev, S., Balandin, S., Koucheryavy, Y. (eds.) NEW2AN/ruSMART/NsCC -2017. LNCS, vol. 10531, pp. 196–205. Springer, Cham (2017). https://doi.org/10.1007/978-3-319-67380-6_18
4. Mach, P., Becvar, Z.: Mobile edge computing: a survey on architecture and computation offloading. IEEE Communications Surveys & Tutorials **19**(3), 1628–1656 (2017)
5. Pirmagomedov, R., Hudoev, I., Shangina, D.: Simulation of medical sensor nanonetwork applications traffic. In: Vishnevskiy, V.M., Samouylov, K.E., Kozyrev, D.V. (eds.) DCCN 2016. CCIS, vol. 678, pp. 430–441. Springer, Cham (2016). https://doi.org/10.1007/978-3-319-51917-3_38

6. Sabella, D., Vaillant, A., Kuure, P., Rauschenbach, U., Giust, F.: Mobile-edge computing architecture: the role of mec in the internet of things. IEEE Consum. Electron. Mag. **5**(4), 84–91 (2016)
7. Cooper, R.B. (ed.): Introduction to Queueing Theory, 2nd edn, p. 347. Elsevier North Holland, New York (1981)
8. Dunaytsev, R., Moltchanov, D., Koucheryavy, Y., Strandberg, O., Flinck, H.: A survey of P2P traffic management approaches: best practices and future directions. Internet Eng. **5**(1), 318–330 (2012)

Network Anomaly Detection Based on WaveNet

Tero Kokkonen$^{(\boxtimes)}$, Samir Puuska, Janne Alatalo, Eppu Heilimo, and Antti Mäkelä

Institute of Information Technology, JAMK University of Applied Sciences, Jyväskylä, Finland
{tero.kokkonen,samir.puuska,janne.alatalo,eppu.heilimo, antti.makela}@jamk.fi

Abstract. Increasing amount of attacks and intrusions against networked systems and data networks requires sensor capability. Data in modern networks, including the Internet, is often encrypted, making classical traffic analysis complicated. In this study, we detect anomalies from encrypted network traffic by developing an anomaly based network intrusion detection system applying neural networks based on the WaveNet architecture. Implementation was tested using dataset collected from a large annual national cyber security exercise. Dataset included both legitimate and malicious traffic containing modern, complex attacks and intrusions. The performance results indicated that our model is suitable for detecting encrypted malicious traffic from the datasets.

Keywords: Intrusion detection · Anomaly detection · WaveNet · Convolutional neural networks

1 Introduction

Intrusion detection systems (IDS) are divided into two categories: anomaly-based detection (anomaly detection) and signature-based detection (misuse detection). Anomaly-based-detection can be applied without pre-recorded signatures for unknown attack patterns and even for encrypted network traffic, however the weakness for anomaly detection is the high amount of false positive detections [3,13].

Machine learning techniques have recently been applied successfully to network anomaly detection and classification [6]. Bitton and Shabtai in [1] have studied machine learning based IDS for Remote Desktop Protocols (RPD). Different machine learning techniques have been applied, e.g. Wiewel and Yang used Variational Autoencoder in their study [28], Chen et al. used Convolutional Autoencoder [2] while Long Short-Term Memory (LSTM) and Gated Recurrent Unit methods are used in the paper [6]. Paper [23] presents technique for increasing detection accuracy with feedback.

O. Galinina et al. (Eds.): NEW2AN 2019/ruSMART 2019, LNCS 11660, pp. 424–433, 2019.
https://doi.org/10.1007/978-3-030-30859-9_36

In our earlier study [19], we used Haar wavelet transforms and Adversarial Autoencoders (AA) [10] for implementing unsupervised network anomaly detection based IDS. Our earlier model, described in [19], had reasonable good operational characteristics; in this study we strived to improve it using alternative modeling approach. As argument of efficiency, numerical results are compared with the earlier results using the same dataset from Finland's National Cyber Security Exercise [12]. Performance characteristics are also accomplished using publicly available reference intrusion detection evaluation dataset (CICIDS2017) [27].

Our study presents state-of-the-art network anomaly detection based intrusion detection system that exploits deep learning method WaveNet [15]. First, in Sect. 2, this paper describes implemented anomaly detection method including feature extraction and analysis method. Then, in Sect. 3, we introduce experimental results for the performance characteristics of our model and finally there are conclusions with found future research topics.

2   Anomaly Detection Method

2.1   Dataset

According to Nevavuori and Kokkonen [14], a network anomaly detection data set must (i) include network traffic data and (ii) host activity data, (iii) multiple scenarios, (iv) be representative of real-world circumstances, and (v) the format of the data must be usable.

Since many publicly available datasets already exist [20], we decided to utilize them in this research. Although notable public datasets, such as the KDD99 [25] and DARPA datasets [7–9] exist and are used in many existing network intrusion detection research, they are very old, and many researches have directed a lot of criticism against them [11,24]. The main problem is that datasets do not include modern threat and attack patterns with required statistical characteristics nor sophisticated and modern architectures [4,14,22,26]. In many datasets the raw data is already processed into network flows losing the information of individual packet timings. Fortunately, in addition to the processed flow data, some datasets include the raw packet captures.

The Intrusion Detection Evaluation Dataset (CICIDS2017) by the Canadian Institute for Cybersecurity [27] is one of the more modern publicly available datasets. Although the dataset was created with a traffic generator, it was modeled after modern real-world network traffic. It includes benign HTTPS network traffic and therefore is suitable for research concerning encrypted communication. Unfortunately, the dataset does not include many TLS based attacks, which form a sizable amount of modern malware control channels.

We decided to use the benign traffic from the CICIDS2017 dataset as clean traffic during the model development and testing, but because the anomalous traffic in the dataset was not large enough, more anomalous traffic was required. We generated additional anomalous traffic in our own environment using Empire

PowerShell post-exploitation agent[1] and Cobalt Strike[2]; both are adversary simulation frameworks that use real-world malware characteristics. A small amount of benign traffic was also generated in the environment. The benign traffic was generated by controlling Windows virtual machine using a scripted bot that operated normal GUI software with virtual mouse and keyboard aided by computer vision. This data was used in the evaluation to make sure that the environments are compatible enough so that our generated benign traffic is not classified as anomalous with the model that is trained with the CICIDS2017 benign data.

In addition to the CICIDS2017 dataset and the self-generated dataset, the final model was also tested with the Finland's National Cyber Security Exercise dataset (FNCSE2018), also used in our previous publication [19]. This dataset was used to get comparable results to our previous research. RGCE Cyber Range (Realistic Global Cyber Environment) is used for research and development or training and exercises. In the RGCE Cyber Range main structures and services of the real Internet are modeled with the realistic user traffic patterns of users. RGCE offers tailored organization environments with real assets [5]. Finland's National Cyber Security Exercise is conducted annually in the RGCE Cyber Range. Network data from the real Cyber Security Exercise conducted in the RGCE Cyber Range includes realistic complex environment and legitimate network traffic mixed with modern attack patterns for testing the capabilities of Intrusion Detection System capability [12]. In this study we were authorized to use the traffic captures from Finland's National Cyber Security Exercise of 2018.

2.2 Feature Extraction

Our research focused on finding the anomalies based on packet timing patterns. This choice was made to accommodate encrypted command and control channels modern malware use. Traditional deep inspection techniques and statistical analyses that utilize payloads are incompatible with modern security landscape, made e.g. decrypting proxies obsolete due to various certificate pinning features. In this project we used a modified version of Suricata IDS software [18] to process the raw packet capture files into parsed network data. The modification in the software allowed the packet timings information to be extracted from packet capture files along with the parsed data.

The CICIDS2017 dataset includes the raw packet captures in addition to labeled processed flow data. Since the processed flow data does not include packet timings, the raw data had to be reprocessed to flow data with the modified Suricata software. The processed flows were then labeled by joining the flows to the CICIDS2017 flow labels by matching flow timestamps, IP addresses and network ports. The result was labeled flows from the CICIDS2017 dataset including packet timings. Because our system used different software for packet capture to network flow conversion from the one used in CICDS2017, the resulting flows did not match exactly, resulting in lost flows. Only the flows that matched correctly

---

[1] https://www.powershellempire.com/.
[2] https://www.cobaltstrike.com/.

between Suricata processed flows and CICDS2017 labeled flows were retained in the dataset. Based on the flow label, the dataset was then split to anomaly and benign flows. All the flows that did not have *benign* label were treated as anomalies. The final processed CICDS2017 dataset included 1,425,742 flows, of which 1,107,695 were labeled as benign flows, and 318,047 flows were labeled as non-benign flows. From the 1,107,695 benign flows, 307,771 were TLS flows. Originally the Suricata processed CICIDS2017 packet capture files included 1,956,363 flows, so 530,621 flows did not find matching flow in the CICIDS2017 flow label files. This can be almost certainly accounted on the poor quality of the flow label files in CICIDS2017 dataset. The files include a duplicate entry for most of the flows and the flow timestamps are recorded in a minute accuracy with an ambiguous 12-h clock format.

The FNCSE2018 dataset and our self generated datasets were processed in the same way. The labels were assigned by hand based on known origin and destination addresses of the attacks. The FNCSE2018 dataset included 715,158 benign TLS flows, and 653 non-benign TLS flows. The self generated dataset included 15,124 benign flows and 7,991 non benign flows.

The resulting flows were then further processed by calculating timing differences between packets. The final features for one packet in a flow were: *packet direction, time difference to next received packet, time difference to next transmitted packet* and *packet size*. The timing differences varied from microseconds to minutes with most of the differences being very small. Because our model required quantization of the input data, the timing differences were scaled with the common logarithm to better utilize the reduced quantization precision. The packet sizes were scaled in similar way for the same reason. This choice is warranted, because in network traffic large delays are often the result of an unrelated problem, and not an inherent feature of the protocol in question. Although many protocols, including malware command channels, may use delays and timers, there usually is no reason to keep using the same flow. Packet sizes follow the same scaling principle, the maximum size being the MTU of the path. Small packet sizes and the variation therein are likely to be indicative of the intrinsic properties of the protocol, unlike the variation near the MTU. This is especially apparent in many malware communication protocols, which often use fixed size binary messages. The aforementioned adversary simulation frameworks also exhibit this phenomenon.

2.3 Multi-feature WaveNet

The network traffic was analyzed with a deep neural network model based on the WaveNet [15] architecture, illustrated in the Fig. 2. WaveNet was chosen as a basis for our model for its capability to directly interface with variable length sequential data. This enables us to feed complete and unreduced sequences to the model. We utilized this trait to predict network traffic connections of varying length packet by packet.

The primary task of the model is to predict the next sample by using prior samples. The core network structure consists of a variation of the WaveNet

architecture configured for multiple features. The modified WaveNet is extended
to utilize two-dimensional dilated causal convolutions; input data is arranged
into a two-dimensional lattice, discrete time steps forming the first dimension
and individual sample features along the other dimension. Dilated convolutions
expand the receptive field of the network exponentially [29], giving the model a
potential to observe long term temporal dependencies. Dilation of convolutions is
only performed along the time axis of the data, as the receptive fields are exceed-
ingly large and thus not optimal for the relatively small fixed length feature
axis. The causality aspect of the convolutions is used to assert an ordered time-
dependency on the input data: predicted samples may only depend on preceding
input samples. We implemented the causality by padding the beginning of the
sequence by the filter size in the first layer and by (filter size−1)×dilation rate in
the subsequent layers, effectively shifting the convolution operations. The causal
layer stack is visualized in Fig. 1.

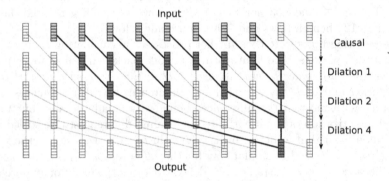

Fig. 1. Visualization of the models two-dimensional dilated causal layers and the first
causal layer.

The input variables are quantized to n bins, continuous and discrete variables
alike, matching the practice used in WaveNet [15] as well as PixelRNN [16]. As
the length of the input data varies with each example, a special end of sequence
value is used to represent sequence termination. The network utilizes a discretized mixture of logistic distributions, as described in PixelCNN++ [21] and
Parallel WaveNet [17]. We found this to perform slightly better when compared
to a more classical soft-max layer.

The individual residual layers follow closely the structure present in WaveNet.
Unlike the WaveNet architecture, we included a dropout layer before each dilated
convolution layer as shown in Fig. 2. Applying dropout inside each residual
layer has been previously explored in PixelCNN++ [21] and Wide Residual
Networks [30].

To distinguish anomalous data from benign data, an anomaly score is quan-
tified from the network outputs with a single forward pass, effectively avoiding
the downside of slow sampling of the WaveNet model. In our approach, we computed the training loss contributions for each sample in the input sequence.

The overall anomaly score of the whole sequence was the mean of these loss values, with samples past the end of sequence marker masked out to account for different length of sequences.

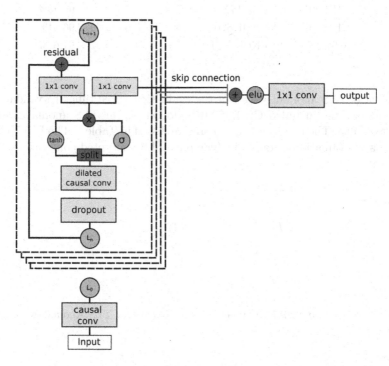

Fig. 2. The architecture is similar to the original WaveNet [15], with the exception of a dropout layer between all dilation layers and exclusive weights between' residual and skip connections.

3   Experimental Results

For the numerical results, we created receiver operating characteristic (ROC) curves by plotting the true positive rate (TPR) to y-axis and false positive rate (FPR) to x-axis. As a comparable score we also calculated the area under curve (AUC) from the ROC.

In order to model an anomaly detector we split the clean data from CICIDS2017 and FNCSE2018 datasets into training and evaluation parts using 80/20 ratio. We took 256 first packets from each flow and trained a model with 9 dilation layers (receptive field of 256), vertical filter size of 3 and horizontal 2, 128 filters each layer for ~15 epochs while evaluating the model using the evaluation part of the dataset to keep the model from over-fitting. During and after the training we ran an evaluation where we included the anomaly data to validate the anomaly detection capability of the model. Since the CICIDS2017

Table 1. Area under curve scores for four different evaluation dataset combinations.

Training dataset	Evaluation dataset	AUC
CICIDS2017	CICIDS2017	97.11%
CICIDS2017	Our TLS anomalies	99.48%
CICIDS2017	CICIDS2017 + Our TLS anomalies	96.81%
FNCSE2018	FNCSE2018	91.61%

dataset lacks TLS anomalies we ran the evaluation three times to validate the model against the included CICIDS2017 anomalies, our TLS anomalies and a mixture of both. The resulting AUC scores are listed in Table 1. The FNCSE2018 training and evaluation datasets include only TLS encrypted connections.

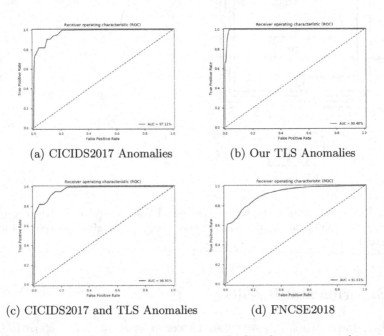

(a) CICIDS2017 Anomalies   (b) Our TLS Anomalies

(c) CICIDS2017 and TLS Anomalies   (d) FNCSE2018

Fig. 3. Receiver operating characteristic curves on the four datasets we used to evaluate the model.

From the results in Fig. 3 we concluded that the model is capable of detecting anomalies in both datasets, while also retaining the capability of detecting anomalous connection with TLS encryption. The model also performs significantly better than our earlier model [19], which had 80% AUC whereas the new model got 91.61% AUC on the same dataset.

4   Conclusion

In this study we applied the WaveNet and PixelCNN models for constructing an IDS based on anomaly detection. For the feature extraction and data processing, an open source software -based data pipeline was constructed. We utilized network data from Finland's National Cyber Security Exercise as well as public reference dataset CICIDS2017. The combined dataset was relatively extensive, although further efforts should be made to include a more diverse selection of applications and web browsing activities.

Results suggest that the machine learning model is suitable for detecting malicious command and control channels from TLS encrypted connections. The model is able to circumvent issues arising from samples of various lengths, and quantize timing and packet size differences into ranges suitable for neural networks.

Future work includes a conditioned WaveNet, variational or adversarial encoder to self-condition the WaveNet, and further testing on possible anomaly scores. Furthermore, visualization methods of found network anomalies should be studied for achieving better situational awareness in operative environments.

Acknowledgment. This research project is funded by MATINE - The Scientific Advisory Board for Defence.

References

1. Bitton, R., Shabtai, A.: A machine learning-based intrusion detection system for securing remote desktop connections to electronic flight bag servers. IEEE Trans. Dependable Secure Comput. 1 (2019). https://doi.org/10.1109/TDSC.2019.2914035

2. Chen, Z., Yeo, C.K., Lee, B.S., Lau, C.T.: Autoencoder-based network anomaly detection. In: 2018 Wireless Telecommunications Symposium (WTS), pp. 1–5, April 2018. https://doi.org/10.1109/WTS.2018.8363930

3. Chiba, Z., Abghour, N., Moussaid, K., Omri, A.E., Rida, M.: A clever approach to develop an efficient deep neural network based IDS for cloud environments using a self-adaptive genetic algorithm. In: 2019 International Conference on Advanced Communication Technologies and Networking (CommNet), pp. 1–9, April 2019. https://doi.org/10.1109/COMMNET.2019.8742390

4. Creech, G., Hu, J.: Generation of a new IDS test dataset: time to retire the KDD collection. In: IEEE Wireless Communications and Networking Conference, WCNC, pp. 4487–4492. IEEE, April 2013. https://doi.org/10.1109/WCNC.2013.6555301

5. JAMK University of Applied Sciences, Institute of Information Technology, JYV-SECTEC: Rgce cyber range. http://www.jyvsectec.fi/en/rgce/. Accessed 26 Apr 2019

6. Li, Z., Rios, A.L.G., Xu, G., Trajković, L.: Machine learning techniques for classifying network anomalies and intrusions. In: 2019 IEEE International Symposium on Circuits and Systems (ISCAS), pp. 1–5, May 2019. https://doi.org/10.1109/ISCAS.2019.8702583

7. Lincoln Laboratory, Massachusetts Institute of Technology: 1998 DARPA Intrusion Detection Evaluation Dataset. https://www.ll.mit.edu/r-d/datasets/1998-darpa-intrusion-detection-evaluation-dataset. Accessed 29 Apr 2019
8. Lincoln Laboratory, Massachusetts Institute of Technology: 1999 DARPA Intrusion Detection Evaluation Dataset. https://www.ll.mit.edu/r-d/datasets/1999-darpa-intrusion-detection-evaluation-dataset. Accessed 29 Apr 2019
9. Lincoln Laboratory, Massachusetts Institute of Technology: 2000 DARPA Intrusion Detection Scenario Specific Datasets. https://www.ll.mit.edu/r-d/datasets/2000-darpa-intrusion-detection-scenario-specific-datasets. Accessed 29 Apr 2019
10. Makhzani, A., Shlens, J., Jaitly, N., Goodfellow, I.: Adversarial autoencoders. In: International Conference on Learning Representations (2016). http://arxiv.org/abs/1511.05644
11. McHugh, J.: Testing intrusion detection systems: a critique of the 1998 and 1999 DARPA intrusion detection system evaluations as performed by Lincoln laboratory. ACM Trans. Inf. Syst. Secur. **3**(4), 262–294 (2000). https://doi.org/10.1145/382912.382923
12. Ministry of Defence Finland: The national cyber security exercises is organised in Jyväskylä - Kansallinen kyberturvallisuusharjoitus kyha18 järjestetään Jyväskylässä, official bulletin 11th of May 2018, May 2018. https://valtioneuvosto.fi/artikkeli/-/asset_publisher/kansallinen-kyberturvallisuusharjoitus-kyha18-jarjestetaan-jyvaskylassa. Accessed 26 Apr 2019
13. Narsingyani, D., Kale, O.: Optimizing false positive in anomaly based intrusion detection using genetic algorithm. In: 2015 IEEE 3rd International Conference on MOOCs, Innovation and Technology in Education (MITE), pp. 72–77, October 2015. https://doi.org/10.1109/MITE.2015.7375291
14. Nevavuori, P., Kokkonen, T.: Requirements for training and evaluation dataset of network and host intrusion detection system. In: Rocha, Á., Adeli, H., Reis, L.P., Costanzo, S. (eds.) WorldCIST'19 2019. AISC, vol. 931, pp. 534–546. Springer, Cham (2019). https://doi.org/10.1007/978-3-030-16184-2_51
15. van den Oord, A., et al.: WaveNet: a generative model for raw audio (2016). https://arxiv.org/pdf/1609.03499.pdf
16. van den Oord, A., Kalchbrenner, N., Kavukcuoglu, K.: Pixel recurrent neural networks. In: Balcan, M.F., Weinberger, K.Q. (eds.) Proceedings of the 33rd International Conference on Machine Learning. Proceedings of Machine Learning Research, vol. 48, pp. 1747–1756. PMLR, New York, 20–22 June 2016. http://proceedings.mlr.press/v48/oord16.html
17. van den Oord, A., et al.: Parallel WaveNet: fast high-fidelity speech synthesis. CoRR abs/1711.10433 (2017). http://arxiv.org/abs/1711.10433
18. Open Information Security Foundation (OISF): Suricata Open Source IDS/IPS/NSM engine. https://suricata-ids.org/. Accessed 7 May 2019
19. Puuska, S., Kokkonen, T., Alatalo, J., Heilimo, E.: Anomaly-based network intrusion detection using wavelets and adversarial autoencoders. In: Lanet, J.-L., Toma, C. (eds.) SECITC 2018. LNCS, vol. 11359, pp. 234–246. Springer, Cham (2019). https://doi.org/10.1007/978-3-030-12942-2_18
20. Ring, M., Wunderlich, S., Scheuring, D., Landes, D., Hotho, A.: A survey of network-based intrusion detection data sets. Comput. Secur. **86**, 147–167 (2019). https://doi.org/10.1016/j.cose.2019.06.005

21. Salimans, T., Karpathy, A., Chen, X., Kingma, D.P.: PixelCNN++: improving the PixelCNN with discretized logistic mixture likelihood and other modifications. In: 5th International Conference on Learning Representations, ICLR 2017, 24–26 April 2017, Toulon, France (2017). https://openreview.net/references/pdf?id=rJuJ1cP_l

22. Shiravi, A., Shiravi, H., Tavallaee, M., Ghorbani, A.A.: Toward developing a systematic approach to generate benchmark datasets for intrusion detection. Comput. Secur. **31**(3), 357–374 (2012). https://doi.org/10.1016/j.cose.2011.12.012

23. Siddiqui, M.A., et al.: Detecting cyber attacks using anomaly detection with explanations and expert feedback. In: ICASSP 2019–2019 IEEE International Conference on Acoustics, Speech and Signal Processing (ICASSP), pp. 2872–2876, May 2019. https://doi.org/10.1109/ICASSP.2019.8683212

24. Tavallaee, M., Bagheri, E., Lu, W., Ghorbani, A.A.: A detailed analysis of the KDD CUP 99 data set. In: Proceedings of the Second IEEE International Conference on Computational Intelligence for Security and Defense Applications, CISDA 2009, pp. 53–58. IEEE Press, Piscataway (2009). http://dl.acm.org/citation.cfm?id=1736481.1736489

25. The University of California Irvine (UCI): KDD Cup 1999 Data. http://kdd.ics.uci.edu/databases/kddcup99/kddcup99.html. Accessed 29 Apr 2019

26. Umer, M.F., Sher, M., Bi, Y.: Flow-based intrusion detection: techniques and challenges. Comput. Secur. **70**, 238–254 (2017). https://doi.org/10.1016/j.cose.2017.05.009

27. University of New Brunswick, Canadian Institute for Cybersecurity: Intrusion Detection Evaluation Dataset (CICIDS 2017). https://www.unb.ca/cic/datasets/ids-2017.html. Accessed 30 Apr 2019

28. Wiewel, F., Yang, B.: Continual learning for anomaly detection with variational autoencoder. In: 2019 IEEE International Conference on Acoustics, Speech and Signal Processing (ICASSP), ICASSP 2019, pp. 3837–3841, May 2019. https://doi.org/10.1109/ICASSP.2019.8682702

29. Yu, F., Koltun, V.: Multi-scale context aggregation by dilated convolutions. CoRR abs/1511.07122 (2016). https://arxiv.org/pdf/1511.07122.pdf

30. Zagoruyko, S., Komodakis, N.: Wide residual networks. In: Richard C. Wilson, E.R.H., Smith, W.A.P. (eds.) Proceedings of the British Machine Vision Conference (BMVC), pp. 87.1–87.12. BMVA Press, September 2016. https://doi.org/10.5244/C.30.87

Steganographic WF5 Method for Weighted Embedding: An Overview and Comparison

Tamara Minaeva[1] , Natalia Voloshina[1] , Sergey Bezzateev[2(✉)] ,
and Vadim Davydov[1]

[1] ITMO University, St. Petersburg, Russia
minaeva-toma-ya@yandex.ru, nataliv@ya.ru, vadimdavydov@outlook.com
[2] Saint-Petersburg State University of Aerospace Instrumentation,
St. Petersburg, Russia
bsv@aanet.ru

Abstract. In this paper, we discuss steganographic methods LSB, F5, and WF5 for embedding information into a container to enable copyright protection. We analyze the difference between selected methods and show how to decrease the container distortion by dividing codewords into several zones of significance using perfect codes in the weighted Hamming metric. Additionally, we propose a number of variants for different container types and test some parameters on several images.

Keywords: Digital watermarking · Error-correcting codes · F5 method · WF5 method

1 Introduction

With the rapid development of information technology, most people's data is stored in the electronic form, which is convenient in terms of accessing the data or identifying the personality including biometric and knowledge-related factors [1]. Simultaneously, security and privacy threats are present in the databases where personal data could be stolen or lured out illegally. Afterward, compromised data could be used for identification under someone else's identity for the most diverse systems and operations.

Stealing identity details and selling it to anyone in the world or uploading to free access is extremely widespread nowadays. According to the CompariTech report [2], the average price of a digital passport scan is only $14.71. Due to the high value of such personal data, it must be protected with exceptional attention. Overall, there are several methods to protect the data, and one of them is digital watermarking, which is widely used today.

In general, many solutions were proposed to mitigate the challenges of digital image privacy, but most of them have limitations related to the amount of embedded information. To overcome these scopes, multi-level embedding into

O. Galinina et al. (Eds.): NEW2AN 2019/ruSMART 2019, LNCS 11660, pp. 434–440, 2019.
https://doi.org/10.1007/978-3-030-30859-9_37

different bit planes allowing to implant practically much more copyright information may be utilized [3,4]. Furthermore, Hammering perfect codes are used to highlight the zones of significance and "low-influenced" zones on the image where we could embed more information with no damage to container distortion. In this paper, we consider new variants of forming workspace in the container using WF5 method [5].

The rest of the paper is organized as follows. In Sect. 2, we highlight the main criteria to opt for the steganographic method for embedding into a static image. Section 3 contains the algorithm of embedding information into the image and potential options for the structure of the image container workspace. In Sect. 4, we analyze our improved method by calculating some important parameters using several test images.

2   Main Criteria for Choosing Steganographic Method

The number of methods utilized for embedding information into the images is vast. Those are divided into many groups, for instance, the group by embedding area (frequency domain, spatial domain, redundancy area) or the group by object format dependency (format dependent and format independent). To opt for the optimal methods for specific tasks, some rules or criteria should be defined. The main suggestions are as following:

1. Embedding of the information must not distort an image to such an extent that this information could be seen by an eye;
2. Method should allow embedding maximum information to have a possibility to improve the security level by encrypting and coding;
3. Method should be extremely difficult or even impossible for a hacker to have access to the information or to modify it.

This study includes five different methods used for embedding information into the spatial image area. These methods are LSB method [6], F5 method [7], weighted WF5 method [8], Kutter-Jordan-Bossen method [9], and adaptive LSB method [10].

First of all, the LSB and Kutter-Jordan-Bossen methods were analyzed. In consequence, it was shown that they are not efficient due to the ease of information extraction from the image. None the less, the disadvantage of these methods is the lack of robustness to other common types of attacks and distortions [11].

After analyzing all methods, based on the previously selected criteria, the most optimal methods for further research are F5 and WF5 that would be considered in Sect. 3.

3   F5 and WF5 Methods for Embedding Information

3.1   F5 Method

F5 method is based on the error correction Hamming code $(7, 4, 3)$ with the parity check matrix H_1 [12]:

$$H_1 = \begin{bmatrix} 1\,0\,0\,1\,1\,0\,1 \\ 0\,1\,0\,1\,0\,1\,1 \\ 0\,0\,1\,0\,1\,1\,1 \end{bmatrix}$$

It is a perfect code that ensures compliance with any non-zero binary vector of length r (which is called syndrome) of a unique vector E (which is called error vector) of length n with the number of ones equal to 1. This feature is used for embedding information as an error vector into the source vector.

The algorithm is the following. To embed information $I = \{I_1, I_2, ..., I_r\}$ to the source vector $B = \{B_1, B_2, ..., B_n\}$ of workspace from the image, the following steps should be done.

1. Calculate a syndrome $S = B \cdot H^T$.
2. Calculate a vector $A = S \oplus I$ (length of S should be equal to length of I).
3. Using matrix H and vector A, find error vector E with length n such that $E \cdot H^T = A$.
4. Finally, calculate a new vector for the image $B' = B \oplus E$. As a result, we add information I of length n by adding errors to the bitstream of an image.
5. Steps are repeated for the whole container unless all information is added.

To extract information I, the syndrome should be calculated:

$$I = B' \cdot H^T$$

Increasing the length of the codeword allows embedding more copyright information into the container (in particular, into an image).

3.2   WF5 Method

The F5 method uses Hamming code for embedding process. It has a feature that a number of errors is controlled. At the same time, the main drawback of the method is an equal weight of distortion on each codeword position, therefore bit errors can occur among all bit levels used for embedding, which lead to more container distortion which is a great restriction for expansion of working space of the container. To correct this deficiency, WF5 method was proposed [4, 5, 8]. The main idea of WF5 is to use the codes that are perfect in the weighted Hamming metric. Here, we consider a case when the error correcting code of length n is perfect in the weighted Hamming metric is used. The WF5 method differs from the usual F5 method that it allows to control bit distortions and "weightedly" embed information into the image. As a consequence, it reduces the level of distortion by dividing the working area of the image into several zones of significance.

Let's consider as an example the case when two zones with weights of distortions $w1$ and $w2$, $w1 = 1, w2 = 2$ exist. In this case, the error-correcting code with length 9 is used, which is a perfect code in the weighted Hamming metric. The codeword of this code is divided into two zones with lengths $n1 = 4$ and $n2 = 5$, $n1 + n2 = 9$ with weights $w1$ and $w2$ respectively. Such code has a minimal distance in the weighted Hamming metric equal to 5 $(d_w = 5)$ and the redundancy level equal to 4. If the vector $c = (c_0, c_1, \ldots, c_8)$ is a codeword of this code, then $(c_0 + c_1 + c_2 + c_3) \cdot w1 + (c_4 + c_5 + \ldots + c_8) \cdot w2 \geq d_w$. Therefore, the parity check matrix H_w for this code is formed in such a way, that the sum of any columns with total weight of their positions less than d_w, is not zero. The parity check matrix, where first four columns have weights equal to $w1 = 1$ and next five columns have weights equal to $w2 = 2$, is shown below.

$$H_w = \begin{bmatrix} 1 & 0 & 0 & 0 & 1 & 1 & 1 & 0 & 1 \\ 0 & 1 & 0 & 0 & 1 & 1 & 0 & 1 & 1 \\ 0 & 0 & 1 & 0 & 1 & 0 & 1 & 1 & 1 \\ 0 & 0 & 0 & 1 & 0 & 1 & 1 & 1 & 1 \end{bmatrix}$$

We already have the error-correcting code in weighted Hamming metric with the length $n = 9$ and the minimal distance $d_w = 5$. Let's show that we obtained a binary code perfect in weighted Hamming metric by parity check matrix H_w. To achieve this goal, it should be proved that the code lies on the weighted Hamming bound

$$2^4 = \binom{15}{0} + \sum_{i=1}^{2} \binom{4}{i} + \binom{5}{1} = 1 + 4 + 6 + 5 = 16. \tag{1}$$

Ordinary Hamming code with the same redundancy has the parity check matrix H_2 with length $n = 15$ and minimal distance $d = 3$.

$$H_2 = \begin{bmatrix} 1 & 0 & 0 & 0 & 1 & 1 & 1 & 0 & 0 & 0 & 1 & 1 & 1 & 0 & 1 \\ 0 & 1 & 0 & 0 & 1 & 0 & 0 & 1 & 1 & 0 & 1 & 1 & 0 & 1 & 1 \\ 0 & 0 & 1 & 0 & 0 & 1 & 0 & 1 & 0 & 1 & 1 & 0 & 1 & 1 & 1 \\ 0 & 0 & 0 & 1 & 0 & 0 & 1 & 0 & 1 & 1 & 0 & 1 & 1 & 1 & 1 \end{bmatrix}$$

Therefore, comparing with the F5 with WF5, we can use error correcting codes with shorter code length and the same number of redundancy symbols. Consequently, matrix dimensions are reduced.

As a part of the study of the WF5 method, the options for forming the workspace with a "weighted" distribution of the container bits were developed. Variants were formed based on the length of the codeword and the use of a different number of bit planes. The codeword was formed from a single pixel of each color component of the initial image. Thus, the working area for one codeword is formed from the bits of three color components of the image pixel. The variants of the working space configuration are shown in Fig. 1.

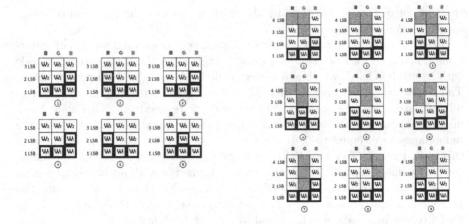

Fig. 1. Variants of forming workspace.

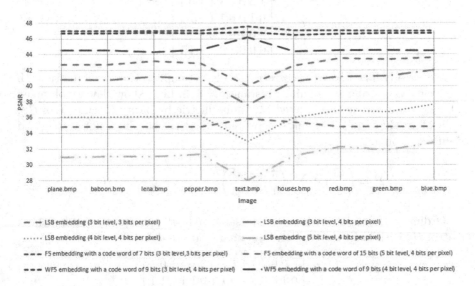

Fig. 2. PSNR of several test images using different steganographic methods and bit levels.

4   Comparison of the Methods and Practical Results

When solving the problem of embedding information volume and increasing the number of bit planes with information embedding, there is a degradation in the image quality. In the image protection preserving the quality of the container after embedding is an important task, therefore methods LSB, F5, and WF5 were tested. The main goal of the tests is to estimate the influence of building multi-level models of the working areas of the image on the degree of distortion after embedding information by various steganographic methods. Figure 2 illustrates the PSNR of the images with the maximum information embedded using a different amount of image bit planes. When the LSB method is used, the values of PSNR are lower than using the F5 and WF5 embedding methods, which means that the visual quality of the resulting image is worse.

Let us consider in more detail the comparison of PSNR values for embedding by the WF5 method using different codeword lengths and different bit planes. As it can be seen from Fig. 3, the visual distortion does not drop significantly as the length of the codeword increases, which means that when the maximum possible embedding volume is increased by 15%, the container distortion does not practically increase even when using four-bit planes. Indeed, the WF5 method with a codeword length of 9 bits with multi-level embedding allows embedding the same amount of information as to when embedding using the F5 method with a codeword length of 15 bits, while reducing the degree of container distortion and using the less number of bit levels.

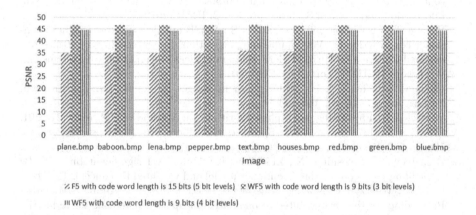

Fig. 3. Embedding using WF5 method with different codeword lengths and bit levels.

5   Conclusion

In this paper, we have shown that a weighted WF5 method is an optimal approach for embedding information into images. It was proven that WF5 with a codeword length of 9 bits is the most optimal method of multi-level embedding, according to the selected criteria. The method allows embedding various information into the spatial domain of an image of various types, which makes possible to completely fill the container with the author's data, thereby protecting the image from unauthorized distribution and use. We have shown new variants of forming workspace in such method, compared it with LSB and F5 embedding methods. Our future research includes studies of the optimal use of other various error-correcting codes for steganographic multi-level embedding, as well as the protection of the embedded author's signature from distortions.

Acknowledgment. This work was partly financially supported by the Russian Foundation for Basic Research in 2017 (grant 17-07-00849-A).

References

1. Ometov, A., Petrov, V., Bezzateev, S., Andreev, S., Koucheryavy, Y., Gerla, M.: Challenges of multi-factor authentication for securing advanced IoT applications. IEEE Netw. **33**(2), 82–88 (2019)
2. Bischoff, P.: Passports on the dark web: how much is yours worth? https://www.comparitech.com/blog/vpn-privacy/passports-on-the-dark-web-how-much-is-yours-worth/. Accessed 17 May 2019
3. Pandey, A., Chopra, J.: Comparison of various steganography techniques using LSB and 2LSB: a review. Int. J. Sci. Res. Eng. Technol. (IJSRET) **6**(5), 522–525 (2017)
4. Bezzateev, S., Voloshina, N., Zhidanov, K.: Steganographic method on weighted container. In: 2012 XIII International Symposium on Problems of Redundancy in Information and Control Systems, pp. 10–12. IEEE (2012)
5. Voloshina, N., Minaeva, T., Bezzateev, S.: MLSB optimal effective weighted container construction for WF5 embedding algorithm. In: Proceedings of 10th International Congress on Ultra Modern Telecommunications and Control Systems and Workshops (ICUMT), pp. 1–6. IEEE (2018)
6. Chan, C.K., Cheng, L.M.: Hiding data in images by simple LSB substitution. Pattern Recogn. **37**(3), 469–474 (2004)
7. Westfeld, A.: F5—a steganographic algorithm. In: Moskowitz, I.S. (ed.) IH 2001. LNCS, vol. 2137, pp. 289–302. Springer, Heidelberg (2001). https://doi.org/10.1007/3-540-45496-9_21
8. Bezzateev, S., Voloshina, N., Zhidanov, K.: Multi-level significant bit (MLSB) embedding based on weighted container model and weighted F5 concept. In: Abraham, A., Wegrzyn-Wolska, K., Hassanien, A.E., Snasel, V., Alimi, A.M. (eds.) Proceedings of the Second International Afro-European Conference for Industrial Advancement AECIA 2015. AISC, vol. 427, pp. 293–303. Springer, Cham (2016). https://doi.org/10.1007/978-3-319-29504-6_29
9. Kutter, M., Petitcolas, F.A.: Fair benchmark for image watermarking systems. In: Security and Watermarking of Multimedia Contents, vol. 3657, pp. 226–240. International Society for Optics and Photonics (1999)
10. Yang, H., Sun, X., Sun, G.: A high-capacity image data hiding scheme using adaptive LSB substitution. Radioengineering **18**(4), 509–516 (2009)
11. Vovk, O., Astrahantsev, A.: Synthesis of optimal steganographic method meeting given criteria. Informatyka, Automatyka, Pomiary w Gospodarce i Ochronie Środowiska (2015)
12. Van Lint, J.: A survey of perfect codes. J. Math. **5**(2), 199–224 (1975)

Modeling of Routing as Resource Distribution in SDN

Alexander Paramonov[1(✉)] and Regina Shamilova[2]

[1] The Bonch-Bruevich Saint-Petersburg State University
of Telecommunications, Saint-Petersburg, Russia
alex-in-spb@yandex.ru
[2] Saint Petersburg National Research University of Information Technologies,
Mechanics and Optics, Saint-Petersburg, Russia

Abstract. This paper presents some results of the routing task formulation and solving for the software defined networks. We consider the routing task as search of the optimal path and the task of the resources distribution. Choosing of the path search method affects on the network resources utilization and quality of service. We consider this task as the optimization problem and consider the objective functions forming in the points of view of both of the tasks. Some examples of solutions were obtained by using of simulation method. Analysis of these results shows difference between the different approaches for the objective function formulation.

Keywords: SDN · Routing · Resource distribution · Optimization

1 Introduction and Related Works

The architecture of the software-defined network (SND) is the basic architecture of the advanced fifth-generation networks (5G) [1]. One of the key properties of such networks is the ability to improve an efficiency of resource utilization and/or quality of service by dynamical management of structural parameters.

In the point of view of the technological implementation the 5G networks are heterogeneous [1, 2] networks, so there are problems of ensuring interaction between the network fragments implemented using various technologies. For example, at the access level, access networks implemented using different standards can be used, which leads to the need to use a vertical handover, which ensures the continuity of services when transferring services from one network to another. The implementation of such functionality is associated with the implementation of criteria and functions of technology selection, as well as with the implementation of methods for selecting the appropriate network configuration to serve the traffic. The given example is only one of the cases when during the provision of the service the network configuration changes, namely the traffic transmission route.

In reality, the need to change the route can be associated not only with changes at the level of the access network, but also with changes in user traffic and the physical configuration of the network itself [3, 4]. In other words, the network should provide a dynamic configuration change (routing) under the influence of some factors, according to

O. Galinina et al. (Eds.): NEW2AN 2019/ruSMART 2019, LNCS 11660, pp. 441–454, 2019.
https://doi.org/10.1007/978-3-030-30859-9_38

the relevant criteria. The configuration usually refers to all network parameters that determine its capabilities in terms of servicing traffic in various directions (routing rules, throughput, traffic priorities, etc.) [5–7]. Further from this set we will consider only the routing rules, naming them simply routing. The definition of routing rules, in general, is based on the knowledge of such entities as network structure, lines (channels) of communication, traffic and quality of service. Depending on the problem being solved, these entities are expressed with specific data and models necessary to solve the problem.

This paper is dedicated to the task of routing traffic to SDN. The tasks of traffic routing in communication networks are well studied and widely covered in the literature [1, 2, 5–7]. However, as a rule, they are solved as optimal route selection problems based on a certain criterion (distance, data delivery delay, throughput, reliability, etc.) [1, 2]. Most of the solutions are based on the use of graph theory algorithms, for example [10].

In our work, we consider the choice of routes as a task of optimal distribution of network resources, taking into account the criteria used.

2   The Model and the Task Definition

Let suppose that the network is described by the graph $G(V, E)$, where $V = \{v_1, v_2 \ldots v_n\}$ the set of vertices and $E = \{e_{i,j}\}$, $i, j = 1 \ldots n$ the set of edges between the vertices, n - the number of vertices in the graph.

An example of a graph model and the conventions used are shown in Fig. 1.

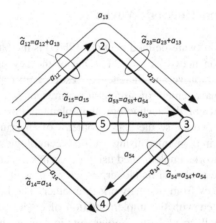

Fig. 1. A network model and legend.

Each of the vertices of the graph (each of the network nodes) is a source of traffic x_i, the part of which $x_{i,j}$ can be transmitted in any direction (i, j), i.e. to each of the network nodes $(j = 1 \ldots n)$, including the source node itself. We suppose that the traffic produced in a node can be represented as a superposition of a finite number of flows, where K_i is the number of traffic flows originated in the i-th node. The same property applies to the traffic of each of the directions, $K_{i,j}$ is the number of traffic flows originated in the direction (i, j). Then for the network all flows can be described as $X = \left\{ x_{i,j}^{(k)} \right\}$, $i,j = 1 \ldots n$; $k = 1 \ldots K$. We assume that the flow, within the framework of this task, is an indivisible unit, although it may have arbitrary characteristics, traffic parameters and requirements for the quality of service.

The edges of the graph may be described by numerical characteristics, reflecting the parameters of communication lines. In general, there may be some weights c_{ij}. Then the matrix $C = \left\{ c_{ij} \right\}$, $i,j = 1 \ldots n$, $e_{i,j} = 1$ describes all the communication lines (edges of the graph).

To characterize the functioning of the network, we introduce the means to describe the functioning of each of the communication lines. To do this, we use the functions that characterize the service of traffic [8–12] with the corresponding edge (line of communication)

$$q_{i,j} = q(\tilde{x}_{i,j}, c_{i,j}) \tag{1}$$

where $\tilde{x}_{ij}$ the traffic parameter in the communication line between nodes i and j is the communication line parameter between the same nodes.

The matrix $Q = \left\{ q_{i,j} \right\}$, $i,j = 1 \ldots n$ contains the characteristics of all communication lines of the network (edges of the graph).

The $\tilde{x}_{ij}$ defined as the result of the mapping of streams on the communication line. The result of this mapping for the network can be written as a set $\tilde{X} = \left\{ \tilde{x}_{i,j} \right\}$, $i,j = 1 \ldots n$.

$$g : X \to \tilde{X} \tag{2}$$

The mapping g is defined by the flow routing rules, i.e. routes for the delivery of traffic between network nodes.

In this graph G, there are many paths $P = \{P_1, P_2 \ldots P_K\}$ between the set of vertices $S = \left\{ v_1^{(S)}, v_2^{(S)} \ldots v_{m_S}^{(S)} \right\}$ and the set of vertices $T = \left\{ v_1^{(T)}, v_1^{(T)} \ldots v_{m_T}^{(T)} \right\}$, which can be used to serve the traffic originated at the vertices of the set S and directed to the vertices of the set T. Each of the paths consists of one or more network elements (graph).

The network operation is in the service (delivery) of traffic between its nodes. To perform this work the network spends a resource, which is the bandwidth of communication lines. The ability of the network to perform this work described by the performance indicators (quality of service traffic) [12, 13]. Of course, using of big resources amount leads to an increase in the quality of service, however, this may lead to the underutilization of these resources. Thus, the organization of traffic service by the communication network may be described both by the quality of the network functioning and the resource using efficiency.

Given this statement, the organization of the logical structure of the network should take into account both the quality requirements of the services provided and the requirements for efficient use of network resources. It is likely that the task of finding a compromise between the amount of resources and quality of service can be formulated as an optimization problem.

The routing problem, under the conditions described above, can be formulated as follows. It is required to find such a mapping of the set X on the set P, for which the optimality criterion is satisfied

$$f : X \to P \tag{3}$$

$$C = true \tag{4}$$

It should be noted, that in this formulation of the problem there is no time reference. Practically, traffic flows in communication networks can most often be considered as stationary only on limited time intervals or under special operating conditions. Therefore, the task of routing traffic in a functioning communication network, like any control problem, should be solved and the predicate basis, i.e. projected baseline data. The task of building forecasts is not included in the scope of this work, therefore, further speaking of the source data, we will assume that they were obtained as predicates of the corresponding parameters.

There are two options for solving this problem. The first, when the elements of the set P (routes) are known and the second, when unknown. Both options should be attributed to the problem of the optimal allocation of resources [11], although the ways of its solution for these options may be different.

3   Objective Function and Method of Optimization

The optimization problem was formulated above is the optimization task for communication network in general. It doesn't take into account the specific features of the network implementation, models and quality of service indicators, traffic characteristics and parameters. Let us restrict the scope of the problem by using specific parameters. We assume that the main characteristic of a communication line is its throughput μ_{ij}, and the traffic between a pair of nodes is its intensity a_{ij} and the traffic arriving on the communication line $\tilde{a}_{ij}$. Note that considering of the communication line, we will mean the network section, which includes both the communication line and node elements that provide data transmission over this line.

By bandwidth, we mean the maximum traffic intensity at which the requirements for quality of service are provided. In this case, the quality of service can be characterized by both the time of packet transmission on the line, the waiting time at the network node (in the buffer), and the probability of packet (frame) losses. In other words, the throughput of the route section (line) is numerically equal to the maximum traffic intensity that can be serviced by it.

Then, instead of matrix C, we will consider the throughput matrix of all communication lines (graph edges).

Along with the throughput we introduce the concept of bandwidth $b_{i,j}$. Here, the bandwidth will be understood as a free throughput resource, i.e. traffic intensity that can be served by the route section (or route) when the quality of service requirements are met. We will assume that if traffic $\tilde{a}_{i,j}$ is transmitted through the route section (i, j), then the section bandwidth is equal to the difference between its throughput and the intensity of the traffic being served $\tilde{b}_{i,j} = \mu_{i,j} - \tilde{a}_{i,j}$. Naturally, that $0 \le \tilde{b}_{i,j} \le \mu_{i,j}$.

The route in the communication network (the path in the graph) P_i is intended to serve the traffic between certain network nodes (the vertices of the corresponding sets S and T). We assume that the quality of the route is determined by the main parameters of the quality of service of the traffic. The value of these parameters depends on the characteristics and parameters of the traffic and the parameters of the sections of the route.

$$\tilde{q}_{i,j} = q(\tilde{a}_{i,j}, \mu_{i,j}) \tag{5}$$

where $\tilde{a}_{i,j}$ traffic intensity in the communication line, $\mu_{i,j}$ the throughput of the route segment.

The choice of route P_i is equivalent to the allocation of network resources to service the traffic. The resource in this task is the capacity of sections of the route, which is necessary to serve the traffic flow.

The problem of searching of paths defined above will be considered as the task of distributing network resources. We consider the solution of the routing problem from two positions: first, from the standpoint of ensuring the quality of service of the traffic in the network and secondly, from the standpoint of the spending of network resources.

In the first case, all network resources are distributed in such a way that provides maximum quality of service of the traffic in the network, and in the second, the quality satisfies certain requirements with minimum consumption of resources. In both cases, the goal is the maximum quality per unit of resources expended, i.e. resource efficiency. But in the second case, the goal is to achieve sufficient quality with a minimum amount of resources. Both approaches can take place in the tasks of planning and operating a communication network.

The objective function, in the first case, will be the maximum of the functional q, describing the dependence of the quality of service on the volume of resources and traffic (provided that its higher value corresponds to a higher quality level). The limitation is the invariance of the total amount of allocated resources, which is equal to the entire volume of network resources, in this case, the bandwidth M.

$$P = \arg \max_{P}\{q(P, A)\}, \quad \sum_{i=1}^{K} \mu_i = M \tag{6}$$

In the second case, the objective function is the minimum of the total volume of resources, and the limitation is the equality of the functional q to the specified value of the quality indicator

$$P = \arg \min_{P}\left\{ \sum_{i=1}^{K} b_i \right\}, \quad q(P, A) = q_0 \tag{7}$$

The choice of the functional $q(P, A)$ determines the further solution of the optimization problem. As mentioned above, its meaning should be directly or indirectly related to the quality of service. Consider further some possible options.

First, as a functional, a model can be used that describes the dependence of quality indicators on traffic parameters and the capacity of a route section, which establishes a direct connection between parameters. Secondly, it can be a bandwidth value (available bandwidth value). This parameter establishes an indirect connection with the quality of service.

Consider the case when the functionality $q(P, A)$ is expressed by the model of a queuing system [17], which establishes a relationship between traffic, throughput and delivery time in the considered section of the route. Let us make the assumption that the average delivery delay on the route segment can be determined as $T(\tilde{a}_k, \mu_k)$ the fact that the delays on different sections are random and independent. We also assume that the degree of impact of the route on the quality of the functioning of the network is proportional to the traffic served by this route.

Taking into account the degree of impact of each of the routes, the objective function (6) can be written as

$$P = \arg\min_{P} \left\{ \sum_{i=1}^{n} \sum_{j=1}^{n} \frac{\tilde{a}_{ij}}{\tilde{a}_{\Sigma}} \sum_{e_k \in P_{ij}} T(\tilde{a}_k, b_k) \right\}, \quad \sum_{i=1}^{K} b_i = M, \quad i,j = 1\ldots n \quad (8)$$

Where $\tilde{a}_{\Sigma} = \sum_{i=1}^{m} \tilde{a}_i$, where m is the number of communication lines.

The expression (8) is just one of the possible formulations of the optimization problem, constructed with respect to temporary indicators of the quality of service. Replacing the desired extremum by the minimum in (8) is explained by the fact that the QoS model used gives an estimate of the magnitude of the delay, and the quality of service increases with its decrease.

The formulation of the problem in the form of (7), assuming that $T(\tilde{a}_k, \mu_k)$ they are convex functions, and also that it is impossible to fulfill the conditions of the restrictions under strict equality, generally, we do

$$P = \arg\min_{P} \left\{ d\left(\sum_{i=1}^{n} \sum_{j=1}^{n} \frac{\tilde{a}_{ij}}{\tilde{a}_{\Sigma}} \sum_{e_k \in P_{ij}} T(\tilde{a}_k, b_k) - T_0 \right) \right\}, \quad \sum_{i=1}^{K} b_i \leq M, \quad i,j = 1\ldots n \quad (9)$$

where $d(a, b)$ is a function expressing the difference between the values a and b.

As a function of $d(a, b)$, the modulus or squared difference can be chosen. The choice of the difference square makes it possible to obtain more stable solutions to the optimization problem and, therefore, is preferable.

Expression (9) means choosing such a set of routes P, at which the maximum proximity of the average delay to the value of T_0 is reached, provided that the maximum resource costs do not exceed M. The disadvantages of this formulation include the fact that the target value of T_0 reflects only the average delays, and the standard is usually set for flow or direction.

Given this specificity of requirements for each of the directions, as well as when choosing the square of the difference as $d(a, b)$, the objective function can be formulated as

$$P = \arg\min_{\mathbf{P}} \sum_{i=1}^{n} \sum_{j=1}^{n} \left(\sum_{e_k \in P_{ij}} T(\tilde{a}_k, b_k) - T_{ij}^{(0)} \right)^2, \quad \sum_{i=1}^{K} b_i \leq M, \quad i,j = 1\ldots n \qquad (10)$$

According to (10), a route will be chosen for servicing the traffic flow, the bandwidth of which is closest to that which provides the required amount of delay.

Further, we will consider the statements of problem (8) and (10) as two main (in this paper) alternatives.

It is easy to show that when describing the delay by waiting time in the buffer and transmission over the communication line when using the classical QoS models for this, task (8) is equivalent to the task of choosing a route with the highest bandwidth, and task (10) is a route with a given bandwidth.

It was noted above that there may be a different set of source data for the solution. The most important factor for choosing a solution is that situations are possible when the set of possible routes is known and when it needs to be found. Next, we consider the choice of solving problems in those and other conditions.

(a) Known routes
Solving problems in formulation(8) and (10) with a known set of routes P can be obtained by exhaustive search, i.e. enumeration of all options for the distribution of traffic flows available to them routes. As a result, a distribution is chosen that allows reaching the extremum of the objective function. If in the direction (i, j) there are m flows and K_{ij} routes, then the total number of options will be K_{ij}^m options. Already with numbers of the order of 10, the number of calculations may be too large to solve practical problems.

To reduce the number of calculations, we suggest using the dynamic programming method. Theoretically, the number of paths between the vertices of the graph can be very large, however, in the practical implementation of the network, this number is limited by various factors and from the point of view of computational complexity, this problem is, in most cases may be solved.

(b) Unknown routes
In the case when the set P is unknown, problems (8) and (10) need to be solved by searching routes that satisfy certain conditions. To construct such conditions, the corresponding functionals selected for the objective functions can be used. As a search method can be used well-known algorithms based on finding the shortest path.

Let us further consider the use of algorithms based on the use of a "triple" operation [15] of the form:

$$r_{i,j} = \min\{r_{i,j}, \ r_{i,k} + r_{k,j}\} \qquad (11)$$

where $r_{i,j}$ is the minimized parameter of the route segment.

When solving a problem of the form (8), a triple operation will look like

$$T(\tilde{a}_{i,j}, \mu_{i,j}) = \min\{T(\tilde{a}_{i,j}, \mu_{i,j}), \quad T(\tilde{a}_{i,k}, \mu_{k,j}) + T(\tilde{a}_{k,j}, \mu_{k,j})\} \tag{12}$$

If the selected delay model in (12) is a convex function, then as a result of performing this operation, for all $i, j = 1 \ldots n$ a set of routes with the minimum value of the selected metric will be found, in this case, routes providing the minimum delay.

The use of triple operations in solving the problem of the form (10) leads to the search for routes in which the delay in each of the sections is as close as possible to the specified value T_0. However, this is not the purpose of routing, which according to (10) is to provide a delay on the route as a whole. In this case, to solve the problem, it is more convenient to use the characteristic of the route sections with the bandwidth $b_{i,j}$. Assuming that the bandwidth of the route is determined by the section with the lowest bandwidth, the triple operation can be written as

$$\tilde{b}_{i,j} = \arg\min\left\{ (\tilde{b}_{i,j} - b_0)^2, \; [\min(\tilde{b}_{i,k}, \tilde{b}_{k,j}) - b_0]^2 \right\} \tag{13}$$

The result of its application will be the choice of routes with a bandwidth that may be close to the required value of b_0.

Note that the problem in the form of (8) can also be solved on the basis of the data on the bandwidth. In this case, its solution will be the choice of a route with a maximum bandwidth.

It will be natural to assume that the results of choosing routes for different problem statements and different conditions for their solution will differ. Let us further analyze the results of solving the considered routing problems.

4   Analysis and Simulation of Routing Methods

To evaluate the solution of the routing problem, one should choose parameters that can describe the quality of the solution of the routing problem. We will proceed from the fact that the main criterion for evaluating the functioning of a communication network is the quality of service. Consequently, the parameters that can characterize the solution of the problem can be the parameters of quality of service, as well as parameters reflecting the use of network resources (efficiency).

Of course, as a result of the solution, a mapping of the generated traffic to multiple routes will be obtained. In this case, the delivery time in different routes may be different. We will assume that the solution of the routing problem can be considered as a random distribution of traffic flows along the network routes, which provides some average delay. However, the average delivery time is not a sufficiently complete characteristic of the solution.

Therefore, we describe the distribution of flows (or routes) by the magnitude of the delay. For this we will describe the probability density of the delivery time in the resulting solution. An example of comparing solutions using probability density functions is shown in Fig. 2.

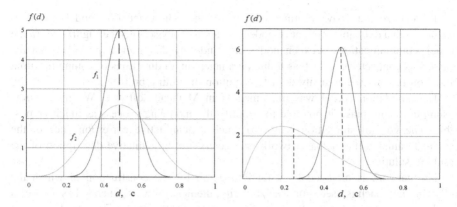

Fig. 2. Probability density of the solutions.

As can be seen from Fig. 2, solutions can have different mean values, variances and distribution forms. The form of the probability density function gives the mostly full view of the set of solutions obtained.

Also, an indicative parameter may be the quantile of the distribution over a given probability value, for example, the share of routes, the delivery delay of which exceeds a certain specified value d_0.

The distributions shown in Fig. 3 can correspond to the case when the objective function is formulated with respect to a single target value, i.e. when quality of service requirements are the same for all streams and directions, but in reality this is not the case. As noted above, different flows may have different requirements for the quality of service, i.e. to the parameters of delay and packet loss. The form of distribution in this case may differ significantly from the examples given.

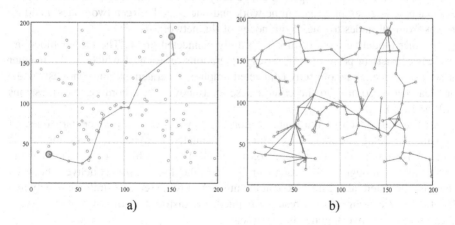

a)   b)

Fig. 3. Illustration of network model and route selection.

To analyze the above routing methods, a simulation network model was constructed with a definable number of nodes and a randomly selected configuration. In the model, communications between nodes are randomly determined (provided that the network is connected), as well as traffic parameters in all directions of communication. Based on this model, an analysis of the solution of routing problems was carried out.

The simulation model was implemented in Mathcad software. We made some assumptions that make it possible to simplify the modeling process as much as possible. The goal of simulation is not so much a quantitative correspondence of the obtained values to the actual network parameters, but a qualitative description of the resulting solution.

The network is modeled by a random graph for the construction of which the following approach is used. For greater clarity, the network was generated as a set of n graph vertices, i.e. points with random coordinates on the plane. The coordinates are random numbers with a uniform distribution law, selected in a given range of values. In this model, we made the assumption that the vertices of the graph have an edge between them, if the distance between them does not exceed R. Such a model, in fact, describes a wireless network model, the communication zone of the nodes of which is described by a circle with radius R.

Of course, the approach to obtaining a random graph may be different. Each of the graph vertices is assigned a random variable of the traffic produced in it, this traffic is distributed in the network between other nodes in proportion to the traffic produced in them. This assumption is made on the basis of considerations of the dependence of the distribution coefficients of traffic on the value of traffic produced in a node.

After all nodes have been generated, all shortest paths are searched for in the resulting graph; for this, the Floyd algorithm has been implemented, which allows finding the shortest paths between all pairs of graph vertices in one pass. If not all vertices of the graph can be reached, then either their number or connection radius R changes, and the process is repeated until a fully connected graph is obtained. In Fig. 3. An example is given of a random network and one route between two nodes a and all routes from all nodes to one of the nodes of the network b.

Further modeling is performed for a fully connected graph. The further modeling process consists in performing a series of iterations on each of which, in the manner described above, a random fully connected graph is generated and the "shortest" paths are searched according to the criteria described above, i.e. according to expressions (12) and (13).

In this case, each of the route sections is considered as an elementary queuing system, the delay introduced by which is calculated analytically, according to the chosen simplest model and depends on the intensity of the traffic route that the route serves and its throughput. For each pair of nodes, the delay in packet delivery between them is calculated taking into account the number of route sections, traffic intensity and bandwidth. According to the results obtained, the distribution of the packet delivery delay between network nodes is constructed.

Figure 4 shows a histogram of the routing solution obtained as a result of a simulation experiment for a model of a network of 20 nodes, containing 400 directions and the same number of routes. The optimization was performed according to the formulation of the problem (10), with target values of the delay of 50 ms and 30 ms.

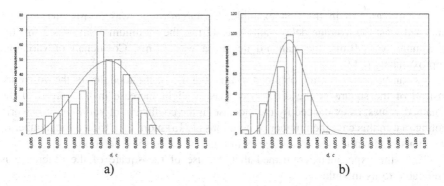

a)   b)

Fig. 4. Distribution of routes by the magnitude of the delay for two target values (a - 50 ms, b - 30 ms).

The graph also shows the approximation of the histogram γ-distribution. In the first case, the result is: an average delay of 43 ms, a minimum delay of 8 ms, a maximum of 75 ms, a standard deviation of 15 ms.

In the second case, the received: the average delay of 26 ms, the minimum delay of 4 ms, the maximum 46 ms, the standard deviation of 9 ms.

From the above results it can be seen that a large share of routes (more than 60%) has a delay not exceeding the target value.

The share of routes in which the delay exceeds the target value was 33.5% and 34.5% in the first and second cases, respectively.

The coefficient of variation, in both cases, approximately 0.31.

In both experiments, the obtained histogram can be quite accurately described by the γ-distribution with the corresponding parameters.

In Fig. 5 for comparison, the optimization result obtained when estimating not the square, but the modulus of the difference with the target value.

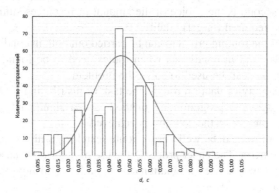

Fig. 5. Solution in estimating the modulus of proximity to the target value.

In this case, as in the first example, the target value was 50 ms, the solution obtained was: the average delay value was 42 ms, the minimum delay was 5 ms, the maximum was 90 ms, the standard deviation was 15 ms. Coefficient of variation, approximately 0.37.

As can be seen from the obtained results, the use of the modulus of the difference instead of the square of the difference gives a slight increase in the spread of the obtained values, as evidenced by the increase in the coefficient of variation. It is worth noting that in this case the number of iterations (1713) more than tripled the number of iterations required to find a solution in the first example (549 iterations).

Thus, the experiment confirmed that the use of the square of the difference is preferable to its module.

5 Conclusions

Effective solution of routing problems directly affects the quality of functioning and efficiency of the communication network, therefore, the choice of methods for their solution is an actual problem in promising communication networks.

The task of routing traffic may be regarded as the task of allocating network resources, i.e. as the task of optimal allocation of resources. In this case, the network resource is the capacity of communication lines, which is allocated to serve the traffic flows between network nodes.

The quality of the solution of the routing task, both the problem of optimal allocation of resources, can be described both from the standpoint of the quality of service of the traffic (quality of the network) and from the standpoint of efficient use of network resources.

The result of solving the problem of optimal choice of routes depends on its formulation (problem statement) and the method of its solution. The paper shows that when setting a task two main approaches can be used, which can be briefly formulated as: the first is the allocation of all available resources so that the quality of service is maximal and the second is the choice of the amount of resources sufficient to service the traffic with the required (target) quality.

The solution of the routing problem can be carried out both in conditions when data on routes is known, and in conditions when routes need to be found. Depending on this, different methods can be used to solve the problem.

With the availability of data on routes, the task of distributing traffic flows can be solved using the dynamic programming method, and a significant reduction in the computational complexity of the solution is achieved.

In the absence of route data, well-known route search algorithms can be used based on their selection criteria. The paper proposes the use of algorithms based on the use of a triple operation, as well as the formulation of criteria for choosing routes for different problem statements.

References

1. Shu, Z., et al.: Traffic Engineering in Software-Defined Networking: Measurement and Management. Special Section on Green Communications and Networking for 5G Wireless, 21 June 2016. Date of current version 7 July 2016. https://doi.org/10.1109/access.2016.2582748

2. Meneses, F., Silva, R., Santos, D., Corujo, D., Aguiar, R.: Using SDN and Slicing for Data Offloading over Heterogeneous Networks Supporting non-3GPP Access, September 2018. https://doi.org/10.1109/pimrc.2018.8580969

3. Kirichek, R., Vladyko, A., Paramonov, A., Koucheryavy, A.: Software-defined architecture for flying ubiquitous sensor networking. In: 19th International Conference on Advanced Communication Technology (ICACT), pp. 158–162 (2017)

4. Makolkina, M., Paramonov, A., Vladyko, A., Dunaytsev, R., Kirichek, R., Koucheryavy, A.: The use of UAVs, SDN, and augmented reality for VANET applications. In: 3rd International Conference on Artificial Intelligence and Industrial Engineering (AIIE 2017) DEStech Transactions on Computer Science and Engineering, pp. 364–368 (2017)

5. Mahmood, O.A., Paramonov, A.: Optimization of routes in the internet of things. In: Galinina, O., Andreev, S., Balandin, S., Koucheryavy, Y. (eds.) NEW2AN/ruSMART - 2018. LNCS, vol. 11118, pp. 584–593. Springer, Cham (2018). https://doi.org/10.1007/978-3-030-01168-0_52

6. Dao, N., Koucheryavy, A., Paramonov, A.: Analysis of routes in the network based on a swarm of UAVs. Information Science and Applications (ICISA) 2016. LNEE, vol. 376, pp. 1261–1271. Springer, Singapore (2016). https://doi.org/10.1007/978-981-10-0557-2_119

7. Vybornova, A., Paramonov, A., Koucheryavy, A.: Analysis of the packet path lengths in the swarms for flying ubiquitous sensor networks. In: Vishnevskiy V., Samouylov K., Kozyrev D. (eds.) DCCN 2016. CCIS, vol. 678, pp. 361–368. Springer, Cham (2016). https://doi.org/10.1007/978-3-319-51917-3_32

8. Usmanov, O., Muthanna, A., Paramonov, A.: Analysis of SDN traffic using full-scale modeling. In: 10th International Congress on Ultra Modern Telecommunications and Control Systems and Workshops (ICUMT) (2018)

9. Paramonov, A., Koucheryavy, A.: M2M traffic models and flow types in case of mass event detection. In: Balandin, S., Andreev, S., Koucheryavy, Y. (eds.) NEW2AN 2014. LNCS, vol. 8638, pp. 294–300. Springer, Cham (2014). https://doi.org/10.1007/978-3-319-10353-2_25

10. Volkov, A., Khakimov, A., Muthanna, A., Kirichek, R., Vladyko, A., Koucheryavy, A.: Interaction of the IoT traffic generated by a smart city segment with SDN core network. In: Koucheryavy, Y., Mamatas, L., Matta, I., Ometov, A., Papadimitriou, P. (eds.) WWIC 2017. LNCS, vol. 10372, pp. 115–126. Springer, Cham (2017). https://doi.org/10.1007/978-3-319-61382-6_10

11. Makolkina, M., Paramonov, A., Koucheryavy, A.: Resource allocation for the provision of augmented reality service. In: Galinina, O., Andreev, S., Balandin, S., Koucheryavy, Y. (eds.) NEW2AN/ruSMART -2018. LNCS, vol. 11118, pp. 441–455. Springer, Cham (2018). https://doi.org/10.1007/978-3-030-01168-0_40

12. Muthanna, A., Khakimov, A., Gudkova, I., Paramonov, A., Vladyko, A., Kirichek, R.: Openflow switch buffer configuration method. In: Proceedings of the International Conference on Future Networks and Distributed Systems, ICFNDS 2017 (2017)

13. Muthanna, A., et al.: SDN multi-controller networks with load balanced. In: The 2nd International Conference on Future Networks and Distributed Systems (ICFNDS) Conference Proceedings (2018)
14. Iversen, V.B.: Teletraffic Engineering and Network Planning. DTU Course 34340. http://www.com.dtu.dk/teletraffic/Technical. University of Denmark (2009)
15. Christofides, N.: Graph Theory. An Algorithmic Approach. Academic Press, London (1975)

Survey of Cyber Security Awareness in Health, Social Services and Regional Government in South Ostrobothnia, Finland

Tero Haukilehto[1(✉)] and Jari Hautamäki[2]

[1] Hospital District of South Ostrobothnia, Hanneksenrinne 7,
60220 Seinäjoki, Finland
tero.haukilehto@epshp.fi
[2] JAMK University of Applied Sciences, Piippukatu 2,
40100 Jyväskylä, Finland
jari.hautamaki@jamk.fi

Abstract. As the health, social services and regional government reform set great expectations for the new technology and the savings it brings, the importance of cyber security increases. When training cyber security for the personnel of the South Ostrobothnia Hospital District (EPSHP), the shortcomings in cyber security awareness emerged. Because improving the awareness is the easiest, fastest and cheapest way to improve cyber security level in organizations, the current level of cyber security awareness was seen valuable to measure in organizations under the reform in the region of South Ostrobothnia.

The study investigated the current level of cyber security awareness and the reasons affecting it. Cyber security awareness studied with three different surveys. The first two surveys organized as a part of cyber security lessons for the personnel of the Hospital District of South Ostrobothnia. The third survey conducted as an internet survey for all organizations involved in the reform in South Ostrobothnia. A total of over 1,200 responses to the questionnaires analyzed using material-based content analysis. The results enabled to create an overall view of the current level of cyber security awareness in the organizations, the coverage of the education and the reasons affecting them.

According the results, the cyber security awareness and education are lacking among the personnel and management in the target organization. The overall cyber security awareness should be improved in all target organizations.

Keywords: Cyber security · Information security · Data protection ·
Awareness · Education · Health ·
Social services and regional government reform ·
Hospital District of South Ostrobothnia

1 Introduction

The Health, social services and regional government reform and Cyber security awareness are both current topics that concern a great amount of organizations and people in Finland. Whereas the importance of cyber security awareness is becoming

© Springer Nature Switzerland AG 2019
O. Galinina et al. (Eds.): NEW2AN 2019/ruSMART 2019, LNCS 11660, pp. 455–466, 2019.
https://doi.org/10.1007/978-3-030-30859-9_39

clearer, the upcoming reform will even increase the value of the two for the whole society.

Cyber security protects and enables many of the modern services; however, still the terms cyber, and cyber security are often seen in the news, especially when something bad happens. For example, a big data breach comes to public attention, hackers have caused a distributed denial of service attack, or just a system error causes some trouble in a service that one is so used to think is available all the time in everyday life.

In fact, many of the modern conveniences such as electricity [1], water supply [2], transportation [3] and finance depend on cyber environment. Another vital service for everyone is healthcare. In the beginning of 2010 decade healthcare organizations reported rising of attacks from 20% to 40% [4].

To get for example an appointment, treatment, prescription and medicine, does not only depend on the availability of nursing staff but a vast amount of systems and services running on many different computers connected to each other on several locations. Cyber-security risk event influence to the health care institution's reputation and is becoming increasingly complex. It often due to the evolution of network technology where new technology are integrated to existing digital environment like connected Wi-Fi devices for medical devices and electronic databases [5]. Systems from air ventilation to lighting and from heating to fire alarm detectors can be dependent on the cyber environment as well, where one single point of failure in this chain can lead into a situation where also the treatment process is disturbed. Large number of systems and access points in a physical building outside the health care systems cause additional risk which enhanced needs to threat prevention, detection, and mitigation activities [6].

To secure these operations and services, the cyber security level must be high in the organizations. It has been recognized that several threats to the organization's digital information systems are due to the behavior of system users [7]. The easiest, fastest and cheapest way to improve cyber security level in an organization is to improve the cyber security awareness.

In an ideal world, the users would know everything related to cyber security and there would be no need to improve cyber security awareness. In the real world, no one knows everything about cyber security; the ever-changing ecosystems, their vulnerabilities and effect on other systems or even the processes behind the systems they are using every day. This produces a fundamental problem, namely, if one cannot know everything about cyber security, what should they at least know, and how could this knowledge be reached? In addition, without knowing the current level of cyber security awareness, the improvement process is difficult. The highly skilled employees in cyber security are required [8].

The main objective for this research was to clarify what the current level of cyber security awareness and education is among employees working in organizations that are to join in the upcoming Health, social services and regional government reform in South Ostrobothnia, and why the awareness is at this level. Secondly, this study aims to study how the cyber security awareness could be improved in the selected organizations and what the elements are that should be taken care of in this process.

This knowledge can be used to form guidelines in improving cyber security awareness inside organizations from strategic to operational planning. The second objective will create an overall picture of the starting level of cyber security awareness for the upcoming organization or organizations and form a base where the organization can continue improving its cyber security.

The infrastructure providing the most vital functions for a society is called critical infrastructure [9]. This kind of infrastructure has several different sectors for example, the US Department of Homeland security identifies 16 different critical infrastructure sectors, healthcare sector being one of them [10]. All these sectors have in common that they are essential for the society and if one or more is damaged, it could impact the whole society negatively.

A simplified structure of critical infrastructure is shown below (see Fig. 1). Although the structure is simple, it shows how strongly today's society depends on the electricity network. Without electricity there are no data networks nor the services using it. If there is electricity but no data network, then again the services built on it do not work [11].

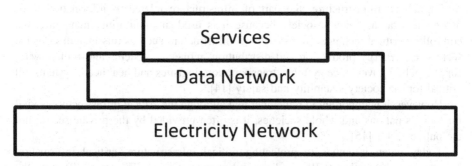

Fig. 1. Simplified structure of critical infrastructure [11]

In order to build a data network, first a reliable electricity network providing electricity is needed. Only with these two bottom layers electricity and data network functioning the digital services to help people's everyday lives can be created and maintained.

Critical infrastructure sectors also depend on each other, and healthcare is no exception. These dependencies are both positive and negative. On the positive side, the sectors support one another; however, on the negative side, they are in need of services the others provide. It is easy to see that without running water or electricity the healthcare sector is in big trouble [12].

Still, troubles for healthcare can exist without a direct sector correlation. For example, if a healthcare sector is damaged, it will not have a wide effect on transportation systems. However, if the transportation systems are damaged and accidents take place, the healthcare sector is also affected.

In real life, here is an example with a cyber security issue: Patient information system is running improperly due to the cyber security interruption, which causes no effect on transportation systems sector; in comparison, the transportation systems are running improperly due to cyber security issue and healthcare must take care of the possible casualties. The damage will increase more rapidly if it hits both or even more of these sectors at the same time or in the short run.

Clearly, such important infrastructure needs good protection. Securing the nation's critical infrastructure from threats including national public healthcare in a digitalized world is not a new idea. In the US, the term critical infrastructure protection (CIP) was defined in the 1990s, although the importance of infrastructure had been noted 30 years prior to this, and it has been studied ever since [13].

Along the digitalization, the terms and definitions evolved during that time also in Finland. According to Kananen [14], the former chief executive officer of Finnish National Emergency Supply Agency, Finland has produced several studies about the resilience of the society's digital systems for decades. Before these studies were connected to the term cyber security, they were known as ICT security, information system security or data network security, just to name a few. However, that time they did not receive the public attention they are receiving today, and some of those studies were classified as confidential material and were never published [14].

Digitalized infrastructure, the part of infrastructure related to information technology is vital in today's society because it is used in monitoring, managing and controlling critical technical infrastructure. Infrastructure such as this includes diverse sectors, e.g. energy production and distribution, monetary systems and traffic, water supply, public governance systems, transportation, logistics and healthcare systems, all critical for the society's stability and safety [14].

However, it is important to remember that although the critical infrastructure is vital for today's nations and whole societies, it is often provided by the private sector, not the public sector [15].

Today when it comes to protecting critical infrastructure, critical information infrastructure protection (CIIP) is discussed more and more. The term CIP, with the extra letter "I", refers to infrastructure sectors and altogether all sectors connected via information infrastructure and therefore, also dependent on it. Consequently, the information infrastructure can be seen as one of the critical information sectors itself [16].

As the Global Forum on Cyber Expertise suggests in their paper The GFCE-MERIDIAN Good Practice Guide on Critical Information Infrastructure Protection for governmental policy-makers, the nations should start to determine the set of possible critical information infrastructure. The paper also suggests that the nations should be prepared for critical information infrastructure crises [17].

Presently, in the world where sectors dependent on information infrastructure from water supply to healthcare could suffer in a wide scale and produce notable damage for the whole society just because of an IT problem. In addition, the digitalized world and critical information infrastructure have produced another challenge: the dependency on cyber.

2   Cyber Security Surveys

In this research three surveys were executed.

2.1   First Survey

The first survey was conducted during the first half of 2017 with cyber security awareness lectures for the employees of Hospital District of South Ostrobothnia. The attendees consisted mostly of nursing staff and their supervisors at the Seinäjoki central hospital. In addition, lectures were held for supervisors training attendees, hospital supporters, mental health personnel and executive groups working in the Hospital District in Seinäjoki City area. After an approximately one-hour lesson the attendees were given the survey forms to be filled in and returned directly back to the lecturer.

The questions of the first survey form consisted of grading the lecture, its importance and usefulness in work and off work as well as giving a grade for the lecturer and for the presentation material. This survey was to give data to show if these kinds of lectures are needed and wanted by the participants and to point out what the weakest links are in the lecture. What is more, these results could be used to prove the importance and effectiveness of the lectures.

2.2   Second Survey

The second survey was a continuum for the first survey and was conducted during the second half of 2017 during the cyber security awareness lectures for the employees of the Hospital District of South Ostrobothnia. After a successful first survey with over hundred respondents, the results were analyzed. As seen in the average numeric results of the first survey (see Fig. 2), the average grades were similar to each other varying only by 0.5 points from 4.4 to 4.9 on scale 1 to 5 with median value of all grades being excellent 5. Therefore, after the 153 answers were tabulated, continuing with the same survey form was thought not to produce any new information.

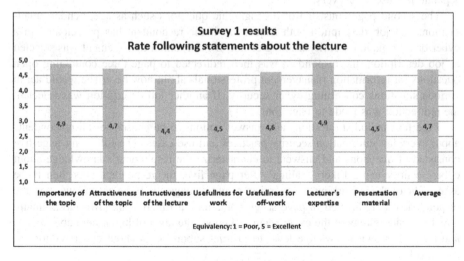

Fig. 2. Rate following statements about the lecture

Because of the results, a change of approach seemed necessarily in order to gather deeper knowledge about the feelings and thoughts of the users about the subject. A new survey form was created with open questions and grading the personal and organizational level of information security. This survey form was then given to the participants after the lectures.

2.3   Third Survey Conducted Online

The third survey had a more ambitious plan than the two previous surveys; to gather data of cyber security knowledge from employees and their supervisors of all organizations that are to join in the upcoming health, social services and regional government reform in South Ostrobothnia.

Gathering data via lectures for the focus group could have taken considerably long time from one person and conducting this kind of survey with paper was estimated not to be feasible, which is why an online survey was created to be sent to the participant organizations. A great opportunity to reach all organizations in the area with improved value and profile arose when the CIO (chief information officer) from the Hospital District of South Ostrobothnia was able to contribute to The Head of Change from Regional Council of South Ostrobothnia to forward the survey to the target organizations.

The survey consisted of eight pages total. The first page was a welcoming page with a welcoming text, the current regional logo for the reform and personal contact information, reminder that the data gathered via the survey will be anonymized and no personalized information will be shown in the final data. The first page included a selection of organizations to be chosen by the respondent. An approximation duration for answering the survey was mentioned as well, based on the average duration of ten separate test users' answers.

The second page consisted of demographic questions such as age, gender, qualification, and job description with a question if the respondent has participated in a cyber or information security training or not. If information management was selected as job description, the respondent was then redirected to page three containing open questions for information management professionals about how he or she would act in various situations concerning cyber security. If another job description was selected, the redirection was made to page four.

Page four included questions and answer options to be chosen for all respondents about everyday information security practices and use cases. The following page five consisted of questions and answers to characterize the respondent's knowledge about cyber security using Likert scaling. After page five, the respondent was then to be asked if he or she is working as supervisor or not on page six. If the answer was yes, the respondent was redirected to page seven containing a Likert scale [18] questionnaire how he or she felt about the cyber security knowledge level of his or her subordinates and his or her own knowledge level to instruct subordinates about cyber security in their work.

The final page eight included questions for all respondents to rate their own knowledge level about cyber security using Likert scaling with a question if they have read the information security policy of their current organization and how they would like to get information concerning cyber security. The last two questions were about how the respondents felt about their own confidential information is secured in the current working organization and how it could be secured in the future organization after the reform. Above the send-button at the bottom of the page, a blank text-box was shown to the respondents with a possibility to give feedback about the survey before submitting the answers.

The survey was published, a test group of ten persons tested the survey to gather feedback. This feedback was essential to ensure the proper operation of the survey and was used to improve the survey's functions, instructions, questions and to find out the average duration the survey will take.

3   Results

The first survey form was answered by 153 participants. The second survey form was answered by 195 participants. In this survey answerer's rated of information security level in own and whole EPSHP organizations (see Fig. 3).

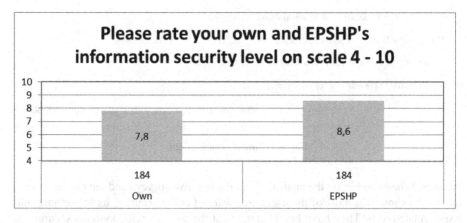

Fig. 3. Rate of information security level in own and whole EPSHP organizations

The third survey form was answered by 881 respondents totally. Besides the large number of answers, a great result was that respondents from all target organizations participated in the online survey (see Fig. 4).

The results were analyzed by using data-driven content analysis. Data-driven content analysis [19] can be used to form interpretations from the content and to increase the value of the gathered information [20]. The results are discussed and analyzed in the following chapter.

3.1   Key Observations and Recommendations

It is often said that the user is the weakest link in the cyber security. Yet, according to the results, the weakest link in cyber security awareness is actually the organization, not

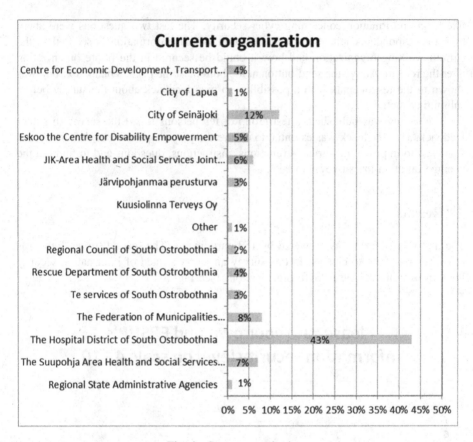

Fig. 4. Current organizations

the user. When combining the analysis from the first two surveys and the online survey, the results show that most of the users are aware of cyber security, its importance and issues related to it. They have knowledge about the use of digital systems securely in theory and they are able to make observations and improvement suggestions from the used infrastructure as well as their own and others' behavior regarding cyber security. The users know that their knowledge about the cyber security should be improved and they want it as well.

Still, the organization has not taken care of improving the cyber security awareness; neither do their employees have enough knowledge for their jobs. What is more, the management does not have enough knowledge to instruct their subordinates in cyber security. It can be questioned if the current level of cyber security awareness and education comply with laws and regulations for personnel security.

There is a contradiction between the feel of personal cyber security level and the answers concerning the actual cyber security awareness. According to the results, the respondent's own cyber security level is felt to be high even as most of the respondents have not participated in any cyber security lecture or training. In addition, many of the respondents answered that they have not read their organization's information security instructions, and most of the respondents are not sure where they could even find them, signaling lack of active cyber security culture in the organizations.

The reasons behind the low amount of people that have participated in cyber security training or lectures are found in the overall level of knowledge. The results signal that the organizations are not committed to improve their cyber security awareness. This is because the importance and meaning of the cyber security awareness to the overall cybersecurity level and to the organization's operation have not been realized. Yet, the results do not include signs about keeping the employee's cyber security awareness level low intentionally.

Based on the results, the prediction for the future is that if the cyber security awareness stays on the current level and will not be improved the lack of the awareness will be causing problems in the target organization's operation sooner or later. In critical infrastructure sector these kind of events can have devastating consequences affecting large amount of valuable information, people and services.

These scenarios include losing control of large amount valuable information or losing confidence to an organization operation or public sector. In the worst case scenario, the patient safety is affected and the lack of cyber security awareness will be risking someone's health indirectly or directly.

What is more, the importance of cyber security awareness is likely to be increased than decreases in the future. An organization or an individual with high cyber security awareness is likely to be more secure against the cyber threats of tomorrow as well.

The lower the level of cyber security awareness is the higher the risks are. Currently, the cyber security awareness is not on acceptable level and the risks are too high. Unfortunately, until the lack of cyber security training and education have proven to have serious consequences such as to be blamed on dangerous situations the resources are not likely to be spent on improving cyber security awareness.

As recommendations, the lacks in cyber security awareness should be taken seriously and the organizations should take immediate actions to improve the cyber security awareness and cyber security related personnel security. Improving the overall cyber security awareness requires that the heads of the organizations have understood the meaning and importance of cyber security awareness to the whole operation of the organizations and they are committed to improve it; i.e., the level of cyber security awareness is improved first among the heads of the organizations and the management level.

After the organization is committed to improve the cyber security awareness, the road map to educate the employees should be created. This road map should include a plan for continuous education and testing the skills and knowledge of all current and future employees in the organization. Participating in the education and testing should be mandatory with no exceptions. Employees from trainees to managers should provide a proof of a basic understanding about cyber security before any access to the organization's digital ecosystem is granted. Advanced education and training should be organized related to the employees' work and tasks. Participating in education and training should be documented and revisable.

Educating employees with lectures about cyber security is one considerable way to improve cyber security awareness in the organizations. The participants experienced the cyber security lectures given as important, educative and instructive as well as useful in work and off-work.

4 Conclusions

This research studied improving cyber security awareness in the selected organizations. The target organizations were located in the region of South Ostrobothnia in Finland; and they were to join in the upcoming Health, social services and regional government reform.

The main objective was to study what the level of cybersecurity awareness and education is in the target organizations and to find reasons affecting them. The results included positive and negative findings.

As positive results, cyber security is seen as important, and the respondents trust their organizations' cyber security and think that the organizations' cyber security is at a high level. In addition, the respondent's own cyber security level is perceived to be high. Most of the respondents have information about the basics of cyber security and they know how to use digital systems and devices securely and avoid common issues in theory.

Yet, according to the results, the overall knowledge is not at a sufficient level in the organizations. The knowledge lacks among the personnel as well as among the management. This analysis is supported by the data gathered; most of the personnel or management have not participated in any training or lectures related to cyber security, many subordinates have answered that they do not have the sufficient knowledge about cyber security for their work, and most of the superiors replied they do not have the sufficient knowledge about cyber security to give guidance to their subordinates in cyber security for their jobs.

Cyber resilience in the target organizations is low and instructions for working in case of computer malfunction or when a critical service for the work is not available are unknown for many. Most of the employees have not read the organization's information security instructions and a significant number of respondents does not know where to find them.

The second objective for this research was to study how the cyber security awareness could be improved in the selected organizations. According to the results, the employees are most willing to participate in cyber security training or lectures onsite.

The onsite lectures given were seen as important and educational by the participants. The results show that persons who had participated in cyber security lectures had better know-how and confidence in cyber security compared to persons who had not participated in any cyber security lectures or training.

The amount of education received is low and lacks behind the level in municipal and governmental sectors in Finland. As a conclusion, cyber security awareness has not been taken care of and should be improved in all target organizations.

References

1. Onyeji, I.: Cyber security and critical energy infrastructure. Electr. J. **27**(2), 52–60 (2014)
2. Clark, R.M.: Protecting water and wastewater utilities from cyber-physical threats. Water Environ. J. **32**(3), 384–391 (2018)
3. Kure, H.I.: An integrated cyber security risk management approach for a cyber-physical system. Appl. Sci. **8**(6), 898 (2018)
4. Parwani, A.V.: Healthcare industry steps up security as cyber attacks increase. Med. Lab. Obs. **49**(11), 56 (2017)
5. Diana, S.C.: The information confidentiality and cyber security in medical institutions. Ann. Univ. Oradea: Econ. Sci. **25**(1), 855–864 (2015)
6. Qi, J.: Demand response and smart buildings: a survey of control, communication, and cyber-physical security. ACM Trans. Cyber-Phys. Syst. **1**(4), 1–25 (2017)
7. Parsons, K.: Determining employee awareness using the human aspects of information security questionnaire (HAIS-Q). Comput. Secur. **42**(C), 165–176 (2014)
8. Lehto, M.: Cyber security education and research in the finland's universities and universities of applied sciences. Int. J. Cyber Warfare Terrorism (IJCWT) **6**(2), 15–31 (2016)
9. Critical Infrastructure, European Comission. https://ec.europa.eu/home-affairs/what-we-do/policies/crisis-and-terrorism/critical-infrastructure_en. Accessed 20 May 2019
10. Homeland, S.: Healthcare and Public Health Sector-Specific Plan. https://www.dhs.gov/sites/default/files/publications/nipp-ssp-healthcare-public-health-2015-508.pdf. Accessed 19 May 2019
11. Lehto, M.: Suomen kyberturvallisuuden nykytila, tavoitetila ja tarvittavat toimenpiteet tavoitetilan saavuttamiseksi. https://tietokayttoon.fi/documents/10616/3866814/30_Suomen+kyberturvallisuuden+nykytila%2C+tavoitetila+ja+tarvittavat+toimenpiteet+tavoitetilan+saavuttamiseksi_.pdf/372d2fd4-5d11-4991-862c-c9ebfc2b3213?version=1.0. Accessed 19 May 2019
12. Janius, R.: Development of a disaster action plan for hospitals in Malaysia pertaining to critical engineering infrastructure risk analysis. Int. J. Disaster Risk Reduct. **21**(C), 168–175 (2017)
13. Lewis, T.G.: Critical Infrastructure Protection in Homeland Security: Defending a Networked Nation, 2nd edn, p. 3. Wiley, Hoboken (2006)
14. Kananen, I.: Suomen huoltovarmuus, Finland's Emergency supply, Docendo, Jyväskylä, pp. 138, 270–271 (2015)
15. Mattioli, R., Levy-Bencheton, C.: Methodologies for the identification of Critical Information Infrastructure, European Union Agency for Network and Information Security. https://www.enisa.europa.eu/publications/methodologies-for-the-identification-of-ciis. Accessed 9 May 2019
16. Personick, S.D., Patterson, C.A.: Critical Information Infrastructure Protection and the Law: An Overview of Key Issues. National Academies Press, Washington (2003). ProQuest Ebook Central, p. 9
17. Luiijf, E., van Schie, T., van Ruijven, A.H.: The GFCE-MERIDIAN good practice guide on critical information infrastructure protection for governmental policy-makers. https://www.thegfce.com/initiatives/c/critical-information-infrastructure-protection-initiative/documents/reports/2017/10/22/the-gfce-meridian-good-practice-guide-on-critical-information-infrastructure-protection-for-governmental-policy-makers. Accessed 5 May 2019
18. Ivanov, O.A.: Likert-scale questionnaires as an educational tool in teaching discrete mathematics. Int. J. Math. Educ. Sci. Technol. **49**(7), 1110–1118 (2018)

19. Schwartz, H.A.: Data-driven content analysis of social media: a systematic overview of automated methods. Ann. Am. Acad. Polit. Soc. Sci. **659**(1), 78–94 (2015)
20. Laadullisen aineiston analyysi ja tulkinta, KAMK University of Applied Sciences. https://www.kamk.fi/fi/opari/Opinnaytetyopakki/Teoreettinen-materiaali/Tukimateriaali/Laadullisen-analyysi-ja-tulkinta. Accessed 21 Aug 2018

Data Delivery Algorithm for Latency Sensitive IoT Application

Omar Abdulkareem Mahmood[1,2], Ammar Muthanna[2,3(✉)],
and Alexander Paramonov[2]

[1] Department of Communications Engineering, College of Engineering,
University of Diyala, Diyala, Iraq
mahmood_omar@list.ru
[2] The Bonch-Bruevich Saint Petersburg State University
of Telecommunications, Saint Petersburg, Russia
ammarexpress@gmail.com, alex-in-spb@yandex.ru
[3] Peoples' Friendship University of Russia (RUDN University),
6 Miklukho-Maklaya St, Moscow 117198, Russia

Abstract. Internet of things represents the third generation of the Internet that is expected to connect billions of heterogeneous devices in a smart way. This large number of connected devices puts high constraints on the system structure and design. In this paper we analyze the features of the IoT networking. The aim of this paper is the Internet of Things organization, as well as the modeling of the message delivery process in conditions of low density of users and in case of low mobile networks coverage. The target network is considered as a network with moving nodes that can perform the functions of data transporting between points with different geographic coordinates. A proposed system in this paper allows to implement an approach for building of such network.

Keywords: Internet of Things · IoT · DTN · MANET

1 Introduction

By 2020 it is expected that the number of connected devices will be higher than 50 billion devices [1]. This enormous number put constraints on communication systems responsible for such collaborations. With great development in Internet of things devices manufacturing and the massive devises exploded in the market such as wearable devices, smart phones and haptic devices; design challenges such as traffic volume and connectivity become critical points in designing next generation networks. The data traffic in 2020 is expected to be 200 times higher than that in 2010 and in 2030 will be 20,000 times higher than that of 2010. The Internet of Things (IoT) is one of the perspective directions for the infocommunication systems development [2, 3].

To enable this massive number of devices and to handle the dramatic increase in traffic new technologies such as MEC, D2D, NFV, SDN and AI should be deployed [4, 5].

Regarding to the IoT conditions, data sources may only be available from their customers, and the volume of data transferred may be relatively low, which is not

© Springer Nature Switzerland AG 2019
O. Galinina et al. (Eds.): NEW2AN 2019/ruSMART 2019, LNCS 11660, pp. 467–480, 2019.
https://doi.org/10.1007/978-3-030-30859-9_40

enough to develop a full-fledged telecommunication system [6]. Data sources can be both automotive and ground, as well as fixed or moving objects. If this is not important, how the IoT device cannot communicate with each other, then the only way to provide communication is the mobility of objects that can act as "couriers" in the delivery of information.

In this paper we show that a promising communication system of the Internet of Things should be a complex "intelligent" system with different telecommunication methods.

A communication system should have a cognitive management system that takes into account external conditions and network operation experience in the previous period of time, which adapts to external conditions by choosing a certain set of scenarios. These scenarios correspond to typical situations that may arise during operation. Naturally, there is a scenario for the normal functioning of the system, which we did not consider in this paper. We believe that under normal operating conditions, the available communication facilities fully satisfy the requirements for data transmission between system elements.

First scenario is in case of high concentration of users and traffic, which is typical for cities with high density of devices, as well as emergency situations. This scenario involves the use of available communication technologies for unloading traffic, for example, from a mobile network (5G) by organizing clusters, transits and boundary calculations in a limited problem area.

The second one is designed to increase the probability of data delivery when there is no coverage of mobile networks or broadband access, and there is also a low density of mobile devices, which makes it impossible to deliver data quickly enough. This situation may be characteristic of areas remote from populated areas.

In this paper we propose models and approaches, which can improve the efficiency of the Internet of Things by increasing the availability of information about its elements.

2   Related Works

In [12] authors proposed a new routing protocol to improve delivery rate and optimize delivery delay with low overhead in DTN for Internet of Thing applications.

In [13, 14] authors select nodes with physical location and mobility behavior. Where routing protocols calculate transmission reliability.

Authors in [15] proposed routing protocol, which consider the social energy of encountering nodes and is in favor of the node with a higher social energy in its or the destination's social community. Authors in [16] introduce DTN routing without a priori mobility knowledge and provide approximation algorithms with theoretical guarantees that can be applied to cases where the number of hops allowed in the routing process is arbitrary.

In [17] authors survey how to use of DTN solutions in Internet of things applications to overcome connectivity problems considering the opportunities and challenges for each technology. There were introduced solutions that enable delay tolerant Internet of Things.

3   System Structure

The peculiarity of IoT is the Internet of things can end up in a wide variety of conditions: as in the conditions of a super-high density network, for example, in conditions of a city; and in a low-density network with limited connectivity, for example, in conditions of low population density and lack of coverage by wireless networks [7, 8]. This can be supplemented by the possibility of a sudden change in operating conditions due to various types of incidents (emergency situations, emergencies, etc.). In those and other conditions, the IoT network must provide data delivery. Under certain conditions, it may be the only means of information delivery.

In addition, different operating conditions or changes in them can lead to a change in the priorities of the IoT [9]. For example, in the normal mode, the target parameters can be such as message time and energy consumption, but in the case of an emergency and a threat to people's lives, energy consumption should no longer be considered as a target.

In other words, the IoT network should be built according to the principles that at the right time provide maximum target characteristics, also depending on the external conditions and the purpose of the network. In particular, under certain conditions, it may be advisable to build an IoT network as a network with acceptable delays.

As noted above, considering IoT as a DTN leads us to consider the principles of building a MANET (Mobile Ad Hoc Network), however, a feature of IoT is that it is not an abstract ad hoc network, but a network with parameters in which it does not otherwise function can.

This "mode" of the network is forced, due to the "desire" of the network to maintain its functionality in "any" conditions. These are conditions where a relatively small number of network nodes can "contact" among themselves. This may be due to the influence of various factors (obstacles and interference to the propagation of signals between nodes) or low network density (remote areas, etc.).

Imagine transporting data from node to node by some hypothetical protocol, Fig. 1. The message is transmitted from node s to node d, not only the nodes included in the delivery route, but also other neighboring nodes, which were not included in the route for various reasons, can be involved. This is understandable, because at the time of sending a message, information about network connectivity and other routing information may be missing. Therefore, depending on the logic of the routing protocol in the network, additional traffic will be generated.

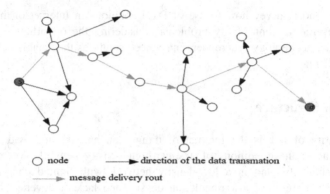

○ node   ——▶ direction of the data transmation
         ——▶ message delivery rout

Fig. 1. Example of data delivery in DTN

Network connectivity and node latency. To describe a network model with an unstable configuration, a random graph model is often used [10]. In many works, the Erdős-Rényi theorem [11] is used to describe networks, which allows us to estimate the network connectivity and the average length of a route through the number of graph nodes (network nodes). In this model, the "threshold" value of the connectivity probability of a node can be determined, the excess of which brings the network into a connected state and a decrease, vice versa. This value is defined as

$$p_0 = \frac{\ln n}{n} \tag{1}$$

Where n- number of the network nodes

The probability of node connectivity for a 2D model, assuming that the node's communication zone is a circle of radius R, can be defined as

$$p_n = \frac{1}{n} \pi R^2 \rho \tag{2}$$

Where ρ - the density of network nodes, $1/m^2$

According to the Erdős-Rényi theorem, if $p_n > p_0$, then the network is more connected, and if $p_n < p_0$ the network is not connected. This formulation suggests that near values there is a phase transition of the network from a disconnected state to a connected state.

We will consider the probability of node connectivity as the probability of its contact with other nodes in the time interval τ. Under the contact we understand the presence of any other network node in the communication zone.

If, per unit of time, λ nodes fall into the communication zone, then the probability of a node's connectivity, taking into account time, can be defined as

$$\hat{p}_n = \frac{1}{n} \lambda \tau \tag{3}$$

The definition of λ may be specific for each specific case, depending on the nature of the movement of network nodes.

The presence of nodes in the communication zone during λ is equivalent to the fact that they will cross the border of the communication zone of the node and will be in it and will stay in the communication zone for a period of time sufficient for the exchange of messages. In this model, we will assume that this time is negligible compared to the waiting time for delivery.

Consider a simplified model of motion when the node under consideration moves in some direction at the speed v (or it is stationary, and all other nodes move, which is equivalent).

Then

$$\lambda(v) = 2R\rho v, \tag{4}$$

Where v - average speed of the node m/s

We will consider the nodes falling into the communication zone as a stream of events with intensity (4), then the probability of falling into the communication zone k nodes can be determined by Poisson's law as

$$p_k(\tau) = \frac{(\lambda\tau)^k}{k!} e^{-\lambda\tau} \tag{5}$$

Where λ is determined according to (4).

This expression, in fact, determines the dependence of the probability of a node's connectivity on time τ given that in this case it will be defined as k/n.

The probability that the number of nodes in a communication zone over time will be at least k will be defined as

$$p_{>k}(\tau) = 1 - \sum_{i=1}^{k} p_k(\tau) \tag{6}$$

Where $p_k(\tau)$ is determined according to (5).

Expression (6), in fact, is the probability distribution of the waiting time until a given probability of connectivity of the node (a certain value of k0) is reached. This dependence is shown in Fig. 2.

The average of the waiting time will be

$$\bar{\tau} = \frac{k_0}{\lambda} \tag{7}$$

For the route of traffic passing through g network nodes, the average time will be, respectively

$$\bar{\tau}_P = g\frac{k_0}{\lambda} \tag{8}$$

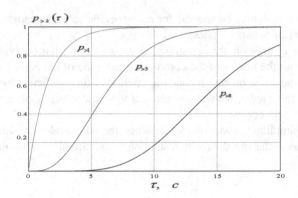

Fig. 2. Dependence of the probability of connectivity of a node with moving nodes from time

The average time values obtained according to (7) and (8) reflect the time required before the onset of a given level of connectivity, defined by the threshold value p0, according to (1). In this model, we will interpret it as the time required to deliver the message to the DTN, with a given probability, which is determined on the basis of the Erdos-Rényi theorem. With sufficient confidence, we can assume that if the probability of a node is connected twice the threshold probability, then the probability of connectivity is so close to one that it can be taken for it in most practical cases.

Route Length and Delivery Time. In the case we focused on at the beginning, the peculiarity of the IoT network is that the density of nodes in the service area is low, i.e. the random graph should be considered strongly sparse. Under such conditions, as shown in [3]. The Erdos-Rényi theorem gives inflated estimates of the route length. In the same paper it is shown that for sparse graph is more adequate evaluation is the use of Bollobas – Riordan theorem.

So in the usual graph (Erdos-Rényi model), the average path length is $\tilde{g}(n) = \ln(n)$, while for the Bollobas-Riordan model, this value is

$$g(n) = \frac{\ln(n)}{\ln(\ln(n))},\tag{9}$$

Figure 3 shows the difference between the two models.

In Fig. 3 above, the average path length in a discharged graph is significantly shorter, and its length (in the number of transits) grows very slowly with increasing number of nodes. Even in a network of several billion nodes, it does not exceed 8 (8 at approximately 100 billion).

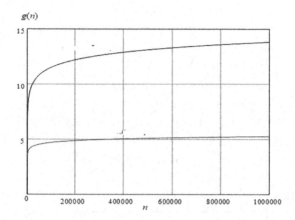

Fig. 3. Dependence of the average path length (in the number of transits) on the number of network nodes

According to the obtained above model for delivery time, taking into account the expected length of the route, you can record

$$\bar{\tau}_P = g(n)\frac{k_0}{\lambda} = \frac{\ln(n)}{\ln(\ln(n))}\frac{k_0}{\lambda} \tag{10}$$

We estimate the value of k0 based on the condition of ensuring connectivity (1), taking into account a certain "reserve" η » .

$$k_0 = \eta p_0 n = \eta \ln(n) \tag{11}$$

As a result, the average delivery time in the interpretation adopted in this model can be estimated as

$$\bar{\tau}_P = \eta\frac{\ln(n)^2}{\lambda \ln(\ln(n))} = \eta\frac{\ln(n)^2}{2R\rho v \ln(\ln(n))} \tag{12}$$

Where - R is the node coupling radius (m), ρ is the density of nodes in the service zone (1/m2), v - average speed of the node m/s.

The value of the coefficient η is selected on the basis of the required "margin" in terms of the probability of connectivity. For most practical applications, $\eta \geq 1$.

Figure 4 shows the results of estimating the average delivery time depending on the average speed of movement of the network nodes for a network consisting of different numbers of nodes.

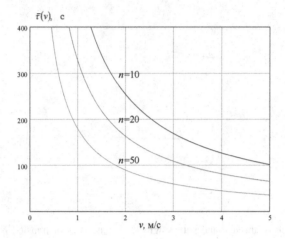

Fig. 4. Dependence of the average path length (in the number of transits) on the number of network nodes

The figure shows that the average delivery time decreases with the growth of the average speed of the nodes and the network capacity (number of nodes) with fixed service area sizes (200 × 200 m), coefficient $\eta = 2$, connection radius R = 50 m.

The obtained results allow us to describe a network with moving nodes as networks with permissible delays and to estimate the value of these delays and their dependence on the average speed of the network nodes movement and the density of nodes.

It should be noted that according to the results (7) given in Fig. 2 any network with arbitrarily low density of nodes allows to provide arbitrarily high connectivity, assuming unlimited time delayed message. This result is expected, since in this model, for a sufficiently long period of time, the necessary number of nodes will probably fall into the communication zone.

The results obtained should be an illustration of the dependencies between the main parameters of the DTN. In specific cases, the implementation of the network, peculiarities of movement should be taken into account.

4   Data Delivery Method in DTN

The model proposed above makes it possible to describe DTN in the general case, in terms of its potential capabilities (within the framework of accepted assumptions). In particular, the model assumes that the network implements methods for finding message delivery routes. These models are indifferent to the types of methods and algorithms for choosing a route. However, these algorithms are important for the functioning of the network and largely determine its effectiveness. In this case, efficiency can be understood as the proportion of useful traffic transmitted in the network and the time of message delivery in relation to a certain hypothetical value obtained, for example, on the basis of the considered model.

As noted above, the operating conditions of Internet of Things devices can be different, both in a high or ultra-high density network and in a low density network. For example, cities are characterized by a high density of devices, while in areas with low population it is likely that low density of Internet of things should be expected. There is also the largest amount of communication available in cities: wireless and broadband access networks, road infrastructure networks, VANET (Vehicle Ad Hoc Network). In conditions of high density, problems arise with the possibility of traffic congestion of networks. To solve these problems, various methods of managing traffic and network configuration can be used.

The opposite problem in low-density conditions is to ensure the data delivery, often in the face of network connectivity problems. In this case, mobile devices (Internet - things, smartphones or other devices) can serve as message carriers. To organize a network, it is necessary to be able to select a message carrier that can deliver it to the recipient or the next transit point. The choice of carrier (transit node) could be made on the basis of some data about the network, while the selection criterion may be to ensure an improvement in its performance.

In general, the choice can be either completely random or fully defined, based on some data. Since the network is built on moving nodes with a random movement, it is likely that the method of selecting transit nodes should also take these accidents into account. The network must be able to identify the state, i.e. choosing a control strategy under conditions of high density of nodes, under normal conditions or under conditions of low density, i.e. it must support the various available technologies and have a cognitive management system that adapts to environmental conditions.

5   Data Delivery on the Network with Allowable Losses

When driving in the absence of coverage of wireless networks and low network density, mobile devices can act as message carriers, i.e. make their physical movement to a certain point from which they can be transferred to the recipient or the next transit node of the network moving in the absence of coverage of wireless communication networks and low-density network. Naturally this possibility depends on the routes of movement of the devices. We can assume that the routes are predetermined, i.e. stored in the device's memory. For example, public transport routes or routes in the memory of navigation systems, unmanned vehicles control systems, etc. In this case, the transmitting node may select the mobile device that follows the most appropriate route, if there is an alternative choice.

If the routes are not predetermined, the choice can be made on the basis of statistical data on possible routes. For example, the control or monitoring system of each mobile device, based on the collection of statistics for a long time, may assess the probability of a mobile device being located in a particular territory, for example, as shown in Fig. 5.

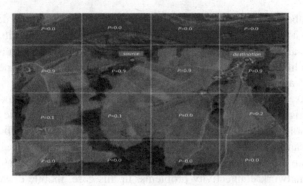

Fig. 5. An example of collecting statistics on routes

All possible delivery area is divided into slices (slice) and for each of the slices is calculated the probability of passing through the route. Most people (with smartphones), personal vehicles and regular vehicles on a daily basis are the same (or similar) routes. Therefore, the daily update of statistics will provide a fairly accurate idea of their possible routes. For example, if n is the number of days of observations, and n0 is the number of days when the car passed through the i-th slice, then

$$p_i = \frac{n_0}{n} \tag{13}$$

Determines the probability that the route will pass through this slice. On this principle, more accurate statistics can be built, for example, taking into account the time, which will determine the likely direction of movement of the vehicle.

Naturally, in such a network message delivery can't be guaranteed. However, the source can also have statistics on the probability of passing and oncoming vehicles and estimate the necessary number of sending messages in order to ensure a sufficient probability of delivery.

$$p_d = 1 - \prod_{i=1}^{k} (1 - p_i) \tag{14}$$

If the delivery probabilities for each vehicle are the same pi = p, then

$$p_d = 1 - (1 - p)^k \tag{15}$$

Where p - the probability of delivery by car, k is the number of shipments.
Then the required number of transmission message will be

$$k(p) = \frac{\lg(1 - p_d)}{\lg(1 - p)} \tag{16}$$

Figure 6 shows the dependence of the number of attempts on the probability of delivery with one attempt p and the required probability of delivery pd.

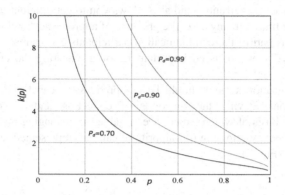

Fig. 6. Dependence of the number of attempts to send the probability of delivery in one attempt (p) and the required probability of delivery pd

The graphs show that the required number of attempts, with the probability of success of one attempt of about 0.5 is from 2 to 6 attempts, which is quite a realistic number.

It is possible more complex routing of messages, with several jumps, in the case when each of the slices are characterized by the probability of delivery to other slices. In other words, when a territory is described by a matrix r × r, where r is the number of slices

$$\mathbf{R} = \{p_{i,j}\}, \quad i,j = 1\ldots r \tag{17}$$

Then the route with the maximum probability of delivery can be found by one of the methods of finding the shortest paths.

To use the search method that uses the additive criterion of the path length, the matrix (6) should be replaced by a matrix composed of the logarithms of the corresponding probabilities

$$\hat{\mathbf{R}} = \{-\lg p_{i,j}\}, \quad i,j = 1\ldots r \tag{18}$$

Thus, promising IoT networks can operate in low density as networks with acceptable delays and losses, also the probability of message delivery can be determined with sufficient probability. This system can be called cognitive, because the information about the routes of its participants is continuously updated on the basis of statistics, i.e. the experience of its work. Next, we consider a possible method of controlling such a system.

6 Management Method of Cognitive Communication System

As shown above, under different conditions, the operation of a DTN can require different methods of organization, and also allows you to get different indicators of the quality of service traffic. In any case, the presence of appropriate "intelligent" means of communication as part of the communication participants can increase the probability of message delivery, which is an important factor in ensuring the efficiency of the system.

The communication system should be comprehensive, capable of analyzing the external conditions, having the functions of collecting and analyzing statistics (training), on the basis of which experience is gained in managing this system.

In general, the system control algorithm can be represented as a simplified algorithm below (Fig. 7).

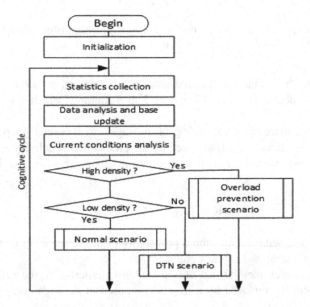

Fig. 7. IoT cognitive management algorithm

The algorithm above is an example of the implementation of communication system management and is an endless cycle of data collection and replenishment of experience, which are used when choosing one of the network operation scenarios. In this case, there are three scenarios: normal - corresponding to the normal situation with normal user density and traffic intensity; scenario of preventing overloads is selected in case of high density of users or overloads (possible overloads) of available communication channels; network scenario with allowable losses is selected in case of lack of coverage of mobile networks or other means of communication.

7 Conclusion

The proposed models and approaches can improve the efficiency of the Internet of Things by increasing the availability of information about its elements.

The paper results and the selection model for describing IoT shows that on a network, even with a very large number of nodes (if we compare it, for example, with the number of people on the Earth), the route between nodes (on average) does not exceed 8 transits.

We proposed a new architecture for organize Internet of Things networks and provided modeling for message delivery process in for areas with low density of users and in case of low mobile networks coverage.

Acknowledgement. The publication has been prepared with the support of the "RUDN University Program 5-100".

References

1. Recommendation ITU-R M.2083: IMT Vision—Framework and overall objectives of the future development of IMT for 2020 and beyond, September 2015
2. RecomendationY.2066 Common requirements of the Internet of things. Telecommunication Standardization Sector of ITU, Geneva (2014)
3. RecomendationY.2068 Functional framework and capabilities of the Internet of things. Telecommunication Standardization Sector of ITU, Geneva (2015)
4. Andreev, S., Pyattaev, A., Johnsson, K., Galinina, O., Koucheryavy, Y.: Cellular traffic offloading onto network-assisted device-to-device connections. IEEE Commun. Mag. **52**(4), 20–31 (2014)
5. Muthanna, A., et al.: Analytical evaluation of D2D connectivity potential in 5G wireless systems. In: Galinina, O., Balandin, S., Koucheryavy, Y. (eds.) NEW2AN/ruSMART -2016. LNCS, vol. 9870, pp. 395–403. Springer, Cham (2016). https://doi.org/10.1007/978-3-319-46301-8_33
6. Farhan, L., Kharel, R., Kaiwartya, O., Hammoudeh, M., Adebisi, B.: Towards green computing for Internet of things: energy oriented path and message scheduling approach. Sustain. Cities Soc. **38**, 195–204 (2018)
7. Ateya, A., Muthanna, A., Koucheryavy, A.: 5G framework based on multi-level edge computing with D2D enabled communication. In: Advanced Communication Technology (ICACT) (2018)
8. Andreev, S., Moltchanov, D., Galinina, O., Pyattaev, A., Ometov, A., Koucheryavy, Y.: Network-assisted device-to-device connectivity: contemporary vision and open challenges. In: Proceedings of 21st European Wireless Conference European Wireless (2015)
9. Hammoudeh, M., Newman, R.: Adaptive routing in wireless sensor networks: QoS optimisation for enhanced application performance. Inf. Fusion **22**, 3–15 (2015)
10. Erdős, P., Rényi, A.: On the evolution of random graphs. Publ. Math. Inst. Hungar. Acad. Sci. **5**, 17–61 (1960)
11. Erdős, P., Rényi, A.: On the evolution of random graphs. Bull. Inst. Int. Statist. Tokyo. **38**, 343–347 (1961)

12. Mao, Y., Zhou, C., Ling, Y., Lloret, J.: An optimized probabilistic delay tolerant network (DTN) routing protocol based on scheduling mechanism for internet of things (IoT). Sensors **19**, 243 (2019)

13. Lebrun, J., Chuah, C.N., Ghosal, D., Zhang, M.: Knowledge-based opportunistic forwarding in vehicular wireless ad hoc networks. In: Proceedings of the 2005 IEEE 61st Vehicular Technology Conference; Stockholm, Sweden. 30 May – 1 June 2005, pp. 2289–2293 (2005)

14. Leguay, J., Friedman, T., Conan, V.: Evaluating Mobility Pattern Space Routing for DTNs. arXiv. cs/0511102 12 (2006)

15. Basilico, N., Cesana, M., Gatti, N.: Algorithms to find two-hop routing policies in multiclass delay tolerant network. IEEE Trans. Wirel. Commun. **15**, 4017–4031 (2016). https://doi.org/10.1109/TWC.2016.2532859

16. Li, F., Jiang, H., Wang, Y., Hashang, L., Cheng, Y.: SEBAR: social energy based routing scheme for mobile social delay tolerant networks. IEEE Trans. Veh. Technol. **66**, 7195–7206 (2017). https://doi.org/10.1109/TVT.2017.2653843

17. Benhamida, F.Z., Bouabdellah, A., Challal, Y.: Using delay tolerant network for the Internet of things: opportunities and challenges. In: Conference: 2017 8th International Conference on Information and Communication Systems (ICICS) (2017)

Development of the Mechanism of Assessing Cyber Risks in the Internet of Things Projects

Sergei Grishunin[1]([⊠]), Svetlana Suloeva[1]([⊠]), Tatiana Nekrasova[1]([⊠]), and Alexandra Egorova[2]([⊠])

[1] St. Petersburg State Polytechnic University, St. Petersburg, Russia
sergei.v.grishunin@gmail.com, emm@spbstu.ru,
dean@fem.spbstu.ru
[2] Moscow State University, Moscow, Russia
alxegorova@gmail.com

Abstract. We developed the mechanism of assessing cyber risks for Internet of Things (IoT) projects. The relevance of this topic is explained by growing sophistication of cyber-attacks, the speed of new threats emergence and increasing damage from the attacks. The paper addresses decreasing efficiencies of existing mechanisms of cyber risk assessment and fills the research gaps in this area. Results include development of the mechanism's concept, its block diagram, the specification and description of its comprising tools and the case study. Unlike peers, the mechanism provided holistic approach to cyber risk assessment; integrated and coordinated all related activities and tools. It simulated the confidence interval of project return on investments (ROI) and showing the chances to go above risk appetite. It makes cyber risk assessment dynamic, iterative, responsive to changes in cyber environment. These advantages let us conclude that the mechanism should have a significant scientific and practical use.

Keywords: Internet of Things · Cyber risks · Cybersecurity · Risk controlling

1 Introduction

The Internet of Things (IoT) provides a wide avenue for innovations from the industrial use to healthcare and consumer. However, IoT can create significant risks for developers and users. The number and frequency of IoT attacks has been increasing while the direct and indirect damages have been rising. A single infected device can open the entire company ecosystem for attack, with potential disruptions ranging from individual privacy breaches to massive breakdown of public systems and threat to human life.

In such an environment, the procedures of cyber risk assessment must be integrated and coordinated in the single mechanism, which relies on combination of methods and tools and provide the range of likely monetary losses from the cybercrime during a given period. Such mechanism should ensure timely identification and assessment of threats, anticipation of likely new threats as well as the development and implementation of risk mitigation decisions. However, our research shows that many existing

© Springer Nature Switzerland AG 2019
O. Galinina et al. (Eds.): NEW2AN 2019/ruSMART 2019, LNCS 11660, pp. 481–494, 2019.
https://doi.org/10.1007/978-3-030-30859-9_41

mechanisms such as operationally critical threat-asset-vulnerability evaluation (OCTAVE) or cyber value at risk (CyVAR) only partially serve these purposes.

We closed the gaps in the research and developed the mechanism of cyber risk assessment in IoT projects. The novelty of the paper is driven by the advantages of the mechanism over its peers. Unlike peers, which estimate a single point estimate of risk impact, the mechanism simulates the confidential interval for project's ROI and shows the chances to go beyond risk appetite. It includes tools and methods that allow estimating risk frequencies with few data points. The mechanism provides the holistic approach for cyber risk assessment, integrates and co-ordinates all the activities of cyber risk assessment. It makes cyber risk assessment dynamic, iterative, responsive to changes in cyber environment.

In Sects. 2 and 3 we present the outlook for IoT and explain advantages of risk controlling application in cyber risk management. Section four presents the literature review in the area and identifies the research gaps. In Sects. 5 and 6 the mechanism of cyber risk assessment is developed and a case study is provided. In Sect. 7 the advantages of the mechanism are discussed and conclusions are formulated.

2 Internet of Things: Outlook and Challenges

The Internet of Things (IoT) is a suite of technologies and applications that equip devices and locations to generate data and information and to connect those devices and locations for instant data analysis and, ideally, "smart" action [2]. IoT implies physical objects being able to utilize the Internet to communicate data about their conditions, position or other attributes. In IoT, information and communication industries have merged together and formed info-communication space [3].

The number of IoT-connected devices will grow at compound annual rate of 15% to reach 31 billion units by 2020 with the estimated market value of $1.1 trillion [2]. It is fueled by (1) declining prices for bandwidth, data storage and computing; and (2) growing usage of augmented intelligence; and penetration of industrial robots. Industrial IoT units (such as devices for the conditions-based monitoring and predictive maintenance of capital assets) will capture of around 50% of global IoT spending [2]. The consumer, health or public services devices will take a share of 25% each [2].

Along with these opportunities, this industry is characterized by significant challenges that could stop investing in the sector. They include (1) lack of infrastructure to manage devices; (2) threat of users' non-acceptance of devices; (3) poor vendor execution, (4) operational problems; or (5) lack of regulation [2]. The most dangerous threats is the growing number of cyber-attacks on IoT units; the cyber-crime alone costs nations more than $1 billion globally [5–7]. Examples include (1) the distributed denial of services (DDOS) attacks; (2) data and identity theft; (3) reconnaissance attacks; (4) man in the middle penetration: (5) Trojans and viruses; and others [1]. Growing complexity, interconnection and pervasiveness of IoT expose these devices to new type of hazards that existing risk management methods are neither designed to anticipate nor predict [5, 8]. These challenges are forcing IoT developers, vendors and users to reconsider the approaches to cyber risk management [8] and switch to newest systems such as the project risk controlling [9, 11].

3   Risk Controlling in IoT Development Projects

Risk controlling (RC) is a goals-oriented set of methods, processes and tools for risk management in IoT development projects, the integral part of investment controlling [9, 11, 19]. RC provides the architecture (infrastructure and processes) of risk management while the project managers applying this infrastructure to particular risks make risk-informed decisions. The functions of RC are listed in [19]. Advantages of RC over the commonly applied integrated risk management are (1) fostering risk governance; (2) integration of risk management into the decision-making at all stages of the project; (3) co-ordination of all risk management activities; and (4) application of tools with low risk tolerance and increased focus on quantitative assessment of risks [9, 19].

In cyber security, RC is aimed to reduce the risk that the users of IoT solution fail to achieve the target return on investments (ROI) due to losses from the cybercrime [11]. The more complex the IoT project is the higher the gap can be between the realized and the target ROI (Fig. 1).

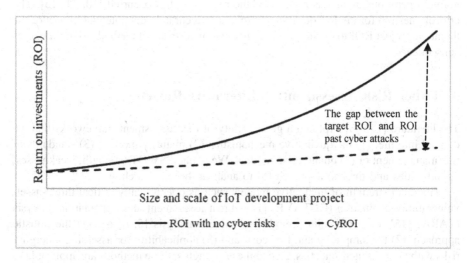

Fig. 1. Impact of cyber-attacks on IoT project ROI

This is underpinned by the increasing with the scale the attractiveness of the device to the attackers and the growing sophistication of the attacks resulting as increase in cost of controls and remediation. [10] To reflect the cyber exposures we developed the cyber ROI (CyROI) metric that measured the effectiveness of investments into IoT given cybercrime and related controls. For one-year horizon:

$$CyROI = \frac{(B - CL \times ME - C_{cs}) - (I_{IoT} + I_s)}{I_{IoT} + I_s} \tag{1}$$

Where, B – customer's benefits from application of IoT device; CL – losses from the cybercrime; ME – mitigation ratio, given the cybersecurity solution; I_{IoT} –

customer's investments into IoT device; I_s – customer's investments into development of cybersecurity solution; C_{cs} – maintenance costs of cybersecurity solution.

Consequently, for lengthy development projects with duration of more than one year, calculation of CyROI involves discounting for project's weighted average cost of capital (WACC):

$$CyROI = \frac{\sum_{t=1}^{T} \frac{(B_t - CL_t \times ME - C_{cs,t})}{(1 + WACC)^t} - (I_{IoT} + I_s)}{I_{IoT} + I_s} \tag{2}$$

Where, t –the period of analysis; T – total number of periods.

To reduce this gap, IoT developers need to optimize the relationship among the benefits from adoption of technology; (2) residual losses from cyberattacks given control system; and (3) investments into control and remediation systems (cybersecurity solutions). Risk controlling system should (1) prevent and anticipate threats before they take hold; (2) monitor and neutralize risks already in play; and (3) restore normal operations as fast as possible if the risk event has occurred [10, 11, 13]. The critical question for the IoT developer is to assess accurately the potential impact of the threat on project ROI to decide on cost-effective measures and methods to minimize its consequences.

4   Cyber Risk Assessment: A Literature Review

The literature [1, 5–8] explored a great variety of risk assessment frameworks that we divided into the: (1) the qualitative mechanisms; (2) maturity models; (3) standards of risk management or (4) quantitative models. We applied SWOT (strength, weaknesses, opportunities and threats) approach [5] to analyze these researches.

The main strength of qualitative models such as operationally critical threat-asset-vulnerability evaluation (OCTAVE) [14], threat assessment and remediation analysis (TARA) [15], or cyber failure mode and effects analysis [5, 6] are (1) the holistic approach; (2) the simplicity and low cost; and (3) applicability for assessing emerging risks with no or limited statistics. Due to these strengths, these methods are applicable to small and medium IoT projects. Their weaknesses include: (1) qualitative interpretation of risk probability and impact: (2) usage of single point estimates; (3) simplification of the correlations among risks and calculation of aggregate exposure; and (4) absence of linkage between the risk impact and project targets [18]. The resulting threats are risk-acceptance inconsistency, range compression or centering bias [16]. The opportunities for improvements include (1) extensions to quantitative risk assessment; or (2) adding fuzzy logic that improves the integration of the opinions given by experts [4, 17]. Still, these improvements are not sufficient to mitigate weaknesses.

The strengths of risk management maturity models (RMM) such as the capability maturity model integrated or Exostar [5] is providing assessment of maturity of IoT cyber risk management system and identification of the gaps. Their disadvantage is the focus on pointing out vulnerabilities without assessing the magnitude of exposures in these weak spots and the impact on project's targets. The opportunity for RMMs is to be integrated with other approaches.

The strengths of information security standards such as ISO 27001 or NIST [12, 20] are that they (1) are auditable and widely recognized international standards in cybersecurity; and (2) they provide the holistic framework for organizing cyber risk management [5, 12]. However, they do not provide detailed models and tools for risk assessment and the guidance how the models can be applied in risk-oriented decision-making.

Lastly, the stochastic quantitative models (SQM) such as cyber value at risk (CyVAR) use probability theory to estimate the confidence interval of likely losses from cybercrime during the given timeframe [5]. The strengths of SQM are that they provide the quantitative assessment of losses with simulation of a very large number of scenarios. The limitations of existing SQM approaches [1, 5] are that they (1) provide only risk assessment tool but not a holistic risk assessment mechanism; (2) rarely access the impact of losses on project's targets; (3) may result in threats of overlooking of the emerging risks due to lack of data [5]. The opportunity for SQMs is development into full-scale mechanisms of risk management and risk assessment.

To summarize, the existing risk assessment frameworks in cyber area are constrained by a number of critical limitations. We will close these gaps in the research by development of the risk assessment mechanism based on principles of risk controlling.

5 Development of the Mechanism of IoT Cyber Risk Assessment

The developed block diagram of the mechanism is presented at Fig. 2. The prerequisite for applying the mechanism is the cyber risk controlling system in the project company. Such system can be built on the base of ISO 27000 or NIST [5, 20]. In addition, a clear understanding of the company's business drivers, security considerations as well as legal, regulatory and contractual requirements specific to its use of a particular IoT technology is required.

5.1 Mechanism Inputs and Risks Identification Step

The first inputs of the mechanism are business characteristics of IoT device that are necessary for further criticality and vulnerability assessment and calculation of CyROI. The second input of the mechanism is the assessment of the criticality of IoT asset for the project company. We applied additive-multiplicative scoring model for this purpose [21]. The model profiles IoT asset by critical factors in the several dimensions. They have influence on company's key business processes and operations, outsiders (customers, suppliers) and personnel. The others include (1) ties and interdependence with other critical information assets; (2) direct and indirect cost of IoT failure (including reputation, regulatory impact and goodwill); (3) cost of information loss and its recovery; (4) time and cost of asset return to normal operations; and (5) investments into IoT rehabilitation. Behind these, confidentiality, integrity and availability must be also considered [12].

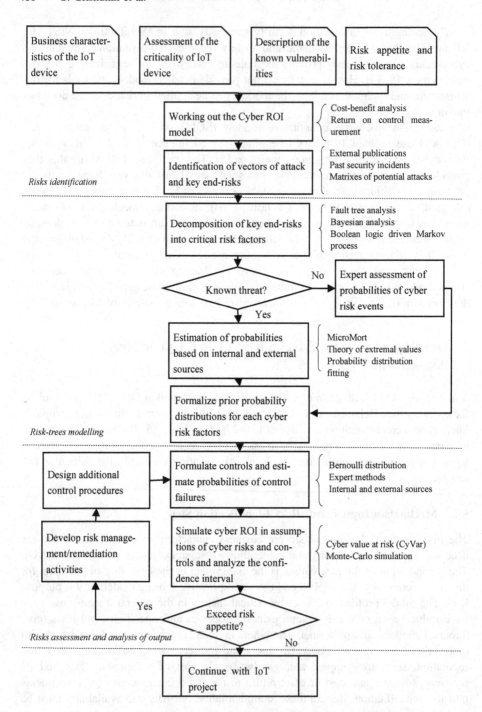

Fig. 2. The block diagram of the developed risk assessment mechanism

The second input is the results of IoT vulnerability assessment. It is done with application of RMM and vulnerability scanning software [5]. The former benchmarks the maturity of IoT cyber security model against the standards [12, 20], allowing identifying the weaknesses in the security system in general, and the specific areas for scanning and helps to decide on areas of improvement. The scanning software identifies the actual vulnerabilities in the weak spots and helps assessing their severity. The last inputs are the risk capacity ($CyROI_T$) and risk appetite ($CyROI_R$) thresholds. The former is the total cyber risk that the company can bear while risk capacity is the level of cyber risks that it can accept. There are characterized by the probabilities of achievements - γ or δ respectively. We set γ and δ at 90% and 95% respectively. After all inputs are collected, CyROI model is worked out with using the formulas (1, 2) and approach [4].

In the next step, the potential vectors of attack and end-risks are identified for each selected vulnerability. This analysis starts from identification of the generic vectors. The sources for this are (1) the publications of reputable organizations such as Verizon's Data Breach Investigation Report, Symantec or FireEye reports [21]; (2) the analysis of the internal or external databases of the past cyber incidents; or (3) expert knowledge. Then, the vectors are customized for the particular IoT and the matrices of potential attacks are formed [21]. The result is the set of end-risks $\{R_i\}_{i \in 1,N}$.

5.2 The Modelling of Risks Trees and Probabilities of Risk Factors

At this stage, each end-risk is decomposed into the key risk factors and for the latter the probability distributions are identified. This is done by the bowtie tree [19] (Fig. 3). The right side of the diagram is the CyROI model developed at the previous stage. The left side of the diagram is the causal network, where the ending node is the end-risk, the leave nodes represent the most credible risk factors (C_i) and conditions to achieve the malefactor's objectives, and the bottom nodes are the initial sets of the cyber-attack.

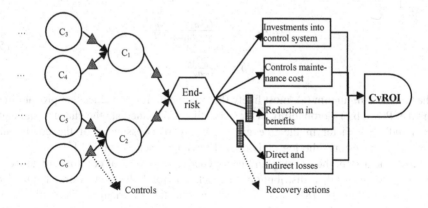

Fig. 3. Example of bowtie diagram

To build the casual network the fault tree analysis (FTA) is applied [17]. This can be done with construction of Bayesian network.

$$P(C_1...C_M) = \prod_{j=1}^{M} P(C_j/pa(C_j)) \qquad (3)$$

Where M – the number of risk factor; C_j – j-th risk factor, $P(C_1...C_M)$ – joint probability distribution of all risk factors and $P(C_j/pa(C_j))$ – conditional probability of j-th risk factor given ancestors of C_j; j- the number of risk factor. For bottom risk factors in the tree, the probability distribution function is determined. In rare cases, there are data of sufficient depth and length ($\geq$ 5 years) and there are expectations that the risks will repeat in the future. In this case, distribution fitting technique [5] is applied. The commonly applied distributions in this case are Poisson, Weibull, lognormal or distributions from theory of extreme values.

Often there are very limited internal information about past cyber events or there are some evidence from the reputable sources (such as Cybersecurity, Ventures, Kaspersky Lab, Verizon, Symantec or others), partners, customers or peers. In this case, the MicroMort approach is used [5] and beta probability distribution is applied [18].

$$P(X/\alpha, \beta) = \frac{x^{\alpha-1}(1-x)^{\beta-1}}{B(\alpha,\beta)}$$
$$B(\alpha, \beta) = \int_0^1 t^{\alpha-1}(1-t)^{\beta-1}dt \qquad (4)$$

Where: α – the number of evidences in which the cyber threat was detected in the period, β – the number of evidences in which cyber threat were not detected, $P(X/\alpha, \beta)$ the probability of cyber threat,

In "zero-day exploit" situation when the new vulnerability is discovered and no data are available, the probability distribution is determined by experts. In this case, the beta-PERT (program evaluation and review technique) distribution is applied:

$$P(X/a, b.c) = \frac{(x-a)^{\alpha-1}(c-x)^{\beta-1}}{B(\alpha,\beta)(c-a)^{\alpha+\beta-1}}$$
$$\alpha = \frac{4b+c-5a}{c-a} \qquad (5)$$
$$\beta = \frac{5c-a-4b}{c-a}$$

Where, a,b,c – minimum (a), most likely (b) and maximum (c) values that probabilities can take; $P(X/a,b,c)$ – probability of the cyber threat. To calculate the chances (a,b,c) of the "child" risk factor in the tree given the expert assessments of the conditional chances of "ancestors" the log odds ratio (LOR) approach is applied [18].

At the next step, the existing and new cyber security critical controls are added to the tree and probability distributions of each control failure are worked out. The Bernoulli distribution [18] in case of proper statistics or triangle distribution with parameters assessed by experts (in absence of statistics) is the most common in this case. Finally, the correlations between end-risks should be established. These are calculated from the past statistics (if available) or evaluated by experts.

5.3   Risk Assessment and Analysis of Output

Once the previous steps are completed, the Monte Carlo simulation is run. It can be done in MS Excel with installed @risk modelling engine. The outcome of the simulation is (1) a range of possible values of CyROI given risks; and (2) CyROI descriptive statistics. The analysis of the outcome includes assessment of (1) expected variance of CyROI from planned ROI; (2) the most probable value of CyROI; (3) what end-risks and risk factors contributed the most to the deviation of outcome. If the lower bound of δ-confidence interval of CyROI is below $CyROI_R$, than the possible cyber losses are not acceptable for the company and the IoT project should be abandoned or sent back for rework and remediation of vulnerabilities. If the lower bound of γ-confidence interval of CyROI is above $CyROI_R$ but below $CyROI_T$ than the possible cyber losses are above the level the company is willing to tolerate. Additional control procedures should be introduced, the reliability of the existing control should be increased and remediation measures against vulnerabilities should be performed. After these measures are taken the Monte-Carlo simulation is run again to ensure that the range conditions are met. Analysis of the output helps to determine: (1) what reserves should be maintained in case of realization of adverse scenarios; (2) what are the key risk areas to concentrate attention; (3) what is the most optimistic and pessimistic scenarios of CyROI; and (4) what contingency plans need to be developed.

5.4   Advantages and Novelties of the Mechanism

The mechanism has important advantages over its peers such as RMM, OCTAVE or CyVAR (Table 1).

Table 1. Advantages of the mechanism over its peers

Peers	Developed mechanism
Qualitative assessment. Risk are single-points estimates with fix-value assumptions	Quantitative risk assessment. generates confidential interval of CyROI
Do not predict the chance of achievement of project's target ROI given cyber risks	Predicts the chance of achievement of project's target ROI given cyber risks
Weak analysis of the impact of each risk factor on potential variances of ROI	Shows the impact of each risk factor on potential variances of CyROI
Limited number of scenarios	Monte-Carlo simulation
Do not estimate the chances of going beyond the risk appetite	Estimates the chances of going beyond the risk appetite and risk tolerance
Difficulties in calculation of aggregated exposures	Calculates aggregated risk exposures given correlations among risks
Weak co-ordination and integration of risk assessment procedures	Integrates and co-ordinates all processes, activities and tools of cyber risk assessment
Do not model the probabilities of control failures	Model the probabilities of control failures
Require the data of sufficient depth and length to quantify the probabilities	Applies MicroMort approach allowing to quantify probabilities with few data points
Often do not provide the holistic approach of risk analysis	Provides the holistic approach of risk analysis

6   The Case Study Example for IoT Cyber Risk Assessment

The case study examines application of the mechanism in the IoT project company that develops preventive equipment maintenance system for the smartphone assembly plant. The project team concluded that the device will be exposed to DDOS attacks, remote malware code execution and related theft of intellectual property and hardware attacks (destruction of sensors). The resulting bowtie diagram is presented in Fig. 4.

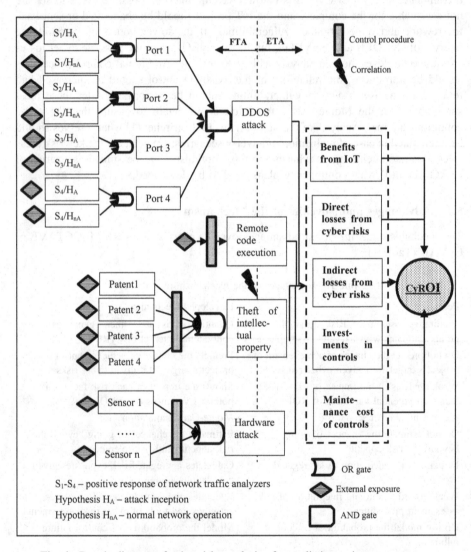

Fig. 4. Bowtie diagram of cyber risks analysis of a predictive maintenance system

To perform the calculations, the project team made following inputs to the mechanism (Table 2).

Table 2. Inputs for the risk assessment mechanism

End-risks	Modelling of probabilities			Modelling of impact
	Prior probabilities of risks	Posterior probabilities of risks	Probabilities of control failures	Probability of impact values
DDOS attack	Beta distribution with inputs from Gartner report	Bayesian network	Bayesian network	Triangle distributions with expert inputs
Remote code execution	Triangle distribution with expert inputs	–	Bernoulli distribution, inputs from internal statistics	Lognormal distribution, inputs from internal statistics
Theft of intellectual assets	Triangle distribution with expert inputs	Application of Boolean algebra	Bernoulli distribution, inputs from internal statistics	Triangle distribution with expert inputs
Hardware physical attack	PERT distribution	Log odds ratio	Bernoulli distribution, inputs from internal statistics	PERT distribution, inputs from internal statistics

The resulted risk CyROI probability analysis charts following the Monte-Carlo simulation are at Fig. 5.

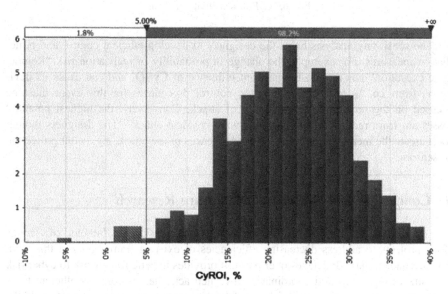

Fig. 5. CyROI probability analysis

Chart demonstrates that despite of potential cyber risk impact there is 98% probability that the value of CyROI will be above 5%. Thus, the developed system satisfies the customers' requirement of not to exceed risk appetite. Moreover, chart demonstrates that planned value of CyROI (10% and more) is not reached with 5% probability and exceeded with 95% probability.

CyROI sensitivity analysis shows the each end-risk influence on CyROI is at Fig. 6.

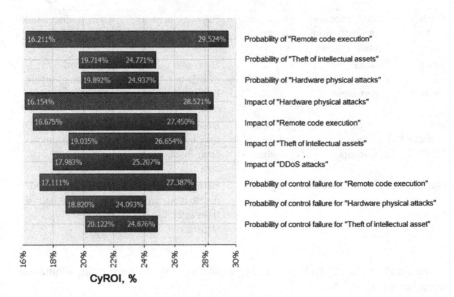

Fig. 6. CyROI sensitivity analysis

The sensitivity analysis helps the designers to develop efficient control and remediation measures. For example, the change in probability of realization risk "Remote code execution" has the most significant influence of CyROI and can cause its deviations from 16.2% to 29.2%. Thus, the control procedures for this event must be focused on the reduction of probability of attack. Conversely, the highest potential losses are from realization of risk "Hardware physical attack". The designers though must create the measures reducing the consequences of the attack, e.g. better protection of sensors.

7   Conclusion and Directions for Future Research

We developed the mechanism for assessing cyber risks for Internet of Things (IoT) projects to address decreasing efficiencies of existing frameworks in this area. The mechanism has advantages over peers. It provides holistic framework to cyber risk assessment; integrates and coordinates all related activities. It contains efficient tools and methods that quantify cyber risks, analyze their impact on project's target, build the distribution of project ROI and analyze the chances of going beyond risk appetite. These advantages make cyber risk assessment dynamic, iterative, responsive to changes in cyber environment and emerging of new threats. The directions of future research will be further elaboration of mechanism's tools such as (1) fitting appropriate probability distributions for different types of cyber risks by the available data; (2) enhancing models for identification attack vectors and evaluation of vulnerabilities; and (3) developing the advanced models of expert opinion calibration and probabilities assessment.

References

1. Abomhara, M., Koien, G.: Cyber security and internet of things: vulnerabilities, threats, intruders and attacks. J. Cyber Secur. **4**, 65–68 (2015)
2. Deloitte Inside. The Internet of Things. A technical primer (2018). https://www2.deloitte.com/insights/us/en/focus/Internet of Things/technical-primer.html. Accessed 2 Mar 2019
3. Glukhov, V., Balashova, E.: Economics and Management in Info-Communication: Tutorial. Piter SPb, St. Petersburg (2012)
4. Grichounine, S.: Developing the mechanism of qualitative risk assessment in strategic controlling. SPbSPU J. Econ. **10**(2), 64–74 (2017)
5. Radanliev, P., et al.: Future developments in cyber risk assessment for the Internet of things. Comput. Ind. **102**, 14–22 (2018)
6. Ralston, P.A.S., Graham, J.H., Hieb, J.L.: Cyber security risk assessment for SCADA and DCS networks. ISA Trans. **46**, 583–594 (2007)
7. Cherdantseva, Y., Burnap, P., et al.: A review of cyber security risk assessment methods for SCADA systems. Comput. Secur. **56**, 1–27 (2016)
8. Nurse, S., Greese, S., De Roure, D.C.: Security risk assessment in internet of things systems. IT Prof. **19**(5), 20–26 (2017)
9. Grishunin, S., Mukhanova, N., Suloeva, S.: Development of concept of risk controlling for industrial enterprise. Organ. Prod. **26**(1), 45–46 (2018)
10. Antonucci, D.: The cyber risk handbook: creating and measuring effective cyber-security capabilities. Wiley, Hoboken (2017)
11. Filko, S., Filko, I.: Risk Controlling of Information Security. Accounting, Analysis and Audit: Theoretical and Practical Problems. SSAU 16, pp. 123–127 (2016)
12. ISO/IEC 27005:2013.: Information technology - security techniques - information security risk management. International Organization for Standardization (2005)
13. Abie, H., Balashingham, I: Risk-based adaptive security for smart IoT in e-health. In: Proceedings of the 7th Conference on Body Area Networks, Oslo, pp. 269–275 (2002)
14. Caralli, R., Stevens, J., Young, L., Wilson, W.: Introducing OCTAVE: Improving the Information Security Risk Assessment Process. Hansom AFB, MA (2007)
15. Wynn, J., et al.: Threat assessment and remediation analysis methodology, Bedford (2011)
16. Thomas, P., Bickel, J., Bratvold, R.: The risk of using risk matrices. SPE Econ. Manag. **6**, 56–66 (2013)
17. Gusmao, A., Poleto, T., Silva, M., Silva, L.: Cybersecurity risk analysis model using fault tree analysis and fuzzy decision theory. Int. J. Inf. Manag. **43**(6), 248–260 (2018)
18. Hubbard, D., Seiersen, R.: How to measure Anything in Cybersecurity Risk. Wiley, Hoboken (2016)

19. Grishunin, S., Suloeva, S., NekrasovaT, T.: Development of the mechanism of risk-adjusted scheduling and cost budgeting of R&D projects in telecommunications. In: Galinina, O., Andreev, S., Balandin, S., Koucheryavy, Y. (eds.) NEW2AN 2018, ruSMART 2018. LNCS, vol. 11118, pp. 456–470. Springer, Cham (2018). https://doi.org/10.1007/978-3-030-01168-0_41
20. Framework for improving critical infrastructure cybersecurity. National Institute of Standards and Technology (2018)
21. Kotenko, I., Chechulin, A.: A cyber attack modeling and impact assessment framework. In: 5th Conference on Cyber Conflict Proceedings, pp. 1–24. IEEE, Tallinn (2013)

Engineering and Architecture Building of 5G Network for Business Model of High Level Mobile Virtual Network Operator

Valery Tikhvinskiy[1,2](✉), Sergey Terentyev[2](✉),
Altay Aitmagambetov[3](✉), and Bolat Nurgozhin[4](✉)

[1] Moscow Technical University of Communications and Informatics (MTUCI),
Moscow, Russia
vtniir@mail.ru
[2] JSC National R&D Institute of Technologies and Communications (NIITC),
Moscow, Russia
s.ter@mail.ru
[3] JSC International Information Technology University (IITU),
Almaty, Kazakhstan
altayzf@mail.ru
[4] JSC Almaty University of Power Engineering and Telecommunications
(AUES), Almaty, Kazakhstan
binl5aues@gmail.com

Abstract. The Authors have analyzed features of mobile virtual network operators (MVNO) building following levels: high, low and reseller for different business model taking into account utilization new technological and architectural possibilities of 5G networks. During the research work topology and structure of MVNO high level based on 5G network functions were designed. The proposed structure of MVNO based on sharing 5G network resource: 5G MOCN (5G Multi-Operator Core Network) и 5G GWCN (5G Gateway Core Network). In the aim of seamless radio coverage realization by multi standard access network were considered issues of MVNO-network integration with base stations 4G/5G. Joint utilization of numeration and addressing for mobile network operator (MNO) and High level MVNO were systematized in the article. The principles of search and selection by a subscriber device of a network of a high-level virtual mobile operator, the principle of selecting network functions (NF Select) depending on network slices are considered.

Keywords: 5G · 5G MOCN · 5G GWCN · gNB · MNO · NG-RAN · Full MVNO

1 Introduction

The emergence of a new 5G standard of mobile communications confronts virtual operators with the task of assessing the possibility of creating MVNO networks of various levels based on the elements of architecture and protocols of 5G networks. The use of new technologies in 5G networks, the service-oriented network architecture and

© Springer Nature Switzerland AG 2019
O. Galinina et al. (Eds.): NEW2AN 2019/ruSMART 2019, LNCS 11660, pp. 495–504, 2019.
https://doi.org/10.1007/978-3-030-30859-9_42

the reduction of infrastructure dependency due to the virtualization of network functions and network capabilities, moreover the formation of network layers (slices) for each type of services and network orchestration create new additional capabilities for MVNO operators that were not available in the virtual mobile networks of previous generations.

Researching and designing of new variants of MVNO networks architecture of various levels based on the elements of architecture and protocols of 5G networks will open up new business opportunities for MVNO operators and provide high flexibility for the introduction of new services in the mobile communications market.

2   Characteristics of MVNO Business Models

Modern virtual operators are characterized by several business models [1]: the high-level virtual operator (Full MVNO), the low-level virtual operator (Light MVNO), the reseller. The main difference lies in the availability of its own core network, subsystems of operation and billing, platforms for the provision of Value Added Services (VAS), service promotion network (Fig. 1).

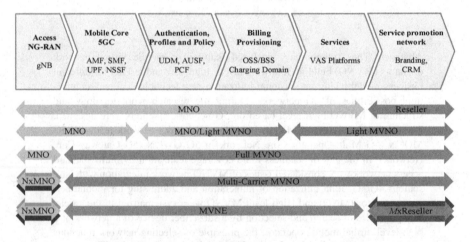

Fig. 1. Business model classification for virtual operator.

The least complicated and costly business model is the model of a reseller, which does not have its own infrastructure and carries out only branding, subscriber services and promotion of its services on the market.

The most complex and costly business model is the high-level virtual operator model, which has its own infrastructure in except for the NG-RAN radio access network.

The high-level virtual operator (Multi-Carrier MVNO) can have simultaneous connection to several radio access networks of various MNO operators. The intermediate trade-off business model is the low-level virtual operator model, which, in contrast to the reseller, also has its own Value-Added Services (VAS) platforms. In addition, the low-level virtual operator can have its own 5G virtual network functions (NF), such as Unified Database Function (UDM), Authentication Function (AUSF) and Policy Management Function (PCF) [1]. The business models of MVNE network infrastructure are also known in the telecommunications market. They are specialized in creating and marketing the resellers. The infrastructure (platform) of the MVNE virtual operator is essentially a layer between the mobile network operator (MNO) and the resellers.

3 MOCN Architecture Variants for Virtual Operator Network

The example of a 5G network architecture constructing of a high-level virtual operator that has its own core network can be an architecture based on network resource sharing technology. In accordance with the 3GPP technical specification [1], only one resource sharing technology is defined for the 5G network. This is the technology of sharing 5G radio access network (NG-RAN) simultaneously by several operators, which is called 5G MOCN (5G Multi-Operator Core Network).

The scheme of such a sharing network of a mobile operator (MNO) and a high-level virtual operator (Full MVNO) based on the 5G MOCN technology is shown in Fig. 2.

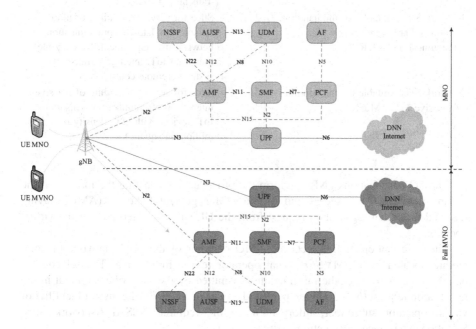

Fig. 2. Sharing MNO and Full MVNO network scheme based on MOCN technology.

This variant of the network construction involves the joint use of both the hardware and the software resources of gNB base stations, NG-RAN radio access network and radio resources by several MNO and MVNO operators. The radio resources of the NG-RAN radio access network are distributed between the MNO and MVNO operators on the basis of the agreed internal policy and Service Level Agreement (SLA).

If the high-level virtual operator Full MVNO has its own mobile network code MNCMNO, the gNBs of the Broadcast System Information of BCCH logical channel transmit this code along with the MNCMNO code of the mobile network operator (MNO) in each cell.

User terminals (UE) of the virtual operator decide to register in the network or perform cell reselection or handover procedures depending on the current state (RM- DEREGISTERED, RM-REGISTERED/CM-IDLE, RM-REGISTERED/CM-CONNECTED) when the code is detected in the network identifiers (PLMN-ID = MCC‖MNCMNO) transmitted by the PBCCH broadcast channel.

Table 1. Relationship between standardized SST values and 5G key business models.

Slice/Service type (SST)	SST Value	Description
Network Slice suitable for the handling of 5G enhanced Mobile Broadband – eMBB	1	Slice for service providing of high speed data transmitting, for example, streaming high-quality video, large files transmitting, etc.
Network Slice suitable for the handling of ultra- reliable low latency communications – URLLC	2	Slice for service providing of ultra-reliable low latency communication between user equipment, for example, industrial IoT, intelligent transport systems, remote control systems
Network Slice suitable for the handling of massive IoT – MIoT	3	Slice for service providing of massive interaction big numbers of sensors or IoT devices with high density over 1 million on square km

The 5G base stations gNBs perform the procedure of selecting the module of the serving network Core Access and Mobility Management Function (AMF), having received the initial registration request from the UE subscriber terminal of the virtual operator.

The base station gNB selects the NMO operator of the AMF network function module located in the MVNO virtual operator responsibility area. The selection is based on the analysis of the information transmitted by the subscriber terminal in the registration request (5G-GUTI temporary identifier or SUPI/IMSI constant identifier of virtual operator subscriber, information on the required NSSAI (Network Slice Selection Assistance Information) network layers.

4   Identification Features for Virtual Operator Subscriber

IMSI and GUTI identifiers used to identification the subscriber in the 5G network have a new designation 5G-GUTI and SUPI/IMSI and contain the network operator code MNCMVNO of the virtual operator.

The 5G-GUTI identifier also contains the global GUAMI identifier of the network access control function (AMF) of the virtual operator. After gNB base station has selected the module that realizing the AMF network function in the virtual operator's area of responsibility, the standard procedure described in Technical Specification [2] is performed and all signaling and traffic of these subscribers are routed to the network of Full MVNO.

A distinctive feature of 5G networks is the technical realization of the software network functions (UDM, AMF, SMF, UPF, PCF, AUSF, NSSF) of the 5G Core network using a set of virtual network functions VNF. Virtual network functions VNF, in fact, are software (programs) and virtual machines (VMs), deployed on the servers of modern data centers (DC). Modern applications of the virtual infrastructure of the data center make it easy to deploy and manage such VMs. Thus, a high-level virtual operator can have one or more sets of network functions of the 5G Core and each set of which is responsible for the implementation of a certain type of service: UDM, AMF, SMF, UPF, PCF, AUSF and NSSF. According to 3GPP terminology, these sets are defined as Network Slices of 5G. The relationship between parameter SST (Slice/Service type) and key business models eMBB, uRLCC and MIoT are given in Table 1 [2].

MNO base stations gNBs have to support all Network Slices realized by Full MVNO. 5G Core can supports big number of Network Slices (over 100). While 5G network can support large number of slices (hundreds), the UE need not support more than 8 slices simultaneously [3].

With slicing, it is possible for Mobile Network Operators (MNO) to consider customers as belonging to different tenant types with each having different service requirements that govern in terms of what slice types each tenant is eligible to use.

Base stations gNBs of MNO can consider each Network Slice as a network infrastructure which based on Service Level Agreement (SLA) and subscriptions. Radio resources of NG-RAN could be allocated as statistically between different Network Slices or allocated as to use by several Network Slices jointly.

Full MVNO can either obtain an own numbering or rent it from the mobile network operator MNO. Developed requirements to numbering of Full MVNO in depends on availability of MNCMVNO for Full MVNO to shown in Table 2.

An important of network aspect of developing high-level virtual operator Full MVNO is realization of program entity for network charging function (CHF) with automated traffic accounting system. The new charging function (CHF) was introduced in the 5G system architecture in Release 15. The interaction of 5G Core architecture with the CHF module of the Full MVNO and the billing domain is shown in Fig. 3.

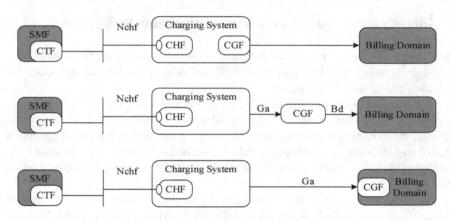

Fig. 3. Interaction architecture of 5G Core of Full MVNO with CHF and Billing Domain.

Table 2. Requirements to Identification and Numbering of Full MVNO.

Identifiers structure which used in 5G network	Requirement to availability of MNCMVNO for Full MVNO	
	Yes	No
SUPI (IMSI = MCC ‖ MNC ‖ MSIN) – Subscriber Permanent Identifier	100% of resource belongs to Full MVNO	Numbering resource MSIN has to be assigned between MNO and Full MVNO as shared resource
SUPI (NAI = username @ realm)	100% of resource belongs to Full MVNO	
SUCI (MCC ‖ MNC ‖ Routing Indicator ‖ Protection Scheme Id ‖ Home Network Public Key Id ‖ Concealed SUPI) – Subscription Concealed Identifier	100% of resource belongs to Full MVNO	Routing Indicator resource has to be assigned between MNO and Full MVNO for delivering of signaling data to AUSF and UDM including MVNO subscriber profile
M-TMSI – M - Temporary Mobile Subscriber Identity, which is unique in scope of AMF	100% of resource belongs to Full MVNO	
5G-S-TMSI (AMF Set ID ‖ AMF Pointer ‖ M-TMSI) – 5G S-Temporary Mobile Subscriber Identity, which is unique in scope of AMF Region ID	Numbering resource AMF Set ID ‖ AMF Pointer has to be assigned between MNO and Full MVNO. Numbering allocation needs for avoiding 5G-S-TMSI Identity duplication for MNO and Full MVNO because paging of subscribers in shared NG-RAN based on 5G-S-TMSI	
DNN – Data Network Name	All Numbering resource belong to Full MVNO	

Figure 3 shows that the program module of the trigger function CTF, generating online and offline events of charging is built directly into the network function of managing SMF subscriber sessions. Thus, the monitoring of traffic consumption by subscribers of the network is performed by session management module and the network billing domain.

An analysis of the experience of creating 4G high-level virtual network operators of Full MVNO type in the Russian Federation shows [4] that the option of building a network based on MOCN technology has not been widely used for a number of reasons, such as:

Impossibility of Subscriber traffic registration. In accordance of network architecture this traffic routing by virtual operators from base stations 5G of MNO directly and these gNBs cannot technical ability of Subscriber traffic registration for Full MVNO;

Complexity of MOCN technical realization due to need of interaction adjustment for huge number of MNO base stations with 5G Core of Full MVNO and IP Backhaul.

Simplification of the architecture that was shown in Fig. 2 is the Cloud RAN based on gNBs with high-level CU on the MNO network. However, in this case, simultaneous utilization of gNBs with high-level CU and gNBs with low and middle centralization [5] are not be excluded.

Existing 4G networks are utilized a gateway model GWCN which is most popular model for integration of MNO network and Full MVNO core network. This model has reference point of MNO and Full MVNO networks which located in Core Network. Developed variant of such architecture applied to 5G Core Network is shown in Fig. 4.

In accordance Fig. 4, MNO keeps on traffic of Full MVNO network under control (for example, receiving and analyzing of CDR-files from SMF which is located in its area of responsibility) and in this case technical realization of Full MVNO network become more simplified.

5   GWCN Architecture Variants for Virtual Operator Network

Taking in account that 5G Network deployment can be occur by several scenarios [5], that evolutional scenario is most vital on transition stage when 5G services will have fragmentary of 5G radio coverage and they will concentrate on the places with high demand of these 5G services. The total radio coverage will be provide by ng-eNB base stations of new generation for 4G (LTE Advanced Pro).

Developed variant of sharing MNO and Full MVNO network scheme based on 4G/5G GWCN technology shown in Fig. 5.

As can be seen from Fig. 5 realization of Sharing 4G/5G network scheme based on GWCN assumed utilization of shared network elements (SMF+PGW-C, UDM+HSS, PCF+PCRF) of Control Plane and utilization of shared network elements (UPF+PGW-U) of User Plane as well as 4G Core network and 5G Core. Besides that, MNO has to provide access of Full MVNO to 4G Core elements as MME and SGW. Service Gateway SGW can be realized as common network element which distributes in accordance with CUPS technology.

Network architecture MNO consisting from such elements as AMF, SGW+PGW-C, MME, SGW can be realized by two variants: superposed and dedicated.

Superposed variant of MNO architecture assumes the use of AMF, SGW+PGW-C, MME, SGW network elements in the architecture, which simultaneously serve subscribers of MNO and Full MVNO operators.

Dedicated variant of MNO architecture assumes utilization in MNO architecture of the previously mentioned network elements allocated by the mobile network operator only to Full MVNO as a separate network slice of 5G Core or use it in the form of a dedicated EPC network (for example, a dedicated DECORE core network for the UE Usage Type = "MVNO" containing MME and SGW network modules).

Existing dimension of network slice parameter (Slice/Service type) for realization dedicated variant of 5G Core is proposed to expand to 4 (SST ∈ 1-4) and SST = 4 will determine the network slice containing the network functions AMF and SMF for MVNO services.

Developed schemes present in Figs. 2, 4 and 5 are correct when NG-RAN uses base stations gNBs connected to Core network in accordance of Option 2 of 5G architecture [6, 7]. In case of utilization of other options for 5G architecture deployment (for example: 3/3A, 4/4A, 5, 7/7A) presented schemes have to change for providing of Full MVNO network tasks and it takes additional study network without using the NG-RAN radio access network and to provide 5G services with Value-added Services to the market using these technological capabilities.

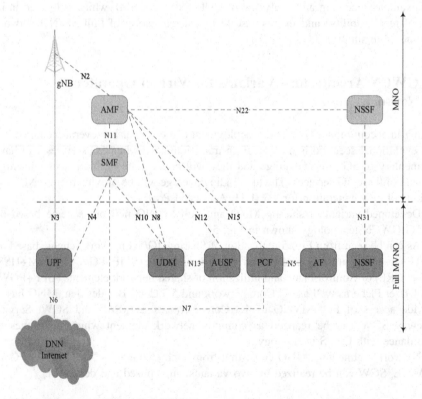

Fig. 4. Sharing MNO and Full MVNO network scheme based on GWCN technology.

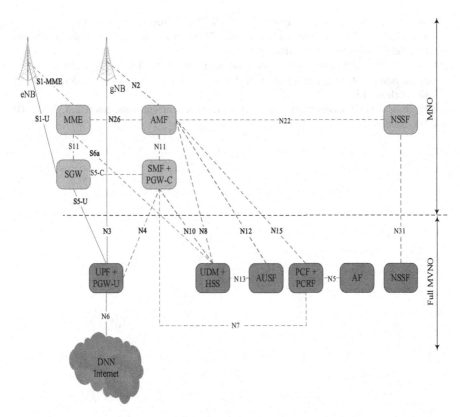

Fig. 5. Sharing MNO and Full MVNO network scheme based on 4G/5G technology.

6   Conclusion

New 5G network technologies allow high-level virtual operators Full MVNO to solve the problem of building a network without using the NG-RAN radio access network and to bring 5G services to the market with additional value using these technological capabilities.

The variants for the virtual network architecture design of Full MVNO which based on the 5G network elements will depend on the options of the network architecture that determine the configuration of the 5G core network.

The structural schemes for 5G Full MVNO deployment developed by the authors on the basis of the GWCN architecture cover the superposed or dedicated variants of the construction of a shared architecture MNO 5G and 4G/5G networks.

References

1. System Architecture for the 5G System, 3GPP TS 23.501, Stage 2, Release 15
2. Procedures for the 5G System, 3GPP TS 23.502, Stage 2, Release 15

3. NR and NG-RAN Overall Description, 3GPP TS 38.300, Stage 2, Release 15
4. J'son & Partners, Report "The market of MVNO in Russia and Europe: main trends and development prospects", June 2016
5. NR; Base Station (BS) radio transmission and reception, 3GPP TS 38.104, Release 15
6. Study on new radio access technology: Radio access architecture and interfaces, 3GPP TS 38.801, Release 15
7. Study on new radio access technology: Radio access architecture and interfaces, 3GPP TR 23.799, Release 15

Development of Infocommunications Services in Russia

Tatyana Nekrasova, Valery Leventsov$^{(\boxtimes)}$, and Vladimir Gluhov

Peter the Great St. Petersburg Polytechnic University, Saint Petersburg, Russia
dean@fem.spbstu.ru, {vleventsov,
vicerector.me}@spbstu.ru

Abstract. We have analyzed the market of infocommunication services in Russia. The number of both mobile Internet users and fixed-line Internet subscribers increased in 2011–2017. In 2017, mobile Internet in Russia was used by 124.83 million people, which is 7.7% higher than in 2016, and 47.7% higher than in 2011; 93% of mobile Internet subscribers are users of broadband Internet access with a declared speed of 256 Kbps. However, even though mobile operators still have large revenues from mobile communications, no serious revenue increase can be reported in the sector. For this reason, Russian telecom companies are beginning to follow the path of digital development, namely, developing financial services, system integration, e-commerce and the Internet of things. Active development of this line of business in the near future will lead to a decrease in the share of telecom services in the revenues, but mobile operators will not suffer much as the decrease will be offset by revenue from new activities. Analysis of the status and specifics of the telecom services market has revealed two issues that merit more detailed consideration: (1) finding the capacity of the telecom services market; (2) forecasting the number of mobile Internet users. An existing method for determining the market capacity is an estimate that takes into account the purchasing activity and the level of demand. For the info-communication market, this is the volume of all expected purchases, namely, the charges for using cellular communication, the Internet and the like for a specific audience for the billing period, for example, one year. We have calculated three types of market capacity using the example of the mobile operator Mega-Fon: the potential, the available and the actual. The estimation of market capacity was performed taking into account the number of consumers of cellular communication services, the average number of services for data transmission per one subscriber and the average cost of one service for a subscriber. The potential of the market (the difference between the potential and available capacity) allows to determine the growth potential. The given data indicate that the telecommunication market is exhausting itself. In order to be successful in the market, the cellular operators need to expand their business by seeking new directions for development. The Gompertz curve was used to determine the predicted number of mobile Internet users. The graph has an S-shape. The number of subscribers is limited. The number is approaching 137 million people. The forecast was carried out up to 2030. Thus, the stage of saturation takes place at present and the Russian market of infocommunications (in terms of the number of subscribers) has already significantly slowed down its growth. Providing high-speed broadband Internet access, fourth (4G) and then fifth (5G) generation mobile communication and introducing digital services can be considered as a strategic objective.

© Springer Nature Switzerland AG 2019
O. Galinina et al. (Eds.): NEW2AN 2019/ruSMART 2019, LNCS 11660, pp. 505–514, 2019.
https://doi.org/10.1007/978-3-030-30859-9_43

Keywords: Infocommunications · Forecast · Market capacity ·
Market potential · Homperts curve · Digital services

1   Introduction

By the end of 2016, mobile phone penetration in Russia had reached 178%, corresponding to 257M users [1]. The demand of mobile services equaled 137M people.
Meanwhile, revenue from the voice services of providers is declining. These types of
services are becoming disadvantageous for users, as they can get the same, and much
more, through mobile internet. The problem for users is that they cannot use ordinary
phones. They need a smart phone which supports at least 3G [2, 3]. It is not difficult to
see what percentage of SIM-cards are used in smart phones and what percentage is
used in ordinary cellphones. According to the calculations of MegaFon and MTS,
smart phones with a 3G network made up 52% and 63% respectively, with cellphones
taking 48% and 37% respectively [4].

Considering that a little less than half of users have cellphones, the revenue from
these cellphone users is fairly stable. Moreover, owners of smart phones use both voice
communication and mobile internet. This indicates a growth of the average revenue per
user (ARPU) in the near future. Furthermore, additional services have yet to become
highly profitable for companies [5]. The reason for this is, firstly, that they are simply
not needed by some users (no interest, no advertising), and secondly, the remaining
users may not be able to use them due to a lack of financial resources.

The number of users who have mobile internet or fixed internet access is presented
in Fig. 1 [6].

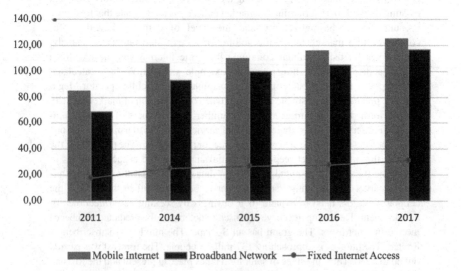

Fig. 1. Number of active internet users by mil./persons

For the period from 2011–2017, there was an increase in mobile internet users as well as in users with fixed internet access. In 2017 mobile internet in Russia was used by 124.83M people - 7.7% higher than in 2016 and 47.7% higher than in 2011. At that, 93% of mobile internet users are users of broadband internet connection with a reported speed of 256 KB/sec. Accordingly, the remaining 7% use cellphones supporting 2G network or lower.

It has been suggested that the number of internet users will increase in up-coming years [7]. There are two reasons for this:

- not all residents of Russia have internet access, mainly due to the lack of base stations. Providers have been gaining new markets, actively building base stations all over Russia;
- introduction of the new 5G network opens up a faster internet with a wider selection of possibilities. This helps attract the attention of new users.

An analysis of the dynamics of creating mobile communications systems, starting with 1G and finishing with 4G, shows that the creation of each new system took place over a 10-year span, specifically in 1970, 1980, 1990, and 2000. There are estimates that a 5G mobile network will be implemented in Russia by 2020 [8].

However, despite network providers still making large revenues from mobile network, there has yet to be a serious growth of revenue for companies [9, 10]. Figure 2 shows the economic performance for the main network providers in Russia for 2017 [11]. An analysis of the data shows that the revenue from mobile communications services is more than 80% of the total revenue for each of the companies while the net profit of the companies is low. MTS had the highest net profit with 127.2M roubles, while MegaFon had the lowest with 5.2M roubles [12]. The reason behind this is the high commercial and management expenses, as well as the high cost of the services.

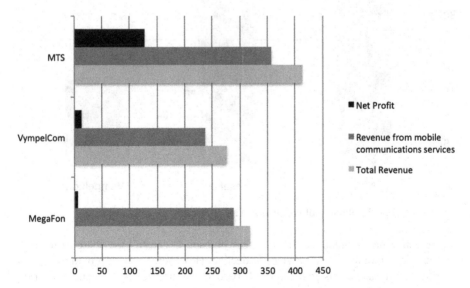

Fig. 2. Economic performance of network providers for 2017 in mil./roubles

Therefore, Russian companies are starting to move towards digital development, i.e., broadening financial services, system integration, e-commerce, and the Internet of things [13].

Network providers started creating financial services in 2017 with an active focus on credit cards with cashback reward programs.

The appearance of new applications like WhatsApp, Telegram, and Viber, which started offering their own video services, immediately impacted the revenues of mobile communications companies. In turn, network providers began creating their own products with similar functions: "MTS Connect" (MTS), eMotion (MegaFon), Veon (VympelCom) [14].

The internet of things market opens up many opportunities for network providers. It offers three sources for profit earning: during data transfer, from sales of devices, and the service they provide itself. These devices have a wide range of activity. They can be installed in homes or in urban infrastructures (e.g., on transport) to control energy consumption, opening and closing doors, heating, security. Special sensors which connect to the internet network analyze and trace the necessary indicators for the users and can also manage them. One important point is that mobile users are starting to use a larger number of SIM-cards for connecting their devices, while their number does not increase. These cards are called M2M. MTS is at the front of connecting such cards (4.5M connections), making up 42% of the total number of cards. MegaFon has 3.85M connections (36%), while VympelCom has 1.92M connections (18%) (see Fig. 3) [15].

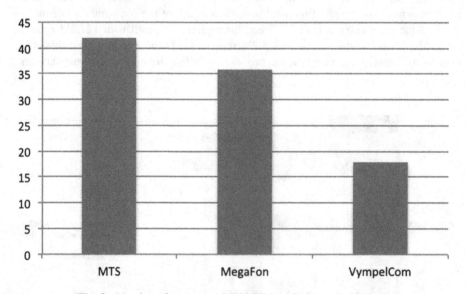

Fig. 3. Number of connected M2M SIM-cards for early 2017, %

The active development of this business division is causing the share of revenue from mobile communications services to decline. However, network providers are not suffering greatly since this reduction will be compensated for by revenue from new activities.

2   Method of Research

An analysis of the condition and specificity of the infocommunications market has shown the need to take a more thorough look at the following two problems:

- determining the capacity of the infocommunications market;
- predicting the number of mobile internet users.

2.1   Determining the Capacity of the Infocommunications Market

When planning and predicting the commercial activities of any organization, it is important to determine the capacity of the market. In its simplest form, the capacity of a market is the maximum number of goods which can be realized in a specific sector of the market over a certain time period [16].

From the existing methods of determining the capacity of a market, we will use assessment, taking into account the buying activity and level of demand [17]. For the infocommunications market, this will be the total of all expected product sales, i.e., payment of network use, internet, etc., by a specific group for an appraisal period (for example, one year).

Using the example of the network provider MegaFon, we can calculate the market capacity in three forms: potential, available, and actual. To determine the potential market capacity, the maximum level of development of demand for the service among users is used, i.e., all potential users know and use services of network communications. The actual market capacity is the portion of the population who currently use services of network communications. When calculating the available market capacity, we define it as the portion of the demand which the network provider is capable of satisfying with the resources they have.

The market capacity is calculated taking into account the number of network service users, the average number of data transfer services to one user, the average cost of one service for one user.

Data from the yearly report of the provider MegaFon along with data from Rosstat were used to calculate the market capacity of these forms (see Table 1) [1, 18].

Table 1. Data for calculating the market capacity of MegeFon

Number of users for calculating market capacity, mil./pers.:	2014	2015	2016	2017
Potential	134.77	137.01	137.03	137.22
Actual	102.97	104.68	104.7	104.85
Available	69.7	74.8	75.6	77.3
(DSU) Average number of services for data transfer for one user (Mb/month)	2,614	3,286	4,286	7,827
(ARPDU) Average cost of services for one user per month (roubles)	321	306	287	272

The market capacity of MegaFon is shown in Table 2.

Table 2. Market capacity of MegaFon, mil./roubles

	2014	2015	2016	2017
Potential capacity	1,357.00	1,653.13	2,022.74	3,505.66
Actual capacity	1,036.88	1,263.16	1,545.58	2,678.67
Available capacity	701.82	902.55	1,115.93	1,974.81

An analysis of the results showed that the network provider MegaFon is not using all of its opportunities. It is quite clear that expanding the number of users is only possible in remote areas. It is not possible to increase market capacity by way of new users in areas where infocommunications is already in full use (e.g., big cities).

Table 3 gives data on the market capacity of mobile communications in Russia over three years.

Table 3. Market capacity of mobile network in Russia, bil./roubles [18]

	2014	2015	2016
Potential	637.457	576.098	551.433
Available	551.521	486.482	465.593

The market potential (difference between potential and available capacity) makes it possible to determine the opportunities of the market. The presented data shows that the mobile network market is fraying. In order to remain on the market, network providers need to expand their business by searching for new areas of development.

2.2 Predicting the Number of Mobile Internet Users

Expanding the market of mobile internet access has presented the task of compiling projected development indicators for this market, specifically: the number of users of infocommunications services, ARPU – average revenue per user for services and the revenue itself of the providers offering theses services.

Statistical data from Rosstat and the Ministry of Economic Development of Russia were used for research and to construct a trend-time series model [19]. Different methods were used to make market projections: direct count, extrapolation of the past on the future, expert methods. The closest situation to reality is the Gompertz curve [20, 21]. The Gompertz curve is an S-shaped curve of growth. Processes which are modeled with this curve have several stages: first a slow growth, then a sharp acceleration followed by another period of slow growth, until, finally, reaching a limit. This model also helps describe the infocommunications market in Russia. The following formula is used to make calculations [20]:

$$Y_t = A + K \times e^{-e^{-b \times (t-m)}} \tag{1}$$

where Yt is the demand at time t; t is the time; A is the asymptotic distribution of demand (A = 0); K is the upper limits of demand; b is the percentage growth of the curve at point Y, when t = m; m is the point of maximum growth at point Y.

Table 4 shows the input for the calculations, specifically the number of users of mobile internet, including broadband internet access [18].

Table 4. Number of users of mobile internet in Russia

Year	Number of users, mil./pers.
2011	84.5
2012	91.2
2013	101.9
2014	105.8
2015	109.9
2016	115.8
2017	124.8

During calculations it was revealed that the upper limit of demand equals 137M people. This includes the total population of Russia in 2017, excluding children under the age of four. The number of users cannot exceed the total population. The average rate at which users of mobile internet increased for the period of 2011–2017 equaled 6.717M people. Calculations gave us a Gompertz curve model of the following form:

$$Y_t = 137 \times e^{-e^{-0,5774 \times (t-2012,89)}} \tag{2}$$

Table 5 shows the predicted number of users of mobile internet, including broadband network, which were gotten by constructing a Gompertz curve model.

In Fig. 4 we see the graph has an S-shaped form, since the number of users is limited. The number is approaching 137M people. Thus, there is currently a stage of saturation and the Russian infocommunications market (in terms of number of users) has already significantly slowed down in growth. Therefore, providers are looking for new ways to increase revenue.

Table 5. Predicted number of users of mobile internet access in Russia

Years	Number of users, mil./pers.	
	Factual data	Gompertz curve
2011	84.5	
2012	91.2	
2013	101.9	
2014	105.8	
2015	109.9	
2016	115.8	
2017	124.8	
2018		130.0218
2019		133.0379
2020		134.7615
2021		135.7389
2022		136.2906
2023		136.6013
2024		136.7761
2025		136.8742
2026		136.9294
2027		136.9604
2028		136.9777
2029		136.9875
2030		136.993

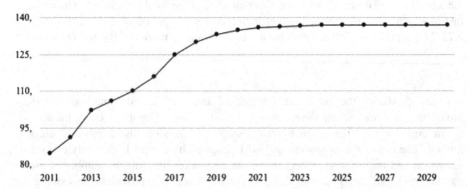

Fig. 4. Predicted number of users of mobile internet in Russia, based on the Gompertz curve model

3 Conclusion

The following main results were obtained during the research:

1. The mobile network market is oversaturated. While making predictions for up to the year 2030, it was determined that the number of users will not increase. Network providers should seek out new areas of business if they wish to stay on the market.
2. New areas of development can be found in digital development and, more importantly, the internet of things.
3. While determining the capacity of the infocommunications services market on the example of MegaFon, it was determined that companies do not fully take advantage of their opportunities. After developing remote areas in the near future, the market will be completely divided among providers.
4. One strategic method is to provide high-speed, broadband internet access, fourth generation mobile network (4G), and later fifth generation, and to introduce digital services.

References

1. Yearly financial report of "MegaFon". (2017). https://corp.megafon.ru/investoram/shareholder/msfo/. Accessed 28 May 2018
2. Krymov, S.M., Leventsov, V.A.: Conceptual bases and tendencies for transforming relations of modern businesses at different stages of development. Russ. Entrepreneurship **18**(22), 3593–3604 (2017). Publisher "Kreativnaya Ekonomika" (Moscow)
3. Kobzev, V.V., Leventsov, V.A., Radaev, A.E.: Procedure for determining the features of transport and warehousing systems in industrial enterprises in a megapolis environment. In: SHS Web of Conferences: 3rd International Conference on Industrial Engineering (ICIE-2017), vol. 35, 6 p. EDP Sciences (2017)
4. Market report of cellphones, smart phones, and tablets. EUROSET.RU: internet resource (2017). https://euroset.ru/corp/pr_information/press_release/14957253. Accessed 14 Apr 2018
5. Financial indicators and areas of growth for the telecommunications services market. News and analytics. http://1234g.ru/novosti/doli-rynka-sotovykh-operatorov-2016/. Accessed 26 May 2018
6. Prokhorov, A.: Capacity of the telecomm market: what are we counting? CNEWS.RU: scientific journal (2017). http://www.cnews.ru/reviews/rossijskij_telekommunikatsionnyj_rynok/articles/rynok_telekommunikatsij_vse_chemto_zanyaty_no_rosta_net/. Accessed 03 Mar 2018
7. Leventsov, V.A., Radaev, A.N., Nikolaevskiy, N.N.: Aspects of the "Industry 4.0" concept in terms of designing production processes. Scientific and technical journal of Saint Petersburg Polytechnic University. Economics. Publisher of the Polytechnic University, Saint Petersburg, vol. 10, No. 1, pp. 19–31 (2017)
8. Growth indicators for the information society. GKS.RU: federal service for state statistics (Rosstat) (2018). http://www.gks.ru/wps/wcm/connect/rosstat_main/rosstat/ru/statistics/publications/plan/. Accessed 29 Apr 2018

9. Abushova, E., Burova, E., Suloeva, S.: Strategic analysis in telecommunication project management system. In: Galinina, O., Balandin, S., Koucheryavy, Y. (eds.) NEW2AN/ ruSMART 2016. LNCS, vol. 9870, pp. 76–84. Springer, Cham (2016). https://doi.org/10. 1007/978-3-319-46301-8_7

10. Grishunin, S., Suloeva, S.: Development of project risk rating for telecommunication company. In: Galinina, O., Balandin, S., Koucheryavy, Y. (eds.) NEW2AN/ruSMART 2016. LNCS, vol. 9870, pp. 752–765. Springer, Cham (2016). https://doi.org/10.1007/978-3-319-46301-8_66

11. Sukharevskaya, A.: Where network providers look for new sources of revenue. VEDOMOSTI.RU: e-newspaper (2018). https://www.vedomosti.ru/technology/articles/ 2018/03/21/754381-sotovie-operatori-istochniki. Accessed 15 May 2018

12. Korolev, I., Beeline, MTS, and MegaFon: who earned more in a year. Numbers. CNEWS. RU: scientific journal (2018). http://www.cnews.ru/news/top/2018-03-20_bilajnmts_i_ megafon_kto_bolshe_zarabotal/. Accessed 11 May 2018

13. Leventsov, V., Radaev, A., Nikolaevskiy, N.: Design issues of information and communication systems for new generation industrial enterprises. In: Galinina, O., Andreev, S., Balandin, S., Koucheryavy, Y. (eds.) NEW2AN/ruSMART/NsCC 2017. LNCS, vol. 10531, pp. 142–150. Springer, Cham (2017). https://doi.org/10.1007/978-3-319-67380-6_13

14. Shmyrova, V., Beeline, MTS, and MegaFon: who earned more in a year. CNEWS.RU: publication on high tech (2017). http://www.cnews.ru/news/top/2018-03-20_bilajnmts_i_ megafon_kto_bolshe_zarabotal/. Accessed 25 May 2018

15. Trosnikova, D.: Automated users require more and more traffic. VEDOMOSTI.RU: e-newspaper. (2018). https://www.vedomosti.ru/technology/articles/2015/09/02/607147-avtomatizirovannie-abonenti-potreblyayut-vse-bolshe-trafika/. Accessed 12 May 2018

16. Noskova, E.: How is market capacity calculated? RUSHBIZ.RU: information on business in Russia (2015). http://rushbiz.ru/startbiz/terms/raschet-emkosti-rynka.html/. Accessed 08 May 2018

17. Matyushina, T.: Main methods for analyzing the market. SRC.RU: business school (2016). https://www.src-master.ru/article26190.html/. Accessed 21 Apr 2018

18. Russian statistical yearbook. 2017: P76 Stat. collection/Rosstat. M., 689 p. (2017)

19. Ministry of Economic Development of the Russian Federation. GOV.RU: Scenario conditions for the long-term prediction of the socio-economic development of the Russian Federation until 2030 (2013). http://economy.gov.ru/minec/activity/sections/macro/prognoz/ doc20130325_06/. Accessed 16 May 2018

20. Volodina, E.E.: Predicting the development of innovation services in infocommunications: textbook for institutes of higher education. Moscow Technical University of Communications, 45 p. (2017)

21. Kuzovkova, T.A., Volodina, E.E.: Main areas of scientific and technical development of infocommunications. Publisher of Moscow Technical University of Communications, 42 p. (2015)

A Concept of Smart Medical Autonomous Distributed System for Diagnostics Based on Machine Learning Technology

Elena Velichko[1(✉)] , Elina Nepomnyashchaya[1] ,
Maxim Baranov[1] , Marina A. Galeeva[1] , Vitalii A. Pavlov[1] ,
Sergey V. Zavjalov[1] , Ekaterina Savchenko[1] ,
Tatiana M. Pervunina[2] , Igor Govorov[2] ,
and Eduard Komlichenko[2]

[1] Peter the Great Saint Petersburg Polytechnic University,
Saint Petersburg, Russia
velichko-spbstu@yandex.ru, elina.nep@gmail.com,
baranovmal993@gmail.com,
{arviana, savchenko-spbstu}@mail.ru,
{pavlov_va, zavyalov_sv}@spbstu.ru
[2] Almazov National Medical Research Centre, Saint Petersburg, Russia
ptm.pervunina@yandex.ru, govorov.igor.med@gmail.com,
e_komlichenko@mail.ru

Abstract. Telemedicine is a promising direction in the development of medical technologies for the interaction of patients with doctors at a distance. In this paper, we consider the use of telemedicine technologies for the development of smart medical autonomous technology. An example of a smart medical autonomous distributed system for diagnostics is also discussed. To develop this system for medical image analysis we review several processing methods and machine learning algorithms. Some examples of medical system processing results are presented.

Keywords: Telemedicine · Medical diagnostics · Image processing

1 Introduction

With each passing year, technologies of communication and information transfer are becoming more and more embedded in our lives. There is virtually no area of human activity today where information transfer systems are not involved. One of these areas is medicine. Due to communication systems and information transfer development, a new direction in medical practice, called telemedicine, has formed.

Telemedicine has been defined as «the use of medical information exchanged from one site to another via electronic communications to improve a patient's clinical health status». Telehealth has historically had a broader definition, encompassing telemedicine's clinical care for patients and tele-education, and tele-research. Telemedicine and telehealth, being commonly used today, can be considered synonymous [1].

© Springer Nature Switzerland AG 2019
O. Galinina et al. (Eds.): NEW2AN 2019/ruSMART 2019, LNCS 11660, pp. 515–524, 2019.
https://doi.org/10.1007/978-3-030-30859-9_44

Telemedicine can be used for tele-education, tele-consultation, tele-practice, and tele-research. Tele-education can be delivered through live interactive AV links, by live streaming video, and stored educational material. Tele-education programs allow physicians to stay current, travel less, obtain free continuing medical education, foster relationships between academic and community-based physicians, and establish public peer groups to learn from academicians [2].

The use of telemedicine as an important part of health care delivery has occurred in a number of systems. Technological developments are increasing the number of tools that can be used for telemedicine while driving down the cost of these tools [3–5].

One of the options for telemedicine is an autonomous system that allows diagnostics even for unqualified personnel. It is becoming possible because of the latest advances in computer technology, which are widely used both in the diagnosis and in therapy. This article considers the concept of the distributed autonomous network for diagnostics based on machine learning methods applied to the realities of the Russian Federation.

The article is organized as follows. In Sect. 2, a reader finds a description of the distributed autonomous network concept for diagnostics in Russian Federation based on machine learning methods. Section 3 describes the machine learning approaches used for medical diagnostics. In conclusion, the most promising areas for further development are identified.

2 A Concept of Smart Medical Distributed System for Diagnostics

Building a system for diagnostics is an incredibly difficult task. In the realities of the Russian Federation, the additional difficulty is due to the multiplicity of time zones, which causes the inability of urgent diagnostics. Based on this fact, it is necessary to build an autonomous system of expert evaluation of the results of surveys and analyzes. For the solution of urgent medical problems, it is necessary to exclude the need for the doctor permanent participation. Periodic verification of diagnostic results by an experienced doctor will still be necessary, but the whole task can be solved by telemedicine methods.

Realization of autonomous diagnostics can be achieved by machine learning methods. Such systems use pre-training on some dataset. Although they can make some diagnostics mistakes, the accuracy of machine training for the diagnostics is superior or comparable to experienced doctors [6, 7]. It is also possible to re-train the working system periodically to update and add the information required for efficient operation. It should be noted that training is a rather complicated and time consuming process. However, the operation time of already trained machine-learning module is almost comparable to the time of information transfer.

The requirement of the system to be distributed implies the absence of a single processing center. This will eliminate the need to build a single powerful data center. This fact will reduce the overhead of building infrastructure and the time for information transmission and processing. This may be a critical factor in some situations. In addition, the distribution of the structure implies the existence of several ways of information transmission through different transmission channels (radio relay communication, optical fiber, satellite communication). Multiple channels make it possible to increase the stability of the system significantly to possible force majeure situations (break of the optical fiber, their failure to reach amplifiers on a satellite, etc.). Thus, the reliability of the information transmission system increases substantially, which is undoubtedly important for medical applications.

An example of building a smart medical autonomous distributed system for diagnostics is shown in Fig. 1. There are several data centers for processing of diagnostic information, and there are always more than one path to each data center. There are also no extended communication lines.

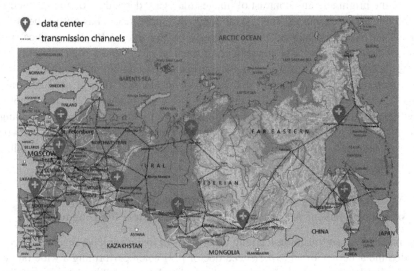

Fig. 1. The example of the design of the smart medical autonomous distributed system for diagnostics

It is possible to note that this concept is consistent with the ideology of building 5-6G systems [8–12]. One of the directions of these systems is to reduce delays in the transmission of information (to less than 1 ms) [13–15]. There is also a constant tendency to increase the speed of transmission [16–19]. In such conditions, the transfer of even large images and video files necessary for diagnosis needs a few seconds. With processing information, diseases will also be diagnosed in few seconds, which is extremely important for medical decision making (especially during surgery, when a patient's life depends on the surgeon's decision).

3   Image Processing for Medical Diagnostics

In this chapter, we will consider machine-learning algorithms for processing medical images using the example of endoscopic images of mucous membranes.

3.1 Endoscopic Inspections

Currently, the problems of diagnosis during endoscopic studies and efficiency evaluation of the internal organs treatment are widely analyzed. An increasing number of diseases and growing mortality from cancer are noted, while these diseases can be detected on time during preventive endoscopic examinations.

At the same time, the analyzes of endoscopic images turns out to be very difficult, mainly because of the variety of textures of the surfaces under study, their substantial heterogeneity and a wide range of scales of processed images. The following features are characteristic of endoscopic images: tissue images can vary significantly during contraction of muscle fibers; images may have blur; images may contain glare and artifacts; the brightness and contrast of images may vary depending on the geometry of the area under consideration. The task of analyzing such images has not yet been fully resolved and is not introduced into the diagnostic practice of Russian medical institutions, which leads to a large percentage of medical errors and a high mortality rate.

The development of software for the automated processing and analysis of diagnostic images based on computer vision and deep learning (neural networks) algorithms seems promising [20]. The literature provides mathematical and graphical output data on the degree of learning of the networks to identify pathological changes in the analysis of medical images [21]. Preliminary results, published in the first works of foreign authors, show a high degree of reliability (up to 95%) of the results of diagnosis of cervical pathologies [6, 7].

3.2 Machine Learning Algorithms for Implementing the Concept of Smart Medical Autonomous Distributed System for Diagnostics

In the past few years, machine-learning algorithms in medicine have reached and even exceeded the accuracy of an experienced doctor [22] when performing a wide range of diagnostic tasks. Examples of such tasks are the detection of melanomas, diabetic retinopathy, cardiovascular risks, and lesions of the mammary gland on mammograms and magnetic resonance studies. It was shown that even one model of deep learning is effective in diagnosing in various medical conditions (for example, in radiology and ophthalmology) [23].

There are three tasks which deep learning solves in relation to medicine: a detection, a segmentation, a classification. We are going to consider each of them.

The main purpose of the detection is to identify the area of interest in the image and select this area with a bounding box. Typically, systems are aimed at identifying the earliest signs of pathology in patients. To solve the problem of detection and recognition, two types of methods are used: based on of classification and regression. The first type can be attributed R-CNN [24], Fast-R-CNN [25], Faster-RCNN [26] to the second YOLO [27], YOLO 2 [28], YOLO 3 [29], RetinaNet [30], SSD [31], and others.

The next task is segmentation. Segmentation is the process of dividing an image into various significant segments (which have similar characteristics) by means of automatic or semi-automatic selection of borders in the image. In medical imaging, these segments are usually commensurate with different classes of tissues, pathologies, organs, or some other biological structure. A typical segmentation network consists of a coder network and a decoder. The encoder extracts high-level features through a convolution operation; the decoder interprets these high-level features using a class mask. Coding is a coding mechanism using pre-trained networks, decoder weights are generated during segmentation network training. The encoder gradually reduces the spatial resolution of the layers by a pooling operation, and the decoder gradually restores the object details and spatial resolution using an inverse operation. It is important to understand how semantic segmentation occurs in convolutional networks. The following architectures can be distinguished: FCN [32], Deconvolutionalnetwork [33], SegNet [34], U-Net [35], PSPNet [36], DeepLab [37], Mask-R-CNN [38], EncNet [39].

The classification of images is a long-standing key problem in the analysis of medical images. For medical images, the task is to classify the image to one of two or more classes. The simplest classifier is the binaryone. In this case, a decision is made about the absence or presence of pathology. In this embodiment, even unqualified personnel can carry out a primary diagnosis and, if necessary, send the patient for further examination to an experienced doctor. The following architectures can be applied to binary and more complex classifiers: VGG [40], AlexNet [41], DarkNet19 [42], as well as DarkNet53 and GoogLeNet. Most convolutional network architectures have the following components in their structure: convolutional layers, activation layers, pooling, dropout regularization or batch normalization [43].

As mentioned above, the complexity of the application of machine learning methods is due to the need for training and periodic retraining. Moreover, significant resources are spent on training. The training time depends directly on the number of layers and filters, the size of the training selection, the power of the hardware platform (video cards, central processors and communication network bandwidth) and the type of problem to be solved. Therefore, this procedure is feasible in a reasonable time only in chief data centers. However, after the learning procedure, an already configured network can be used in other peripheral data centers, which simplifies system deployment, reduces cost, and increases reliability.

3.3   Examples of Processing and Diagnostics

To demonstrate the capabilities of the above concept of an expert diagnostic system, we consider the options for automatic processing of endoscopic images.

Initially, we consider the use of analytic features, such as local binary patterns. Examples of local binary patterns are shown in Fig. 2. It can be seen that the structure in Fig. 2b is more fine-grained. Thus, the construction of a binary classifier is possible; it is only necessary to type many examples for constructing a hypothesis for separating the two classes and testing it.

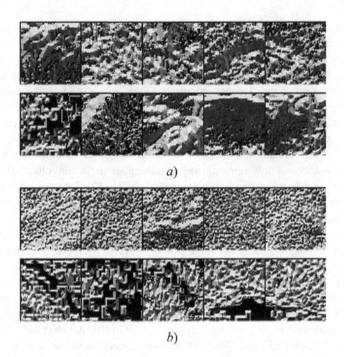

a)

b)

Fig. 2. Examples of local binary patterns for image fragments of "Cancer" (*a*) and "not Cancer" (*b*) classes.

Next, consider the VGG-16 convolutional neural network, which allows you to extract features using the convolution operation automatically. The architecture of VGG-16 is provided in Fig. 3. This network uses convolution filters of size 3×3, which are formed during training.

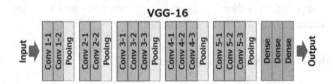

Fig. 3. Processing structure of VGG-16.

In Fig. 4(a) and (c) 2 examples are presented: without pathology and with pathology, respectively. In Fig. 4(b) and (d), visualized feature maps of the final convolution layer for case (a) and (b) are shown.

As can be seen from the analysis of the visualization results, the cases of presence and absence of pathology are different, which allows them to be used in the construction of a smart medical autonomous distributed system for diagnostics.

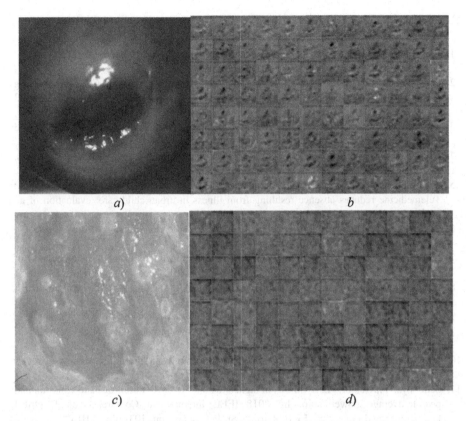

Fig. 4. Examples of processing and visualization of signs of processing endoscopic images using the VGG-16 convolutional neural network.

4 Conclusion

In this paper, we considered a concept of a smart medical system for Russian Federation. We have shown that the described concept, built on 5-6G systems, is able to process large volumes of medical data, in particular, medical images, quickly.

The basic machine learning algorithms for automated diagnostics based on image processing are presented. It is shown that the created neural system successfully copes with the task of recognizing cancerous diseases of the mucous membrane. This technology will create a smart medical autonomous system for diagnostics based on machine learning technology.

The developed concept will be highly demanded by the healthcare system of the Russian Federation for the provision of surgical care to the population.

References

1. Spooner, S.A., Gotlieb, E.M.: Telemedicine: pediatric applications. Pediatrics **113**(6), e639–e643 (2004)
2. González-Espada, W.J., Hall-Barrow, J., Hall, R.W., Burke, B.L., Smith, C.E.: A Achieving success connecting academic and practicing clinicians through telemedicine. Pediatrics **123** (3), 476–483 (2009)
3. Doolittle, G.C., Spaulding, A.O., Williams, A.R.: The decreasing cost of telemedicine and telehealth. Telemed. Health **17**(9), 671–675 (2011)
4. Izquierdo, R., et al.: School-centered telemedicine for children with type 1 diabetes mellitus. Pediatr. **155**(3), 374–379 (2009)
5. McConnochie, K.M., Wood, N.E., Kitzman, H.J., Herendeen, N.E., Roy, J., Roghmann, K.J.: Telemedicine reduces absence resulting from illness in urban child care: evaluation of an innovation. Pediatrics **115**(5), 1273–1282 (2005)
6. Sato, M., et al.: Application of deep learning to the classification of images from colposcopy. Oncol. Lett. **15**(3), 3518–3523 (2018)
7. Fernandes, K., Cardoso, J.S., Fernandes, J.: Automated methods for the decision support of cervical cancer screening using digital colposcopies. IEEE Access **6**, 33910–33927 (2018)
8. Gupta, A., Jha, R.K.: A survey of 5G network: architecture and emerging technologies. IEEE Access **3**, 1206–1232 (2015)
9. Ometov, A., Moltchanov, D., Komarov, M., Volvenko, S.V., Koucheryavy, Y.: Packet level performance assessment of mmWave backhauling technology for 3GPP NR systems. IEEE Access **7**, 9860–9871 (2019)
10. Gapeyenko, M., et al.: On the degree of multi-connectivity in 5G millimeter-wave cellular urban deployments. IEEE Trans. Veh. Technol. **68**(2), 1973–1978 (2019)
11. Sadovaya, Y., Gelgor, A.: Synthesis of signals with a low-level of out-of-band emission and peak-to-average power ratio. In: 2018 IEEE International Conference on Electrical Engineering and Photonics (EExPolytech), St. Petersburg, pp. 103–106 (2018)
12. Gorbunov, S., Rashich A.: BER performance of SEFDM signals in LTE fading channels. In: 2018 41st International Conference on Telecommunications and Signal Processing (TSP), pp. 1–4 (2018)
13. Andreev, S., et al.: Exploring synergy between communications, caching, and computing in 5G-grade deployments. IEEE Commun. Mag. **54**(8), 60–69 (2016)
14. Varga, J., Hilt, A., Rotter, C., Járó, G.: Providing ultra-reliable low latency services for 5G with unattended datacenters. In: 2018 11th International Symposium on Communication Systems, Networks & Digital Signal Processing (CSNDSP), Budapest, pp. 1–4 (2018)
15. Schmoll, R., Pandi, S., Braun, P.J., Fitzek, F.H.P.: Demonstration of VR/AR offloading to Mobile Edge Cloud for low latency 5G gaming application. In: 2018 15th IEEE Annual Consumer Communications & Networking Conference (CCNC), Las Vegas, NV, pp. 1–3 (2018)
16. Ovsyannikova, A.S., Zavjalov, S.V., Volvenko, S.V.: On the joint use of turbo codes and optimal signals with increased symbol rate. In: 2018 10th International Congress on Ultra Modern Telecommunications and Control Systems and Workshops (ICUMT), Moscow, Russia, 1–4 (2018)
17. Gelgor, A., Gorlov, A.: A performance of coded modulation based on optimal Faster-than-Nyquist signals. In: 2017 IEEE International Black Sea Conference on Communications and Networking (BlackSeaCom), Istanbul, pp. 1–5 (2017)

18. Rashich, A., Urvantsev, A.: Pulse-shaped multicarrier signals with nonorthogonal frequency spacing. In: 2018 IEEE International Black Sea Conference on Communications and Networking (BlackSeaCom), Batumi, pp. 1–5 (2018)
19. Ghannam, H., Nopchinda, D., Gavell, M., Zirath, H., Darwazeh, I.: Experimental demonstration of spectrally efficient frequency division multiplexing transmissions at E-Band. IEEE Trans. Microw. Theory Tech. **67**(5), 1911–1923 (2019)
20. Topol, E.J.: High-performance medicine: the convergence of human and artificial intelligence. Nat. Med. **25**(1), 44 (2019)
21. Jha, S., Topol, E.J.: Adapting to artificial intelligence: radiologists and pathologists as information specialists. JAMA **316**(22), 2353–2354 (2016)
22. Esteva, A., et al.: A guide to deep learning in healthcare. Nat. Med. **25**, 24–29 (2019)
23. Kermany, D.S., et al.: Identifying medical diagnoses and treatable diseases by image-based deep learning. Cell **172**, 1122–1131 (2018)
24. Girshick, R., Donahue, J., Darrell, T., Malik, J.: Rich feature hierarchies for accurate object detection and semantic segmentation. In: 2014 IEEE Conference on Computer Vision and Pattern Recognition (CVPR), pp. 580–587 (2014)
25. Girshick, R.: Fast R-CNN. In: Proceedings of the International Conference on Computer Vision (ICCV) (2015)
26. Ren, S., He, K., Girshick, R., Sun, J.: Faster R-CNN: towards real-time object detection with region proposal networks. In: Neural Information Processing Systems (NIPS) (2015)
27. Redmon, J., Divvala, S., Girshick, R., et al.: You only look once: unified, real – time object detection. In: 2016 IEEE Conference on Computer Vision and Pattern Recognition (CVPR) (2016)
28. Redmon, J., Farhadi, A.: YOLO9000: better, faster, stronger. In: 2017 IEEE Conference on Computer Vision and Pattern Recognition (CVPR)
29. Redmon, J., Farhadi, A.: YOLOv3: an incremental improvement, arXiv preprint arXiv:1804.02767 (2018)
30. Lin, T.Y., Goyal, P., Girshick, R., et al.: Focal loss for dense object detection, arXiv preprint arXiv:1708.02002 (2017)
31. Liu, W., Anguelov, D., Erhan, D., Szegedy, C., Reed, S.E.: SSD: single shot multibox detector. In: Leibe, B., Matas, J., Sebe, N., Welling, M. (eds.) ECCV 2016, pp. 21–37. Springer, Cham (2016)
32. Long, J., Shelhamer, E., Darrell, T.: Fully convolutional networks for semantic segmentation. In: 2015 IEEE Conference on Computer Vision and Pattern Recognition (CVPR), Boston, MA, pp. 3431–3440 (2015)
33. Noh, H., Hong, S., Han, B.: Learning deconvolution network for semantic segmentation. In: 2015 IEEE International Conference on Computer Vision (ICCV), Santiago, pp. 1520–1528 (2015)
34. Badrinarayanan, V., Kendall, A., Cipolla, R.: SegNet: a deep convolutional encoder-decoder architecture for image segmentation. IEEE Trans. Pattern Anal. Mach. Intell. **39**(12), 2481–2495 (2017)
35. Ronneberger, O., Fischer, P., Brox, T.: U-Net: convolutional networks for biomedical image segmentation. In: Navab, N., Hornegger, J., Wells, William M., Frangi, Alejandro F. (eds.) MICCAI 2015. LNCS, vol. 9351, pp. 234–241. Springer, Cham (2015). https://doi.org/10.1007/978-3-319-24574-4_28
36. Zhao, H., Shi, J., Qi, X., Wang, X., Jia, J.: Pyramid scene parsing network. In: 2017 IEEE Conference on Computer Vision and Pattern Recognition (CVPR), Honolulu, HI, pp. 6230–6239 (2017)

37. Chen, L., Papandreou, G., Kokkinos, I., Murphy, K., Yuille, A.L.: DeepLab: semantic image segmentation with deep convolutional nets, atrous convolution, and fully connected CRFs. IEEE Trans. Pattern Anal. Mach. Intell. **40**(4), 834–848 (2018)
38. He, K., Gkioxari, G., Dollár, P., Girshick, R.: Mask R-CNN. In: 2017 IEEE International Conference on Computer Vision (ICCV), Venice, pp. 2980–2988 (2017)
39. Zhang, H., et al.: Context encoding for semantic segmentation. In: 2018 IEEE/CVF Conference on Computer Vision and Pattern Recognition, Salt Lake City, UT, pp. 7151–7160 (2018)
40. Simonyan, K., Zisserman, A.: Very deep convolutional networks for large-scale image recognition. arXiv preprint arXiv:1409.1556 (2014)
41. Krizhevsky, A., Sutskever, I., Hinton, G.E.: Imagenet classification with deep convolutional neural networks. In: Advances in Neural Information Processing Systems, 25, pp. 1097–1105. Curran Associates Inc. (2012)
42. Szegedy, C., et al.: Going deeper with convolutions. In: 2015 IEEE Conference on Computer Vision and Pattern Recognition (CVPR), Boston, MA, pp. 1–9 (2015)
43. Musakulova, Z., Mirkin, E., Savchenko, E.: Synthesis of the backpropagation error algorithm for a multilayer neural network with nonlinear synaptic inputs. In: 2018 IEEE International Conference on Electrical Engineering and Photonics (EExPolytech), pp. 131–135 (2018)

New Method for Determining the Probability of Signals Overlapping for the Estimation of the Stability of the Radio Monitoring Systems in a Complex Signal Environment

Alexey S. Podstrigaev[1,2], Andrey V. Smolyakov[1,2],
Vadim V. Davydov[3,4], Nikita S. Myazin[3], Nadya M. Grebenikova[3],
and Roman V. Davydov[3(✉)]

[1] Scientific-Research Institute "Vector" OJSC, St. Petersburg, Russia
[2] Saint Petersburg Electrotechnical University "LETI", St. Petersburg, Russia
[3] Peter the Great Saint-Petersburg Polytechnic University, St. Petersburg, Russia
myazin.n@list.ru, davydovrvv@spbstu.ru
[4] Department of Ecology, All-Russian Research Institute of Phytopathology,
Odintsovo District, B. Vyazyomy, 143050 Moscow Region, Russia

Abstract. The article considers the consequences of the negative influence of signals overlapping from radio emission sources in time. Several examples show that overlapping in time worsens the quality of primary signal processing and, therefore, decreases the stability of radio monitoring systems. A new method of determining the probability of signals overlapping from radio emission sources has been developed. Calculations of the probability of failure of primary processing systems of radio monitoring systems are made.

Keywords: Sources of radio emission · Complex signal environment ·
Radio monitoring · Bearing · Direction finding · Pulse sequence · Signal phase ·
Carrier frequency · Receiving channel · Probability ·
Mathematical expectation · Simulation

1 Introduction

Radio monitoring is carried out in a wide range of frequencies to control the functioning of radio and telecommunication sources of radio emission (SRE), to perform analysis of the electromagnetic environment (EME), and to ensure the electromagnetic compatibility of SRE [1–7]. Analysis of the EME reveals the facts of unauthorized use of the most critical areas of the radio band, allocated for the control of the aircraft landing, satellite communications, radio navigation devices, road radars, etc. [1, 3, 8–16].

On the territory of seaports, airports, and transport hubs of large cities, many SRE are used to solve communication and management problems. They are also widely used during mass events. In such conditions, the problems of detection and selection of signals from different SRE with overlapping of time-frequency parameters should be solved by radio monitoring systems. For this purpose, the parameters of the received

© Springer Nature Switzerland AG 2019
O. Galinina et al. (Eds.): NEW2AN 2019/ruSMART 2019, LNCS 11660, pp. 525–533, 2019.
https://doi.org/10.1007/978-3-030-30859-9_45

signal (frequency, duration, amplitude, arrival time, and direction) are measured at the primary processing stage.

Most civilian and military SRE, such as various radars [12–14], as well as communication systems and switching devices for various purposes [17, 18], use impulse signals with a batch structure. In a complex signal environment (CSE), the impulses of different SRE are inevitably overlapped in time, which can disrupt the normal functioning of primary processing.

2 Problems Arising from the Combination of Impulse Sequences

For the most common today's superheterodyne receiver [19, 20], the signals overlapped in time with close carrier frequencies can interfere with the combination and intermodulation channels of reception. Due to the nonlinear characteristics of the mixers and amplifiers in the receiver, many harmonics are produced at high signal strength. These harmonics can be mistaken for real signals, which increases the probability of false alarms and makes it difficult to process true signals.

Also, when designing microwave systems, it is necessary to consider the periodic nature of the frequency response of bandpass devices. This leads to the fact that there are side channels of signal penetration. These channels correspond to the resonant bands of filters and other band-pass microwave devices. In this case, two effects must be borne in mind.

The first effect is the increase of the probability of a false alarm when a signal enters one of the spurious bands of the input filter. This effect is due to the low attenuation in these bands. Therefore, false detection of the signal entering the spurious bandwidth occurs even at low power. Thus, the probability of false alarm of radio monitoring tools is increased.

The second effect is the decreasing of the probability of correct detection. In general, when using automatic gain control (AGC), a strong signal at the receiver's input causes weak signals to be lost. This is a normal process. However, it is rarely considered that the loss of weak signals is also caused by powerful signals in the spurious bandwidths. For this reason, the probability of correct detection is further decreased.

Another problem that arises when impulses are overlapped is the reduced accuracy of the direction finding. In practice, this is most pronounced when using the phase method of direction finding [21, 22]. Operation in nonlinear mode, as well as AGC system operation and modification of any other parameters of the receiver, lead to changes in the phase response over time. As a result, the jitter of the received signal phase increases, and abrupt phase jumps can occur. This results in a brief increase in the direction of finding an error.

The stability of radio monitoring system in CSE is affected by the width of the instantaneous bandwidth processed. Given the a priori uncertainty of the frequency of the received signal and the need to analyze simultaneously operating SRE scattered over the frequency, it is necessary to have a wide instantaneous reception bandwidth. However, at high power of overlapped impulses, the spurious harmonics generated in the amplifier and mixer are further processed. Their location is almost impossible to predict, and therefore, they cannot be excluded from processing. Thus, in broadband radio monitoring systems, the quality of processing is reduced in CSE.

Therefore, to estimate the stability of the functioning of the primary processing system of a radio monitoring system in a CSE, it is necessary to determine the probability of overlapping in time of the impulses from SRE. The initial phases of the SRE signals are random values. Since the probability of impulses overlapping is a function of the initial phases of the SRE, it is also a random value. Therefore, to estimate the stability of the primary processing in a CSE, it is necessary to determine the mathematical expectation (ME) of this probability.

3 Known Ways to Determine the Probability of Impulses Overlapping Over Time

There are two ways to determine the probability of impulses overlapping in time. The first method is based on a simulation that implements the following sequence of actions:

1. Some combination of initial SRE phases is set.
2. The simultaneous operation is simulated over a long period, and the areas of impulse overlap are determined.
3. The probability of overlapping impulses is determined.
4. Another combination of the initial phases of the SRE is set, steps 2–4 are repeated many times.
5. After iterating over the set of combinations of initial phases of the SRE, the arithmetic mean of the found probability values is calculated.

The time required to perform the simulation of this algorithm can be roughly determined by the formula:

$$T \approx t_0 \cdot n^{N-1}, \tag{1}$$

where n is the number of initial phase values of one SRE, N is the number of SRE under consideration, t_0 is the simulation time at a fixed combination of initial SRE phases. The value t_0 is chosen in such a way that it is much greater than the maximum of the periods of repetition of the impulses of the SRE under consideration to correctly determine the probability of overlapping of these impulses.

According to the expression (1), with the increase of the number of considered SRE, the time T grows as an exponential function. With the increase in accuracy of the simulation, the time T grows as the power function. Thus, the application of simulation for estimation of stability of primary processing tools in difficult signal conditions demands considerable computing resources.

The second method is based on the calculation of the mathematical expectation of the probability of overlapping impulses of two or more SRE using an analytical expression [23]:

$$R = \prod_{j=1}^{N} \left[\int_0^\infty x q_j(x) dx \right], \tag{2}$$

where $q_j(x)$ is a function defined in such a way that $q_j(x)\Delta x$ is the average number of impulses that occur per unit of time in the j-th sequence and have a duration greater than x no more than Δx.

However, the formula (2) obtained by the authors allows determining the probability of simultaneous impulses overlapping only from a certain number of SRE. Impulse overlaps from a certain combination of SRE are not mutually exclusive. This leads to the fact that it is necessary to calculate the probability of combining many events. Therefore, when using (2) to estimate real-life situations, the formula becomes much more complex. Because of this, it is difficult to use it in practice.

There is another method for estimating the probability of impulses overlapping [24]. This method is partly based on the previous one, but it provides a mathematical apparatus for more detailed statistical analysis of overlapping events. Unfortunately, the method is also hardly applicable for cases with more than two pulse trains under consideration.

In the course of the literature review, no other methods for estimating the probability of interest were found. The challenges associated with CSE are relatively new and will significantly complicate the radio monitoring tasks only in the next 5–10 years. This explains the lack of information about this problem in scientific sources.

4   A New Method for Determining the Probability of Signals Overlapping in Time

The proposed method is based on the probabilistic description of impulse sequences. Let us consider the signal environment with M SRE emitting impulse signals with a batch structure. By the previously developed methods [9], the expression for determining the total share of time taken up by the impulses of a certain sequence can be represented by the following expression:

$$R = \int_0^\infty x q(x) dx, \tag{3}$$

where $q(x)$ is such a function that $q(x)\Delta x$ is the average number of impulses, the duration of which is greater than x by Δx, and which occur per unit of time in the described sequence.

The value of R can be interpreted as a probability that at a random moment, the signal level of the sequence under consideration will correspond to the presence of an impulse.

In practice, the statistical parameters of the SRE signal are unknown. However, according to the mode of operation of the SRE, the average period of repetition of the batch T_b, the average period of repetition of impulses in the batch T_i, the average duration of impulses τ and the average number of impulses in the batch n_i are known.

Then the probability of R can be defined as the product of the average occupancy of a sequence of impulse batches by the average occupancy of individual impulse batches:

$$R = \frac{n_i T_i}{T_b} \cdot \frac{\tau}{T_i} = \frac{n_i \tau}{T_b}. \tag{4}$$

In this case, the probability that at a random moment, the signal level measured at the receiver input will correspond to the presence of an impulse only in one of the considered impulse sequences will be equal to:

$$P_1 = \sum_{i=1}^{M} \left[R_i \cdot \prod_{i \neq j}^{M} (1 - R_j) \right]. \tag{5}$$

The probability P_1 can be considered as a probability that at a random moment only one impulse is received at the receiver's input, and there are no overlaps with impulses from other sequences.

The probability that, at a random time, no pulse from the considered M pulse sequences is received at the receiver input is equal to:

$$P_2 = \prod_{i=1}^{M} (1 - R_i). \tag{6}$$

Then the probability that, at a random time, there are no overlaps at the receiver input is equal to:

$$Q_0 = P_1 + P_2. \tag{7}$$

Using expressions (5), (6), (7) the probability of two or more pulses overlapping is determined by the expression:

$$P_0 = 1 - Q_0 = 1 - \left(\sum_{i=1}^{M} \left[R_i \cdot \prod_{i \neq j}^{M} (1 - R_j) \right] + \prod_{i=1}^{M} (1 - R_i) \right). \tag{8}$$

In practice, an algorithm of fast Fourier transform (FFT) is widely used before determining the time-frequency parameters of signals. The algorithm involves the accumulation of a certain number of measurements (samples) before calculations. Therefore, the short-term overlap of pulses may slightly distort the result of the primary processing. Thus, of interest is the probability that of n accumulated counts the overlap will affect at least k. The Bernoulli formula can be used for this:

$$P = \sum_{i=k}^{n} C_n^i \cdot P_0^i \cdot Q_0^{n-i}. \tag{9}$$

In cases when the probability of P_0 is not too close to 0 or 1, and condition $n \cdot P_0 \cdot Q_0 > 10$ is met, the Laplace integral theorem can be used to calculate the probability P.

The resulting expressions can also be modified by limiting the minimum overlap duration d:

$$R'_i = \frac{n_{ii} \cdot (\tau_i - d)}{T_{bi}}. \tag{10}$$

Using (10), modified expressions (5)–(9) may be useful if it is necessary to determine the probability that pulses are overlapped in k consecutive samples and the sampling rate f_s is known. In this case, the minimum overlay duration can be set based on the following expression:

$$d = \frac{k}{f_{OT}}. \tag{11}$$

The time required to calculate P_0 by the formula (8) depends on the number of pulse sequences M:

$$T \sim M \cdot (M - 1) + M. \tag{12}$$

The expression (12) shows that the time of calculating the probability P_0 by the proposed method increases polynomial with the number of considered SRE. When simulating, according to (1), time T grows as a power function with increasing accuracy. Therefore, time costs increase slower compared to simulation with more accurate calculations.

5   Analysis of the Accuracy of Calculation Results Obtained by the New Method

To estimate the accuracy of the calculation of the probability of two or more impulses overlapping, two values of ME of this probability are compared. The first value P_{sim} is obtained as a result of the simulation, the second value P_{an} is obtained using expressions (4)–(9).

The dependence of the error square of the simulation result ΔP on the step of the phase variation $\Delta \varphi$, used during the simulation, is determined by the formula $\Delta P = (P_{\text{сим}} - P_{\text{ан}})^2$. The number of combinations of initial phases, for which the probability of impulses overlapping is calculated in the simulation, depends on the $\Delta \varphi$ value. Therefore, the accuracy of determining the ME of overlapping probability and the time of simulation execution also depend on this value.

As an example, calculations were made for three SRE with the following values of the probability of the SRE impulse presence at a random time: $R_1 = 0.1$; $R_2 = 0.125$; $R_3 = 0.167$. The results of these calculations are shown in Fig. 1.

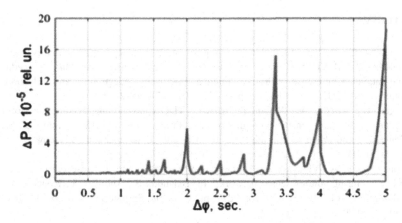

Fig. 1. Dependence of the simulation error square ΔP on the phase variation step $\Delta \varphi$

The analysis of the results presented in Fig. 1 shows that with a decrease in the step of phase variation $\Delta \varphi$, the simulation result approaches the one obtained by the expressions (4)–(9), which confirms the correctness of the proposed method. This allows us to conclude that the developed technique allows determining the probability of failure of primary processing of radio monitoring system, without requiring significant time and computing resources.

We propose the following procedure for using the developed method:

1. Make a complete list of all SRE which may be active in a region where radio monitoring system will be used.
2. Obtain the information about the signal structure of the listed systems (from reference books, manufacturers web-sites or other sources).
3. Calculate all the probabilities R using the expressions (3) or (4).
4. Calculate the probability of interest using the expression (8).

6   Conclusion

The paper shows the consequences of the negative influence of signals overlapping in time. As a result of this overlapping, the quality of the primary signal processing deteriorates and, as a result, the stability of the radio monitoring equipment decreases.

Based on the data on the expected location of SRE and time-frequency parameters of their signals, the developed methodology allows determining the probability of signals overlapping. This makes it possible to predict the errors of signal processing occurring during radio monitoring in various conditions characterized by the presence of a complex signal environment.

Predicting such situations and taking into account possible consequences allows taking preventive measures that increase the efficiency of radio monitoring. In practice, this makes it possible to identify more effectively and faster the location of offenders. These offenders interfere with the normal operation of navigation, communication, and other equipment.

References

1. Mashkov, G., Borisov, E., Fokin, G.: A positioning accuracy experimental evaluation in SDR-based MLAT with joint processing on range measurement. In: Proceedings - International Conference on Radar, Antenna, Microwave, Electronics, and Telecommunications, ICRAMET 2016, Jakarta, pp. 7–12 (2016). 7849572

2. Fokin, G., Kireev, A., Al-Odliari, A.H.A.: TDOA positioning accuracy performance evaluation for arc sensor configuration. In: Proceedings - 2018 Systems of Signals Generating and Processing in the Field of on Board Communications, Moscow, vol. 2018, pp. 1–5 (2018)

3. Koucheryavy, A., Vladyko, A., Kirichek, R.: State of the art and research challenges for public flying ubiquitous sensor networks. In: Balandin, S., Andreev, S., Koucheryavy, Y. (eds.) ruSMART 2015. LNCS, vol. 9247, pp. 299–308. Springer, Cham (2015). https://doi.org/10.1007/978-3-319-23126-6_27

4. Hoang, T., Kirichek, R., Paramonov, A., Koucheryavy, A.: Influence of intentional electromagnetic interference on the functioning of the terrestrial segment of flying ubiquitous sensor network. Information Science and Applications (ICISA) 2016. LNEE, vol. 376, pp. 1249–1259. Springer, Singapore (2016). https://doi.org/10.1007/978-981-10-0557-2_118

5. Borodulin, R.U., Sosunov, B.V., Makarov, S.B.: The principles of antennas constructive synthesis in dissipative media. In: Galinina, O., Andreev, S., Balandin, S., Koucheryavy, Y. (eds.) NEW2AN/ruSMART/NsCC 2017. LNCS, vol. 10531, pp. 455–465. Springer, Cham (2017). https://doi.org/10.1007/978-3-319-67380-6_41

6. Petrov, A.A., Davydov, V.V.: Improvement frequency stability of caesium atomic clock for satellite communication system. In: Balandin, S., Andreev, S., Koucheryavy, Y. (eds.) ruSMART 2015. LNCS, vol. 9247, pp. 739–744. Springer, Cham (2015). https://doi.org/10.1007/978-3-319-23126-6_68

7. Tarasenko, M.Yu., Davydov, V.V., Lenets, V.A., Akulich, N.V., Yalunina, T.R.: Features of use direct and external modulation in fiber optical simulators of a false target for testing radar station. In: Galinina, O., Andreev, S., Balandin, S., Koucheryavy, Y. (eds.) NEW2AN/ruSMART/NsCC 2017. LNCS, vol. 10531, pp. 227–232. Springer, Cham (2017). https://doi.org/10.1007/978-3-319-67380-6_21

8. Fokin, G., Ali, A.-o.A.H: Algorithm for positioning in non-line-of-sight conditions using unmanned aerial vehicles. In: Galinina, O., Andreev, S., Balandin, S., Koucheryavy, Y. (eds.) NEW2AN/ruSMART 2018. LNCS, vol. 11118, pp. 496–508. Springer, Cham (2018). https://doi.org/10.1007/978-3-030-01168-0_44

9. Koucheryavy, A., Bogdanov, I., Paramonov, A.: The mobile sensor network life-time under different spurious flows intrusion. In: Balandin, S., Andreev, S., Koucheryavy, Y. (eds.) NEW2AN/ruSMART 2013. LNCS, vol. 8121, pp. 312–317. Springer, Heidelberg (2013). https://doi.org/10.1007/978-3-642-40316-3_27

10. Petrov, A.A., Davydov, V.V., Myazin, N.S., Kaganovskiy, V.E.: Rubidium atomic clock with improved metrological characteristics for satellite communication system. In: Galinina, O., Andreev, S., Balandin, S., Koucheryavy, Y. (eds.) NEW2AN/ruSMART/NsCC 2017. LNCS, vol. 10531, pp. 561–568. Springer, Cham (2017). https://doi.org/10.1007/978-3-319-67380-6_52

11. Koucheryavy, A.: Networks interoperability. In: Proceedings – International Conference on Advanced Communication Technology, ICACT (Phoenix Park; South Korea), vol. 1, pp. 691–693 (2009). N 4810044

12. Podstrigaev, A.S., Smolyakov, A.V., Davydov, V.V., Myazin, N.S., Slobodyan, M.G.: Features of the development of transceivers for information and communication systems considering the distribution of radar operating frequencies in the frequency range. In: Galinina, O., Andreev, S., Balandin, S., Koucheryavy, Y. (eds.) NEW2AN/ruSMART 2018. LNCS, vol. 11118, pp. 509–515. Springer, Cham (2018). https://doi.org/10.1007/978-3-030-01168-0_45

13. Podstrigaev, A.S.: All-purpose adjuster for microwave microstrip devices. In: Proceedings - CriMiCo 2014 - 2014 24th International Crimean Conference Microwave and Telecommunication Technology (Sevastopol), pp. 896–897 (2014). 6959682

14. Sivers, M., Fokin, G., Dmitriev, P., Kireev, A., Volgushev, D., Hussein Ali, A.-o.A.: Indoor positioning in WiFi and NanoLOC networks. In: Galinina, O., Balandin, S., Koucheryavy, Y. (eds.) NEW2AN/ruSMART 2016. LNCS, vol. 9870, pp. 465–476. Springer, Cham (2016). https://doi.org/10.1007/978-3-319-46301-8_39

15. Pirmagomedov, R., Blinnikov, M., Kirichek, R., Koucheryavy, A.: Wireless nanosensor network with flying gateway. In: Chowdhury, K.R., Di Felice, M., Matta, I., Sheng, B. (eds.) WWIC 2018. LNCS, vol. 10866, pp. 258–268. Springer, Cham (2018). https://doi.org/10.1007/978-3-030-02931-9_21

16. Semenov, V.V., Nikiforov, N.F., Ermak, S.V., Davydov, V.V.: Calculation of stationary magnetic resonance signal in optically oriented atoms induced by a sequence of radio pulses. Sov. J. Commun. Technol. Electron. 36(4), 59–63 (1991)

17. Ryazantsev, L.B., Likhachev, V.P.: Assessment of range and radial velocity of objects of a broadband radar station under conditions of range cell migration. Meas. Tech. 60(11), 1158–1162 (2018)

18. Kovalchukov, R., Moltchanov, D., Samuylov, A., Ometov, A., Andreev, S., Koucheryavy, Y.: Analyzing effects of directionality and random heights in drone-based mmWave communication. IEEE Trans. Veh. Technol. 67(10), 10064–10069 (2018). 8412525

19. Podstrigaev, A.S., Davydov, R.V., Rud, V.Yu., Davydov, V.V.: Features of transmission of intermediate frequency signals over fiber-optical communication system in radar station. In: Galinina, O., Andreev, S., Balandin, S., Koucheryavy, Y. (eds.) NEW2AN/ruSMART 2018. LNCS, vol. 11118, pp. 624–630. Springer, Cham (2018). https://doi.org/10.1007/978-3-030-01168-0_56

20. Davydov, R.V., et al.: Fiber-optic transmission system for the testing of active phased antenna arrays in an anechoic chamber. In: Galinina, O., Andreev, S., Balandin, S., Koucheryavy, Y. (eds.) NEW2AN/ruSMART/NsCC 2017. LNCS, vol. 10531, pp. 177–183. Springer, Cham (2017). https://doi.org/10.1007/978-3-319-67380-6_16

21. Podstrigaev, A.S., Ryazantsev, L.B., Likhachev, V.P.: Technique for tuning microwave strip devices. Meas. Tech. 59(5), 547–550 (2016)

22. Sivers, M., Fokin, G.: LTE positioning accuracy performance evaluation. In: Balandin, S., Andreev, S., Koucheryavy, Y. (eds.) ruSMART 2015. LNCS, vol. 9247, pp. 393–406. Springer, Cham (2015). https://doi.org/10.1007/978-3-319-23126-6_35

23. Self, A.G., Smith, B.G.: Intercept time and its prediction. In: IEE Proceedings. Part F: Communications Radar and Signal Processing. vol. 132, No. 4, pp. 215–222 (1985)

24. Perkins, J.: Probability and Distribution in Time of Pulse Overlap in Periodic Settings, pp. 1–19 (1995)

Signal Transmitting in Pheromone Networks

Maxim Zakharov[1]([⊠]), Ruslan Kirichek[2], Maria Makolkina[2],
and Andrey Koucheryavy[1]

[1] The Bonch-Bruevich Saint - Petersburg State University of Telecommunications,
Moika Embankment 61, St. Petersburg 191186, Russia
zaharov.spbgut@gmail.com, akouch@mail.ru
[2] Peoples' Friendship University of Russia (RUDN University),
6 Miklukho-Maklaya St, Moscow 117198, Russian Federation
kirichek@sut.ru, makolkina@list.ru

Abstract. The authors of this article consider the transfer of data in pheromone networks, which are the application of molecular networks. The authors consider the types of signals that can be used in molecular networks for data transfer (information transfer by moving substances—human and animal pheromones, plant spores and pollen, etc.). The article also discusses the data transmission channel, affecting the channel additive and multiplicative interference. A review of possible interference of natural and man-made nature is made, their possible impact is considered. The article describes the design of a laboratory test bench for conducting an experiment on data transfer by moving a substance, as well as experimental results.

Keywords: Internet of things · Molecular networks ·
Pheromone data transmitting

1 Introduction

The development of science and technology at the beginning of the XXI century led to the emergence of such telecommunication technologies that force us to rethink both the principles of providing telecommunication services and the services themselves. These include ultradense networks with ultra-low latency, Internet skills, tactile Internet, and many other technologies that are planned to be implemented as part of the transition to fifth-generation networks (5G). However, all of the above technologies work with the macro-world objects, and the sources of information in the nanoworld have so far been available only in laboratory conditions [1].

Today, advanced studies of domestic and foreign scientists laid the foundation for an integrated and systematic approach to the implementation of interaction with objects of the nanoworld and the provision of new telecommunication services [2–4]. Formed approaches to the organization of nanoscale based on the

© Springer Nature Switzerland AG 2019
O. Galinina et al. (Eds.): NEW2AN 2019/ruSMART 2019, LNCS 11660, pp. 534–539, 2019.
https://doi.org/10.1007/978-3-030-30859-9_46

most recent, which are able to solve problems at the molecular level. A classification of nanoseconds has been developed that divides them into electromagnetic - in which information is transmitted using radio waves in the terahertz range, and molecular - in which information transfer is used to transmit information. The study of molecular networks is currently quite promising, since this area of knowledge is located at the junction of several disciplines - organic chemistry, biology and telecommunications, and in the future allows us to expand the boundaries of knowledge about the nanoworld around us and learn how to interact with it [5].

2   Data Link in Pheromone Networks

Next, we consider a data link for networks in which information is transmitted by moving volatile organic substances (VOC) - alcohols, pheromones, spores and pollens, etc.

For the beginning let's define what the signal is. Signal (from the Latin. Signum - sign) in the classical theory of communication is called the physical process by which the message is transmitted at a certain distance, while the signal itself is described by a function of time. In molecular networks, pheromones, pollen, plant spores, etc. can be used as a signal to transmit information. Environment is the surrounding area. Any objects of the biosphere can act as a transmitter/receiver: man, animals, plants, mushrooms, etc. At the same time, most of the objects in the biosphere have not only powerful transmitters, but also quite sensitive receivers that allow analyzing information that is continuously coming from the outside world. For example, many predators are able to smell the prey for several kilometers, and plants can attract pollinating insects during flowering. However, a person is deprived of a sensitive receiver - we are able to smell the flowers, but we cannot distinguish the poisonous fruit from the edible one on the basis of the aroma emitted by it [6]. Therefore, it is necessary, based on modern achievements of science and technology, to develop methods for the exchange of information between the objects of the nanoworld and the macrocosm, including using molecular nano-nets and modern technical means (gas analyzers, microspectrometers, etc.).

It should be noted that molecular networks have not only theoretical, and practical application [7]. For example, in the provision and organization of augmented and virtual reality services, various smells are often used, contributing to a deeper user immersion in the surrounding virtual or augmented reality. Molecular networks can also find their application in ways where it is not possible to use traditional methods of information transfer - in medicine, agriculture, industry, etc. However, before considering the possible range of applications of such networks, it is necessary to determine what obstacles may arise in the path of molecular communications.

3 Laboratory Test Bench

For exploring the problem of signal propagation in molecular networks, as well as the effect of interference on the signal, the authors propose a laboratory test bench architecture that transforms volatile organic compounds into digital information for further processing and analysis. Figure 1 shows this model network, which consists of the following nodes: a client collecting environmental indicators and acting as a receiver in molecular networks, an IEEE 802.11n wireless router, a device for intercepting and analyzing traffic, and a server for storing and further analyzing data.

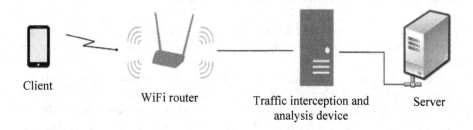

Client   WiFi router   Traffic interception and analysis device   Server

Fig. 1. Laboratory test bench architecture.

For the client's hardware, the NodeMCU ESP8266 was used, which is conveniently used in projects for transmitting a signal to a local network or the Internet via Wi-Fi. To collect environmental data, the CJMCU-811 air quality sensor was chosen, which measures 3 parameters: the concentration of carbon dioxide (CO_2), the concentration of volatile organic compounds (TVOC) and the ambient temperature. The air quality sensor has 2 signal outputs (SDA and SCL), which are connected to the SPI outputs on the board, as shown in Fig. 2.

For the client's software, the Arduino IDE development environment and the Arduino Processing programming language were used. As a server software, the JavaScript programming language (NodeJS) was chosen, which is well suited for implementing server applications and applications for interacting with the database. MongoDB was used as a DBMS. Figure 3 shows the interaction between the client, the server and the DBMS.

The device for intercepting and analyzing traffic is a personal computer with two network interfaces and the Wireshark traffic analyzer installed on it, with which the delays and quality of service (QoS) parameters for the developed model network were obtained. During the experiment 150 measurements were carried out. As a result of data analysis carried out the following values were obtained for each of the above parameters: mean value, standard deviation, confidence deviation and confidence interval. To calculate these indicators, a confidence level of 95% was chosen. The results are shown in Table 1.

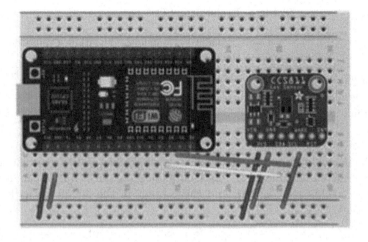

Fig. 2. Connection diagram of peripheral devices to the controller.

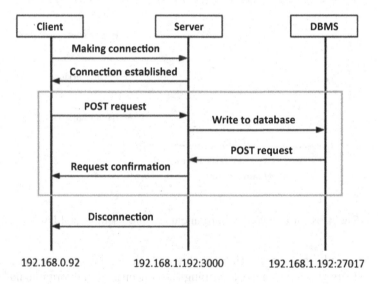

Fig. 3. General scheme of network interaction.

4 Testing the Client Device

On the developed model network, experiments were conducted to analyze the concentration of volatile organic compounds in the environment, using the developed model network. The following substances were used: perfume with pheromones and insecticidal powder from ants, which contains molecules of the substance attracting insects to the powder.

During the experiment, the possibility of data transmission using pheromones was investigated. For this purpose, the sensor was used to collect information

Table 1. Evaluation of QoS parameters.

	Average	Standard deviation	Confidence deviation	Confidence interval
Delays, ms	70,46	14,05	3,89	70,46 ± 3,89
Packet size, bytes	209,34	53,88	8,62	209,34 ± 8,62
Throughput, bytes/s	958,03	432,45	69,21	958,03 ± 69,2

about the normal content of volatile organic substances in the room air (natural noise), the increased content of volatile organic substances due to the use of ant insecticide (technogenic noise), and the content in the air of the useful signal (pheromones) (Fig. 4).

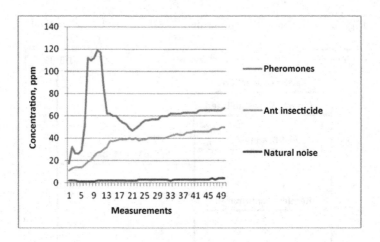

Fig. 4. Natural noise, technogenic noise and useful signal levels.

On the graph you can see that the line, which displays the sensor readings from the effects of pheromones, describing the concentration of organic molecules increases sharply at the beginning, which proves the rapid evaporation of the liquid after its primary application. On the graph you can see that the line, which displays the sensor readings from the effects of ant insecticide, describing the concentration of organic molecules in an isolated environment increases gradually without sudden bursts and has smaller values, as the powdered substance slowly releases volatile compounds and fills the space of the box.

5   Conclusion

The present findings confirm that the software and hardware complex for the analysis of the concentration of organic substances in the environment was developed and successfully tested. The developed model network can be used for further analysis of traffic in molecular networks. The experiments have shown that the useful signal transmitted by pheromones is clearly visible against the background of technogenic and natural noise.

Future research should be devoted to the development of a more sophisticated and isolated from the environmental test bench to conduct cleaner measurements and eliminate the influence of the human factor. It is a question of future research to investigate the most suitable organic matter for the organization of the data receiving and transmitting channel.

Acknowledgements. The publication has been prepared with the support of the "RUDN University Program 5-100" and funded by RFBR according to the research projects No. 12-34-56789 and No. 12-34-56789.

References

1. Farsad, N., Yilmaz, H.B., Eckford, A., Chae, C.-B., Guo, W.: A comprehensive survey of recent advancements in molecular communication. IEEE Commun. Surv. Tut. **18**(3), 1887–1919 (2016)
2. Akyildiz, I.F., Pierobon, M., Balasubramaniam, S., Koucheryavy, Y.: The internet of bio-nano things. IEEE Commun. Mag. **53**(3), 32–40 (2015)
3. Pirmagomedov, R., Hudoev, I., Kirichek, R., Koucheryavy, A., Glushakov, R.: Analysis of delays in medical applications of nanonetworks. In: 8th International Congress on Ultra Modern Telecommunications and Control Systems and Workshops (ICUMT), pp. 80–86 (2016)
4. Pirmagomedov, R., Blinnikov, M., Glushakov, R., Muthanna, A., Kirichek, R., Koucheryavy, A.: Dynamic data packaging protocol for real-time medical applications of nanonetworks. In: Galinina, O., Andreev, S., Balandin, S., Koucheryavy, Y. (eds.) NEW2AN/ruSMART/NsCC -2017. LNCS, vol. 10531, pp. 196–205. Springer, Cham (2017). https://doi.org/10.1007/978-3-319-67380-6_18
5. Unluturk, B.D., Akyildiz, I.F.: An end-to-end model of plant pheromone channel for long range molecular communication. IEEE Trans. Nanobiosci. **16**(1), 11–20 (2017)
6. Bicen, A.O., Akyildiz, I.F.: System-theoretic analysis and least-squares design of microfluidic channels for flow-induced molecular communication. IEEE Trans. Signal Process. **61**(20), 5000–5013 (2013)
7. Pierobon, M., Akyildiz, I.F.: Capacity of a diffusion-based molecular communication system with channel memory and molecular noise. IEEE Trans. Inf. Theory **59**(2), 942–954 (2013)

Integrating Internet of Things with the Digital Object Architecture

Mahmood Al-Bahri[1,2(✉)], Kirichek Ruslan[1,2(✉)],
and Borodin Aleksey[2,3(✉)]

[1] The Bonch-Bruevich Saint-Petersburg State University
of Telecommunications, 22 Prospekt Bolshevikov,
193232 St. Petersburg, Russian Federation
albahri.89@hotmail.com, kirichek@sut.ru
[2] Peoples Friendship University of Russia (RUDN University),
6 Miklukho-Maklaya St., 117198 Moscow, Russian Federation
[3] PJSC "Rostelecom", Moscow, Russia
aleksey.borodin@rt.ru,
https://www.sut.ru/
http://www.rudn.ru/

Abstract. The Internet of Things (IW) is a modern concept for the development of communication networks in the short and long term. The concept implies the integration of devices ("things"), equipped with built-in technologies for interacting with each other or with the external environment, into a single worldwide network. Today, the number of devices connected to the network exceeds the number of all inhabitants of the planet and continues to grow rapidly, which raises the question of assigning each object a unique address to ensure confidentiality and security during data transmission, as well as unique identification of Internet of Things on the global network. The article presents a method for identifying devices and applications of the Internet of Things in heterogeneous communication networks based on the Digital Objects Architecture. It is also proposed to develop methods for installing the DO identifier into devices of the Internet of things with various wireless data transmission modules.

Keywords: DOA · Identification · Internet of Things ·
Wireless communications network technology · Integration

1 Introduction

Today, the Internet of Things (IoT) is the most advanced platform in the framework of the development of digital intelligence in the concept of "smart country" [1]. According to consulting company Gartner, over the next five years, more than 25 billion devices will be present in each of the spheres of human activity, including everyday life and business. Discovering tremendous opportunities, this assumption at the same time makes one worry about the term "Personality" within the concept IoT. Identification plays an important role on the Internet of Things. For example, the ability of hackers to intercept data in communication networks in order to obtain IMEI-numbers of multiple

© Springer Nature Switzerland AG 2019
O. Galinina et al. (Eds.): NEW2AN 2019/ruSMART 2019, LNCS 11660, pp. 540–547, 2019.
https://doi.org/10.1007/978-3-030-30859-9_47

devices and the subsequent clogging of the database by mass sending distorted (broken) messages. Another example is that hackers can use a portable RFID reader to steal personal data of other people's bank cards in public transport, using technologies like PayPass, due to the lack of confirmation of the identity of the owner of the RFID reader [2] (Fig. 1).

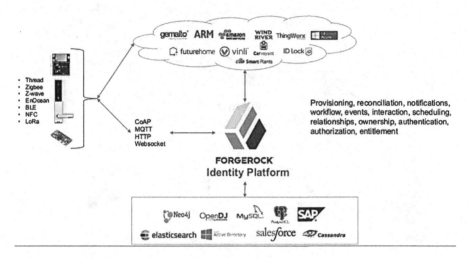

Fig. 1. IoT- Reconcile and integrate

Now, there are many scientific studies focused mainly on the use of communication identifiers, like the IP address and mobile phone number in the IoT. At the same time, identification has a much wider scale and is more appropriate for many applications and entities (subjects) in IoT. In addition to the purposes of identification in the field of communications, research includes the identification of things, as well as, for example, services, users, data, places. Various identification schemes that exist now are already standardized and introduced in the market.

Different types of identifiers are used depending on the application and user requirements. At the heart of the Internet of Things is the interaction between things and users of things through electronic means such as sensors, actuators, and wireless communication.

Things and users must be identified in order to achieve this interaction. Many other entities are also involved in the interaction, while being part of the IoT system, for them identification is also an important aspect. Different entities with associated identifiers in the IoT sphere model are shown in Fig. 2. Various identification schemes already exist, standardized and implemented.

The purpose of this work is to describe the Digital Object Architecture (DOA) and its capabilities to achieve the generalized requirements of the Internet of Things, in particular the use of DOA to solve security issues and compatibility in applications of IoT. At the same time, the paper presents the possibilities of the Digital Object Architecture in overcoming the complex tasks related to the compatibility of the Internet of Things.

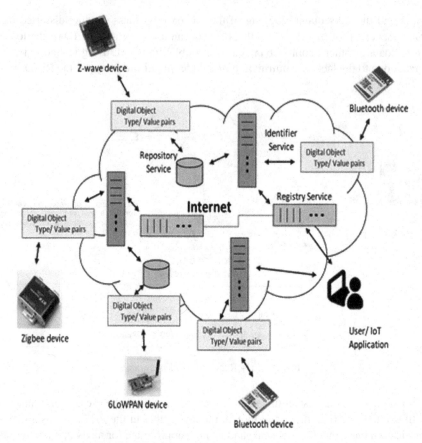

Fig. 2. DOA representation and components [4].

2   Overview of Work on This Subject

As mentioned earlier, our last paper presented a review of DOA, its purpose and motivation. Details of the DOA service components were also presented, including how the components interact with each other. This paper will describe the principles for using DOA capabilities to ensure compatibility in devices and applications of the Internet of Things.

The Digital Object Architecture is presented in the form of an open framework that allows you to openly and safely transfer information within a heterogeneous information system. A minimum set of necessary architectural components, protocols and services has been defined, which together form a structure that implements common information functions and service compatibility. The core of a DOA structure is a concept in which any information represented in digital form can be structured as a digital object, which is assigned a globally unique identifier [3]. This identifier returns the accompanying information of a digital object that exists in the resolution system regardless of changes made to the digital object by the entity receiving the digital object. This accompanying information may include location, metadata, checksums, digital signatures, certificates, public key, etc. The accompanying information components associated with a digital object are treated as attributes of a digital object. A digital object is also used as a representation of IoT entities, for example, on top of TCP/IP networks, regardless of the technologies underlying the system [4]. Internet of Things devices that use various protocols for communication can be connected through private and public networks, or via the Internet, using various protocols, including TCP/IP [5].

The DOA allows any digital information, pre-structured as a digital object, to be safely identified, defined and distributed regardless of the particular system, service or application where the information was created or stored [6]. The DOA framework consists of three basic, fundamental components that, as they are implemented, implement the following services: global identification service, digital object storage service, digital object service register [7].

The global identification service allows you to assign a global identifier to any digital object. This service provides a resolution and administration protocol designed to determine the accompanying information associated with a digital object: storage location, information origin, with the ability to retrieve and control in compliance with the necessary security measures. The identification service must be a distributed system with built-in protection mechanisms to ensure the integrity of the service, its reliability, and the integrity of the stored data. Authentication and confidentiality of operations with stored data, the presence of selective access control for any metadata associated with the identifier are also required. A set of distributed services for storing digital objects facilitates the safe storage, access and distribution of objects using their identifiers. The storage itself is a digital object that can store other objects inside it [8]. A digital object can perform a specific set of actions, including accessing other digital objects, creating new digital objects, etc. The storage of digital objects can be a set of IoT devices, which are also digital objects.

A digital object can have many attributes associated with a real object. Some attributes may describe the nature of the IoT device. In particular, an object may have control attributes that are associated with software, providing direct interaction with the functions of the IoT device, for example, turning the system on or off, obtaining readings of the temperature sensor on the device. In addition, a digital object can also have attributes that determine the availability of basic device attributes, thus determining who can interact with the IoT device using the interface described in the object attributes [9].

The structure of a digital object can be formed as a digital representation of the physical device IoT. The system of components, namely the registry, has the ability to determine how to find and access such entities. Compatibility criteria in terms of the Internet of Things implies the existence of an API, so that digital objects can interact with the devices they are attached to [10]. This approach can be used to achieve specific access controls for the convenience of each repository. On the other hand, the storage can be a digital object providing access to data generated by a separate device of the Internet of Things. The architecture of digital objects does not limit the number of possible storages.

A set of federated digital object registry services allows you to detect any digital object. The registry of digital objects may also provide the ability to search for metadata or simple data in a digital object. Using the registry allows you to detect digital objects according to various criteria [11], for example:

1. search for different types of digital objects metadata records within various registry services,
2. search through different levels of network connection within different registry services,
3. search by different types of data management services,
4. Search by various types of security and access control systems.

The access policy, which includes the authentication and authorization of client requests applied to many federated registry services, must be explicitly defined [12].

3   Interaction and Integration of the Internet of Things with the Digital Object Architecture

To solve a number of tasks posed by the Digital Objects Architecture in the field of monitoring, as well as in smart home systems, it is necessary to analyze the existing issues of integrating the unique identifier DOI into electronic devices of various levels.

As examples of existing systems, the integration of which is possible in real conditions, take the following technologies: Wi-Fi, ZigBee and LoRaWAN. These technologies are used in devices of various levels: for example, Wi-Fi controllers can be simultaneously found in low-power micro-controllers to solve problems of a smart home, and in network cards of devices based on x86 or x64 architecture, which have enormous computing power. Depending on the power of the devices, the available functionality of such devices is determined, however, the basic methods of introducing an identifier into the devices are the same for all (Fig. 3).

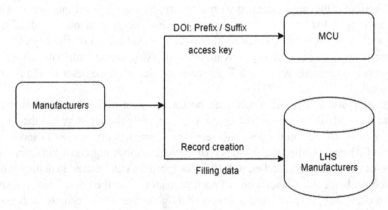

Fig. 3. Introduction of basic information into the device on the microcontroller at the production stage

In this case, at the production stage, each device defined in the global resolution system must have a digital object identifier and a key for accessing the identifier metadata specified by software methods (by analogy with existing identifiers such as MAC or IMEI). Filling in these data should be accompanied by the creation of the corresponding handle-records in the LHS of the device manufacturer.

The microcontroller has its own digital copy in the global resolution system due to the possibility of creating universal methods for identifying Internet devices of Things. Such information stored in a unique domain for each released device can be a version of the available protocols of the device, binding to other identification technologies, accompanying information, or even basic access commands for the device. In general, the access key and DOI must be accessible to control devices. If the controlling device has a full-fledged operating system, this data is accessed using the OS driver; if the microcontroller acts as a control device, data access is performed through the basic commands of the microcontroller. An example of use (Fig. 4) can be the situation when device 1, using an application, writes to the available global device address in the TCP/IP network in the field of the network_address digital object accessible to the device. This allows the interaction of two different devices (1 and 2) without intermediary servers, usually helping to establish a connection. To establish a connection, it is enough to have a DOI device.

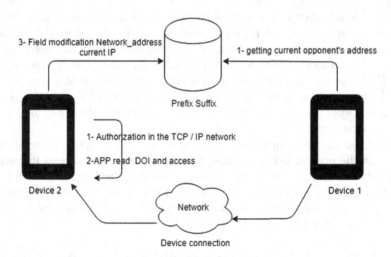

Fig. 4. Implementing digital object fields to store the device's network address

This method is suitable for complex devices containing at least two devices on a microprocessor, examples of which are most modern smartphones. Inability to implement cryptographic functions necessary to modify the data in the system resolution, as well as the lack of direct access methods to the global network makes it impossible to implement such functions on simple devices. At the same time, in the case of using networks for smart home devices using, for example, the ZigBee protocol stack, this example is not realizable, because there is no direct access to the Internet, and as a result, there is no resolution system to the end devices. Implementation of this functionality should be carried out through the gateway software, together with the implementation of the necessary functionality in ZigBee applications.

Additional functionality necessary for implementation on devices simultaneously with the receipt of a DOI and a unique key is the ability to rewrite those to arbitrary while maintaining the original. Device manufacturers are obliged to limit the maximum amount of data contributed by one device to their domain, assuming reassignment of the main (but not the original) digital object identifier and the corresponding access key to impartial, in order to increase the volume of metadata and increase the speed of data access. It is worth noting that it is important to maintain the original DOI, which is one of the possible proofs of the authenticity of devices, with the possibility of its return. This functionality can be used in scenarios of combating counterfeit Internet of Things devices.

4 Conclusions

Now, the system of end-to-end global digital identifier implemented based on the architecture of digital objects has been actively developed. As noted in the paper, the use of an identifier based on DOA will allow to take into account all existing unique identifiers and addresses (for example, MAC, IMEI, ID, IPv4/IPv6, etc.), providing end-to-end identification of devices and applications of the Internet of things without reference to a specific identifier. This will allow the implementation of a global and truly international identification system, as it is implemented with the support of ITU. There are various identification methods that cannot be used by many devices of the Internet of things for a number of objective reasons. At the same time, a very important property is the fixedness of the correlation of the identifier with the actual device of the Internet of things (physical address), as well as universality in the use of the identifier in various industries.

The global identifier allows access to the device and detect it not only for the purposes of the original IoT application, but also by other applications that want to interact with the device. Ownership and access control defined by a digital object can provide secure access to a device in a variety of IoT applications without losing the necessary security systems. The interface that interacts with the device can be detected on the fly, without restriction to the original application.

Acknowledgements. The publication has been prepared with the support of the RUDN University Program 5-100 and funded by RFBR according to the research projects No. 12-34-56789 and No. 12-34-56789.

References

1. Al-Bahri, M., Ateya, A.A., Muthanna, A., et al.: Combating counterfeit for IoT system based on DOA. In: Proceedings of the 2018 10th International Congress on Ultra Modern Telecommunications and Control Systems and Workshops (ICUMT) 2018, St. Petersburg, Russia, 5–9 November 2018, pp. 338–342. IEEE (2018)

2. Al-Bahri, M., Yankovsky, A., Borodin, A., Kirichek, R.: Smart system based on DOA and IoT for products monitoring and anti-counterfeiting. In: 2019 4th MEC International Conference on Big Data and Smart City (ICBDSC), pp. 25–31. IEEE (2019)
3. Al-Bahri, M., Yankovsky, A., Borodin, A., Kirichek, R.: Testbed for identify IoT-devices based on digital object architecture. In: Galinina, O., Andreev, S., Balandin, S., Koucheryavy, Y. (eds.) NEW2AN/ruSMART 2018. LNCS, vol. 11118, pp. 129–137. Springer, Cham (2018). https://doi.org/10.1007/978-3-030-01168-0_12
4. Recommendation ITU-T T.181203: An architecture for IoT interoperability. Geneva: ITU-T (2018). 25
5. Kirichek, R., Pham, V.-D., Kolechkin, A., Al-Bahri, M., Paramonov, A.: Transfer of multimedia data via LoRa. In: Galinina, O., Andreev, S., Balandin, S., Koucheryavy, Y. (eds.) NEW2AN/ruSMART/NsCC 2017. LNCS, vol. 10531, pp. 708–720. Springer, Cham (2017). https://doi.org/10.1007/978-3-319-67380-6_67
6. Muthanna, M.S.A., Abdukodir, K., Ateya, A.A, Al-Bahri, M.: Delay Tolerant Network model based on D2D communication. In: 2019 4th MEC International Conference on Big Data and Smart City (ICBDSC), pp. 1–5. IEEE (2019)
7. Ignatova, L., Khakimov, A., Mahmood, A., Muthanna, A.: Analysis of the Internet of Things devices integration in 5G networks. In: 2017 Systems of Signal Synchronization, Generating and Processing in Telecommunications (SINKHROINFO), pp. 1–4. IEEE (2017)
8. Al-Bahri, M., Ruslan, K., Borodin, A.: The digital object architecture as a basis for identification in the era of the digital economy. elektrosvyaz (2019). No. 1, pp. 12–22
9. Al-Bahri, M.: Method of identification of devices and applications of the internet of things in heterogeneous communication networks based on digital object architecture. elektrosvyaz (2019). No. 4, pp. 41–47
10. Al-Bahri, M., Ruslan, K., Sazonov, D.: A digital object architecture based internet of things devices identification system modeling. In: Proceedings of Educational Institutions of Communication (2019). T. 5, No. 1, pp. 42–47
11. Digital Object Architecture for IoT. http://www.wileyconnect.com/home/2016/11/8/what-governmentsdecided-on-digital-object-architecture-for-iot. Accessed Mar 2018
12. The DONA Foundation. https://dona.net/. Accessed Mar 2018

Industrial Internet of Things Classification and Analysis Performed on a Model Network

V. Kulik[1]([✉]), R. Kirichek[2], and A. Sotnikov[1]

[1] The Bonch-Bruevich Saint-Petersburg State University of Telecommunications,
22 Prospekt Bolshevikov, 193232 St. Petersburg, Russian Federation
vslav.kulik@gmail.com
[2] Peoples' Friendship University of Russia (RUDN University),
6 Miklukho-Maklaya St, 117198 Moscow, Russian Federation
https://www.sut.ru/, http://www.rudn.ru/

Abstract. The authors conduct a study of various types of networking in the Industrial Internet of Things, classify the IIoT traffic based on the previously studied types, analyze the IIoT traffic that received during the industrial equipment work. Based on the analysis, a model network for packet generation is proposed to simulate the operation of industrial equipment in communication networks by the authors.

Keywords: Internet of Things · Industrial Internet of Things · Traffic analysis · Model network

1 Introduction

Modern telecommunication and information technologies are getting new impulses for communication network's development, with occurrence such concepts like the Internet of Things (IoT) and fifth-generation communication network 5G/IMT-2020. Such concepts as IoT and 5G networks associated with various types of human life activities, for example, city infrastructure (smart cities) [1], unmanned vehicles (cars, drones, and others) [2,3], medicine (medical networks, Internet of Nanothings and others) [4,5], buildings (smart house, smart building and others) and etc. Furthermore, IoT concept includes such area as Industrial Internet of Things (IIoT), that contains various types of automated technologies to automate different industrial fields (example: agriculture, textile industry, metallurgical industry, electronics manufacturing, automotive and others) [6].

IIoT is including such aspects of the production cycle as collecting data from end nodes (sensors); controlling industrial devices by actuators; storing

The publication has been prepared with the support of the "RUDN University Program 5-100".

industrial data; data analyzing; security control; analyzing a company's environment; enterprise resource planning; customer relationship management; human resources management; supply chain management; production control; data synchronization and etc.

Those aspects are realized by different software, that using for sending, storing and analyzing IoT data, which based on various types of network communication structures. Thereby, IIoT has various types of network traffic which are generated by different applications. Tasks of traffic classification and traffic separation are the most important problem for the implementation of the generation of the IIoT traffic, that could be used for tasks of IIoT company modeling.

2 Types of the Industrial Internet of Things Traffic

There are a number of tasks, that realization is necessary for the construction of IIoT solutions. These tasks are required a special approach to building a computer appliance (CA) system for IIoT. Different CA systems use various network communication scenarios and network technologies and therefore it is very important to classify IIoT traffic and investigate its properties. This research will be used to develop special software for IIoT traffic generation.

It is necessary to allocate the main sources of IIoT traffic:

- Sensors and actuators. Sensors and actuators are using for automation industrial equipment work and can be connected and controlled with it by the special application (by numerical control for example) or with the special control and management systems (for example SCADA, SAP Hana, OPC UA, and others) [7–9];
- Business applications (CRM, ERM, SCM and others). Business process automation is the most important aspect of a company's work procedure [10,11];
- Open web data. Open source web information may be used to provide work of the different business processes and to optimize the company's work by various analytical algorithms [12];
- Multimedia systems. Video cameras and microphones may be used as source devices by the company's secure systems [13,14];
- Location systems. Those systems may be used for positioning industrial equipment, humans, a vehicle in the company demesne or globally, regionally or worldwide by the special computer appliance systems [15].

Each of those listed traffic types has unique network parameters as latency's between packet reception, middle packet size, maximum and minimum communication channel bandwidth, type of network, transport and application layers, that used for data transmission, presence of various cryptographic protocols and others.

Traffic sources may be working by special unique network connectivity scenario, therefore it is important to allocate the following scenarios of work various type of devices in Industrial Internet of Things:

- Regular. IIoT end-node sends data packets to a distant data collection device (for example database management system, industrial management and control systems, and others), with pre-set data delivery frequency;
- Event-oriented. IIoT end-node sends data packets to distant data collection device when some event has occurred (for example data request by the user or other devices, sending data about sensors or actuators status change and others);
- By schedule. IIoT end-node sends data packets to distant data collection device by the previously defined schedule.

IIoT applications and services are their own requirements for service time. Next types of applications and services should be allocated:

- Real-time. Those application and services have severe message delivery time limits and therefore those type of messages delivered in minimum time;
- Delay-tolerant. Those application and services do not have severe message delivery time limits.

Presented types of IIoT traffic classification is necessary for create computer appliance system of testing local networks to IIoT traffic loading stability.

3   IIoT Traffic Analysis

Investigation of Industrial Internet of Things traffic characteristics should be build performed to build a computer appliance system for IIoT traffic generation by the real industrial system.

As it defined at ITU-T Q.3900 "Methods of testing and model network architecture for NGN technical means testing as applied to public telecommunication networks" model network is a network which simulates the capabilities similar to those available in present telecommunication networks, has a similar architecture and functionality and uses the same telecommunication technical means [16].

Figure 1 is shown the structure of the model network for data collection, which built by using real industrial equipment. Next network elements are shown on this structure:

- Industrial server. This server is the response for collection, storing and processing information from industrial equipment with IP address: 10.30.106.200. Traffic is captured and analyzed on the server;
- Client 1. Client equipment is represented as industrial equipment with an integrated calculation device – numerical control unit with IP address: 10.30.106.203. This equipment is using for collecting data from some quantity of sensors, which were data sources for the created model network;

- Client 2. Client equipment is represented as industrial equipment with an integrated calculation device – numerical control unit with IP address: 10.30.106.204. This equipment is using for collecting data from some quantity of sensors, which were data sources for the created model network;
- Local network. It was built by laboratory crew and based on the network infrastructure of 3D printing prototyping workshop. Its network gateway address is 10.30.106.1 and the netmask is 255.255.255.0.

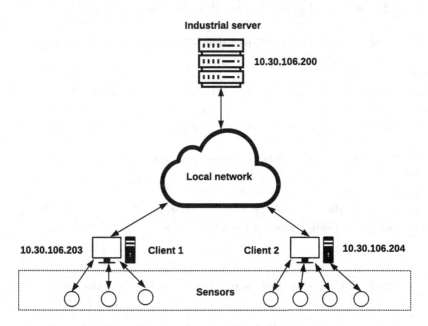

Fig. 1. The structure of model network for data collection built by real industrial equipment.

The collection of industrial sensors and actuators is the traffic source for that system. Traffic was captured while industrial equipment working. Captured traffic is related to event-oriented industrial equipment work scenarios. Investigating industrial equipment is a delay tolerant device because this device is autonomous with optional properties of data collection to the remote industrial equipment management server.

The traffic collection experiment was served 3193 s. While experiment clients 1 and 2 were sent data, which are collected from sensors, that installed on equipment, to the local network's industrial server, which was capture traffic, with different time intervals. Then the server was proceed received information. Traffic was divided into two different groups for analysis:

- Packets that are referred to the procedure of useful data transmission from client 1 to industrial server. The captured packet quantity is 3461. Captured traffic for client 1 have different parameters on different time intervals and therefore those packets were divided by the next groups:

1. All captured packets from client 1 were analyzed;
2. Packets from client 1 that were analyzed in the time interval from 35.94 to 990.38 s;
3. Packets from client 1 that were analyzed in the time interval from 990.70 to 3191.02 s.

- Packets that are referred to the procedure of useful data transmission from client 2 to industrial server. Captured packet quantity is 19681.

Table 1. Generated traffic parameters for packets from the client 1.

Parameters	Values
The confidence interval for time periods between packets arrival (for all), ms	91.19 ± 5.23
The confidence interval for time periods between packets arrival (1st interval), ms	31.62 ± 0.02
The confidence interval for time periods between packets arrival (2nd interval), ms	501.21 ± 3.99
The average bandwidth value (for all), bytes/s	1059.72
The average bandwidth value (1st interval), bytes/s	3417.29
The average bandwidth value (2nd interval), bytes/s	37.23

Table 2. Generated traffic parameters for packets from the client 2.

Parameters	Values
The confidence interval for time periods between packets arrival (for all), ms	16.27 ± 0.22
The average bandwidth value (for all), bytes/s	3519.79

Tables 1 and 2 represent parameters for two groups of captured traffic. Next parameters are selected for investigation: confidence interval for time periods between packets arrival with confidence probability 95% and the average value of the bandwidth.

Figures 2 and 3 show graphics that represented the dependence of the values of the time intervals between package delivery from the current time of the experiment for 1st and 2nd intervals for client 1.

Figure 4 shows graphics that represented the dependence of the values of the time intervals between package delivery from the current time of the experiment for client 2.

It was conducted to analyze the probability distribution for time intervals between package delivery. Each of investigated group was split into 100 equal

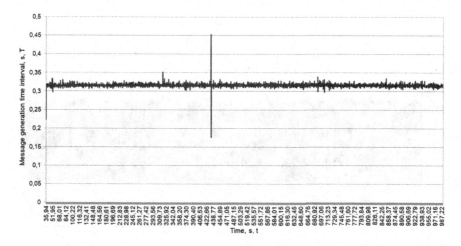

Fig. 2. Time intervals to current time dependency graphics for the 1st interval for client 1.

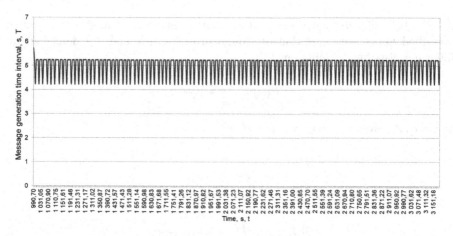

Fig. 3. Time intervals to current time dependency graphics for the 2nd interval for client 1.

observations, after this, it was produced probability calculation of getting the package to the researched group.

1st client intervals were split into three groups that probability percent summarily is 99.80%:

- Group 1. This group includes intervals from 0.295 to 0.339 s. Intervals belong to this group with a probability percent of 87.18%;
- Group 2. This group includes intervals from 4.213000 to 4.213442 s. Intervals belong to this group with a probability percent of 2.92%;
- Group 3. This group includes intervals from 5.218000 to 5.218442 s. Intervals belong to this group with a probability percent of 9.71%.

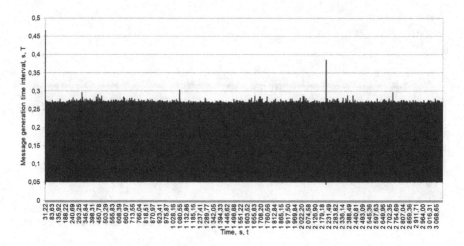

Fig. 4. Time intervals to current time dependency graphics for client 2.

The beta distribution is selected for mathematical modeling of two groups described above with criterion of Kolmogorov-Smirnov and it's described by following distribution function [17]:

$$F(x) = \frac{B_x(\alpha, \beta)}{B(\alpha, \beta)},$$ (1)

where α, β - shapes properties ($\alpha > 0, \beta > 0$).

$$B_x(\alpha, \beta) = \int_0^x t^\alpha (1 - t)^{\beta - 1} dt,$$ (2)

incomplete beta function.

$$B(\alpha, \beta) = \int_0^\infty \frac{t^{\alpha - 1}}{(1 + t)^{\alpha + \beta}} dt,$$ (3)

beta function.

It is selected following objective function based on the generalized reduced gradient method by the least square summary coefficient:

$$K_{lsm}(t) = \sum_1^n [P(t_i) - P(t_i, t_{i+1}]^2,$$ (4)

where P(t) is the probability of getting random value between package delivery from t_i to t_{i+1} for experimental results,

$$P(t_i, t_{i+1}) = F(t_{i+1}) - F(t_i),$$ (5)

is the probability of getting a random value between package delivery for selected distribution law F(t).

Graphic on Fig. 5 shows the ratio between the probability of getting a random value between package delivery and time observations for the 1st group. Following shape values was getting: $\alpha = 64.53$; $\beta = 3089.15$.

Graphic on Fig. 6 shows the ratio between the probability of getting a random value between package delivery and time observations for the 2nd group. Following shape values was getting: $\alpha = 6.56$; $\beta = 446.59$.

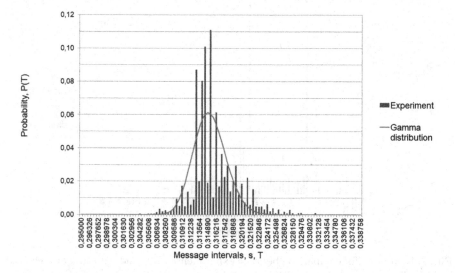

Fig. 5. Histogram for comparison between experimental data and data that are received by beta distribution for group 1.

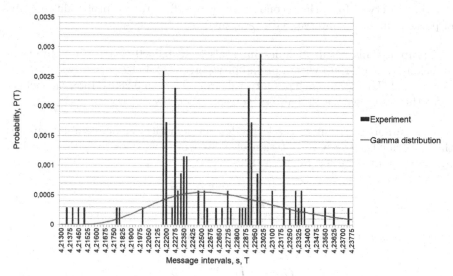

Fig. 6. Histogram for comparison between experimental data and data that are received by beta distribution for group 2 (1).

Graphic on Fig. 7 shows the ratio between the probability of getting a random value between package delivery and time observations for the 2nd group. Following shape values was getting: $\alpha = 6.66$; $\beta = 422.41$.

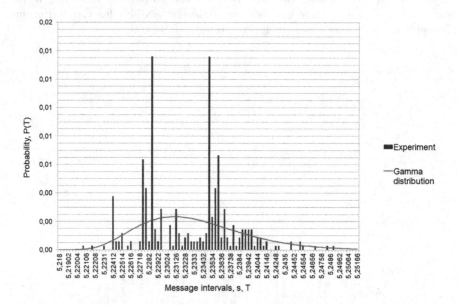

Fig. 7. Histogram for comparison between experimental data and data that are received by beta distribution for group 2 (2).

Time intervals from the second client were split into two probability groups of packets hit to the interval with total hit percent 99.48%:

- Group 1. This group includes intervals from 0.0500 to 0.0563 s. Intervals belong to this group with a probability percent of 49.62%;
- Group 2. This group includes intervals from 0.235 to 0.288 s. Intervals belong to this group with a probability percent of 49.86%.

The gamma distribution is selected for mathematical modeling of two groups that described above with criterion of Kolmogorov-Smirnov and it's described by following distribution function [17]:

$$F(x) = \frac{1}{G(v)} \int_0^{ux} t^{u-1} e^{-t} dt, \tag{6}$$

where u | shape parameter, v | scale parameter ($u > 0$, $v > 0$),

$$G(v) = \int_0^{\infty} t^{v-1} e^{-t} dt, \tag{7}$$

gamma function.

It is selected following objective function based on the generalized reduced gradient method by the least square summary coefficient (4).

Graphic on Fig. 8 shows the ratio between the probability of getting a random value between package delivery and time observations for the 1st group. Following shape and scale values was getting: u = 92.86; v = 0.00002.

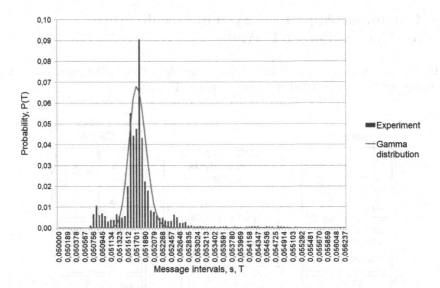

Fig. 8. Histogram for comparison between experimental data and data that are received by gamma distribution for group 1.

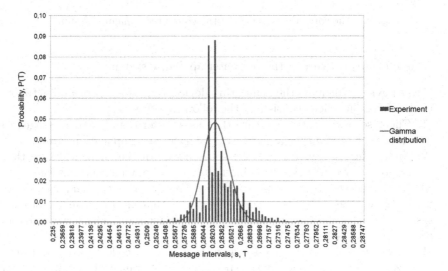

Fig. 9. Histogram for comparison between experimental data and data that are received by gamma distribution for group 2.

Graphic on Fig. 9 shows the ratio between the probability of getting a random value between package delivery and time observations for the 2nd group. Following shape and scale values was getting: u = 165.99; v = 0.00017.

Message stream similar to experimental data can be generated by selected gamma and beta distribution, with the obtained coefficient of shape and scale.

4   Computer Appliance System Architecture for Generation IIoT Traffic

The following computer appliance architecture offers to test the local network's stability for IIoT traffic and it's represented in Fig. 10.

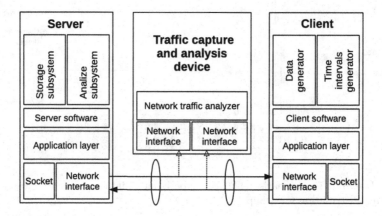

Fig. 10. The software structure of the computer appliance system.

The software structure of the computer appliance system:

- Client device (Client) is the computing device used as a sensor and actuator in the IIoT. The client consists of the following software:
  1. Client software is the software that uses to form and send messages to the server by the special network, transport and application layers.
  2. Time intervals generator is the sub-software of the client software that response for generation time intervals values between sent messages.
  3. Data generator is the sub-software for data generation for messages.
- Server device (Server) is the computing device that used as a server of IIoT system and it's collect, store and analyze incoming data. The server consists of the following software:
  1. Server software is the sub-software that uses to collect, store and analyze incoming messages by the special network, transport and application layers.

2. Storage subsystem is the sub-software of the server software that uses to collect data from messages and for interaction with data management systems.
3. Analyze subsystem is the sub-software of the server software that uses to analyze the client's data.

– Traffic capture and analysis device (TCAD) is the computing device used to capture network traffic while the server and client interact and for data analysis by the special software that calls network traffic analyzer.

The model network that proposes for the testing generation of the IIoT traffic procedure is shown in Fig. 11.

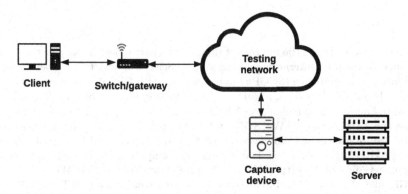

Fig. 11. The model network for testing the generation of IIoT traffic.

This model network consists of the following elements:

– Server is a computer appliance system that realizes the server's functions;
– Client is a computer appliance system that realizes the client's functions;
– Capture device is a computer appliance system that realizes the function of traffic capture and analysis device;
– Switch/gateway is a computer appliance device that uses as local network's switch or as the gateway for connecting client to testing network;
– Testing network is a local network that simulates the architecture of the industry company's local network.

5 Conclusion

Experimental data are roughly corresponded to obtained in mathematical modeling gamma and beta distributions. Obtained coefficients may be used for generation message stream that is similar to the traffic received from industrial equipment.

The model network is developed for investigation questions of the Industrial Internet of Things traffic generation. This network may be used to solve for following tasks: an investigation of various types of the IIoT traffic; IIoT systems network interaction modeling; benchmark testing of industrial company's networks. Received results can be used for planning network architecture of the IIoT systems.

The described model network may be used to develop special software for IIoT traffic generation and generate by its various types of IIoT traffic specified in previously developed classification.

Acknowledgements. The publication has been prepared with the support of the "RUDN University Program 5-100" and funded by RFBR according to the research projects No. 12-34-56789 and No. 12-34-56789.

References

1. Rhee, S.: Catalyzing the internet of things and smart cities: global city teams challenge. In: 2016 1st International Workshop on Science of Smart City Operations and Platforms Engineering (SCOPE) in partnership with Global City Teams Challenge (GCTC) (SCOPE–GCTC) (2016). https://doi.org/10.1109/SCOPE.2016.7515058
2. Makolkina, M., Paramonov, A., Vladyko, A., Dunaytsev, R., Kirichek, R., Koucheryavy A.: The use of UAVs, SDN, and augmented reality for VANET applications. In: 3rd International Conference on Artificial Intelligence and Industrial Engineering (AIIE 2017) "DEStech Transactions on Computer Science and Engineering", pp. 364–368 (2017). https://doi.org/10.12783/dtcse/aiie2017/18244
3. Giyenko, A., Cho, Y.I.: Intelligent UAV in smart cities using IoT. In: 2016 16th International Conference on Control, Automation and Systems (ICCAS-2016), pp. 207–210 (2016). https://doi.org/10.1109/ICCAS.2016.7832322
4. Pirmagomedov, R., Blinnikov, M., Glushakov, R., Muthanna, A., Kirichek, R., Koucheryavy, A.: Dynamic data packaging protocol for real-time medical applications of nanonetworks. In: Galinina, O., Andreev, S., Balandin, S., Koucheryavy, Y. (eds.) NEW2AN/ruSMART/NsCC 2017. LNCS, vol. 10531, pp. 196–205. Springer, Cham (2017). https://doi.org/10.1007/978-3-319-67380-6_18
5. Pirmagomedov, R., Hudoev, I., Kirichek, R., Koucheryavy, A., Glushakov, R.: Analysis of delays in medical applications of nanonetworks. In: 8th International Congress on Ultra Modern Telecommunications and Control Systems and Workshops (ICUMT), pp. 49–55 (2016). https://doi.org/10.1109/ICUMT.2016.7765231
6. Geng, H.: Internet of Things and Data Analytics Handbook, 1st edn., p. 816, pp. 41–81. John Wiley & Sons, Inc. (2017). https://doi.org/10.1002/9781119173601.ch3
7. Tom, R. J., Sankaranarayanan, S.: IoT based SCADA integrated with Fog for power distribution automation. In: 2017 12th Iberian Conference on Information Systems and Technologies (CISTI 2017) (2017). https://doi.org/10.23919/CISTI.2017.7975732
8. Shahzad, A., Kim, Y.G., Elgamoudi, A.: Secure IoT platform for industrial control systems. In: 2017 International Conference on Platform Technology and Service (PlatCon-2017) (2017). https://doi.org/10.1109/PlatCon.2017.7883726

9. Cho, H., Jeong, J.: Implementation and performance analysis of power and cost-reduced OPC UA gateway for industrial IoT platforms. In: 2018 28th International Telecommunication Networks and Applications Conference (ITNAC-2018) (2018). https://doi.org/10.1109/ATNAC.2018.8615377

10. Chamekh, M., Hamdi, M., Asmi, S.E., Kim, T.H.: Secured distributed IoT based supply chain architecture. In: 2018 IEEE 27th International Conference on Enabling Technologies: Infrastructure for Collaborative Enterprises (WETICE-2018), pp. 199–202 (2018). https://doi.org/10.1109/WETICE.2018.00045

11. Drobintsev, P.D., Kotlyarov, V.P., Chernorutsky, I.G., Kotlyarova, L.P., Aleksandrova, O.V.: Approach to adaptive control of technological manufacturing processes of IoT metalworking workshop. In: 2017 XX IEEE International Conference on Soft Computing and Measurements (SCM-2017), pp. 174–176 (2017). https://doi.org/10.1109/SCM.2017.7970530

12. Smit, H., Delamer, I.M.: Service-oriented architectures in industrial automation. In: 2006 4th IEEE International Conference on Industrial Informatics (INDIN-2006) (2006). https://doi.org/10.1109/INDIN.2006.275707

13. Kirichek, R., Golubeva, M., Kulik, V., Koucheryavy, A.: The home network traffic models investigation. In: 18th International Conference on Advanced Communication Technology (ICACT), pp. 97–100. (2016) https://doi.org/10.1109/ICACT.2016.7423288

14. Escudero, J.I., Gonzalo, F., Mejias, M., Parada, M., Luque, J.: Multimedia in the operation of large industrial networks, vol. 3, pp. 1281–1285 (1997). https://doi.org/10.1109/ISIE.1997.648929

15. Macagnano, D., Destino, G., Abreu, G.: Indoor positioning: a key enabling technology for IoT applications. In: 2014 IEEE World Forum on Internet of Things (WF-IoT) (2014). https://doi.org/10.1109/WF-IoT.2014.6803131

16. ITU-T Q.3900. Methods of testing and model network architecture for NGN technical means testing as applied to public telecommunication networks. Telecommunication standartization sector of ITU 2006 (2006). https://www.itu.int/rec/T-REC-Q.3900-200609-I/en

17. Ross, S.M.: Introduction to Probability Models, 11th edn., p. 784. Academic Press, Amsterdam (2014). ISBN: 9780124081215

Mobile Edge Computing for Video Application Migration

Steve Manariyo[1], Dmitry Poluektov[2], Khakimov Abdukodir[1],
Ammar Muthanna[1,2(✉)], and Maria Makolkina[1,2]

[1] St. Petersburg State University of Telecommunication, 22 Prospekt
Bolshevikov, St. Petersburg 193232, Russia
mansteve06@mail.ru, khakimov.a@sdnlab.ru,
ammarexpress@gmail.com, makolkina@list.ru
[2] Peoples' Friendship University of Russia (RUDN University),
6 Miklukho-Maklaya Street, Moscow 117198, Russia
poluektov-ds@rudn.ru

Abstract. Modern telecommunication networks show steady growth in the
digital cable television and IPTV market. In today's telecommunication networks most of the services are related to transmission of video traffic. For
client's loyalty operators are busy looking for new video delivery optimization
methods with the appropriate quality of experience. MEC or Mobile edge
computing offers significant advantages for example, enabling operators to bring
applications and content closer to the network edge i.e. close to the video
content end consumer and other advantages. This is particularly interesting in
video delivery, where latency negatively affect video quality. Users can receive
video content with minimum delays and operators can realize operational and
cost efficiencies while reducing network latency and, ultimately improving the
end consumer's quality of experience. In this paper, a video content application
migration method using edge-computing technology is proposed. The proposed
method is applied in a managed mode over a Software-Defined Networking
(SDN); what improves the efficiency of video traffic delivery. A new concept
"Exchange" is introduced for flexible and automated interaction between video
delivery chain members, i.e. the network operator, the content provider and the
end user.

Keywords: Video traffic · Software-Defined Network · SDN ·
Edge computing · QoE · Exchange

1 Introduction

Today's Video streaming over public or private networks is becoming popular each
end every year. Globally, IP video traffic will represent more than 82% of consumer
traffic by 2022 [1], and not only OTT and IPTV video, but also Internet video
streaming, and video streaming applications are the highest rank of applications consuming Bandwidth in the network. The load on the network has increased, and user's
requirements for video quality are more and more close to the ideal. Assessing video
services evolution and perspective, we can assume that in the near future there will be

© Springer Nature Switzerland AG 2019
O. Galinina et al. (Eds.): NEW2AN 2019/ruSMART 2019, LNCS 11660, pp. 562–571, 2019.
https://doi.org/10.1007/978-3-030-30859-9_49

way more advanced technologies that allow achievement of the maximum realism and image details. Network devices optimization to improve the quality of content and increase data transmission speed continues. Therefore, the development of new network configuration to optimize the transmission of large volumes of video traffic with proper and better quality is necessary. The emerging ETSI standard on Mobile Edge Computing (MEC) plays an important role in this direction [2]. MEC makes new services optimization possible by deploying applications and content at the network edge close to the end user. This can facilitate content dissemination within the access network and users can receive video content with minimum delay and the network operators can realize operational and cost efficiencies while reducing network latency and, ultimately, improving the end user's quality of experience [3, 4]. Software Defined Network (SDN) is used as transport network. SDN has a number of significant advantages when it comes to delivering content. It provides greater automation and orchestration of the network fabric, and allows dynamic configuration of networks and services. SDN allows a more distributed, flexible, and scalable network and is one of the great methods to increase the capacity of any network [5, 6]. The use of mobile edge computing technology in a controlled manner over a software-configured network improves network performance and improves the efficiency of video traffic delivery profiting from the way MEC jointly work with SDN [7].

One of the tasks of telecom operators is the optimization of network devices and communication channels resources utilization. Network device resources stand for RAM usage, CPU utilization, and power consumption by a network element. In today's telecommunications sector there are many solutions of process optimization. In this paper, we focus on using MEC in communication networks for video content applications. We are deeply convinced that operators need to optimize video delivering process to the end user from a remote cloud, since the majority of subscribers assess operator's communication quality from subjective indicators called the quality of Experience of the viewed video. In our work, MEC is used as a temporary allocation, for the time of viewing video content by a group of subscribers. In practice, it is well illustrated in the example of new episodes of famous televisions series that are published every week/day at a certain time, while subscribers are waiting to enjoy their favorite TV show. Before the publication of the new episode start, the source file is sent from the video content provider to the network operator, and then the operator allocates related applications in their MEC hosts where the many video requests are. Now, when the publication begins on the main official site, access to the video content is open on all MEC hosts. Content migration to the edge of the network reduces network load and delays in the network, thereby providing high QoS and the video quality of experience respectively.

2 Related Work

Content delivery network or content distribution network (CDN), often refers to a set of servers with specialized software, which accelerates content delivery to the end user. Servers are located worldwide in such a way that, for site visitors to have a minimal response time. "Content" refers to videos and static web sites elements (that do not require server code or database requests, such as css/js), but content also refers to other things - for example, games on Steam (uses CDN to upload games), operating systems updates, etc. By having the data closer to the end user allows the user to receive data faster. It is logical. What about the reduction in traffic? Caching is a commonly used technique to improve the performance of any application, be it desktop, mobile or web. Nodes-side data caching reduces the number of requests to the "main server". Plus, we can always transfer compressed data from the main server to the CDN node, in Kee-palive connections. CDN node can be configured so that the competing cached requests are not (cannot be) executed in parallel. This allows to save CPU time on the "main server". The combination allows the reduction in the number of requests and traffic to the "main server". The CDN node is also a great place to host a DNS slave server, actually, for the same reasons.

Mobile Edge Computing (MEC) is poised to make Content Delivery Networks up to 40% more efficient for cellular communications service providers as shown in many works. Cost savings will be shared between CSPs and CDN providers the latter improving margins by up to 25%. MEC represents cloud-computing capabilities and an IT service environment at the edge of the mobile network. The edge of the network includes base station infrastructure and data centers close to the radio network, which can extract network context from Radio Access Network (RAN) and process in a distributed manner.

In [8] the authors provided a classification of application models and an investigation of the latest mobile cloud application models. In [9] a summary of challenges for MCC, application partition and offloading technologies, classification of contexts and context management methods is provided. In [10] the authors provided an overview of MCC definition, architecture, and applications, as well as the generic issues and some of existing solutions. In [11], an investigation of existing works on representative platforms and intelligent access schemes of MCC can be found. [12] gives a detailed taxonomy of mobile cloud computing based on the key issues and the approaches to tackle them. In [13], was provided a comprehensive survey of the state of-the-art authentication mechanism in MCC and comparison within the cloud computing is made. Authors in [14] provide a taxonomy of MEC based on different aspects including its characteristics, access technologies, applications, objectives and so on. Some of MEC open issues are identified. In [15], a classification of deployed applications in mobile edge according to MEC technical metrics and the MEC benefits for stakeholders in the network is provided. A discussion of the security threats and challenges in the edge paradigms, as well as the promising solution for each specific challenge can be found in [16]. In [17] highlighted the representative applications and various aspects of research issues in fog computing. In [18], a survey of the new security and privacy challenges in fog computing in addition cloud-computing challenges is given. An investigation of the web caching and prefetching techniques in improving the web performance as well as a classification of caching policies can be found in [19]. A description of advantages and disadvantages of cache replacement strategies can be found in [20].

In this paper we realize a system based on model network, where MEC is used as a temporary allocation, at the time when a group of user is watching video content.

3 Algorithm Description

In our work, we suggest content exchange method for flexible and automated interaction between network operator and the content provider, which was introduced. This method is a kind of organization - regulator, which has neutral legal relations with operators, and regulates the process of content provider's application migration to the operator, ensuring safety, reliability, and quality assessment. It is not free of charge. Our method also performs as an automated financial regulator between the operator and the content provider (O2B) Fig. 1.

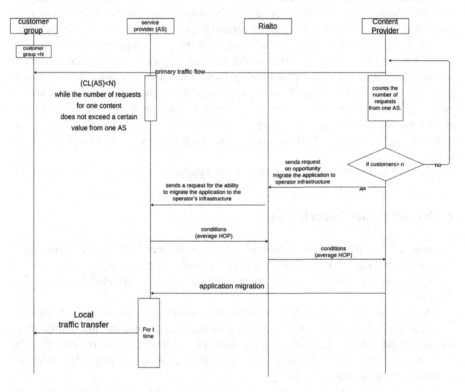

Fig. 1. A flowchart of developed system management algorithm

Figure 1 shows a flowchart of developed system management algorithm
Description of the algorithm:

To improve the quality of video content service, our algorithm allows unloading communication channels.

We assume we have network area where M is the number of users. Weekly on Monday evening at 19:00, a certain video hosting publishes a new episode of the series. In the network area a lot of requests to the show are submitted, i.e. N requests from one network area are simultaneously submitted to one remote server. Therefore, remote video host ask "Exchange" (Rialto) to upload application content to MEC in that network area assuming that QoS degradation is expected during the video hosting period due to high request level. Then "Exchange" in turn asks the last mile provider the permission to place the video hosting container in MEC. In the case of positive response, the video hosting sends the DOCKER container with the configuration file to the MES hosting. Further, the SDN-based network operator creates a logical channel for redirecting remote video hosting requests to the local MEC located on that network area. Subscribers will receive video content from the local "server". When the number of requests to this content decreases to a threshold (N) value, the container is deleted.

There are a lot of operators and many video hosting sites around the world. In order to not make contracts between each service and the b2b operator, that cost money and time, "Exchange" registers operators and video hosting sites in its database and at the same time create a short-term digital agreement between them during service provision time. Each operator and video hosting will have its own unique registration number with a key, and key exchange between the operator and video hosting occurs through an API. The operator and video hosting prescribe conditions and requirements in advance. For example, video hosting may request the disk space volume, hosting duration, RAM characteristics, average value of OpenFlow switches hops and maximum financial cost for the allocation time. ($ per hour). The operator, in turn, can define its capabilities; the MEC host rent cost, the average value of OpenFlow switch hops.

Figure 2 shows the work algorithm of our system.

4 Experiments Description

Communication channel load experiment using traffic generator Traffic generator is used as an application. It is used for a network load qualitative assessment.

In the SDN network, a video content application was installed in one end in Docker.

On the other hand, 50 streams of the same content are simultaneously requested.

As a result, the application has moves (migrates) to the nearest and optimal point for all request sources. Unloading of the network transport channel is ensured, delay tends to minimum, and the necessary QoS level is ensured. Migration rules are described in the algorithm in Fig. 2.

In our research, a experiment stand is assembled and is consisted of operator's network, consisting of OpenFlow OVS switches, a controller and an SDN orchestrator. To get closer to the real conditions, the video content server is placed in a remote hosting as shown in Fig. 3 (Table 1).

In Fig. 4, it can be seen how the communication channel was released after unloading the content application to MEC host. This solution solves problems very well in case of simultaneous http request to one video server. In addition, it can be seen that, at the moment when the application is transferred to the MEC host, traffic is seamlessly downloaded. This was achieved through SDN networks using a P2M logical channel (Table 2).

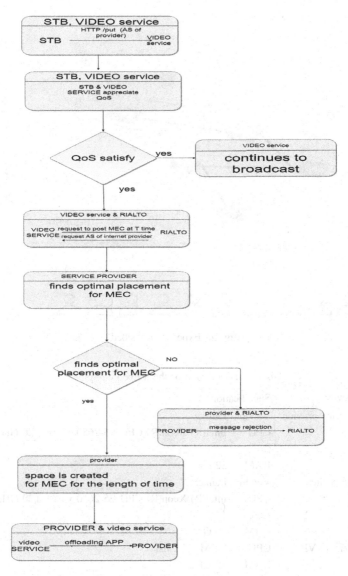

Fig. 2. Algorithm of our system.

$$R_{system}(t) = 1 - \prod_{i=1}^{n}(1 - R_i(t))$$

The reliability of the provided network can be described by the formula of a parallel system, i.e. for a separate operator's network part, the level of reliability increases with the density of located MEC hosts (Fig. 6).

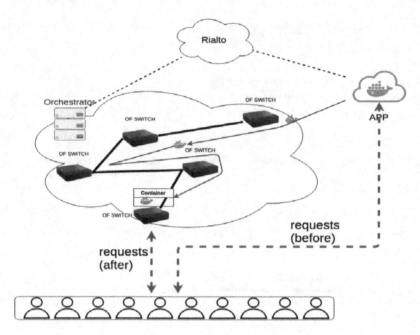

Fig. 3. Experiment testbed

Table 1. Description of the testbed equipment

Device	Specifications	
Service provider	Vendor	Lanner
	CPU	Intel(R) Xeon(R) CPU E5-2650 v4 @ 2.20 GHz
	Core	12
	RAM	32 GB
OF Switch	Vendor	Lanner
	CPU	Intel(R) Xeon(R) CPU E5-2650 v4 @ 2.20 GHz
	Core	12
	RAM	40 GB
Remote VPS	CPU	KVM 3.5 GhZ
	RAM	32 GB
	VIDEO	16 GB

Figure 5 shows how video content system reliability increases in general for the network segment with the number of subscribers of 1000 people, i.e. the more tightly MEC hosts are placed, the more reliability increases.

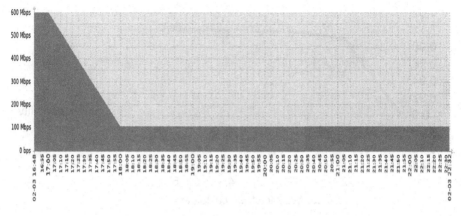

Fig. 4. Research results

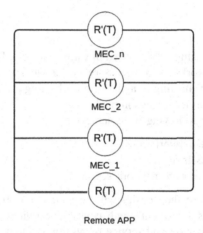

Fig. 5. Representation of our experiment components in parallel system

Table 2. Content description

Video	ULTRA HD
N (query threshold)	50
Protocol	HTTP

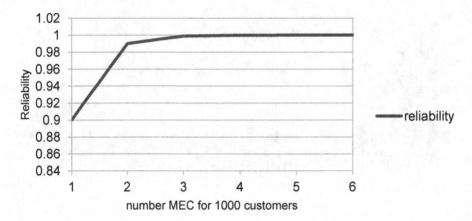

Fig. 6. System reliability of 1000 subscribers' network segment

5 Conclusion

MEC technology is increasingly being used in various areas of IT services. Using MEC in fifth generation networks is a necessary for achieving indispensable network requirements. Using MEC hosting as a platform for hosting a video content service, the difference in provided service was examined.

Deploying MEC, the following were achieved:

- Reduced traffic in the network core
- Improvement of MOS Index
- Increased reliability for each network segment

In 5G networks, where the density will be equal to 1,000,000 users per square kilometer, such decisions are urgently indispensable. Services decentralization lead to quality of service optimization and service reliability and QoE increases.

Acknowledgment. The publication has been prepared with the support of the "RUDN University Program 5-100".

References

1. Cisco: Cisco Visual Networking Index: Forecast and Trends, 2017–2022, White Paper, February 2019
2. McGarry, M.P., et al.: Mobile-edge computing introductory technical whitepaper, White Paper, September 2015
3. Beck, M.T., Feld, S., Fichtner, A., Linnhoff-Popien, C., Schimper, T.: ME-VoLTE: network functions for energy efficient video transcoding at the mobile edge. In: Proceedings of the 18th International Conference on Intelligence in Next Generation Networks (ICIN 2015) (2015)

4. Chiang, M., Zhang, T.: Fog and IoT: an overview of research opportunities. IEEE Internet Things J. **3**, 854–864 (2016)
5. Vladyko, A., Muthanna, A., Kirichek, R.: Comprehensive SDN testing based on model network. In: Galinina, O., Balandin, S., Koucheryavy, Y. (eds.) NEW2AN/ruSMART 2016. LNCS, vol. 9870, pp. 539–549. Springer, Cham (2016). https://doi.org/10.1007/978-3-319-46301-8_45
6. Muhizi, S., Shamshin, G., Muthanna, A., Kirichek, R., Vladyko, A., Koucheryavy, A.: Analysis and performance evaluation of SDN queue model. In: Koucheryavy, Y., Mamatas, L., Matta, I., Ometov, A., Papadimitriou, P. (eds.) WWIC 2017. LNCS, vol. 10372, pp. 26–37. Springer, Cham (2017). https://doi.org/10.1007/978-3-319-61382-6_3
7. Manariyo, S., Khakimov, A., Pyatkina, D., Muthanna, A.: Optimization algorithm for IPTV video service delivery over SDN using MEC technology. In: Galinina, O., Andreev, S., Balandin, S., Koucheryavy, Y. (eds.) NEW2AN/ruSMART 2018. LNCS, vol. 11118, pp. 419–427. Springer, Cham (2018). https://doi.org/10.1007/978-3-030-01168-0_38
8. Ur Rehman Khan, A., Othman, M., Madani, S.A., Khan, S.U.: A survey of mobile cloud computing application models. IEEE Commun. Surv. Tutorials **16**(1), 393–413 (2014). First Quarter
9. Guan, L., Ke, X., Song, M., Song, J.: A survey of research on mobile cloud computing. In: 2011 IEEE/ACIS 10th International Conference on Computer and Information Science (ICIS), Sanya, China, pp. 387–392 (2011)
10. Dinh, H.T., Lee, C., Niyato, D., Wang, P.: A survey of mobile cloud computing: architecture, applications, and approaches. Wirel. Commun. Mob. Comput. **13**(18), 1587–1611 (2013)
11. Fan, X., Cao, J., Mao, H.: A survey of mobile cloud computing. ZTE Commun. **9**(1), 4–8 (2011)
12. Fernando, N., Loke, S.W., Rahayu, W.: Mobile cloud computing: a survey. Future Gener. Comput. Syst. **29**(1), 84–106 (2013)
13. Alizadeh, M., Abolfazli, S., Zamani, M., Baharun, S., Sakurai, K.: Authentication in mobile cloud computing: a survey. J. Network Comput. Appl. **61**, 59–80 (2016)
14. Ahmed, A., Ahmed, E.: A survey on mobile edge computing. In: 2016 10th International Conference on Intelligent Systems and Control (ISCO), Coimbatore, India, pp. 1–8 (2016)
15. Beck, M.T., Werner, M., Feld, S., Schimper, S.: Mobile edge computing: a taxonomy. In: Proceedings of the Sixth International Conference on Advances in Future Internet, Citeseer, pp. 48–54 (2014)
16. Roman, R., Lopez, J., Mambo, M.: Mobile edge computing. In: Fog, et al. (ed.) A Survey and Analysis of Security Threats and Challenges. arXiv preprint arXiv:1602.00484 (2016)
17. Yi, S., Li, C., Li, Q.: A survey of fog computing: concepts, applications and issues. In: Proceedings of the 2015 Workshop on Mobile Big Data, Hangzhou, China, pp. 37–42. ACM (2015)
18. Yi, S., Qin, Z., Li, Q.: Security and privacy issues of fog computing: a survey. In: Xu, K., Zhu, H. (eds.) WASA 2015. LNCS, vol. 9204, pp. 685–695. Springer, Cham (2015). https://doi.org/10.1007/978-3-319-21837-3_67
19. Ali, W., Shamsuddin, S.M., Ismail, A.S.: A survey of Web caching and prefetching. Int. J. Adv. Soft Comput. Appl. **3**(1), 18–44 (2011)
20. Podlipnig, S., Böszörmenyi, L.: A survey of web cache replacement strategies. ACM Comput. Surv. (CSUR) **35**(4), 374–398 (2003)

An Accurate Approximation of Resource Request Distributions in Millimeter Wave 3GPP New Radio Systems

Roman Kovalchukov[1]($\boxtimes$) (iD), Dmitri Moltchanov[1] (iD), Yuliya Gaidamaka[2,3] (iD), and Ekaterina Bobrikova[2] (iD)

[1] Tampere University, Korkeakoulunkatu 10, 33720 Tampere, Finland
rnkovalchukov@sci.pfu.edu.ru
[2] Peoples' Friendship University of Russia (RUDN University),
6 Miklukho-Maklaya St, Moscow 117198, Russian Federation
[3] Federal Research Center "Computer Science and Control" of the Russian Academy of Sciences (FRC CSC RAS), 44-2 Vavilov St, Moscow 119333, Russian Federation

Abstract. The recently standardized millimeter wave-based 3GPP New Radio technology is expected to become an enabler for both enhanced Mobile Broadband (eMBB) and ultra-reliable low latency communication (URLLC) services specified to future 5G systems. One of the first steps in mathematical modeling of such systems is the characterization of the session resource request probability mass function (pmf) as a function of the channel conditions, cell size, application demands, user location and system parameters including modulation and coding schemes employed at the air interface. Unfortunately, this pmf cannot be expressed via elementary functions. In this paper, we develop an accurate approximation of the sought pmf. First, we show that Normal distribution provides a fairly accurate approximation to the cumulative distribution function (CDF) of the signal-to-noise ratio for communication systems operating in the millimeter frequency band, further allowing evaluating the resource request pmf via error function. We also investigate the impact of shadow fading on the resource request pmf.

Keywords: 5G · New Radio · Millimeter-wave · SNR · Shadow fading · Performance evaluation

1 Introduction

The future 5G New Radio (NR) systems are expected to provide three primary services, massive machine-type communications (MTC), enhanced mobile broadband (eMMB) and ultra-reliable low-latency communications (URLLC). New radio NR interface operating in the millimeter frequency range is planned to become enabling technology for the latter two services [1]. The first two phases of NR standardization providing LTE-anchored and standalone NR operations

© Springer Nature Switzerland AG 2019
O. Galinina et al. (Eds.): NEW2AN 2019/ruSMART 2019, LNCS 11660, pp. 572–585, 2019.
https://doi.org/10.1007/978-3-030-30859-9_50

have been completed by 3GPP in December 2017 and August 2018, respectively. The NR standardization efforts are expected to commence by the end of 2020.

In addition to inherent advantages of 5G NR related to the use directional antenna radiation and reception patterns and extremely wide bandwidth, the use of millimeter-wave frequency band (30–100 GHz) brings unique challenges to system designers, e.g., blockage of propagation between communicating entities path that may lead to abrupt fluctuations of the signal-to-noise ratio [2–5]. To provide deployment guidelines for 5G NR network operators and evaluate forthcoming technology under a wide variety of prospective scenarios, researchers currently analyze the performance of NR systems in various deployments. The effects of three-dimensional communications scenarios in 5G NR have been addressed in [6]. In [7,8] the authors have analyzed performance aerial access points operating in the millimeter-wave band. Aiming to improve spectral efficiency and reduce outage probability, the authors [9] have deeply investigated the effect of multi-connectivity option recently proposed by 3GPP. The upper bound on spectral efficiency in the presence of multi-connectivity has been developed in [10].

Most of the performance evaluation studies of 5G NR technology carried out so far concentrated on system aspects characterizing time-averaged user performance using spectral efficiency, achieved rate, and outage probability as the primary metrics of interest. However, the prospective applications of 5G NR include applications generating bandwidth-greedy non-elastic traffic patterns. Thus, in addition to spatial randomness of users request distributions, performance evaluation models need to capture traffic dynamics as well. Recently, these studies started to appear. In [11], the authors developed a framework that jointly captures spatial and session-level traffic dynamics in 5G NR systems in the presence of a 3GPP multi-connectivity option. The authors in [12] proposed a new approach to improve the reliability of the session service process at 5G NR base stations (BS) using the concept of resource reservation. The effects of both multi-connectivity and bandwidth reservation have been studied in [13], where the authors demonstrated that initial selection of NR BS having sufficient amount of resources to handle arriving session provides the positive impact of new and ongoing session drop probabilities. The effect of multi-RAT NR/LTE service process in the street deployment of 5G NR systems has been investigated in [14,15]. Finally, the joint support of multicast and unicast sessions in 5G NR systems has been analyzed in [16].

Accounting for traffic dynamics at the 5G NR air interfaces requires the joint use of queuing theory and stochastic geometry. Due to the randomness of user locations in the service area as well as propagation and environmental dynamics, the queuing models for this type of analysis need to capture random session resource requirements by [17,18]. Thus, the critical part in most of the above mentioned studies is a derivation of probability mass function (pmf) of resources required by a session from NR BS. As one may observe, this pmf is the function of multiple system parameters including antenna gains at transmitting and receiving side, emitted power, interference, the randomness of user location within the service area of interest and propagation environments including the

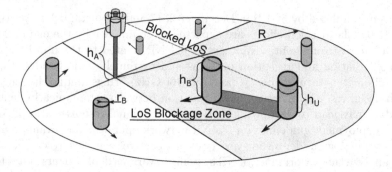

Fig. 1. An illustration of the considered 5G NR cellular deployment.

density of dynamic blockers. As a result, no closed-form expression is available for the sought pmf. Furthermore, the use of various approximations by the authors to simplify the derivations of the pmf in question restrains potential readers from comparing the reported results across the studies.

This study aims to unify the efforts towards an accurate and reliable approximation of pmf of session resource requirements in millimeter-wave 5G NR systems. Using the standardized propagation model, typically assumed coverage of NR BS and random user equipment (UE) distribution we first demonstrate that the signal-to-noise ratio (SNR) perceived at UE can be fairly well approximated by Normal distribution. This fact allows expressing the pmf of session resource requirements in terms of well-known error function drastically reducing the computational efforts. Our numerical results confirm that the proposed model provides an accurate approximation for the session request distribution. Finally, we investigate the effect of shadow fading on the considered pmf.

The paper is organized as follows. Section 2 describes the system model. The develop approximation for session resource request pmf in Sect. 3. Numerical illustration of the proposed approximation is provided in Sect. 4. The last section concludes the paper.

2   System Model

Deployment. Figure 1 illustrates the considered deployment. We assume a single NR BS with a certain coverage area with radius r_A (Table 1) around it. In practice, this is achieved by using several antenna arrays, each service its sector. In what follows, we consider a single sector. The coverage of NR BS is determined by the propagation model specified below, cell-edge outage probability p_C, and the set of modulation and coding schemes for 5G NR specified in [19]. UE is assumed to be randomly and uniformly distributed in the service area of an NR BS sector. The height of UE and NR BS are assumed to be h_U and h_A, respectively.

In our scenario, similarly to [4], we also assume dynamic blockage by the mobile crowd around UE. The spatial density of blockers is assumed to be λ_B.

Table 1. Notation used in the paper.

Parameter	Definition
f_c	Carrier frequency
λ_B	Users density
h_A	NR BS height
r_A	Radius of coverage area
h_U	UE height
h_B	Height of blockers
r_B	Blocker radius
ζ	Path loss exponent
N_0	Thermal noise
L_B	Loss blockage loss
P_T	NR BS transmit power
C_O	Control channel overhead
C_L	Cable losses
M_I	Interference margin
$M_{S,nB}, M_{S,B}$	Shadow fading margins in non-blocked and blocked states
$\sigma_{S,B}, \sigma_{S,nB}$	STD of fading in LoS blocked and non-blocked states
N_F	Noise figure
p_C	Cell edge coverage probability
T	Session rate
K_B, K_U	Number of planar antenna elements at NR BS and UE
ω_B, ω_U	Antenna directivities at NR BS and UE
G_B, G_U	NR BS transmit and UE receive antenna gains
$F_X(x), f_X(x)$	CDF and pdf of random variable X
$S_{\min}$	SNR outage threshold
S_i, S	SNR with NR BS i and overall SNR
$p_{B,i}(x), p_{B,i}$	Distance-dependent/independent blockage probabilities
s_j	SNR margins
m_j	Probability of choosing MCS j
e_j	Spectral efficiency of MCS j
$\mathrm{erfc}(\cdot)$	Complementary error function

Blockers move around the area according to random direction mobility model (RDM, [20]). The flux of blockers across the cell boundary is assumed to be constant; i.e., the density of blockers is homogeneous. Blockers are modeled as cylinders with constant base radius r_B and constant height h_B.

Propagation Model and SNR. The SNR at UE can be written as

$$P_R(y) = \frac{P_T G_B G_U}{L_{dB}(y) N_0 C_O C_L M_I N_F M_S},$$ (1)

where P_T is the NR BS transmit power, G_B and G_U are the antenna gains at the NR BS and UE sides, respectively, y is the three-dimensional (3D) distance between the UE and the NR BS, $L_{dB}(y)$ is the propagation loss in decibels, C_O is the control channel overhead , C_L is the cable losses, M_I is the interference margin, N_F is the noise figure, M_S is the shadow fading margin.

Table 2. CQI, MCS and SNR mapping for 3GPP NR.

CQI	MCS	Spectral efficiency	SNR in dB
0	Out of range		
1	QPSK, 78/1024	0.15237	−9.478
2	QPSK, 120/1024	0.2344	−6.658
3	QPSK, 193/1024	0.377	−4.098
4	QPSK, 308/1024	0.6016	−1.798
5	QPSK, 449/1024	0.877	0.399
6	QPSK, 602/1024	1.1758	2.424
7	16QAM, 378/1024	1.4766	4.489
8	16QAM, 490/1024	1.9141	6.367
9	16QAM, 616/1024	2.4063	8.456
10	16QAM, 466/1024	2.7305	10.266
11	16QAM, 567/1024	3.3223	12.218
12	16QAM, 666/1024	3.9023	14.122
13	16QAM, 772/1024	4.5234	15.849
14	16QAM, 873/1024	5.1152	17.786
15	16QAM, 948/1024	5.5547	19.809

We capture interference from adjacent NR BSs via interference margin M_I. For a given NR BS deployment density, one may estimate the interference margin using stochastic geometry-based models [6,7,21]. The effect of shadow fading is accounted for using shadow fading margins, $M_{S,B}$, $M_{S,nB}$ for LoS blocked, and non-blocked states provided in [22].

The LoS path between the UE and the NR BS might be temporarily occluded by moving users. Depending on the current link state (LoS blocked or non-blocked) as well as the distance between the NR BS and the UE, the running session employs an appropriate MCS specified in TR 38.211 to maintain reliable data transmission [19]. We also utilize the 3GPP urban micro (UMi) street canyon model specified in TR 38.901 with blockage enhancements that provide path loss for a certain separation distance with and without blockage [22]. Particularly, the path loss is

$$L_{dB}(y) = \begin{cases} 32.4 + 21\log(y) + 20\log f_c, & \text{non-blocked,} \\ 52.4 + 21\log(y) + 20\log f_c, & \text{blocked,} \end{cases} \tag{2}$$

where y is the 3D distance, f_c is the carrier frequency in GHz.

Session Resource Requirements. We assume that the session requires constant bitrate R. Technically, to determine pmf of resources required from NR BS to serve a session with bitrate R, we have to know the CQI and MCS values as well as SNR to CQI mapping. As these parameters are usually vendor-specific, in our study, we use MCS mappings from [23] provided in Table 2.

Denote by s_j, $j = 1, 2, .., K$, the SNR margins of the NR MCS schemes, where K is the MCS number, and by m_j the probability that the UE session is assigned to MCS j. We have

$$m_j = \Pr\{s_j < s < s_{j+1}\} = W_S(s_{j+1}) - W_S(s_j), \tag{3}$$

where $F_S(x)$ is the CDF of SNR S.

Once m_j, $j = 1, 2, \ldots, K$, are available, the probabilities that user will request i resources for a session with rate R is provided as

$$p_i = \sum_{\forall j: e_j \in \left[\frac{R}{i W_{\text{PRB}}}, \frac{R}{(i-1) W_{\text{PRB}}}\right)} m_j, \tag{4}$$

where e_j is a spectral efficiency of j-th CQI and W_{PRB} is a bandwidth of the primary resource block (PRB).

3 The Proposed Methodology

In this section, we develop the approximation for pmf of session resource requirements in the considered scenario. First, we determine the maximum coverage area of NR BS such that cell edge UE experiences no more than a fraction of time p_C in an outage. Next, we develop approximation for SNR that simultaneously accounts for random UE location in the service area and shadow fading. Finally, we derive the pmf of the session resource requirements.

3.1 NR BS Coverage

We first determine maximum coverage of the deployment area, r_A, such that no UEs experience outage with any of NR BS located on the circumference. Let $S_{\min}$ be the SNR outage threshold, i.e., $S_{\min}$ is the lower bound of the SNR range corresponding to the lowest MCS [19]. Using the propagation model for LoS blockage state, we have the following relation

$$S_{\min} = \frac{P_T G_B G_U}{N_0 C_O C_L M_I N_F M_{S,B}} (r_A + [h_A - h_U]^2)^{-\zeta/2}, \tag{5}$$

where ζ (Table 3) is the path loss exponent, h_A and h_U are the heights of NR BS and UE, P_T is the NR BS transmit power, G_B and G_U are the NR BS transmit and the UE receive antenna gains, N_0 is the thermal noise, C_O is the control channel overhead, C_L is the cable losses, M_I is the interference margin, N_F is the noise figure, $M_{S,B}$ is the fading margin in LoS blocked state.

Solving (5) with respect to r_A, we obtain

$$r_A = \sqrt{\left(\frac{P_T G_B G_U}{N_0 C_O C_L M_I N_F M_{S,B} S_{\min}}\right)^{\zeta/2} + (h_A - h_U)^2}, \tag{6}$$

where $M_{S,B}$ is computed as follows

$$M_{S,B} = \sqrt{2}\sigma_{S,B} \text{erfc}^{-1}(2p_C), \tag{7}$$

where $\text{erfc}^{-1}(\cdot)$ is the inverse complementary error function, p_C is the cell edge coverage probability, and $\sigma_{S,B}$ is standard deviation (STD) of shadow fading distribution for LoS blocked state, which is provided in [22].

3.2   SNR CDF Approximation

We now proceed deriving SNR CDF. Observe that in the considered model, the randomness of SNR is due to two factors, UE location, and shadow fading. For accounting for shadow fading, that follows Lognormal distribution (i.e., Normal distribution in the decibel scale), it is easier to operate in decibel scale.

To derive SNR CDF approximation we first obtain CDF of the 3D distance between UE and NR BS assuming that the position of UE is uniformly distributed within the coverage zone. Recall, that the 2D distance is distributed according to probability density function (pdf) $w_R(x) = 2x/2r_A$ [24]. Now, 3D distance can be expressed as a function of 2D distance using $\phi_D(r) = \sqrt{(h_A - h_U)^2 + r^2}$. The 3D distance can be found using the random variable (RV) transformation technique [25]. Particularly, recall that pdf of a RV Y, $w(y)$, expressed as function $y = \phi(x)$ of another RV X with pdf $f(x)$ is

$$w(y) = \sum_{\forall i} f(\psi_i(y)) \left| \frac{d\psi_i'(y)}{dy} \right|, \tag{8}$$

where $x = \psi_i(y) = \phi^{-1}(x)$ is the inverse functions.

Substituting $w_r(x)$ and $\phi_r(x)$ into (8) we arrive at

$$W_d(x) = \begin{cases} 1 & \sqrt{d_E^2 + h_A^2 - 2h_A h_U + h_U^2} \le x, \\ \frac{2h_A h_U - h_A^2 + h_U^2 + x^2}{d_E^2}, & h_A - h_U < x < \sqrt{d_E^2 + h_A^2 - 2h_A h_U + h_U^2}, \\ 0 & \text{elsewhere.} \end{cases} \tag{9}$$

The inverse of the SNR in decibel without shadow fading can be found using the same technique, where SNR in decibels is a function of 3D distance $\phi_{SNR,dB}(d) = 10 \log_{10}\left(Ad^{-\zeta}\right)$, where d is a 3D distance between UE and NR BS, ζ is a pathloss exponent and A is a term representing all gains and losses except propagation losses and fluctuations due to shadow fading. Substituting $\phi_{SNR,dB}(x)$ and $W_d(x)$ into (8) we get SNR CDF:

$$
W_{S^{dB}}(x) =
\begin{cases}
1 - \dfrac{10^{-\frac{x}{5\zeta}}A^{2/\Gamma} - (h_A - h_U)^2}{d_E^2} & x > 10\log 10\left[AB^{-\zeta/2}\right], \\
0 & \text{elsewhere,}
\end{cases}
\tag{10}
$$

where $B = d_E^2 + (h_A - h_U)^2$.

Now, recalling that shadow fading is characterized by Lognormal distribution in linear scale leading to Normal distribution in the decibel scale, the random variable specifying the SNR distribution can be written as

$$
S_{SF} = S^{dB} + Norm(0, \sigma_{SF}). \tag{11}
$$

Finally, we determine the SNR CDF as a convolution of $W_S(y)$ and the probability density function of a normal distribution with zero mean and standard deviation σ, i.e.,

$$
W_{S_{SF}}(y) = \int_{-\infty}^{\infty} W_S(y + u)\frac{e^{-\frac{u^2}{2\sigma^2}}}{\sqrt{2\pi}\sigma}\,du. \tag{12}
$$

Unfortunately, the latter cannot be evaluated in closed-form by using random variables transformation technique but can be represented in terms of error functions as follows:

$$
\begin{aligned}
W_{S_{SF}}(x) = \frac{1}{2d_E^2}\Bigg[& A^{2/\gamma}10^{-\frac{x}{5\zeta}}e^{\frac{\sigma^2\log^2(10)}{50\gamma^2}} \\
& \Bigg[\mathrm{erf}\left(\frac{50\zeta\log A - 25\gamma^2\log B + \sigma^2\log^2 10 - 5\zeta x\log 10}{5\sqrt{2}\gamma\sigma\log(10)}\right) \\
& -\mathrm{erf}\left(\frac{50\zeta(\log A - \gamma\log(h_A - h_U)) + \sigma_S^2\log^2 10 - 5\zeta x\log(10)}{5\sqrt{2}\gamma\sigma\log(10)}\right)+\Bigg] \\
& + \left(d_E^2 + (h_A - h_U)^2\right)\mathrm{erf}\left(\frac{-10\log A + 5\zeta\log B + x\log 10}{\sqrt{2}\sigma\log(10)}\right) - (h_A - h_U)^2 \\
& \times\mathrm{erf}\left(\frac{\sqrt{2}(-10\log A + 10\zeta\log(h_A - h_U) + x\log 10)}{\sigma\log 100}\right) + d_E^2\Bigg],
\end{aligned}
\tag{13}
$$

where $B = d_E^2 + (h_A - h_U)^2$, $\mathrm{erf}(\cdot)$ is the error function.

Including the blockage induced losses L_B into A and using $\sigma_{S,B}$ and $\sigma_{S,nB}$ into (13), we can obtain two SNR CDFs $W_{S_{nB}}$ and W_{S_B} for non-blocked LoS and blocked LoS conditions.

To determine SNR S_i, we also need the blockage probability. Observe that with the specified RDM mobility model the fraction of time UE located at the 2D distance x from NR BS is in blocked conditions coincides with the blockage probability provided in [26],

$$p_B(x) = 1 - e^{-2\lambda_B r_B \left[x \frac{h_B - h_U}{h_A - h_U} + r_B\right]}. \tag{14}$$

leading to the following weighted blockage probability with the NR BS

$$p_B = \int_0^{r_A} p_B(x) w_D(x) dx. \tag{15}$$

The final result for SNR CDF accounting for shadow fading and blockage is

$$W_S(x) = P_B W_{S_B(x)} + (1 - P_B) W_{S_{nB}(x)}. \tag{16}$$

Once SNR CDF is obtained, session resource requirements pmf can be obtained using (3) and (4). Observe that it is expressed in terms of error functions.

Table 3. System parameters.

Parameter	Value
Carrier frequency, f_c	28 GHz
Transmit power, P_T	23 dBm
UE receive antenna gain G_U	5.57 dBi
NR BS transmit gain G_B	20.58 dBi
LoS blockage loss, L_B	20 dB
NR BS height, h_A	4 m
UE height, h_U	1.5 m
Blocker height, h_B	1.7 m
Blocker radius, r_B	0.3 m
User density, λ_B	0.2 users/m^2
Session rate, R	2 Mbps
Control channel overhead, C_O	1 dB
Cable losses, C_L	2 dB
Interference margin, M_I	3 dB
Thermal noise, N_0	−174 dBm/Hz
Noise figure, N_F	7 dB
Noise figure, W_{PRB}	1.44 Mhz
Min SNR, S_{min}	−9.478 dB
Cell edge coverage probability, p_C	0.01, 0.05, 0.1
Radius of coverage area, r_A	65, 119, 165 m
Standard deviation of shadow fading, $\sigma_{S,B}, \sigma_{S,nB}$	4, 8.2 dB

4   Numerical Results

In this section, we report our numerical results. We first demonstrate that the resulting SNR CDF closely follow Normal distribution. Then, we proceed illustrating the effect of shadow fading on session resource requirements pmf. System parameters used in this section is shown in Fig. 3.

4.1   SNR Approximation

We first start assessing the proposed Normal approximation to SNR CDF. Figure 2 illustrates SNR CDFs with and without shadow fading as well as their approximations by a weighted sum of two Normal distributions, similar to (16), with parameters $\mu = E[S^{dB}]$ and $\sigma = \sqrt{\sigma^2_{SF,\cdot} + \sigma^2_{S^{dB}}}$ for LoS and nLoS cases. To assess the closeness of original and approximating distributions, we use the notion of statistical distance. Particularly, we apply the Kolmogorov Statistic (K-S) also shown in Fig. 2. Analyzing the resulting values of K-S statistic, the proposed approximation is extremely close to the original SNR CDF with shadow fading. Furthermore, as one may observe, the approximation becomes better standard deviation of shadow fading increases, e.g., in nLoS case.

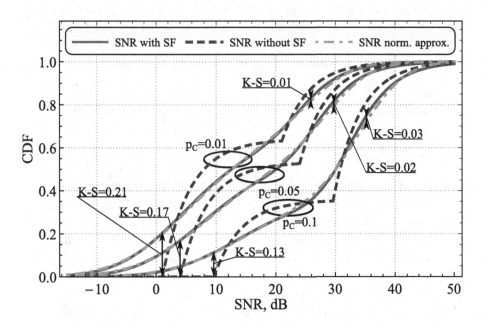

Fig. 2. SNR CDF with/without shadow fading and their approximations.

Note that the Gaussian distribution of shadow fading partially explains the suitability of Normal approximation and partially due to other random effects involved in SNR CDF, e.g., blockage, random UE location.

The second critical observation is that SNR CDF without shadow fading, also shown in Fig. 2 drastically deviates from the one accounting for this effect. Thus, excluding the effects of shadow fading from the model characterizing resource requirements pmf may lead to drastic errors in the predicted system and user performance metrics. Recalling the non-linear mapping of SNR into MCS schemes dropping the effects of shadow fading leads to overly optimistic results.

4.2   Resource Request Approximation

We now proceed highlighting the effect of shadow fading on resource request pmf. Observe that its effect manifests itself in two ways: (i) it affects the coverage area of NR BS, r_A, and (ii) once r_A is determined the shadow fading affects the number of requested resources directly by introducing another source of uncertainty in addition to distance-induced path losses.

The comparison of resource requirements pmfs with and without shadow fading is provided in Fig. 3 for $R = 2$ Mbps and in Fig. 4 for $R = 5$ Mbps. In both cases, the fraction of outage time for cell-edge UE in nLoS state is set to $p_C = 0.1$. As one may observe, the difference is rather significant and may drastically affect the absolute values of the performance metrics of interest in 5G NR system analysis. Note that not only the form of the distribution changes but its moments as well. The mean resource requirements with shadow fading taken into account are now higher compared to the model without this effect.

Table 4. Summary of approximations for $R = 5$ Mbps.

p_C	0.01	0.05	0.1
Mean SNR	27.016	17.8982	12.7958
Mean SNR no SF	27.016	17.8982	12.7958
Mean SNR Approximation	27.016	17.8982	12.7958
STD SNR	12.0514	12.7718	12.7065
STD SNR no SF	10.538	10.9461	10.6159
STD SNR Approximation	12.0513	12.7718	12.7065
Mean Resource Requirement	1.42256	2.37408	3.27115
Mean Resource Requirement no SF	1.22362	1.69017	2.21262
Mean Resource Requirement Approximation	1.43419	2.35948	3.20999
SDT Resource Requirement	2.71887	11.7379	20.8665
STD Resource Requirement no SF	0.17362	0.64415	1.29765
STD Resource Requirement Approximation	3.03627	11.9472	20.5599

Table 4 provides the mean and standard deviation of SNR with and without shadow fading and the proposed approximation for SNR using Normal distribution. It also illustrates the mean and standard deviation of resulting session resource requirements, including original pmf and its approximation.

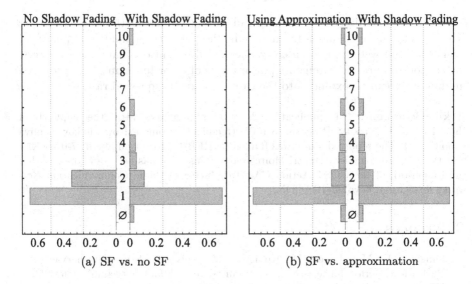

Fig. 3. Comparison of pmfs of resource requirements for session rate 2 Mbps.

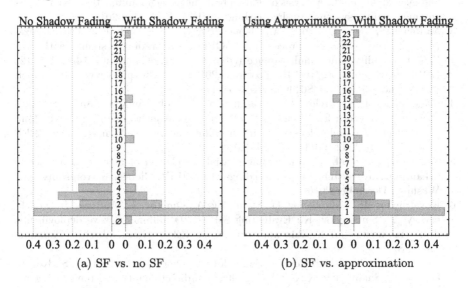

Fig. 4. Comparison of pmfs of resource requirements for session rate 5 Mbps.

5 Conclusion

Characterizing session request requirements is an essential step in assessing user-level performance provided by forthcoming 5G NR systems. In this study, to derive pmf of resources requested by a session from NR BS we have proposed a unified methodology that accounts for 3GPP propagation model, random location of UE within the coverage area of NR BS and other environmental impairments including shadow fading and blockage of LoS path between UE and NR BS. We further demonstrated that in the presence of shadow fading SNR CDF could be well approximated by Normal distribution that allows expressing session resource requirements pdf in terms of error functions. The proposed methodology can be extended to the case of random requested rate R.

Acknowledgment. The publication has been prepared with the support of the "RUDN University Program 5-100" (recipients Roman Kovalchukov, Yuliya Gaidamaka). The reported study was funded by RFBR, project numbers 17-07-00845, and 18-07-00576 (recipient Dmitri Moltchanov). This work has been developed within the framework of the COST Action CA15104, Inclusive Radio Communication Networks for 5G and beyond (IRACON).

References

1. Ometov, A., Moltchanov, D., Komarov, M., Volvenko, S.V., Koucheryavy, Y.: Packet level performance assessment of mmWave backhauling technology for 3GPP NR Systems. IEEE Access **7**, 9860–9871 (2019)
2. Gapeyenko, M., et al.: Analysis of human body blockage in millimeter-wave wireless communications systems. In: Proceedings of IEEE ICC, May 2016
3. Moltchanov, D., Ometov, A., Koucharyavy, Y.: Analytical characterization of the blockage process in 3GPP new radio systems with trilateral mobility and multi-connectivity. Comput. Commun. **146**, 110–120 (2019). https://doi.org/10.1016/j.comcom.2019.07.010, http://www.sciencedirect.com/science/article/pii/S0140366418309411
4. Gapeyenko, M., Samuylov, A., Gerasimenko, M., Moltchanov, D., Singh, S., Akdeniz, M.R., Aryafar, E., Himayat, N., Andreev, S., Koucheryavy, Y.: On the temporal effects of mobile blockers in urban millimeter-wave cellular scenarios. IEEE Trans. Veh. Technol. **66**(11), 10124–10138 (2017)
5. Samuylov, A., et al.: Characterizing spatial correlation of blockage statistics in urban mmWave systems. In: Proceedings of the IEEE Globecom Workshops (GC Wkshps), December 2016
6. Kovalchukov, R., Moltchanov, D., Samuylov, A., Ometov, A., Andreev, S., Koucheryavy, Y., Samouylov, K.: Evaluating SIR in 3D millimeter-wave deployments: direct modeling and feasible approximations. IEEE Trans. Wireless Commun. **18**(2), 879–896 (2018)
7. Kovalchukov, R., Moltchanov, D., Samuylov, A., Ometov, A., Andreev, S., Koucheryavy, Y., Samouylov, K.: Analyzing effects of directionality and random heights in drone-based mmWave communication. IEEE Trans. Veh. Technol. **67**(10), 10064–10069 (2018)
8. Gapeyenko, M., Petrov, V., Moltchanov, D., Andreev, S., Himayat, N., Koucheryavy, Y.: Flexible and reliable UAV-assisted backhaul operation in 5G mmWave cellular networks. IEEE J. Sel. Areas Commun. **36**(11), 2486–2496 (2018)
9. Gapeyenko, M., Petrov, V., Moltchanov, D., Akdeniz, M.R., Andreev, S., Himayat, N., Koucheryavy, Y.: On the degree of multi-connectivity in 5G millimeter-wave cellular urban deployments. IEEE Trans. Veh. Technol. **68**(2), 1973–1978 (2018)
10. Moltchanov, D., Ometov, A., Andreev, S., Koucheryavy, Y.: Upper bound on capacity of 5G mmWave cellular with multi-connectivity capabilities. Electron. Lett. **54**(11), 724–726 (2018)

11. Petrov, V., et al.: Dynamic multi-connectivity performance in ultra-dense urban mmWave deployments. IEEE J. Sel. Areas Commun. **35**(9), 2038–2055 (2017)

12. Moltchanov, D., Samuylov, A., Petrov, V., Gapeyenko, M., Himayat, N., Andreev, S., Koucheryavy, Y.: Improving session continuity with bandwidth reservation in mmWave communications. IEEE Wireless Commun. Lett. **8**(1), 105–108 (2018)

13. Kovalchukov, R., et al.: Improved session continuity in 5G NR with joint use of multi-connectivity and guard bandwidth. In: 2018 IEEE Global Communications Conference (GLOBECOM), pp. 1–7. IEEE (2018)

14. Petrov, V., Lema, M.A., Gapeyenko, M., Antonakoglou, K., Moltchanov, D., Sardis, F., Samuylov, A., Andreev, S., Koucheryavy, Y., Dohler, M.: Achieving end-to-end reliability of mission-critical traffic in softwarized 5G networks. IEEE J. Sel. Areas Commun. **36**(3), 485–501 (2018)

15. Begishev, V., Samuylov, A., Moltchanov, D., Machnev, E., Koucheryavy, Y., Samouylov, K.: Connectivity properties of vehicles in street deployment of 3GPP NR systems. In: 2018 IEEE Globecom Workshops (GC Wkshps), pp. 1–7. IEEE (2018)

16. Samuylov, A., Moltchanov, D., Krupko, A., Kovalchukov, R., Moskaleva, F., Gaidamaka, Y.: Performance analysis of mixture of unicast and multicast sessions in 5G NR systems. In: 2018 10th International Congress on Ultra Modern Telecommunications and Control Systems and Workshops (ICUMT), pp. 1–7. IEEE (2018)

17. Samouylov, K., Naumov, V., Sopin, E., Gudkova, I., Shorgin, S.: Sojourn time analysis for processor sharing loss system with unreliable server. In: Wittevrongel, S., Phung-Duc, T. (eds.) ASMTA 2016. LNCS, vol. 9845, pp. 284–297. Springer, Cham (2016). https://doi.org/10.1007/978-3-319-43904-4_20

18. Naumov, V., Samouylov, K.: Analysis of multi-resource loss system with state-dependent arrival and service rates. Prob. Eng. Inf. Sci. **31**(4), 413–419 (2017)

19. 3GPP. NR; Physical channels and modulation (Release 15), 3GPP TR 38.211, December 2017

20. Nain, P., Towsley, D., Liu, B., Liu, Z.: Properties of random direction models. In: Proceedings of the IEEE 24th Annual Joint Conference of the IEEE Computer and Communications Societies (INFOCOM), March 2005

21. Petrov, V., Komarov, M., Moltchanov, D., Jornet, J.M., Koucheryavy, Y.: Interference and SINR in millimeter wave and terahertz communication systems with blocking and directional antennas. IEEE Trans. Wireless Commun. **16**(3), 1791–1808 (2017)

22. 3GPP. Study on channel model for frequencies from 0.5 to 100 GHz (Release 14), 3GPP TR 38.901 version 15.0.0, July 2018

23. Fan, J., Yin, Q., Li, G.Y., Peng, B., Zhu, X.: MCS selection for throughput improvement in downlink LTE systems. In: 2011 Proceedings of 20th International Conference on Computer Communications and Networks (ICCCN), pp. 1–5. IEEE (2011)

24. Moltchanov, D.: Distance distributions in random networks. Elsevier Ad Hoc Netw. **10**, 1146–1166 (2012)

25. Ross, S.M.: Introduction to Probability Models, 10th edn. Academic Press Inc, Orlando (2010)

26. Gerasimenko, M., Moltchanov, D., Gapeyenko, M., Andreev, S., Koucheryavy, Y.: Capacity of Multiconnectivity mmWave Systems With Dynamic Blockage and Directional Antennas. IEEE Trans. Veh. Technol. **68**(4), 3534–3549 (2019)

Numerical Study of the Consensus Degree Between Social Network Users in the Group Decision Making Process

Olga Chukhno[1(✉)], Nadezhda Chukhno[1(✉)], Anna Gaidamaka[1(✉)],
Konstantin Samouylov[1,3(✉)], and Enrique Herrera-Viedma[2(✉)]

[1] Peoples' Friendship University of Russia (RUDN University),
Moscow, Russia
olgachukhno95@gmail.com, nvchukhno@gmail.com,
aagajdamaka@sci.pfu.edu.ru, samuylov-ke@rudn.ru
[2] University of Granada, Granada, Spain
viedma@decsai.ugr.es
[3] Federal Research Center "Computer Science and Control" of the Russian
Academy of Sciences, Moscow, Russia

Abstract. The introduction of Web 2.0 and Web 3.0 has changed not only the available web technologies, but also the ways in which users interact. The growing ubiquity of Internet access and a variety of mobile devices have allowed people to choose the most attractive tools and services. The conditions of the created environment are well suited for conducting group decision making processes: a large number of users participate in the network, each of whom has his own interests, knowledge and experience. Despite the huge technological leap, there are still problems to be solved. First, in social networks, people communicate and express opinions with the help of words, while traditional methods of group decision making operate with exact numbers. Experts are required to provide estimates in terms of qualitative aspects. Secondly, it is not enough to find a joint solution for all experts, it is also necessary to reach an acceptable level of consensus. The purpose of this work is to conduct a numerical analysis of the group decision making process in social networks, using user publications as ratings. The paper also proposes a format for conducting a process of reaching consensus and its analysis, using the advantages and features of social networks.

Keywords: Group decision making · Social networks · Sentiment analysis · Clustering · Consensus measures

1 Introduction

Currently we can observe the increasing complexity of the socio-economic environment. For this reason, group decision making processes are becoming more and more popular. Many organizations and companies have moved from a decision taken at the sole discretion to forming expert groups involved in the decision making process.

© Springer Nature Switzerland AG 2019
O. Galinina et al. (Eds.): NEW2AN 2019/ruSMART 2019, LNCS 11660, pp. 586–598, 2019.
https://doi.org/10.1007/978-3-030-30859-9_51

At the same time in the modern world social networks are becoming an increasingly important component of the global network. Online communities are an excellent platform for communication and discussion of problems, news, etc. [1–3]. For this reason, recently there has been an interest in the study of social networks in the field of group decision making [4, 5].

During the group decision making (GDM), information aggregation processes and consensus assessment play a special role. During the aggregation of information, the opinions of all experts are taken into account, and the consensus assessment is responsible for analyzing the degree of expert agreement. Obviously, the consensus assessment process is an important component in group decision making, since only with a high level of expert agreement can the results be considered reliable [6, 7].

In addition, one should pay attention to another feature: people express opinions with the help of words, while computers operate with numbers. An algorithm is needed to help the system translate users' verbal statements into numbers that the computer is able to understand.

One solution is the format of presenting preferences by experts regarding alternatives: for example, linguistic expressions [5, 8] or fuzzy preferences relations [9–11] are often used in the literature. Many GDM approaches suggest that all experts use the same scale to present information about preferences. However, there is a possible scenario in which each participant in the group decision making process is able to choose the granularity of a set of linguistic expressions [5, 12].

This paper focuses on group decision making problems, in which experts should express their preferences on qualitative aspects that cannot be assessed using values. In these cases, the most appropriate method is sentiment analysis, namely its application to the messages of users. Thus, experts will be able to freely communicate and share their opinions without worrying about giving preference to the system.

In the work a numerical experiment based on the method of [2] is conducted. The experiment examines the consensus degree of company employees in the decision making process regarding investment in various areas of the company's development.

The rest part is organized as follows: Sect. 2 offers basic information about the process of group decision making, namely, an introduction to the GDM field, clustering and sentiment analysis. Section 3 presents a theoretical model of the group decision making process. A numerical experiment conducted under real-life conditions is demonstrated in Sect. 4. In Sect. 5, the advantages and disadvantages of the proposed method are analyzed. Finally, the main conclusions are pointed out.

2 Precursive Remarks

2.1 Group Decision Making Process

A process can be considered as a group decision making process if (1) there is some problem that has several solutions; (2) there are 2 or more experts with their own experience, knowledge and opinions about the problem; (3) it is necessary to find a solution which share all experts. In an ideal scenario, a consensus decision should be reached. However, complete consensus is often unattainable, due to some differences inherent in the level of knowledge and personal interests of the participants of the GDM process. In addition, experts deal with vague or biased information and are forced to express their opinions on qualitative aspects.

The GDM process consists of 4 steps:

(1) Providing preferences. Participants are asked to answer the question using the suggested response options. They need to evaluate which of the proposed options are the most preferable in their opinion. One of the most popular methods for this stage is the pairwise comparison of alternatives [13]. This method is convenient for experts since they compare pairs of alternatives, without being distracted by all the options. Parallel assessments help to focus only on two alternatives, simultaneously reducing the expert's uncertainty and leading to high consistency.

(2) Aggregation of the information received from all experts. During this phase, the individual preferences of experts are combined into a single collective assessment matrix, which reflects the opinion of all experts.

(3) Consensus calculation. This step verifies the validity and reliability of the GDM process results. Only in the case when the level of consensus among experts is quite high (which means that the experts agree with each other) the results can be approved. Otherwise, it is worth holding an additional circle of discussion in order to bring together the opinions of experts.

(4) Exploitation of the information obtained in step 2. This step forms the final ranked list of alternatives and determines the decision of all experts.

2.2 Social Networks

In social networks, any network user can communicate with other users by entering a name, for example, @username. This feature is now included in almost every online community, for example, Twitter, Instagram, Facebook, etc. In addition, these platforms have another feature - hashtags. Words or phrases prefixed with # are the primary way to group publications on specific topics, which makes it easier to search for information on a specific subject. Using hashtags in social networks is also useful in the context of group decision making. With a single hashtag, all process-related information can be collected and shared with participants.

2.3 Sentiment Analysis

Sentiment analysis helps to understand the attitude of the author to objects in the text by identifying emotionally colored vocabulary. This method is well suited to the GDM process in social networks, where messages and publications of users are employed. Applying the sentiment analysis to reviews about alternatives written by experts, it becomes possible to identify some properties and determine the opinion of people [14].

Sentiment analysis of natural language processing (NLP) creates systems that identify and extract emotionally colored words. In addition to defining opinions directly, one can also evaluate their polarity (positive or negative feelings), the subject of discussion, as well as identify an expert associated with a specific message. Currently, the sentiment analysis is a subject of interest for many researchers and has many practical applications [15, 16].

In our work with the help of sentiment analysis, expert opinions were translated into a format of preference values.

2.4   Cluster Analysis Concepts

Cluster analysis divides objects into groups based on information that describes objects and their relationships. The goal is to put objects with similar features in one group. The greater the similarity within the group and the greater the differences between the groups, the better or clearer is clustering [17].

Dividing a set of objects into groups called clusters is the task of cluster analysis. In the example with experts, cluster analysis allows you to identify clusters that are grouped by a certain attribute. In the example described below gender and age of the expert were taken into account while splitting.

We chose the K-means method in order to divide the experts into clusters. Clustering using the K-means algorithm refers to methods based on the prototype—some object defining the cluster. Thus, a group is formed from a set of objects that are closer to a given prototype than to any other. In the case of K-means, the prototype is termed as the centroid, a point that is the center of the cluster.

At the first step of the algorithm, K initial centroids are selected. The parameter K - the number of clusters - is set by the user. Each point is then assigned to the nearest centroid. The set of points which were assigned to some centroid forms a cluster. After the centroid of each cluster is updated based on the points assigned to the cluster. The assignment and update steps are repeated until no point is re-assigned to a new cluster or the centroids remain the same.

A distinctive feature of the K-means algorithm is that the user himself sets the number of clusters into which the data set must be divided.

3   Theoretical Model of the Group Decision Making Process

Definition 1. The classic GDM process scenario describes a situation in which there is some problem that needs to be solved. From now on, $\mathcal{X} = \{x_1, \ldots, x_M\}$ is a set of alternatives and $\mathcal{E} = \{e_1, \ldots, e_K\}$ is a set of experts, who give their preferences to the system. The purpose of this process is to determine a ranked list of alternatives.

3.1   Fuzzy Preference Relations

In many situations of group decision making it is assumed that each expert transmits an assessment of each pair of alternatives to the system through a fuzzy preference relationship.

Definition 2. The fuzzy preference relation on the set of alternatives X is the set on the Cartesian product $X \times X$, i.e., it is characterized as

$$\mu_{P_k} : X \times X \rightarrow [0, 1], (x_i, x_j) \rightarrow p_{ij}(k)_{i,j \in 1,...,M} \in [0, 1]. \tag{1}$$

Note that p_{ij} is interpreted as the degree of preference for alternative x_i over alternative x_j. Thus, $p_{ij} = 0.5$ indicates that x_i and x_j are the same; $p_{ij} = 1$ that x_i is superior to x_j. Moreover, $p_{ij} = 1 - p_{ji}$.

3.2   Carrying Out the Process of Comparing Alternatives

To carry out this process it is necessary:

Identify the hashtag of the group decision process: this hashtag will represent the group decision process, for example, #GDM.

Create a list of keywords for all alternatives: a set of keywords $W = \{W_1, ..., W_M\}$ must be associated with each alternative, where M is the number of alternatives. Here, the meaning of W_i forms a set of words that are associated with the alternative x_i.

Provide an opportunity for experts to speak out: This step depends entirely on the experts, their desire and interest in obtaining a solution.

3.3   Extracting Expert Texts

In traditional group decision making methods, experts often provide information to the system using a specific format. Therefore, they get in an uncomfortable situation and cannot express themselves, as they want. To solve this problem in the work it is proposed to use the sentiment analysis to the reports of experts [2]. Thanks to this method experts can use words or phrases instead of exact numerical values while evaluating alternatives.

When performing this process, texts (messages) written by users of the social network are retrieved, i.e. expert messages containing the hashtag are retrieved and the set of texts $A = \{A_1, ..., A_K\}$ associated with the set of experts $\mathcal{E} = \{e_1, ..., e_K\}$ is determined, where K is the number of experts participating in the GDM process.

3.4   Calculation of Expert Preferences

Group decision making methods operate with numbers, while experts use words to express their opinions. There is a need to convert expert texts into a set of preferences. Using sentiment analysis, we are able to determine whether the experts have positive, negative or neutral feelings on two specific alternatives, using three lists of words: PL, NPL and SL, respectively. Each of the lists contains an expression used to compare two elements and the degree of preference of one alternative over the other. Note that a comparative expression from the SL list is assigned an average grade.

Next, it is necessary to calculate the degree of preference of each pair of alternatives: if a comparative expression refers to PL, then the value of preference corresponds to the value of comparative expression, if NPL, then the value of preference corresponds confirms to a negative value. Finally, if the comparative expression belongs to SL, then $p_{ij} = 0$.

To release of negative numbers, there can be realized to change the range and express all the information using the interval $[0, 2 \cdot g]$ (the previously obtained representation generates values that belong to the interval $[-g, g]$).

Definition 3. The value of the preference $p_{ij}(k)$ of expert e_k regarding a pair of alternatives x_i and x_j can be defined as

$$p_{ij_{[0,2 \cdot g]}}(k) = p_{ij_{[-g,g]}}(k) + g. \tag{2}$$

Normalizing of assessments: in the paper we propose to use normalized assessments on the interval $[0,1]$.

Definition 4. The normalized value of the preference $p_{ij}(k)$ of expert e_k for a pair of alternatives x_i and x_j can be calculated from

$$\tilde{p}_{ij}(k) = \frac{p_{ij_{[0,2 \cdot g]}}(k)}{2 \cdot g}. \tag{3}$$

3.5 Consensus Calculation

This step is necessary to measure the quantification of the agreement between experts. Only in the case when the level of consensus is acceptable, the solution of the problem can be considered final.

The approach to assessing the achieved consensus used in the work is based on preferences drawn from the discussion texts [2]. The results obtained are determined from the similarity of the values of preferences provided by each expert.

Definition 5. Consensus for a pair of alternatives x_i and x_j measures the degree of agreement between experts e_k and e_l and is defined as

$$sm_{ij}(e_k, e_l) = s(\tilde{p}_{ij}(k), \tilde{p}_{ij}(l)) = 1 - |\tilde{p}_{ij}(k) - \tilde{p}_{ij}(l)|. \tag{4}$$

Definition 6. Consensus among all experts between two pairs of alternatives x_i and x_j is calculated as

$$cp_{ij} = \frac{\sum\limits_{k=1,k<l}^{K-1} sm_{ij}(e_k, e_l)}{\binom{K}{2}}. \tag{5}$$

Definition 7. Consensus at the alternative level is found from the formula

$$ca_i = \frac{\sum\limits_{j=1,i\neq j}^{M} (cp_{ij} + cp_{ji})}{2(M-1)}. \tag{6}$$

Definition 8. Global group decision making process consensus is

$$GCP = \frac{\sum\limits_{i=1}^{M} ca_i}{M}. \tag{7}$$

3.6 Collective Value Calculation

The values of the preferences of all experts are combined into a single collective matrix, which represents information about the overall opinion of the participants in the GDM process. In our work, we use average as an aggregation operator.

Definition 9. The elements of the matrix of collective preferences $C = (c_{ij})_{i,j\in 1,...,M}$ are calculated as follows

$$c_{ij} = \frac{\sum\limits_{k=1}^{K} \tilde{p}_{ij}(k)}{K}. \tag{8}$$

3.7 Creating a Rating of Alternatives

Using the matrix of collective preferences C, we calculate the rating of alternatives.

The ranking is determined from two degrees of choice: Quantifier Guided Dominance Degree, *GDD*, and Quantifier Guided Non Dominance Degree, *GNDD* [18].

Definition 10. GDD operator calculates the degree in which an alternative x_i dominates the rest. GDD operator expression is defined as

$$GDD_i = \sum_{j=1}^{M} C_{ij}, i = 1, \ldots, M. \tag{9}$$

Definition 11. GNDD calculates the degree in which an alternative x_i is not dominated by the rest. GNDD operator is calculated as

$$GNDD_i = \sum_{j=1}^{M} (1 - \max\{c_{ji} - c_{ij}, 0\}), i = 1, \ldots, M. \tag{10}$$

Definition 12. The final ranking values can be obtained from

$$RV_i = \frac{(GDD_i + GNDD_i)}{2}, i = 1, \ldots, M. \tag{11}$$

Thus, information about alternatives is converted to a rating.

4   Case Study and Numerical Analysis

This section presents the experiment of the application of the group decision making process under real conditions. A tiny company, whose managers have free money, stands in the way of implementing the company's further development program. In this regard, it is necessary to solve in which of the directions for the development would be better to invest money. The company has a private corporate portal for all employees, where it is proposed to carry out the GDM process using the above method. Employees of the company are invited to leave comments regarding pairwise comparison of alternatives, while indicating the hashtag in the general chat of the company's employees.

The problem is to choose the best option for investing money: a set of alternatives $X = \{x_1, x_2, x_3, x_4, x_5, x_6, x_7\}$ are described in detail in Table 1. Forty experts $\mathcal{E} = \{e_1, e_2, e_3, \ldots, e_{40}\}$ will provide their opinions in the company chat.

Above all, a set of keywords $W = \{W_1, \ldots, W_7\}$ must be determined for each alternative in order to use them to find alternatives in expert texts. Column 3 of Table 1 shows a set of keywords for each alternative.

When all the necessary parameters have determined, it is time to perform an analysis of the texts provided by the experts in the general chat. Table 2 shows examples of proposals used by experts. To convert them into preferences values, it is necessary to compare the values of expressions associated with each comparative word (Table 3). The values of the expressions are given in the interval [0, 5].

Given that we have a large number of experts in GDM process, we cannot demonstrate information received from experts. The rating of alternatives and the results of the consensus analysis are presented in Table 4.

Table 1. Description of alternatives.

x_i	Description	W_i
x_1	Staff development	Employee, staff, qualifications
x_2	Lease/purchase of new premises	Office, premises, accommodation, rent, purchase
x_3	New equipment	Equipment, facilities, procurement
x_4	Investing in the development of new product	Product, offer, creation, release
x_5	Promotion and advertisement	Advertising, marketing, promotion, popularity
x_6	Website and blog development	Website, blog, web page, content site
x_7	Keeping money	Saving, not invest, keeping, safety, and preservation

Table 2. Sentences and preference values.

Sentence	p_{ij}	$\tilde{p}_{ij}$
Staff development is **needed more** than the purchase of new equipment	$p_{13} = 4$	$\tilde{p}_{13} = 0.9$
Product development **is slightly higher** to renting a new office	$p_{42} = 2$	$\tilde{p}_{42} = 0.7$
It is much more important to invest in advertising than in web page	$p_{56} = 5$	$\tilde{p}_{56} = 1$
It is **equally** useful to invest money in a new product and purchase facilities	$p_{43} = 0$	$\tilde{p}_{43} = 0.5$
It would be **worse** to spend money on promotion than to release a new product	$p_{54} = -4$	$\tilde{p}_{54} = 0.1$
Renting a new office is **not as important** as purchasing new procurement	$p_{23} = -2$	$\tilde{p}_{23} = 0.3$
Advertising is **unprofitable** compared to the creation of the site	$p_{56} = -3$	$\tilde{p}_{56} = 0.2$
Renting a new accommodation is **better** than saving money	$p_{27} = 4$	$\tilde{p}_{27} = 0.9$
I think that it is **much worse** not to invest money than to raise the staff qualifications	$p_{71} = -5$	$\tilde{p}_{71} = 0$
Buying equipment is a **little better** than not invest	$p_{37} = 3$	$\tilde{p}_{37} = 0.8$
No matter where to invest money in blog or advertising	$p_{56} = 0$	$\tilde{p}_{56} = 0.5$
Creating a new offer is **not as important** as conducting training for employees	$p_{41} = -2$	$\tilde{p}_{41} = 0.3$
To release a new product will be **better** than saving money	$p_{47} = 4$	$\tilde{p}_{47} = 0.9$

Note that the degree of agreement among experts is 0.8061, which fully satisfies the requirements stated by the firm's managers. Therefore, the decision could be regarding as having been finally, and the results are reliable.

Thus, the company's employees collectively identified the following ranked list of alternatives for investing money (see Fig. 1): $x_1, x_4, x_3, x_5, x_2, x_6, x_7$.

Table 3. Comparative terms with values.

Comparative terms	Value
much more important	5
is needed more/better	4
little better	3
slightly higher/slightly exceeds	2
equally/the same/identically/no matter	–
not important	2
unprofitable	3
worse	4
much worse	5

Table 4. The results of the group decision making process.

x_i	GDD_i	$GNDD_i$	RV_i	Rating	ca_i
x_1	4.6275	7	5.8138	1	0.8376
x_2	3.3275	6.0675	4.6975	5	0.7784
x_3	3.9275	6.65	5.2888	3	0.8348
x_4	4.14	6.735	5.4375	2	0.8144
x_5	3.4125	6.395	4.9038	4	0.7866
x_6	3.305	6.085	4.695	6	0.7716
x_7	1.7875	3.56	2.6738	7	0.8193
Global consensus GCP					0.8061

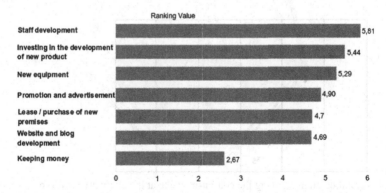

Fig. 1. Ranking of alternatives.

The next task of the study is to divide the experts into an optimal number of groups using k-means clustering. Given such parameters as gender and age, we obtain the following results:

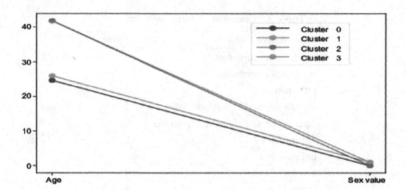

Fig. 2. K-means clustering results.

As it is obvious from the Fig. 2, the experts are divided into 4 groups: men aged about 25 years (cluster 3), men aged about 40 (cluster 0), women aged about 25 (cluster 2) and women aged about 40 (cluster 1) (Fig. 3) and Table 5.

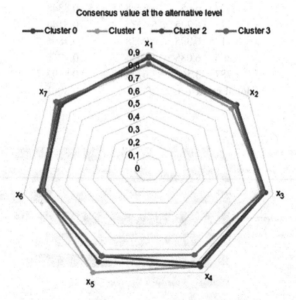

Fig. 3. Diamond diagram showing the consensus value at the alternative level for clusters 0, 1, 2, 3.

Table 5. Global consensus values.

Cluster number	Cluster 0	Cluster 1	Cluster 2	Cluster 3
GCP	0.7996	0.8076	0.8206	0.7852

The results of the experiment showed that the division of experts into clusters depending on gender and age gave a good result in terms of increasing the consistency of assessments in most groups (except for cluster 2). This decision is quite expected, since in matters of implementing the company's development strategy, each person has his own point of view. However, it can be noted that men and women of the older age group are more unanimous in their opinion.

The noteworthy feature here is that the results obtained are based on the subjective opinions of specific people and may not be applicable to any company.

5   Main Findings and Discussions

The novelty of the numerical experiment presented in Sect. 4 is that an example of such a scale was first carried out in real conditions. The procedures of sentiment analysis were carried out. Thanks to this, experts can freely communicate and discuss problems in social networks without worrying about how to present their preferences to the system.

The exploitation of online communities for decision making by a group of experts has the following advantages.

1. Social networks allow to monitor fully all actions occurring during the process. Namely, now it is possible to find out which message belongs to a specific expert, at what time the message was sent (which is important in cases where the expert changes his point of view).
2. Since the provided opinions and discussion texts are stored on the social network, it is easy to access and analyze them using sentiment analysis procedures.
3. Thanks to social networks, discussion processes can be conducted at any time from any location.

6   Conclusions

We live in the era of the development of technologies Web 2.0 and Web 3.0, where a large number of users interact in real time and share opinions and knowledge. Therefore, in connection with the widespread use of social networks, it is necessary to develop decision making mechanisms that take into account the views of users in web communities.

The paper demonstrates a model of the GDM process in which experts can express their opinions through everyday communication and actions in a network. A real experiment of group decision making was carried out, extracting expert assessments by sentiment analysis. Clustering methods are used to assess the degree of agreement among experts in small groups.

This work is the first attempt to implement a large-scale experiment of group decision making process in a social network.

Acknowledgement. The publication has been prepared with the support of the "RUDN University Program 5-100" (N. Chukhno - review and editing; O. Chukhno - original draft preparation, examples; K. Samouylov - supervision and project administration, conceptualization). The reported study was funded by RFBR, project numbers 18-00-01555 (18-00-01685) and 19-07-0093.

References

1. Internet world stats - internet users (2019). https://www.internetworldstats.com/. Accessed 21 May 2019
2. Morente-Molinera, J.A., Kou, G., Samuylov, K., Ureña, R., Herrera-Viedma, E.: Carrying out consensual Group Decision Making processes under social networks using sentiment analysis over comparative expressions. Knowl. Based Syst. **165**, 335–345 (2019)
3. Scott, J.: Social Network Analysis. Sage, Thousand Oaks (2017)
4. Dong, Y., et al.: Consensus reaching in social network group decision making: research paradigms and challenges. Knowl. Based Syst. **162**, 3–13 (2018)
5. Chukhno, N., Samouylov, K., Chukhno, O., Gaidamaka, A., Herrera-Viedma, E.: A new ranking method of alternatives for group decision making in social networks. In: International Congress on Ultra Modern Telecommunications and Control Systems and Workshops, November 2018, no. 8631258 (2019)
6. Liu, F., Wu, Y.H., Pedrycz, W.: A modified consensus model in group decision making with an allocation of information granularity. IEEE Trans. Fuzzy Syst. **26**, 3182–3187 (2018)
7. Wu, J., Xiong, R., Chiclana, F.: Uniform trust propagation and aggregation methods for group decision making in social network with four tuple information. Knowl. Based Syst. **96**, 29–39 (2016)
8. Ureña, R., Chiclana, F., Fujita, H., Herrera-Viedma, E.: Confidence-consistency driven group decision making approach with incomplete intuitionistic preference relations. Knowl. Based Syst. **9**, 86–96 (2015)
9. Zadeh, L.: Fuzzy logic = computing with words. IEEE Trans. Fuzzy Syst. **4**(2), 103–111 (1996)
10. Torra, V.: Hesitant fuzzy sets. Int. J. Intell. Syst. **25**(6), 529–539 (2010)
11. Szmidt, E., Kacprzyk, J., Bujnowski, P.: How to measure the amount of knowledge conveyed by Atanassov's intuitionistic fuzzy sets. Inf. Sci. **257**, 276–285 (2014)
12. Xu, Z.S., Wang, H.: Managing multi-granularity linguistic information in qualitative group decision making: an overview. Granul. Comput. **1**(1), 21–35 (2016)
13. Chukhno, N.V., Chukhno, O.V., Gudkova, I.A., Samouylov, K.E.: Using the Gini coefficient to calculate the degree of consensus in group decision making process. In: CEUR Workshop Proceedings, pp. 97–105 (2018)
14. Saif, H., He, Y., Fernandez, M., Alani, H.: Contextual semantics for sentiment analysis of twitter. Inf. Process. Manage. **52**, 5–19 (2016)
15. Taboada, M., Brooke, J., Tofiloski, M., Voll, K., Stede, M.: Lexicon-based methods for sentiment analysis. Comput. Linguist. **37**(2), 267–307 (2011)
16. Pang, B., Lee, L.: Opinion mining and sentiment analysis. Found. Trends Inf. Retrieval **2**(1–2), 1–135 (2008)
17. Tan, P., Steinbach, M., Kumar, V.: Introduction to Data Mining, p. 769. Addison-Wesley, Boston (2006)
18. Morente-Molinera, J.A., Wikstrom, R., Herrera-Viedma, E., Carlsson, C.: A linguistic mobile decision support system based on fuzzy ontology to facilitate knowledge mobilization. Decis. Support Syst. **81**, 66–75 (2016)

Joint Device-to-Device and MBSFN Transmission for eMBB Service Delivery in 5G NR Networks

Federica Rinaldi[1], Olga Vikhrova[1,2(✉)], Sara Pizzi[1], Antonio Iera[1], Antonella Molinaro[1], and Giuseppe Araniti[1]

[1] DIIES Department, University Mediterranea of Reggio Calabria, Reggio Calabria, Italy
{federica.rinaldi,olga.vikhrova,sara.pizzi,antonio.iera, antonella.molinaro,araniti}@unirc.it
[2] Peoples' Friendship University of Russia (RUDN University), Moscow, Russian Federation
vikhrova-og@rudn.ru

Abstract. Next to come 5G New Radio (NR) radio access technology is foreseen to support a massive number of "resource-hungry" connections and provide high-quality services. Multimedia Broadcast/Multicast Service over Single Frequency Network (MBSFN) for NR, expected in forthcoming 3GPP releases, will enable the simultaneous delivery of the same content over the multiple 5G cells synchronized in time. In this paper, we show that Device-to-Device (D2D) communications can improve the MBSFN network coverage, data rate, and latency for the future 5G use cases. More specifically, we propose a new D2D-aided MBSFN area formation algorithm, which foresees that in such an area the content can be delivered through either MBSFN or D2D transmissions. Achieved simulation results testify that our proposed algorithm is able to improve the system Aggregate Data Rate (ADR) and, at the same time, to satisfy the user's requirements.

Keywords: 5G · eMBB · Dynamic MBSFN Area formation · D2D

1 Introduction

The fifth-generation (5G) radio access technology, named as "New Radio" (NR), enables the provision of innovative advanced services [1].

A primary service classification includes three macro-categories. *Enhanced Mobile Broadband (eMBB)* supports larger data volume and higher data rate than today's mobile broadband services. *Massive Machine Type Communications (mMTC)* focus on the support of low energy consuming connections among

---

The publication has been prepared with the support of the "RUDN University Program 5-100".

O. Galinina et al. (Eds.): NEW2AN 2019/ruSMART 2019, LNCS 11660, pp. 599–609, 2019.
https://doi.org/10.1007/978-3-030-30859-9_52

a huge number of low-cost, low-power, and long-life devices. *Ultra-Reliable and Low-Latency Communications (URLLC)* allow a bidirectional communication among devices requiring low latency and high network reliability. From the combination of the above categories, further use cases can be obtained. As an example, virtual augmented reality can be seen as the fusion of eMBB and URLLC.

Moreover, the exponential growth in the number of devices requesting new services requires that TELCO operators jointly face several challenges, such as radio resource management, network scalability, and flexibility [2].

On one hand, an interesting means to boost the system capacity and to improve users' perceived video quality in 5G NR networks is Device-to-Device (D2D) communications [3]. Indeed, D2D guarantees that high-performing links are set up among users in mutual proximity. On the other hand, to satisfy the ever-demanding requests for resource-hungry applications and to guarantee the network scalability, in forthcoming 3GPP releases, the 5G NR technology is expected to provide eMBB services in single frequency mode by supporting Multimedia Broadcast/Multicast Service over Single Frequency Network (MBSFN) [4]. Thanks to the MBSFN protocol, adjacent cells (i.e., gNB) of a MBSFN Area are synchronized in time to perform the so-called MBSFN transmission [5], that is a simulcast technique for the delivery of the same service over the same radio resources.

The solution traditionally utilized for the MBSFN Area formation follows a fixed approach where coordinated gNBs deliver the same eMBB content to all the interested users with the most robust modulation and coding scheme. As a consequence, the system performance is negatively affected by users experiencing adverse channel conditions. To overcome the limitations of the static approach, authors in [6,7] proposed a dynamic MBSFN Area formation, known as Single Content Fusion (SCF), where users experiencing good channel qualities are served through the MBSFN Area while users with poor channel conditions are served via unicast links. As defined in the Long Term Evolution (LTE) Release 11 [8], SCF foresees a fixed radio resource allocation, where up to 60% of the available radio resources are allocated for multicast/broadcast transmissions, and the remaining 40% for unicast connections. To better understand which radio resources SCF allocates to the unicast users kept out of the MBSFN Area, in [9] a performance evaluation has been done for two cases: *in-band*, where unicast users left out of the MBSFN Area exploit the radio resources allocated to the MBSFN transmission; and, *out-band*, where unicast users excluded from the MBSFN transmission exploit the radio resources destined to the unicast traffic in background.

In the 3GPP Release 15 [4], the 5G technology supports the coexistence of multicast/broadcast and unicast services by allowing for both static and dynamic radio resource allocation. Moreover, 5G foresees that up to 100% of available radio resource can be allocated for the multicast/broadcast service delivery.

In this paper, we propose a novel algorithm for the dynamic formation of MBSFN Area assisted by D2D communications, hereinafter referred to as D2D-MBSFN scheme, which is aimed at boosting the system aggregate data rate (ADR) under the constraint to satisfy all UEs' requests. We compare our proposed algorithm with the SCF scheme through extensive simulation campaigns.

The rest of the paper is organized as follows. Background and motivations of our work are reported in Sect. 2. Section 3 describes our proposed D2D-aided MBSFN area formation algorithm. Simulations and results of the performance analysis are discussed in Sect. 5. Conclusive remarks are given in the last section.

2   Background and Motivations

Evolved-Multimedia Broadcast and Multicast Services (eMBMS) is an evolution of the broadcast technology for the cellular systems. It was designed to provide broadband services over a single cell via Single-Cell Point-to-Multipoint (SC-PTM) connections or over multiple cells arranged to perform a MBSFN transmission.

MBSFN related data and signalling traffic are carried in dedicated logical, transport, and physical channels. Counting procedure allows the network to choose between unicast and MBSFN transmission modes according to the number of UEs interested in MBMS service. In 3GPP Release 13, SC-PTM transmission mode was introduced to broadcast MBMS data on a per-cell basis with the aim of improving network spectral efficiency.

The eMBMS architecture is based on two main logical entities. The *Broadcast/Multicast Service Center* (BM-SC) supports various eMBMS user services, sets up the eMBMS session and initiates content delivery, and the *MBMS Gateway* MBMS-GW forwards MBMS traffic to the downstream nodes using IP multicast distribution. To enable broadcasting and multicasting in the forthcoming 5G system, the 5G-Xcast project [10] proposes three possible architectures, where the BM-SC and MBMS-GW are split into control plane function (CXF) and user plane function (UXF) [11].

One of the main issues for multicast/broadcast service delivery concerns the choice of transmission parameters. There are different Radio Resource Management (RRM) techniques to determine the transmission parameters for the group-based communications. The Conventional Multicast Scheme (CMS) and the Opportunistic Multicast Scheme (OMS) are both single-rate approaches. The main idea of CMS is to broadcast the content to all relevant UEs with the lowest modulation and coding scheme (MCS). It guarantees that all UEs will be able to decode the transmission but at the expense of the poor network spectral efficiency. In OMS the data transmission is scheduled each Transmission Time Interval (TTI) only to the set of UEs with the best channel conditions. OMS achieves long-term fairness but suffers from short-term unfairness. Multicast Subgrouping (MS) is a multi-rate approach which splits UEs into several subgroups with different MCSs. MS offers a good trade-off between fairness and throughput.

The problem of MBSFN Area Formation relates to the issues on how to group eMBMS-enabled cells and how to broadcast the content.

The SCF scheme provides a dynamic method for MBSFN Areas formation based on the content preferences. First, SCF creates single-content MBSFN areas by grouping cells with similar content interests. Then, it merges the overlapping single-content MBSFN areas into multi-content MBSFN areas to maximize the overall throughput. In detail, SCF chooses a better MCS level for the MBSFN transmission and serves the users with poor channel conditions via unicast link.

3 The Proposed D2D-Aided MBSFN Area Formation Algorithm

The idea of D2D-MBSFN scheme is to exploit both MBSFN and D2D transmissions to improve the system performance and the Quality of User Experience (QoE). The algorithm consists of the following steps:

1. *Channel Status Indicator (CSI) collection.* All users send their channel status feedback to the gNBs.
2. *MBSFN Configuration.* Based on the collected CSI, the gNB selects the appropriate transmission parameters, i.e, the lowest supported MCS (the most robust modulation).
3. *D2D configuration.* The gNB iteratively increases the MCS level of the multicast transmission. If all UEs in the area support the selected MCS, the eMBB content will be delivered through the MBSFN Area; otherwise, it will be delivered by means of the D2D communications towards the out-of-service UEs. All relevant gNBs select appropriate forwarding nodes (FNs) among the UEs in the MBSFN Area and verify if all the out-of-service UEs are reached by the FNs via D2D links.

The algorithm stops either when at least one UE is not able to receive the eMBB service or when the ADR after n iterations is lower than the ADR after the $n-1$ iterations. Figure 1 shows how the proposed algorithm works.

4 System Model

We consider a 5G NR system and MBSFN network where all gNBs are synchronized in time and simultaneously transmit data over the same frequency. Transmissions are scheduled in frames, each of which consists of 10 subframes. A subframe of 1 ms can be composed of a different number of slots according to the NR numerology [12]. The radio spectrum is managed in terms of Resource Blocks (RBs). One RB comprises 12 consecutive subcarriers. NR numerology also supports different subcarrier spacings. UEs are interested in an eMBB service, e.g. Video on Demand (VoD). Therefore, for the wideband service like VoD, we consider the numerology $\mu = 0$ with SCS of 15 KHz and TTI of 1 ms.

Let C denote a set of gNBs deployed inside the Synchronization Area, which may include one or more MBSFN Areas. All MBSFN Areas belong to a set M.

The content can be delivered towards UEs in two modes: *(i) cellular mode*, when UEs receive data directly from the gNB, and *(ii) D2D mode*, when UEs receive data via D2D links from the forwarding UE. Let U be a set of UEs receiving the service through the MBSFN transmission. We denote D as a set of UEs served by D2D communications and R as a set of D2D-relays.

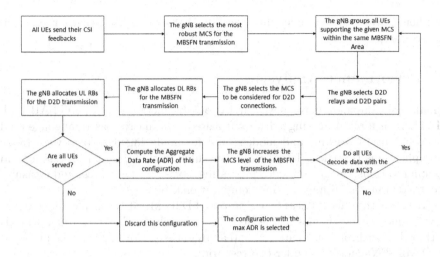

Fig. 1. Flowchart of the proposed D2D-aided MBSFN area formation algorithm.

Each gNB collects the CSI of all devices within the area and selects such an MSC to satisfy all UEs. A scheduler allocates radio resources every TTI. Let $\mathcal{RB}_m$ stand for the set of available RBs for the MBSFN Area $m \in \mathcal{M}$, while $\mathcal{RB}_{b,m}$ and $\mathcal{RB}_{D2D,m}$ correspond to the sets of dedicated RBs to the MBSFN and D2D transmissions, respectively.

We consider that D2D communications take place in uplink subframes in order to exploit unused radio resources. Thus, the number of RBs allocated to D2D communications must be less or equal to the number of available RBs in a given MBSFN Area:

$$| \mathcal{RB}_{D2D,m} | \leq | \mathcal{RB}_m^{\mathrm{UL}} |, \quad \forall m \in \mathcal{M} \qquad (1)$$

The overall ADR of MBSFN transmission is a sum of individual rates of the UEs from the set $\mathcal{U}$ with the respect to the number of allocated RBs:

$$\mathcal{ADR}_{B,m} = \sum_{u \in \mathcal{U}} Rate(\mathcal{U}) \times | \mathcal{RB}_{b,m} |, \forall m \in \mathcal{M} \qquad (2)$$

In a similar way, we define the overall ADR of D2D connections:

$$\mathcal{ADR}_{D2D,m} = \sum_{u \in \mathcal{D}} Rate(\mathcal{D}) \times | \mathcal{RB}_{D2D,m} |, \forall m \in \mathcal{M} \qquad (3)$$

Finally, the ADR over all MBSFN Areas is given by:

$$\mathcal{ADR} = \sum_{m \in \mathcal{M}} \left(\mathcal{ADR}_{B,m} + \mathcal{ADR}_{D2D,m} \right) \qquad (4)$$

MBSFN Area formation problem for the D2D-MBSFN algorithm can be formulated as:

$$\underset{\mathcal{RB}}{\arg \max} \, \mathcal{ADR}$$
$$\text{subject to} (1), (4) \qquad (5)$$

and heuristically solved under the condition that all UEs will receive the broadcast content.

5   Performance Analysis

Simulative campaigns have been carried out by means of the MATLAB tool. MATLAB is a programming software featured by a matrix-based language used for designing algorithms and models, and for analyzing data of a wide range of applications (i.e., deep learning and machine learning, signal processing and communications, image and video processing, control systems, test and measurement, computational finance, and computational biology).

To assess the effectiveness of our proposed D2D-MBSFN scheme, we compare it with one of the schemes in the literature, namely SCF. We consider 10 cells within the synchronization area [5] and analyze the performance of the proposed D2D-MBSFN algorithm under two scenarios:

- *Scenario 1.* The number of users per cell is set to 500, while the channel bandwidth varies from 3 to 20 MHz.
- *Scenario 2.* The channel bandwidth is fixed to 10 MHz and the number of users per cell varies from 200 to 600.

UEs are randomly deployed in a cell. We consider a 2 GHz carrier frequency and we set the numerology $\mu = 0$ and the subcarrier spacing equals to 15 kHz. The rest of simulation settings are reported in Table 1.

The percentage of radio resource assigned to the MBSFN transmission is set to 60% (i.e., 60 RBs of the available 100 RBs in the case of a channel bandwidth of 20 MHz) in order to perform a fair comparison between D2D-MBSFN and SCF. We assume that both SCF and the proposed D2D-MBSFN algorithm operate in in-band mode. According to SCF, the considered 60% of the available RBs are split between the MBSFN and unicast transmissions (i.e., again 60% and 40%, respectively). D2D-MBSFN performs the MBSFN transmission during the downlink subframes, and the D2D communications during the uplink subframes by exploiting all 60% of the available RBs for both MBSFN and D2D.

Each simulation has been run several times to achieve the most reliable results with 95% confidence intervals.

The system performance has been evaluated in terms of the following metrics:

- *Aggregate Data Rate (ADR)* is the sum of UEs data rates.
- *Spectral Efficiency* is the data rate normalized by the bandwidth.
- *Average Throughput* is the average data rate experienced by all users.

Figure 2 shows the ADR for D2D-MBSFN and SCF algorithms when the number of UEs per cell is set to 500 and the channel bandwidth varies from 3 to 20 MHz. D2D-aided MBSFN algorithm significantly outperforms SCF at each bandwidth configuration. As expected, both algorithms improve ADR when the channel bandwidth gets wider. Indeed, D2D-MBSFN and SCF increase the ADR from 3.94 Gbps to 34.25 Gbps and from 1.38 Gbps to 20.3 Gbps, respectively.

Table 1. Main simulation assumptions

Parameter	Value
Cell layout	Hexagonal grid, 10 cells
Inter Site Distance	500 m
Pathloss model	$128.1 + 37.6\, log_{10}(R)$, R in kilometers
gNB transmit power	46 dBm
D2D node Tx power	23 dBm
gNB antenna gain	15 dBi
UE antenna gain	0 dBi
gNB noise figure	5 dB
UE noise figure	9 dB
Carrier frequency	2 GHz
Scheduling Frame	10 ms
RB size	12 sub-carrier
μ	0
Sub-carrier spacing	15 kHz
TTI	1 ms
BLER target	1%

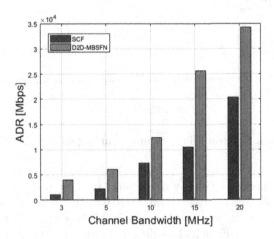

Fig. 2. Aggregate Data Rate under varying channel bandwidth.

However, as we can see in Fig. 3 when both SCF and D2D-MBSFN utilize the same bandwidth but the number of UEs per cell less than 400, the ADR in case of SCF scheme is higher than that of our proposal. It is because the higher the number of UEs, the higher the probability to find users with poorest channel conditions. Due to the cumulative nature of ADR, the metric, in general, is increasing together with the increasing number of UEs per cell.

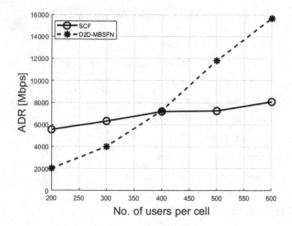

Fig. 3. Aggregate Data Rate under varying number of UEs.

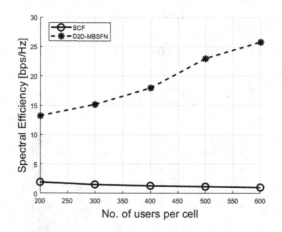

Fig. 4. Spectral Efficiency under varying number of UEs.

In Fig. 4, the spectral efficiency is evaluated in the case of simulation scenario 2. As the number of UEs increases from 200 to 600, the spectral efficiency of D2D-MBSFN significantly grows since less robust MCSs can be utilized for serving MBSFN users. In fact, UEs in bad conditions are served via D2D communications which, in addition, can take advantage over the utilization of all available RBs. On the contrary, the performance of SCF worsens under an increasing number of users since the presence of a higher number of UEs perceiving bad channel qualities forces the selection of more robust MCSs. As a result, the spectral efficiency is considerably (on average 15-fold) higher for our proposed D2D-MBSFN algorithm with respect to SCF.

As shown in Fig. 5, D2D-MBSFN considerably improves the spectral efficiency with respect to SCF in the simulation scenario 1 as well.

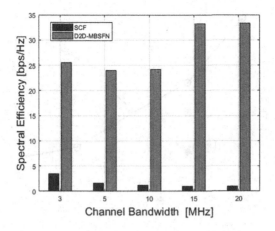

Fig. 5. Spectral Efficiency under varying channel bandwidth.

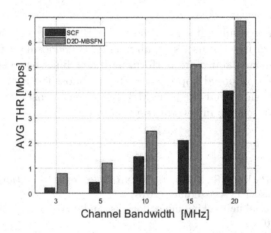

Fig. 6. Throughput under varying channel bandwidth.

In D2D-MBSFN, cell edge UEs benefit from the D2D transmissions in single frequency mode with respect to the unicast connections in SCF. The data rates of such UEs significantly increase leveraging high performing D2D links, Fig. 6. The average data rate of UEs exploiting SCF drops as the number of UEs per cell increases (see Fig. 7) while the metric continues to grow even when more devices are served through the D2D connections.

6   Conclusions

In this paper, we proposed a novel MBSFN area formation algorithm assisted by D2D communications that aim at improving the system Aggregated Data Rate and the system spectral efficiency thanks to the establishment of connections

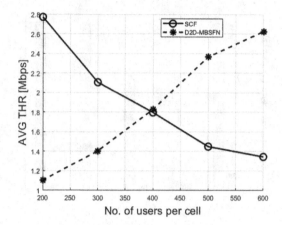

Fig. 7. Throughput under varying number of users per cell.

between nodes in proximity. It selects the best MBSFN Area configuration under the condition to satisfy the user's requirements. By choosing a proper MCS for MBSFN transmissions and more performing direct links between MBSFN UEs and edge UEs, the proposed approach is able to boost the average throughput. Simulation results show that the D2D-MBSFN scheme considerably outperforms the reference SCF scheme in several test conditions.

References

1. Dahlman, E., Parkvall, S., Sköld, J.: 5G NR: The Next Generation Wireless Access Technology. Elsevier, Amsterdam (2018)
2. Cisco: Cisco visual networking index: forecast and methodology, 2016–2021. White paper, June 2017
3. Asadi, A., Wang, Q., Mancuso, V.: A survey on device-to-device communication in cellular networks. IEEE Commun. Surv. Tutor. **16**(4), 1801–1819 (2014)
4. 3GPP, TS 38.913: 5G; Study on Scenarios and Requirements for Next Generation Access Technologies. Rel. 15 (2018)
5. 3GPP, TS 36.300: Evolved Universal Terrestrial Radio Access (E-UTRA) and Evolved Universal Terrestrial Radio Access Network (E-UTRAN). Rel. 9 (2009)
6. Borgiattino, C., Casetti, C., Chiasserini, C.F., Malandrino, F.: Efficient area formation for LTE broadcasting. In: IEEE International Conference on Sensing, Communication, and Networking (SECON), June 2015
7. Casetti, C., Chiasserini, C.F., Malandrino, F., Borgiattino, C.: Area formation and content assignment for LTE broadcasting. Comput. Netw. **126**, 174–186 (2017)
8. Lecompte, D., Gabin, F.: Evolved multimedia broadcast/multicast service (eMBMS) in LTE-advanced: overview and Rel-11 enhancements. IEEE Commun. Mag. **50**, 11 (2012)
9. Rinaldi, F., Scopelliti, P., Iera, A., Molinaro, A., Araniti, G.: Delivering multimedia services in MBSFN areas over 5G networks: a performance analysis. In: IEEE International Symposium on Broadband Multimedia Systems and Broadcasting, Valencia (2018)

10. 5GPPP: 5G-Xcast. https://5g-xcast.eu
11. Estevan, C.B., et al.: Broadcast and Multicast Communication Enablers for the Fifth Generation of Wireless Systems; Mobile Core Network. 5G-Xcast Project, Deliverable 4.1, May 2018
12. 3GPP TS 38.211: NR; Physical channels and modulation, Release 15 (2018)

Calculation of Packet Jitter
for Correlated Traffic

Igor Kartashevskiy[✉] and Marina Buranova

Povolzhskiy State University of Telecommunications and Informatics,
443010 23 L. Tolstoy, Samara, Russia
ivk@psuti.ru

Abstract. Here we present a jitter calculation for correlated traffic. As a model, we consider a general-type queuing system, where the distribution of packet interarrival time and service time are approximated by hyperexponential distributions. We use a method that allows us to consider correlation properties through the parameters of these distributions. The resulting expression for jitter allows to calculate its average value with known or estimated statistical characteristics of random interarrival time, service time and packet waiting time for a general type queuing system serving correlated traffic. We show the calculation of packet jitter for real IPTV traffic both with pronounced correlation properties and without correlation. The dependence of the packet jitter value on the utilization coefficient is also given. We made the comparison of the results of analytical evaluation of packet jitter with the results of simulation conducted in the ns2 system, where a two-dimensional sequence was captured on the real network and consisted of packet generation times and their lengths.

Keywords: Jitter · Correlated queue · Hyperexponential distribution

1 Introduction

The packet delay variation (jitter) at the input of a network device significantly affects the quality of service (QoS) in IP networks, especially for real-time applications such as videoconferencing, VoIP, video streaming, and can affect more than delay and packet loss.

The analysis of packet jitter is simply carried out by the with methods of the queuing theory [1] when traffic is represented in the *M/M/1* system where the interarrival time and service time (packet duration) are the sequences of independent random variables with exponential probability density functions (PDF). It is well known that IP network traffic has fractal properties, and we have to consider the autocorrelation in mentioned sequences and heavy-tailed distributions for instantaneous intervals [2]. In other words, traffic should be represented in a *G/G/1* system with correlated time sequences.

There are a lot of works dedicated to the calculation of jitter for non-Poisson traffic models. An expression for the jitter density function was obtained in [3]. The calculation of jitter for various distributions (lognormal, Pareto, Gamma and truncated normal), used as the source for the systems G/M/1 and G/D/1, is also given. However,

© Springer Nature Switzerland AG 2019
O. Galinina et al. (Eds.): NEW2AN 2019/ruSMART 2019, LNCS 11660, pp. 610–620, 2019.
https://doi.org/10.1007/978-3-030-30859-9_53

these results were obtained without taking into account the autocorrelation of the interarrival time and the service time. Authors presents extensive analytical model for the delay jitter in the single queue single server queuing system in [4]. They considered only correlated service time intervals and found that for different values of utilization coefficient and service statistics for the queue, the parameters affecting the delay jitter change accordingly. Authors in [5] describes condition under which the jitter buffer at the receiver is to be operated for minimum output variance in a Time Division Multiplexing over IP framework, to achieve better voice quality. They model the receiver jitter buffer as a correlated M/G/1 queueing system with correlation between the interarrival and the service times.

There are different approaches to modeling the correlated traffic. A well-known approach [6], based on the use of BMAP (Batch Markovian Arrival Process), where requests are received under the control of some irreducible Markov chain with continuous time and finite space of states. This model allows to achieve any predetermined value of the coefficient of variation and correlation coefficient. That is, to simulate a stream with the parameters obtained by statistical processing of the results of monitoring the real stream. However, approximating a real flow with a BMAP is quite challenging.

Here we consider another possible way to simulate the $G/G/1$ queuing system with both correlated interarrival time and correlated service time using $H_2/H_2/1$ approximation when traffic is modeled by the renewal process. The parameters of renewal process are determined by statistical parameters of the observed traffic. Obviously, this option operates with an equivalent Markov system $(H_2/H_2/1)$ and, for its implementation, requires knowledge of the second-order statistical characteristics of real traffic.

2   The Model of a Queuing System with Correlated Traffic

We can perform an analysis of the general-type queuing system with pronounced correlation properties of the interarrival time and service time using hyperexponential distributions. This requires an approximation of the system $G/G/1$ by the system $H_2/H_2/1$. In this case, the PDF for the H_2 distribution written as:

$$w_H(\tau) = p \cdot \alpha_1 e^{-\alpha_1 \tau} + (1-p) \cdot \alpha_2 e^{-\alpha_2 \tau}, \tag{1}$$

with parameters p, α_1, α_2 that must be defined considering the correlation function and coefficient of variation of initial sequence.

Consider the sequence of correlated time intervals $X_i, i = 1, 2, \ldots$ with the rate λ_x and other parameters: m_x is the mean, $\sigma_x^2 = Var\{X_n\}$ is the variance, $R_x(k) = \frac{E\{[X_i - m_x] \cdot [X_{i+k} - m_x]\}}{\sigma_x^2}$, $k = 0, 1, \ldots$ is the correlation coefficient, $C_x^2 = \sigma_x^2/m_x^2$ is the coefficient of variation and $I_x = C^2(X)\left(1 + 2\sum_{j=1}^{\infty} R_x(j)\right)$ is the dispersion index.

Replace the sequence of random variables X_i, $i = 1, 2, \ldots$ with the renewal process H [7, 8]. For H we have $m_H = m_x$ (or $\lambda_H = \lambda_x$) and the squared coefficient of variation is equal to the dispersion index of the original sequence, i.e. $C_H^2 = I_x$.

Equating the number of events $N_X(t)$ in the sequence X_i, $i = 1, 2, \ldots$ to the number of events in the renewal stream $N_H(t)$ in the interval $(0, t]$, i.e. $N_X(t) = N_H(t)$, $t \to \infty$ we can show [9] that $E(N_X(t)) \to \lambda_x t$ and:

$$E(N_H(t)) \to \lambda_H t + A_H(t),$$

where:

$$A_H(t) = A_H(1 - e^{-\alpha_H t}),$$

and decrement α_H together with the parameter A_H describe the variation of the average value $E(N_H(t))$ over time.

For any renewal process, it is true [8] that:

$$C_H^2 = 1 + 2A_H.$$

If a mixture of two Poisson processes with probabilities p and $(1 - p)$ and parameters α_1 and α_2 with density (1) is selected as the renewal process, then the parameters of the renewal process are [9]:

$$
\begin{cases}
\alpha_1 = \frac{1}{2}\left(\lambda_x + \alpha_H(1 + A_H) \pm \sqrt{[\lambda_x + \alpha_H(1 + A_H)]^2 - 4\lambda_x \alpha_H} \right) \\
\alpha_2 = \lambda_x + \alpha_H(1 + A_H) - \alpha_1 \\
p = \frac{\lambda_x + \alpha_H A_H - \alpha_2}{|\alpha_1 - \alpha_2|}
\end{cases}
\tag{2}
$$

So, A_H and α_H define the properties of the renewal process.

Determining the values of these parameters is quite difficult. It was shown in [8] that the values of A_H and α_H can be determined from the observed traffic sample. It's fair for A_H:

$$A_H = \frac{C_x^2 - 1}{2} + C_x^2 \sum_{i=1}^{\infty} R_x(i), \tag{3}$$

and the α_H can be approximately calculated with the expression:

$$\alpha_H \approx \frac{2\lambda_x}{(2A_H + 1)[1 - R^2(1)]}, \tag{4}$$

where the factor $[1 - R^2(1)]$ in the denominator refines the behavior of $E(N_H(t))$ for small t [5].

Thus, this approach does not require knowledge of the initial one-dimensional probability density functions of interarrival and service time and allows us to simulate the *G/G/1* queuing system with correlated traffic by approximating it with a queuing system with hyperexponential distributions.

3   The Expression for Packet Jitter

In accordance with the recommendation of the IETF [10], jitter is a random variable, defined as

$$J_{i+1} = |T_{i+1} - T_i|.$$

Here T_i is the delay time of the *i*-th packet in the network node, and defined as $T_i = W_i + Q_i$, where W_i is the *i*-th packet waiting time in the queue and Q_i is the time it is serviced.

It's obvious that:

$$J_{i+1} = |T_{i+1} - T_i| = |W_{i+1} + Q_{i+1} - (W_i + Q_i)|$$

Further, following the works [3, 11], we will use Lindley assumption [1], which namely that the $(i+1)$ packet will not wait in the queue when the condition $V_{i+1} \geq T_i$ is satisfied, where V_{i+1} is the time interval between the arrival of the $(i+1)$ and *i*-th packet. Then:

$$W_{i+1} = \begin{cases} 0 & \text{when } V_{i+1} \geq T \\ W_i + Q_i - V_{i+1} & \text{when } V_{i+1} < T \end{cases}$$

And then:

$$J_{i+1} = \begin{cases} |Q_{i+1} - T_i| & \text{when } V_{i+1} \geq T \\ |Q_{i+1} - V_{i+1}| & \text{when } V_{i+1} < T \end{cases} \tag{5}$$

Considering the random variables T_i, Q_i and V_i as independent among themselves and non-correlated in the structure of each sequence, the index *i* in the corresponding PDFs can be discarded and the following notation can be entered: $f_T(x)$ is PDF of a random variable T, $f_V(y)$ is a PDF of a random variable V, $f_Q(z)$ is a PDF of a random variable Q.

Non-correlated random variables within each sequence will be provided by introducing appropriate approximations based on the use of renewal processes.

In accordance with (5) we can write:

$$P(J_{i+1} = \omega) = \iiint\limits_{x,y,z} P(V_{i+1} = y, \ Q_{i+1} = z, \ T_i = x) dx dz dy =$$

$$= \iiint\limits_{x,y,z} P(V_{i+1} = y) \cdot P(Q_{i+1} = z) \cdot P(T_i = x) dx dz dy. \tag{6}$$

The expression (6) takes into account the independence of random variables T_i, Q_i and V_i. Areas of integration X, Y, Z are selected considering the possible values of random variables determined similar to condition (5):

$$\omega = \begin{cases} |z - x| & y \geq x \\ |z - y| & y < x \end{cases}$$

For PDFs from the expression (6) it follows that [3]:

$$f_{J_{i+1}}(\omega) = \int_0^\infty f_{V_{i+1}}(y) \int_0^\infty f_{Q_{i+1}}(z)[f_T(z - \omega)U(y - x) + f_T(z - y)U(x - y)]\, dzdy$$

where $U(x) = \begin{cases} 1, & x \geq 0 \\ 0, & x > 0 \end{cases}$.

The density $f_{J_{i+1}}(\omega)$ allows us to find the average value $E(J_{i+1})$:

$$E(J_{i+1}) = \int_0^\infty \omega f_{J_{i+1}}(\omega)\, d\omega =$$
$$= \int_0^\infty f_{V_{i+1}}(y) \int_0^\infty f_{Q_{i+1}}(z) \int_0^\infty [\omega f_{T_i}(z - \omega)U(y - x) + \omega f_{T_i}(z - y)U(x - y)]dxdzdy \tag{7}$$

Consider $\omega = |z - x|$, when $y \geq x$, and $\omega = |z - y|$, when $y < x$, then the integral in (7) can be rewritten in the form, omitting the index i:

$$\int_0^\infty [\omega f_{T_i}(z - \omega)U(y - x) + \omega f_{T_i}(z - y)U(x - y)]dx$$
$$= \int_0^y |z - x|f_T(x)dx + |z - y| \int_y^\infty f_T(x)dx$$

Now we can finally write expression for the average jitter:

$$J = E(|T_{i+1} - T_i|) = \int_0^\infty f_V(y) \int_0^\infty f_Q(z) \left[\int_0^y |z - x|f_T(x)dx + |z - y| \int_y^\infty f_T(x)dx \right] dzdy. \tag{8}$$

Expression (8) allows us to calculate the average jitter value with known or estimated PDFs of random interarrival, service and packet waiting time for a general type queuing system (such as $G/G/1$) serving correlated traffic.

We use hyperexponential approximations of densities $f_V(\cdot)$ for interarrival time and $f_Q(\cdot)$ for service time. It was shown in [12], that using a hyperexponential representation for densities $f_V(\cdot)$ and $f_Q(\cdot)$ leads to the hyperexponential form of the PDF of the waiting time in the queue $f_W(\cdot)$. However, to simplify the calculations here we assume that $f_W(\cdot)$ is the exponential distribution and has the form $f_W(\tau) = \delta e^{-\delta\tau}$. For example, this assumption is exactly fulfilled for the system $G/M/1$, where the parameter δ is defined as $\delta = \mu(1 - \xi)$, where ξ is the root of the equation $\xi = \Lambda_V(\mu - \mu\xi)$, where Λ_V is the Laplace transform for $f_V(\cdot)$, μ is the mean service time in the $G/M/1$ system [1].

So, we accept the following representation for PDFs $f_V(\cdot)$ and $f_Q(\cdot)$:

$$f_V(\tau) = p\gamma_1 e^{-\gamma_1\tau} + (1 - p)\gamma_2 e^{-\gamma_2\tau},$$

$$f_Q(\tau) = q\mu_1 e^{-\mu_1\tau} + (1 - q)\mu_2 e^{-\mu_2\tau},$$

where parameters (p, γ_1, γ_2) and (q, μ_1, μ_2) are determined by the mean and variance of the initial distribution with regard to the corresponding correlation function of the sequence of time intervals.

Since the packet delay time T_i in the system is $T_i = W_i + Q_i$, the PDF $f_T(\cdot)$ is determined by the convolution of the distributions (if the condition of independence of random variables W and Q is satisfied):

$$f_T(y) = \int_0^\infty f_W(u)f_Q(y - u)\, du = \frac{\delta q\mu_1}{|\delta - \mu_1|}e^{-\mu_1 y} + \frac{\delta(1 - q)\mu_2}{|\delta - \mu_2|}e^{-\mu_2 y}$$

Now, all the initial densities in (8) are defined, so we can perform a threefold integration, considering the obvious relation:

$$\int_0^y |z - x|f_T(x)dx = \int_0^z (z - x)f_T(x)dx + \int_z^y (x - z)f_T(x)dx$$

The result will look like:

$$
\begin{aligned}
J = {} & A\left[\mu_1\mu_2 q + \mu_2^2 + \mu_1^2(1 - q)\right] + B\left[\mu_1\mu_2(1 - q) + q\mu_2^2 + \mu_1^2\right] - \\
& -\frac{A}{\mu_1}\left[\frac{p\gamma_1}{\mu_1 + \gamma_1} + \frac{(1 - p)\gamma_2}{\mu_1 + \gamma_2}\right] - \frac{B}{\mu_2}\left[\frac{p\gamma_1}{\mu_2 + \mu_1} + \frac{(1 - p)\gamma_2}{\mu_2 + \gamma_2}\right] + \\
& +\frac{2qA}{\mu_1}\left[\frac{p\gamma_1}{2\mu_1 + \gamma_1} + \frac{(1 - p)\gamma_2}{2\mu_1 + \gamma_2}\right] + \frac{2(1 - q)B}{\mu_2}\left[\frac{p\gamma_1}{2\mu_2 + \mu_1} + \frac{(1 - p)\gamma_2}{2\mu_2 + \gamma_2}\right] + \\
& +2\left[\frac{A(1 - q)}{\mu_2} + \frac{Bq}{\mu_1 + \gamma_2}\right] \cdot \left[\frac{p\gamma_1}{\mu_1 + \mu_2 + \gamma_1} + \frac{(1 - p)\gamma_2}{\mu_1 + \mu_2 + \gamma_2}\right],
\end{aligned}
\tag{9}
$$

where $A = \frac{\delta q}{|\delta - \mu_1|}$, $B = \frac{\delta(1 - q)}{|\delta - \mu_2|}$.

For calculations using expression (9), we have to know all the parameters of the used hyperexponential distributions.

4 Calculation of Packet Jitter for Captured IPTV Traffic

We take the real data characterizing IPTV traffic at the access level where for the observed time intervals between packets have the following mean value, variance and coefficient of variation.

$$m_x = 0,0049, \quad \sigma_x^2 = 0,00014, \quad C_x^2 = \frac{\sigma_x^2}{m_x^2} = 6 \tag{10}$$

In this case, the correlation function of the sequence of time intervals between packets has the form of Fig. 1(a).

For service intervals, the same parameters has values:

$$m_y = 0,0041, \quad \sigma_y^2 = 0,00046, \quad C_y^2 = \frac{\sigma_y^2}{m_y^2} = 2,76 \tag{11}$$

and the corresponding correlation function is shown in Fig. 1(b).

As follows from the values C_x^2 and C_y^2 both distributions interarrival and service time belong to the type of distributions with heavy tails. The utilization factor, defined as $\rho = m_y/m_x$, for the considered system $G/G/1$ with the received parameters (10) and (11) takes the value $\rho = 0,84$.

Next, we approximate the distribution $w(x)$ characterizing the interarrival time by the distribution $f_V(\tau) = p\gamma_1 e^{-\gamma_1 \tau} + (1-p)\gamma_2 e^{-\gamma_2 \tau}$, and the distribution $w(y)$ of the service time intervals by distribution $f_Q(\tau) = q\mu_1 e^{-\mu_1 \tau} + (1-q)\mu_2 e^{-\mu_2 \tau}$.

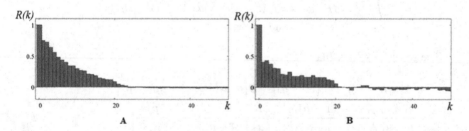

Fig. 1. Graph of correlation coefficients: (a) inter-arrival time; (b) IPTV traffic packet sizes

To calculate the values of the parameters (p, γ_1, γ_2) and (q, μ_1, μ_2), we use the expressions (2)–(4).

Simplifying the calculations, we present the sequences $R_x(j)$, $j = 1, 2, \ldots$ (Fig. 1 (a)) and $R_y(j), j = 1, 2, \ldots$ (Fig. 1(b)) as arithmetic progressions with a step of 0.05 starting from $R_x(1) = 0,75$ and $R_y(1) = 0,45$, respectively. In this case, the sum of the correlation coefficients will take the values $\sum_{j=1}^{15} R_x(j) = 5,625$ and $\sum_{j=1}^{9} R_y(j) = 2,025$.

Parameters of the renewal processes V and Q are obtained in the form:

$$A_V = 36.25; \quad \alpha_V = 12,7; \quad \gamma_1 = 3,85; \quad \gamma_2 = 673,3; \quad p = 0,013;$$

$$A_Q = 6,47; \quad \alpha_Q = 43,88; \quad \mu_1 = 19,4; \quad \mu_2 = 552,3; \quad q = 0,045.$$

Estimated parameters (p, γ_1, γ_2) and (q, μ_1, μ_2) allows us to calculate packet jitter, according to the expression (9).

The parameter δ value required for (9) can be approximately estimated as follows. Authors in [13] show the Laplace transform for PDF of waiting time for $H_2/H_2/1$ queuing system as $W*(s) = \frac{s_1 s_2 (s+\mu_1)(s+\mu_2)}{\mu_1 \mu_2 (s+s_1)(s+s_2)}$ and expression for average waiting time in queue as $\delta = 1/s_1 + 1/s_2 - 1/\mu_1 - 1/\mu_2$, where s_1 and s_2 are negative real parts of the roots of cubic equation $s^3 - c_2 s^2 - c_1 s - c_0 = 0$. Coefficients for this equation are:

$$c_0 = a_0 b_1 - a_1 b_0 - a_0(\mu_1 + \mu_2) + b_0(\gamma_1 + \gamma_2),$$

$$c_1 = -a_1 b_1 - a_0 - b_0 + (\gamma_1 + \gamma_2)(\mu_1 + \mu_2),$$

$$c_2 = \gamma_1 + \gamma_2 - \mu_1 - \mu_2,$$

and intermediate parameters are: $a_0 = \gamma_1 \gamma_2$, $a_1 = p\gamma_1 + (1-p)\gamma_2$, $b_0 = \mu_1 \mu_2$, $b_1 = q\mu_1 + (1-q)\mu_2$.

Now we can calculate the value of J with the expression (9), substituting into it the values (p, γ_1, γ_2), (q, μ_1, μ_2) and δ. The calculation gives the value $J = 0,887$ (conditional units of time).

It is of interest to compare the obtained result with the result of jitter analysis in case when there are no correlation in interarrival and service time while maintaining the distributions $w(x)$ and $w(y)$ with the same parameters. This situation is easily modeled by setting zero values for the sums of the correlation coefficients $\sum_{j=1}^{15} R_x(j) = 0$ and $\sum_{j=1}^{9} R_y(j) = 0$. The latter assumption will lead us to the following coefficients for approximating hyperexponential distributions

$$A_V = 2,5; \quad \alpha_V = 68,02; \quad \gamma_1 = 34,01; \quad \gamma_2 = 408,14; \quad p = 0,09; \atop A_Q = 0,88; \quad \alpha_Q = 176,74; \quad \mu_1 = 88,68; \quad \mu_2 = 486,1; \quad q = 0,78. \tag{12}$$

The value of δ must also be recalculated.

Using the values of the coefficients from relations (12) to calculate the jitter according to (9) we have $J = 0,082$ (conditional units of time).

Comparison of two jitter values $J = 0,887$ and $J = 0,082$ with a rather large load ($\rho = 0,84$) suggests that, if there are obvious correlation properties for time intervals that determine the nature of traffic, the average value of jitter is much larger for correlated traffic in the queuing system. In the specifically considered case, jitter is more than 10 times.

5   The Impact of Utilization on Jitter

A more complete picture of jitter variation for correlated traffic in the *G/G/1* system gives dependence $J(\rho)$. The presented method of analysis allows us to do this quite simply.

We will variate the load $\rho = \frac{m_y}{m_x}$ by changing the parameter m_x, preserving the properties of the distributions $w(x)$ and $w(y)$, i.e. keeping the values of the coefficients of variation C_y^2 and C_x^2. We leave the type of correlation functions unchanged. The change of m_x will require recalculation of the entire set of coefficients $(A_V, \alpha_V, \gamma_1, \gamma_2, p)$. In connection with the change of ρ, it is necessary to recalculate the parameter δ.

The dependency $J(\rho)$ is shown at Fig. 2, where the red line shows the result for correlated traffic, and the blue line is for non-correlated traffic.

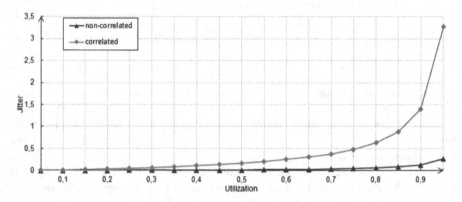

Fig. 2. Dependence of jitter on load factor (Color figure online)

6   Simulation

Compare the results of the analytical evaluation of jitter and the results of simulation.

Simulation was implemented in the ns2, allowing us to generate and process real traffic flows. A two-dimensional sequence registered on a real network and consisting of time intervals between packets and intervals of their processing is used as a processed stream. The simulation model of traffic processing implemented according to the Fig. 3. Two streams were selected for processing: first with characteristics corresponding to parameters (10) and (11) with the correlation coefficients from Fig. 1, the second with characteristics (10) and (11), but without any correlation coefficients.

The results of simulation presented at Fig. 4. Analytical results presented for comparison by dashed lines.

It follows from Fig. 4 that the results of simulation are consistent with the analytical results. This indicates the adequacy and sufficiently high accuracy of the analytical model. The greatest discrepancy between the analytical results and the simulation results is manifested only for the utilization tending to 1 for traffic without correlation.

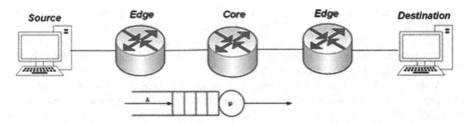

Fig. 3. Simulation scheme in ns2

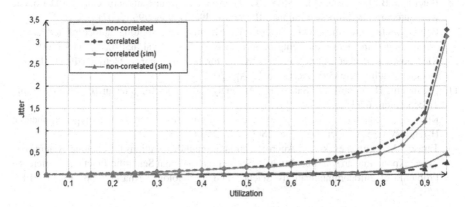

Fig. 4. Simulation results

7 Conclusion

So, within the framework of the considered approximations with the use of hyperexponential approximations, the presented analysis technique makes it possible to estimate jitter for the G/G/1 queuing system with given correlation characteristics for interarrival time and service time. At the same time, we have not used the specific type of marginal distributions for the G/G/1 queuing system, and all possible distributions are described by two characteristics—the mean time and the coefficient of variation. This is a consequence of using the system $H_2/H_2/1$ as an approximation for G/G/1. This is a simplified solution. The use of real distributions (instead of hyperexponential) in the final expression (8) for the average value of jitter eliminates the possibility of obtaining an analytical result.

References

1. Kleinrock, L.: Queueing Systems, volume I: Theory. Wiley, New York (1975)
2. Chakraborty, D., Ashir, A., Suganuma, T., Manseld Keeni, G., Roy, T.K., Shiratori, N.: Self-similar and fractal nature of Internet traffic. Int. J. Netw. Manag. **14**(2), 119–129 (2004)
3. Dbira, H., Girard, A., Sanso, B.: Calculation of packet jitter for non-poisson traffic. Ann. Telecommun. **71**(5–6), 223–237 (2016)

4. Hammad, K., Moubayed, A., Shami, A., Primak, S.: Analytical approximation of packet delay jitter in simple queues. IEEE Wirel. Commun. Lett. **5**(6), 564–567 (2016)
5. Seshasayee, U.R., Rathinam M.: Correlated M/G/1 queue modelling of jitter buffer in TDMoIP. In: 8th Advanced International Conference on Telecommunications, AICT 2012, Stuttgart, pp. 191–196. IARIA (2012)
6. Vishnevskii, V.M., Dudin, A.N.: Queueing systems with correlated arrival flows and their applications to modeling telecommunication networks. Autom. Remote Control **78**(8), 1361–1403 (2017)
7. Jagerman, D.L., Balcioğlu, B., Altiok, T., Melamed, B.: Mean waiting time approximations in the G/G/1 queue. Queueing Syst. **46**(3–4), 481–506 (2004)
8. Balcioğlu, B., Jagerman, D.L., Altiok, T.: Approximate mean waiting time in a GI/D/1 queue with autocorrelated times to failures. IIE Trans. **39**(10), 985–996 (2007)
9. Balcioğlu, B., Jagerman, D.L., Altiok, T.: Merging and splitting autocorrelated arrival processes and impact on queueing performance. Perform. Eval. **65**(9), 653–669 (2008)
10. RFC 3393 IP Packet Delay Variation Metric for IP Performance Metrics (IPPM). https://tools.ietf.org/html/rfc3393. Accessed 12 Apr 2019
11. Dahmouni, H., Girard, A., Sanso, B.: An analytical model for jitter in IP networks. Ann. Telecommun. **67**(1–2), 81–90 (2012)
12. Keilson, J., Machihara, F.: Hyperexponential waiting time structure in hyperexponential $H_k/H_l/1$ system. J. Oper. Soc. Jpn. **28**(3), 242–250 (1985)
13. Tarasov, V., Kartashevskiy, I.: Approximation of input distributions for queuing system with hyper-exponential arrival time. In: 2015 Second International Scientific-Practical Conference Problems of Infocommunications Science and Technology, Kharkiv, pp. 15–17. IEEE (2015)

Modeling and Performance Analysis of Elastic Traffic with Minimum Rate Guarantee Transmission Under Network Slicing

Anastasiya Vlaskina[1], Nikita Polyakov[1], and Irina Gudkova[1,2(✉)]

[1] Peoples' Friendship University of Russia (RUDN University),
6 Miklukho-Maklaya St, Moscow 117198, Russian Federation
vlaskina.anastasia@yandex.ru, goto97@mail.ru
[2] Institute of Informatics Problems, Federal Research Center "Computer Science and Control" of the Russian Academy of Sciences, 44-2 Vavilova St., Moscow 119333, Russian Federation
gudkova-ia@rudn.ru

Abstract. The technology of network slicing allows network resources to be distributed among virtual mobile operators in the context of a general increase in the amount of information transmitted. It is important to guarantee the quality of service requirements for user delays. For effective resource sharing, a mathematical and simulation model of a queuing system with elastic traffic, finite queue, and a finite number of sources has been developed. A numerical analysis of such characteristics as the average time and the average number of requests in the system, the share of resource utilization was performed.

Keywords: Network slicing · 5G · Simulation · Queuing system

1 Introduction

Recent studies suggest that fifth-generation (5G) mobile networks will support a large number of users requesting different services, each of which has different requirements for quality of service (Quality of Service, QoS) [1]. The usual division of radio resources leads to an inefficient use of resources with relatively high costs, which are no longer acceptable. Instead, operators are more interested in flexible solutions [2].

The concept Network Slicing is intended for the design, separation, and organizing virtual network resources, taking into account the various uses [3].

The publication has been prepared with the support of the "RUDN University Program 5-100" (recipients A.V., I.G., mathematical model development). The reported study was funded by RFBR, project numbers 18-00-01555(18-00-01685) and No. 18-07-00576 (recipient I.G., simulation model development and numerical analysis).

© Springer Nature Switzerland AG 2019
O. Galinina et al. (Eds.): NEW2AN 2019/ruSMART 2019, LNCS 11660, pp. 621–634, 2019.
https://doi.org/10.1007/978-3-030-30859-9_54

In other words, one physical network is divided into several virtual networks, each of which is designed and optimized for a specific requirement and/or a specific request/service [4,5]. This technology allows operators to effectively share their radio resources with several virtual operators who do not own any physical infrastructure, but each of which has different requirements [6]. Consequently, virtual operators will be able to support many varied services provided in 5G networks [7].

Thus, in the conditions of limited radio resources, the problem of the effective distribution between the various services of various virtual operators arises [8]. At the same time, it is important to guarantee the QoS and requirements for delay to users [9,10].

Based on the set of services and requirements, VNOs may have different Service Level Agreement (SLA), but these SLAs can usually be divided into three types [11]:

- Guaranteed Bitrate (GB). VNOs are assigned minimum and maximum data rates, regardless of the network status. With this type of service, the VNO service is assigned a minimum and maximum guaranteed speed. Such a distribution leads to the fact that subscribers always get good QoE values, but at the same time such services are more expensive.
- Best effort with minimum Guaranteed (BG). For VNO, only the minimum guaranteed data rate is assigned. Thus, with the lowest cost there is a minimum-guaranteed speed [12].
- Best Effort (BE). In this case, there are no guaranteed speeds for servicing VNO subscribers. When allocating resources, subscribers are allocated as much as is currently available. A significant drawback is the resource starvation for the operators in the loaded hours. At the same time, this type of service is the least expensive in monetary terms.

In this paper, an integrated approach was used: a simulation model for loading a file with a minimum guaranteed transfer rate was built, a numerical analysis was performed for various input data, a mathematical model of a queuing system with elastic traffic [13,14], patient for delays, and a finite number of sources was constructed, formulas for calculating such characteristics as the average time and the average number of requests in the system, the share of resource utilization and a comparative analysis of the results was held.

2   Network Slicing and Resource Allocation Problem

2.1   Network Operators and Services

The overall structure of the considered cellular network represents the following. Initially, the network is divided into two large infrastructures:

- base operator, i.e. wireless service provider who owns and controls all the items required for sale and provision of services to the end user, including distribution of radio frequency spectrum. In addition to selling services under its own brand, base operator can also sell the access to network services to mobile virtual network operators.

- virtual operator (or operator), i.e. wireless service provider who does not own a physical wireless infrastructure, through which he provides services to his clients. The virtual operator enters into an agreement with the underlying operator to obtain access to its network services.

Consequently, the network has a base operator that has its own radio frequency resource of volume C. The volume of the resource characterizes the technical capability to provide services, i.e. subscriber service speed. At the same time, the base operator provides access for N virtual operators to its cellular network. A virtual operator, based on its resources, can provide subscribers with all $\mathcal{M} = \{1, 2, ..., M\}$ services, $M = 19$. Depending on the service provided, the minimum and maximum data transfer rates can be set, b_m^{min}, b_m^{max} i.e. the upper and lower limits that the user can use, respectively, where m is the service number from the set of all services, $m \in \mathcal{M}$.

All services based on the type of traffic can be divided into two classes:

- streaming services, $M_s \in \mathcal{M}$, when the service is characterized by duration $\mu_m^{-1}, m \in M_s$;
- and elastic traffic services, $M_e \in \mathcal{M}$, when service is characterized by size s_m, $m \in M_e$, $M_s \cup M_e$.

In addition, each n - operator provides users with its own set of services, $M_n \subseteq \mathcal{M}$, and (n, m) - service number for n - operator.

Virtual operators provide their services to their customers and users. The number of users having access to the m- service of n- operator is denoted by K_{nm}, then the total number of users of the virtual operator is $K_n = \sum\limits_{m \in M_n} K_{nm}$, $n = 1..N$.

2.2   Resource Allocation and Optimization Problem

Thus, the resource C of the base operator is divided between virtual operators and their services. At the same time, the distribution of resources among users within the same service occurs equally under the conditions of existing possibilities and limitations.

Therefore, the distribution procedure can be formulated as an optimization problem as follows:

$$\max_{\mathbf{C}} f(\mathbf{C}) = \sum_{n=1}^{N} \sum_{m \in M_n} a_{nm} C_{nm}$$

$$s.t: \begin{cases} \sum\limits_{n=1}^{N} \sum\limits_{m \in M_n} a_{nm} C_{nm} \leq C \\ 0 \leq C_{nm}^{min} \leq C_{nm} \leq C_{nm}^{max} \leq C \end{cases} \tag{1}$$

where

- C_{nm}: the amount of resource allocated for the m - service n - operator, $m \in M_n$ ($C_n = \sum\limits_{m \in M_n} C_{nm}$: the amount of the resource allocated for the n - operator);
- a_{nm}: the priority distribution of resources between the services of different operators, $0 \le a_{nm} \le 1$ ($\sum\limits_{n=1}^{N} \sum\limits_{m \in M_n} a_{nm} = 1$);
- C_{nm}^{min}, C_{nm}^{max}: minimum and maximum amounts of allocated resource;
- C: the amount of resource of the basic operator.

3   Mathematical Model

3.1   System Model

As a model under study, we investigated the access scheme for a single operator service with the type of service SLA - BG. We were faced with the task of analyzing, within a single virtual operator with elastic traffic and the minimum guaranteed data transfer rate, how the service takes place, taking into account delays. For the network, we introduce the following assumptions: $N = 1$ virtual operator that requests resources from the base operator; virtual operator provides its users with a $M_n = M = 1$ service; service will has priority $a_m = a = 1$; K virtual operator users.

We describe these scenarios in the form of a queuing system. The system contains a resource C over which elastic data blocks are transmitted. At the input a stream of requests comes with the intensity $0 < \lambda_k < \infty$, $k = \overline{1, K}$ (ε_k for second case) and the average block length is $0 < \theta_k < \infty$, $k = \overline{1, K}$. Because the service is of type BG, then there may be delays in the system, which leads to the appearance of a length queue r. If the serving resource C cannot be divided equally between the users with the minimum guaranteed rate b, the request enters the queue r. Consequently, the maximum number of requests that can be serviced simultaneously on a resource is $\lfloor \frac{C}{b} \rfloor = N$. Due to impatience, users leave the system with intensity $0 < \gamma_k < \infty$, $k = \overline{1, K}$.

3.2   Two Models of User Behaviour

Under these assumptions, two models of user behavior and the corresponding traffic models were considered. In the first case (model with variable number of users, *Model 1*), each user behaves as follows: user sends a request to download a file, downloads it and disappears from the system. Disappearance can be associated with cell withdrawal, service change, and completion service. In the second case (model with a fixed number of users, *Model 2*), the behavior of each user is determined according to the following principle: first, user sends a request to download a file, loads it, then waits, then loads the file again, etc. As it can be seen from the description, the first case is described by the Poisson flow of the first kind, the second - by the Poisson flow of the second kind.

This system is described by a one-dimensional random process $N(t) \in \{1, ..., \lfloor \frac{C}{b} \rfloor\}$ — this is the number of requests in the system in $t \ge 0$.

3.3   Model with Variable Number of Users

We consider in detail the *Model 1*.

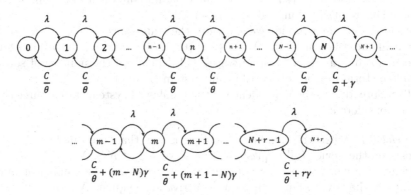

Fig. 1. Conversion intensity diagram for *Model 1*.

Further, we present the state-space representation of the system denoted by $X := \{n \in \{0, ..., N, ...(N + r)\}\}$. For this model, the chart of the transitions can be presented (Fig. 1), in which states are encoded by the number of applications that currently in the system, and the diagram is a birth-and-death process.

According to the obtained diagram the system of global balance equations are given by (2). Solving the system (3) of local balance equations taking into account the normalization condition $\sum_{i=0}^{N+r} p_i = 1$ we obtain probability distributions for model with variable number of users (4).

$$\begin{cases} \lambda p_0 = \frac{C}{\theta} p_1 \\ \left(\lambda + \frac{C}{\theta}\right) p_n = \lambda p_{n-1} + \frac{C}{\theta} p_{n+1}, \quad n = \overline{1, (N-1)} \\ \left(\lambda + \frac{C}{\theta} + (m - N)\gamma\right) p_m = \lambda p_{m-1} + \left(\frac{C}{\theta} + (m + 1 - N)\gamma\right) p_{m+1}, \\ m = \overline{N, (N+r-1)} \\ \left(\frac{C}{\theta} + r\gamma\right) p_{N+r} = \lambda p_{N+r-1} \end{cases} \qquad (2)$$

$$\begin{cases} \lambda p_{n-1} = \frac{C}{\theta} p_n, \quad n = \overline{1, N} \\ \lambda p_{m-1} = \left(\frac{C}{\theta} + (m - N)\gamma\right) p_m, \quad m = \overline{(N+1), (N+r)} \end{cases} \qquad (3)$$

$$p_n = \begin{cases} \left(\frac{\lambda \theta}{C}\right)^n p_0, & n \in \overline{1, N} \\ \left(\frac{\theta}{C}\right)^N \dfrac{\lambda^n}{\prod_{i=1}^{n-N}\left(\frac{C}{\theta} + i\gamma\right)} p_0, & n \in \overline{(N+1), (N+r)} \end{cases},$$

where (4)

$$p_0 = \left(\sum_{n=0}^{N} \left(\frac{\lambda \theta}{C}\right)^n + \left(\frac{\theta}{C}\right)^N \sum_{n=N+1}^{N+r} \frac{\lambda^n}{\prod_{i=1}^{n-N}\left(\frac{C}{\theta} + i\gamma\right)} \right)^{-1}$$

3.4   Model with a Fixed Number of Users

Now consider the situation when there is not one source in the system, but K sources. It should be noted that the source cannot submit a new request until the previous request submitted by it is served.

As in the previous model, we can enter the state-space representation of the system in the following form $X := \{n \in \{0, ..., N, ... \min{(K, N + r)}\}\}$. At the same time, the number of system states will depend on the number of sources K and the total number of places in the system $(N + r)$.

Therefore, due to the dependence of the number of system states, three cases can be considered:

- *Model 2.1.*: when the number of sources is less than the number of empty seats on the device (all requests will be served), $0 < K \leq N$;
- *Model 2.2.*: when the number of sources is greater than the number of empty seats on the device, but their requests arrive in the queue $N < K \leq (N + r)$;
- *Model 2.3.*: when the number of sources is greater than the number of empty seats on the device and in the queue $K > (N + r)$;

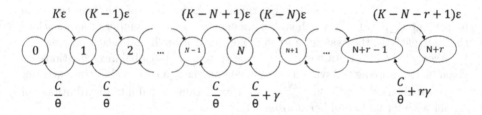

Fig. 2. Conversion intensity diagram for *Model 2.3*.

As in the *Model 1*, to obtain a system of linear algebraic equations for stationary probabilities, we construct the charts of the transitions for all cases, assuming that only one event can occur at a time (receipt of an application in the QS or completion of the application). Here we give the most common case *Model 2.3.* (Fig. 2) for $K > (N + r)$, for other models, the equations are derived similarly. For the other two cases, the number of states will be equal to the number of sources K. Consequently, for *Model 2.1.*, the absence of a queue of requests is taken into account.

Below are the systems of global balance equations and local balance equations (5, 6) for *Model 2.3*. Solving equations for three models, we find that the probability distribution differs only in the boundary values. Therefore, the decision will take the form 7.

$$
\begin{cases}
K\varepsilon p_0 = \frac{C}{\theta} p_1 \\
\left((K-n)\varepsilon + \frac{C}{\theta}\right) p_n = (K-n+1)\varepsilon p_{n-1} + \frac{C}{\theta} p_{n+1}, \quad n = \overline{1, (N-1)} \\
\left((K-n)\varepsilon + \frac{C}{\theta} + (n-N)\gamma\right) p_n = (K-n+1)\varepsilon p_{n-1} + \\
\quad + \left(\frac{C}{\theta} + (n+1-N)\gamma\right) p_{n+1}, \quad n = \overline{N, (N+r-1)} \\
\left(\frac{C}{\theta} + r\gamma\right) p_{N+r} = (K-N-r+1)\varepsilon p_{N+r-1}
\end{cases}
\tag{5}
$$

$$
\begin{cases}
(K-n+1)\varepsilon p_{n-1} = \frac{C}{\theta} p_n, \quad n = \overline{1, N} \\
(K-n+1)\varepsilon p_{n-1} = \left(\frac{C}{\theta} + (n-N)\gamma\right) p_n, \quad n = \overline{(N+1), (N+r)}
\end{cases}
\tag{6}
$$

$$
p_n =
\begin{cases}
\left(\frac{\varepsilon\theta}{C}\right)^n A_K^n \, p_0, & n = \overline{1, \min(N, K)} \\
\left(\frac{\theta}{C}\right)^N \dfrac{\varepsilon^n}{\prod_{i=1}^{n-N}\left(\frac{C}{\theta}+i\gamma\right)} A_K^n p_0, & n = \overline{N+1, \min(N+r, K)}
\end{cases},
$$

where

$$
p_0 = \left(\sum_{n=0}^{\min(N,K)} \left(\frac{\varepsilon\theta}{C}\right)^n A_K^n + \left(\frac{\theta}{C}\right)^N \sum_{n=1}^{N \min(r, K-N)} \frac{\varepsilon^n}{\prod_{i=1}^{n-N}\left(\frac{C}{\theta}+i\gamma\right)} A_K^n \right)^{-1}
\tag{7}
$$

$$
A_K^n = \frac{K!}{(K-n)!}.
$$

3.5 Performance Measures for Model with Variable Number of Users

Analysis of QS is necessary for the effective organization of the system. So, it is necessary to calculate the parameters that characterize the efficiency of a queuing system. Having found the probability distribution of the model (4) with variable number of users (*Model 1*), we compute its performance measures:

– the probability that the system p_0;
– utilization factor $UTIL$ (8);
– blocking probabilities π (9);
– mean time of requests in the queue W_q, on the device W_{ser} and in the system W_{sys} (10–12);
– mean number of requests in the queue L_q, on the device L_{ser} and in the system L_{sys} (13–15).

The mean time of requests in the queue and mean number of requests on the device can be obtained by the Little formula.

$$
UTIL = 1 - p_0,
\tag{8}
$$

$$
\pi = p_{N+r} = \left(\frac{\theta}{C}\right)^N \frac{\lambda^{N+r}}{\prod_{i=1}^{r}\left(\frac{C}{\theta}+i\gamma\right)} p_0 \, ,
\tag{9}
$$

$$W_{\mathrm{q}} = \frac{L_{\mathrm{q}}}{\lambda} = \left(\frac{\theta}{C}\right)^N p_0 \sum_{i=1}^{r} i \cdot \frac{\lambda^{N+i-1}}{\prod_{j=1}^{i}\left(\frac{C}{\theta}+j\gamma\right)}, \tag{10}$$

$$W_{\mathrm{ser}} = \frac{L_{\mathrm{ser}}}{\lambda} = p_0 \left(\sum_{i=1}^{N} i \cdot \left(\frac{\theta}{C}\right)^i \lambda^{i-1} + N \cdot \left(\frac{\theta}{C}\right)^N \sum_{i=1}^{r} \frac{\lambda^{N+i-1}}{\prod_{j=1}^{i}\left(\frac{C}{\theta}+j\gamma\right)}\right), \tag{11}$$

$$W_{\mathrm{sys}} = L_{\mathrm{ser}} + L_{\mathrm{q}}, \tag{12}$$

$$L_{\mathrm{q}} = \left(\frac{\theta}{C}\right)^N p_0 \sum_{i=1}^{r} i \cdot \frac{\lambda^{N+i}}{\prod_{j=1}^{i}\left(\frac{C}{\theta}+j\gamma\right)}, \tag{13}$$

$$L_{\mathrm{ser}} = p_0 \left(\sum_{i=1}^{N} i \cdot \left(\frac{\lambda\theta}{C}\right)^i + N \cdot \left(\frac{\theta}{C}\right)^N \sum_{i=1}^{r} \frac{\lambda^{N+i}}{\prod_{j=1}^{i}\left(\frac{C}{\theta}+j\gamma\right)}\right), \tag{14}$$

$$L_{\mathrm{sys}} = L_{\mathrm{ser}} + L_{\mathrm{q}},$$

$$L_{\mathrm{sys}} = p_0 \left(\sum_{n=0}^{N} n \cdot \left(\frac{\lambda\theta}{C}\right)^n + \left(\frac{\theta}{C}\right)^N \sum_{n=N+1}^{N+r} n \cdot \frac{\lambda^n}{\prod_{i=1}^{n-N}\left(\frac{C}{\theta}+i\gamma\right)}\right). \tag{15}$$

3.6   Performance Measures for Model with Fixed Number of Users

Similarly to *Model 1* and on the basis of the obtained probability distributions (7), we obtain formulas for calculating the performance indicators of the QS with multiple sources, Table 1. For *Model 2.1.* ($0 < K \leq N$), where the number of sources is less than the maximum number of requests that the device can process

Table 1. Performance indicators.

Param.	$0 < K \leq N$	$N < K \leq (N+r)$	$K > (N+r)$
π	0	0	p_{N+r}
L_{ser}	$\sum_{i=0}^{K} i \cdot p_i$	$\sum_{i=0}^{N} i \cdot p_i + N \cdot \sum_{i=1}^{K-N} p_{N+i}$	$\sum_{i=0}^{N} i \cdot p_i + N \cdot \sum_{i=1}^{r} p_{N+i}$
L_q	–	$\sum_{i=1}^{K-N} i \cdot p_{N+i}$	$\sum_{i=1}^{r} i \cdot p_{N+i}$
L_{sys}	$\sum_{i=0}^{K} i \cdot p_i$	$\sum_{i=0}^{K} i \cdot p_i$	$\sum_{i=0}^{N+r} i \cdot p_i$
W_{ser}	$\dfrac{\sum_{i=0}^{K} i \cdot p_i}{\sum_{n=0}^{K-1}(K-n)\varepsilon p_n}$	$\dfrac{\sum_{i=0}^{N} i \cdot p_i + N \cdot \sum_{i=1}^{K-N} p_{N+i}}{\sum_{n=0}^{K-1}(K-n)\varepsilon p_n}$	$\dfrac{\sum_{i=0}^{N} i \cdot p_i + N \cdot \sum_{i=1}^{r} p_{N+i}}{\sum_{n=0}^{N+r-1}(K-n)\varepsilon p_n}$
W_q	–	$\dfrac{\sum_{i=1}^{K-N} i \cdot p_{N+i}}{\sum_{n=0}^{K-1}(K-n)\varepsilon p_n}$	$\dfrac{\sum_{i=1}^{r} i \cdot p_{N+i}}{\sum_{n=0}^{N+r-1}(K-n)\varepsilon p_n}$
W_{sys}	$\dfrac{\sum_{i=1}^{K} i \cdot p_i}{\sum_{n=0}^{K-1}(K-n)\varepsilon p_n}$	$\dfrac{\sum_{i=1}^{K} i \cdot p_i}{\sum_{n=0}^{K-1}(K-n)\varepsilon p_n}$	$\dfrac{\sum_{i=1}^{N+r} i \cdot p_i}{\sum_{n=0}^{N+r-1}(K-n)\varepsilon p_n}$

at the same time, the probability of loss (blocking) of the request is zero $\pi = 0$ -
absolutely all received requests will be served, moreover, immediately (without
waiting). Consequently, there is no need to calculate the characteristics of the
queue (because there is no queue), but we can calculate the mean number/time
of requests on the device L_{ser} and in the system L_{sys}.

4 Numerical Analysis

A simulation model is a program written in the Java programming language.
This is a model of discrete event type. The following events are highlighted: the
receipt of the request in the system, the completion of the maintenance of the
request, and the departure of the request from the queue due to impatience.

With the help of a simulation model, several experiments were conducted.
We begin with a numerical analysis for the data source from (Table 2).

Denote a model with a variable number of users as System I, and a model
with a fixed number of users as System II.

4.1 Example 1. Comparison of Systems and File Length
Distributions

In example 1, we consider the graphs of performance indicators, compiled from
the results of the simulation and presented in Fig. 3. After going through all the
combinations of System I, System II and the file length distributions we got 4
options for each graph. It is worth noting that for System I, the number of users

Table 2. The data source for examples.

Param.	Characteristic	Ex. 1 [15]	Ex. 2	Ex. 3
K	Number of users	[1; 350]	50	50
b	Minimum guaranteed speed [Mbps]	0,384	0,384	1
C	Resource [Mbps]	16,76	16,76 and 3500	[1;30]
ϵ	Upstream rate for System II Exponential distribution	0,0056	[0,01;20]	0,0056
λ	Input flow rate for System I Exponential distribution, $\lambda = \epsilon K$	[0,0056;1,96]	0.28	0.28
θ	File size [MB] Exponential distribution and truncated log-normal distribution with parameters $\mu = 14.45$, $\omega = 0.35$	2	2	2
γ	Leaving impatience	0,000001	0,000001	0,000001
r	Queue length	20	20	20

is in fact variable, because it is believed that each file transfer request comes from a new subscriber. The behavior of the indicators on the graphs is such that we can conclude that the exponential law is an estimate from above for a truncated log-normal. Since we have a mathematical model, we can conclude that it is possible to use it to get an estimate from above.

From the point of view of modeling and computational complexity, the Poisson flow of the first kind is simpler than the Poisson flow of the second kind. Thus, the value of this experiment lies in the fact that we can find a moment in which the Poisson flow of the second kind can be replaced by a Poisson flow of the first kind without loss of accuracy of calculations. For the initial data selected for this numerical analysis, such a pattern is observed when the value $K = 350$ or more. A detailed study of all the circumstances of such a change of flows may be the task of further research.

4.2   Example 2. Analysis of the Probability-Time Characteristics of the Virtual Operator Traffic

In Example 2, we will look at the simulation results only for System II - we will build graphs for the probability-time characteristics for the source data from Table 2 and analyze them.

At Fig. 4(a, b) the simulation results are provided, with the resource size $C = 16.76$ [Mbps] such source data is indicated in the source [15].

However, it is easy to note that in this case, the total time for submission services exceeds 40 s, with a file size of 2 MB. With even a minor the increase in the intensity of receipt of requests the system is quickly filled, and a queue is formed.

To get a delay closer to reality was simulation was carried out with the resource size $C = 3500$ [Mbps] at Fig. 4(c, d). For such initial data, the value of the average time in the system ranges from 8.27 to 156.9 ms and has almost a direct dependence on the intensity of receipt of requests. The queue at the selected intensities is not formed.

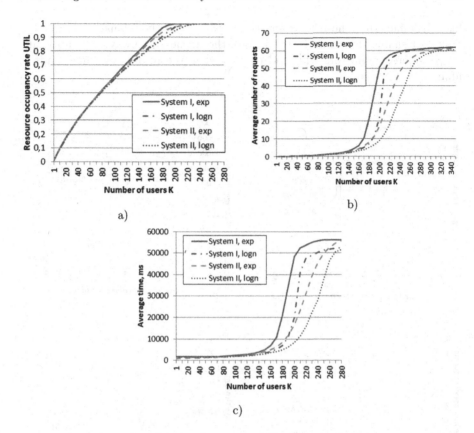

Fig. 3. (a) Graphs of the dependence of the occupation rate of the resource on the number of users for different systems and file length distributions. (b) Graphs of the average number of requests in the system of the number of users for different systems and file length distributions. (c) Graphs of the average time on the number of users for different systems and file length distributions.

4.3 Example 3. Analysis of the Impact of the Amount of Resources on the Probability-Time Characteristics of Traffic

Last consider example 3. With the original data from Tab. 2 consider the dependence of the characteristics of the resource size. The results of this experiment will have the feature of the system in question elastic traffic.

Consider the graphs in Fig. 5. On the first two graphs, we observe a sharp decrease in the number of requests and waiting time in the queue, as well as a jump in the number of requests on the device with resource sizes of $3-4$ [Mbps]. In this interval, the intensity of receipt of requests ceases to be critical for the system. With a further increase in the volume of the resource, the number of requests on the device, as well as the time they are serviced, gradually decreases due to the fact that the rate of service increases, and the intensity of receipt of requests remains the same. In the future, the queue will always be empty, and the considered indicators will tend to zero when the resource volume tends to infinity.

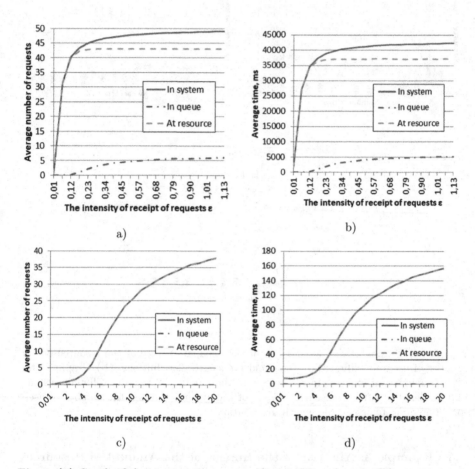

Fig. 4. (a) Graph of the average time versus the intensity of receipt of requests ε, at $C = 16.76$ [Mbps]. **(b)** Graph of the average number of requests from the intensity of receipt of requests ε, at $C = 16.76$ [Mbps]. **(c)** Graph of the average time from the intensity of receipt of requests ε, at $C = 3500$ [Mbps]. **(d)** Graph of the average number of requests from the intensity of receipt of requests ε, at $C = 3500$ [Mbps].

Referring to the graph of $UTIL$ resource occupancy rate versus resource size (Fig. 5c) for this indicator, we observe a regular increase in model time spans when the system was completely empty, which again is explained by the specific increase in the service rate of each individual request.

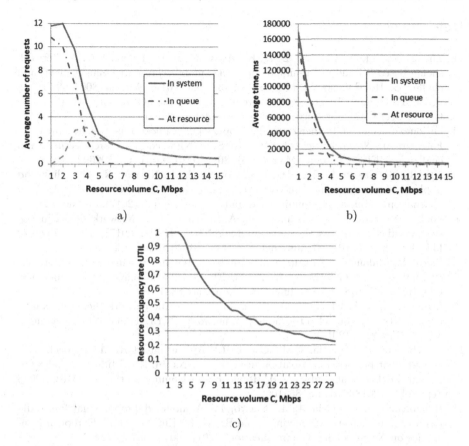

Fig. 5. (a) Graph of the average number of requests from the amount of resource C. (b) Graph of average time versus resource size C. (c) Graph of $UTIL$ resource occupancy rate versus resource size C.

5   Conclusion

Network Slicing, the most important 5G technology, allows to maximize the efficiency of resource allocation between virtual operators. We have proposed the task of optimizing the use of resources in the conditions of the use of NS.

In this paper, we proposed and analyzed two virtual operator traffic service systems based on two different user behavior patterns. For each mathematical model, probability distributions and formulas for performance indicators were obtained, and simulation models of discrete event type were constructed. Our results show that (i) two mathematical models are interchangeable in certain cases, (ii) each of them (with an increase in the number of virtual operators) is consistent for modeling network slicing and solving an optimization problem.

References

1. Zhang, H., Liu, N., Chu, X., Long, K., Aghvami, A.-H., Leung, V.C.M.: Network slicing based 5G and future mobile networks: mobility, resource management, and challenges. IEEE Commun. Mag. **55**(8), 138–145 (2017). Art. no. 8004168
2. 3GPP TS 23.501 V15.4.0 - System architecture for the 5G System (5GS)
3. GB999 User Guide for Network Slice Management R18.5.1
4. Ordonez-Lucena, J., Ameigeiras, P., Lopez, D., Ramos-Munoz, J.J., Lorca, J., Folgueira, J.: Network Slicing for 5G with SDN/NFV: concepts, architectures, and challenges. IEEE Commun. Mag. **55**(5), 80–87 (2017). art. no. 7926921
5. Khatibi, S., Caeiro, L., Ferreira, L.S., Correia, L.M., Nikaein, N.: Modelling and implementation of virtual radio resources management for 5G Cloud RAN. Eurasip Journal on Wireless Communications and Networking (1) (2017), art. no. 128
6. Foukas, X., Patounas, G., Elmokashfi, A., Marina, M.K.: Network slicing in 5G: survey and challenges. IEEE Commun. Mag. **55**(5), 94–100 (2017). art. no. 7926923
7. ITU-T Rec. Y.0.3101 - Requirements of the IMT-2020 network, Jan 2018
8. Lieto, A., Malanchini, I., Capone, A.: enabling dynamic resource sharing for slice customization in 5G networks. In: 2018 IEEE Global Communications Conference, GLOBECOM 2018 - Proceedings (2019). art. no. 8647249
9. Lee, Y.L., Loo, J., Chuah, T.C., Wang, L.-C.: Dynamic network slicing for multi-tenant heterogeneous cloud radio access networks. IEEE Trans. Wirel. Commun. **17**(4), 2146–2161 (2018)
10. Rouzbehani, B., Correia, L.M., Caeiro, L.: An optimised RRM approach with multi-tenant performance isolation in virtual RANs. In: IEEE International Symposium on Personal, Indoor and Mobile Radio Communications, PIMRC, 2018 September (2018). art. no. 8581050
11. Rouzbehani, B., Correia, L.M., Caeiro, L.: A modified proportional fair radio resource management scheme in virtual RANs. In: EuCNC 2017 - European Conference on Networks and Communications (2017). art. no. 7980724
12. Caballero, P., Banchs, A., De Veciana, G., Costa-Perez, X., Azcorra, A.: Network slicing for guaranteed rate services: admission control and resource allocation games. IEEE Trans. Wirel. Commun. **17**(10), 6419–6432 (2018). art. no. 8424561
13. Samouylov, K.E., Gudkova, I.A.: Recursive computation for a multi-rate model with elastic traffic and minimum rate guarantees. In: 2010 International Congress on Ultra Modern Telecommunications and Control Systems and Workshops, ICUMT 2010, pp. 1065–1072 (2010). art. no. 5676509
14. Gudkova, I.A., Markova, E.V., Abaev, P.O., Antonova, V.M.: Analytical modelling and simulation of admission control scheme for non-real time services in LTE networks. In: Proceedings - 29th European Conference on Modelling and Simulation, ECMS 2015, pp. 689–695 (2015)
15. Khatibi, S.: Radio Resource Management Strategies in Virtual Networks. Thesis approved in public session to obtain Ph.D. degree in Electrical and Computer Engineering - 2016

Probability Model for Performance Analysis of Joint URLLC and eMBB Transmission in 5G Networks

Elena Makeeva[1], Nikita Polyakov[1], Petr Kharin[1], and Irina Gudkova[1,2](✉)

[1] Peoples' Friendship University of Russia (RUDN University),
6 Miklukho-Maklaya St, Moscow 117198, Russian Federation
len16730637@yandex.ru, goto97@mail.ru, gruzavjeg@mail.ru,
gudkova-ia@rudn.ru
[2] Institute of Informatics Problems, Federal Research Center,
"Computer Science and Control" of the Russian Academy of Sciences,
44-2 Vavilova St., Moscow 119333, Russian Federation

Abstract. 5G technology is designed to resolve issues such as the growth of mobile traffic, the increase in the number of devices connected to the network, reducing delays for the implementation of new services and the lack of a frequency spectrum. The main services in future 5G networks are enhanced Mobile Broadband (eMBB), massive Machine Type Communications (mMTC) and Ultra Reliable Low Latency Communications (URLLC).

In this paper, we consider two cases of joint eMBB and URLLC transmissions. In the first case, when all resources are occupied, the blocking of new requests happens, and in the second case, it is possible to reduce the transmission speed of the eMBB application and accept the new URLLC request for service.

Keywords: eMMB · URLLC · Joint transmission · Multiple access · NOMA · OMA · Slots · Mini-slots · Queuing systems

1 Introduction

Modern 5G systems are supposed to support both enhanced mobile broadband traffic (eMBB) and Ultra Reliable Low Latency Communication traffic (URLLC). On the one hand, broadband traffic eMBB has a bandwidth of 200–350 Mbit/s with an average delay (several milliseconds). URLLC traffic, on the other hand, requires extremely low delays (1 ms or less) with very high reliability - probability of successful delivery is 0,99999.

The publication has been prepared with the support of the "RUDN University Program 5-100" (recipients P.Kh., I.G., mathematical model development). The reported study was funded by RFBR, project numbers 18-00-01555(18-00-01685) and No. 17-07-00845 (recipient I.G., simulation model development and numerical analysis).

O. Galinina et al. (Eds.): NEW2AN 2019/ruSMART 2019, LNCS 11660, pp. 635–648, 2019.
https://doi.org/10.1007/978-3-030-30859-9_55

URLLC and eMBB topic is gaining increasingly research attention. First group of researchers is connected with network slicing. Authors of the paper [5] propose a risk-sensitive based formulation to allocate resources to the incoming URLLC traffic while minimizing the risk of the eMBB transmission and ensuring URLLC reliability. In article [6], authors introduced the first comprehensive tutorial for integrated mmWave, in which envisioned integrated design will enable wireless networks to achieve URLLC along with eMBB. In article [8], authors define the main usage scenarios and related technical requirements. In paper [9], authors study the potential advantages of non-orthogonal sharing of RAN resources in uplink communications for eMBB, mMTC and URLLC devices. Paper [10] provides a comparative study on the performance of different modulation options for orthogonal frequency division multiplexing (OFDM) in terms of their spectral efficiency, reliability and e.t.c. In paper [11], authors propose a novel method that can fit the low latency and high spectral efficiency requirements of future 5G wireless services by eliminating the need for inserting CP between successive OFDM symbols. Comparing OMA and NOMA is also a popular topic for research within the 5G framework. In paper [12], a novel solution is proposed that enables the non-orthogonal coexistence of URLLC and eMBB services by processing URLLC traffic at the edge nodes, while eMBB communications are handled centrally at a cloud processor. Paper [13] provides an information-theoretic perspective in the performance of URLLC and eMBB traffic under both OMA and NOMA.

We consider the following structure of the joint transmission of URLLC and eMBB traffic: as in cellular systems, time is divided into slots. In order to ensure low latency for URLLC, the slots are divided into minislots as shown in Fig. 1. In NR the one slot can consider 7 or 14 OFDM symbols ($\leq$60 kHz subcarrier spacing) or 14 OFDM symbols ($\geq$120 kHz subcarrier spacing). The duration of slot can be varied in dependence of subcarrier spacing from 1000 ms and less. There are two methods of Multiple Access: Non-Orthogonal Multiple Access (NOMA) and Orthogonal Multiple Access (OMA). NOMA is a multiple access method used in a 5G cellular wireless network. The main function of NOMA is to serve several UEs (user devices) using one 5G-NB (Node B or base station). It serves several users with the same time-frequency resources. The OMA technology uses OFDMA method. In the OFDMA method, BS (base station) shares its resources by transmitting to the UE (user equipment) at different times and frequencies. OFDMA allocates subchannels and time slots to users

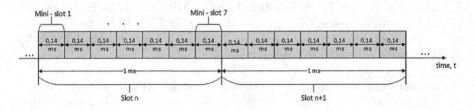

Fig. 1. Structure of frame

based on the desired bandwidth or data rate. In this paper, we consider the case of OMA technology and streaming traffic in first model and the case of NOMA technology and streaming traffic in the second. Let us introduce these two technologies with streaming eMBB traffic using the following scenario: Two devices with eMBB traffic and one device with URLLC traffic that arrived during the first slot, second mini-slot, send their service requests to the same base station, see Fig. 2.

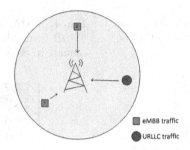

Fig. 2. The example of devices location to illustrate multiple access methods

Firstly, we consider the OMA technology: eMBB traffic requests are distributed by the scheduler to the resource units of each slot, when a URLLC request is received, it can be assigned to any free resource unit of any subsequent minislot, due to the delay requirements. Problems arise in the case when all resource units are engaged in the transmission of eMBB traffic data. Then the transfer of data of any device are have to be blocked and start to transfer data of URLLC request using the freed resource units. But not all released resource blocks will be used for the new URLLC traffic transmission. In the case of NOMA technology, the division occurs not only in time and frequency, but also in power. The scheduler distributes eMBB requests to resource units and there can be vacant resource units, then incoming URLLC requests can be assigned to them. In this case, the allocation of resources is no different from the orthogonal multiple access. The overlay of two traffics will occur when all resource units are occupied. When two requests for eMBB traffic are received, the scheduler distributes these requests to resource units, for example as illustrated in Fig. 3. During the servicing of scheduled eMBB requests in some minislot, a URLLC request occurs. Then, if there are free resource units, the URLLC request that appears will be scheduled on any free resource unit in both cases, as shown in Fig. 4. Otherwise, in case of OMA technology the session of any eMBB traffic device is blocked and a URLLC request will be scheduled on its resource units, as illustrated in Fig. 5. And in case of NOMA technology the URLLC request will be overlay on eMBB request, for example as shown in Fig. 6.

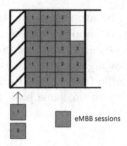

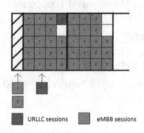

Fig. 3. OMA/NOMA technology and eMBB Streaming Traffic: before URLLC request arrives

Fig. 4. OMA/NOMA technology and eMBB Streaming Traffic: after URLLC request receives

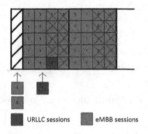

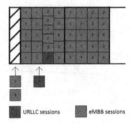

Fig. 5. OMA technology and eMBB Streaming Traffic: blocking eMBB session

Fig. 6. NOMA technology and eMBB Streaming Traffic: overlay URLLC request on eMBB session

2   System Model

In this section, we formulate a the system model of joint eMBB and URLLC traffic transmission, which allows us to consider two possible options for resource allocation planning: there is one base station with devices of two traffic types that try to establish a data connection.

We consider two systems, each of which represents a frame with an initial number of slots - N, the duration of which is 1 ms. Each slot is divided into an equal number of mini-slots - b. Also, in each slot $d(d < b)$ mini-slots are allocated for transferring d URLLC sessions with one current eMBB session simultaneously. Sessions of both classes arrive according to Poisson processes with intensities λ_m and λ_u, respectively. For simplicity, we assume that each session of the same type of traffic has specific requirements for the number of mini slots occupied. So, each session of URLLC traffic takes 1 mini-slot, and each session of eMBB - b mini-slots, i.e. 1 slot. Therefore, the maximum number of eMBB sessions is N, and the URLLC sessions - $C = N \cdot b$. And the maximum number of active URLLC sessions in one slot with the current active session eMBB $= d$. The average duration of URLLC/eMBB sessions is assumed to be exponentially distributed: μ_u^{-1} and μ_m^{-1}, respectively. $\mu_u^{-1} << \mu_m^{-1}$. Let's denote the corresponding offered loads: $\rho_m = \lambda_m/\mu_u$ and $\rho_u = \lambda_u/\mu_u$.

All the main parameters of the systems used in this work are presented in Table 1.

Table 1. System parameters

Parameter	Description
λ_m	eMBB session arrival rate
λ_u	URLLC session arrival rate
μ^{-1}	Average URLLC/ eMBB session duration
n	Number of established (active) sessions
N	Total number of slots
b	Number of mini-slots in one slot
$C = N \cdot b$	Total number of mini-slots
d	Max number of active URLLC session in one slot with an ongoing active eMBB session

3 Model for eMBB Sessions Interruption

Let us describe the process of servicing applications for two eMBB and URLLC traffic as follows: due to limited channel: C is the maximum transmission rate, and because of the requirements for the number of resource units for each request due on type of traffic, the state space can be written as follows:

$$N = \{(n_m, n_u) : n_m \geq 0,\ n_u \geq 0,\ bn_m + n_u \leq C\} \tag{1}$$

where n_m - number of eMBB sessions in the system, n_u - number of URLLC sessions in the system. Scheme of the model is shown in Fig. 7.

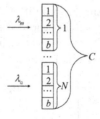

Fig. 7. Scheme of the model

In the case of an eMBB request arriving, the following cases are possible:

- If there is at least one free resource block (b resource units in the block) in the system, then the new eMBB request is accepted to service.
- If there is not free block in the system, the new eMBB request is blocked.

Also in the case of an URLLC request arriving, the following cases are possible:

- There is at least one free resource unit in the system, then the new URLLC request is accepted to service.
- There is not free resource unit and at least one eMBB application is served in the system, then the transfer of the eMBB application is interrupted, b resource units are released, and the new URLLC request is accepted to service.
- There is not free resource unit and eMBB request is not served in the system, then the new URLLC request is blocked.

Thus, possible transitions for an arbitrary state $(n_m, n_u) \in N$ are as follows:

$$
\begin{cases}
a(n_m, n_u)(n_m + 1, n_u) = \lambda_m, \\
a(n_m, n_u)(n_m, n_u + 1) = \lambda_u, \\
a(n_m, n_u)(n_m - 1, n_u + 1) = \lambda_u, \\
a(n_m, n_u)(n_m - 1, n_u) = n_m \mu_m, \\
a(n_m, n_u)(n_m, n_u - 1) = n_u \mu_u.
\end{cases}
\tag{2}
$$

The general structure of the state transition diagram of a defined model is shown in the Fig. 8 while the detailed transition rates for internal states and the diagram for all states with the appropriate conditions are shown in the Figs. 9 and 10 respectively. And the example of the state spaces with marked blocking states is shown in Fig. 11.

We describe the system of equations as follows

$$
\begin{aligned}
&p(n_m, n_u)[\lambda_m I(bn_m + n_u + b \le C) + \lambda_u [I(bn_m + n_u + 1 \le C) \\
&+ I(bn_m + n_u + 1 > C,\ n_u + 1 \le C)] + n_u \mu_u I(n_u > 0) \\
&+ n_m \mu_m I(n_m > 0) = p(n_m + 1, n_u)(n_m + 1)\mu_m I(bn_m + n_u + b \le C) \\
&+ p(n_m, n_u + 1)(n_u + 1)\mu_u I(bn_m + n_u + 1 \le C) + p(n_m, n_u - 1) \\
&\cdot \lambda_u I(n_u > 0) + p(n_m - 1, n_u)\lambda_m I(n_m > 0)
\end{aligned}
\tag{3}
$$

where $(p(n_m, n_u))_{(n_m, n_u) \in N}$ is the probability distribution of the steady state. This distribution can be found by solving the system of equilibrium Eq. 3, written in the following form $p \cdot A = 0$, $p \cdot 1^T = 1$, where the $a((n_m, n_u), (n'_m, n'_u))$-elements of a transposed infinitely small generator A are defined as

$$
a\left((n_m, n_u)\left(n'_m, n'_u\right)\right) =
\begin{cases}
\lambda_m, & n'_m = n_m + 1,\ n'_u = n_u; \\
\lambda_u, & n'_m = n_m,\ n'_u = n_u + 1, \\
\quad or \\
& n'_m = n_m - 1,\ n_m > 0, \\
& n'_u = C - (bn_m - 1); \\
n_m \mu_m, & n'_m = n_m - 1,\ n'_u = n_u; \\
n_u \mu_u, & n'_m = n_m,\ n'_u = n_u - 1; \\
0, & \text{otherwise.}
\end{cases}
\tag{4}
$$

Finding the probability distribution of the stationary state $(p(n_m, n_u))_{(n_m, n_u) \in N}$ we can calculate the performance indicators of the system under consideration as follows:

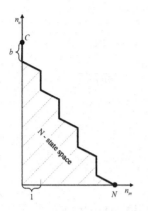

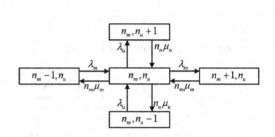

Fig. 8. State transition diagram in general

Fig. 9. Central state diagram

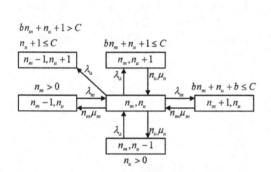

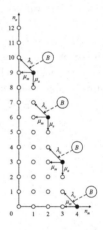

Fig. 10. Central state diagram with block state

Fig. 11. Example of the state spaces

- Average number of eMBB/URLLC sessions:

$$\overline{n_m} = \sum_{n_m=1}^{N} n_m \cdot \sum_{n_u=0}^{C-bn_m} p(n_m, n_u); \tag{5}$$

$$\overline{n_u} = \sum_{n_u=1}^{C} n_u \cdot \sum_{n_m=0}^{\lfloor \frac{C-n_u}{b} \rfloor} p(n_m, n_u). \tag{6}$$

- Blocking probability of the eMBB/URLLC requests:

$$B_M = \sum_{n_u=0}^{C} p\left(\left\lfloor \frac{C-n_u}{b} \right\rfloor, n_u\right); \tag{7}$$

$$B_U = p(0, C).$$ (8)

– The probability of interruption of eMBB request's service:

$$I = \sum_{n_m=1}^{N-1} p(n_m, C - bn_m) \frac{\lambda_u}{\lambda_u + n_m \mu_m + n_u \mu_u} \frac{1}{n_m} \\ + p(N, C - bN) \frac{\lambda_u}{\lambda_u + n_m \mu_m} \frac{1}{n_m}.$$ (9)

– Coefficient of system utilization:

$$UTIL = \frac{1}{C} \sum_{n_m=0}^{N} \sum_{n_u=0}^{C-bn_m} (bn_m + n_u) p(n_m, n_u).$$ (10)

4   Model for eMBB Session Bit Rate Degradation

In the case of the second model, it is possible to impose a URLLC request on an already distributed eMBB request if there are not free resource units, and at least one eMBB request is serviced in the system. In this case, the transfer of the eMBB device does not stop, but there will be a reduction in the rate of its transfer on the transmission rate of the URLLC application. The maximum number of requests to the URLLC in one block can be d, $d < b$ pieces, thus, the minimum transmission rate of eMBB requests - $b' = b - d$ Mbps. Scheme of the model is shown in Fig. 12.

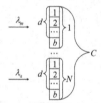

Fig. 12. Scheme of the model

Therefore, compared with the first model, the state space has the following form:

$$N = \{(n_m, n_u) : n_m \geq 0, \ n_u \geq 0, \ (b - d)n_m + n_u \leq C, \ n_m \leq N\}.$$ (11)

In the case when the eMBB request arrives, the same cases are possible as in the first model:

– If there is at least one free resource block (b resource units in the block) in the system, then the new eMBB request is accepted to service.
– If there is not free block in the system, the new eMBB request is blocked.

And in the case when the URLLC application arrives, the following cases are possible:

– There is at least one free resource unit in the system, then the new URLLC request is accepted to service.
– There is not free resource unit, and at least one eMBB request is serviced in the system, then a new URLLC application is accepted for servicing, while the transfer rate of an existing eMBB application is reduced by 1 Mbps.
– There is not free resource unit in the system, and eMBB application isn't served in the system, then the new application URLLC is blocked.

Thus, possible transitions for an arbitrary state $(n_m, n_u) \in N$ are as follows:

$$
\begin{cases}
a(n_m, n_u)(n_m + 1, n_u) = \lambda_m, \\
a(n_m, n_u)(n_m, n_u + 1) = \lambda_u, \\
a(n_m, n_u)(n_m - 1, n_u + 1) = \lambda_u, \\
a(n_m, n_u)(n_m - 1, n_u) = n_m \mu_m, \\
a(n_m, n_u)(n_m, n_u - 1) = n_u \mu_u.
\end{cases}
\tag{12}
$$

The general structure of the state transition diagram of a defined model is shown in the Fig. 13 while the detailed transition rates for internal states and the diagram for all states with the appropriate conditions are shown in the Figs. 14 and 15 respectively. And the example of the state spaces with marked blocking states is shown in Fig. 11.

Let's describe the system of equations as follows:

$$
\begin{aligned}
&p(n_m, n_u)[\lambda_m I(bn_m + n_u + b \leq C) + \lambda_u[I((b-d)n_m + n_u + 1 \leq C) \\
&+ I((b-d)n_m + n_u + 1 > C,\ n_u + 1 \leq C)] + n_u \mu_u I(n_u > 0) \\
&+ n_m \mu_m I(n_m > 0)] = p(n_m + 1, n_u)(n_m + 1)\mu_m I(bn_m + n_u + b \leq C) \\
&+ p(n_m, n_u + 1)(n_u + 1)\mu_u \cdot I((b-d)n_m + n_u + 1 \leq C) + p(n_m, n_u - 1) \\
&\cdot \lambda_u I(n_u > 0) + p(n_m - 1, n_u)\lambda_m I(n_m > 0).
\end{aligned}
\tag{13}
$$

And the elements $a((n_m, n_u), (n'_m, n'_u))$ of a transposed infinitely small generator A are defined as

$$
a\left((n_m, n_u)\left(n'_m, n'_u\right)\right) =
\begin{cases}
\lambda_m, & n'_m = n_m + 1,\ n'_u = n_u; \\
\lambda_u, & n'_m = n_m,\ n'_u = n_u + 1, \\
& or \\
& n'_m = n_m - 1,\ n_m > 0, \\
& n'_u = C - ((b-d)n_m - 1); \\
n_m \mu_m, & n'_m = n_m - 1,\ n'_u = n_u; \\
n_u \mu_u, & n'_m = n_m,\ n'_u = n_u - 1; \\
0, & \text{otherwise.}
\end{cases}
\tag{14}
$$

We find the probability distribution of the stationary state $(p(n_m, n_u))_{(n_m, n_u)in N}$ solving the system of Eq. 13 and calculate the following characteristics (Fig. 16):

– Average number of eMBB/URLLC sessions:

$$
\overline{n_m} = \sum_{n_m=1}^{N} n_m \cdot \sum_{n_u=0}^{C-(b-d)n_m} p(n_m, n_u);
\tag{15}
$$

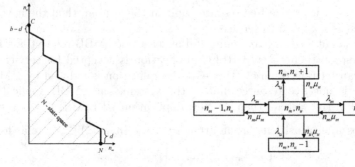

Fig. 13. State transition diagram in general

Fig. 14. Central state diagram

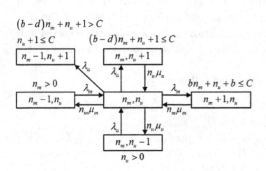

Fig. 15. Central state diagram with block state

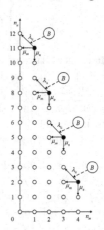

Fig. 16. Example of the state spaces

$$\overline{n_u} = \sum_{n_u=1}^{C} n_u \cdot \sum_{n_m=0}^{\left\lfloor \frac{C-n_u}{b-d} \right\rfloor} p(n_m, n_u). \qquad (16)$$

– Blocking probability of the eMBB/URLLC sessions:

$$B_M = \sum_{n_u=0}^{C} p\left(\left\lfloor \frac{C-n_u}{b-d} \right\rfloor, n_u\right); \qquad (17)$$

$$B_U = p(0, C). \qquad (18)$$

– The probability of interruption of eMBB request's service:

$$I = \sum_{n_m=1}^{N} p(n_m, C-(b-d)n_m)\frac{\lambda_u}{\lambda_u + n_m\mu_m + n_u\mu_u}\frac{1}{n_m}. \qquad (19)$$

– The average eMBB sessions transfer rate:

$$\bar{b} = \frac{1}{\sum_{(n_m,n_u) \in N_+ \cup N_-} p(n_m,n_u)}$$
$$\times \left[\sum_{(n_m,n_u) \in N_+} b p(n_m,n_u) + \sum_{(n_m,n_u) \in N_-} \frac{C-n_u}{n_m} p(n_m,n_u) \right], \qquad (20)$$

where $N_+ = (n_m, n_u) \in N : n_m > 0, bn_m + n_u \leq C$ and
$N_- = (n_m, n_u) \in N : n_m > 0, bn_m + n_u > C$.

– Coefficient of system utilization:

$$UTIL == \frac{1}{C} \left(\left(\sum_{n_m=1}^{N} \sum_{n_u}^{C-n_m} (bn_m + n_u) p(n_m,n_u) \right) + 1 \right). \qquad (21)$$

5 Numerical Analysis

In this section, we present the results of a numerical analysis of the two systems mentioned above. We refer to the first system and consider the dependence of such quantities as: the average number of eMBB and URLLC sessions in the system, blocking probability of the eMBB/URLLC sessions. We consider the channel with the subcarrier 120 kHz with 10 slots. The duration of each slot = 1 ms. Since the subcarrier equals to 120 kHz, the number of mini slots is 14, the duration of each is 0.066 ms. The radio channel under consideration consists of 15 resource blocks (N), each of which has 12 resource units (b). For the numerical analysis, we will use two scenarios: The source data for the two scenarios are shown in Table 2.

Table 2. System parameters for numerical analysis

Parameter	Scenario 1	Scenario 2
λ_m	0–1000	100, 200, 300
λ_u	400, 500, 600	0–1000
μ_m^{-1}	1	1
μ_u^{-1}	3^{-1}	3^{-1}
N	15	15
b	12	12
d	6	6

First, consider the effect of the arrival rates of the URLLC and eMBB sessions on the average on the number of eMBB sessions in the system, shown in Figs. 17 and 18 respectively. As we can see, a much larger impact on the average number of requests is exerted by the intensity of the URLLC session arrival. Figures 19 and 20 let us understand that the average number of URLLC sessions in the system does not depend on the arrival rate of eMBB sessions. These results are expected, as we initially know that the URLLC sessions are prioritized relative to eMBB.

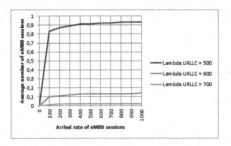

Fig. 17. Scenario 1: Average number of eMBB sessions

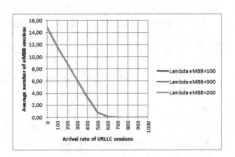

Fig. 18. Scenario 2: Average number of eMBB sessions

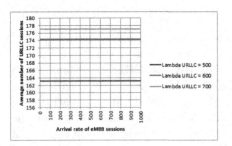

Fig. 19. Scenario 1: Average number of URLLC sessions

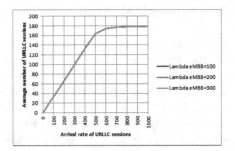

Fig. 20. Scenario 2: Average number of URLLC sessions

Next proceed to consider the dependence the blocking probability of the eMBB sessions on the arrival rates of the URLLC and eMBB sessions. The results of the numerical analysis are shown in Figs. 21 and 22, respectively. According to these results, it can be concluded that with an increase in the two arrival rates, the blocking probability of the eMBB sessions grows, however, it is worth noting that with an increase in the arrival rate of eMBB sessions, the blocking probability increases much faster than with an increase in the arrival rate of URLLC sessions. This conclusion can be explained by the fact that one eMBB session requires more resources for data transmission than for one URLLC session. Let us analyze the dependence of the probability of blocking URLLC sessions on changes in the intensity of the arrival of eMBB/URLLC sessions, looking at the illustrated results in Figs. 23 and 24. As we can see these results indicate that there is no effect of the arrival rate of eMBB sessions. This phenomenon is also evident due to the priority of URLLC sessions.

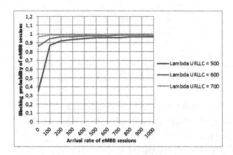

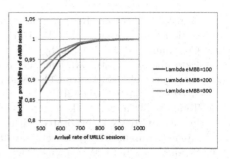

Fig. 21. Scenario 1: Blocking probability of eMBB sessions

Fig. 22. Scenario 2: Blocking probability of eMBB sessions

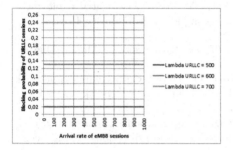

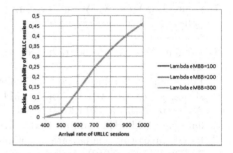

Fig. 23. Scenario 1: Blocking probability of URLLC sessions

Fig. 24. Scenario 2: Blocking probability of URLLC sessions

6   Conclusion

In this paper, we have introduced two system models that provide joint transfer of sessions of different types of traffics: eMBB and URLLC. We described the structure of determining the basic characteristics of these systems, witch determine the appropriate efficiency of the system model. After conducting a numerical analysis of the first system, we concluded that the intensity of the eMBB sessions revenues does not have any effect on the average number and on the probability of blocking URLLC sessions, explained it by the priority of URLLC sessions.

References

1. Gerasimenko, M., et al.: Adaptive resource management strategy in practical multi-radio heterogeneous networks. IEEE Access **5**, 219–235 (2017)
2. Anand, A., de Veciana, G., Shakkottai, S.: Joint scheduling of URLLC and eMBB traffic in 5G wireless networks. In: IEEE INFOCOM 2018 - IEEE Conference on Computer Communications, October 2018. https://doi.org/10.1109/INFOCOM.2018.8486430

3. Begishev, V., et al.: Resource allocation and sharing for heterogeneous data collection over conventional 3GPP LTE and emerging NB-IoT technologies. Comput. Commun. **120**, 93–101 (2018)
4. Ekterina, M., Dmitri, M., Irina, G., Konstantin, S., Koucharyavy, Y.: Performance assessment of QoS-Aware LTE sessions offloading onto LAA/WiFi systems. IEEE Access **7**, 36300–36311 (2019)
5. Tang, J., Shim, B., Quek, T.Q.S.: Service multiplexing and revenue maximization in sliced C-RAN incorporated with URLLC and multicast eMBB. IEEE J. Sel. Areas Commun. **37**(4), 881–895 (2019). Article no. 8638932
6. Semiari, O., Saad, W., Bennis, M., Debbah, M.: Integrated millimeter wave and sub-6 GHz wireless networks: a roadmap for joint mobile broadband and ultra-reliable low-latency communications. IEEE Wireless Commun. **26**(2), 109–115 (2019). Article no. 8642794
7. Guan, W., Wen, X., Wang, L., Lu, Z., Shen, Y.: A service-oriented deployment policy of end-to-end network slicing based on complex network theory. IEEE Access **6**, 19691–19701 (2018)
8. Soldani, D.: 5G beyond radio access: a flatter sliced network. Mondo Digitale **17**(74)
9. Popovski, P., Trillingsgaard, K.F., Simeone, O., Durisi, G.: 5G wireless network slicing for eMBB, URLLC, and mMTC: a communication-theoretic view. IEEE Access **6**, 55765–55779 (2018). Article no. 8476595
10. Jaradat, A.M., Hamamreh, J.M., Arslan, H.: Modulation options for OFDM-based waveforms: classification, comparison, and future directions. IEEE Access **7**, 17263–17278 (2019). Article no. 8631007
11. Hamamreh, J.M., Ankarali, Z.E., Arslan, H.: CP-Less OFDM with alignment signals for enhancing spectral efficiency, reducing latency, and improving PHY security of 5G services. IEEE Access **6**, 63649–63663 (2018). Article no. 8501913
12. Kassab, R., Simeone, O., Popovski, P., Islam, T.: Non-orthogonal multiplexing of ultra-reliable and broadband services in fog-radio architectures. IEEE Access **7**, 13035–13049 (2019). Article no. 8612914
13. Matera, A., Kassab, R., Simeone, O., Spagnolini, U.: Non-orthogonal eMBB-URLLC radio access for cloud radio access networks with analog fronthauling. Entropy **20**(9), 661 (2018)
14. Lu, F., Cheng, L., Xu, M., Wang, J., Shen, S., Chang, G.-K.: Orthogonal and sparse chirp division multiplexing for MMW fiber-wireless integrated systems. IEEE Photon. Technol. Lett. **29**(16), 1316–1319 (2017). Article no. 7964782

Optimization of Shaping Pulse by Spectral Mask to Enhance DVB-S2

Phuoc Nguyen Tan Hoang[(✉)] and Aleksandr Gelgor

Peter the Great St. Petersburg Polytechnic University, St. Petersburg, Russia
nguentan.hf@edu.spbstu.ru, agelgor@spbstu.ru

Abstract. This paper proposes new formulation and solution of optimization problems to synthesis bandwidth-efficient signals. Different types of optimization problems such as maximizing free Euclidean distance and minimizing partial correlation coefficients with and without capacity constraint are considered to obtain optimal pulses. Spectral mask is used as additional constraint for optimization problem instead of using bandwidth calculation. Spectral mask templates and other system parameters are taken from DVB-S2 standard. The optimization problems are solved for multicomponent signals which are the kind of partial response signals with finite pulses. BCJR demodulator is used in order to deal with intentional intersymbol interference (ISI). In presence of LDPC and BCH coding, the system with proposed optimal finite pulses provides 10% bandwidth efficiency gain and suffers energy losses of 0.1 dB and 0.25 dB compared to the system with RRC-pulses for QPSK and 8PSK respectively.

Keywords: Bandwidth efficiency · Faster-than-Nyquist signaling · Partial response signaling · Optimal finite pulses · Multi-Component signals · DVB-S2

1 Introduction

In the development of new data transmission systems, it is crucial to provide electro-magnetic compatibility. To some extent, it means that a spectral mask is required to minimize the interference between neighboring frequency bands. Communication systems developers obviously should try to use the whole provided spectrum-time resources. Ideally, the signal spectrum is supposed to be matched with the system spectral mask. However, the task of choosing appropriate pulse shape, which consequently defines signal spectrum, is usually solved taking into account only some numerical characteristics of the spectral mask instead of the entire mask. In this case, the requirements on spectral mask are satisfied by using additional signal filtering at transmitter. Consequently, some characteristics of transmitted signals, unfortunately, may be changed.

An increasing amount of literature is devoted to the topic of searching for new bandwidth-efficient signal forms. In these publications, the objective is to improve bandwidth efficiency γ, which is defined as a ratio of bit rate R to signal bandwidth W calculated by some given criterion.

© Springer Nature Switzerland AG 2019
O. Galinina et al. (Eds.): NEW2AN 2019/ruSMART 2019, LNCS 11660, pp. 649–660, 2019.
https://doi.org/10.1007/978-3-030-30859-9_56

There are two main known types of bandwidth-efficient single-carrier signals for systems operating with AWGN. The first one is Faster-than-Nyquist (FTN) signals proposed in [1] by Mazo. The idea of FTN is that shaping pulses are placed closer than symbol period T, which ensures ISI-free condition at receiver with matched filtering. In this case, bandwidth efficiency gain is achieved by increasing the bit rate R while the bandwidth remains unchanged. The second type suggests a transition from full response signaling (FRS) to partial response signaling (PRS) with optimal pulses. Probably the most effective and popular solution for searching optimal pulses for PRS were presented in [2, 3]. The idea of spectral efficiency gain lies on the fact that signal spectrum is narrowed thank to special shape of the optimal pulse. In contrast to the previous case, the bit rate R remains constant while the bandwidth W is reduced. In general, these outlined types of signals are fundamentally identical to one another since they both introduce intentional ISI which may lead to energy loss.

To deal with ISI, advanced detection algorithms were adopted such as the Viterbi [4] and BCJR [5]. Implementation of these algorithms is difficult because of their computational complexity. In [6, 7], sub-optimal M-Viterbi and M-BCJR were proposed. Nowadays, thanks to the development of electronics components, these sub-optimal algorithms seem to be realizable. Compared to the Viterbi algorithm, BCJR algorithm is more preferable since it provides soft decisions about coded symbols to improve the efficiency of forward error correction decoders.

In different studies, there are different criteria used to calculate signal bandwidth. In [8, 9], for instance, bandwidth is calculated by the level of out-of-band emissions. In [3, 10, 11], the authors calculate bandwidth containing a certain amount of signal power (for example 99% or 99.9%). Obviously, the best criterion for bandwidth calculation doesn't exist, since different parameters are considered to be important in different situations.

However, using these above-mentioned criteria is convenient for some reasons. First of all, this approach provides the universality of obtained results, i.e. they are not bound to any specific system. On the other hand, the utilization of different criteria is dictated by methods used for searching for new signals. It is especially important when it comes to solving optimization problems. While one criterion of bandwidth definition provides an opportunity to formulate linear optimization problem with the guaranty of finding global extremum, another criterion leads to non-linear optimization problem.

This study proposes a new approach to setting the problem of searching bandwidth-efficient signals. Initially, an existing telecommunication system or an existing spectral mask of projected system is selected. Then it is stated the problem of searching for the signals providing maximal data rate with a fixed signal/noise ratio in compliance with the requirements of given spectral mask. In other words, the bandwidth value W is the same for all possible signals. To put concrete numbers in further calculations, we propose to define the bandwidth as $W = 1/T$, where T is symbol period when an ensemble of full response signals is used.

Under this approach, our aim now seems to be the realization of pulse shaping filter. This task is usually not considered when solving optimization problems. In [3], for instance, searching optimal discrete impulses is implemented, but the realization of these impulses remains something unclear for equipment developers. In [3], the authors only point out that it is possible to use an interpolating filter with impulse characteristic from the family of RRC-pulses. However, RRC-pulses are not finite. Consequently, they are truncated and weighted by a window function in practice. In this case, of course, spectrum characteristics and BER performance of the signals may change.

The idea of using spectral mask in the pulse optimization instead of using the bandwidth values was proposed in [14]. In [15], spectral mask taken from 802.11 g WiFi standard is used as constraint on optimization to reduce the effect of spectrum broadening caused by precoding process. However, in these studies, the spectral mask is applied only for $fT \in [0, 1/2]$. It means that we lose control over spectrum at frequencies above $1/2T$.

In [9], we proposed multicomponent signals (MCS) as some kind of PRS signals. The distinguishing feature of MCS is the use of finite shaping pulses. Based on the above, it is clearly convenient for our approach to formulate the problem of shaping pulse synthesis taking into account spectral mask.

In this work, we take the DVB-S2 standard [16], one of the most widespread satellite digital television broadcast standard in the world, as an example of current system. Therefore, the aim of this study is to synthesize the pulses for spectrally efficient MCS signals providing minimal energy loss when using spectral masks from the DVB-S2 standard.

The rest of this study is divided as follows. In Sect. 1, the multicomponent signals and their characteristics are examined. In Sect. 2, the optimization problems for pulse synthesis are formulated and the noise immunity of synthesized pulses is examined by simulation method. Then, Sect. 3 describes a simulation model of the DVB-S2 system and provides performance comparison for the cases of RRC-pulses and proposed optimal finite pulses. Finally, conclusions are outlined in the last section.

2 Multi-Component Signals

The idea of developing MCS signals [9] is to decompose a signal into a sum of components whose adjacent finite pulses do not overlap (i.e. without ISI). In general, components can be transmitted in different frequencies, using different constellation and even different pulses. A baseband multicomponent signal with L components and the same pulse in each component can be represented as follows:

$$y_L(t) = \sum_{p=1}^{L} y_{L,p}(t) = \sum_{p=1}^{L} \sum_{k} \frac{1}{\sqrt{L}} C_{r,p,k} a\left(\frac{t - \Delta t_p - kLT}{L}\right) \exp(j2\pi\Delta f_p t), \quad (1)$$

where L is the number of components; $C_{r,p,k}$ is modulated symbol transmitted in k-th time period of p-th component ("time period of a component" refers to an interval of duration LT); $a(t)$ is a finite pulse occupying the time interval T: $[-T/2, T/2]$ or $[0, T]$; Δt_p is a relative delay of time period between different components; Δf_p is a frequency shift of component p relative to the central frequency of the multicomponent signal. The argument of the pulse is designed to allow spraining of the original pulse to L times. The factor $1/\sqrt{L}$ is needed to make pulse energy independent from the number of components. It is convenient in solving the optimization problem.

Notice that when we choose $\Delta t_p = 0$, $\Delta f_p = (p - L/2)/T$ and rectangular pulse shape, the Eq. (1) corresponds to OFDM signals. If we use the frequency shift $\Delta f_p = (p - L/2)/T$, where $0 < \alpha < 1$, then we have the equation for popular multi-frequency bandwidth-efficient signals SEFDM.

Now it is supposed that all components of MCS signal are transmitted on the same carrier frequency, i.e. $\Delta f_p = 0$, and pulses of components are shifted from each other uniformly by $\Delta t_p = (p - 1)T$, then traditional single-carrier signals are formed. For example, to obtain signals, which are used in DVB-S2, we need to use RRC-family pulses as shaping pulses and increase the number of components to infinity $L \to \infty$ taking into account the infinity of the pulses. Further in this paper, by MCS signals we mean the MCS signals with $\Delta f_p = 0$, $\Delta t_p = (p - 1)T$ and the same signal constellation for each component.

It can be shown in [9] that the spectrum of single-carrier multicomponent signal with symmetrical signal constellations in components is defined as follows:

$$G(f) = \frac{LZ}{T}|F_a(Lf)|^2 \tag{2}$$

where $F_a(f)$ is the spectrum of shaping pulse $a(t)$. The factor Z depend on signal constellation and can be omitted because usually it is used normalized spectrum. From (2), it is clear that the spectrum of MCS signal mainly depends on shape of the pulse.

Now we consider the correlation properties of the proposed signal. In [9], we proposed partial correlation coefficients (PC) as a measure of ISI. The normalized PC between the signal from k-th time interval of p-th component and the signal from l-th time interval of d-th component is defined as follows:

$$PC_{p,d}^{(k,l)} = (1/E_a)C_{r_p}^{(k)}C_{q_d}^{(l)*}\int a_p^{(k)}(t,L)a_d^{(l)}(t,L)dt \tag{3}$$

For signal constellations with all symbols symmetrically lying on the unit circle, the maximum of absolute partial correlation coefficients (MaxPC) can be computed as follows:

$$MaxPC = \max_{C,p,d,k,l}\left\{\left|PC_{p,d}^{(k,l)}\right|\right\} = \max_{2 \leq d \leq L}\left\{\left|(1/E_a)\int a(t)a(t - \Delta t_d/L)dt\right|\right\}, \tag{4}$$

where E_a is the pulse energy.

Along with MaxPC, free Euclidean distance (d_{free}) can be used as a criterion in optimization process. Squared free Euclidean distance is the minimum of all possible squared Euclidean distances d_n^2 and formulated as follows:

$$d_{free}^2 = \min_n\{d_n^2\} = \int_{-\infty}^{+\infty}|e_n(t)|^2dt, \tag{5}$$

where $e_n(t)$ is the difference signal between two signals $y^{(1)}(t)$ and $y^{(2)}(t)$, which are generated from two symbol sequences $(C^{(1)}[0], C^{(1)}[1],..., C^{(1)}[L - 1])$ and $(C^{(2)}[0],$

$C^{(2)}[1],\ldots, C^{(2)}[L - 1])$. In general, squared Euclidean distances should be considered over all data sequences, but in accordance with [3], we consider the sequences with the length of L symbols. We define n-th symbol difference sequence $(\Delta C_n[0], \Delta C_n[1],\ldots, \Delta C_n[L - 1])$ as the difference between any two symbol sequences with the length of L. More detailed examination of difference signal $e_n(t)$ in relationship with symbol difference sequences and pulse $a(t)$ is provided in [3].

Having partial correlation coefficients, squared Euclidean distances for all symbol difference sequences $(\Delta C_n[0], \Delta C_n[1],\ldots, \Delta C_n[L - 1])$ can be computed as follows:

$$d_n^2 = \sum_k b^*_{\Delta C_n}[k]b_a[k] = b_{\Delta C_n}[0] + 2\sum_{k=1}^{L-1} b^*_{\Delta C_n}[k]PC_{1,1}^{1,k+1}, \tag{6}$$

where: symbol b represents autocorrelation function; symbol $*$ represents complex conjugate operation. Notice that the autocorrelation function of pulse $b_a[k]$ is nothing other than the values of partial correlation coefficients of MCS signals.

The idea of using d_{free} as a metric of noise immunity proposed in [3] unfortunately has some disadvantages. Firstly, maximization of free Euclidean distance guarantees the best noise immunity only asymptotically, i.e. in the condition of low bit error rate (BER). In other words, when the system works in low signal/noise ratio conditions, d_{free} no longer remains the best metric. This is confirmed in the next sections of this paper where BER curves of different pulses are considered. Secondly, it is required an exhausted searching process for pulses with high L. For example, for QPSK constellation, it is hard to consider $2 \cdot 3^{L-1}$ conditions with the length of pulse $L > 14$.

3   Optimization Problem for Shaping Pulse

3.1   Formulation of Optimization Problems

In previous works, we presented different results for various types of optimization problems. In [9] and [10], we used maximum group correlation (MGC) and partial correlation (PC), which indirectly linked with noise immunity of the signal, as additional constraints. In [11] we followed another way to achieve better noise immunity by maximizing free Euclidean distance.

In this work, we propose to mix above-described ideas with a new type of constraint for optimization problem. As optimization criterion, we can use maximizing free Euclidean distance in (5) or minimizing the maximum of absolute partial correlation coefficients in (4). As additional constraint, we propose to use a spectral mask from the DVB-S2 standard to control the signal spectrum. The DVB-S2 standard establishes 6 values of roll-off factors β (parameter of RRC-pulses): 0.35, 0.25, 0.2, 0.15, 0.1 and 0.05. For each value of roll-off factor a spectral mask is defined. Figure 1 illustrates masks for $\beta = 0.05$ and $\beta = 0.35$.

We also consider the idea, proposed in [15], to approach the ultimate capacity of the mask. The study supposed that decreasing the spectrum gap between the normalized signal spectrum $G(f)$ and the spectral mask $Mask(f)$ may help to reach the ultimate capacity of the mask. However, there is a compromise between decreasing the

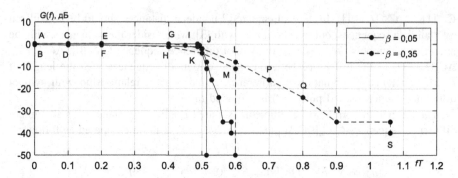

Fig. 1. Spectral masks of DVB-S2 system for $\beta = 0.05$ and $\beta = 0.35$

spectrum gap and increasing d_{free}. In this work, we control spectrum gap by using it as a constraint of the optimization problem. Further, in this paper, we refer this constraint on capacity as "capacity constraint".

The idea of achieving spectral efficiency gain by using MCS signals is to reduce the symbol period defined by a standard, i.e. to move from the value of T to τT $(0 < \tau < 1)$, while the spectrum of signal still fits into the given spectral mask. However, when setting and solving the optimization problem, the value of symbol period is assumed to be equal to T, so as constraints for the spectrum of MCS signals we use a mask compressed by $1/\tau$.

As in [9, 10, 12, 13], in this study we use the decomposition of pulse in truncated Fourier series:

$$a(t) = \frac{c_0}{2} + \sum_{k=1}^{K-1} \left(c_k \cos\left(\frac{2\pi}{T} kt\right) + s_k \sin\left(\frac{2\pi}{T} kt\right) \right), \tag{7}$$

where $(2K-1)$ is the number of sought coefficients of this decomposition. When using (7), the spectrum of MCS signal can be represented as follows [9]:

$$G(f) = \frac{LZT}{\pi^2} \left[\left(\frac{c_0}{2} \frac{\sin(\pi fLT)}{fLT} + \sum_{k=1}^{K-1} \sin(\pi fLT) \frac{(-1)^k c_k fLT}{(fLT)^2 - k^2} \right)^2 + \left(\sum_{k=1}^{K-1} \sin(\pi fLT) \frac{(-1)^k s_k k}{(fLT)^2 - k^2} \right)^2 \right], \tag{8}$$

where Z is defined by signal constellation. In this work, $Z = 1$ since we consider QPSK and 8PSK.

The form of signal spectrum is expressed through sought coefficients of the decomposition in truncated Fourier series. Thus, we can formulate constraints for the signal spectrum due to spectral mask:

$$G(f_i T) \leq Mask(\tau f_i T), \ i = 0, 1, \ldots, N_f - 1, \tag{9}$$

where $Mask(f)$ is a spectral mask as a continuous function of frequency. Notice that in our experiment the spectral constraints are only applied for a finite set of frequencies $f_i \in$ H. We choice $N_i = 300$ points which are uniformly distributed on considered interval $0 \leq fT \leq 6$. After obtaining the pulse, verification for 10^5 points can be used. If the verification fails, we can add more points and repeat the experiment. In our experiments, the verification always successfully passed. It shows that the number of 300 points is acceptable.

The above-mentioned capacity constraint, which related to spectrum gap, can be formulated as:

$$\sum_{f_i} |Mask(\tau f_i T) - G(f_i T)| \leq P, \ i = 0, 1, \ldots, N_f - 1, \tag{10}$$

where P is a constant playing a role as a given threshold.

Finally, the optimization problem of maximizing free Euclidean distance with constraint on spectrum (9) and capacity constraint (10) can be expressed as follows:

$$a(t) = \arg\left(max\left(\min_n\{d_n^2\}\right)\right). \tag{11}$$

The optimization problem of minimizing the maximum of absolute partial correlation coefficients with the same constraints (9) and (10) is defined as follows:

$$a(t) = \arg(\min(MaxPC)). \tag{12}$$

These non-linear optimization problems (11) and (12) seeking a point that minimizes (or maximizes) the maximum (or minimum) of a set of objective functions with non-linear constraints are solved by using function *fminimax* in MATLAB.

3.2   Results of Solving Optimization Problems

In this work, we consider following parameters: the length of pulse $L = 8$; spectral mask with roll-off factor $\beta = 0.35$; the mask is compressed by $1/0.9$ times (i.e. $\tau = 0.9$); signal constellation is QPSK.

After solving the proposed optimization problems, we obtain four pulses. Pulse #1 and pulse #2 are solutions of the problem (11) using d_{free} with and without capacity constraint respectively. They both have squared free Euclidean distance value of 2 that potentially corresponds to the lossless detection. Pulse #3 and pulse #4 are obtained from the problem (12) using $MaxPC$ with and without capacity constraint respectively. However, because of capacity constraint, pulse #3 has the value of $MaxPC$ smaller than pulse #4 has (the values of $MaxPC$ for pulse #3 and pulse #4 are 0.15 and 0.06 respectively, note that $MaxPC$ equal to 0 corresponds to FRS). Figure 2 shows pulse #4, its spectrum and compressed spectral mask as an example. For evaluation of noise immunity for obtained pulses, we use a simulation model with QPSK modulation and AWGN channel. True BCJR algorithm is considered in demodulator.

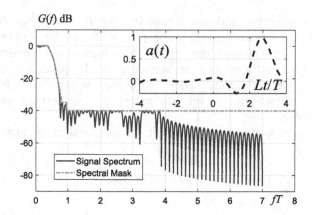

Fig. 2. Pulse #4, its spectrum and compressed spectral mask for $\beta = 0.35$

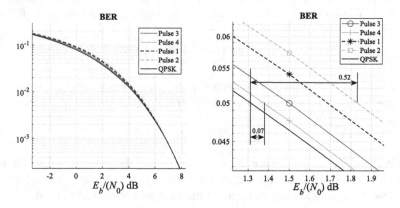

Fig. 3. BER for QPSK with four pulses in large scale (left) and closer view (right)

Figure 3 shows BER curves for four pulses. As seen in the left figure with large scale, all pulses have almost the same noise immunity and their BER curves are close to theoretical BER for QPSK. It is clear that for $E_b/N_0 > 7$ dB there is no difference between these pulses in terms of noise immunity. However, the right figure shows that in the area of low E_b/N_0 the pulse #4 is the best since it suffers an energy loss of about 0.07 dB (for BER = 0.05) instead of 0.52 dB for the pulse #2.

In DVB-S2 standard, thanks to outer coding (BCH) and inner coding (LDPC) the demodulator can perform in low signal/noise ratios. As we mentioned above, in this condition there is a significant difference in performance of pulses. We applied these obtained optimal pulses into the model of DVB-S2 described in next sections. The simulation results showed that for coded data pulse #4 requires less energy consumptions compared to other pulses (#1, #2 and #3). Thus, we choose pulse #4 as the optimal pulse with intentional ISI for shaping filters, which is used in simulation model in next sections.

4   Simulation Model

4.1   Explanation of Simulation Model

System DVB-S2 exploits flexible combinations of signal constellation types (QPSK, 8PSK, 16APSK, 32APSK), error correction coding (BCH, LDPC) and other options to provide different application areas such as broadcast, interactive and professional services. For broadcast application, it is usually recommended to use QPSK and 8PSK with the normal FECFRAME length of 64800 bits. Modulated symbols are sent to RRC filters with different roll-off factors to make free-ISI condition in the receiver.

In this study, we consider the possibility of the improvement in bandwidth efficiency with energy loss as small as possible by using the MCS with optimal pulses instead of signals with RRC-pulses. A simplified simulation model is presented in Fig. 4. In the transmitter, there is only one change in the block "Shaping Filter" where RRC-pulses are replaced by finite optimal pulses. The AWGN channel is utilized in this study. BCJR demodulator is implemented on the receiver side to deal with intentional ISI and replace the matched filter and the demapper which are used in conventional scheme. BCJR demodulator sends soft decisions about transmitted bits to deinterleaver block and then to LDPC and BCH decoders.

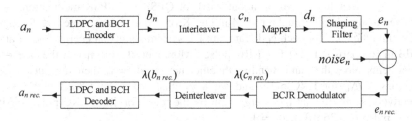

Fig. 4. Simulation model of DVB-S2 using MCS signals with optimal pulses

For comparison, results will be presented in the plane of energy consumption E_b/N_0 and bandwidth efficiency β_F as follows:

$$\beta_F = (R_{Code} * R_{Mod})/W, \tag{13}$$

where: E_b/N_0 is the energy per information bit to noise one-sided power spectral density ratio, which is needed to obtain the useful transport packet error rate (PER) at PER $= 5 \times 10^{-4}$ (this value of PER is compared to BER $= 10^{-7}$); R_{Code} is the code rate of LDPC plus BCH construction; R_{Mod} equals to 2 for QPSK and to 3 for 8PSK; the bandwidth W is computed as $(1 + \beta)/T$.

As mentioned above, for QPSK we use the pulse #4 which derived from the optimization problem using the mask $\beta = 0.35$ and the compression factor $\tau = 0.9$. It means that we fix the bandwidth efficiency gain at 10%. The question now is how serious the energy loss is when using the proposed pulse with ISI.

The study in [17] proposes that it is possible to gain additional bandwidth efficiency when the intentional ISI is exploited simultaneously with the increase of the size of signal constellation. Therefore, we also analyze the case of 8PSK. We use the same approach to synthesis optimal pulse as for pulse #4. However, because of the computational complexity of BCJR algorithm, we limit the length of pulse at $L = 4T$. It makes the demodulator more realizable but causes more ISI than in the case of pulse #4 since the value of *MaxPC* for new pulse is higher.

4.2 Simulation Results

The simulation results are obtained by Monte-Carlo method in MATLAB for two types of pulses (RRC and MCS optimal pulses), QPSK and 8PSK modulations and LDPC code rates from 1/4 to 9/10. All simulation parameters are chosen from the DVB-S2 standard. For sufficient statistics, each point of PER curves is calculated until the condition of 1000 packet errors is satisfied.

In Fig. 5, the performance of orthogonal RRC-pulse (two solid lines) for different code rates is presented. Each point corresponds to one LDPC code rate. From left to right, for QPSK the code rates [1/4 1/3 2/5 1/2 3/5 2/3 3/4 4/5 5/6 8/9 9/10] are used and for 8PSK the code rates [3/5 2/3 3/4 5/6 8/9 9/10] are used. It is clear that the DVB-S2 standard provides a large variety of choices for different signal/noise conditions from E_b/N_0 lower than 0.5 dB to about 7 dB for QPSK and 8PSK modulations.

When using MCS optimal pulse #4 in QPSK (two dash lines), there is 10% gain in bandwidth efficiency for each code rate, but the system suffers an energy loss of about 0.1 dB in comparison to ISI-free RRC-pulses. Moreover, if we consider the case when two systems have the same bandwidth efficiency at $\beta_F = 1$, then the energy consumptions E_b/N_0 for RRC-pulse and MCS optimal pulse are 1.8 dB and 1.44 dB respectively. It means that the energy gain by the transition from RRC-pulse to MCS optimal pulse is 0.36 dB at $\beta_F = 1$.

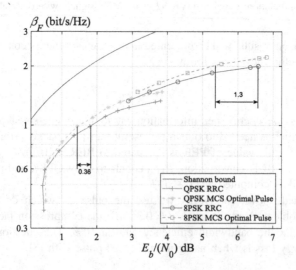

Fig. 5. Performance of signals with RRC-pulse and MCS optimal pulse for DVB-S2 system

As seen in Fig. 5, 8PSK with MCS optimal pulse has bandwidth efficiency gain of 10% in comparison with RRC, but the energy loss is higher than in the case of QPSK for the same code rate. The average value of the energy loss for all code rates is about 0.25 dB. It can be explained by the fact that the pulse used for 8PSK is shorter ($L = 4$) and consequently introduces more ISI level. In the case of fixing bandwidth efficiency at 2 (bit/s/Hz), significant energy gain (about 1.3 dB) is provided by using optimal pulses for 8PSK.

5 Conclusion

This paper presents formulation and solutions of optimization problems of MCS shaping pulse synthesis. The results show that taking into account given spectral masks it is possible to improve bandwidth efficiency and minimize energy loss caused by the transition from FRS to PRS. We consider different types of optimization problems namely maximizing free Euclidean distance and minimizing absolute partial correlation coefficients with and without the capacity constraint. Among obtained optimal pulses, the pulse which is obtained from the problem of minimizing absolute partial correlation coefficients without the capacity constraint is the best in terms of noise immunity under AWGN channel conditions.

A model of the DVB-S2 system is described to evaluate the performance of obtained optimal pulses for QPSK and 8PSK constellations in presence of LDPC and BCH coding. The results show that using QPSK with MCS optimal pulse leads to either 0.36 dB energy gain without bandwidth efficiency reduction or 10% gain in bandwidth efficiency with 0.1 dB of energy loss compared to the case of RRC-pulses. The results also demonstrate that for 8PSK with MCS optimal pulse in order to gain 10% bandwidth efficiency the system suffers about 0.25 dB energy loss. The energy loss for 8PSK (or higher modulation orders of constellations) can be reduced by decreasing ISI level of optimal pulses by using longer pulses and exploiting sub-optimal demodulation algorithms that remain a considerable challenge for further researches.

Acknowledgements. This work was supported by the Ministry of Education and Science of the Russian Federation (the state contract #8.2880.2017/ПЧ). The simulation results were obtained with the use of computational resources of the Supercomputing Center of Peter the Great St. Petersburg Polytechnic University.

References

1. Mazo, J.E.: Faster-than-Nyquist signaling. Bell Syst. Tech. J. **54**(8), 1451–1462 (1975)
2. Liveris, D., Georghiades, C.N.: Exploiting faster-than-Nyquist signaling. IEEE Trans. Commun. **51**(9), 1502–1511 (2003)
3. Said, A., Anderson, J.B.: Bandwidth-efficient coded modulation with optimized linear partial-response signals. IEEE Trans. Inform. Theory **44**(2), 701–713 (1998)
4. Forney, G.D.: The viterbi algorithm. Proc. IEEE **61**(3), 268–278 (1973)

5. Bahl, L.R., et al.: Optimal decoding of linear codes for minimizing symbol rate. IEEE Trans. Inf. Theory **IT-20**, 284–287 (1974)
6. Anderson, J.B.: Limited search trellis decoding of convolutional codes. IEEE Trans. Inf. Theory **35**, 944–955 (1989)
7. Franz, V., Anderson, J.: Concatenated decoding with a reduced search BCJR algorithm. IEEE J. Sel. Areas Commun. **16**(2), 186–195 (1998)
8. Zavjalov, S., Volvenko, S., Makarov, S.: A method for increasing the spectral and energy efficiency SEFDM signals. IEEE Commun. Lett. **20**, 2382–2385 (2016)
9. Gelgor, A., Gorlov, A., Popov, E.: On the synthesis of optimal finite pulses for bandwidth and energy efficient single-carrier modulation. In: Balandin, S., Andreev, S., Koucheryavy, Y. (eds.) ruSMART 2015. LNCS, vol. 9247, pp. 655–668. Springer, Cham (2015). https://doi.org/10.1007/978-3-319-23126-6_59
10. Gorlov, A., Gelgor, A., Nguyen, V.P.: Root-raised cosine versus optimal finite pulses for faster-than-Nyquist generation. In: Galinina, O., Balandin, S., Koucheryavy, Y. (eds.) NEW2AN/ruSMART -2016. LNCS, vol. 9870, pp. 628–640. Springer, Cham (2016). https://doi.org/10.1007/978-3-319-46301-8_54
11. Gelgor, A., Gorlov, A.: A performance of coded modulation based on optimal faster-than-Nyquist signals. In: 2017 IEEE International Conference BlackSeaCom, pp. 1–5 (2017)
12. Gelgor, A., Gorlov, A., Popov, E.: Improving energy efficiency of partial response signals by using coded modulation. In: IEEE International Conference BlackSeaCom, pp. 1–5 (2015)
13. Gelgor, A., Gorlov, A., Popov, E.: Multicomponent signals for bandwidth-efficient single-carrier modulation. In: IEEE International Conference BlackSeaCom, pp. 1–5 (2015)
14. Davidson, T.N., Lu, Z.-Q., Wong, K.M.: Orthogonal pulse shape design via semidefinite programming. IEEE Trans. Signal Process. **48**, 1433–1445 (2000)
15. Wen, S.: Optimal precoding based spectrum compression for faster-than-Nyquist signaling. In: 2018 IEEE International Symposium BMSB, pp. 1–5 (2018)
16. ETSI EN 302 307-1 v1.4.1 (2014-11): Digital Video Broadcasting (DVB); Second generation framing structure, channel coding and modulation systems for Broadcasting, Interactive Services, News Gathering and other broadband satellite applications (DVB-S2)
17. Nguyen, V.P., Gorlov, A., Gelgor, A.: An Intentional Introduction of ISI Combined with Signal Constellation Size Increase for Extra Gain in Bandwidth Efficiency. In: Galinina, O., Andreev, S., Balandin, S., Koucheryavy, Y. (eds.) NEW2AN/ruSMART/NsCC -2017. LNCS, vol. 10531, pp. 644–652. Springer, Cham (2017). https://doi.org/10.1007/978-3-319-67380-6_61

BER Performance Improvement for Optimal FTN Signals with Increased Signal Constellation Size

Anna S. Ovsyannikova$^{(\boxtimes)}$ ⓘ, Sergey V. Zavjalov ⓘ,
and Sergey B. Makarov ⓘ

Peter the Great St. Petersburg Polytechnic University, St. Petersburg, Russia
anny-ov97@mail.ru, zavyalov_sv@spbstu.ru,
makarov@cee.spbstu.ru

Abstract. Using signals with pulse amplitude modulation (PAM) provides a higher transmission rate in the same frequency bandwidth. However, decreasing frequency bandwidth in order to improve spectral efficiency causes degradation of bit error rate (BER) performance. To minimize energy losses faster than Nyquist (FTN) signaling on the basis of RRC pulses or optimal pulses may be applied. In this work, the possibility of improving BER performance of coherent bit-by-bit detection for optimal FTN signals with increased size of PAM signal constellation is considered. The optimization problem is solved according to the specified size of signal constellation (M). Comparison between RRC pulses and obtained optimal pulses in time and frequency domain is made. It is shown that signals with PAM based on obtained optimal pulses provide an energy gain up to 7 dB regarding signals with RRC pulses for M = 64 at error probability 10^{-4}.

Keywords: Optimal signals · Faster than Nyquist signaling ·
Pulse amplitude modulation · BER performance · Optimization problem ·
Root raised cosine

1 Introduction

Application of signals with pulse amplitude modulation (PAM) allows increasing the transmission rate in fixed frequency bandwidth ΔF significantly [1, 2]. However, attempts to decrease ΔF and therefore to improve spectral efficiency $R/\Delta F$ (R – symbol rate) of such signals generally lead to degradation of bit error rate (BER) performance. It can be explained by the fact that decreasing ΔF results in a longer duration of signals $s(t)$ and intersymbol interference (ISI) takes place. The level of ISI becomes greater when spectral efficiency tends to the Nyquist limit or exceeds it. For binary systems, the properties of such faster than Nyquist (FTN) signals are considered in details and estimations of their BER performance in the additive white Gaussian noise (AWGN) channel with power spectral density $N_0/2$ are made [3–6].

Random FTN signals provide transmission of binary data with the rate $R = 1/\xi T$ ($0 < \xi \leq 1$) higher than the Nyquist limit (for $\Delta F = 1/2T$) [7–9]. A random sequence of N single pulses $s(t)$ with duration $T_s = LT$ ($L = 2, 3, \ldots$) with energy E_s may be written as follows:

© Springer Nature Switzerland AG 2019
O. Galinina et al. (Eds.): NEW2AN 2019/ruSMART 2019, LNCS 11660, pp. 661–669, 2019.
https://doi.org/10.1007/978-3-030-30859-9_57

$$y(t) = \sqrt{E_s/T} \sum_{n=-N/2}^{N/2} c_j^{(n)} s(t - n\xi T) \tag{1}$$

For FTN signals with the volume of channel alphabet M values of $c_j^{(n)}$ in (1) are defined by the next formula:

$$c_j^{(n)} = \frac{M - 2j + 1}{M - 1}, j = 1 \ldots M.$$

For example, if M = 2, then $j = 1, 2$ and $c_1^{(n)} = 1$; $c_2^{(n)} = -1$. For M = 4 $j = 1, 2, 3, 4$ and $c_1^{(n)} = 1$; $c_2^{(n)} = 0.33$; $c_3^{(n)} = -0.33$; $c_4^{(n)} = -1$. The values $c_j^{(n)}$ are identically distributed in the area from -1 to $+1$. For FTN signals with PAM, values $c_j^{(n)}$ are equiprobable for each n.

Random sequences (1) of $s(t)$ may be formed on the basis of root-raised-cosine (RRC) pulses [5, 9, 10] which duration T_s is much longer than transmission time of the channel symbol. In this case, energy losses are quite small.

Another way to reduce these losses is the application of optimal FTN signals $s_{opt}(t)$ [11–15] which allow increasing symbol rate of transmission with minimal energy losses. The feature of such signals is controllable ISI level. Using these signals with binary channel alphabet (M = 2) provides at least doubled symbol rate with energy losses no more than 0.2 dB [11, 16–19].

In this work, it is proposed to consider the possibility of improving BER performance of coherent bit-by-bit detection for optimal FTN signals with increased size of PAM signal constellation (up to M = 64).

2 Optimal FTN Signals with PAM

If multi-level PAM is used, the problem of the searching for optimal shape $s(t) = s_{opt}(t)$ with minimal ISI level arises. It is especially important when the pulses with maximum and minimum amplitudes interchange. To solve this problem it is necessary to add an extra constraint to the optimization procedure [11, 13, 18, 19]. Note that optimization problem may be transformed into the problem of searching for the minimum of a function of many variables, as it was shown in [11, 18, 19]. This transformation assumes representation of even function $s_{opt}(t)$ in terms of limited Fourier series:

$$s_{opt}(t) = \frac{s_0}{2} + \sum_{k=1}^{m-1} s_k \cos\left(\frac{2\pi}{T} kt\right). \tag{2}$$

Then the problem of searching for the global minimum of the function of m variables (2) may be written by the next expression:

$$\min_{\{s_k\}_{k=0}^{m-1}} J\left(\{s_k\}_{k=0}^{m-1}\right), J\left(\{s_k\}_{k=0}^{m-1}\right) = T_s/2 \sum_{k=0}^{m-1} (2\pi k/T_s)^{2n} s_k^2. \quad (3)$$

The condition of minimization ISI at the time interval $(0, T_s)$ is provided by the constraint on the coefficient of mutual correlation of transmitted signals $s_{opt}(t)$. Averaged BER performance is determined by BER performance of each realization. Let us consider it in more detail. From the point of view of BER performance, the worst signal sequence (1) or the worst part of such signal sequence consists of interchanging signals with the minimum and maximum possible amplitudes. For instance, for M = 4 it is the sequence of pulses with coefficients $c_1^{(n)} = 1$, $c_2^{(n+1)} = 0,33$, $c_1^{(n+2)} = 1$, $c_1^{(n+3)} = 0,33$ and so on. Let c_{min} to be equal to the minimum absolute value of all possible c_j. Then the constraint on the coefficient of mutual correlation looks as follows:

$$\max_{n=1..([L/\xi]-1)} \left\{ (M-1)^2 \int_{n\xi T}^{LT} c_{min} s_{opt}(t) c_j s_{opt}(t - n\xi T)dt \right\} \leq K_0 \quad (4)$$

Due to the constraint (4) the average error probability of random signal (1) detection may be reduced, especially for realizations with a great number of interchanging symbols with maximum and minimum values. The constraint (4) includes the dependency of optimization results on the parameter of PAM signal constellation. The volume of channel alphabet M determines c_j which takes part in (1).

In Fig. 1 the process of solving the optimization problem is illustrated. Rectangular pulse shape with duration $T_s = T$ and correlation coefficient 0.5 is used as an initial approximation. The energy spectrum of signals with such pulse shape decays with the

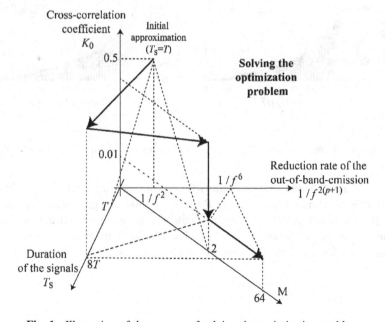

Fig. 1. Illustration of the process of solving the optimization problem

rate $1/f^2$ in the area of out-of-band emissions. The original volume of channel alphabet is M = 2.

At the first stage, optimal pulse shape with increased duration (e.g., $T_s = 8T$) needs to be found. The solution obtained at each stage is used as an initial approximation at the next stage. During the second step specified rate of out-of-band emissions (e.g., $1/f^6$) is being achieved. Then the constraint on the coefficient of mutual correlation (4) should be satisfied. In Fig. 1 correlation coefficient $K_0 = 0.01$ providing energy losses no more than 0.2 dB was chosen. At the last step of optimization, the volume of channel alphabet (M = 4, M = 16, ..., M = 64) for which requirement of K_0 should be met is taken into account. The result of calculations represents the set of expansion coefficients (2) of even function $s_{opt}(t)$ into limited Fourier series.

3 Time and Spectral Characteristics

As a result of solving optimization problem (3) time and spectral characteristics of pulses (Figs. 2 and 3) with duration $T_s = 8T$ were obtained via modeling.

Figure 2(a) shows the shape of $s_{opt}(t)$ (bold line) for reduction rate of out-of-band emissions proportional to $1/f6$ (a constraint on the coefficient of mutual correlation $K_0 = 0.01$ for M = 4). The samples of RRC pulse with roll-off factor $\alpha = 0.3$ are given for comparison. It can be seen that these values almost match the values of $s_{opt}(t)$. In Fig. 2 (b) similar graphics for M = 64 are presented. Now we can note the difference between the values of RRC pulse and $s_{opt}(t)$.

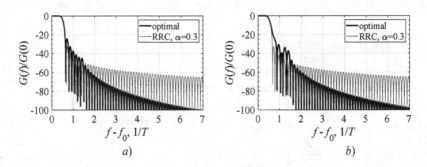

Fig. 3. Shapes of energy spectra $G(f)$ of random sequences of optimal FTN pulses with PAM

In Fig. 3 normalized energy spectra $G(f)/G(0)$ of random sequences (1) with pulse duration $T_s = 8T$ transmitted at the rate $R = 1/T$ are shown. Figure 3(a) corresponds to pulses illustrated in Fig. 2(a). It is easy to note that the energy spectrum of a random sequence of optimal FTN pulses with PAM has a higher reduction rate of out-of-band emissions than RRC pulses. The spectra for M = 64 may be found in Fig. 3(b).

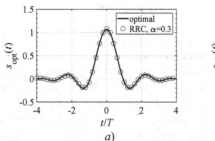 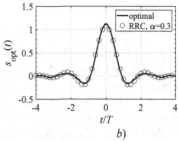

$$a) \qquad\qquad b)$$

Fig. 2. Optimal FTN pulse shapes for bit rate $R = 1/T$

Table 1 includes the values of occupied frequency bandwidth containing 99% of signal energy $\Delta F_{99\%}$ and bandwidth defined for the level of energy spectrum −60 dB ΔF_{-60dB} as well as the same parameters for a random sequence of signals with RRC pulses.

Table 1. Values of $\Delta F_{99\%}$ and $\Delta F_{-60\ dB}$

	$\Delta F_{99\%}$, $1/T$	$\Delta F_{-60\ dB}$, $1/T$
RRC pulse	1.13	8
$s_{opt}(t)$ for M = 4	1.12	3.7
$s_{opt}(t)$ for M = 64	1.19	4.1

Optimal FTN pulses and RRC pulses have close values of occupied frequency bandwidth containing 99% of energy. The difference does not exceed 1% for the volumes of channel alphabet from M = 4 to M = 64. At the same time, we can conclude that optimal FTN signals with PAM have occupied frequency bandwidth ΔF_{-60dB} defined by the level of energy spectrum −60 dB about two times smaller than signals with RRC pulses.

4   BER Performance

To detect optimal FTN signals with PAM coherent bit-by-bit detection algorithm providing high performance in conditions of controllable ISI may be used [16, 18, 19]. According to this algorithm, the symbol $c_j^{(n)*}$ is chosen between all possible symbols by minimization of the next expression:

$$c_j^{(n)*} = \min_k \left\{ \int_{n\xi T}^{n\xi T + LT} \left[c_k^{(n)} s_{opt}(t - n\xi T) - (y(t) + awgn(t)) \right] dt \right\}, \qquad (5)$$

where $awgn(t)$ is AWGN realization, $y(t)$ is FTN signal (1).

Realization of the algorithm (5) assumes the formation of the set of reference pulses (1) for all possible values of channel symbols $\left\{c_j^{(n)}\right\}_{j=1}^{M}$. At each time interval with duration $T_s = LT$ corresponding to transmission of one symbol the Euclidean distance between an obtained mixture of the desired signal with AWGN and reference signals. The decision is made in favor of the symbol minimizing the Euclidean distance.

Let us estimate BER performance of optimal FTN signals with the help of simulation modeling of the algorithm (5) in Matlab system (Fig. 4). The base block of the model is the block of initialization of simulation parameters such as pulse duration T_s, shape $s_{opt}(t)$, signal-to-noise ratio E_b/N_0 (E_b is averaged bit energy) and volume of channel alphabet M. AWGN channel is used as a transmission channel. After signal detection error probability is calculated. Each value of error probability p for a specific value of E_b/N_0 is determined by 10^6 transmitted symbols of channel alphabet.

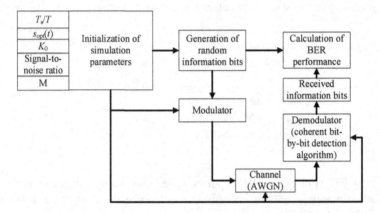

Fig. 4. The flowchart of simulation modeling for analyzing BER performance

The results of simulation modeling (dependency of error probability on the ratio E_b/N_0) are presented in Fig. 5. The error probabilities are obtained for optimal FTN signals (solid curves) with the volume of channel alphabet M = 4, 16 and 64. The dashed curves relate to BER performance of signals on the basis of RRC pulses with roll-off factor $\alpha = 0.3$. When the signal-to-noise ratio is small (up to 8 dB), the error probabilities of these signals match each other. For an increased volume of channel alphabet (M = 64), energy gain provided by optimal FTN signals regarding the signals with RRC pulses takes place.

This energy gain reaches about 2 dB for error probability $p = 10^{-2}$, 4 dB for $p = 10^{-3}$ and 7 dB for $p = 10^{-4}$. We can suppose that the tendency to increase energy gain by increasing the volume of channel alphabet will be kept.

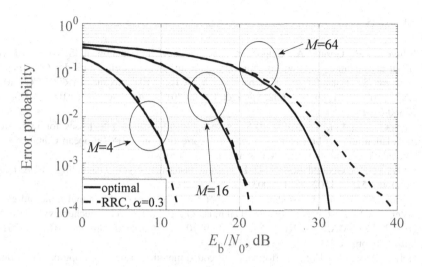

Fig. 5. BER performance (rate $R = 1/T$)

5 Conclusion

In this paper, the possibility of improving BER performance of coherent bit-by-bit detection for optimal FTN signals with increased size of PAM signal constellation (up to M = 64) is considered. It is shown that:

- With the increase in the volume of channel alphabet for optimal FTN signals with PAM up to the value M = 64 the gain comparing to the signals with RRC pulses [8–10] in the area of $E_b/N_0 = 30....40$ dB reaches about 7 dB for error probability $p = 10^{-4}$.
- For small values M = 4–16 sequences of optimal FTN pulses and RRC pulses provide almost identical BER performance.
- Optimal FTN pulses and RRC pulses have close values of occupied frequency bandwidth containing 99% of energy. The difference is not more than 1% for M = 4...16.

Acknowledgement. The results of the work were obtained under the State contract № 8.2880.2017/ПЧ with Ministry of Education and Science of the Russian Federation and used computational resources of Peter the Great Saint-Petersburg Polytechnic University Supercomputing Center (http://www.scc.spbstu.ru).

References

1. Nguyen, V.P., Gorlov, A., Gelgor, A.: An intentional introduction of ISI combined with signal constellation size increase for extra gain in bandwidth efficiency. In: Galinina, O., Andreev, S., Balandin, S., Koucheryavy, Y. (eds.) NEW2AN/ruSMART/NsCC -2017. LNCS, vol. 10531, pp. 644–652. Springer, Cham (2017). https://doi.org/10.1007/978-3-319-67380-6_61
2. Alvarado, A., Lei, Y., Millar, D.S.: Achievable information rate losses for high order modulation and hard-decision forward error correction. In: 2018 European Conference on Optical Communication (ECOC), Rome, pp. 1–3 (2018)
3. Liveris, A.D., Georghiades, C.N.: Exploiting faster-than-Nyquist signaling. IEEE Trans. Commun. **51**(9), 1502–1511 (2003)
4. Gorlov, A., Gelgor, A., Nguyen, V.P.: Root-raised cosine versus optimal finite pulses for faster-than-Nyquist generation. In: Galinina, O., Balandin, S., Koucheryavy, Y. (eds.) NEW2AN/ruSMART -2016. LNCS, vol. 9870, pp. 628–640. Springer, Cham (2016). https://doi.org/10.1007/978-3-319-46301-8_54
5. Gelgor, A., Gorlov, A.: A performance of coded modulation based on optimal faster-than-Nyquist signals. In: 2017 IEEE International Black Sea Conference on Communications and Networking (BlackSeaCom), Istanbul, pp. 1–5 (2017)
6. Rashich, A., Urvantsev, A.: Pulse-shaped multicarrier signals with nonorthogonal frequency spacing. In: 2018 IEEE International Black Sea Conference on Communications and Networking (BlackSeaCom), Batumi, pp. 1–5 (2018)
7. Mazo, J.E.: Faster-than-Nyquist signaling. Bell Syst. Tech. J. **54**, 1451–1462 (1975)
8. Le, C., Schellmann, M., Fuhrwerk, M., Peissig, J.: On the practical benefits of faster-than-Nyquist signaling. In: 2014 International Conference on Advanced Technologies for Communications (ATC 2014), Hanoi, pp. 208–213 (2014)
9. Anderson, J.B., Rusek, F., Owall, V.: Faster-than-Nyquist signaling. Proc. IEEE **101**(8), 1817–1830 (2013)
10. Gelgor, A., Gorlov, A., Nguyen, V.P.: The design and performance of SEFDM with the Sinc-to-RRC modification of subcarriers spectrums. In: 2016 International Conference on Advanced Technologies for Communications (ATC), Hanoi, pp. 65–69 (2016)
11. Zavjalov, S.V., Volvenko, S.V., Makarov, S.B.: A Method for increasing the spectral and energy efficiency SEFDM signals. IEEE Commun. Lett. **20**(12), 2382–2385 (2016)
12. Sadovaya, Y., Gelgor, A.: Synthesis of signals with a low-level of out-of-band emission and peak-to-average power ratio. In: 2018 IEEE International Conference on Electrical Engineering and Photonics (EExPolytech), St. Petersburg, pp. 103–106 (2018)
13. Gelgor, A., Gorlov, A., Nguyen, V.P.: Performance analysis of SEFDM with optimal subcarriers spectrum shapes. In: 2017 IEEE International Black Sea Conference on Communications and Networking (BlackSeaCom), Istanbul, pp. 1–5 (2017)
14. Rashich, A., Kislitsyn, A., Gorbunov, S.: Trellis demodulator for pulse shaped OFDM. In: 2018 IEEE International Black Sea Conference on Communications and Networking (BlackSeaCom), Batumi, pp. 1–5 (2018)
15. Makarov, S.B., Ovsyannikova, A.S., Lavrenyuk, I.I., Zavjalov, S.V., Volvenko, S.V.: Distributions of probability of power values for random sequences of optimal FTN signals. In: 2018 International Symposium on Consumer Technologies (ISCT), St. Petersburg, pp. 57–59 (2018)

16. Zavjalov, S.V., Ovsyannikova, A.S., Lavrenyuk, I.I., Volvenko, S.V., Makarov, S.B.: Application of optimal finite-length signals for overcoming "Nyquist limit". In: Galinina, O., Andreev, S., Balandin, S., Koucheryavy, Y. (eds.) NEW2AN/ruSMART -2018. LNCS, vol. 11118, pp. 172–180. Springer, Cham (2018). https://doi.org/10.1007/978-3-030-01168-0_16
17. Fan, J., Guo, S., Zhou, X., Ren, Y., Li, G.Y., Chen, X.: Faster-than-Nyquist signaling: an overview. IEEE Access **5**, 1925–1940 (2017)
18. Ovsyannikova, A.S., Zavjalov, S.V., Makarov, S.B., Volvenko, S.V.: Choosing parameters of optimal signals with restriction on correlation coefficient. In: Galinina, O., Andreev, S., Balandin, S., Koucheryavy, Y. (eds.) NEW2AN/ruSMART/NsCC -2017. LNCS, vol. 10531, pp. 619–628. Springer, Cham (2017). https://doi.org/10.1007/978-3-319-67380-6_58
19. Ovsyannikova, A.S., Zavjalov, S.V., Makarov, S.B., Volvenko, S.V., Quang, T.L.: Spectral and energy efficiency of optimal signals with increased duration, providing overcoming "Nyquist barrier". In: Galinina, O., Andreev, S., Balandin, S., Koucheryavy, Y. (eds.) NEW2AN/ruSMART/NsCC -2017. LNCS, vol. 10531, pp. 607–618. Springer, Cham (2017). https://doi.org/10.1007/978-3-319-67380-6_57

The Efficiency of Detection Algorithms for Optimal FTN Signals

Sergey V. Zavjalov⬤, Anna S. Ovsyannikova$^{(\boxtimes)}$⬤,
Ilya I. Lavrenyuk⬤, and Sergey V. Volvenko⬤

Peter the Great St. Petersburg Polytechnic University, St. Petersburg, Russia
zavyalov_sv@spbstu.ru, anny-ov97@mail.ru,
knaiser@mail.ru, volk@cee.spbstu.ru

Abstract. Optimal signals represent the branch of FTN signaling and may be used for spectral efficiency improvement. Such signals may be detected with a simple coherent bit-by-bit algorithm if the transmission rate is equal to the one defined in the optimization problem. In this case, bit-error rate (BER) performance is close to the theoretical one. Extra increase in symbol rate leads to the necessity of application of more complex detection algorithms. In this paper, different detection algorithm including Viterbi algorithm and maximum likelihood sequence estimations are compared. It is shown that for an 11% increase in the symbol rate algorithm of partial enumeration with 2 iterations may be used instead of the Viterbi algorithm because provides almost the same BER performance with less computational complexity.

Keywords: Faster than Nyquist signaling · Optimal signals ·
Optimization problem · Detection algorithms · MLSE · Viterbi algorithm

1 Introduction

Faster than Nyquist (FTN) signaling is one of the most efficient techniques providing spectral efficiency improvement. The basis of FTN was laid down in Mazo work [1]. The difference between this technology and spectrally efficient frequency division multiplexing (SEFDM) [2–8] consists in an attempt to maximize the symbol rate of transmission. In this case, the problem of time-domain interference of signals has a top priority. Lately, this technology is becoming more and more popular because of 5G systems and Internet of Things (IoT) which require constant improving spectral efficiency $R/\Delta F$ (R – symbol rate, ΔF – occupied frequency bandwidth). In addition to this, it is necessary to avoid energy losses. There are many pieces of research in this area [9–14]. These papers are devoted to the application of error-correcting coding in order to deal with interference, constructing complex demodulation algorithms for such signals and so on [15, 16]. However, another approach to FTN signals forming also needs to be mentioned. This approach is associated with the application of smoothed envelope shape obtained as a solution of optimization problem [17–20]. During solving optimization problem limitations on reduction rate of out-of-band emissions and bit-error-rate (BER) performance at a given symbol rate are set. The limitation on BER performance is expressed numerically by the coefficient of mutual correlation. Due to

© Springer Nature Switzerland AG 2019
O. Galinina et al. (Eds.): NEW2AN 2019/ruSMART 2019, LNCS 11660, pp. 670–680, 2019.
https://doi.org/10.1007/978-3-030-30859-9_58

this limitation, such optimal signals may be detected with the use of a simple coherent bit-by-bit detection algorithm. At this time, envelope shapes with duration $T_s = 2T...$ 16T (T is the duration of initial symbols with rectangular shape transferred at symbol rate $1/T$) providing energy losses regarding the theoretical BER performance no more than 0.5 dB at doubled symbol rate $2/T$ have been obtained. The value of T_s is increased with a view to reducing occupied frequency bandwidth ΔF. However, it is rather difficult to achieve significant reducing of bandwidth for signals with $T_s >$ 16T. This fact is caused by a limitation on the fixed reduction rate of out-of-band emissions used during the optimization procedure. According to the equivalent of the law of energy conservation, the main petal of the spectrum cannot be reduced considerably. It should be noted that extra increase in symbol rate (more than $2/T$ defined in optimization procedure) leads to the necessity of application of more complex detection algorithms. At the moment there are no researches devoted to this problem.

This paper is dedicated to the issues of efficient using of optimal signals with different detection algorithms (including maximum likelihood sequence estimation – MLSE). We use AWGN channel to determine the boundary for BER performance. Application of all other channels leads to EBR performance degradation. Thus, this article presents the limits of applicability of optimal signals by spectral-energy efficiency.

The paper consists of three sections. In the first section, different detection algorithms for optimal signals are considered. The second section contains a description of the used simulation model. In the third section, BER performance and computational complexity of corresponding detection algorithms are discussed.

2 Detection Algorithms for Optimal Signals

2.1 Optimal Coherent Detection Algorithm

The j-th realization of optimal FTN signals with duration T_s, envelope shape $a(t)$, carrier frequency f_0, amplitude A_0 and symbol rate $R = 1/\xi T$ (ξ – time acceleration factor) may be written as shown below:

$$s_j(t) = A_0 \sum_{k=0}^{Q-1} a(t - k/R) d_{j,h}^{(k)} \cos\left(2\pi f_0 t + \varphi_j^{(k)}\right), \tag{1}$$

where $d_{j,h}^{(k)}$ is the k-th modulation symbol, $\varphi_j(k)$ – the initial phase of k-th symbol, Q – number of modulation symbols. The value of $d_{j,h}^{(k)}$ depends on index $h = 1, 2, ...,$ m (m is the volume of channel alphabet). For BPSK modulation $m = 2$, $d_{j,1}^{(k)} = 1$, $d_{j,2}^{(k)} = -1$, $\varphi_j(k) = 0$.

The form of the signal at the receiver input:

$$x(t) = s_j(t) + n(t) = A_0 \sum_{k=0}^{Q-1} a(t - k/R) d_{j,h}^{(k)} \cos(2\pi f_0 t) + n(t), \tag{2}$$

where $n(t)$ is the average power spectral density of additive white Gaussian noise (AWGN).

It is known that the receiver task is to associate a set of modulation symbol in (1) with the input realization of the process $x(t)$ in (2). There is a one-to-one mapping between modulation symbols and information bits.

Let us make an assumption that prior probabilities of occurrence of modulation symbols are equal. Then optimal detection algorithm which represents maximum likelihood estimation (MLE) may be expressed by the next formula:

$$\int_{-T_s/2}^{T_s/2} x(t)s_j(t)dt - \frac{1}{2}E_j > \int_{-T_s/2}^{T_s/2} x(t)s_k(t)dt - \frac{1}{2}E_k, \ k \neq j, \tag{3}$$

where E_j, E_k are energies of corresponding realizations of signals (1). It means that if condition (3) is fulfilled, then the decision is made in favor of j-th realization of modulation symbols. To implement this algorithm a large amount of resources is required (the number of correlators is equal to m^Q), that is why it is often impossible.

2.2 Coherent Bit-by-Bit Detection Algorithm

Let us consider optimal algorithm of coherent bit-by-bit detection for signal transmitting k-th modulation symbol [21]:

$$\Delta_{pr} = \frac{\exp\left\{\frac{2}{N_0}\int_{-T_s/2+k/R}^{T_s/2+k/R} x(t)s_{y,1}^{(k)}(t)dt - \frac{1}{N_0}\int_{-T_s/2+k/R}^{T_s/2+k/R} \left(s_{y,1}^{(k)}(t)\right)^2 dt\right\}}{\exp\left\{\frac{2}{N_0}\int_{-T_s/2+k/R}^{T_s/2+k/R} x(t)s_{y,2}^{(k)}(t)dt - \frac{1}{N_0}\int_{-T_s/2+k/R}^{T_s/2+k/R} \left(s_{y,2}^{(k)}(t)\right)^2 dt\right\}},$$

where $s_{y,1}^{(k)}(t) = y_j^{(-)}(t) + s_{j,1}^{(k)}(t) + y_j^{(+)}(t), s_{y,2}^{(k)}(t) = y_j^{(-)}(t) + s_{j,2}^{(k)}(t) + y_j^{(+)}(t),$

$$s_{j,h}^{(k)}(t) = a(t - k/R)d_{j,h}^{(k)}\cos\left(2\pi f_0 t + \varphi^{(k)}\right), h = 1, 2; \ d_{j,1}^{(k)} = -d_{j,2}^{(k)},$$

$$y_j^{(-)}(t) = A_0 \sum_{i=0}^{k-1} a(t - i/R)d_{j,h}^{(i)}\cos\left(2\pi f_0 t + \varphi^{(k)}\right),$$

$$y_j^{(+)}(t) = A_0 \sum_{i=k+1}^{Q} a(t - i/R)d_{j,h}^{(i)}\cos\left(2\pi f_0 t + \varphi^{(k)}\right).$$

In this way, we can get optimal algorithm of bit-by-bit detection (Fig. 1) by averaging numerator and denominator over all possible combinations of $y_j^{(-)}(t)$ and $y_j^{(+)}(t)$ in (3). Nevertheless, the implementation of this algorithm is computationally difficult. It is necessary to look at simpler algorithms.

For example, take a look at coherent bit-by-bit detection algorithm at each k-th interval. In the case of BPSK modulation ($d_{j,1}^{(k)} = -d_{j,2}^{(k)}$):

$$\int_{-T_s/2+k/R}^{T_s/2+k/R} x(t)a(t-k/R)\cos\left(2\pi f_0 t + \varphi^{(k)}\right)dt \underset{d_{j,2}^{(k)}}{\overset{d_{j,1}^{(k)}}{\gtrless}} 0. \tag{4}$$

The disadvantage of this algorithm is that all signals from the previous and following clock intervals represent additional noise. Therefore, a further increase in symbol rate leads to degradation of BER performance.

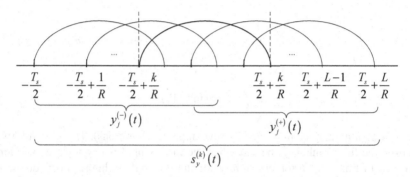

Fig. 1. Illustration of optimal coherent bit-by-bit detection algorithm.

To eliminate this drawback suppose that till the k-th interval all symbols have been detected right. Then following nonlinear bit-by-bit detection algorithm with interference compensation may be obtained:

$$\int_{-T_s/2+k/R}^{T_s/2+k/R} \left[x(t) - y_j^{*(-)}(t)\right]a(t-k/R)\cos\left(2\pi f_0 t + \varphi^{(k)}\right)dt \underset{d_{j,2}^{(k)}}{\overset{d_{j,1}^{(k)}}{\gtrless}} 0, \tag{5}$$

$$y_j^{*(-)}(t) = A_0 \sum_{i=0}^{k-1} a(t-i/R)d_{j,h}^{*(i)}\cos\left(2\pi f_0 t + \varphi^{(k)}\right),$$

where $d_{j,h}^{*(i)}$ – estimations of earlier detected symbols. Indeed, if previous symbols were detected wrong, BER performance worsens.

2.3   Coherent Detection Algorithm Based on Partial Enumeration

Another algorithm worth paying attention to is based on partial enumeration in relation to a sequence of optimal signals. At the first stage signal representing the first symbol of the sequence is formed. There are m different modulation symbols $d_{j,h}^0$ for the volume of channel alphabet m. The decision $d_{j,h}^{*(0)}$ is made in favor of the $d_{j,h}^0$ which minimizes the next function:

$$
W_{0,h} = \int_{-T_s/2}^{T_s/2+(Q-1)/R} \left| x(t) - a(t)d_{j,h}^{(0)} \cos(2\pi f_0 t) \right| dt.
$$

At each subsequent stage, the decisions made at previous steps are taken into account. Thus, at k-th stage, the estimations $d_{j,h}^{*(i)}, i = 0 \ldots k - 1$ of previous $(k - 1)$ symbols are obtained, and the function to be minimized looks as follows:

$$
W_{k,h} = \int_{-T_s/2}^{T_s/2+(Q-1)/R} \left| x(t) - \left[\sum_{i=0}^{k-1} a(t - i/R)d_{j,h}^{*(i)} \cos(2\pi f_0 t) + a(t - k/R)d_{j,h}^{(k)} \cos(2\pi f_0 t) \right] \right| dt
$$

$$
d_{j,h}^{*(k)} = \arg \min_h W_{k,h}. \tag{6}
$$

This algorithm may be improved by several passes (iterations). The straightforward pass from the first symbol to the last when the first symbol is unknown is mentioned above (6). In this case, error probability for first symbols is higher than for the last symbols. The next pass may be done in reverse direction from the last symbol to the first one. It is logical to assume that further passes will not have a significant influence on BER performance. To develop this algorithm some preassigned symbol sequence in the beginning and at the end of information sequence may be inserted.

Let us estimate the computational complexity of the algorithm with one pass (iteration). At each k-th step, it is required to keep $(k - 1)$ previous decisions and make m comparisons. For demodulation of Q information bits, mQ comparisons need to be made and Q decisions obtained at each step need to be kept. When two passes (iterations) are done, the number of comparisons increases by two times.

2.4   Viterbi Algorithm and Maximum Likelihood Sequence Estimation (MLSE)

Except for "simple" processing algorithms it is also worth to check the efficiency of optimal signals detection with the use of Viterbi algorithm [22] and maximum likelihood sequence estimation (MLSE). The complexity of Viterbi algorithm grows exponentially with an increase in the depth of intersymbol interference L. "Depth" means the number of previous symbols which influence on the current one at clock interval. At each clock interval $m \cdot m^L = m^{L+1}$ Euclidean distances should be calculated. Therefore, to demodulate the sequence consisting of Q information symbols Qm^{L+1} comparisons have to be done [23].

3 Simulation Model

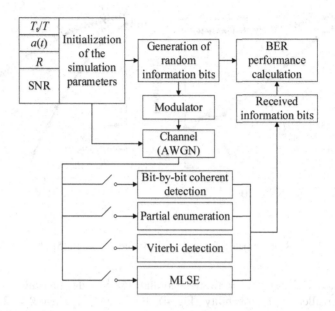

Fig. 2. Simulation modeling flowchart.

The efficiency of detection algorithms for optimal FTN signals is estimated with simulation modeling in Matlab (Fig. 2). This simulation model is similar to the one used in [19]. In this work, optimal signals with parameters $T_s = 2T$, $8T$, $16T$ and $n = 2$, $K_0 = 0.01$, $R = 2/T$ were chosen. These signals provide energy losses comparing to the theoretical BER performance no more than 0.5 dB when coherent bit-by-bit detection algorithm is applied (error probability is equal to 10^{-3}). In the same conditions signals with $T_s = 2T$ have energy losses about 1 dB because of their small duration.

Let us check the possibility of signal transmission at higher rates with the application of different detection algorithms.

4 Results and Discussions

Figure 3a) illustrates results corresponding to coherent bit-by-bit detection algorithm (4). Increasing R even by 11% relatively to original value leads to significant degradation of BER performance: energy losses for error probability 10^{-3} vary from 5 dB for signals with $T_s = 2T$ to more than 10 dB for signals with $T_s = 16T$ (Fig. 3a). The higher the duration T_s, the greater the error probability. This fact can be explained by the increase in interference from adjacent signals. Extra increasing R does not make sense for this detection algorithm. In case of the algorithm (5), the rate may be increased up to 25% with energy losses regarding the theoretical BER performance about 4.5 dB for $T_s = 2T$ and 8 dB for $T_s = 16T$ (Fig. 3b).

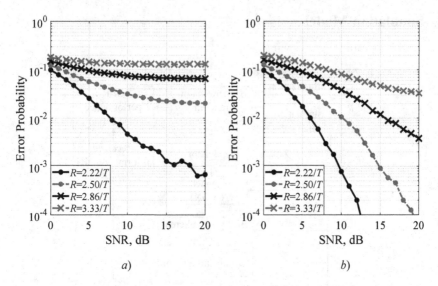

Fig. 3. BER performance for algorithm (4) and algorithm (5) ($T_s = 16T$).

For a fixed value of SNR = 10 dB algorithm (5) provides transmission at higher rates with smaller error probability (Fig. 4). For example, when $R = 2.2/T$, error probability for the algorithm (5) is an order of magnitude less than for algorithm (4). At the same time, there is a stronger dependency of the algorithm (5) on signal duration (at $R = 2.2/T$ error probability for $T_s = 16T$ is twice as much as for $T_s = 2T$).

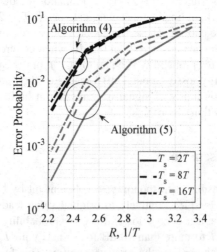

Fig. 4. Error probability vs. symbol rate for different T_s and algorithms (4), (5). SNR = 10 dB.

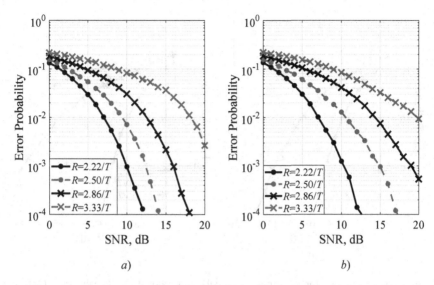

Fig. 5. BER performance for algorithm (6): (1 iteration for $T_s = 2T$ (a), $T_s = 8T$ (b)).

Comparing to the algorithm of partial enumeration (6) using a state diagram with one iteration, algorithm (5) provides about 1 dB gain for error probability 10^{-3} (Fig. 5).

Using two iterations gives much better results for short signal duration and a little increase in the symbol rate. So, when the symbol rate is 11% higher than original, signals with duration $2T$ and $8T$ have energy losses regarding the theoretical BER performance about 1 dB (for error probability equal to 10^{-3}). However, an attempt to increase the rate up to $2.5/T$ for signals with $T_s = 8T$ does not allow to achieve a significant gain in energy with respect to mentioned algorithms (Fig. 6).

For a greater increase in symbol rate ($R = 3.33/T$) the number of iterations does not influence on error probability. Nevertheless, for smaller R using two iterations instead of one allows reducing error probability by up to 2.5 times. In terms of reducing error probability, a further increase in the number of iterations is inefficient (Fig. 7).

It was shown that it is possible for optimal signals to increase the symbol rate additionally using summarized detection of the signal sequence. Note that there are no losses in the minimal Euclidean distance [4, 20]. The work [20] presents estimations of minimal Euclidean distance depending on symbol rate: for $T_s = 8T$ and $R = 2/T$ limit symbol rate is equal to $2.5/T$. It means that BER performance does not degrade until the symbol rate is less than or equal to $2.5/T$. Figure 6(b) in [20] shows BER curves of optimal binary signals with increased symbol rates for summarized detection of the signal sequence. In particular, it can be seen that BER performance corresponds to the theoretical one for symbol rate $2.5/T$.

MLSE algorithm is optimal for summarized detection of a signal sequence but computationally inefficient. With increasing packet length, the complexity of the algorithm grows exponentially. At the same time, the receiver has to keep in memory m^Q possible sequences for the MLSE algorithm. Therefore, it makes sense to use this algorithm in systems with short packet length (tens or hundreds of bits). In practice,

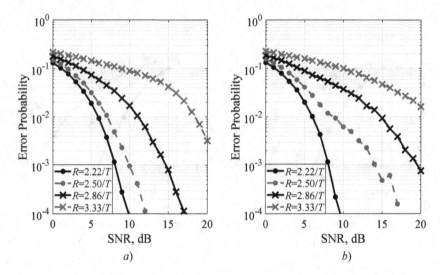

Fig. 6. BER performance for algorithm (6): partial enumeration using state diagrams (2 iteration for $T_s = 2T$ (a), $T_s = 8T$ (b)).

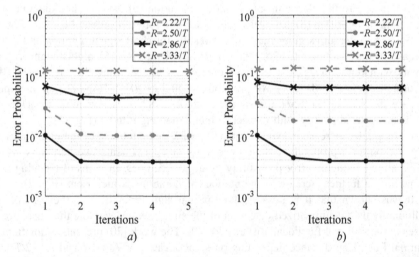

Fig. 7. Error probability vs. number of iterations for fixed SNR = 7 dB and $T_s = 2T$ (a), $T_s = 8T$ (b).

computationally efficient algorithms like Viterbi algorithm are used. In Fig. 8(a) the results of detection of optimal signal sequences with different symbol rates are given. Obviously, there is an energy loss comparing to the algorithm of MLSE (Fig. 8(b)). Note that algorithm of partial enumeration with 2 iterations allows to get almost the same BER performance as Viterbi algorithm provides, if symbol rate is increased no more than by 11% (the difference is less than 0.5 dB for error probability 10^{-3}). Besides, this algorithm is characterized by less computational complexity. Further increase in symbol rate results in much greater losses (more than 6 dB).

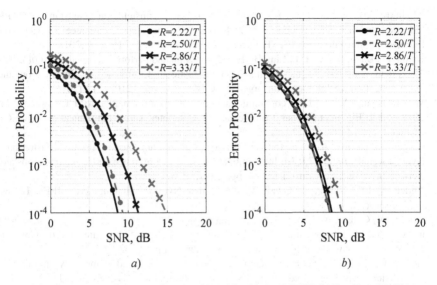

Fig. 8. BER performance for algorithm Viterbi (*a*), MLSE (*b*) for $T_s = 8T$ (*b*).

Acknowledgement. The results of the work were obtained under the grant of the President of the Russian Federation for state support of young Russian scientists (agreement MK-1571.2019.8 №. 075-15-2019-1155) and used computational resources of Peter the Great Saint-Petersburg Polytechnic University Supercomputing Center (http://www.scc.spbstu.ru).

References

1. Mazo, J.E.: Faster-than-Nyquist signaling. Bell Syst. Tech. J. **54**, 1451–1462 (1975)
2. Ghannam, H., Darwazeh, I.: SEFDM over satellite systems with advanced interference cancellation. IET Commun. **12**(11), 59–66 (2018)
3. Darwazeh, I., Ghannam, H., Xu, T.: The first 15 years of SEFDM: a brief survey. In: 2018 11th International Symposium on Communication Systems, Networks and Digital Signal Processing (CSNDSP), Budapest, pp. 1–7 (2018)
4. Vasilyev, D., Rashich, A.: SEFDM-signals Euclidean distance analysis. In: 2018 IEEE International Conference on Electrical Engineering and Photonics (EExPolytech), St. Petersburg, pp. 75–78 (2018)
5. Rashich, A., Urvantsev, A.: Pulse-shaped multicarrier signals with nonorthogonal frequency spacing. In: 2018 IEEE International Black Sea Conference on Communications and Networking (BlackSeaCom), Batumi, pp. 1–5 (2018)
6. Gelgor, A., Gorlov, A., Nguyen, V.P.: Performance analysis of SEFDM with optimal subcarriers spectrum shapes. In: 2017 IEEE International Black Sea Conference on Communications and Networking (BlackSeaCom), Istanbul, pp. 1–5 (2017)
7. Gelgor, A., Gorlov, A., Nguyen, V.P.: The design and performance of SEFDM with the Sinc-to-RRC modification of subcarriers spectrums. In: 2016 International Conference on Advanced Technologies for Communications (ATC), Hanoi, pp. 65–69 (2016)

8. Kholmov, M., Fadeev, D.: The effectiveness of active constellation extension for PAPR reduction in SEFDM systems. In: 2018 IEEE International Conference on Electrical Engineering and Photonics (EExPolytech), St. Petersburg, pp. 116–118 (2018)

9. Kwon, H., Baek, M., Yun, J., Lim, H., Hur, N.: Design and performance evaluation of DVB-S2 system with FTN signaling. In: 2016 International Conference on Information and Communication Technology Convergence (ICTC), Jeju, pp. 1210–1212 (2016)

10. Wu, Z., Huang, X.: A LDPC convolutional code optimization method for FTN systems. In: 2017 3rd IEEE International Conference on Computer and Communications (ICCC), Chengdu, pp. 1469–1473 (2017)

11. Gelgor, A., Gorlov, A.: A performance of coded modulation based on optimal faster-than-Nyquist signals. In: 2017 IEEE International Black Sea Conference on Communications and Networking (BlackSeaCom), Istanbul, pp. 1–5 (2017)

12. Anderson, J.B., Rusek, F., Owall, V.: Faster-than-Nyquist signaling. Proc. IEEE **101**(8), 1817–1830 (2013)

13. Gorlov, A., Gelgor, A., Popov, E.: Improving energy efficiency of partial response signals by using coded modulation. In: 2015 IEEE International Black Sea Conference on Communications and Networking (BlackSeaCom), Constanta, pp. 58–62 (2015)

14. Zhang, G., Wei, Y., Shen, Y., Guo, M., Nie, S.: A reduced complexity interference cancellation technique based on matrix decomposition for FTN signaling. In: 2016 SAI Computing Conference (SAI), London, pp. 622–625 (2016)

15. Le, C., Schellmann, M., Fuhrwerk, M., Peissig, J.: On the practical benefits of faster-than-Nyquist signaling. In: 2014 International Conference on Advanced Technologies for Communications (ATC 2014), Hanoi, pp. 208–213 (2014)

16. Liang, X., Liu, A., Wang, K., Zhang, Q., Peng, S.: Symbol-by-symbol detection for faster-than-Nyquist signaling aided with frequency-domain precoding. In: 2016 6th International Conference on Electronics Information and Emergency Communication (ICEIEC), Beijing, pp. 14–17 (2016)

17. Zavjalov, S.V., Volvenko, S.V., Makarov, S.B.: A method for increasing the spectral and energy efficiency SEFDM signals. IEEE Commun. Lett. **20**(12), 2382–2385 (2016)

18. Sadovaya, Y., Gelgor, A.: Synthesis of signals with a low-level of out-of-band emission and peak-to-average power ratio. In: 2018 IEEE International Conference on Electrical Engineering and Photonics (EExPolytech), St. Petersburg, pp. 103–106 (2018)

19. Zavjalov, S.V., Ovsyannikova, A.S., Volvenko, S.V.: On the necessary accuracy of representation of optimal signals. In: Galinina, O., Andreev, S., Balandin, S., Koucheryavy, Y. (eds.) NEW2AN/ruSMART -2018. LNCS, vol. 11118, pp. 153–161. Springer, Cham (2018). https://doi.org/10.1007/978-3-030-01168-0_14

20. Zavjalov, S.V., Ovsyannikova, A.S., Lavrenyuk, I.I., Volvenko, S.V., Makarov, S.B.: Application of optimal finite-length signals for overcoming "Nyquist limit". In: Galinina, O., Andreev, S., Balandin, S., Koucheryavy, Y. (eds.) NEW2AN/ruSMART -2018. LNCS, vol. 11118, pp. 172–180. Springer, Cham (2018). https://doi.org/10.1007/978-3-030-01168-0_16

21. Tsikin, I.A., Shcherbinina, E.A.: Algorithms of GNSS signal processing based on the generalized maximum likelihood criterion for attitude determination. In: 2018 25th Saint Petersburg International Conference on Integrated Navigation Systems (ICINS), St. Petersburg, pp. 1–4 (2018)

22. Viterbi, A.: Error bounds for convolutional codes and an asymptotically optimum decoding algorithm. IEEE Trans. Inf. Theory **13**(2), 260–269 (1967)

23. Proakis, J.G.: Digital Communications, 4th edn. McGraw-Hill, New York (2001)

The Effectiveness of Application of Multi-frequency Signals Under Conditions of Amplitude Limitation

Dac Cu Nguyen, Sergey V. Zavjalov[ID],
and Anna S. Ovsyannikova[(⊠)][ID]

Peter the Great St. Petersburg Polytechnic University, St. Petersburg, Russia
daccu91.spb@gmail.com, zavyalov_sv@spbstu.ru,
anny-ov97@mail.ru

Abstract. Simulation modeling of information transmission using spectrally efficient multi-frequency signals with an amplifier and a limiter on transmission was held. The possibility of increasing the average radiation power while limiting the peak-to-average power ratio (PAPR) is shown. However, with a more significant limitation on the PAPR, a greater interference between signals from adjacent subcarriers begins to appear. Thus, it is possible to search for an effective value for PAPR limiting. In this paper, cases of multi-frequency signals are considered with a different value of the frequency separation between subcarriers, and the case of the simultaneous use of a smoothed shape of envelopes is affected.

Keywords: Nonorthogonal multi-frequency spectrally efficient frequency division multiplexing · Peak-to-average power ratio · Amplitude limitation · SEFDM · PAPR

1 Introduction

At the present time, an increase in spectral efficiency under conditions of limited frequency resources is one of the important areas of research [1–4]. Among the well-known methods of increasing the spectral efficiency of signals, we can pay attention to the generalization of technology with orthogonal frequency division multiplexing (OFDM) for the case of non-orthogonal frequency separation between subcarriers - spectrally efficient frequency division multiplexing (SEFDM) [5–8].

We consider baseband SEFDM-symbols with duration T of the following form [9–11]:

$$s(t) = \sum_{k=1}^{N} C_k \cos(2\pi k \Delta f t), t \in [0; T].$$

It is a multi-frequency signal with N subcarriers, complex modulation symbols C_k on each subcarrier (BPSK, QPSK, QAM-16, etc.) with no cyclic prefix. The interval between adjacent subcarriers is Δf. For SEFDM, $\Delta f = \alpha/T$, where $\alpha < 1$. For OFDM

© Springer Nature Switzerland AG 2019
O. Galinina et al. (Eds.): NEW2AN 2019/ruSMART 2019, LNCS 11660, pp. 681–687, 2019.
https://doi.org/10.1007/978-3-030-30859-9_59

signals, $\alpha = 1$, that is, $\Delta f_{OFDM} = 1/T$. The cost of improving the spectral efficiency of SEFDM signals is the degradation of BER performance due to the interference of signals from neighboring subcarrier frequencies because the signals are not orthogonal.

SEFDM signals have higher spectral efficiency compared to OFDM signals and occupy a lower bandwidth, but have a high PAPR [12–14]. This leads to the need to limit the value of PAPR to increase the efficiency of the amplifier stages of transceiver devices.

Due to PAPR limiting the average power begins to grow, which can lead to an increase in the communication distance or a decrease in the probability of error at a fixed distance between the transmitter and receiver and a fixed value of signal-to-noise ratio. However, with a significant limitation of the PAPR, a greater interference between signals from adjacent subcarriers begins to appear [15]. As a result, deterioration in bit error rate (BER) performance takes place. Thus, the purpose of this paper is to find the effective value of the PAPR restriction to ensure BER performance of the reception at a given level.

The rest of the paper is organized as follows. In Sect. 2 we will consider simulation model with preamplifier and limiter on the transmission to study the effect of reducing the PAPR on the quality of information transmission. Simulation results are presented in Sect. 3. Finally, in Sect. 4 we will draw conclusions about the effectiveness of this method and the necessary parameters.

2   Simulation Model

This model (Fig. 1), which was built in the Matlab system, includes an information source, a modulator of SEFDM signals, a block of transmitter model, calculation blocks of spectral characteristics and PAPR, a block simulating a transmission channel with additive white Gaussian noise (AWGN) and block of receiver model.

In the information source block, a pseudo-random sequence of zeros and ones of a given volume is formed depending on the number of used subcarriers and equiprobable symbols. At each point for limiting the PAPR, at least $2 \cdot 10^7$ information bits were transmitted.

The transmitter model includes a SEFDM modulator, a preamplifier, a limiter, and a power amplifier. A pseudo-random sequence received from an information source is fed to the modulator input. Modulation type of BPSK or QPSK is performed. The number of subcarrier frequencies is equal to 200. The formed SEFDM symbols from the modulator output are fed to the input of the preamplifier to increase the average radiation power. Then, the received signal is sent to the limiter to limit the signal amplitude to the highest amplitude of the output signal of SEFDM modulator. This means that the average power of the signal increases, while the peak power of the signal does not change, leading to a PAPR reduction. The average value of the PAPR is considered, and in this paper, 10^4 iterations of averaging are done to calculate the average PAPR value. The procedure for limiting the PAPR of SEFDM symbols is shown in Fig. 2. As long as level A, which determines the value of ΔPAPR, does not exceed the value of A_{lim}, the output value U_{out} linearly depends on the input U_{in}. Starting with A_{lim}, U_{out} stops changing.

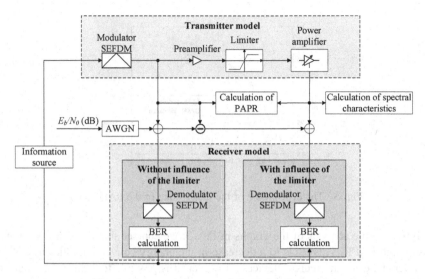

Fig. 1. Structural diagram of the simulation model

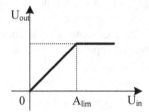

Fig. 2. Limiter characteristic

The amplified signal is transmitted to the calculation blocks of spectral characteristics and PAPR values. At the same time, the received SEFDM signal in the limiter is transmitted via the communication channel, where the AWGN is added, with signal-to-noise ratio per bit E_b/N_0 (E_b is the signal energy per bit and N_0 is spectral noise density) ranging from 0 to 10 dB. Moreover, the noise is added at the same level as the SEFDM signal without amplitude limitation.

The receiver model includes demodulators of SEFDM signals, and blocks for BER performance calculating. The receiver model performs the procedures for detecting and demodulating SEFDM signals received from the communication channel. An element-by-element algorithm is used as a detection algorithm.

3   Results and Discussions

Let's consider the values of the PAPR in the BPSK and QPSK modulation for different values of the frequency separation between the subcarriers. Figure 3 shows the dependence of the mean value of the PAPR on α in the BPSK and QPSK modulation.

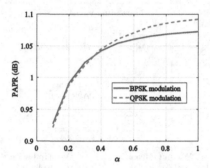

Fig. 3. The dependence of the mean value of the PAPR on α

Based on the analysis of the graphs in Fig. 3, we can see that the SEFDM signals have a high PAPR. This leads to a decrease in the energy conversion efficiency of the amplifier stages of transceiver devices. However, it is possible to reduce the values of the PAPR, as will be discussed further.

Figures 4 and 5(a) show the types of normalized instantaneous power $p(t)/\max(p(t))$ and spectra of SEFDM signals $|S(f)|^2/|S(0)|^2$ at different values of PAPR limit and $\alpha = 0.7$. Based on the analysis of the graphs in Fig. 5, it can be noted that with a more stringent restriction on the PAPR, significant interference between signals from adjacent frequency subcarriers begins to appear due to an increase in out-of-band emission of SEFDM signals.

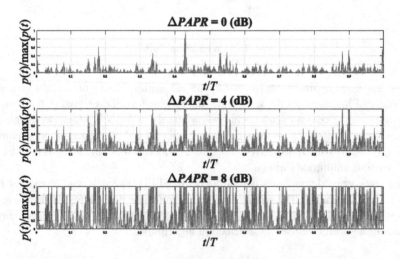

Fig. 4. Normalized instantaneous power of SEFDM signals at different values of the PAPR limitation

In Fig. 6 the dependence of the average PAPR and average power P(t) value on the value of the ΔPAPR limit is given. It is easy to notice that a higher value of ΔPAPR

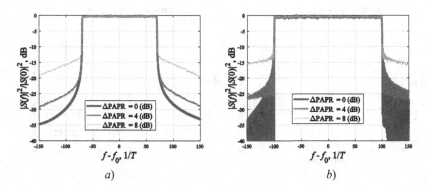

Fig. 5. The spectra of SEFDM (*a*) and OFDM (*b*) signals for different values of the PAPR limitation

results in smaller average PAPR values. This is due to an increase in the average power value shown in Fig. 6(*b*).

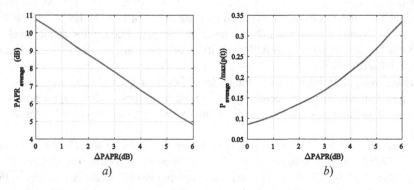

Fig. 6. Dependence of the average PAPR (*a*) and average power P(*t*) value (*b*) on the ΔPAPR limit

Figure 7(*a*) and (*b*) illustrates the dependency of BER performance of SEFDM signals on the PAPR limitation level for the cases of BPSK (7, *a*) and QPSK (7, *b*) modulation, signal-to-noise ratio $E_b/N_0 = 7$ dB and different values of α. In both cases, it is easy to see that increasing spectral efficiency of SEFDM signals (i.e. decreasing α) leads to a greater number of errors. When α = 1, the received OFDM signal has the best BER performance.

In addition, in both cases, it can be seen that with an increase in the ΔPAPR limit value, the BER value decreases and reaches a minimum of $5 \cdot 10^{-7}$ at the 4.5 dB point for BPSK, $6 \cdot 10^{-7}$ at the 4.5 dB point for QPSK modulation when α = 1. When ΔPAPR ≥ 5 dB, the BER performance of the reception starts to deteriorate. This means that the effective PAPR limit value is 4.5–5 dB.

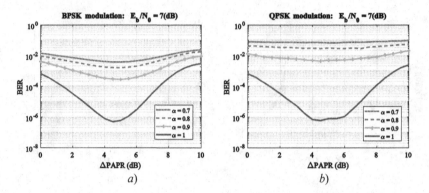

Fig. 7. BER of receiving SEFDM signals depending on the ΔPAPR limit

4 Conclusions

To conclude, there is a certain optimal value of the PAPR, at which the BER is minimal when a scheme with a preamplifier and a limiter is used. In this scheme, the effective PAPR limit value is 4.5–5 dB with $E_b/N_0 = 7$ dB. However, at the same time, the band of occupied frequencies expands due to an increase in out-of-band emission of SEFDM signals. For example, with $\alpha = 0.7$, the band of occupied frequencies at the level of 20 dB reaches 206.9 ($1/T$) for ΔPAPR = 4 dB, while it reaches only 167.9 ($1/T$) for the signal itself (ΔPAPR = 0 dB).

Taking everything into account, the trade-off between spectral efficiency and BER performance should be made according to the specific purposes. Some applications of this method include satellite communications, 5–6G systems, industrial internet of things.

The results of the work were obtained under the grant of the President of the Russian Federation for state support of young Russian scientists (agreement MK-1571.2019.8 №075-15-2019-1155) and used computational resources of Peter the Great Saint-Petersburg Polytechnic University Supercomputing Center (http://www. scc.spbstu.ru).

References

1. Luo, F.-L., Zhang, C. (eds.): Signal Processing for 5G: Algorithms and Implementations. Wiley, Hoboken (2016)
2. Proakis, J.G.: Digital Communications. McGraw-Hill, New York (1995)
3. Ometov, A., Moltchanov, D., Komarov, M., Volvenko, S.V., Koucheryavy, Y.: Packet level performance assessment of mmWave backhauling technology for 3GPP NR systems. IEEE Access 7, 9860–9871 (2019)
4. Sadovaya, Y., Gelgor, A.: Synthesis of signals with a low-level of out-of-band emission and peak-to-average power ratio. In: 2018 IEEE International Conference on Electrical Engineering and Photonics (EExPolytech), St. Petersburg, pp. 103–106 (2018)

5. Zavjalov, S.V., Ovsyannikova, A.S., Lavrenyuk, I.I., Volvenko, S.V., Makarov, S.B.: Application of optimal finite-length signals for overcoming "Nyquist limit". In: Galinina, O., Andreev, S., Balandin, S., Koucheryavy, Y. (eds.) NEW2AN/ruSMART -2018. LNCS, vol. 11118, pp. 172–180. Springer, Cham (2018). https://doi.org/10.1007/978-3-030-01168-0_16

6. Isam, S., Darwazeh, I.: Characterizing the intercarrier interference of non-orthogonal spectrally efficient FDM system. In: 2012 8th International Symposium on Communication Systems, Networks & Digital Signal Processing (CSNDSP), pp. 1–5, July 2012

7. Gelgor, A., Gorlov, A., Nguyen, V.P.: Performance analysis of SEFDM with optimal subcarriers spectrum shapes. In: Black Sea Conference on Communications and Networking (BlackSeaCom), pp. 1–5, June 2017

8. Gelgor, A., Gorlov, A., Nguyen, V.P.: The design and performance of SEFDM with the Sinc-to-RRC modification of subcarriers spectrums. In: 2016 International Conference on Advanced Technologies for Communications (ATC), Hanoi, pp. 65–69 (2016)

9. Kislitsyn, A.B., Rashich, A.V., Tan, N.N.: Generation of SEFDM-signals using FFT/IFFT. In: Balandin, S., Andreev, S., Koucheryavy, Y. (eds.) NEW2AN 2014. LNCS, vol. 8638, pp. 488–501. Springer, Cham (2014). https://doi.org/10.1007/978-3-319-10353-2_44

10. Isam, S., Darwazeh, I.: Simple DSP-IDFT techniques for generating spectrally efficient FDM signals. In: 7th International Symposium on Communication Systems Networks and Digital Signal Processing (CSNDSP), pp. 20–24, July 2010

11. Rashich, A., Kislitsyn, A., Fadeev, D., Ngoc Nguyen, T.: FFT-based trellis receiver for SEFDM signals, In: 2016 IEEE Global Communications Conference (GLOBECOM), Washington, DC, pp. 1–6 (2016)

12. Gelgor, A., Gorlov, A., Popov, E.: Multicomponent signals for bandwidth-efficient single-carrier modulation. In: 2015 IEEE International Black Sea Conference on Communications and Networking (BlackSeaCom), Constanta, pp. 19–23 (2015)

13. Antonov, E.O., Rashich, A.V., Fadeev, D.K., Tan, N.: Reduced complexity tone reservation peak-to-average power ratio reduction algorithm for SEFDM signals. In: 2016 39th International Conference on Telecommunications and Signal Processing (TSP), Vienna, pp. 445–448 (2016)

14. Kholmov, M., Fadeev, D.: The effectiveness of active constellation extension for PAPR reduction in SEFDM systems. In: 2018 IEEE International Conference on Electrical Engineering and Photonics (EExPolytech), St. Petersburg, pp. 116–118 (2018)

15. Fadeev, D.K., Rashich, A.V.: Optimal input power backoff of a nonlinear power amplifier for SEFDM system. In: Balandin, S., Andreev, S., Koucheryavy, Y. (eds.) ruSMART 2015. LNCS, vol. 9247, pp. 669–678. Springer, Cham (2015). https://doi.org/10.1007/978-3-319-23126-6_60

BER Analysis in Dual Hop Differential Amplify-and-Forward Relaying Systems with Selection Combining Using M-ary Phase-Shift Keying over Nakagami-m Fading Channels

Mamoun F. Al-Mistarihi[1(✉)], Arwa S. Aqel[1], and Khalid A. Darabkh[2]

[1] Electrical Engineering Department,
Jordan University of Science and Technology, Irbid, Jordan
mistarihi@just.edu.jo, arwa.aqell@yahoo.com
[2] Computer Engineering Department, The University of Jordan, Amman, Jordan
k.darabkeh@ju.edu.jo

Abstract. In this work, a dual-hop relaying system using the differential amplify-and-forward relaying scheme along with post-detection selection combining techniques is proposed. This technique used previously received samples to estimate current fading gain in a branch in order to accomplish a cooperative diversity at cases that exclude the need of having explicit channel estimation. A closed-form expression for the exact end-to-end bit error rate of the communication system, assuming M-ary phase-shift keying constellation over independent non-identical Nakagami-m fading channels, is derived. Different fading parameters were used to validate the correctness of our end-to-end bit error rate tight expression.

Keywords: Differential encoding · Amplify-and-forward relaying · Nakagami-m fading channel · Post detection · Selection combining

1 Introduction

In some of wireless networks, the source and the destination nodes can be communicating together directly [1–7]. But with the presence of multipath fading and channel impairment, the complete communication becomes necessary which is investigated by using cooperative relaying nodes between the source and destination nodes [8–12].

Cooperative Relaying schemes are mainly found in three types: amplify-and-forward (AF), decode-and-forward (DF) and Compress-and-Forward (CF) [13–16]. In the AF protocol, the relay amplifies the received signal from the source by the amplification factor which is often constrained by the source-relay and the source-destination links, then it forwards the amplified signal to the destination [17, 18]. While, in the DF protocol, the relay decodes and re-encodes the received signal from the source, then it forwards it to the destination [19]. In the CF protocol, the relay compresses the received signal from the source and forwards it to the destination taking

© Springer Nature Switzerland AG 2019
O. Galinina et al. (Eds.): NEW2AN 2019/ruSMART 2019, LNCS 11660, pp. 688–699, 2019.
https://doi.org/10.1007/978-3-030-30859-9_60

into account the exclusion of being decoded. With cooperative networks, the capacity and throughput are improved, and the network's coverage area is expanded. In this paper, the AF protocol has been considered and implemented due to its simplicity.

Multiple signal copies are received at the destination through multipath fading and to improve the system performance, cooperative diversity is applied [20]. Actually, the most common combining diversity techniques are Selection Combining (SC), Equal Gain Combining (EGC), and Maximum Ratio Combining (MRC) [21]. SC is the simplest technique due to the its low cost and complexity. With SC technique the strongest received signal is selected.

The detection process is either a "Pre-Detection" or "Post-Detection" [22]. In this paper, Post Detection Selection Combining technique is applied on the system due to its low implementation cost by removing the co-phasing operation. The combining process here is done before the detection process.

The average bit error rate (BER) analysis for multihop relaying systems with the consideration of multipath fading channels and various channel conditions was introduced in [23] in which the authors assumed that all the relays are using the AF generative scheme of relaying. However, the model uses the Moment Generation Function (MGF) approach to compute the BER and assumes that the combining scheme at the destination is the maximal ratio combining (MRC). A closed-form expression has been derived for the average symbol error probability considering M-ary quadrature amplitude modulation in [24]. In addition, a closed form expression of the exact end-to-end symbol error probability of a scaled selection combining based a multi relay cooperative diversity environment with DF relaying scheme has been derived for flat Rayleigh fading considering M-ary phase shift modulation in [25] and given an iterative scheme to calculate the optimum value of the scale factor as well. Not only to this end, but the cooperative diversity has been investigated over AF relaying systems considering many combining techniques [26]. SC analysis has been evaluated considering Nakagami-m fading and Weibull fading channels in [27] and [28], respectively.

In this paper, a dual-hop relaying system using differential amplify-and-forward scheme with the post-detection selection combining reception scheme has been proposed. This technique used previously received samples to estimate current fading gain in a branch in order to accomplish a cooperative diversity at cases that exclude the need of having explicit channel estimation. It is noteworthy to mention that the Signal-to-noise ratio (SNR) is very popular in literature and basically used in different contexts [29–33]. In this context and at the destination, the SNR is computed for each link and the highest signal is selected for non-coherent detection. The exact end-to-end bit error rate performance of the proposed system is investigated. Closed form expression of the system BER is derived and simulation results are presented to validate the analytical ones.

The remainder of this paper is structured as follows. Section 2 presents our proposed system model. In Sect. 3, the exact end-to-end bit error rate performance is evaluated. Section 4 discusses some plotted simulation results. In Sect. 5, a conclusion about this paper is presented.

2 System Model

Consider a cooperative diversity network as shown in Fig. 1, which has a source node (S) transmitting to a destination node (D) through two branches. In the first branch, the source transmits its data to a cooperating AF relay node (R) which amplifies the source signals and forwards the amplified signals to the destination. In the second branch, the source transmits its data directly to the destination. Each node is assumed to be equipped with a single antenna and operates in half-duplex mode. The communication in this proposed model is a kind of virtual selection combining by selecting a best path.

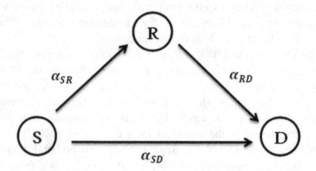

Fig. 1. System model

All types of links like source-relay, relay-destination, and source-destination paths, are characterized of being spatially uncorrelated and mutually independent. They communicate over a flat Nakagami-m fading channel. As seen in figure, the channel envelopes between S-R, R-D, and S-D, are given by α_{SR}, α_{RD}, and α_{SD} respectively. It is assumed that the additive white Gaussian noise (AWGN) terms of all the links (S-R, R-D, S-D) have a zero mean and unit variance.

When the differential encoding technique is applied on each symbol, the encoder output symbol becomes highly sensitive to the difference between current and prior output symbols. Therefore, the encoded symbol at instant time k with considering $x[0] = 1$ as an initial condition can be represented as:

$$x[k] = v[k]x[k - 1] \tag{1}$$

where, $v[k]$ is encoder's input symbol at instant time k.

For the proposed system model, the transmission of the symbol by the source occurs over two time slots. In the first time slot, the source transmits encoded signal to the relay and the destination. The received signals at the relay and at the destination are, respectively, as follows:

$$y_{SR}[k] = \sqrt{P_s}\alpha_{SR}[k]x[k] + n_{SR}[k] \tag{2}$$

$$y_{SD}[k] = \sqrt{P_s}\alpha_{SD}[k]x[k] + n_{SD}[k] \tag{3}$$

where, P_s is the average power per symbol for the source while n_{SR} and n_{SD} refer both to the AWGN with zero mean and unit variance for the relay and destination, respectively.

In the second time slot, the AF relay amplifies the received signal from the source by multiplying it with a factor that is dependent upon the received signal's power then it forwards it to the destination. To maintain the average transmit power by the relay P_r in our proposed model, the gain factor is assumed to be:

$$G^2 = \frac{P_r}{P_s + 1} \tag{4}$$

where, P_r is the relay mean power per symbol and "1" denotes to the noise variance.

Therefore, the received signal at the relay is amplified by G before re-transmitting it to the destination. It is as follows:

$$y_{RD}[k] = G\alpha_{RD}[k]y_{SR}[k] + n_{RD}[k] \tag{5}$$

Applying (2) into (5) to get

$$y_{RD}[k] = \sqrt{P_s}G\alpha_{SR}[k]\alpha_{RD}[k]x[k] + G\alpha_{RD}[k]n_{SR}[k] + n_{RD}[k] \tag{6}$$

The signal to noise ratio SNR is obtained by dividing the received signal power over the received noise power. Thus, the received SNR at the relay from the source is signified as:

$$\gamma_{SR} = P_s|\alpha_{SR}[k]|^2 \tag{7}$$

As far as, the end-to-end received SNR at the destination from the relay is signified as:

$$\gamma_{RD} = \frac{P_s G^2|\alpha_{SR}[k]|^2|\alpha_{RD}[k]|^2}{G^2|\alpha_{RD}[k]|^2 + 1} \tag{8}$$

As far as, the end-to-end received SNR at the destination from the source is signified as:

$$\gamma_{SD} = P_s|\alpha_{SD}[k]|^2 \tag{9}$$

3 Bit Error Rate Analysis

In this section, we analyze the error performance of the proposed model for M-ary phase shift keying modulation schemes. To analyze the end-to-end bit error rate, we have to study the error performance of the two branches; the source to-destination link and the source-to-relay-to-destination link. The end-to-end bit error rate is the sum of all error sources due to the usage of these two branches, which can be expressed as:

$$P_e = P_e(\gamma_{SD} > \gamma_{RD}) + P_e(\gamma_{SD} < \gamma_{RD}) \tag{10}$$

where $P_e(\gamma_{SD} > \gamma_{RD})$ and $P_e(\gamma_{SD} < \gamma_{RD})$ are the bit error probabilities of the source-to-destination link and the source-to-relay-to-destination link, respectively.

For MPSK constellation, the conditional error probability in a AWGN environment $P_e(\gamma)$, which is conditioned on γ is given as:

$$P_e(\gamma) = Q\left(\sqrt{2\gamma}\,sin\left(\frac{\pi}{M}\right)\right) \tag{11}$$

where $Q(\cdot)$ is the Gaussian Q-function and γ is a general notation that represents the instantaneous SNR of the S-R, R-D, and S-D links which is symbolized by γ_{SR}, γ_{RD} and γ_{SD} respectively.

The distribution of the channel envelopes between S-R, R-D, and S-D links is Nakagami-m. Therefore, the power variable γ is a gamma distribution [34]. The PDF of the instantaneous SNRs γ_{SR}, γ_{RD}, and γ_{SD} is expressed as:

$$f_{\gamma_{SR}}(\gamma, m_{SR}, P_s) = \frac{1}{\Gamma(m_{SR})}\left(\frac{m_{SR}}{P_s}\right)^{m_{SR}}\gamma^{m_{SR}-1}e^{-\frac{m_{SR}}{P_s}\gamma} \tag{12}$$

$$f_{\gamma_{RD}}(\gamma, m_{RD}, c) = \frac{1}{\Gamma(m_{RD})}\left(\frac{m_{RD}}{c}\right)^{m_{RD}}\gamma^{m_{RD}-1}e^{-\frac{m_{RD}}{c}\gamma} \tag{13}$$

and

$$f_{\gamma_{SD}}(\gamma, m_{SD}, P_s) = \frac{1}{\Gamma(m_{SD})}\left(\frac{m_{SD}}{P_s}\right)^{m_{SD}}\gamma^{m_{SD}-1}e^{-\frac{m_{SD}}{P_s}\gamma} \tag{14}$$

where, m_{SR}, m_{RD}, and m_{SD} are the parameters of Nakagami-m fading related to the S-R, R-D, and S-D links, respectively, and $c = \frac{P_s G^2 |\alpha_{RD}[k]|^2}{G^2 |\alpha_{RD}[k]|^2 + 1}$.

3.1   Error Analysis of Source-to-Destination Link

The average error probability of the source-to-destination link over a flat fading channel is obtained by averaging the conditional error probability of the S-D link in a AWGN environment conditioned on the instantaneous SNR γ_{SD} which is given in (11) over the statistics of γ_{SD}, under the condition that the instantaneous SNR of the S-D link γ_{SD} is higher than the instantaneous SNR of the R-D link γ_{RD}. That means that the BER $P_{e(\gamma_{SD} > \gamma_{RD})}$ of the S-D link can be obtained by:

$$P_e(\gamma_{SD} > \gamma_{RD}) = \int_{u=0}^{\infty}\int_{v=0}^{u} P_e(u) f_{\gamma_{SD}}(u) f_{\gamma_{RD}}(v)\,dv\,du \tag{15}$$

Substituting (11), (13) and (14) into (15) and after some mathematical manipulations and using [35, 3.381 (1)] and [35, 8.352 (6)], $P_{e(\gamma_{SD} > \gamma_{RD})}$ can be expressed as:

$$
\begin{aligned}
P_e(\gamma_{SD} > \gamma_{RD}) = &\frac{(m_{RD} - 1)!}{\pi \, \Gamma(m_{SD})\Gamma(m_{RD})} \left(\frac{m_{SD}}{P_s}\right)^{m_{SD}} \\
&\times \left[\frac{\Gamma(m_{SD})}{\left(\frac{m_{SD}}{P_s}\right)^{m_{SD}}} \left[\frac{\pi}{2}\left(1 - \sqrt{\frac{P_s sin^2 \frac{\pi}{M}}{m_{SD} + P_s sin^2 \frac{\pi}{M}}}\right)\right]^{m_{SD}} - \sum_{m=0}^{m_{RD}-1} \frac{\left(\frac{m_{RD}}{c}\right)^m}{m!} \right. \\
&\left. \times \frac{\Gamma(m_{SD} + m)}{\left(\frac{m_{SD}}{P_s} + \frac{m_{RD}}{c}\right)^{m_{SD}+m}} \left[\frac{\pi}{2}\left(1 - \sqrt{\frac{P_s \, c \, sin^2 \frac{\pi}{M}}{m_{SD} \, c + m_{RD} \, Ps + P_s \, c \, sin^2 \frac{\pi}{M}}}\right)\right]^{m_{SD}+m} \right]
\end{aligned}
\tag{16}
$$

where $\Gamma(\cdot)$ is the gamma function.

3.2   Error Analysis of Source-to-Relay-to-Destination Link

The source transmits the signal to the relay which amplifies it and retransmits it to the destination. Therefore, the probability of obtaining a correct detected signal at the destination due to source-to-relay-to-destination link conditioned on the instantaneous SNRs γ_{SR}, γ_{RD} can be expressed as:

$$
P_c(\gamma_{SR}, \gamma_{RD}) = (1 - P_e(\gamma_{SR})) + (1 - P_e(\gamma_{RD}))
\tag{17}
$$

The conditional error probability of the S-R-D link conditioned on the instantaneous SNRs γ_{SR}, γ_{RD} can be obtained by:

$$
P_e(\gamma_{SR}, \gamma_{RD}) = 1 - P_c(\gamma_{SR}, \gamma_{RD}) = P_e(\gamma_{SR}) + P_e(\gamma_{RD}) - P_e(\gamma_{SR})P_e(\gamma_{RD})
\tag{18}
$$

By averaging (18) over the instantaneous SNR γ_{SR}, we obtain:

$$
P_e(\overline{\gamma_{SR}}, \gamma_{RD}) = P_{e,avg}(\overline{\gamma_{SR}}) + \left(1 - P_{e,avg}(\overline{\gamma_{SR}})\right) P_e(\gamma_{RD})
\tag{19}
$$

where:

$$
P_{e,avg}(\overline{\gamma_{SR}}) = E[P_e(\gamma_{SR})] = \int_0^\infty P_e(\gamma_{SR})f_{\gamma_{SR}}(\gamma_{SR})d\gamma_{SR}
\tag{20}
$$

Substituting (11) and (12) into (20) and after some mathematical manipulations and using [35, 3.381 (4)] $P_{e,avg}(\overline{\gamma_{SR}})$ becomes:

$$
P_{e,avg}(\overline{\gamma_{SR}}) = \frac{1}{\pi}\left[\frac{\pi}{2}\left(1 - \sqrt{\frac{P_s sin^2 \frac{\pi}{M}}{m_{SR} + P_s sin^2 \frac{\pi}{M}}}\right)\right]^{m_{SR}}
\tag{21}
$$

By averaging the conditional error probability of the S-R-D link conditioned on the instantaneous SNRs γ_{SR}, γ_{RD} which is given in (19) over the statistics of γ_{RD}, under the condition that the instantaneous SNR of the R-D link γ_{RD} is higher than the instantaneous SNR of the S-D link γ_{SD}, The average error probability of the source-to-relay-to-destination link $P_{e(\gamma_{SD} < \gamma_{RD})}$ can be obtained:

$$P_e(\gamma_{SD} < \gamma_{RD}) = \int_{v=0}^{\infty} \int_{u=0}^{v} P_e^{\cdot}(\overline{\gamma_{SR}}, v) f_{\gamma_{SD}}(u) f_{\gamma_{RD}}(v) \, du \, dv \tag{22}$$

Substituting (19) and (14) into (22) and after some mathematical manipulations and using [35, 8.352 (6)], $P_{e(\gamma_{SD} < \gamma_{RD})}$ can be expressed as:

$$P_e(\gamma_{SD} < \gamma_{RD}) = \frac{(m_{SD} - 1)!}{\Gamma(m_{SD})\Gamma(m_{RD})} \left(\frac{m_{RD}}{c}\right)^{m_{RD}} \left[\frac{P_{e,avg}(\overline{\gamma_{SR}})\Gamma(m_{RD})}{\left(\frac{m_{RD}}{c}\right)^{m_{RD}}} + \frac{(1 - P_{e,avg}(\overline{\gamma_{SR}}))\Gamma(m_{RD})}{\pi\left(\frac{m_{RD}}{c}\right)^{m_{RD}}} \right.$$

$$\times \left[\frac{\pi}{2}\left(1 - \sqrt{\frac{c \sin^2 \frac{\pi}{M}}{m_{RD} + c \sin^2 \frac{\pi}{M}}}\right) \right]^{m_{RD}}$$

$$-P_{e,avg}(\overline{\gamma_{SR}}) \sum_{n=0}^{m_{SD}-1} \frac{\left(\frac{m_{SD}}{P_s}\right)^n}{n!} \frac{\Gamma(m_{RD} + n)}{\left(\frac{m_{SD}}{P_s} + \frac{m_{RD}}{c}\right)^{m_{RD} + n}}$$

$$-\frac{(1 - P_{e,avg}(\overline{\gamma_{SR}}))}{\pi} \sum_{n=0}^{m_{SD}-1} \frac{\left(\frac{m_{SD}}{P_s}\right)^n}{n!} \frac{\Gamma(m_{RD} + n)}{\left(\frac{m_{SD}}{P_s} + \frac{m_{RD}}{c}\right)^{m_{RD} + n}}$$

$$\times \left[\frac{\pi}{2}\left(1 - \sqrt{\frac{P_s \, c \sin^2 \frac{\pi}{M}}{m_{SD} \, c + m_{RD} P_s + P_s \, c \sin^2 \frac{\pi}{M}}}\right) \right]^{m_{RD} + n} \right] \tag{23}$$

3.3   End-to-End BER

Previously, we mentioned that the end-to-end bit error rate is the sum of all error probabilities due to the usage of the source-to-destination link and the source-to-relay-to-destination link. Hence, we used the bit error probability expressions of the S-D and S-R-D links which is computed in the previous subsections.

By substituting (16) and (23) in (10), the exact end-to-end BER can be expressed as:

$$P_e = \frac{(m_{RD} - 1)!}{\pi \, \Gamma(m_{SD})\Gamma(m_{RD})} \left(\frac{m_{SD}}{P_s}\right)^{m_{SD}} \times \left[\frac{\Gamma(m_{SD})}{\left(\frac{m_{SD}}{P_s}\right)^{m_{SD}}} \left[\frac{\pi}{2}\left(1 - \sqrt{\frac{P_s \, sin^2 \frac{\pi}{M}}{m_{SD} + P_s \, sin^2 \frac{\pi}{M}}}\right)\right]^{m_{SD}}\right.$$

$$\left. - \sum_{m=0}^{m_{RD}-1} \frac{\left(\frac{m_{RD}}{c}\right)^m}{m!} \frac{\Gamma(m_{SD} + m)}{\left(\frac{m_{SD}}{P_s} + \frac{m_{RD}}{c}\right)^{m_{SD}+m}} \left[\frac{\pi}{2}\left(1 - \sqrt{\frac{P_s \, c \, sin^2 \frac{\pi}{M}}{m_{SD} \, c \, + m_{RD} \, Ps + P_s \, c \, sin^2 \frac{\pi}{M}}}\right)\right]^{m_{SD}+m}\right]$$

$$+ \frac{(m_{SD} - 1)!}{\Gamma(m_{SD})\Gamma(m_{RD})} \left(\frac{m_{RD}}{c}\right)^{m_{RD}} \left[\frac{P_{e,avg}(\overline{\gamma_{SR}})\Gamma(m_{RD})}{\left(\frac{m_{RD}}{c}\right)^{m_{RD}}} + \frac{\left(1 - P_{e,avg}(\overline{\gamma_{SR}})\right)\Gamma(m_{RD})}{\pi\left(\frac{m_{RD}}{c}\right)^{m_{RD}}}\right]$$

$$\times \left[\frac{\pi}{2}\left(1 - \sqrt{\frac{c \, sin^2 \frac{\pi}{M}}{m_{RD} + c \, sin^2 \frac{\pi}{M}}}\right)\right]^{m_{RD}} - P_{e,avg}(\overline{\gamma_{SR}}) \sum_{n=0}^{m_{SD}-1} \frac{\left(\frac{m_{SD}}{P_s}\right)^n}{n!} \frac{\Gamma(m_{RD} + n)}{\left(\frac{m_{SD}}{P_s} + \frac{m_{RD}}{c}\right)^{m_{RD}+n}}$$

$$- \frac{\left(1 - P_{e,avg}(\overline{\gamma_{SR}})\right)}{\pi} \sum_{n=0}^{m_{SD}-1} \frac{\left(\frac{m_{SD}}{P_s}\right)^n}{n!} \frac{\Gamma(m_{RD} + n)}{\left(\frac{m_{SD}}{P_s} + \frac{m_{RD}}{c}\right)^{m_{RD}+n}}$$

$$\times \left[\frac{\pi}{2}\left(1 - \sqrt{\frac{P_s \, c \, sin^2 \frac{\pi}{M}}{m_{SD} \, c + m_{RD} \, Ps + P_s \, c \, sin^2 \frac{\pi}{M}}}\right)\right]^{m_{RD}+n}$$

$$\tag{24}$$

4 Numerical Results

To illustrate the efficiency of our proposed model, numerical results of the bit error rate performance P_e are shown. The impact of changing several parameters is investigated such as the Nakagami-m parameters m_{SR}, m_{RD} and m_{SD} of S-R, R-D, and S-D links, respectively. Additionally, the impact of modulation index of MPSK is considered like BPSK ($M = 2$), QPSK ($M = 4$) and 8-PSK ($M = 8$).

Figure 2 depicts the performance of BER P_e versus P_s when $m_{SD} = m_{SR} = m_{RD} = 3$ and $P_r = P_s$ with various modulation schemes; BPSK, QPSK and 8-PSK. Comparison of the BER curves shows that BPSK has the better performance of the proposed system than others. Also, QPSK is better than 8-PSK. For example, at $P_e = 10^{-3}$, the performance with BPSK offers a mean source power per symbol P_s of about 3 dB over QPSK and about 8 dB over 8-PSK. Additionally, by increasing the power for the same modulation scheme, the performance will be improved.

Figure 3 shows the performance of BER P_e versus P_s for BPSK and QPSK considering non-identical (S-D) and (R-D) channels. Assuming that $m_{SR} = 2$ and $\frac{P_s}{P_r} = 60$ dB. It is clear that BPSK has the better performance of the proposed system than QPSK.

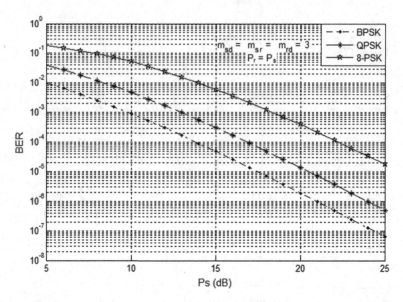

Fig. 2. BER versus the mean source power per symbol P_s with $m_{SD} = m_{SR} = m_{RD} = 3$ and $P_r = P_s$ considering various modulation schemes; BPSK, QPSK and 8-PSK.

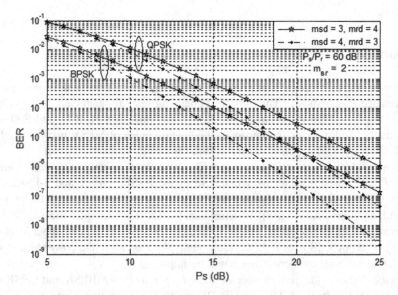

Fig. 3. BER versus the mean source power per symbol P_s for BPSK and QPSK considering non-identical (S-D) and (R-D) channels with $m_{SR} = 2$ and $\frac{P_s}{P_r} = 60$ dB

5 Conclusion

This paper analyzes the performance of a dual-hop differential amplify-and-forward relaying systems considering post detection selection combining techniques. Closed-form expression for the exact end-to-end BER of the proposed model is derived assuming flat Nakagami-m fading channels with M-ary phase shift keying modulation schemes.

References

1. Khalifeh, A.F., AlFasfous, N., Theodory, R., Giha, S., Darabkh, K.A.: An experimental evaluation and prototyping for visible light communication. Comput. Electr. Eng. **72**, 248–265 (2018)
2. Darabkh, K.A., Khalifeh, A., Naser, M., Al-Qaralleh, E.: New arriving process for convolutional codes with adaptive behavior. In: Proceedings of IEEE/SSD 2012 Multi-Conference on Systems, Signals, and Devices, Chemnitz, Germany, pp. 1–6 (2012)
3. Darabkh, K.A., Al-Jdayeh, L.: AEA-FCP: an adaptive energy-aware fixed clustering protocol for data dissemination in wireless sensor networks. Pers. Ubiquit. Comput. (2019). https://doi.org/10.1007/s00779-019-01233-0
4. Al-Zubi, R., Darabkh, K.A., Hawa, M., Jafar, I.F.: RSP-WRAN: resource sharing protocol for inter/intra WRAN communications. High Speed Networks **24**(1), 31–47 (2018)
5. Darabkh, K.A., Abu-Jaradeh, B.: Buffering study over intermediate hops including packet retransmission. In: Proceedings of IEEE International Conference on Multimedia Computing and Information Technology (MCIT-2010), Sharjah, U.A.E, pp. 45–48 (2010)
6. Darabkh, K.A., Abu-Jaradeh, B.: Bounded fano decoders over intermediate hops excluding packet retransmission. In: Proceedings of IEEE 24th International Conference on Advanced Information Networking and Applications (AINA 2010), Perth, Australia, pp. 299–303 (2010)
7. Darabkh, K.A., Pan, W.D.: Queueing simulation for fano decoders with finite buffer capacity. In: Proceedings of the 9th Communications and Networking Simulation Symposium Conference (CNSS 2006), Huntsville, Alabama (2006)
8. Hlayel, M.M., Hayajneh, A.M., Al-Mistarihi, M.F., Shurman, M., Darabkh, K.A.: Closed-form expression of bit error rate in dual-hop dual-branch mixed relaying cooperative networks with best-path selection over rayleigh fading channels. In: Proceedings of the 2014 IEEE International Multi-Conference on Systems, Signals & Devices, Conference on Communication & Signal Processing, 1–4, Castelldefels-Barcelona, Spain (2014)
9. Darabkh, K.A., Ibeid, H., Jafar, I.F., Al-Zubi, R.T.: A generic buffer occupancy expression for stop-and-wait hybrid automatic repeat request protocol over unstable channels. Telecommun. Syst. **63**(2), 205–221 (2016)
10. Khan, I., Rajatheva, N., Tanoli, S.A., Jan, S.: Performance analysis of cooperative network over Nakagami and Rician fading channels. Int. J. Commun. Syst. **27**, 11, 2703–2722 (2014). https://doi.org/10.1002/dac.2500
11. Al-Mistarihi, M.F., Mohaisen, R., Darabkh, K.A.: Closed-form expression for BER in relay-based df cooperative diversity systems over Nakagami-m fading channels with non-identical interferers. In: Galinina, O., Andreev, S., Balandin, S., Koucheryavy, Y. (eds.) NEW2AN/ruSMART 2019. LNCS, vol. 11660, pp. 700–709. Springer, Cham (2019)

12. Al-Mistarihi, M.F., Mohaisen, R., Darabkh, K.A.: On the performance of relay-based decode and forward cooperative diversity systems over rayleigh fading channels with non-identical interferers. IET Commun. (2019). https://doi.org/10.1049/iet-com.2019.0129

13. Al-Mistarihi, M.F., Mohaisen, R., Sharaqa, A, Shurman, M.M., Darabkh, K.A.: Performance evaluation of multiuser diversity in multiuser two-hop cooperative multi-relay wireless networks using maximal ratio combining over Rayleigh fading channels. Int. J. Commun. Syst. **28**(1), 71–90 (2013)

14. Al-Mistarihi, M.F., Mohaisen, R., Darabkh, K.A.: BER analysis in relay-based DF cooperative diversity systems over rayleigh fading channels with non-identical interferers near the destination. In: Proceedings of the 2nd International Conference on Advanced Communication Technologies and Networking (CommNet 2019), Rabat, Morocco, pp. 1–5 (2019)

15. Bai, Z., Dong, P., Gao, S.: An opportunistic relaying-based incremental hybrid DAF cooperative system. Int. J. Commun. Syst. **30**, 3 (2017). https://doi.org/10.1002/dac.2979

16. Al-Zoubi, S., Mohaisen, R., Al-Mistarihi, M.F., Khatalin, S.M., Khodeir, M.A.: On the outage probability of DF relay selection cooperative wireless networks over Nakagami-m fading channels. In: Proceedings of the IEEE 7th International Conference on Information and Communication Systems, ICICS 2016, Jordan University of Science and Technology, Irbid, Jordan (2016)

17. Beaulieu, C.N., Soliman, S.S.: Exact analysis of multihop amplify-and-forward relaying systems over general fading links. IEEE Trans. Commun. **60**(8), 2123–2134 (2012)

18. Seyfi, M., Muhaidat, S., Jie, L.: Amplify-and-forward selection cooperation over rayleigh fading channels with imperfect CSI. IEEE Trans. Wireless Commun. **2**(1), 199–209 (2012)

19. Rui, X., Hou, L., Zhou, L.: Decode-and-forward with full-duplex relaying. Int. J. Commun. Syst. **25**(2), 270–275 (2012)

20. Wang, R., et al.: Joint relay selection and resource allocation in cooperative device-to-device communications. AEU – Int. J. Electron. Commun. **73**, 50–58 (2017). https://doi.org/10.1016/j.aeue.2016.12.023

21. Eng, T., Kong, N., Milstein, L.B.: Comparison of diversity combining techniques for Rayleigh-fading channels. IEEE Trans. Commun. **44**(9), 1117–1129 (1996). https://doi.org/10.1109/26.536918

22. Sendonaris, A., Erkip, E., Aazhang, B.: Increasing uplink capacity via user cooperation diversity. In: Proceedings of IEEE International Symposium on Information Theory, Cambridge, MA, USA (1998). https://doi.org/10.1109/isit.1998.708750

23. Forghani, A.H., Ikki, S.S., Aissa, S.: Novel approach for approximating the performance of multi-hop multi-branch relaying over rayleigh fading channels. In: Proceedings of 2011 IFIP Wireless Days (WD), Niagara Falls, ON (2011)

24. Asghari, V., Maaref, A., Aissa, S.: Symbol error probability analysis for multihop relaying over Nakagami fading channels. In: Proceedings of IEEE Wireless Communications and Networking Conference (WCNC 2010), Sydney, NSW, Australia, pp. 18–21 (2010). https://doi.org/10.1109/WCNC.2010.5506230

25. Selvaraj, M.D., Mallik, R.: Scaled selection combining based cooperative diversity system with decode and forward relaying. IEEE Trans. Vehicular Technol. **59**, 4388–4399 (2010). https://doi.org/10.1109/tvt.2010.2065819

26. Guan, Z.-J., Zhang, W.-J., Zhou, X.-L.: Performance analysis of multi-antenna relay communication systems with MRC. Int. J. Commun. Syst. **25**(11), 1505–1512 (2012). https://doi.org/10.1002/dac.2334

27. Chu, S.I.: Performance of amplify-and-forward cooperative diversity networks with generalized selection combining over Nakagami-m fading channels. IEEE Commun. Lett. **16**(5), 634–637 (2012). https://doi.org/10.1109/lcomm.2012.031212.112443

28. Lei, Y., Cheng, W., Zeng, Z.: Performance analysis of selection combining for amplify-and-forward cooperative diversity networks over Weibull fading channels. In: Proceedings of the IEEE International Conference on Communications Technology and Applications, ICCTA 2009, Beijing, China, pp. 648–651 (2009). https://doi.org/10.1109/iccomta.2009.5349120

29. Darabkh, K.A., Aygun, R.S.: Quality of service evaluation of error control for TCP/IP-based systems in packet switching ATM networks. In: Proceedings of 2006 International Conference on Internet Computing (ICOMP 2006), Las Vegas, Nevada, pp. 243–248 (2006)

30. Yaseen, H.A., Alsalamin, M., Jarwan, A., Al-Mistarihi, M.F., Darabkh, K.A.: A secure energy-aware adaptive watermarking system for wireless image sensor networks. In: Proceedings of the 15th IEEE Multi-conference on Systems, Signals, and Devices (SSD 2018), Hammamet, Tunisia, pp. 12–16 (2018)

31. Darabkh, K.A., Pan, W.D.: Stationary queue-size distribution for variable complexity sequential decoders with large timeout. In: Proceedings of the 44th ACM Southeast Conference, Melbourne, Florida, pp. 331–336 (2006)

32. Darabkh, K.A., El-Yabroudi, M.Z., El-Mousa, A.H.: BPA-CRP: a balanced power-aware clustering and routing protocol for wireless sensor networks. Ad Hoc Netw. **82**, 155–171 (2019)

33. Darabkh, K.A., Aygun, R.S.: Quality of service and performance evaluation of congestion control for multimedia networking. In: Proceedings of 2006 International Conference on Internet Computing (ICOMP 2006), Las Vegas, Nevada, pp. 217–223 (2006)

34. Chen, S., Wang, W., Zhang, X.: Ergodic and outage capacity analysis of cooperative diversity systems under rayleigh fading channels. In: Proceedings of the IEEE International Conference on Communications Workshops, IEEE ICC Workshops 2009, Dresden, Germany (2009). https://doi.org/10.1109/iccw.2009.5208077

35. Zwillinger, D. (ed.): Table of integrals, series, and products, Academic Press, Elsevier, 8th edn. (2014). ISBN-10: 0123849330

Closed-Form Expression for BER in Relay-Based DF Cooperative Diversity Systems Over Nakagami-m Fading Channels with Non-identical Interferers

Mamoun F. Al-Mistarihi[1(⊠)], Rami Mohaisen[1],
and Khalid A. Darabkh[2]

[1] Electrical Engineering Department, Jordan University of Science
and Technology, Irbid, Jordan
mistarihi@just.edu.jo, ramifayezl@hotmail.com
[2] Computer Engineering Department, The University of Jordan, Amman, Jordan
k.darabkeh@ju.edu.jo

Abstract. The deficiencies of regular cooperative relaying schemes were the main reason behind the development of Incremental Relaying (IR). Fixed relaying is one of the regular cooperative relaying schemes and it relies on using the relay node to help in transmitting the signal of the source towards the destination despite the channel's condition. However, adaptive relaying methods allocate the channel resources efficiently; thus, such methods have drawn the attention of researchers in recent years. In this study, we analyze a two-hop Decode-and-Forward (DF) IR system's performance via Nakagami-m fading channels with the existence of the several L distinguishable interferers placed close to the destination which diminishes the overall performance of the system due to the co-channel interference. Tight formula for the Bit Error Rate (BER) is drawn. The assumptions are consolidated by numerical calculations.

Keywords: Nakagami-m channel · Incremental relaying ·
Decode-and-forward · Signal-to-interference-plus-noise ratio

1 Introduction

New generations of wireless communications are required to hold up several multimedia functions; thus, they need to provide high data rate transmissions [1–3]. This type of wireless communications arises various demands on the planning of link budget and the designing of systems [4, 5]. Obviously, providing high data rates with wide coverage area can't be reached only by direct transmissions [6–10]. Therefore, splitting the source-destination's distance into several hops (i.e. multi-hop relaying) can be considered as a good solution to the coverage problem. This solution is considered because of the non-linear connection between the propagation loss and the distance, which reduces the end-to-end attenuation and relaxes the link budget. Recently, traditional relaying is now applied in wireless and mobile networks, instead of microwave links and satellite relays [1].

© Springer Nature Switzerland AG 2019
O. Galinina et al. (Eds.): NEW2AN 2019/ruSMART 2019, LNCS 11660, pp. 700–709, 2019.
https://doi.org/10.1007/978-3-030-30859-9_61

Interestingly, distributed spatial diversity is considered to achieve high data rates and overcome the impairments of wireless channels. Therefore, cooperative diversity methods are used to support spatial diversity, enlarge the covered area and lengthen the life of the battery. These methods are implemented by delivering the source's signal to the destination depending on intermediate nodes [11]. Generally, using a relay, the pattern of transmission has two phases. First, the source transmits a packet of information through the source-relay and source-destination links, i.e., achieving direct link to the destination. Second, the source and the relay will separately transfer the information packet towards the destination using a pre- described multiple-access method.

In cooperative diversity [12], the creation of a virtual antenna array is done by employing the antennas of the distributed nodes in the network. This is very helpful when it comes to combating multipath fading, especially when the nodes can't uphold several antennas due to size or any other limitation. The propagation of error in the relay-destination link is a major part of the disadvantages of cooperative protocols with digital relaying. In order to overcome the propagation of error, the destination should know the source-relay's Channel State Information (CSI) [12], or the selection relaying schemes should be employed, i.e., the relay would adaptively forward the source's signal based on its quality [13]. The latter solution is considered bandwidth inefficient in half-duplex relaying. In cooperative diversity [12], the creation of a virtual antenna array is done by employing the antennas of the distributed nodes in the network. This is very helpful when it comes to combating multipath fading, especially when the nodes can't uphold several antennas due to size or any other limitation.

For example, authors of [14–18] have explained different cooperative diversity protocols and techniques, such as the Amplify-and-Forward (AF) and the DF relaying methods. However, the work in [18] have examined the performance of AF relaying via Rayleigh fading channels by means of OP and symbol error probability (SEP).

The problem facing the fixed relaying methods (i.e., underutilizing the channel resources) can be mitigated by employing Incremental Relaying (IR) systems, which optimizes the performance by utilizing the relay just when required. For instance, the work in [19], have discussed a one-relay DF-IR system via a Rayleigh fading channel, in which the formulas of the BER and the OP were derived.

In this work we examine the DF-IR system's performance via Nakagami-m fading channel with the existence of several L distinguishable interferers close to the destination, which lead to degrading the system's overall performance by causing co-channel interference. The destination-interferers fading channels are presumed to be the summation of distinguishable and independent Rayleigh random variables.

This paper is structured as follows: Sect. 2 provides the system model. Section 3 illustrates how the protocol's performance is analyzed and the procedure of deriving an expression of the bit error rate bearing in mind that results and discussion are provided in Sect. 4. The work is finally summarized in Sect. 5.

2 System Model

As Fig. 1 illustrates, a channel with independent Nakagami-m coefficients $h_{S,D}$, $h_{R,D}$ and $h_{S,R}$ holds the source-destination, the relay-destination and the source-relay link's communications. A negative feedback is going to be transmitted by the destination towards the relay and the source if the source's signal wasn't received properly in the first time slot. This feedback is going to be request the signal of the source from the relay. Then, in the second time slot, both received signals (i.e., from the relay and the source) are going to be combined by the destination using MRC; in order to improve the reception of the signal. A unity variance Additive White Gaussian Noise (AWGN) is presumed to be at the destination. Only a single antenna is available at each node and there exist several L interferers close to the destination, which cause co-channel interference. The destination-interferers links are presumed to be distinguishable and independent Rayleigh random variables and their sum represent the channel holding these connections. The multiple access technique used in this work is time division multiple access.

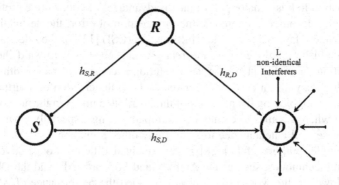

Fig. 1. Description of cooperative diversity network with the consideration of having multiple interferers to the destination node

The source will broadcast its signal within the first time slot. Afterwards, the destination will determine if it needs the relay's help in the second time slot or not. If not (i.e., a successful Direct Transmission (DT)), then a positive feedback will be transmitted through the relay-destination and the source-destination paths requesting another signal from the source in the second time slot.

Defining the above events, the signals received at the destination and the relay in the first time slot are:

$$y_{S,D}(t_1) = h_{S,D}\sqrt{E_s}x(t_1) + n_1(t_1) + I(t_1) \tag{1}$$

$$y_{S,R}(t_1) = h_{S,R}\sqrt{E_s}x(t_1) + n_2(t_1) \tag{2}$$

where $y_{S,D}(t_1)$ and $y_{S,R}(t_1)$ declare the two signals received at the destination and the relay and E_s is the energy of the source signal. $x(t_1)$ is the signal of the source in the first time slot. $n_1(t_1)$, and $n_2(t_1)$ are the AWGN symbols. $I(t_1)$ represents the effect of the interferers close to the destination in the first time slot.

The destination will send a negative feedback to the relay and the source if the DT wasn't successful (i.e., the source-destination path is in outage), this negative feedback will request the help of the relay, which by its turn is going to decode the signal of the source, re-encode it, then transmit it to the destination in the second time slot.

Describing the previous events, the signal received by the destination from the relay in the second time slot is:

$$y_{R,D}(t_2) = h_{R,D}\sqrt{E_s}x_r(t_2) + n_3(t_2) + I(t_2) \tag{3}$$

where $x_r(t_2)$ is the relay's re-encoded signal, $I(t_2)$ is the effect of the interferers close to the destination within the second time slot.

The Signal-to-Noise Ratio (SNR) is widely used in the literature and in different contexts [20–24]. Interestingly, the destination's definition of the source-destination path's outage relies on comparing the path's SNR versus γ_o, which represents the minimum threshold that allows the destination to properly handle the source's signal from the DT.

3 Performance Analysis

In our analysis, we will bypass the noise's effect at the destination; as this effect is not worthy of comparison with the co-channel interference resulted from the interferers close to the destination. In other words, we will use SIR instead of Signal-to-Interference-and-Noise-Ratio (SINR).

3.1 PDF and CDF of SIR's $\eta_{S,D}$ and $\eta_{R,D}$

Provided that the source-destination path's SNR $\gamma_{S,D}$ and the relay-destination path's SNR $\gamma_{R,D}$ are exponentially distributed and the destination-interferers fading channels SNR ($\gamma_b = \sum_{k=1}^{L} \gamma_k$) is presumed to be the summation of distinguishable and independent exponential random variables.

The SIR's ($\eta_{S,D}$) PDF in the source-destination's link is calculated as the following:

$$f_{\eta_{S,D}}(\eta_{S,D}) = \frac{(m)!}{\Gamma(m)} \sum_{k=1}^{L} \pi_k \left[\frac{\lambda_{S,D}\eta_{S,D}^{m-1}}{(\eta_{S,D} + \lambda_{S,D})^{m+1}} \right] \tag{4}$$

where $\lambda_{S,D} = \left(\frac{\gamma_{S,D}}{m\gamma_{k,1}} \right)$, 1 is used to refer to the first time slot, $\pi_k = \prod_{\substack{i=1 \\ i \neq k}}^{L} \frac{\gamma_k}{\gamma_k - \gamma_i}$, L is the number of interferers. $\overline{\gamma_k}$ is the average SNR of each destination interferer's channel.

The SIR's ($\eta_{R,D}$) PDF in the relay-destination's link is calculated as the following:

$$f_{\eta_{R,D}}(\eta_{R,D}) = \frac{(m)!}{\Gamma(m)} \sum_{k=1}^{L} \pi_k \left[\frac{\lambda_{R,D}\eta_{R,D}^{m-1}}{(\eta_{R,D} + \lambda_{R,D})^{m+1}} \right] \tag{5}$$

where $\lambda_{R,D} = \left(\frac{\bar{\gamma}_{R,D}}{m\bar{\gamma}_{k,2}}\right)$, 2 is used to refer to the second time slot.

The SIR's ($\eta_{S,D}$) CDF in the source-destination's link is calculated as the following:

$$F_{\eta_{S,D}}(\eta_{S,D}) = \frac{(m-1)!}{\Gamma(m)} \sum_{k=1}^{L} \pi_k \left(\frac{\eta_{S,D}}{\lambda_{S,D}}\right)^m {}_2^F 1\left(m+1, m; m+1; -\frac{\eta_{S,D}}{\lambda_{S,D}}\right) \quad (6)$$

The SIR's ($\eta_{R,D}$) CDF in the relay-destination's link is provided as the following:

$$F_{\eta_{R,D}}(\eta_{R,D}) = \frac{(m-1)!}{\Gamma(m)} \sum_{k=1}^{L} \pi_k \left(\frac{\eta_{R,D}}{\lambda_{R,D}}\right)^m {}_2^F 1\left(m+1, m; m+1; -\frac{\eta_{R,D}}{\lambda_{R,D}}\right) \quad (7)$$

where ${}_2^F 1(a, b; c; z)$ is the Gaussian Hypergeometric function [25].

3.2 Bit Error Rate Analysis

The BER of the incremental relaying system is designated as:

$$p_e = \Pr(\eta_{S,D} \leq \gamma_o) \times P_{div}(e) + \left(1 - \Pr(\eta_{S,D} \leq \gamma_o)\right) \times P_{direct}(e) \quad (8)$$

The above equation contains both of the average unconditional BER cases, obviously, they rely on whether the relay is going to help or not. For instance, the second part of Eq. (8) is used for the BER when the destination relies only on the source-destination path. However, the first part of Eq. (8) is used for the BER of the relay-destination path, in which the destination is going to use MRC to combine both signals.

$P_{div}(e)$ is the average MRC error probability in the combined diversity communication of the source-destination and relay-destination paths. $P_{direct}(e)$ is the destination's error probability when the DT is successful (i.e., only in the source-destination path).

$P_{direct}(e)$ can be expressed as:

$$P_{direct}(e) = \int_0^\infty P_{direct}(e|\eta) f_{\eta_{S,D}}(\eta|\eta_{S,D} > \gamma_o) d\eta_{S,D} \quad (9)$$

where $P_{direct}(e|\eta)$ is the conditional error probability which equals $\alpha Q(\sqrt{\beta \eta_{S,D}})$ with the constellation parameters α and β. For instance, for Binary Phase Shift Keying (BPSK), $\alpha = 1$ and $\beta = 2$, for M-PSK, $\alpha = 1$ and $\beta = 2\sin\left(\frac{\pi}{M}\right)^2$ And for M-QAM, $\alpha = 4$ and $\beta = \frac{3}{(M-1)}$ and $Q(x) = \frac{1}{\sqrt{2\pi}} \int_x^\infty e^{\left(-\frac{y^2}{2}\right)} dy$ is the Gaussian Q- function. $f_{\eta_{S,D}}(\eta|\eta_{S,D} > \gamma_o)$ is the conditional PDF of $\eta_{S,D}$ given that $\eta_{S,D}$ is above the threshold γ_o to indicate that the DT is successful which is given as:

$$f_{\eta_{S,D}}\left(\eta|\eta_{S,D} > \gamma_o\right) = \begin{cases} 0 & ,\eta \leq \gamma_o \\ \rho \sum_{k=1}^{L} \pi_k \lambda_{S,D} \left(\frac{1}{\eta_{S,D}+\lambda_{S,D}}\right)^2 & ,\eta \geq \gamma_o \end{cases} \tag{10}$$

where $\rho = \dfrac{(m)!}{\left[\Gamma(m)-(m-1)!\sum_{k=1}^{L} \pi_k \left(\frac{\gamma_o}{\lambda_{S,D}}\right)^m {}_2F_1\left(m+1,m;m+1;-\frac{\gamma_o}{\lambda_{S,D}}\right)\right]}$.

The approximated expression of $P_{direct}(e)$ can be found by substituting Eq. (10) into Eq. (9) and using the Prony estimation of the Q-function with [25, (3.353.1)], and it is given as:

$$P_{direct}(e) = \alpha\rho \sum_{i=1}^{2} A_i \sum_{k=1}^{L} \pi_k \lambda_{S,D} \sum_{n=0}^{\infty} \frac{(-a_i\beta)^n}{n!}$$

$$\times \left[\frac{e^{-(n+m)\gamma_o}}{m!}\sum_{j=1}^{m} \frac{(j-1)!(-n-m)^{m-j}}{(\gamma_o+\lambda_{S,D})^j} - \frac{(-n-m)^m}{m!}e^{(n+m)\lambda_{S,D}}\right. \tag{11}$$

$$\left. \times E_i\left[-(n+m)(\gamma_o+\lambda_{S,D})\right]\right]$$

where $E_i[x]$ is the exponential integral function [25, (8.211.1)] and the terms A_i and a_i are the Prony estimation parameters.

The expression of the error probability $P_{div}(e)$, when the relay is helping and the destination is using MRC, is given by:

$$P_{div}(e) = p_{SR}(e)p_x(e) + (1 - p_{SR}(e))p_{com}(e) \tag{12}$$

where $p_{SR}(e)$ represents the relay's error probability in the source-relay path and $p_x(e)$ is the destination's error probability when the relay decodes the signal unsuccessfully and it is limited by 0.5 as mentioned in [17]. $p_{com}(e)$ is the destination's error probability with correct decoding by the relay. The term $(1 - p_{SR}(e))$ declares that the decoding at the relay was successful, which means that spatial diversity is achieved at the destination when it uses MRC to combine both received signals.

$p_{SR}(e)$ is expressed as [1]:

$$p_{SR}(e) = \begin{cases} \frac{1}{2}\left[1 - \mu\left(\frac{\alpha^2\overline{\gamma_{S,R}}}{2m}\right)\sum_{k=0}^{m-1} \binom{2k}{k}\left(\frac{1-\mu^2\left(\frac{\alpha^2\overline{\gamma_{S,R}}}{2m}\right)}{4}\right)^k\right] & m \text{ is integer} \\[4mm] \frac{1}{2\sqrt{\pi}} \frac{\sqrt{\frac{\alpha^2\overline{\gamma_{S,R}}}{2m}}}{\left(1+\frac{\alpha^2\overline{\gamma_{S,R}}}{2m}\right)^{m+(1/2)}} \frac{\Gamma(m+\frac{1}{2})}{\Gamma(m+1)} \times {}_2F_1\left(1,m+\frac{1}{2};m+1;\left(\frac{m}{m+\frac{\alpha^2\overline{\gamma_{S,R}}}{2}}\right)\right) & m \text{ is noninteger} \end{cases}$$

$$\tag{13}$$

where $\mu\left(\frac{\alpha^2\overline{\gamma_{S,R}}}{2m}\right) \triangleq \sqrt{\frac{\alpha^2\overline{\gamma_{S,R}}/2}{m+\alpha^2\overline{\gamma_{S,R}}/2}}$, ${}_2F_1(a,b;c;z)$ is the Gaussian Hypergeometric function [25] and $\Gamma(x)$ is the Gamma function [25], $\overline{\gamma_{S,R}}$ is the relay's average SNR in the source-relay link.

$p_{com}(e)$ is calculated by:

$$p_{com}(e) = \alpha \int_0^\infty f_x(x|\eta_{S,D} \leq \gamma_0) Q\left(\sqrt{\beta x}\right) dx \qquad (14)$$

where $x = \eta_{S,D} + \eta_{R,D}$. Since $\eta_{S,D}$ is not comparable with $\eta_{R,D}$ at the MRC; thus, we will create a violation to work simply on $\eta_{R,D}$ and therefore, $p_{com}(e)$ and after some mathematical manipulations will be given as:

$$p_{com}(e) \approx \frac{\alpha m!}{\sqrt{2\pi}\Gamma(1+m)\Gamma(m)} \left(\frac{2}{\beta}\right)^m \sum_{k=1}^L \pi_k(\lambda_{R,D})^{-m} G_{3,2}^{1,3}\left(\frac{2}{\beta\lambda_{R,D}} \Big| \begin{array}{c} -m, 1-m, \frac{1}{2}-m \\ 0, -m \end{array}\right)$$

$$(15)$$

where $G_{p,q}^{m,n}\left(z\Big|\begin{array}{c}a\\b\end{array}\right)$ is the Meijer G function [26].

Equations (15), (13) and (11) can be substituted into Eq. (8) to get the expression of the BER. However, more details about obtaining Eqs. (4), (5), (6), (7), (11), and (15) will be part of a journal article.

4 Numerical Results

In this section, the DF-IR system's performance via Nakagami-m fading channel with distinguishable interferers close to the destination is analyzed. The impact of varying L along with γ_o and interferers γ's on the system performance is elaborated by showing which parameter has more effect on the performance.

Figure 2 describes the influences of increasing L on the BER of the system for BPSK utilizing m = 3, γ_o = 10 dB, γ_1 = 7 dB, γ_2 = 10 dB, and γ_3 = 13 dB. From this

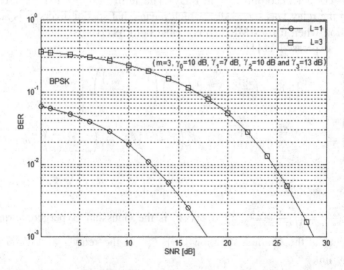

Fig. 2. BER for BPSK modulation for various numbers of interferers (L = 1, 3), m = 3, γ_o = 10 dB, γ_1 = 7 dB, γ_2 = 10 dB, and γ_3 = 13 dB

figure, as L decreases, the system's BER is going to be severely enhanced in reflection to degrading the co-channel interference. To illustrate this, we say, for a 10^{-2} BER, a system with (L = 3) is worse than a system with (L = 1) by (17 dB).

Figure 3 shows the BER versus SNR of various m values with three interferers and $\gamma_o = 10$ dB, $\gamma_1 = 7$ dB, $\gamma_2 = 10$ dB, and $\gamma_3 = 13$ dB. As seen in the figure, if the fading *severity parameter* (m) values increase, the system's BER becomes better. To illustrate this, we say, for a 10^{-2} BER, a system with (m = 4) surpasses another one with (m = 2) by (2 dB).

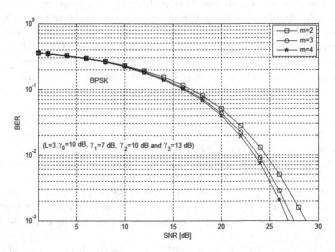

Fig. 3. BER for various m values and, (L = 3, $\gamma_o = 10$ dB, $\gamma_1 = 7$ dB, $\gamma_2 = 10$ dB, and $\gamma_3 = 13$ dB)

5 Conclusion

The DF-IR system's performance via Nakagami-m fading channel with existence of distinguishable interferers close to the destination was analyzed. The analysis clarified the effects of both the co-channel interference and the alternation of the threshold γ_o on the system's BER. Spatial diversity is granted in IR systems and the channel's bandwidth is used efficiently. A derivation of the BER formula was provided in this work. Results have shown that if the Nakagami-m's order decreases and interferes number increases, then the system's BER will get diminished. Additionally, there is a great effect of γ_o and the interferers γ's on the system's performance.

References

1. Hasna, M.O., Alouini, M.-S.: End-to-end performance of transmission systems with relays over Rayleigh-fading channels. IEEE Trans. Wireless Commun. **2**(6), 1126–1131 (2003)
2. Darabkh, K.A., El-Yabroudi, M.Z., El-Mousa, A.H.: BPA-CRP: a balanced power-aware clustering and routing protocol for wireless sensor networks. Ad Hoc Netw. **82**, 155–171 (2019)

3. Darabkh, K.A., Ibeid, H., Jafar, I.F., Al-Zubi, R.T.: A generic buffer occupancy expression for stop-and-wait hybrid automatic repeat request protocol over unstable channels. Telecommun. Syst. **63**(2), 205–221 (2016)

4. Darabkh, K.A.: Fast and upper bounded Fano decoding algorithm: queuing analysis. Trans. Emerg. Telecommun. Technol. **28**(1), 1–12 (2017)

5. Darabkh, K.A.: Evaluation of channel adaptive access point system with Fano decoding. Int. J. Comput. Math. **88**(5), 916–937 (2011)

6. Darabkh, K.A., Khalifeh, A., Naser, M., Al-Qaralleh, E.: New arriving process for convolutional codes with adaptive behavior. In: Proceedings of IEEE/SSD 2012 Multi-conference on Systems, Signals, and Devices, Chemnitz, Germany, pp. 1–6 (2012)

7. Darabkh, K.A., Abu-Jaradeh, B.: Buffering study over intermediate hops including packet retransmission. In: Proceedings of IEEE International Conference on Multimedia Computing and Information Technology (MCIT-2010), Sharjah, U.A.E, pp. 45–48 (2010)

8. Darabkh, K.A., Abu-Jaradeh, B.: Bounded Fano decoders over intermediate hops excluding packet retransmission. In: Proceedings of IEEE 24th International Conference on Advanced Information Networking and Applications (AINA 2010), Perth, Australia, pp. 299–303 (2010)

9. Darabkh, K.A., Pan, W.D.: Queueing simulation for Fano decoders with finite buffer capacity. In: Proceedings of the 9th Communications and Networking Simulation Symposium Conference (CNSS 2006), Huntsville, Alabama (2006)

10. Hasna, M.O., Alouini, M.-S.: Outage probability of multihop transmission over Nakagami fading channels. IEEE Commun. Lett. **7**(5), 216–218 (2003)

11. Fitzek, F.H.P., Katz, M.D.: Cooperation in Wireless Networks: Principles and Applications. Springer, Cham (2006). https://doi.org/10.1007/1-4020-4711-8. ISBN 978-1-4020-4711-4

12. Simon, M.K., Alouini, M.-S.: Digital Communication Over Fading Channels. Wiley, New York (2000)

13. Laneman, J.N., Tse, D.N.C., Wornell, G.W.L.: Cooperative diversity in wireless networks: efficient protocols and outage behavior. IEEE Trans. Inf. Theory **50**, 3062–3080 (2004)

14. Al-Mistarihi, M.F., Aqel, A., Darabkh, K.A.: BER analysis in dual hop differential amplify-and-forward relaying systems with selection combining using M-ary phase-shift keying over Nakagami-m fading channels. In: Galinina, O., et al., (eds.) NEW2AN 2019/ruSMART 2019, LNCS, vol. 11660, pp. 688–699. Springer, Cham (2019)

15. Al-Mistarihi, M.F., Mohaisen, R., Sharaqa, A., Shurman, M.M., Darabkh, K.A.: Performance evaluation of multiuser diversity in multiuser two-hop cooperative multi-relay wireless networks using maximal ratio combining over Rayleigh fading channels. Int. J. Commun Syst **28**(1), 71–90 (2015). https://doi.org/10.1002/dac.2640

16. Al-Mistarihi, M.F., Mohaisen, R., Darabkh, K.A.: BER analysis in relay-based DF cooperative diversity systems over Rayleigh fading channels with non-identical interferers near the destination. In: Proceedings of the 2nd International Conference on Advanced Communication Technologies and Networking (CommNet 2019), Rabat, Morocco, pp. 1–5 (2019)

17. Hlayel, M.M., Hayajneh, A.M., Al-Mistarihi, M.F., Shurman, M., Darabkh, K.A.: Closed-form expression of bit error rate in dual-hop dual-branch mixed relaying cooperative networks with best-path selection over Rayleigh fading channels. In: Proceedings of the 2014 IEEE International Multi-conference on Systems, Signals & Devices, Conference on Communication & Signal Processing, Castelldefels-Barcelona, Spain, pp. 1–4 (2014)

18. Ikki, S., Ahmed, M.H.: Performance analysis of cooperative diversity wireless networks over Nakagami-m fading channel. IEEE Commun. Lett. **11**(4), 334–336 (2007)

19. Ikki, S., Ahmad, M.H.: Performance analysis of decode-and-forward incremental relaying cooperative-diversity networks over Rayleigh fading channels. In: Proceedings IEEE Technology Conference-Spring (VTC-Spring 2009), Barcelona, Spain (2009)
20. Yaseen, H.A., Alsalamin, M., Jarwan, A., Al-Mistarihi, M.F., Darabkh, K.A.: A secure energy-aware adaptive watermarking system for wireless image sensor networks. In: Proceedings of the 15th IEEE Multi-conference on Systems, Signals, and Devices (SSD 2018), Hammamet, Tunisia, pp. 12–16 (2018)
21. Darabkh, K.A., Pan, W.D.: Stationary queue-size distribution for variable complexity sequential decoders with large timeout. In: Proceedings of the 44th ACM Southeast Conference, Melbourne, Florida, pp. 331–336 (2006)
22. Darabkh, K.A., Aygun, R.S.: Quality of service and performance evaluation of congestion control for multimedia networking. In: Proceedings of 2006 International Conference on Internet Computing (ICOMP 2006), Las Vegas, Nevada, pp. 217–223 (2006)
23. Darabkh, K.A., Aygun, R.S.: Quality of service evaluation of error control for TCP/IP-based systems in packet switching ATM networks. In: Proceedings of 2006 International Conference on Internet Computing (ICOMP 2006), Las Vegas, Nevada, pp. 243–248 (2006)
24. Darabkh, K.A., Al-Jdayeh, L.: AEA-FCP: an adaptive energy-aware fixed clustering protocol for data dissemination in wireless sensor networks. Pers. Ubiquit. Comput. (2019). https://doi.org/10.1007/s00779-019-01233-0
25. Gradshteyn, I.S., Ryzhik, I.M.: Table of Integrals, Series, and Products, 7th edn. Academic Press, Cambridge (2007)
26. Miejer's G-Function, The wolfram functions site. http://functions.wolfram.com/07.34.21.0011.01

A New Scheme for Transmitting Heterodyne Signals Based on a Fiber-Optical Transmission System for Receiving Antenna Devices of Radar Stations and Communication Systems

Angelina V. Moroz[1]([⊠]), Roman V. Davydov[1],
and Vadim V. Davydov[1,2,3]

[1] Peter the Great Saint Petersburg Polytechnic University,
St. Petersburg 195251, Russia
`moroz.com3844@gmail.com`
[2] The Bonch-Bruevich Saint - Petersburg State University
of Telecommunications, Saint Petersburg 193232, Russia
[3] Department of Ecology, All-Russian Research Institute of Phytopathology,
B.Vyazyomy, Odintsovo District, Moscow Region 143050, Russia

Abstract. The article discusses the problems arising from the modernization of active phased antenna arrays (for example, an increase in the number of transceiver active elements, a decrease in the weight and size of the antenna system, etc.). It has been substantiated that the most rational solution of these problems is the use of fiber-optical communication systems for transmitting heterodyne signals. A new design of the transmit-receive module with a fiber-optic transmission system for an active phased antenna array has been developed. The results of experimental investigations are presented.

Keywords: Radar · Active phased array antenna · Interference · Fiber-optical transmission system

1 Introduction

The principle of receiving signals and their subsequent primary processing both in radar stations (radar) and in communication systems located on mobile objects do not differ significantly [1–7]. For their more convenient processing in most cases, superheterodyne radios are used. The operation of superheterodyne receivers is based on the principle of converting a received signal into a fixed signal of an intermediate frequency with subsequent amplification. The main advantage of a superheterodyne receiver is the absence of the need to re-adjust it to different frequencies [6, 8–13].

Operation of radar and communication systems, especially on an aircraft, is carried out in conditions of high density of interference of various kinds. Therefore, communication channels and controls are protected by special screens (passive and active) [1, 5, 7, 8, 13–18]. The degree of protection of communication channels and control is provided depending on the functional tasks that must be solved using various signals

© Springer Nature Switzerland AG 2019
O. Galinina et al. (Eds.): NEW2AN 2019/ruSMART 2019, LNCS 11660, pp. 710–718, 2019.
https://doi.org/10.1007/978-3-030-30859-9_62

[9, 13, 16]. The use of combined methods of protection of communication channels does not significantly reduce the weight and dimensions of coaxial cables.

It should also be noted that very high demands are made on the local oscillators for the stability of the frequency and amplitude, as well as the spectral purity of the harmonic oscillations. In the transmission channel, these parameters of the heterodyne signal in different operating conditions of the radar (temperature differences, the presence of interference of various kinds, etc.) should not be changed [1, 5, 8, 14, 19]. Therefore, each communication channel or control uses its own superheterodyne receiver with its own screening and protection systems. This significantly increases the cost of the transmission system, especially with a large number of channels (for example, 256 channels are used in the active phased antenna array (APAA)), as well as its size and weight.

Especially there are many problems with the preservation of the parameters of the heterodyne signal in APAA, located on the flying object, operating in the S and X bands. In these designs APAA very high density of electronic elements and feeder paths. In this case, in addition to external interference, interference related to interference from adjacent channels, etc., begins to play a significant role. The most sensitive to interference in APAA and aircraft communication systems are the feeder paths of superheterodyne receivers. Their "vulnerable" spots are covered with absorbing materials to reduce the effects of induced interference. Experience operating APAA and communication systems shows that this is a temporary solution to the problem. Absorbent coatings for various reasons fail, lose their integrity during repairs, etc.

Currently, objects that need to be detected using radar are manufactured using new absorbing coatings. Therefore, the modernization of the APAA in operation is required. In addition, the flow of information between the aircraft and ground stations, as well as satellites, etc., is constantly increasing. For effective information protection in a complex signal environment (MTR), it is advisable for each flow to allocate its receiving channel [4, 6, 7, 14, 19, 20]. With an increase in communication channels from 8 to 16, the problem of placing additional equipment on an aircraft can be solved. When upgrading APAA, it is necessary to increase the number of transceiver elements in its design to 512 or 1024, while maintaining the weight and size characteristics of the antenna. There are problems with this. The size of transceiver elements can be reduced by retaining their characteristics. To do this, use new materials in their manufacture. But finding additional space in the design of an APAA to accommodate additional feeder paths, especially superheterodyne receivers, is an extremely difficult task.

One of the solutions to this problem is considered in our article. We propose to replace the AFAR feeder paths transmitting microwave signals to fiber-optical transmission systems (FOTS), which are successfully used to transmit microwave signals in various radars [12, 13, 15, 17, 19–26]. Optical fiber is not affected by high-frequency interference in the SSO and has low noise characteristics when transmitting signals, has high flexibility and low mass.

In the case of the high tracts work efficiency a developed by us on the basis of FOTS in the radar, the boundaries of its application can be extended. This development can used in various communication systems, navigation, radio monitoring, etc. [2–5, 10, 11, 15–17]. In these systems, superheterodyne receivers are used to transmit signals from receiving antennas.

2   Features of the Transmission of Heterodyne Signals on a Fiber-Optic System and Its Design

In the case of heterodyne signals for transmission in FOTS radar APAA necessary in its design to take into account a number of features and calculate the following parameters: phase noise, the signal/noise transmission path temporal stability and temperature stability.

The values of the frequencies of heterodyne signals are small (no more than 100 MHz) compared with the frequency of the radar emission (more than 8 GHz). Therefore, nonlinear effects in an optical fiber and dispersion losses can be neglected. This is one of the features of the transmission of these signals to APAA.

In addition, another distinctive feature of the transmission of heterodyne signals is the small length of the optical fiber L (not more than 15 m). This circumstance also makes it possible to neglect the previously noted nonlinear effects. It should be noted that at such distances L the requirements for the optical transmitting module and photodetector are not high in terms of input power, conversion loss, speed, conversion coefficient when transmitting these signals at frequencies less than 100 MHz compared to when FOTS transmits microwave signals from 40 GHz frequency [6, 8, 12, 15, 17].

The use of fiber optic lines allowed us to develop a new scheme for transmitting heterodyne signals to receiving modules that are each connected to an APAA receiving and transmitting element. In Fig. 1 shows the structural scheme developed by us.

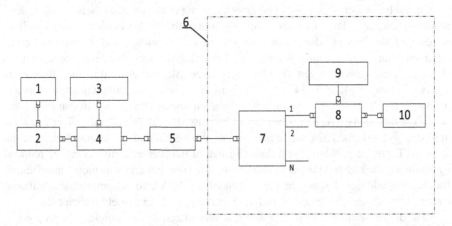

Fig. 1. The structural diagram of a fiber-optic transmission system: 1 - local oscillator power supply; 2 is a local oscillator; 3 - power driver; 4 - transmitting laser module; 5 - optical isolator; 6 - block housing APAA; 7 - optical divider; 8- receiving optical module; 9 - power driver; 10 - receiving and transmitting module APAA.

The proposed design of the FOTS allows you to generate a signal using a single local oscillator. Next, convert its optical modulated signal, which can be divided by an optical divider into a specified number of transceiver modules. The only condition for the implementation of this transmission is the choice of optical transmitting (with a given output power level) and receiving (minimum optical input power) modules that use photodiodes on heterostructures [27, 28]. This is due to the fact that with a large number of N channels (optical divider 1/N) information may be lost.

With this construction of a FOTS with an optical divider, the intrinsic noise factor of a channel on a Kn line becomes an important characteristic. When transmitting heterodyne signals, it limits the number of channels into which the optical signal can be divided. The measurements performed showed that Kn in the channel of the developed FOTS is less than 4 dBm. The intrinsic noise of the feeder path for transmitting the heterodyne signal to APAA is not lower than 6 dBm. This shows that the new design not only significantly increases the free space in the APAA unit, but also improves the characteristics of the transmission channel by more than 30%.

In addition, we found that in the newly developed FOTS design, the dependence of the signal power at the system output on the signal power at its input is linear from – 128 dBm to –22 dBm. This allows us to conclude that the dynamic range (DR) developed by FOTS with an optical divider exceeds 105 dBm. This value of DR is sufficient for transmitting heterodyne signals via the FOTS. Standard DR feeder paths is not more than 110 dBm. One of the main characteristics of the heterodyne communication channel in APAA has not deteriorated.

It should be noted that the channels of the transceiver modules, where the heterodyne signal enters, are optically isolated. This eliminates possible interferences from one another, and also does not require additional screening, as in feeder paths. The latter also frees up space for the introduction of additional elements in the design of APAA.

The use of the developed FOTS in the design of APAA imposes a number of features on the construction of the receiving-transmitting module (RTM) scheme. In Fig. 2 shows the structural diagram of the receiving-transmitting module developed by us.

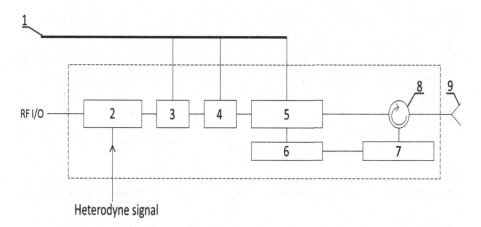

Fig. 2. The structural diagram of the receiving-transmitting module: 1- control bus; 2 - mixer; 3 - phase shifter; 4 - attenuator; 5 - switch; 6 - low noise amplifier; 7 - receiver protection device; 8 - circulator; 9 - receiving and transmitting element APAA.

The main feature of the new scheme is that the phase shifter and attenuator developed by us are located in one receiving and transmitting channel, when in other designs they are spaced apart along separate channels. The second feature is the use of a mixer in RTM. This allows you to combine the channel of reception and transmission in one path, which significantly saves space in the design of APAA and reduces its cost. A heterodyne signal arrives at the mixer input from the output of the FOTS. The mixer also receives a signal from the receiving antenna reflected from the target.

This allows the mixer to generate an intermediate frequency signal, which contains information about the movement of the target. This signal is transmitted to radar processing devices.

3   The Results of Experimental Studies of the Fiber-Optic Transmission System and Their Discussion

The most important characteristic that needs to be investigated when transmitting a heterodyne signal via a FOTS is the dependence of power on frequency. Let us determine the presence of distortions in the spectrum of the signal at the output of the FOTS [8, 9, 15, 17, 29, 30]. On the laboratory model of FOTS developed by us, possible distortions in the spectrum when transmitting a heterodyne signal via FOTS at an operating frequency of 10 MHz were investigated. In Fig. 3, one of the results of these studies is presented as an example.

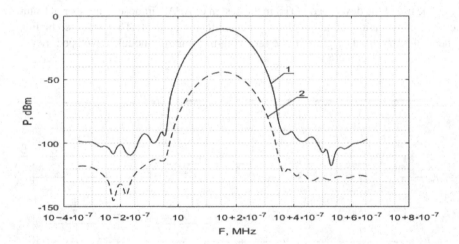

Fig. 3. The spectra of a heterodyne signal at the input (graph 1) and output (graph 2) FOTS.

Comparison of the obtained spectra (Fig. 3) shows the high efficiency of the transmission of the heterodyne signal at a carrier frequency of 10 MHz over FOTS. Distortions in the spectrum are present only on the lateral components, which does not affect the accuracy of determining the distance to the target in the radar.

Since the developed FOTS is intended for APAA, which are located on the aircraft, it will be operated in different temperature conditions. A change in ambient temperature causes both a change in the refractive index of the fiber and additional lengthening of the fiber due to thermal expansion or contraction. This leads to a change in the phase of the light and, accordingly, to a change in the phase of modulation of the radiation transmitted through the fiber. Therefore, an experimental assessment was made of the temperature drift of the modulation phase during propagation in the fiber. In Fig. 4 shows the experimental dependence of the phase shift of the modulation of light $\Delta\varphi m$ on the ambient temperature T.

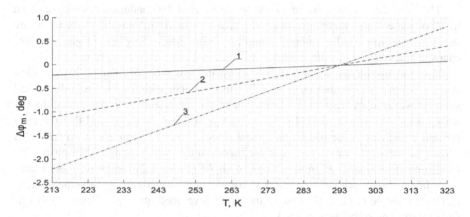

Fig. 4. The phase shift modulation of the medium temperature. Graphs 1, 2 and 3 correspond to the frequency of the heterodyne signal in MHz: 10; 50; 100.

The research results showed that the temperature dependence of the modulation phase change for G.657 fiber is no more than 2° in the selected temperature range from 213 to 323 K for various frequencies of the heterodyne signal.

It is not advisable to investigate the change in the dynamic range of the FOTS from frequency, because the frequency of the heterodyne signal for different systems of APAA does not exceed 120 MHz. In the developed of FOTS, a direct modulated laser module is used to convert the analog microwave signal into an optical signal. The dynamic range of the FOTS in this case will be determined of the characteristics of the transmitting and receiving module, which depend mainly on temperature. Including vibrational amplitude-frequency characteristics of FOTS. Good thermal stabilization means constant dynamic range.

4   Conclusion

The obtained results showed that the FOTS design developed by us can be successfully operated as part of a radar with an APAA. It has been established that the design solutions implemented on the basis of the conducted research in the manufacture of FOTS are justified.

The use of the FOTS design developed in APAA made it possible to implement a new construction of the RTM. This allowed for a more compact placement of a large number of RTMs in APAA compared with the case when feeder paths were used. In addition, the number of used local oscillators in the new design of APAA decreased by an order of magnitude compared with the case of using feeder paths.

Since optical dividers and fiber do not need to be protected from electromagnetic radiation, a part of the channel for transmitting a heterodyne signal with these elements can be placed behind the APAA housing from its rear part (the fiber length of 5 m allows you to do this). In this case, additional space is made available in the APAA housing.

The technical solutions we proposed on the basis of the conducted research allowed us to increase the number of transceiver elements in APAA to 1024 and to ensure reliable operation of the antenna complex while maintaining its weight and size characteristics within specified limits.

It should also be noted that with an increase in the number of channels in the optical divider, Kn in each channel increases. This fact together with a decrease in the optical signal power in each channel is a limitation on the increase in N in the optical divider during the development of FOTS. As a result of research, it was found that the most optimal at present is the use of an optical divider with 10 channels in APAA. This allows the use of a total of 103 local oscillators in the APAA design. Previously, 256 local oscillators were used in the design of an APAA with 256 transceiver elements.

When using feeder paths for aircraft communication systems, 16 superheterodyne receivers must be used. In the case of using the developed fiber optic design, only one superheterodyne receiver is needed.

References

1. Podstrigaev, A.S., Ryazantsev, L.B., Lukashev, V.P.: Technique for tuning microwave strip devices. Meas. Tech. **59**(5), 547–550 (2016)
2. Makolkina, M., Pham, V.D., Kirichek, R., Gogol, A., Koucheryavy, A.: Interaction of AR and IoT applications on the basis of hierarchical cloud services. In: Galinina, O., Andreev, S., Balandin, S., Koucheryavy, Y. (eds.) NEW2AN/ruSMART -2018. LNCS, vol. 11118, pp. 547–559. Springer, Cham (2018). https://doi.org/10.1007/978-3-030-01168-0_49
3. Ateya, A.A., Muthanna, A., Vybornova, A., Darya, P., Koucheryavy, A.: Energy - aware offloading algorithm for multi-level cloud based 5G system. In: Galinina, O., Andreev, S., Balandin, S., Koucheryavy, Y. (eds.) NEW2AN/ruSMART -2018. LNCS, vol. 11118, pp. 355–370. Springer, Cham (2018). https://doi.org/10.1007/978-3-030-01168-0_33
4. Mashkov, G., Borisov, E., Fokin, G.: A positioning accuracy experimental evaluation in SDR-based MLAT with joint processing on range measurement. In: Proceedings - International Conference on Radar, Antenna, Microwave, Electronics, and Telecommuni-cations, ICRAMET 2016 (Jakarta), 7849572, pp. 7–12 (2016)
5. Sivers, M., Fokin, G., Dmitriev, P., Kireev, A., Volgushev, D., Hussein Ali, A.-O.A.: Indoor positioning in WiFi and NanoLOC networks. In: Galinina, O., Balandin, S., Koucheryavy, Y. (eds.) NEW2AN/ruSMART -2016. LNCS, vol. 9870, pp. 465–476. Springer, Cham (2016). https://doi.org/10.1007/978-3-319-46301-8_39

6. Podstrigaev, A.S., Smolyakov, A.V., Davydov, V.V., Myazin, N.S., Slobodyan, M.G.: Features of the development of transceivers for information and communication systems considering the distribution of radar operating frequencies in the frequency range. In: Galinina, O., Andreev, S., Balandin, S., Koucheryavy, Y. (eds.) NEW2AN/ruSMART - 2018. LNCS, vol. 11118, pp. 509–515. Springer, Cham (2018). https://doi.org/10.1007/978-3-030-01168-0_45

7. Davydov, R.V., et al.: Fiber-optic transmission system for the testing of active phased antenna arrays in an anechoic chamber. In: Galinina, O., Andreev, S., Balandin, S., Koucheryavy, Y. (eds.) NEW2AN/ruSMART/NsCC -2017. LNCS, vol. 10531, pp. 177–183. Springer, Cham (2017). https://doi.org/10.1007/978-3-319-67380-6_16

8. Lenets, V.A., Tarasenko, M.Yu., Davydov, V.V., Rodugina, N.S., Moroz, A.V.: New method for testing of antenna phased array in X frequency range. J. Phys: Conf. Ser. 1038(1), 012037 (2018)

9. Lavrov, A.P., Molodyakov, S.A.: A method for measurement of the pulse arrival time of the radio emission of pulsars in a wideband optoelectronic processor. Meas. Tech. 59(10), 1025–1033 (2017)

10. Borodulin, R.U., Sosunov, B.V., Makarov, S.B.: The principles of antennas constructive synthesis in dissipative media. In: Galinina, O., Andreev, S., Balandin, S., Koucheryavy, Y. (eds.) NEW2AN/ruSMART/NsCC -2017. LNCS, vol. 10531, pp. 455–465. Springer, Cham (2017). https://doi.org/10.1007/978-3-319-67380-6_41

11. Koucheryavy, A., Vladyko, A., Kirichek, R.: State of the art and research challenges for public flying ubiquitous sensor networks. In: Balandin, S., Andreev, S., Koucheryavy, Y. (eds.) ruSMART 2015. LNCS, vol. 9247, pp. 299–308. Springer, Cham (2015). https://doi.org/10.1007/978-3-319-23126-6_27

12. Podstrigaev, A.S., Davydov, R.V., Rud, V.Yu., Davydov, V.V.: Features of transmission of intermediate frequency signals over fiber-optical communication system in radar station. In: Galinina, O., Andreev, S., Balandin, S., Koucheryavy, Y. (eds.) NEW2AN/ruSMART - 2018. LNCS, vol. 11118, pp. 624–630. Springer, Cham (2018). https://doi.org/10.1007/978-3-030-01168-0_56

13. Belkin, M.E., Sigov, A.S.: Some trend in super-high frequency optoelectronics. J. Commun. Technol. Electron. 54(8), 655–658 (2009)

14. Fadeenko, V.B., Kuts, V.A., Vasiliev, D.A., Davydov, V.V.: New design of fiber-optic communication line for the transmission of microwave signals in the X-band. J. Phys: Conf. Ser. 1135(1), 012053 (2018)

15. Ivanov, S.I., Lavrov, A.P., Saenko, I.I.: Model of photonic beamformer for microwave phased array antenna. In: Galinina, O., Andreev, S., Balandin, S., Koucheryavy, Y. (eds.) NEW2AN/ruSMART/NsCC -2017. LNCS, vol. 10531, pp. 482–489. Springer, Cham (2017). https://doi.org/10.1007/978-3-319-67380-6_44

16. Tarasenko, M.Yu., Davydov, V.V., Lenets, V.A., Akulich, N.V., Yalunina, T.R.: Features of use direct and external modulation in fiber optical simulators of a false target for testing radar station. In: Galinina, O., Andreev, S., Balandin, S., Koucheryavy, Y. (eds.) NEW2AN/ruSMART/NsCC -2017. LNCS, vol. 10531, pp. 227–232. Springer, Cham (2017). https://doi.org/10.1007/978-3-319-67380-6_21

17. Ivanov, S.I., Lavrov, A.P., Saenko, I.I.: Application of microwave photonics components for ultrawideband antenna array beamforming. In: Galinina, O., Balandin, S., Koucheryavy, Y. (eds.) NEW2AN/ruSMART -2016. LNCS, vol. 9870, pp. 670–679. Springer, Cham (2016). https://doi.org/10.1007/978-3-319-46301-8_58

18. Petrov, A.A., Davydov, V.V., Myazin, N.S., Kaganovskiy, V.E.: Rubidium atomic clock with improved metrological characteristics for satellite communication system. In: Galinina, O., Andreev, S., Balandin, S., Koucheryavy, Y. (eds.) NEW2AN/ruSMART/NsCC -2017. LNCS, vol. 10531, pp. 561–568. Springer, Cham (2017). https://doi.org/10.1007/978-3-319-67380-6_52

19. Filatov, D.L., Galichina, A.A., Vysoczky, M.G., Yalunina, T.R., Davydov, V.V., Rud, V.Yu.: Features of transmission at analog intermediate frequency signals on fiber – optical communication lines in radar station. J. Phys: Conf. Ser. **917**(8), 082005 (2017)

20. Petrov, A.A., Davydov, V.V., Grebenikova, N.M.: Some directions of quantum frequency standard modernization for telecommunication systems. In: Galinina, O., Andreev, S., Balandin, S., Koucheryavy, Y. (eds.) NEW2AN/ruSMART -2018. LNCS, vol. 11118, pp. 641–648. Springer, Cham (2018). https://doi.org/10.1007/978-3-030-01168-0_58

21. Shlyagin, M., Prokofiev, A.V., Pleshakov, I.V., Agruzov, P.M., Nepomnyashchaya, E.K., Velichko, E.N.: Magnetic fluid analysis by optical fiber method. In: Proceedings - International Conference Laser Optics 2018, ICLO 2018, 8435459, pp. 407 (2018)

22. Kiesewetter, D., Malyugin, V., Makarov, S., Korotkov, K., Ming, D., Wei, X.: Application of the optical fibers in the system of determining the distance of jump at ski springboard. In: Proceedings – 2016 Advances in Wireless and Optical Communications, RTUWO 2016, 7821845, pp. 5–8 (2017)

23. Davydov, V.V., et al.: Fiber-optics system for the radar station work control. In: Balandin, S., Andreev, S., Koucheryavy, Y. (eds.) ruSMART 2015. LNCS, vol. 9247, pp. 712–721. Springer, Cham (2015). https://doi.org/10.1007/978-3-319-23126-6_65

24. Grebenikova, N.M., Davydov, V.V., Moroz, A.V., Bylina, M.S., Kuzmin, M.S.: Remote control of the quality and safety of the production of liquid products with using fiber-optic communication lines of the Internet. IOP Conf. Ser.: Mater. Sci. Eng. **497**, 012109 (2019)

25. Davydov, V.V., Ermak, S.V., Karseev, A.U., Nepomnyashchaya, E.K., Petrov, A.A., Velichko, E.N.: Fiber-optic super-high-frequency signal transmission system for sea-based radar station. In: Balandin, S., Andreev, S., Koucheryavy, Y. (eds.) NEW2AN 2014. LNCS, vol. 8638, pp. 694–702. Springer, Cham (2014). https://doi.org/10.1007/978-3-319-10353-2_65

26. Petrov, A.A., Davydov, V.V.: Improvement frequency stability of caesium atomic clock for satellite communication system. In: Balandin, S., Andreev, S., Koucheryavy, Y. (eds.) ruSMART 2015. LNCS, vol. 9247, pp. 739–744. Springer, Cham (2015). https://doi.org/10.1007/978-3-319-23126-6_68

27. Myazin, N.S., Smirnov, K.J., Davydov, V.V., Logunov, S.E.: Spectral characteristics of InP photocathode with a surface grid electrode. J. Phys: Conf. Ser. **929**(1), 012080 (2017)

28. Grebenikova, N.M., Smirnov, K.J., Artemiev, V.V., Davydov, V.V., Kruzhalov, S.V.: The universal optical method for condition control of flowing medium. J. Phys: Conf. Ser. **1038**(1), 012089 (2018)

29. Ateya, A.A., Muthanna, A., Gudkova, I., Abuarqoub, A., Vybornova, A., Koucheryavy, A.: Development of intelligent core network for tactile internet and future smart systems. J. Sens. Actuator Netw. **7**(1), 7 (2018)

30. Ivanov, S.I., Lavrov, A.P., Saenko, I.I., Filatov, D.L.: Chirped fiber grating beamformer for linear phased array antenna. In: Galinina, O., Andreev, S., Balandin, S., Koucheryavy, Y. (eds.) NEW2AN/ruSMART -2018. LNCS, vol. 11118, pp. 594–604. Springer, Cham (2018). https://doi.org/10.1007/978-3-030-01168-0_53

Simulation of Simplex Acousto-Optic Channel on Few-Mode Optical Fiber

Vladimir A. Burdin[(⊠)] and Olga Yu. Gubareva

Povolzhskiy State University of Telecommunications and Informatics,
Samara, Russia
burdin@psati.ru

Abstract. The design of an acousto-optic-fiber simplex communication channel based on a two-mode optical fiber is proposed. The model of this channel has been described and the simulation of data transmission over such channel has been executed. The simulation results are presented. As examples the dependencies of bit error rate from a cumulative differential mode delay in the fiber optic link. They showed that, for optimal reception, the accumulated differential mode delay should lie within 1.1–1.5 rad. Measures are proposed to ensure the required value of a differential mode delay at the receiver input.

Keywords: Distributive acoustic sensor · Few-mode optical fiber ·
Acousto-optic-fiber channel · Differential mode delay · Bit error rate

1 Introduction

Acousto-optic effect is usefully employed for designing of various fiber-optic different purpose devices such as frequency converters, mode convectors, filters and modulating equipment, different physical quantities sensors [1]. In the late 1970s of the past century there was shown the acoustic sound detection capability with the help of fiber-optic sensor [2, 3]. Compared to usual electronic microphone fiber-optic acoustic sensors have broad passband, electromagnetic and radio frequency interference immunity, remote probing capacity, direct connection to optical transmission network and more [4]. There have been executed a large number of studies leading to designing of fiber-optic acoustic sensors and technologies connected with them. They are the sensors on Fabry-Perot interferometers and Mach-Zehnder interferometers, fiber Bragg gratings sensors, tunable laser sensors and so on. However multidrop acoustic detection configurations on such sensors are limited about distributed probing realization capacities.

For the last three decades there have become widely used distributed acoustic sensors (DAS) based on optic fiber and Rayleigh scattering which are defined by small fiber size, electromagnetic immunity, great perimeter catchment and broad passband [5, 6]. The distributed fiber-optic sensors are used everywhere where it is necessary to reveal several cases in long-range different places. Such sensor systems are employed for condition and safety control of extended objects in different areas including transport, oil-and-gas sites, process control systems. DAS market in on the rise and is expected that by 2025 it will have overshot by 2 billion dollars [6]. It should be noted

© Springer Nature Switzerland AG 2019
O. Galinina et al. (Eds.): NEW2AN 2019/ruSMART 2019, LNCS 11660, pp. 719–726, 2019.
https://doi.org/10.1007/978-3-030-30859-9_63

that the realization of the distributed fiber-optic sensors based on reflectometric methods is not cheap and requires quite complex schemes and devices usage.

While for a number of applications localizing the case on distributed object is not needed but all it takes is only eliciting fact of its existence and parameters changing estimation [7, 8]. For this it is enough to make a distributed acousto-optic-fiber communication channel vibro-acoustic transmitter of which can be in an arbitrary point on the fiber length [9–11]. Such distributed acousto-optic-fiber channel can be used for organization of simplex audio communication from explosion danger areas when audio signal transmitter doesn't need to be connected to the definite cable line point. The same approach is also employed for audio information retrieval [12]. As a rule, the ways under consideration as well as reflectometric ones are based on back-scattered and returned optic signals usage [9, 10, 12].

In the study there is provided the way of designing of an acousto-optic-fiber simplex communication channel based on a two-mode optical fiber and there are shown the results of transmitting low-speed signals in audio spectrum in it. The employment of multimode optic fibers for the distributed acousto-optic sensors is also known [13, 14].

However the unknown grades are received by speckle analysis results that require quite complex equipment usage and large-scale computation efforts. The proposed solution is based on that the two-mode optical fiber is in fact a Mach-Zehnder interferometer which is successfully employed for acousto-optic modulating equipment and sensors designing [1, 4, 15, 16].

2   Model of Acousto-Optic Channel on Few-Mode Optical Fiber

The proposed general scheme of a simplex acousto-optical communication channel based on a few-mode optical fiber is shown in Fig. 1.

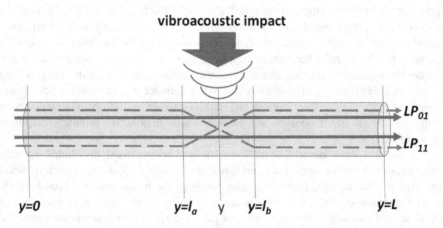

Fig. 1. General scheme of a simplex acousto-optical communication channel based on a few-mode optical fiber.

It is supposed that acousto-optic channel is meant to transmit low-speed signals in audio spectrum. There is proposed to use two-mode optical fiber where launched modes LP_{01} and LP_{11} are sustained. Generically there are propagated six modes in such optical fiber [17–19]. They are orthogonally polarized modes components LP_{01}, LP_{11a}, LP_{11b}. However, as differential modal delay between the orthogonally polarized modes is negligible in comparison with the differential modal delay between the modes LP_{01} and LP_{11}, the orthogonally polarized modes can be believed degenerated. For the same reason modes LP_{11a} and LP_{11b} can also be believed degenerated. That way, in further modelling we will look at propagation in the few-mode optical fiber only of the two modes LP_{01} and LP_{11}.

Each of the modes transmits desired signal and crosstalk caused by the modes coupling. Let us assign to parameters of the mode LP_{01} transmitted by this mode optical signal further identifying it as mode 1. Accordingly, the mode LP_{11} we will assign as mode 2 and its parameters and the optical signal parameters transmitted in it we will assign index 2. The optical signal of the first mode can be written as:

$$s_1 = s_{01} + s_{c1}, \tag{1}$$

where s_{01} – is a desired signal transmitted by mode 1;

s_{c1} – crosstalk transmitted by mode 1.

As the desired signal is transmitted with low speed we will neglect the fiber chromatic dispersion and the optical fiber nonlinearity. Then the going to optical fiber far end signal of mode 1 can be presented as:

$$s_{01} = A_1 \exp^{-\alpha_1 L} \cos(\omega t - \beta_1 L + \delta\beta_1 \cdot \Delta l), \tag{2}$$

where A_1 – is optical emission amplitude of the first mode:

α_1 – specific damping in the first mode optical fiber:

β_1 – phase coefficient in the first mode optical fiber:

$\delta\beta_1$ – the first mode phase coefficient changes in response to acousto-vibrational action on the optical fiber:

$\Delta l = l_b - l_a$ – fiber part length where the optical fiber experiences acousto-vibrational action:

L – optical fiber length:

ω – circular frequency:

t – time.

In regular optical fiber different azimuthal order number modes are not coupled with each other [17–19]. Accordingly for the modes LP_{01} and LP_{11} intermode coupling in regular fiber is equal to zero. However there is no ideal and fiber-optical cabel line has irregularities. They can be divided into internal and joint ones. The joint irregularities occur in a fiber jointings of optical cable delivery lengths. The internal irregularities in their turn are divided into in-fiber and intracable ones. The in-fiber irregularities are conditioned by profile parameters fluctuations of optical fiber refraction index and its core diameter. The intracable ones are mainly caused by excess fiber length in the cable that leads to the fiber bend. We will believe that the crosstalk level is quite low and the reverse transitions can be disregarded. Then for the local irregularity located at the fiber length in the point with the coordinate y and with the part length dy, the crosstalk for the mode 1 can be approximately written as:

$$s_{c1y} = A_2 \cdot \exp^{-\alpha_1 y} \cdot \exp^{-\alpha_2(L-y)} \cdot C(y) \cdot \cos[\omega t - \beta_1 y - \beta_2(L-y) + \delta\varphi_1(y)]dy, \quad (3)$$

$$\delta\varphi_1(y) = \begin{cases} \delta\beta_1\Delta l, & y < l_a \\ \delta\beta_2(y - l_a) + \delta\beta_1(l_b - y), & l_a < y < l_b, \\ \delta\beta_2\Delta l, & y > l_b \end{cases} \quad (4)$$

$$C(y) = \begin{cases} C_{0y}(y), & y < l_a, y > l_b \\ C_{ay}(y), & l_a < y < l_b \end{cases}.$$

Here the coefficient $C(y)$ has regard to the modes connection dependence on the vibro-acoustic action [20–25].

Going forwards we will restrict ourselves with one delivery length analysis of the optical fiber that is on an average 4–6 km long [26]. This allows to exclude form our consideration the joint irregularities and to neglect the loss. We will suppose that the intermodes coupling along the optical fiber length are equally distributed:

$$C(y) = \begin{cases} C_0, & y < l_a, y > l_b \\ C_a, & l_a < y < l_b \end{cases}.$$

Where C_0, C_a – are constant.

Then the far-end crosstalk from all the fiber-optical line irregularities for the mode 1can be defined through integration:

$$s_{c1} = A_2 \cdot \int_0^L C(y) \cos[\omega t - \beta_2 y - \beta_1(L-y) + \delta\varphi_1(y)]dy, \quad (5)$$

From (5) with regard to (4) we get:

$$s_{c1} = \sum_{j=1}^3 s_{c1}^{(j)}, \quad (6)$$

$$s_{c1}^{(1)} = A_2 C_0 \cdot \int_0^{l_a} \cos[\omega t - \Delta\beta y - \beta_1 L + \delta\beta_1\Delta l]dy, \quad (7)$$

$$s_{c1}^{(2)} = A_2 C_0 \cdot \int_{l_b}^{L} \cos[\omega t - \Delta\beta y - \beta_1 L + \delta\beta_2 \Delta l] dy, \qquad (8)$$

$$s_{c1}^{(3)} = A_2 C_a \cdot \int_{l_a}^{l_b} \cos[\omega t - \Delta\beta y - \beta_1 L + (\delta\beta_2 - \delta\beta_1)y + \delta\beta_1 l_b - \delta\beta_2 l_a] dy, \qquad (9)$$

Here $\Delta\beta = \beta_2 - \beta_1$. By integration (7)–(9), we get:

$$s_{c1}^{(1)} = -2\frac{A_2 C_0}{\Delta\beta} \cdot \sin\frac{\Delta\beta L}{2} \sin\left[\omega t - \frac{\Delta\beta L}{2} - \beta_1 L + \delta\beta_1 \Delta l\right], \qquad (10)$$

$$s_{c1}^{(2)} = -2\frac{A_2 C_0}{\Delta\beta} \cdot \sin\frac{\Delta\beta L}{2} \sin\left[\omega t - \frac{\Delta\beta L}{2} - \beta_1 L + \delta\beta_2 \Delta l\right], \qquad (11)$$

$$\begin{aligned} s_{c1}^{(3)} &= \frac{A_2 C_a}{\Delta\beta + \delta\beta_{12}} \cdot \sin[\omega t - \beta_2 L + \delta\beta_{12} L + \delta\beta_1 l_b - \delta\beta_2 l_a] \\ &- \frac{A_2 C_a}{\Delta\beta + \delta\beta_{12}} \cdot \sin(\omega t - \beta_2 L + \delta\beta_1 l_b - \delta\beta_2 l_a) \end{aligned} \qquad (12)$$

Here $\delta\beta_{12} = \delta\beta_2 - \delta\beta_1$.

Far-end optic fiber signal of mode 2 is analogously described by the formulas (2), (6), (10)–(12) where you only need to change the positions the inferior indexes 1 and 2.

If both modes at far end are provided on square detector, we will get signal of the form:

$$s_R = (s_{01} + s_{02})^2 + 2(s_{01} + s_{02})(s_{c1} + s_{c2}) + (s_{c1} + s_{c2})^2, \qquad (13)$$

With uniform mode excitation at the input when A1 = A2 = A, neglecting the crosstalk as it was stated above, setting (2), (6), (10)–(12) with the respect to inferior indexes, filtering high-frequency components and excluding constant component we get the formula for selected by the receiver the low-frequency envelope minding the modes coupling. The formula appears as follows:

$$\begin{aligned} I_{LF} = &\; A^2 \cos(\Delta\beta L - \delta\beta_{12}\Delta l) \\ &+ \frac{A^2 C_a}{\Delta\beta + \delta\beta_{12}} \{\sin(\delta\beta_{12} l_b) \cos(\delta\beta_{12} l_b) + \sin[(\Delta\beta + \delta\beta_{12})L] \cos(\delta\beta_{12} l_a)\} \\ &+ \frac{A^2 C_0^2}{2\Delta\beta^2} \sin^2(\Delta\beta L) \cos\left(\frac{\delta\beta_{12}\Delta l}{2}\right) \\ &+ \frac{A^2 C_a^2}{4(\Delta\beta + \delta\beta_{12})^2} \{\cos(\delta\beta_{12} L) + \cos[(2\Delta\beta - \delta\beta_{12})L + 2\delta\beta_{12}\Delta l]\} \\ &+ \frac{A^2 C_a^2}{4(\Delta\beta + \delta\beta_{12})^2} \{\cos[2\Delta\beta L + 2\delta\beta_{12}\Delta l] + \cos[2(\Delta\beta - \delta\beta_{12})L + 2\delta\beta_{12}\Delta l]\} \end{aligned} \qquad , \quad (14)$$

3 Results of Simulations

For illustrative purposes there has been studied fiber optic transmission line including 1 length of optic cable with two-mode optic fiber. The optic fiber length in the cable is $L = 5000$ m. It was believed that two modes LP_{01} and LP_{11} were generated on the near end of the transmitter line in optic fiber by coherent optical oscillation source (laser). Therewith there were satisfied uniform mode excitation conditions. At fiber-optic transmission line far end both modes are transmitted to square detector, at output of which there is produced low-frequency signal with the help of a filter. There is provided some vibro-acoustic action on the 10 m length cable optical fiber. The distance from the near end to the beginning of the part $l_a = 1000$ m. The differential modal delay was taken equal to $\Delta\beta = 4.5$ ps/m. The average intermode coupling parameter value was taken equal to $C_0 = 10^{-4}$ 1/m. As a result of the vibro-acoustic action the connection between the modes increases and on the part where the action takes place it is taken equal to $C_a = 0.1$ 1/m. The vibro-acoustic action source produces impulses codechain with the period of 1.0 кhz. The maximal difference values of the mode delay changes of the modes under consideration under vibro-acoustic action were chosen in the range from $5\cdot10^{-4}$ to $10\cdot10^{-4}$ rad/m. Low-frequency signal at the square receiver output was produced using the described above model. The examples of the got as a result of simulation the bit error rate (BER) dependencies upon the accumulated differential modal delay latency at the link end for some values of vibro-acoustic action intensities are shown in Fig. 2.

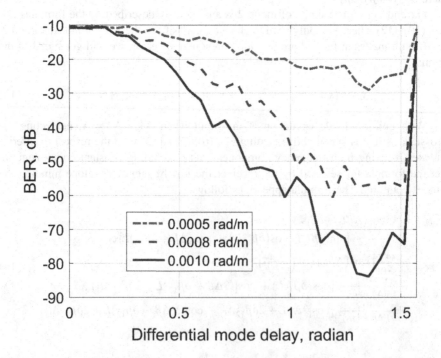

Fig. 2. Dependencies of BER from the cumulative differential mode delay at a far end of fiber.

As the simulation results have shown for the acoustic signal quality reception in the under consideration acousto-optic-fiber channel there is some optimal area of the resulting differential modal delay at the link end. For all the studied vibro-acoustic action intensity values the interval is within 1.1–1.5 rad. In common case this can be provided if to divide the modes at the optical fiber input with mode dividing multi-plexor (MDM) and to let one of the modes before sending at the square detector input through delay line. But more simpler do it by changing the optical fiber length by using stretching or heating a portion of the fiber.

4 Conclusion

There is proposed the design of an acousto-optic-fiber simplex communication channel based on a two-mode optical fiber. The model of this channel has been described and the simulation of data transmission over such channel has been executed. There have been shown the simulation results which showed that for optimal reception, the accumulated differential mode delay should lie within 1.1–1.5 rad. Measures are proposed to ensure the required value of a differential mode delay at the receiver input.

References

1. Pohl, A.A.P., et al.: Advances and new applications using the acousto-optic effect in optical fibers. Photonic Sens. **3**(1), 1–25 (2013)
2. Cole, J.H., Johnson, R.L., Bhuta, P.G.: Fiber-optic detection of sound. J. Acoust. Soc. Am. **62**(5), 1136–1138 (1977)
3. Bucaro, J.A., Hickman, T.R.: Measurement of sensitivity of optical fibers for acoustic detection. Appl. Opt. **18**(6), 938–940 (1979)
4. Teixeira, J.G.V., Leite, I.T., Silva, S., Frazao, O.: Advanced fiber-optic acoustic sensors. Photonic Sens. **4**(5), 198–208 (2014)
5. Wu, Y., Gan, J., Li, Q., Zhang, Z., Heng, X., Yang, Z.: Distributed fiber voice sensor based on phase-sensitive optical time-domain reflectometry. IEEE Photonics J. **7**(6), 6803810 (2015)
6. Muanenda, Y.: Recent advances in distributed acoustic sensing based on phase-sensitive optical time domain reflectometry. Hindawi J. Sens. **23**(3897873), 1–16 (2018)
7. Vazhdaev, K.: Acoustooptic devices and their using in the units and information-measuring systems. Pet. Eng. **10**(1), 148–151 (2012)
8. Bercy, A., et al.: In-line extraction of an ultrastable frequency signal over an optical fiber link. J. Opt. Soc. Am. B **31**(4), 678–685 (2014)
9. Udd, E., Morrell, M.M.: Single fiber Sagnac sensing system. US Patent 6,459,486 (2002)
10. Healey, P., Sicora, E.S.R.: Communicating or reproducing an audible sound. US Patent 8,000,609 (2011)
11. Burdin, V.A., Gubareva, O.Yu.: Simulation of simplex acousto-optical data transmission on fiber optic link. In: Andreev, V., Bourdine, A., Burdin, V., Morozov, O., Sultanov, A. (eds.) SPIE Proceedings Optical Technologies in Telecommunications 2017, 10774, 107740B (2018)
12. Grishachev, V.V.: Detecting threats of acoustic information leakage through fiber optic communications. J. Inf. Secur. **3**, 149–155 (2012)

13. Ha, W., Lee, S., Jung, Y., Kim, J.K., Oh, K.: Acousto-optic control of speckle contrast in multimode fibers with a cylindrical piezoelectric transducer oscillating in the radial direction. Opt. Express **17**(20), 17536–17546 (2009)
14. Gorbachev, O.V.: Optical cables speckle interferometric examination. Photonics **36**(6), 20–24 (2012)
15. Bucaro, J.A., Dardy, H.D., Carome, E.F.: Fiber-optic hydrophone. J. Acoust. Soc. Am. **62**(5), 1302–1304 (1977)
16. Hwang, J.-H., Seon, S., Park, C.-S.: Position estimation of sound source using three optical mach-zehnder acoustic sensor array. Curr. Opt. Photonics **1**(6), 573–578 (2017)
17. Snyder, A.W., Love, J.D.: Optical Waveguide Theory. Radio i svyaz Publisher, Moscow (1987)
18. Black, R.J., Gagnon, L.: Optical Waveguide Modes. The McGraw-Hill Companies Publisher, New York (2010)
19. Kokubun, Y.: Rigorous mode theory and analysis of few-mode fibers. Jpn. J. Appl. Phys. **57**, 08PA05 (2018)
20. Taylor, H.: Bending effects in optical fibers. J. Lightwave Technol. **2**(5), 617–628 (1984)
21. Engan, H.E., Kim, B.Y., Blake, J.N., Shaw, H.J.: Propagation and optical interaction of guided acoustic waves in two-mode optical fibers. J. Lightwave Technol. **6**(3), 428–436 (1988)
22. Zhao, J., Liu, X.: Fiber acousto-optic mode coupling between the higher-order modes with adjacent azimuthal numbers. Opt. Lett. **31**(11), 1609–1611 (2006)
23. Park, H.S., Song, K.Y.: Acousto-optic resonant coupling of three spatial modes in an optical fiber. Opt. Express **22**(2), 1990–1996 (2014)
24. Alcusa-Saez, E., Diez, A., Andres, M.V.: Accurate and broadband characterization of fewmode optical fibers using acousto-optic coupling. In: Zhang, X.-C. (eds.) CONFERENCE 2015, ECOC, vol. 40, pp. 689–692. OSA Publishing, California (2015)
25. Alcusa-Sáez, E., Díez, A., Andrés, M.V.: Accurate mode characterization of two-mode optical fibers by in-fiber acousto-optics. Opt. Express **24**(5), 4899 (2016)
26. Mahlke, G., Gossing, P.: Fiber Optic Cable. Corning Cable Systems Publisher, Novosibirsk (2001)

Broad-Band Fiber Optic Link with a Stand-Alone Remote External Modulator for Antenna Remoting and 5G Wireless Network Applications

Aleksei Petrov[1], Elena Velichko[1], Vladimir Lebedev[2], Igor Ilichev[2], Peter Agruzov[2], Mikhail Parfenov[1,2], Andrei Varlamov[1,2], and Aleksandr Shamrai[1,2(✉)]

[1] Peter the Great Saint Petersburg Polytechnic University, Saint Petersburg, Russia
[2] Ioffe Institute, Saint Petersburg, Russia
velichko-spbstu@yandex.ru, achamrai@mail.ioffe.ru

Abstract. A broad band fiber optic link with a stand-alone remote external modulator at the transmission terminal and laser source and photo detector at the receiving terminal have been demonstrated. The standard single mode fiber without polarization maintaining was used to deliver an optical radiation to the polarization sensitive lithium niobate external modulator. An original scheme with depolarizing laser sources was used. One component of the linear polarization was used for high frequency modulation and the orthogonal component was converted into electricity and feed the modulator bias control electronic scheme. The using of the balanced detection scheme for the increasing spurious-free dynamic range (SFDR) higher than 110 dB/Hz$^{2/3}$ was proposed.

Keywords: Microwave photonics · Optical fiber communication · Fiber-optic links · External modulator · Antenna remoting · 5G systems

1 Introduction

The near to zero attenuation and dispersion properties of the telecommunication grade single mode optical fiber render it an attractive medium for microwave and millimeter-wave transmission to the distances from a few meters to many kilometers.

One of the applications is so called antenna remoting [1, 2] the technology which used in radio astronomy and radar applications where microwave and millimeter-wave signals collected by multiple antennas to be processed in one location.

The other application is radio over fiber (RoF) technique [3, 4] which became of a special interest with the emerging of 5G wireless networks. Due to the strong air-link attenuation at high frequencies the typical communication distance shorten to only a few meters to tens of meters. The RoF allows to increase the area of coverage and to offer undisrupted services across different networks by distributing millimetre-wave signals over optical fibers to different access nodes. In a 5G system baseband signals convert to the millimetre-wave frequency band and are transmitted over fiber optic

© Springer Nature Switzerland AG 2019
O. Galinina et al. (Eds.): NEW2AN 2019/ruSMART 2019, LNCS 11660, pp. 727–733, 2019.
https://doi.org/10.1007/978-3-030-30859-9_64

links from the central station to the base stations [5]. The base stations in turn receive millimetre-wave signals which are transmitted to the central station (head end) for the processing.

This paper describes a new configuration of the broad band fiber optic link with stand-alone remote external optical modulator which is very promising for the antenna remoting and 5G wireless network applications.

2 Fiber Optic Link Configuration

2.1 Optical Scheme

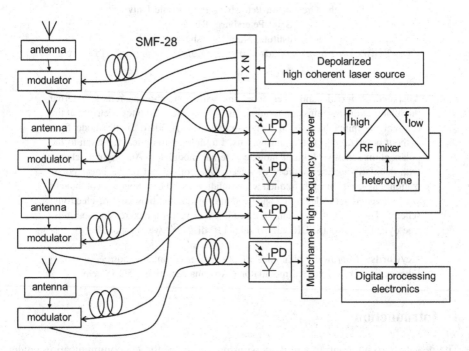

Fig. 1. Configuration of a broad band fiber optic link for antenna remoting and 5G wireless network applications.

The scheme of the proposed fiber optic link is show on the Fig. 1. All complicated and power consuming equipment such as electronics for generation and processing of high frequency signals as well as laser sources are included in the central station. The base station is made most simple with low power consumption. It consist a broad band high efficient photo detector connected with the transmitting antenna and a broad band integrated optical modulator with an electronic system for the bias control which modulate the laser radiation obtained from the central station by a high frequency signal from the receiving antenna. Transmitting and receiving channels can share one and the same optical fiber by using wavelength division multiplexing (WDM) or some other multiplexing technique. We are concentrated on the receiving channel with remote external modulator.

2.2 Integrated Optical Modulator

The lithium niobate integrated optical Mach-Zehnder modulator with dual output [6] was fabricated in the Laboratory of Quantum Electronics at Ioffe Institute. The titanium in-diffused optical waveguides [7] form the modulator optical scheme (Fig. 2). An original technique for the precise photorefractive trimming of the X type waveguide directional coupler [8] was used for exact balancing of modulator outputs. The travelling wave push pull 50-ohm electrodes provide broad band modulation with frequency bandwidth up to 20 GHz [9].

The modulator was set to the quadrature by specially designed electronic system which equalized constant optical power on the different modulator outputs by application of the control DC voltage to the bias electrodes. The power consumption of the bias control system was minimized lower than 5 mW by the choice of the electronic components and by an operation algorithm.

The modulator can be used in the two regimes. If only one modulator output directed back to the central station where it is detected by broad band photo diode we have standard external modulation with direct detection link configuration. Using two modulator outputs with π shifted high frequency modulation the balanced detection can be realized [10] which allows a suppression of RIN and other synphase noises and especially interesting for low signal applications such as radio astronomy.

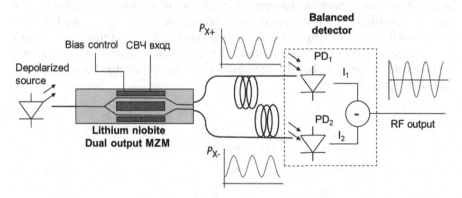

Fig. 2. Lithium niobate integrated optical Mach-Zehnder modulator with dual output in the balanced detection link configuration. The link consist of depolarized source (1550 nm, 200 mW), dual output Mach-Zehnder modulator (20 GHz, Uπ = 5 V), balanced photodetetor (5 GHz, IDC = 60 mA).

2.3 Depolarized Laser Source

The lithium niobate integrated optical modulator is a polarization sensitive device. It requires linear input polarization directed parallel to the chip surface. In order to deliver the laser irradiation to the remote modulator through the standard single mode (SMF 28) fiber a depolarized laser source was used. The problem to produce depolarized light from highly coherent telecommunication laser diode was solved by combining with a fiber optic polarization combiner/splitter light from two DFB lasers diodes whose linear polarization states are perpendicular to each other and adjusting their output power and wavelength [11]. The optical power of the two DFB LD should be exactly equal at the output of the fiber optic polarization combiner/splitter and wavelength difference should large enough to ensure that beat note and mixing intermodulation terms appears out of the frequency range of the system. Thus two DFB LD from 100 GHz ITU grid with a high output power 100 mW and a low noise (RIN < −155 dB/Hz) were used.

At the base station a similar fiber optic polarization combiner/splitter was used to separate and direct required linear polarization state to the integrated optical modulator which contain thin film plasmon polariton polarizer [12] for ensure high (more than 40 dB) polarization extinction. The orthogonal polarization not wasted and directed to a semiconductor photovoltaic power converter with 30% conversion efficiency thus the bias control electronics was powered by incoming light.

2.4   Balanced Detector

A balanced detector (Fig. 3a) was assembled from two broad band photodiode chips. The diodes with 30 mA saturation current were taken for operation with rather high constant optical power. It gave a tradeoff with the diode response bandwidth which was 10 GHz. The diodes were chosen with the equal response efficiency and similar spectral characteristics (Fig. 3b) to ensure high synphase noise suppression (around 20 dB). Note that in balanced scheme the bandwidth of response is reduced by two times in comparison with responses of each photo diode separately. On the Fig. 3b is also shown the graph of synphase noise suppression which demonstrates performance of the balanced assembly. Observed falls of the suppression efficiency with the growth of the signal frequency attributed to the mismatch of efficient electric length of the two arms of the balanced detector.

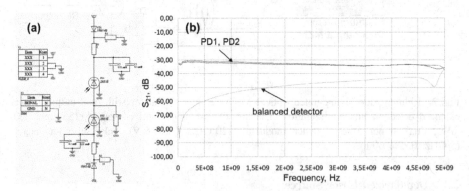

Fig. 3. (a) The scheme of balanced photodetector. (b) Spectral dependences of separate photodiodes and a synphase noise suppression demonstration.

3   Results of the Fiber Optic Link Testing

Proposed fiber optic link was tested in both standard (single arm) and balanced (dual arm) regimes. Very high characteristics were demonstrated. Experimental dependences of key technical characteristics (gain G, noise figure NF and spurious free dynamic range SFDR$_3$) are shown on the Fig. 4. Rather high gain was associated with high power of laser diodes. Note that the gain on 6 dB higher for the balanced configuration

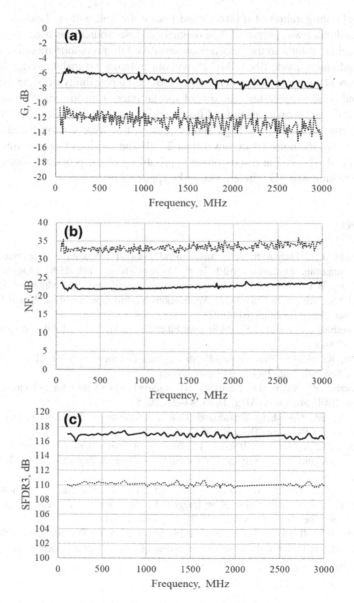

Fig. 4. Experimental results of the fiber optic link testing: (a) gain spectral dependence, (b) noise figure spectral dependence and (c) spurious free dynamic range of third order.

due to sum of signals amplitudes in the two arms. The noise figure drops on 10 dB due to gain growth and RIN suppression. Achieved spurious free dynamic range was about 110 dB/Hz$^{2/3}$ for standard link and growth up to the 117 dB/Hz$^{2/3}$ for balanced configuration.

4 Conclusion

An original configuration of a broad band fiber optic link with a stand-alone remote external modulator was proposed. The depolarized laser source was used for delivering of a high coherent light to the polarization sensitive lithium niobite modulator through the standard single mode fiber SMF 28 without polarization maintaining. The Mach-Zehnder modulator with dual optical output provided a possibility of balanced detector scheme and noise suppression which is of special interest for low signal transmission. The balanced detector was assembled from two identical photodiodes with rather high saturation current of 30 mA. High link characteristics were demonstrated: the gain about of −6 dB, the noise figure lower than 25 dB and spurious free dynamic range as high as 117 dB/Hz$^{2/3}$. The results prove applicability of the proposed approaches for antenna remoting and 5G wireless networks applications.

References

1. Ackerman, E.I., Daryoush, A.S.: Broad-band external modulation fiber-optic links for antenna-remoting applications. IEEE Trans. Microw. Theory Tech. **45**(8), 1436–1442 (1997)
2. Glomb Jr., W.L.: Fiber optic links for antenna remoting. In: Proceedings of SPIE 1703, Optical Technology for Microwave Applications VI and Optoelectronic Signal Processing for Phased-Array Antennas III (1992)
3. Al-Raweshidy, H., Komaki, S.: Radio over Fiber Technologies for Mobile Communications. Artech (2002)
4. Llorente, R., et al.: Ultra-wideband radio signals distribution in FTTH networks. IEEE Photon. Technol. Lett. **20**(11), 945–947 (2008)
5. Waterhouse, R., Novak, D.: Realizing 5G: microwave photonics for 5G mobile wireless systems. IEEE Microwav. Mag. **16**(8), 84–92 (2015)
6. Chen, A., Murphy, E.J.: Broadband Optical Modulators: Science, Technology, and Applications. CRC Press, Boca Raton (2012)
7. Parfenov, M., Agruzov, P., Il'ichev, I., Shamray, A.: Simulation of Ti-indiffused lithium niobate waveguides and analysis of their mode structure. J. Phys: Conf. Ser. **741**, 012141 (2016)
8. Parfenov, M.V., Tronev, A.V., Il'ichev, I.V., Agruzov, P.M., Shamrai, A.V.: Photorefractive correction of the coupling ratio of an integrated optical directional X-coupler on a lithium niobate substrate. Tech. Phys. Lett. **45**(3), 187–189 (2019)
9. Lebedev, V.V., Il'ichev, I.V., Agruzov, P.M., Shamray, A.V.: The influence of the current-carrying electrode material on the characteristics of integral optical microwave modulators. Tech. Phys. Lett. **40**(9), 743–746 (2014)

10. Urick, V.J., Williams, K.J., McKinney, J.D.: Fundamentals of Microwave Photonics. Wiley, Hoboken (2015)
11. Burns, W.K., Moeller, R.P., Bulmer, C.H., Greenblatt, A.S.: Depolarized source for fiber-optic applications. Opt. Lett. **16**, 381–383 (1991)
12. Il'ichev, I.V., Toguzov, N.V., Shamray, A.V.: Plasmon-polariton polarizers on the surface of single-mode channel optical waveguides in lithium niobite. Tech. Phys. Lett. **35**, 831–833 (2009)

Interfering Molecular Communication by Rotating Magnetic Fields

Puhalsky Yan[1,2], Vorobyov Nikolay[1], Pirmagomedov Rustam[3(✉)],
Loskutov Svyatoslav[2], Yakubovskaya Alla[2], and Tolmachev Sergey[2]

[1] All-Russia Research Institute for Agricultural Microbiology, Sh. Podbelskogo 3,
Saint-Petersburg-Pushkin 196608, Russian Federation
arriam2008@yandex.ru
[2] Research-and-production association "BioEcoTech" LTD,
Vozrozhdenia Street, 15, Saint-Petersburg 198188, Russian Federation
info@bioecotech.ru
[3] Peoples' Friendship University of Russia (RUDN University),
6 Miklukho-Maklaya Street, Moscow 117198, Russian Federation
prya.spb@gmail.com

Abstract. This paper presents an experimental study of the interference caused by rotating magnetic fields in interspecies molecular communication. During the experiment, we considered microbial-plant biosystem interaction under the effect of magnetic fields. This study has been one of the first attempts to examine the interference of electromagnetic field to molecular communication. The results of the experiment demonstrate that molecular communication channels can be sensitive to the influence of magnetic fields. Particularly, the experimental results reported in the paper indicate that molecular communication was blocked by magnetic fields.

Keywords: Internet of nano-things · Molecular communication ·
Nanonetworks · Interference

1 Introduction

Recent progress in nanotechnology has had a significant impact in many areas of science. Due to nanotechnology, telecommunication networks can be deployed at the nano level (named as Nanonetworks) promising a new technological era [1]. In addition to electromagnetic communication, recent research efforts were focused on bio-inspired types of information transfer, such as molecular communication. Molecular communication is a widespread type of communication used within an organism; it is considered the primary enabler for data exchange inside the organism, including the gathering of information about the overall state of wellbeing [2].

The publication has been prepared with the support of the "RUDN University Program 5-100".

O. Galinina et al. (Eds.): NEW2AN 2019/ruSMART 2019, LNCS 11660, pp. 734–743, 2019.
https://doi.org/10.1007/978-3-030-30859-9_65

Molecular communication theory has been well developed during the last few years. Recent works developed channel models, protocols and coding schemes for intercellular ionic and molecular exchange [3–6], bacterial networks [7,8], molecular motors [9] and pheromone communication [10,11]. The low reliability of links is a common disadvantage of all these types of molecular communication. According to simulations, the reliability of the molecular channel can be significantly improved by using distributed receivers which collaboratively determine a transmitter's signal [12–15] and by implementing novel detecting techniques [16–18]. These research outcomes allow for developing advanced models of molecular communication.

Most of the proposed models for molecular communication are based on traditional telecommunication paradigms [19–21], and consider molecular channels as a chemically isolated system, which is impossible beyond the laboratory. In vivo, the molecular communication is interacting with a pre-existing biochemical system [22]. This interaction may cause two side effects: (i) the molecular channel's performance may be reduced; (ii) the molecular channel may have a negative impact on the biochemical systems of the organism.

To avoid these adverse effects of artificial molecular communication, technologies enabling their coexistence with the environment are required. Technically, to enable coexistence, it should be ensured that communication particles (e.g., molecules, ions) do not cause toxicity within the organism [23] and that they are resilient to external impacts (e.g., biochemical processes of an organism, electromagnetic waves and fields). The framework for the coexistence between molecular communication and biochemical processes in organisms is suggested in [22]. This framework is based on the theory of chemical reaction networks [24], which account for concentrations of different molecules in the organism evolving over time. However, there is a lack of research considering the coexistence of molecular communication and external electromagnetic fields.

Electromagnetic fields and particularly geometric features of electromagnetic fields may also impact on molecular communication. Because biological objects demonstrate a high level of sensitivity to these [25].

In this paper, we performed an experimental study of the interference caused by rotating magnetic fields to interspecies molecular communication. Particularly, we considered microbial-plant biosystem features under the conditions of hydroponic culture and the effect of magnetic fields. In the considered system, molecular communication is performed by the natural interaction between rhizospheric bacteria; the changes in this communication lead to changes in the phytohormonal balance (homeostasis) of plants. Thus, during the experiment, we are evaluating the effect introduced by a magnetic field measuring the homeostasis of plants.

2   Experiment Design

For the experimentation the field pea (*Pisum sativum*) was utilized. The sprouted seeds were planted in the openings of platforms installed in sterile plastic vessels (OS140box, Duchefa Biochemie) containing a solution with a nutrient medium [26]. Chemical composition of the solution was as follows (ml/l):

- $Ca(NO_3)_2 \times 4H_2O(100\,\text{mM}) - 0.6$
- $K_2HPO_4 \times 3H_2O(600\,\text{mM}) - 0.6$
- $MgSO_4 \times 7H_2O(400\,\text{mM}) - 0.6$
- $CaCl_2 \times 2H_2O(100\,\text{mM}) - 0.6$
- $KCl(400\,\text{mM}) - 0.6$
- $KNO_3(1\,\text{M}) - 0.6$
- $FeC_4H_4O_6 \times 2.5H_2O(2\,\text{mM}) - 6.0$

The initial acidity (pH) of the nutrient solution was 5.5, the degree of mineralization (ppm0.5) was 158, and the redox potential (Eh) was +231.

During the experiment four options of plant growing conditions were utilized:

- Option 1 (control sample) – the plants developed in the absence of the bacteria and magnetic field.
- Option 2 – the plants were exposed to biological factor only.
- Option 3 – the plants were exposed to physical factor only (magnetic field).
- Option 4 – the plants were exposed to biological and physical factors simultaneously.

In each option, 40 plants were grown (4 vessels of 10 plants each). The general scheme of the experiment is shown in Fig. 1.

The Biological Factor. The biological factor (FB) represented by *Sphingomonas sp. K1B* bacteria which synthesized an increased number of signal molecules - auxins - in the nutrient solution (the density of bacterial cells in the nutrient solution at the beginning of the experiment was 1.7×10^5 CFU/ml). This type of bacteria has intensive molecular signaling with the plant roots [27].

The Physical Factor. The physical factor (FM) represents rotating magnetic field of the electromagnetic generator, whose level of magnetic induction did not exceed 10 mT. The magnetic field of the generator is formed utilizing two magnetic components superposition. The magnetic induction vector of the first component is constant in terms of the magnitude and direction, and the magnetic induction vector of the second component is constant in terms of the magnitude and variable in terms of the direction (it rotates cyclically, like an hour hand, in a plane perpendicular to the first vector).

Equipment. Analysis of the experimental results was performed using the high-performance liquid chromatography system UPLC Waters ACQUITY H-class ("Waters," USA). Organic acids were separated by the Waters ACQUITY CSH C18 UPLC column with an ultraviolet detector ($l = 220\,\text{nm}$).

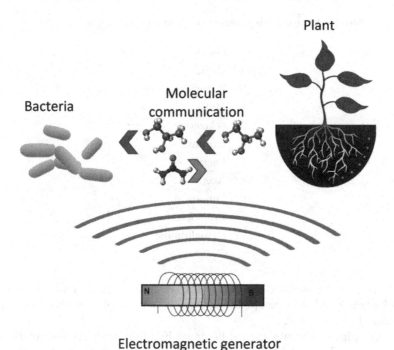

Fig. 1. A general scheme of the experiment.

3   Results and Discussion

At the end of the experiment (it lasted 12 days), the weight of roots and shoots of the plants were measured (Fig. 2). We considered weight as an integrated indicator of a plant development.

The pH of the nutrient solution, end titer of bacteria (CFU), mineralization of the nutrient solution (ppm) and redox potential (Eh) in each vessel was measured (Table 2).

At the end of the experiment the composition of low molecular root exudations (exometabolites), as well as components of signal exchange [27] – were measured. Total list of components, which are determined in the plants presented in Table 1.

According to the measurements of root exometabolites, portraits of a fractal organization (Fig. 3) were computed. These portraits rely on the concept of fractals, and coherence indexes (structuredness), which reflect the degree of consistency of the primary component I under the influence of different factors. The calculation was carried out according to (1).

$$I = N_{PFG}/(N_{EX} - 1) \qquad (1)$$

Table 1. Organic acids determined during exudation by the plants.

1	Oxalic
2	Citric
3	Pyruvic
4	Malic
5	t-Aconitic
6	Succinic
7	Lactic
8	Acetic
9	Fumaric
10	Propionic
11	Pyroglutamic

where N_{PFG} is the number of primary fractal groups (PFG) found on the portrait of plant exudation components; N_{EX} is the number of exudates in the spectrum of organic acids.

In our opinion, the observed differences in the growth rates of plant roots and shoots can be explained by the impact of the magnetic field on the communication between plants and bacteria. Particularly, the field effect (i) signal transmission along signal molecular chains (MSCs) and (ii) by changing the flow of substances in trophic chains (TCs).

Table 2. Microbiological and biochemical characteristics of nutrient solutions after two weeks of hydroponic pea plants cultivation.

Option	pH	ppm (0.5)	Eh	Concetration of bacteria	
				$\times 10^6$ CFU/ml	$\times 10^7$ CFU/g in roots
No. 1	5.54 ± 0.02	63.5 ± 0.1	209.0 ± 0.1	–	–
No. 2	5.02 ± 0.09	115.1 ± 0.1	248.0 ± 0.1	6.4 ± 0.1	11.5 ± 0.1
No. 3	5.90 ± 0.01	33.5 ± 0.1	178.0 ± 0.1	–	–
No. 4	5.35 ± 0.07	32.5 ± 0.1	189.0 ± 0.1	10.0 ± 0.1	5.2 ± 0.1

A decrease in the pH of nutrient solutions can occur due to an increased exudation of organic acids from plants to solutions [28]. The higher the amount of exudate released by the plants to the nutrient solution is, the lower the solution pH is, the fewer organic resources are sent into plants for their own needs, and the lower the rate of shoot development is. A strong correlation between the pH values and the weight of shoots in the experiment options ($r = 0.999$) indirectly confirms the mutual regulation of nutrient flows in microbe-plant

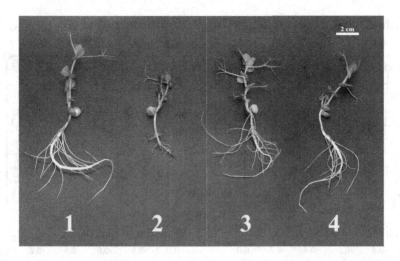

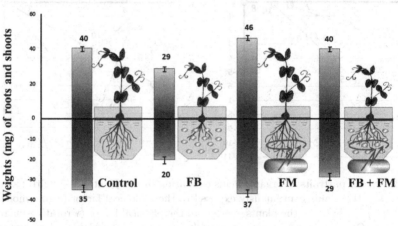

Fig. 2. The plants after experiment: 1 – control sample; 2 – the plants continuously exposed to the biological factor (signal molecules of bacteria); 3 – the plants exposed to the physical factor (a rotating magnetic field); 4 – the plants simultaneously exposed to the biological and physical factors.

biological systems. Such interactions are related to TC, in which control is reduced to a change in the ratio of the nutrient flows between bacteria and plants.

According to experiment option No. 2 only the biological factor influences the plants, and this factor causes a significant reduction in the rate of root system growth as compared to control option No. 1. The additional effect of physical factors in option No. 4 neutralizes the effect of the biological factor. As a result, the rate of roots growth in option No. 4 does not differ from the rate of roots growth in control option No. 1. It is possible to block the biological factor only if there is an element of the molecular chain of control of processes

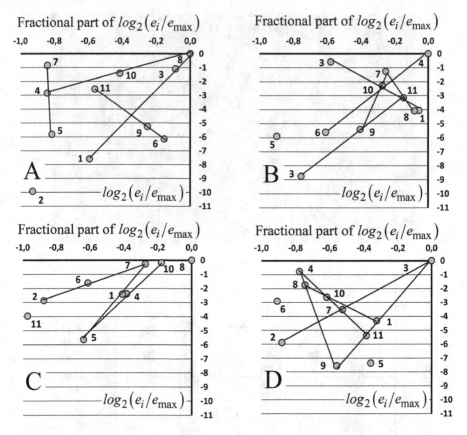

Fig. 3. Fractal portraits of organic acids exudation in the plants. A – control sample, 4 PFG; B – the plants continuously exposed to the biological factor (signal molecules of bacteria), 6 PFG; C – the plants exposed to the physical factor (a rotating magnetic field), 3 PFG; D – the plants simultaneously exposed to the biological and physical factors, 7 PFG.

in the root system sensitive to the influence of rotating magnetic fields (physical factor). Only signal molecules can be such an element. They propagate along the MSC and are sensitive to the blocking action of the physical factor. In the case of the second type of process control in plants, signal molecules are not used and, therefore, this signal path cannot be blocked by external magnetic fields. However, according to the degree of solution mineralization (ppm 0.5), there is a slight increase in the consumption of nutrients by the plants in options No. 3 and 4.

Primary fractal groups (PFG) of organic acids marked on the fractal portraits show different degrees of organization in the synthesis of organic acids by the plants. The index of exudation fractal organization (EFO) of organic acids is 0.4 in the control sample. Under the influence of only the magnetic field (in the absence of bacteria) the EFO index increases to 0.6. In the absence of a magnetic effect on plants and in the presence of bacteria in the nutrient medium, the index of EFO decreases to 0.3. Under conditions of a combined effect of bacteria and the magnetic field on plants the index of EFO is 0.7. So, bacteria disrupt the level of exudation organization, which leads to their chaotic release, and the magnetic field within certain limits restores the consistency of these processes. The higher EFO is, the more efficient functional properties of the plant-microbial community are.

Based on the experiment results, we can conclude that the magnetic field affects the molecular communication between the plants and bacteria, which leads to changes in the development of the plants. Particularly, the interaction of bacteria and plants through TCs under these conditions changes the ratio of substance flows in the nutrient solution in the form of increased exudation.

4   Conclusion

Recent works on nanonetworks elaborate a concept of cognitive molecular communication. This concept focused on reducing interference between molecular communication and other biochemical processes existing in the same environment. In this paper, we demonstrate that molecular communication channels can be also sensitive to the influence of magnetic fields. Particularly, the experimental results reported in the paper indicate that molecular communication can be blocked by magnetic fields. Thus, interference of molecular communication with electromagnetic fields also should be considered developing an application of molecular communication.

To the best of our knowledge, the present study has been one of the first attempts to examine the interference of electromagnetic field to molecular communication. The blockage of molecular communication observed during the experiment can be explained by changes in the spatial configuration of signal molecules and noncompliance of their forms with the forms of plant cell receptors according to the key-lock scheme. The further study on the demonstrated effect requires additional equipment for the confirmation control of organic molecules, as well as frequency scanning of biological objects by magnetic fields. Due to lack of this equipment, a correlation between properties of the magnetic field and changes in the molecular channel it is difficult to define precisely. However, the reported results open new challenges related to enabling reliable molecular communication under the influence of the magnetic field.

References

1. Akyildiz, I.F., Brunetti, F., Blázquez, C.: Nanonetworks: a new communication paradigm. Comput. Networks **52**(12), 2260–2279 (2008)
2. Kirichek, R., Pirmagomedov, R., Glushakov, R., Koucheryavy, A.: Live substance in cyberspace - biodriver system. In: Proceedings of 18th International Conference on Advanced Communication Technology (ICACT), IEEE, pp. 274–278 (2016)

3. Barros, M.T., Balasubramaniam, S., Jennings, B., Koucheryavy, Y.: Transmission protocols for calcium-signaling-based molecular communications in deformable cellular tissue. IEEE Trans. Nanotechnol. **13**(4), 779–788 (2014)
4. Barros, M.T.: Ca2+-signaling-based molecular communication systems: Design and future research directions. Nano Commun. Networks **11**, 103–113 (2017)
5. Akdeniz, B.C., Turgut, N.A., Yilmaz, H.B., Chae, C.-B., Tugcu, T., Pusane, A.E.: Molecular Signal Modeling of a Partially Counting Absorbing Spherical Receiver. arXiv preprint arXiv:1712.07435 (2017)
6. Huang, Y., Wen, M., Yang, L.-L., Chae, C.-B., Ji, F.: Spatial Modulation for Molecular Communication. arXiv preprint arXiv:1807.01468 (2018)
7. Gregori, M., Akyildiz, I.F.: A new nanonetwork architecture using flagellated bacteria and catalytic nanomotors. IEEE J. Sel. Areas Commun. **28**(4), 612–619 (2010)
8. Balasubramaniam, S., et al.: Multi-hop conjugation based bacteria nanonetworks. IEEE Trans. Nanobiosci. **12**(1), 47–59 (2013)
9. Chahibi, Y., Akyildiz, I.F., Balasingham, I.: Propagation modeling and analysis of molecular motors in molecular communication. IEEE Trans. Nanobiosci. **15**(8), 917–927 (2016)
10. Enomoto, A., Moore, M.J., Nakano, T., Suda, T.: Stochastic cargo transport by molecular motors in molecular communication. In: Proceedings of International Conference on Communications (ICC), IEEE, pp. 6142–6145 (2012)
11. Unluturk, B.D., Akyildiz, I.F.: An end-to-end model of plant pheromone channel for long range molecular communication. IEEE Trans. Nanobiosci. **16**(1), 11–20 (2017)
12. Fang, Y., Noel, A., Yang, N., Eckford, A.W., Kennedy, R.A.: Distributed cooperative detection for multi-receiver molecular communication. In: Proceedings of Global Communications Conference (GLOBECOM), IEEE, pp. 1–7 (2016)
13. Fang, Y., et al.: Convex optimization of distributed cooperative detection in multi-receiver molecular communication. IEEE Trans. Mol. Biol. Multi-Scale Commun. **3**(3), 166–182 (2017)
14. Fang, Y., Noel, A., Yang, N., Eckford, A.W., Kennedy, R.A.: Maximum likelihood detection for cooperative molecular communication. In: 2018 IEEE International Conference on Communications (ICC), IEEE, pp. 1–7 (2018)
15. Chae, C.-B., Bon-Hong, K., Cho, Y.J.: Method for receiving data in MIMO molecular communication system. 30 November 2017. US Patent App. 15/605,596
16. Noel, A., Eckford, A.W.: Asynchronous peak detection for demodulation in molecular communication. In: Proceedings of International Conference on Communications (ICC), IEEE, pp. 1–6 (2017)
17. Furubayashi, T., Sakatani, Y., Nakano, T., Eckford, A., Ichihashi, N.: Design and wet-laboratory implementation of reliable end-to-end molecular communication. Wireless Networks **24**(5), 1809–1819 (2018)
18. Mai, T.C., Egan, M., Duong, T.Q., Di Renzo, M.: Event detection in molecular communication networks with anomalous diffusion. IEEE Commun. Lett. **21**(6), 1249–1252 (2017)
19. Pierobon, M., Akyildiz, I.F.: A physical end-to-end model for molecular communication in nanonetworks. IEEE J. Sel. Areas Commun. **28**(4), 602–611 (2010)
20. Kuran, M., Tugcu, T., Edis, B.: Calcium signaling: overview and research directions of a molecular communication paradigm. IEEE Wireless Commun. **19**(5), 20–27 (2012)
21. Guo, W., Deng, Y., Yilmaz, H.B., Farsad, N., Elkashlan, M., Eckford, A., Nallanathan, A., Chae, C.-B.: SMIET: simultaneous molecular information and energy transfer. IEEE Wireless Commun. **25**(1), 106–113 (2018)

22. Egan, M., Mai, T.C., Duong, T.Q., Di Renzo, M.: Coexistence in molecular communications. Nano Commun. Networks **16**, 37–44 (2018)
23. Chahibi, Y., Akyildiz, I.F.: Molecular communication noise and capacity analysis for particulate drug delivery systems. IEEE Trans. Commun. **62**(11), 3891–3903 (2014)
24. Feinberg, M.: Chemical reaction network structure and the stability of complex isothermal reactors-I. The deficiency zero and deficiency one theorems. Chem. Eng. Sci. **42**(10), 2229–2268 (1987)
25. Hong, F.T.: Magnetic field effects on biomolecules, cells, and living organisms. Biosystems **36**(3), 187–229 (1995)
26. Belimov, A.A., Dodd, I.C., Safronova, V.I., Malkov, N.V., Davies, W.J., Tikhonovich, I.A.: The cadmium-tolerant pea (Pisum sativum L.) mutant sgecdt is more sensitive to mercury: assessing plant water relations. J. Exp. Bot. **66**(8), 2359–2369 (2015)
27. Bais, H.P., Weir, T.L., Perry, L.G., Gilroy, S., Vivanco, J.M.: The role of root exudates in rhizosphere interactions with plants and other organisms. Annu. Rev. Plant Biol. **57**, 233–266 (2006)
28. Lu, W., Cao, Y., Zhang, F.: Role of root-exuded organic acids in mobilization of soil phosphorus and micronutrients. Chin. J. Appl. Ecol. **10**(3), 124–127 (1999)

Fiber – Optical System for Governance and Control of Work for Nuclear Power Stations of Low Power

Nikita S. Myazin[1], Valentin I. Dudkin[2], Nadya M. Grebenikova[1],
Roman V. Davydov[1(✉)], Vadim V. Davydov[1,3], Vasiliy Yu. Rud'[3],
and Alexey S. Podstrigaev[4,5]

[1] Peter the Great Saint Petersburg Polytechnic University,
St Petersburg 195251, Russia
davydovrv@spbstu.ru
[2] The Bonch-Bruevich Saint - Petersburg State University
of Telecommunications, Saint Petersburg 193232, Russia
[3] Department of Ecology, All-Russian Research Institute of Phytopathology,
B. Vyazyomy, Odintsovo District, 143050 Moscow Region, Russia
[4] Scientific-Research Institute Vector OJSC, St. Petersburg, Russia
[5] Saint Petersburg Electrotechnical University "LETI", St. Petersburg, Russia

Abstract. Features of control systems and various systems for monitoring the parameters of nuclear power plants using fiber - optical communication systems are considered. A system of control and monitoring of the values of various physical parameters for a low-power nuclear power plant with a Brest-OD-300 reactor has been developed. Various fiber-optic sensors are presented that are integrated without additional devices into a single monitoring and control system. The measurement results are presented. The necessity of continuing development and research in this direction has been substantiated.

Keywords: Fiber-optical transmission system · Nuclear power plant ·
Optical sensor · Remote control · Transmitting laser module

1 Introduction

Currently, the urgent task is to create various systems and complexes that can provide reliable control of equipment in continuous mode at a nuclear power plant (NPP) [1–6]. In addition, they must provide continuous diagnostics of equipment, both of the entire station, and of its individual nodes in automatic mode. The greatest difficulties arise in the development of these systems for nuclear power plants of low power. This is due to the specifics of its operation. In some cases, these stations are located on moving objects with severe space constraints with a small number of personnel to maintain it [2, 4–6].

It should also be noted that the management and control systems must operate reliably, both in normal operation and in a complex emergency situation (one or several parameters reach or exceed the maximum permissible values). Experience in operating various nuclear power plants has shown that monitoring and diagnostic systems for

© Springer Nature Switzerland AG 2019
O. Galinina et al. (Eds.): NEW2AN 2019/ruSMART 2019, LNCS 11660, pp. 744–756, 2019.
https://doi.org/10.1007/978-3-030-30859-9_66

nuclear power plants, implemented on the basis of electronic devices, often fail for various reasons. Especially, in the NPP of low power due to the compact placement of all systems. The greatest negative impact on electronic devices and their connecting elements (connectors, cables, etc.) in this case have powerful surges in voltage and current. As well as the sudden operation of key elements of the relay protection transformer substations. Therefore, it is impractical to use mobile devices, cable communication lines with analog signals and digital systems [7–14] in such a complex electromagnetic environment.

In addition, the spark pulse voltage disrupts some of the sensors and creates large interferences in the operation of control systems [15–18]. False commands appear, etc. In addition, electronic components are quite sensitive to temperature, vibration, as well as to an increase pressure and humidity in the area of their location. Modern methods of protection of electronics cannot provide full compensation for the influence of these factors. This can lead to the fact that control over the operation of the reactor installation can be lost.

Fiber-optic transmission systems (FOTS) for transmitting information for various purposes are immune to electromagnetic interference [12, 17–22]. Especially to those that occur during abrupt voltage surges. This allows devices using an optical fiber to realize the galvanic isolation between monitoring and diagnostic systems and to implement reliable remote monitoring. This is very important when servicing hazardous facilities. In addition, modern technologies for the manufacture of optical fibers from chemical inert quartz glass have made fiber optic lines more resistant to the effects of various adverse factors [21–23]. This allows them to be successfully used at nuclear power plants during the operation of the plant in extreme conditions.

Therefore, the development of control systems and monitoring parameters of low-power NPPs based on FOTS is an important task. One of its possible solutions is presented in our work for NPPs with the Brest-OD-300 reactor. For NPPs with other types of reactors, it will be necessary to make improvements and changes considering the specific design of each station.

2 Features of Operation of Fiber-Optical Systems at Nuclear Power Plants of Low Power

Considering the specifics of operating conditions of reactors at nuclear power plants of low power, especially at mobile objects, we established a number of features that should be taken into account when developing fiber-optic control and monitoring systems:

- These control and monitoring systems should function in some cases in automatic mode for more than 24 months during NPP operation without the attendants access to the reactor;
- Long-term exposure to temperature, overpressure, vibration and increased radiation levels can simultaneously occur on FOTS;
- Systems must ensure the speed and efficiency of information transfer in large volumes in order to make correct and informed decisions, especially at the final

stage of the life cycle of reactor installations. This is quite a difficult period in the operation of the station. There are several negative factors: a high degree of wear and tear of the main equipment, a change in the generations of staff (they did not participate in the start-up of the reactor), etc. Without operational data on the NPP life cycle, it is quite difficult to decommission it [1, 2].

In Fig. 1 is a schematic diagram of the Brest-OD-300 reactor in section. It indicates the main blocks and nodes in which the values of various physical quantities are measured. A number of reactor units receive control commands. Reliable information on their performance should be received from these units and other nodes to the central computer of the NPP.

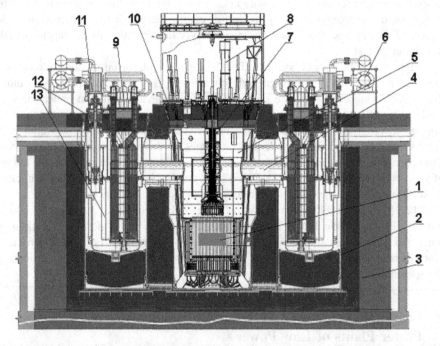

Fig. 1. The design of the reactor Brest-OD-300: 1 - active zone; 2 - case block; 3 - reactor shaft; 4 - manifold pipeline; 5 - core basket; 6 - cooling system; 7 - measuring column; 8 - in-core reloading machine; 9 - steam generator; 10 - overlap with manholes; 11 - pumping system for pumping feed water and coolant; 12—steam generator conversion unit; 13 - filter.

Difficult placement of blocks and other nodes in the reactor design, temperature differences and the presence of a large number of interferences require the development of a specific configuration of optical control and monitoring systems. Studies have shown that it is most appropriate to use separate channels: for measurements, for controlling parameters, for transmitting control commands and confirming their execution.

The advantage of the systems based on optical fibers developed by us is that optical sensors (primary converters) and can be in extreme operating conditions of the reactor. And the means of processing optical signals can be placed on the central panel of the nuclear power plant (in a protected room). The distance of 300–500 m for an optical signal that propagates through fiber optic lines does not pose any problems [18–23]. The attenuation of the optical signal power even with extreme effects on the optical fiber of negative factors will allow to process the signals and obtain data on the measured value (for example, temperature T in the core system, coolant flow, etc.).

In addition, previous studies have shown that optical fibers used in FOTS must have a protective sheath. This shell should ensure tightness during repeated washing with decontamination solutions, withstand at least 200–300 heating and cooling cycles. In addition, the shell must withstand high temperatures up to 573 K for at least one hour (in the emergency mode).

Currently, there are a lot of different optical fibers with different protective coatings. Since the structure of the reactor for low-power nuclear power plants is compact enough, there will be a lot of sharp turns and bends when laying the fiber. In this situation, it will also determine the choice of optical fiber. Table 1 presents several variants of such fibers with different coatings.

Table 1. Characteristics of optical fibers.

Coating material	Operating temperature, K	External diameter, μm	Manufacturing company
Acrylate	233–358	245	Corning (USA), OFS (USA), Draka (Chez)
Mid-Temp Dual acrylate	233–433	245	AFL (Verrilont, USA), Draka (Chez)
Silicone/Mid-temp acrylate	233–433	245	AFL (Verrilont, USA), Draka (Chez), HTPI (Canada)
Silikone/PFA	233–473	25–750	AFL (Verrilont, USA), Draka (Chez), FiberLogix (UK)
Polyimide	223–703	155–255	AFL (Verrilont, USA), Draka (Chez), FiberLogix (UK), IVG (Germany)

The data in Table 1 allow the use of various optical fibers in the development of a fiber-optic system regarding operating temperatures. This makes it possible in some cases more effectively to lay the fiber optic line in areas with sharp bends.

3 The Fiber-Optical Transmission System and Its Operation

The use of optical fiber and optical sensors significantly reduced the area occupied by the measuring elements and information transmission lines. This made it possible to significantly increase the number of channels, both for transmitting information and control commands, and for measuring. In Fig. 2 the block diagram of the FOTS developed by us is presented.

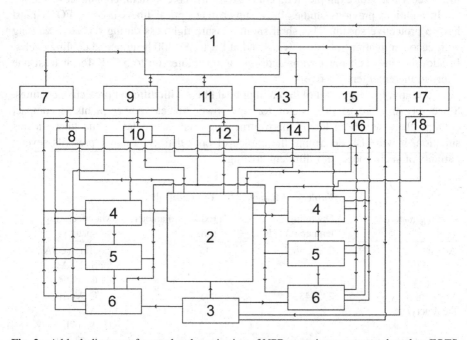

Fig. 2. A block diagram of control and monitoring of NPP operating parameters based on FOTS and optical sensors: 1 - NPP central computer, 2 - reactor, 3 - reactor emergency shutdown system, 4 - turbine, 5 - electrical energy converter, 6 - transformer substation with key elements, 7 - flow rate and feedwater flow control unit, 8 - fiber optic flow control sensors, 9 - pressure control unit in various reactor systems, 10 - fiber optic pressure sensors, 11 - magnet control unit, 12 - fiber-optic induction sensors of the magnetic field, 13 - control unit of the reactor and substation systems to compensate for the voltage of the electric field, 14 - fiber optic sensor of the electric field strength, 15 - temperature control unit, 16 - fiber-optic temperature sensors, 17 - ionizing radiation control unit, 18 - fiber-optic sensors for changing the level of ionizing radiation.

The use of FOTS and optical sensors also made it possible to exclude various transducers and amplifiers from the design of measuring devices. There are numerous interferences on these devices. In addition, they are in the radial zone of the reactor. Therefore, they are subject to various influences and require additional protection [1–6, 15].

The new principle of placement of control units and measuring systems, as well as the use of FOTS allowed to significantly increase the number of channels, both for measurement and for information transfer. Each channel uses a single-mode fiber, which is more resistant to various influences than the multimode used previously in nuclear power plants [22, 23]. The presence of additional free space in the system for placing communication lines allows creating two additional backup channels for each active channel. This significantly increases the reliability of the entire system, unlike the previously developed ones. In addition, the use of FOTS allowed the use of rectangular pulses of various durations to transmit commands to various control systems. In previously used analog systems, pulse shape distortion occurred. In Fig. 3, as an example, one variant of the transmission of the control signal over the fiber-optic fiber and a shielded copper cable developed by us is presented.

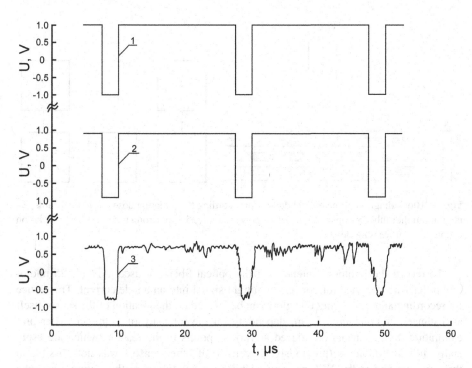

Fig. 3. The form of the pulses of the output signal: 1 - generator 3; 2 - at the exit of FOTS (input oscilloscope 9); 3 - at the exit of a copper cable.

Analysis of the results shows the absence of distortion of the pulse shape in the control signal at the inputs of the control systems of the reactor. This allows the control to change the pulse duration within the required limits, and to keep the specified pulse level until the confirmation signal about the command execution is received. This is a new method of control of command execution compared to the previously used.

The main measuring elements of the FOTS developed by us are fiber-optic sensors. Currently, the conceptual foundations of many fiber-optic sensors are worked out quite well. Various industrial samples of fiber-optic sensors for measuring temperature, pressure, humidity, electric field strength and magnetic field induction are produced [23]. The accuracy of measurements of physical parameters and the reliability of these sensors correspond to the concept developed by us FOTS. Therefore, in this paper we will consider only optical sensors for measuring radiation dose and coolant flow and feed water, the development of which was directly involved by the authors. The remaining types of optical sensors used in the developed system are standard.

To determine the dose of radiation from γ - radiation, a method was developed based on recording the intensity of radiation-induced luminescence in quartz fibers when exposed to external sources of γ - radiation. In Fig. 4 shows the scheme of γ-radiation dose recording using an opticalcable, which consists of 85 fibers.

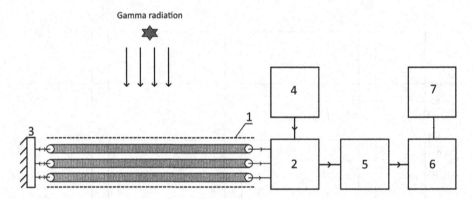

Fig. 4. Block diagram of the optical device for recording the radiation dose. 1 - optical cable, 2 - photomultiplier tube, 3 - opaque mirror, 4 - power supply, 5 - spectrometric amplifier, 6 - photon counter, 7 - processing device.

To record the luminous intensity of the optical fiber, we used a TNFT 25 photomultiplier with a device for converting a light signal into an anode current. The device for recording and processing this glow can be located on the location of the sensor itself at distances of more than 50 m, depending on the intensity of γ-quanta. After precalibration, these sensors are placed at various points in the reactor, which are especially difficult to access (previously, placement of other sensors was not possible at these points) and at the NPP premises. Under the conditions of the station's extreme operating conditions, the radiation dose measurement error is increased to 4% due to the heating of the optical cable. This measurement accuracy is sufficient in this mode of operation of the station to assess the situation and make decisions.

To measure the flow rate of the coolant and feed water, we have developed a design of a correlation flow meter using an annular fiber-optic detector for detecting γ-quanta. In Fig. 5 presents its structural diagram.

The coolant in the reactor core is irradiated with γ - radiation. After leaving the core, the coolant (for example, liquid sodium or lead melt with bismuth), while flowing through the pipeline, emits γ-quanta. The intensity of γ-quanta radiation as it flows

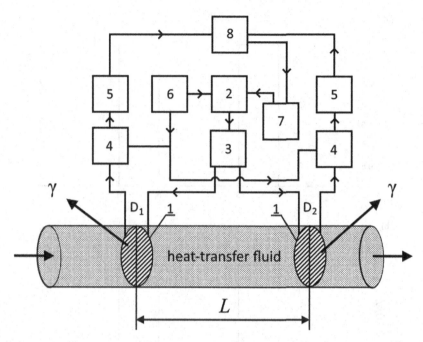

Fig. 5. Block diagram of an optical device for measuring the flow of a liquid medium in a pipeline. 1 - ring fiber-optical system, 2 - laser transmitting module, 3 - optical divider, 4 - photoreceiver module, 5 - analog-digital converter, 6 - power supply, 7 - pulse generator, 8 - processing device and control.

through the pipeline decreases in proportion to a certain dependence (for each coolant it is determined experimentally). When exposed to γ-quanta with a single-mode optical fiber, the intensity of the laser radiation signal that propagates through it varies. From two ring sensors located at a certain distance L (Fig. 5), changes in the intensity of laser radiation are recorded. The FOTS developed by us allows you to locate a large number of these optical pairs of sensors in different parts of the pipeline. Measurements of the intensity change are used to determine the attenuation dependence. This dependence determines the flow time of the coolant and then the flow.

Experiments have shown that for measurements it is most appropriate to use laser radiation in the form of short pulses with a duration of 0.1 ms to 10 ms (depending on the flow velocity of the coolant and its composition). We also found that when using 10 pairs of fiber-optic ring sensors with L = 10 m, it is better to place the sensors at a distance of 20 m from each other. In Fig. 6, as an example, is presented the work of a fiber optic sensor developed by us. In the section D_1 and D_2, a different number of γ - quanta acts on the optical fiber wound on the pipeline (Fig. 5). The fiber transparency factor changes in zones D_1 and D_2. The laser radiation transmitted through these zones has a different intensity (Fig. 6 (b) and (c)). The value of U2 > U1.

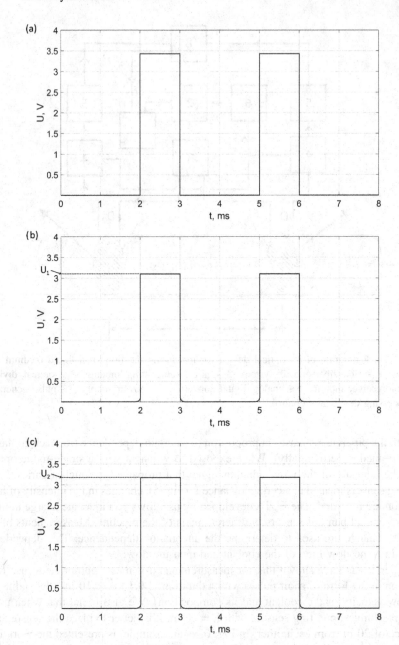

Fig. 6. The dependence of the intensity of laser radiation (amplitude of pulses) on the number of γ-quanta emitted by the current liquid medium: (a) - pulses at the output of an optical divider; (b) - impulses in the D1 zone; (c) - impulses in the D2 zone.

The error in determining the flow rate of the coolant q in the primary circuit of the reactor using the device we developed is less than 2%. This error is obtained using NMR flow meters, which allow measurements with an accuracy of less than 1% [24–27]. Using the NMR flowmeter-relaxometer, the values of the measured coolant flow rates were checked using a ring fiber optic sensor. It should be noted that the NMR flow meter, in contrast to the proposed optical sensor, is a very large instrument that requires special maintenance [24–27]. Only one such device can be placed in the design of a low-power nuclear reactor. This is not enough for effective control.

It should be noted that high power ionizing radiation has a negative effect on losses in an optical fiber. With high radiation power from the coolant, optical loss in the fiber can be up to 10–15 dB/km. The length of the optical fiber in sensor 1 (Fig. 5) is about 100 m. This allows, in addition to the flow rate q, to carry out a rough control of the power of the ionizing radiation of the coolant. The power of ionizing radiation coolant depends on the reliability of the reactor.

Consider the case when the power of ionizing radiation leads to optical losses in the fiber of more than 1000 dB/km, does not make sense. With such a power of radioactive radiation from the coolant, which he acquired in the reactor zone, automatic control systems of nuclear power plants identify this situation as a reactor accident. The reactor is transferred to the "shutdown" mode with locking in a special dome or mine.

Ring optical sensors for measuring the flow of feed water work on the same principle as for measuring the q coolant. The change in the intensity of γ-quanta is recorded at shorter L. Since γ-radiation emitted during the decay of ^{16}N nuclei, which are contained in the feed water, is detected. Also, several pairs of sensors are used for measurements.

The results of determining the flow of feed water showed that the measurement error q does not exceed 2%. As q decreases, the measurement error decreases. The same trend is present when measuring q coolant. This is due to the peculiarities of the principle of operation of the ring fiber-optic detector.

4 Conclusion

The results obtained on the basis of the conducted experiments and data on the operation of the experimental design of FOTS developed by us as part of a low-power nuclear power plant showed that our proposed system for controlling and monitoring the operation of NPPs with FOTS works more reliably and efficiently than previously used systems with analog control and measurement channels. Using FOTS allows to use a large number of complex signal processing systems. With analog signal processing, this will lead to an increase in the number of modules used and the reliability of the entire system will decrease significantly. The use of FOTS allows two additional backup channels to be placed on one operating optical communication or control channel. The reliability of the system is increased three times compared with the use of analog channels.

Based on the studies, it was found that using the system developed by us, it is possible to control various physical parameters of NPPs in the following ranges: temperature from 283 to 673 K, pressure up to 25 MPa, vibration from 20 to 2000 Hz, humidity 1–100%, induction magnetic field B from 0.2 μT to 0.015 mlT, electric field intensity E from 0.2 kV/m from 2 MV/m, radiation dose rate from 10–10 to 1 Gy (gray)/s, coolant flow rate up to 50 m³/h, consumption feed water up to 200 m³/h with an error not exceeding 1.5–2%. This allows you to more effectively control the operation of low-power nuclear power plants, especially at the final stage of the reactor life cycle.

The approaches proposed in the FOTS developed by us to solving problems of managing NPP operation and monitoring the implementation of various commands (for example, to change the technological operation modes of the cooling system, etc.) increase the accuracy and reliability of monitoring the operation of its main equipment. This, in turn, increases the efficiency of production of electric energy at nuclear power plants of low power.

The methods of solving the problems of managing and controlling the operation of low-power NPPs presented in this paper can be used at other nuclear facilities after modifications to the specifics of the object.

References

1. Klinov, D.A., Gulevich, A.V., Kagramanyan, V.S., Dekusar, V.M., Usanov, V.D.: Development of sodium-cooled fast reactors under modern conditions: challenges and stimuli. At. Energy **125**(3), 143–148 (2019)
2. Ignatiev, V.V., et al.: Molten-salt reactor for nuclear fuel cycle closure on all actinides. At. Energy **125**(5), 279–283 (2019)
3. Krapivtsev, V.G., Solonin, V.I.: Model studies of interloop coolant mixing in VVER-1000 in-reactor pressure channel. At. Energy **125**(5), 307–317 (2019)
4. Davydov, V.V., Dudkin, V.I., Karseev, A.U.: Nuclear magnetic flowmeter – spectrometer with fiber – optical communication line in cooling systems of atomic energy plants. Opt. Mem. Neural Netw. (Inf. Opt.) **22**(2), 112–117 (2013)
5. Davydov, V.V., Velichko, E.N., Dudkin, V.I., Karseev, A.Yu.: A nutation nuclear-magnetic teslameter for measuring weak magnetic field. Meas. Tech. **57**(6), 684–689 (2014)
6. Davydov, V.V., Dudkin, V.I., Karseev, A.Yu.: Fiber – optic communication line for the nmr signals transmission in the control systems of the ships atomic power plants work. Opt. Mem. Neural Netw. (Inf. Opt.) **23**(4), 259–264 (2014)
7. Makolkina, M., Pham, V.D., Kirichek, R., Gogol, A., Koucheryavy, A.: Interaction of AR and IoT applications on the basis of hierarchical cloud services. In: Galinina, O., Andreev, S., Balandin, S., Koucheryavy, Y. (eds.) NEW2AN/ruSMART -2018. LNCS, vol. 11118, pp. 547–559. Springer, Cham (2018). https://doi.org/10.1007/978-3-030-01168-0_49
8. Ateya, A.A., Muthanna, A., Vybornova, A., Darya, P., Koucheryavy, A.: Energy - aware offloading algorithm for multi-level cloud based 5 g system. In: Galinina, O., Andreev, S., Balandin, S., Koucheryavy, Y. (eds.) NEW2AN/ruSMART -2018. LNCS, vol. 11118, pp. 355–370. Springer, Cham (2018). https://doi.org/10.1007/978-3-030-01168-0_33
9. Koucheryavy, A., Vladyko, A., Kirichek, R.: State of the art and research challenges for public flying ubiquitous sensor networks. In: Balandin, S., Andreev, S., Koucheryavy, Y. (eds.) ruSMART 2015. LNCS, vol. 9247, pp. 299–308. Springer, Cham (2015). https://doi.org/10.1007/978-3-319-23126-6_27

10. Podstrigaev, A.S., Davydov, R.V., Rud, V.Yu., Davydov, V.V.: Features of transmission of intermediate frequency signals over fiber-optical communication system in radar station. In: Galinina, O., Andreev, S., Balandin, S., Koucheryavy, Y. (eds.) NEW2AN/ruSMART 2018. LNCS, vol. 11118, pp. 624–630. Springer, Cham (2018). https://doi.org/10.1007/978-3-030-01168-0_56

11. Ateya, A.A., Muthanna, A., Gudkova, I., Abuarqoub, A., Vybornova, A., Koucheryavy, A.: Development of intelligent core network for tactile internet and future smart systems. J. Sens. Actuator Netw. 7(1), 7 (2018)

12. Kiesewetter, D., Malyugin, V., Makarov, S., Korotkov, K., Ming, D., Wei, X.: Application of the optical fibers in the system of determining the distance of jump at ski springboard. In: Proceedings – 2016 Advances in Wireless and Optical Communications, RTUWO 2016, pp. 5–8 (2017). 7821845

13. Prokofiev, A., Nepomnyashchaya, E., Pleshakov, I., Kuzmin, Y., Velichko, E., Aksenov, E.: Study of specific features of laser radiation scattering by aggregates of nanoparticles in ferrofluids used for optoelectronic communication systems. In: Galinina, O., Balandin, S., Koucheryavy, Y. (eds.) NEW2AN/ruSMART -2016. LNCS, vol. 9870, pp. 680–689. Springer, Cham (2016). https://doi.org/10.1007/978-3-319-46301-8_59

14. Bystrov, V.V., Likhachev, V.P., Ryazantsev, L.B.: Experimental check of the coherence of radiolocation signals from objects with nonlinear electrical properties. Meas. Tech. 57(9), 1073–1076 (2014)

15. Davydov, V.V., Dudkin, V.I., Karseev, AYu.: Fiber – optic imitator of accident situation for verification of work of control systems of atomic energy plants on ships. Opt. Mem. Neural Netw. (Inf. Opt.) 23(3), 170–176 (2014)

16. Davydov, V.V., Dudkin, V.I., Velichko, E.N., Karseev, AYu.: Fiber-optic system for simulating accidents in the cooling circuits of a nuclear power plant. J. Opt. Technol. (A Translation of Opticheskii Zhurnal) 82(3), 132–135 (2015)

17. Davydov, V.V., Ermak, S.V., Karseev, A.U., Nepomnyashchaya, E.K., Petrov, A.A., Velichko, E.N.: Fiber-optic super-high-frequency signal transmission system for sea-based radar station. In: Balandin, S., Andreev, S., Koucheryavy, Y. (eds.) NEW2AN 2014. LNCS, vol. 8638, pp. 694–702. Springer, Cham (2014). https://doi.org/10.1007/978-3-319-10353-2_65

18. Ivanov, S.I., Lavrov, A.P., Saenko, I.I.: Application of microwave photonics components for ultrawideband antenna array beamforming. In: Galinina, O., Balandin, S., Koucheryavy, Y. (eds.) NEW2AN/ruSMART -2016. LNCS, vol. 9870, pp. 670–679. Springer, Cham (2016). https://doi.org/10.1007/978-3-319-46301-8_58

19. Davydov, V.V., Sharova, N.V., Fedorova, E.V., Gilshteyn, E.P., Malanin, K.Yu., Fedotov, I. V., Vologdin, V.A., Karseev, A.Yu.: Fiber-optics system for the radar station work control. In: Balandin, S., Andreev, S., Koucheryavy, Y. (eds.) ruSMART 2015. LNCS, vol. 9247, pp. 712–721. Springer, Cham (2015). https://doi.org/10.1007/978-3-319-23126-6_65

20. Ermolaev, A.N., Krishpents, G.P., Davydov, V.V., Vysoczkiy, M.G.: Compensation of chromatic and polarization mode dispersion in fiber-optic communication lines in microwave signals transmittion. J. Phys: Conf. Ser. 741(1), 012071 (2016)

21. Belkin, M.E., Sigov, A.S.: Some trend in super-high frequency optoelectronics. J. Commun. Technol. Electron. 54(8), 655–658 (2009)

22. Friman, R.K.: Fiber-Optic Communication Systems, p. 496. Wiley-Inter Science (2012)

23. Agrawal, G.P.: Light Wave Technology: Telecommunication Systems, p. 480. NJ, Wiley-Inter Science (2014)

24. Davydov, V.V.: Control of the longitudinal relaxation time T_1 of a flowing liquid in NMR flowmeters. Russ. Phys. J. 42(9), 822–825 (1999)

25. Davydov, V.V., Dudkin, V.I., Karseev, AYu.: A compact nuclear magnetic relaxometer for the express monitoring of the state of liquid and viscous media. Meas. Tech. **57**(8), 912–918 (2014)

26. Davydov, V.V., Dudkin, V.I., Karseev, AYu.: Governance of the nutation line contour in nuclear-magnetic flowmeters. Russ. Phys. J. **58**(2), 146–152 (2015)

27. Davydov, V.V., Dudkin, V.I., Karseev, AYu., Vologdin, V.A.: Special features in application of nuclear magnetic spectroscopy to study flows of liquid media. J. Appl. Spectrosc. **82**(6), 1013–1019 (2016)

Author Index

Printed in the United States
By Bookmasters